Metals
	Metalloids
	Nonmetals

			13 3A	14 4A	15 5A	16 6A	17 7A	8A 2 He 4.00 helium
			5 B 10.81 boron	6 C 12.01 carbon	7 N 14.01 nitrogen	8 O 16.00 oxygen	9 F 19.00 fluorine	10 Ne 20.18 neon
10 8B	11 1B	12 2B	13 Al 26.98 aluminum	14 Si 28.09 silicon	15 P 30.97 phosphorus	16 S 32.06 sulfur	17 Cl 35.45 chlorine	18 Ar 39.95 argon
28 Ni 58.69 nickel	29 Cu 63.55 copper	30 Zn 65.39 zinc	31 Ga 69.72 gallium	32 Ge 72.63 germanium	33 As 74.92 arsenic	34 Se 78.97 selenium	35 Br 79.90 bromine	36 Kr 83.80 krypton
46 Pd 106.42 palladium	47 Ag 107.87 silver	48 Cd 112.41 cadmium	49 In 114.82 indium	50 Sn 118.71 tin	51 Sb 121.75 antimony	52 Te 127.60 tellurium	53 I 126.90 iodine	54 Xe 131.29 xenon
78 Pt 195.08 platinum	79 Au 196.97 gold	80 Hg 200.59 mercury	81 Tl 204.38 thallium	82 Pb 207.2 lead	83 Bi 208.98 bismuth	84 Po (209) polonium	85 At (210) astatine	86 Rn (222) radon
110 Ds (281) darmstadtium	111 Rg (280) roentgenium	112 Cn (285) copernicium	113 Nh (284) nihonium	114 Fl (289) flerovium	115 Mc (289) moscovium	116 Lv (293) livermorium	117 Ts (294) tennessine	118 Og (294) oganesson

64 Gd 157.25 gadolinium	65 Tb 158.93 terbium	66 Dy 162.50 dysprosium	67 Ho 164.93 holmium	68 Er 167.26 erbium	69 Tm 168.93 thulium	70 Yb 173.04 ytterbium	71 Lu 174.97 lutetium
96 Cm (247) curium	97 Bk (247) berkelium	98 Cf (251) californium	99 Es (252) einsteinium	100 Fm (257) fermium	101 Md (258) mendelevium	102 No (259) nobelium	103 Lr (260) lawrencium

INTRODUCTORY
CHEMISTRY
ESSENTIALS

SIXTH EDITION

Nivaldo J. Tro

Westmont College

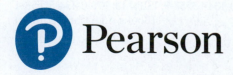Pearson

330 Hudson Street, NY NY 10013

Director, Courseware Portfolio Management: Jeanne Zalesky
Courseware Portfolio Specialist: Scott Dustan
Courseware Director, Content Development: Jennifer Hart
Development Editor: Erin Mulligan
Courseware Analyst: Coleen Morrison
Portfolio Management Assistant: Lindsey Pruett
VP, Product Strategy & Development: Lauren Fogel
Content Producer: Chandrika Madhavan
Managing Producer: Kristen Flathman
Content Producer: Jackie Jakob
Director, Production & Digital Studio: Katie Foley
Senior Mastering Media Producer: Jayne Sportelli

Rights and Permissions Manager: Ben Ferrini
Rights and Permissions Management: Cenveo Publisher Services
Photo Researcher: Eric Schrader
Production Management and Composition: codeMantra
Illustrator: Precision Graphics
Design Managers: Marilyn Perry, Maria Guglielmo Walsh
Cover and Interior Designer: Gary Hespenheide
Contributing Illustrators: Lachina
Manufacturing Buyer: Maura Zaldivar-Garcia
Product Marketer: Elizabeth Bell
Cover Art: Quade Paul

Credits and acknowledgments borrowed from other sources and reproduced, with permission, in this textbook appear on the appropriate page within the text or on page C-1.

Library of Congress Cataloging-in-Publication Data

Names: Tro, Nivaldo J.
Title: Introductory chemistry essentials / Nivaldo J. Tro, Westmont College.
Description: Sixth edition. | New York, NY : Pearson, [2018] | Includes index.
Identifiers: LCCN 2016045461 | ISBN 9780134291802 (hardcover) | ISBN 9780134555584 (a la carte)
Subjects: LCSH: Chemistry—Textbooks.
Classification: LCC QD33.2 .T763 2018 | DDC 540—dc23
LC record available at https://lccn.loc.gov/2016045461

ISBN 10: 0-134-29180-8; ISBN 13: 978-0-134-29180-2 (Student edition)
ISBN 10: 0-134-55558-9; ISBN 13: 978-0-134-55558-4
(Books A La Carte edition)

www.pearsonhighered.com

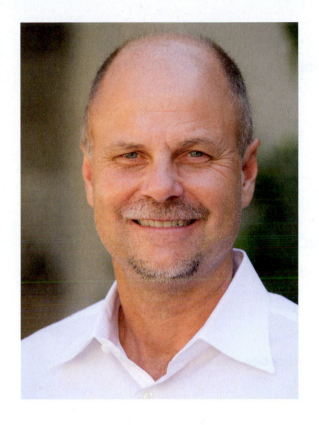

Nivaldo Tro is a professor of chemistry at Westmont College in Santa Barbara, California, where he has been a faculty member since 1990. He received his Ph.D. in chemistry from Stanford University for work on developing and using optical techniques to study the adsorption and desorption of molecules to and from surfaces in ultrahigh vacuum. He then went on to the University of California at Berkeley, where he did postdoctoral research on ultrafast reaction dynamics in solution. Since coming to Westmont, Professor Tro has been awarded grants from the American Chemical Society Petroleum Research Fund, from the Research Corporation, and from the National Science Foundation to study the dynamics of various processes occurring in thin adlayer films adsorbed on dielectric surfaces. He has been honored as Westmont's Outstanding Teacher of the Year three times and has also received the college's Outstanding Researcher of the Year award. Professor Tro lives in Santa Barbara with his wife, Ann, and their four children, Michael, Ali, Kyle, and Kaden. In his leisure time, Professor Tro enjoys mountain biking, surfing, and being outdoors with his family.

To Annie

Brief Contents

Contents

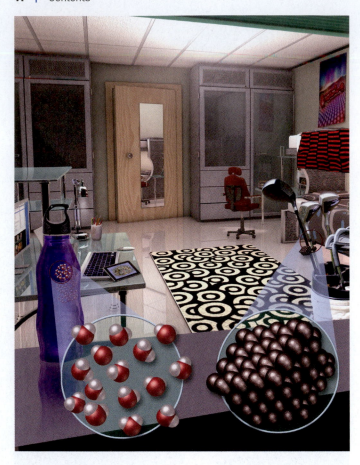

4 Atoms and Elements 98

3 Matter and Energy 60

5 Molecules and Compounds 132

6 Chemical Composition   168

8 Quantities in Chemical Reactions 248

7 Chemical Reactions 206

9 Electrons in Atoms and the Periodic Table 284

10 Chemical Bonding 324

11 Gases 358

13 Solutions   444

12 Liquids, Solids, and Intermolecular Forces   408

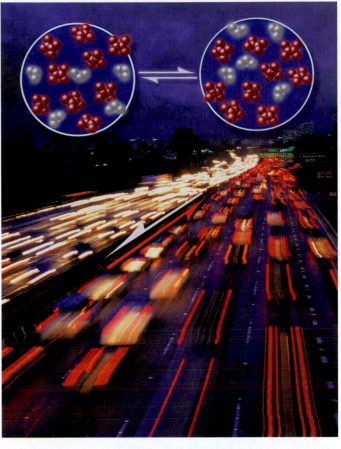

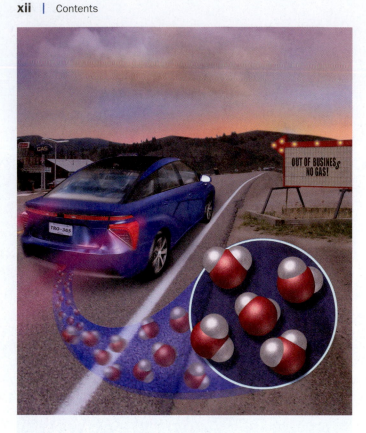

■ Three-Column Problem Solving Strategies

Interactive Media Contents

Key Concept Videos

PEARSON eText 2.0 — The eText 2.0 icon indicates that this feature is embedded and interactive in the eText.

Interactive Worked Examples

PEARSON eText 2.0

To the Student

This book is for *you*, and every text feature is meant to help you learn and succeed in your chemistry course. I wrote this book with two main goals for you in mind: to see chemistry as you never have before and to develop the problem-solving skills you need to succeed in chemistry.

I want you to experience chemistry in a new way. I have written each chapter to show you that chemistry is not just something that happens in a laboratory; chemistry surrounds you at every moment. Several outstanding artists have helped me to develop photographs and art that will help you visualize the molecular world. From the opening example to the closing chapter, you will *see* chemistry. My hope is that when you finish this course, you will think differently about your world because you understand the molecular interactions that underlie everything around you.

My second goal is for you to develop problem-solving skills. No one succeeds in chemistry—or in life, really—without the ability to solve problems. I can't give you a one-size-fits-all formula for problem solving, but I can and do give you strategies that will help you develop the *chemical intuition* you need to understand chemical reasoning.

Look for several recurring features throughout this book designed to help you master problem solving. The most important ones are: (1) a four-step process (Sort, Strategize, Solve, and Check) designed to help you learn how to develop a problem-solving approach; (2) the solution map, a visual aid that helps you navigate your way through problems; (3) two-column Examples, in which the left column explains in clear and simple language the purpose of each step of the solution shown in the right column; and (4) three-column Examples, which describe a problem-solving procedure while demonstrating how it is applied to two different Examples. In addition, the For More Practice feature at the end of each worked Example directs you to the end-of-chapter Problems that provide more opportunity to practice the skill(s) covered in each Example. In addition, Interactive Worked Examples are digital versions of select worked Examples from the text that help you break down problems using the book's "Sort, Strategize, Solve, and Check" technique. Interactive Worked Examples can be found in the eText 2.0 and can be accessed directly at: https://media.pearsoncmg.com/ph/esm/esm_tro_intro_6/media/index.html.

Recent research has demonstrated that you will do better on your exams if you take a multiple-choice pre-exam before your actual exam. At the end of each chapter, you will find a Self-Assessment Quiz to help you check your understanding of the material in that chapter. You can string these together to make a pre-exam. For example, if your exam covers Chapters 5–7, complete the Self-Assessment Quizzes for those chapters as part of your preparation for the exam. The questions you miss on the quiz will reveal the areas you need to spend the most time studying. Studies show that if you do this, you will do better on the actual exam.

Lastly, I hope this book leaves you with the knowledge that chemistry is *not* reserved only for those with some superhuman intelligence level. With the right amount of effort and some clear guidance, anyone can master chemistry, including you.

Sincerely,

Nivaldo J. Tro
tro@westmont.edu

To the Instructor

I thank all of you who have used any of the first five editions of *Introductory Chemistry*—you have made this book the best-selling book in its market, and for that I am extremely grateful. The preparation of the sixth edition has enabled me to continue to refine the book to meet its fundamental purpose: teaching chemical skills in the context of relevance.

Introductory Chemistry is designed for a one-semester, college-level, introductory or preparatory chemistry course. Students taking this course need to develop problem-solving skills—but they also must see *why* these skills are important to them and to their world. *Introductory Chemistry* extends chemistry from the laboratory to the student's world. It motivates students to learn chemistry by demonstrating the role it plays in their daily lives.

This is a visual book. Wherever possible, I use images to help communicate the subject. In developing chemical principles, for example, I worked with several artists to develop multipart images that show the connection between everyday processes visible to the eye and the molecular interactions responsible for those processes. This art has been further refined and improved in the sixth edition, making the visual impact sharper and more targeted to student learning. For example, I have continued to expand and refine a hierarchical system of labeling in many of the images: the white-boxed labels are the most important, the tan boxes are second in importance, and the unboxed labels are the third most important. In many cases, this system allows information to be placed closest to its point of relevance, instead of being lumped together in the caption. In addition, this allows me to treat related labels and annotations within an image in the same way, so that the relationships between them are immediately evident. My intent is to create an art program that teaches and presents complex information clearly and concisely. Many of the illustrations showing molecular depictions of a real-world object or process have three parts: macroscopic (what we can see with our eyes); molecular and atomic (space-filling models that depict what the molecules and atoms are doing); and symbolic (how chemists represent the molecular and atomic world). Students can begin to see the connections between the macroscopic world, the molecular world, and the representation of the molecular world with symbols and formulas.

The problem-solving pedagogy employs four steps as it has done in the previous five editions: Sort, Strategize, Solve, and Check. This four-step procedure guides students as they learn chemical problem solving. Students will also encounter extensive flowcharts throughout the book, allowing them to better visualize the organization of chemical ideas and concepts.

Throughout the worked Examples in this book, I use a *two-* or *three-column* layout in which students learn a general procedure for solving problems of a particular type as they see this procedure applied to one or two worked Examples. In this format, the *explanation* of how to solve a problem is placed directly beside the actual steps in the *solution* of the problem. Many of you have told me that you use a similar technique in lecture and office hours. Since students have specifically asked for connections between worked Examples and end-of-chapter Problems, I include a For More Practice feature at the end of each worked Example that lists the end-of-chapter review Examples and end-of-chapter Problems that provide additional opportunities to practice the skill(s) covered in the example. Also in this edition, we have 39 Interactive Worked Examples, which can be accessed in the eText 2.0.

A successful feature of previous editions is the Conceptual Checkpoints, a series of short questions that students can use to test their mastery of key concepts as they read through a chapter. For this edition, all Conceptual Checkpoints are embedded in eText 2.0. Emphasizing understanding rather than calculation, they are designed to encourage active learning even while reading. Your continued embrace of this feature prompted me to add more of these to the sixth edition.

In this edition, I have also added a new category of End-of-Chapter Questions called *Data Interpretation and Analysis*. These questions present real data in real-life situations and ask students to analyze and interpret that data. They are designed to give students much needed practice in reading graphs, understanding tables, and making data-driven decisions.

In my own teaching, I have been influenced by two studies published in the last few years. The first one is a mega analysis of the effect of active learning on student learning in STEM disciplines.[1] In this study, Freeman and his coworkers convincingly demonstrate that students learn better when they are active in the process. The second study focuses on the effect of multiple-choice pretests on student exam performance.[2] Here, Pyburn and his coworkers show that students who take a multiple-choice pretest do better on exams than those who do not. Even more interesting, the enhancement is greater for lower performing students. In my courses, I have implemented both active learning and multiple-choice pretesting with good results. In my books, I have developed tools to allow you to incorporate these techniques as well.

To help you with active learning, I have added 12 Key Concept Videos to the media package that accompanies this book. These three- to five-minute videos each introduce a key concept from the chapter. They are themselves interactive because every video has an embedded question posed to the student to test understanding. In addition, there are 19 new Interactive Worked Examples adding to a total of 39 new and revised Interactive Worked Example videos in the media package. This means that you now have a library of 31 new interactive videos and a total of 51 new and revised interactive videos to enhance your course.

In my courses, I use these videos in conjunction with the book to implement a *before, during, after* strategy for my students. My goal is simple: *Engage students in active learning before class, during class, and after class.* To that end, I assign a video *before* most class sessions. All Key Concept Videos and Interactive Worked Examples are embedded and interactive in eText 2.0, allowing students to review and test their understanding in real-time. The video introduces students to a concept or problem that I will cover in the lecture. *During* class, I expand on the concept or problem using *Learning Catalytics*™ to question my students. Instead of simply passively listening to a lecture, they are interacting with the concepts through questions that I pose. Sometimes I ask my students to answer individually, other times in pairs or even groups. This approach has changed my classroom. Students engage in the material in new ways. They are actively learning and have to think and process and interact. Finally, *after* class, I give them another assignment, usually a short follow-up question or problem. At this point, they must apply what they have learned to solve a problem.

To help you with multiple-choice pretesting, each chapter contains a Self-Assessment Quiz. Like the Conceptual Checkpoints and the videos, these quizzes are embedded in eText 2.0. These quizzes are designed so that students can test themselves on the core concepts and skills of each chapter. I encourage my students to use these quizzes as they prepare for exams. For example, if my exam covers Chapters 5–8, I assign the quizzes for those chapters for credit (you can do this in MasteringChemistry™). Students then get a pretest on the core material that will be on the exam.

My goal with this new edition is to continue to help you make learning a more active (rather than passive) process for your students. I hope the tools that I have provided here continue to aid you in teaching your students better and more effectively. Please feel free to email me with any questions or comments you might have. I look forward to hearing from you as you use this book in your course.

Sincerely,

Nivaldo J. Tro

tro@westmont.edu

[1] Freeman, Scott; Eddy, Sarah L.; McDonough, Miles; Smith, Michelle K.; Okoroafor, Nnadozie; Jordt, Hannah; and Wenderoth, Mary Pat; Active learning increases student performance in science, engineering, and mathematics, 2014, Proc. Natl. Acad. Sci.

[2] Daniel T. Pyburn, Samuel Pazicni, Victor A. Benassi, and Elizabeth M. Tappin *J. Chem. Educ.*, 2014, 91 (12), pp. 2045–2057.

What's New in This Edition?

The book has been extensively revised and contains more small changes than can be detailed here. The most significant changes to the book and its supplements are listed below.

- I have added a new category of end-of-chapter questions called *Data Interpretation and Analysis*. These questions present real data in real-life situations and ask students to analyze that data. They give students much needed practice in reading graphs, digesting tables, and making data-driven decisions. A new section (Section 1.4), including a new in-chapter worked Example (Example 1.4), introduces these skills.

- There are 12 new Key Concept Videos and 19 new Interactive Worked Examples to accompany the book. That means there are 31 new videos and 51 total new and revised interactive videos to accompany the material in the sixth edition. All Key Concept Videos and Interactive Worked Examples are embedded and interactive in eText 2.0, allowing students to review and test their understanding in real-time. These tools are designed to help professors engage their students in active learning. Recent research has conclusively demonstrated that students learn better when they are active in the learning process, as opposed to passively listening and simply taking in content. The Key Concept Videos are brief (three to five minutes), and each introduces and explains a key concept from a chapter. The student does not just passively listen to the video; the video stops in the middle and poses a question to the student. The student must answer the question before the video continues. Each video also includes a follow-up question that is assignable in MasteringChemistry™. The Interactive Worked Examples are similar in concept, but instead of explaining a key concept, they walk the student through one of the in-chapter worked examples from the book. Like the Key Concept Videos, Interactive Worked Examples stop in the middle and force the student to interact by completing a step in the example. The examples also have a follow-up question that is assignable in MasteringChemistry™. The power of interactivity to make connections in problem solving is immense. I did not quite realize this power until we started making the Interactive Worked Examples and I saw how I could use the animations to make connections that are just not possible on the static page.

- All chapter-ending Self-Assesment Quizzes are embedded in eText 2.0.

- I have added 13 new Conceptual Checkpoint questions throughout the book. For this edition, all Conceptual Checkpoints are embedded in eText 2.0.

- I have updated the data throughout the book to reflect the most recent measurements and developments available. I changed the half-life of carbon-14 to 5715 years in Table 17.2 and throughout Chapter 17 to reflect the current accepted value, and I also added new information about *thermoluminescent dosimeters* (and deleted the information on film badge dosimeters) to Section 17.4. Other updates include changes to Figure 8.2, *Climate change*; Section 10.1, *Bonding Models and AIDS Drugs*; Table 11.5, *Changes in Pollutant Levels for Major U.S. Cities, 1980–2014*; the *Chemistry in the Environment* box in Section 12.8, *Water: A Remarkable Molecule*; and Section 17.8, *Nuclear Power: Using Fission to Generate Electricity*.

- Several chapter-opening sections and (or) the corresponding art, including Sections 1.1, 2.1, 12.1, and 16.1, have been replaced or significantly modified.
- I added a new section (Section 2.8) and new worked example (Example 2.12) as well as new end-of-chapter Problems to address conversions involving quantities with combined units such as mL/kg or km/hr.
- I have extensively modified the art program to move information from the captions and into the art itself. This allows relevant information to be placed right where it is most needed and makes the art a more accessible study and review tool. I have modified 70 figures in this way.
- I have modified end-of-chapter Problems that were showing low levels of student success when assigned in MasteringChemistry™.
- I have added temporary symbols for elements 113, 115, 117, and 118 (Uut, Uup, Uus, and Uuo, respectively) to all periodic tables.
- In all chapters, chapter text was edited for clarity and to limit use of passive voice and extraneous words and phrases.

Teaching Principles

The development of basic chemical principles—such as those of atomic structure, chemical bonding, chemical reactions, and the gas laws—is one of the main goals of this text. Students must acquire a firm grasp of these principles in order to succeed in the general chemistry sequence or the chemistry courses that support the allied health curriculum. To that end, the book integrates qualitative and quantitative material and proceeds from concrete concepts to more abstract ones.

Organization of the Text

The main divergence in topic ordering among instructors teaching introductory and preparatory chemistry courses is the placement of electronic structure and chemical bonding. Should these topics come early, at the point where models for the atom are being discussed? Or should they come later, after the student has been exposed to chemical compounds and chemical reactions? Early placement gives students a theoretical framework within which they can understand compounds and reactions. However, it also presents students with abstract models before they understand why they are necessary. I have chosen a later placement; nonetheless, I know that every course is unique and that each instructor chooses to cover topics in his or her own way. Consequently, I have written each chapter for maximum flexibility in topic ordering. In addition, the book is offered in two formats. The full version, *Introductory Chemistry*, contains 19 chapters, including organic chemistry and biochemistry. The shorter version, *Introductory Chemistry Essentials*, contains 17 chapters and omits these topics.

Print and Media Resources

Instructor and Student Supplements	
0134564057 / 9780134564050	Instructor Resource Manual with Complete Solutions
013455342X / 9780134553429	TestGen Test Bank (Download only)
0134553446 / 9780134553443	Instructor Resource Materials (Download only)
0134553411 / 9780134553412	Study Guide
0134564065 / 9780134564067	Student Selected Solutions Manual
0134555554 / 9780134555553	Modified MasteringChemistry™ with Pearson eText–Instant Access
0134555570 / 9780134555577	MasteringChemistry™ with Pearson eText–Instant Access

Acknowledgments

This book has been a group effort, and I am grateful for all of those who helped me. First and foremost, I would like to thank my editor Scott Dustan. I have known Scott for many years and in various roles, and am grateful to have him as my editor. I appreciate his straightforward style, constant support, and commitment to my work. I am also in a continual state of awe and gratitude to Erin Mulligan, my development editor and friend. Thanks, Erin, for all your outstanding help and advice. Thanks also to Jackie Jakob, media editor extraordinaire. Jackie is the force behind the media elements that accompany this book, and I am grateful for her vision, guidance, and friendship. Thanks also to Jennifer Hart, with whom I have now worked for over a decade. Thanks Jennifer for your constant attention, guidance, and wisdom on all of my projects at Pearson. I am also grateful for Jeanne Zalesky, Adam Jaworski, Paul Corey and the rest of Pearson leadership. You have supported my projects and my vision from the beginning, and I am privileged to work with you.

I would also like to thank Elizabeth Ellsworth, my marketing manager, whose creativity in describing and promoting the book is without equal. I am also grateful to Coleen Morrison, whose help with editing and manuscript preparation was invaluable. Thanks also to the MasteringChemistry™ team who continue to provide and promote the best online homework system on the planet. I also appreciate the expertise and professionalism of my copy editor, Betty Pessagno, as well as the skill and diligence of Francesca Monaco and her colleagues at codeMantra. I am a picky author, and they always accommodated my seemingly endless requests. Thank you, Francesca. Thanks as well to my content producer, Chandrika Madhavan and the rest of the Pearson editorial and production team—they are part of a first-class operation. This text has benefited immeasurably from their talents and hard work. I owe a special debt of gratitude to Quade Paul, who continues to make my ideas come alive in his chapter-opener and cover art.

I am grateful for the assistance of my colleagues, Allan Nishimura, David Marten, Stephen Contakes, Kristi Lazar, Carrie Hill, Michael Everest, Amanda Silberstein, and Heidi Henes-Vanbergen, who have supported me in my department while I worked on this book. I owe a special debt of gratitude to Michael Tro. He has been helping me with manuscript preparation, proofreading, organizing art manuscripts, and tracking changes in end-of-chapter material for the past six years. Michael has been reliable, accurate, and invaluable. Thanks Mikee! I also owe a special thanks to my colleagues Michael Everest and Tom Greenbowe, who collaborated with me in creating some of the end of chapter questions.

I am grateful to those who have given so much to me personally while writing this book. First on that list is my wife, Ann. Her patience and love for me are beyond description. I also thank my children, Michael, Ali, Kyle, and Kaden, whose smiling faces and love of life always inspire me. I come from a large Cuban family, whose closeness and support most people would envy. Thanks to my parents, Nivaldo and Sara; my siblings, Sarita, Mary, and Jorge; my siblings-in-law, Jeff, Nachy, Karen, and John; my nephews and nieces, Germain, Danny, Lisette, Sara, and Kenny. These are the people with whom I celebrate life.

Lastly, I am indebted to the many reviewers, listed next, whose ideas are found throughout this book. They have corrected me, inspired me, and sharpened my thinking on how best to teach this subject we call chemistry. I deeply appreciate their commitment to this project.

Reviewers of the 6th Edition

Premilla Arasasingham
El Camino College

Crystal Bendenbaugh
Southeastern University

Charles Carraher
Florida Atlantic University

Cassidy Dobson
St. Cloud University

David Futoma
Roger Williams University

Galen George
Santa Rosa Junior College

Marcia Gillette
Indiana University Kokomo

Ganna Lyubartseva
Southern Arkansas University

Helen Motokane
El Camino College

David Rodgers
North Central Michigan College

Mu Zheng
Tennessee State University

6th Edition Accuracy Reviewers

Kelly Befus
Anoka-Ramsey Community College

Stevenson Flemer
University of Vermont

Lance Lund
Anoka-Ramsey Community College

Tanea Reed
Eastern Kentucky University

Jennifer Zabzydar
Palomar College

Reviewers of the 5th Edition

Scott Bunge
Kent State University

Ebru Buyuktanir
Stark State College

Claire Cohen
University of Toledo

Robert Culp
California State University — Fresno

Rosa Davila
College of Southern Idaho

Alyse Dilts
Harrisburg Area Community College

Sylvia Esjornson
Southwestern Oklahoma State University

Jennifer Firestine
Lindenwood University

Kathy Flynn
College of the Canyons

Sara Harvey
Los Angeles Pierce College

Michael Hauser
St. Louis Community College — Meramec

Edward Lee
Texas Tech University

Craig McClure
University of Alabama — Birmingham

Virginia Miller
Montgomery College

Michael Rodgers
Southeast Missouri State University

Janice Webster
Ivy Tech Community College —Terre Haute

James Zubricky
University of Toledo

5th Edition Accuracy Reviewers

Alyse Dilts
Harrisburg Area Community College

Stevenson Flemer Jr.
University of Vermont

Connie Lee
Montgomery County Community College

Lance Lund
Anoka-Ramsey Community College

Kent McCorkle
Fresno City College

Reviewers of the 4th Edition

Jeffrey Allison
Austin Community College

Mikhail V. Barybin
The University of Kansas

Lara Baxley
California Polytechnic State University

Kelly Befus
Anoka-Ramsey Community College

Joseph Bergman
Illinois Central College

Simon Bott
University of Houston

Carmela Byrnes
MiraCosta College

Carmela Magliocchi Brynes
MiraCosta College

Guy Dadson
Fullerton College

Maria Cecilia D. de Mesa
Baylor University

Brian G. Dixon
Massachusetts Maritime Academy

Timothy Dudley
Villanova University

Jeannine Eddleton
Virginia Tech

Ron Erickson
University of Iowa

Donna Friedman
St. Louis Community College—Florissant Valley

Luther D. Giddings
Salt Lake Community College

Marcus Giotto
Quinsigamond Community College

Melodie Graber
Oakton Community College

Maru Grant
Ohlone College

Jerod Gross
Roanoke Benson High School

Tammy S. Gummersheimer
Schenectady County Community College

Tamara E. Hanna
Texas Tech University

Michael A. Hauser
St. Louis Community College

Bruce E. Hodson
Baylor University

Donald R. Jones
Lincoln Land Community College

Martha R. Kellner
Westminster College

Farkhondeh Khalili
Massachusetts Bay Community College

Margaret Kiminsky
Monroe Community College

Rebecca Krystyniak
Saint Cloud State

Chuck Laland
Black Hawk College

Richard Lavallee
Santa Monica College

Laurie Leblanc
Cuyamaca College

Nancy Lee
MiraCosta College

Vicki MacMurdo
Anoka-Ramsey Community College

Jack F. McKenna
St. Cloud State University

Virginia Miller
Montgomery College

Geoff Mitchell
Washington International School

Meg Osterby
Western Technical College

John Petty
University of South Alabama

Jason Serin
Glendale Community College

Steven Socol
McHenry Community College

Youngju Sohn
Florida Institute of Technology

Jie Song
University of Michigan—Flint

Clarissa Sorenson-Unruh
Central New Mexico Community College

David Vanderlinden
Des Moines Area Community College

Vidyullata C. Waghulde
St. Louis Community—Meramec

Reviewers of the 3rd Edition

Benjamin Arrowood
Ohio University

Joe Bergman
Illinois Central College

Timothy Dudley
Villanova University

Sharlene J. Dzugan
University of Cumberlands

Thomas Dzugan
University of Cumberlands

Donna G. Friedman
St. Louis Community College

Erick Fuoco
Daley College

Melodie A. Graber
Oakton Community College

Michael A. Hauser
St. Louis Community College, Meramec Campus

Martha R. Joseph
Westminster College

Timothy Kreider
University of Medicine & Dentistry of New Jersey

Laurie Leblanc
Grossmont College

Carol A. Martinez
Central New Mexico Community College

Kresimir Rupnik
Louisiana State University

Kathleen Thrush Shaginaw
Particular Solutions, Inc.

Pong (David) Shieh
Wharton College

Mary Sohn
Florida Tech

Kurt Allen Teets
Okaloosa-Walton College

John Thurston
University of Iowa

Anthony P. Toste
Missouri State University

Carrie Woodcock
Eastern Michigan University

Reviewers of the 2nd Edition

David S. Ballantine, Jr.
Northern Illinois University

Colin Bateman
Brevard Community College

Michele Berkey
San Juan College

Steven R. Boone
Central Missouri State University

Morris Bramlett
University of Arkansas—Monticello

Bryan E. Breyfogle
Southwest Missouri State University

Frank Carey
Wharton County Junior College

Robbey C. Culp
Fresno City College

Michelle Driessen
University of Minnesota—Minneapolis

Donna G. Friedman
St. Louis Community College—Florissant Valley

Crystal Gambino
Manatee Community College

Steve Gunther
Albuquerque Technical Vocational Institute

Michael Hauser
St. Louis Community College—Meramec

Newton P. Hillard, Jr.
Eastern New Mexico University

Carl A. Hoeger
University of California—San Diego

Donna K. Howell
Angelo State University

Nichole Jackson
Odessa College

T. G. Jackson
University of South Alabama

Donald R. Jones
Lincoln Land Community College

Kirk Kawagoe
Fresno City College

Roy Kennedy
Massachusetts Bay Community College

Blake Key
Northwestern Michigan College

Rebecca A. Krystyniak
St. Cloud State University

Laurie LeBlanc
Cuyamaca College

Ronald C. Marks
Warner Southern College

Carol A. Martinez
Albuquerque Technical Vocational Institute

Charles Michael McCallum
University of the Pacific

Robin McCann
Shippensburg University

Victor Ryzhov
Northern Illinois University

Theodore Sakano
Rockland Community College

Deborah G. Simon
Santa Fe Community College

Mary Sohn
Florida Institute of Technology

Peter-John Stanskas
San Bernardino Valley College

James G. Tarter
College of Southern Idaho

Ruth M. Topich
Virginia Commonwealth University

Eric L. Trump
Emporia State University

Mary Urban
College of Lake County

Richard Watt
University of New Mexico

Lynne Zeman
Kirkwood Community College

Reviewers of the 1st Edition

Lori Allen
University of Wisconsin—Parkside

Laura Andersson
Big Bend Community College

Danny R. Bedgood
Arizona State University

Christine V. Bilicki
Pasadena City College

Warren Bosch
Elgin Community College

Bryan E. Breyfogle
Southwest Missouri State University

Carl J. Carrano
Southwest Texas State University

Donald C. Davis
College of Lake County

Donna G. Friedman
St. Louis Community College at Florissant Valley

Leslie Wo-Mei Fung
Loyola University of Chicago

Dwayne Gergens
San Diego Mesa College

George Goth
Skyline College

Jan Gryko
Jacksonville State University

Roy Kennedy
Massachusetts Bay Community College

C. Michael McCallum
University of the Pacific

Kathy Mitchell
St. Petersburg Junior College

Bill Nickels
Schoolcraft College

Bob Perkins
Kwantlen University College

Mark Porter
Texas Tech University

Caryn Prudenté
University of Southern Maine

Rill Ann Reuter
Winona State University

Connie M. Roberts
Henderson State University

Jeffery A. Schneider
SUNY—Oswego

Kim D. Summerhays
University of San Francisco

Ronald H. Takata
Honolulu Community College

Calvin D. Tormanen
Central Michigan University

Eric L. Trump
Emporia State University

Help students develop 21st-century skills to succeed in chemistry courses, future careers, and beyond.

Nivaldo Tro's approach introduces students to 21st-century skills, encouraging them to think critically when they encounter complex information and real-world problems.

NEW! Data Interpretation and Analysis Questions at the end of each chapter allow students to work with real data to develop 21st-century problem-solving skills. These questions ask students to sort, analyze and interpret actual data from real-life situations. Students practice reading graphs, digesting tables, and making data-driven decisions.

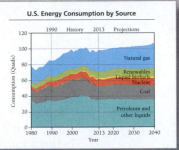

Data Interpretation and Analysis

124. The graph at right shows U.S. energy consumption by source from 1980 to 2040 (based on projections). The consumption is measured in quadrillion BTUs or quads (1 quad = 1.055×10^{18} J).
 (a) What were the three largest sources of U.S. energy in 2013 in descending order? What total percent of U.S. energy do these three sources provide?
 (b) What percent of total U.S. energy is provided by renewables in 2013?
 (c) Which two sources of U.S. energy decline as a percentage of total energy use between 1989 and 2040 (based on projections)?
 (d) How much U.S. energy (in joules) was produced by nuclear power in 1990?

A new section (Section 1.4), which includes a new in-chapter worked example (Example 1.1), introduces data interpretation and analysis skills and emphasizes their importance in student success.

All Data Interpretation and Analysis Questions are assignable in MasteringChemistry™

1.4 Analyzing and Interpreting Data

▶ Identify patterns in data and interpret graphs.

We just learned how early scientists such as Lavoisier and Dalton saw patterns in a series of related measurements. Sets of measurements constitute scientific *data*, and learning to analyze and interpret data is an important scientific skill.

Identifying Patterns in Data

Suppose you are an early chemist trying to understand the composition of water. You know that water is composed of the elements hydrogen and oxygen. You do several experiments in which you decompose different samples of water into hydrogen and oxygen, and you get the following results:

Sample	Mass of Water Sample	Mass of Hydrogen Formed	Mass of Oxygen Formed
A	20.0 g	2.2 g	17.8 g
B	50.0 g	5.6 g	44.4 g
C	100.0 g	11.1 g	88.9 g

Do you notice any patterns in this data? The first and easiest pattern to see is that the sum of the masses of oxygen and hydrogen always sums to the mass of the water sample. For example, for the first water sample, 2.2 g hydrogen + 17.8 g oxygen = 20.0 g water. The same is true for the other samples. Another pattern, which is a bit more difficult to see, is that the ratio of the masses of oxygen and hydrogen is the same for each sample.

Sample	Mass of Hydrogen Formed	Mass of Oxygen Formed	Mass Oxygen Mass Hydrogen
A	2.2 g	17.8 g	8.1
B	5.6 g	44.4 g	7.9
C	11.1 g	88.9 g	8.01

The ratio is 8—the small variations are due to experimental error, which is common in all measurements and observations.

1.4 Analyzing and Interpreting Data | 9

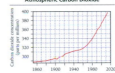

▲ FIGURE 1.5 Atmospheric carbon dioxide levels from 1860 to present.

Seeing patterns in data is a creative process that requires you to not just merely tabulate laboratory measurements, but to see relationships that may not always be obvious. The best scientists see patterns that others have missed. As you learn to interpret data in this course, be creative and try looking at data in new ways.

Interpreting Graphs

Data is often visualized using graphs or images, and scientists must constantly analyze and interpret graphs. For example, the graph in ◀ **FIGURE 1.5** shows the concentration of carbon dioxide in Earth's atmosphere as a function of time. Carbon dioxide is a greenhouse gas that has been rising as result of the burning of fossil fuels (such as gasoline and coal). When you look at a graph such as this one, you should first examine the x and y axes and make sure you understand what each axis represents. You should also examine the numerical range of the axes. In Figure 1.5, the y axis does not begin at zero in order to better display the change that is occurring. How would this graph look different if the y axis began at zero instead of at 290? Notice also that, in this graph, the increase in carbon dioxide has not been constant over time. The rate of increase—represented by the slope of the line—has intensified since about 1960.

EXAMPLE 1.1   Interpreting Graphs

Examine the graph in Figure 1.5 and answer each question.

(a) What was the concentration of carbon dioxide in 1960?
(b) What was the concentration in 2000?
(c) How much did the concentration increase between 1960 and 2000?
(d) What is the average rate of increase over this time?
(e) If the average rate of increase remains constant, estimate the carbon dioxide concentration in 2030.

SOLUTION

a) What was the concentration of carbon dioxide in 1960?

To determine the concentration of carbon dioxide in 1960, draw a vertical line at the year 1960. At the point where the vertical line intersects the carbon dioxide concentration curve, draw a horizontal line. The point where the horizontal line intercepts the y axis represents the concentration in 1960. So, the concentration in 1960 was 318 ppm.

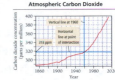

b) What was the concentration in 2000?

Apply the same procedure as in part a, but now shift the vertical line to the year 2000. The concentration in the year 2000 was 370 ppm.

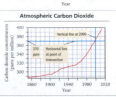

continued on page 10 ▶

Students build a **framework** for solving problems.

Nivaldo Tro's unique problem-solving technique, "Sort, Strategize, Solve, and Check," teaches students how to successfully approach, set up, and solve the problems they encounter in their introductory chemistry course. Solution maps visually walk students through problems and help them learn how to organize and use given information to successfully solve problems.

> **Two- and three-column example** formats help students break down the steps of each problem and learn and practice problem-solving techniques they can apply in other assignments.

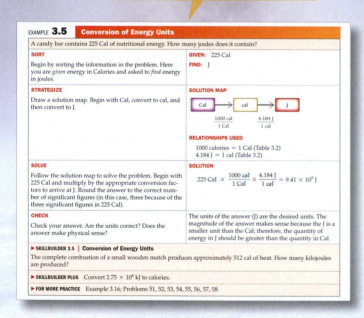

EXAMPLE **3.5**   **Conversion of Energy Units**

A candy bar contains 225 Cal of nutritional energy. How many joules does it contain?

SORT
Begin by sorting the information in the problem. Here you are *given* energy in Calories and asked to *find* energy in joules.

GIVEN: 225 Cal
FIND: J

STRATEGIZE
Draw a solution map. Begin with Cal, convert to cal, and then convert to J.

SOLUTION MAP

Cal → cal → J

$$\frac{1000 \text{ cal}}{1 \text{ Cal}} \qquad \frac{4.184 \text{ J}}{1 \text{ cal}}$$

RELATIONSHIPS USED
1000 calories = 1 Cal (Table 3.2)
4.184 J = 1 cal (Table 3.2)

SOLVE
Follow the solution map to solve the problem. Begin with 225 Cal and multiply by the appropriate conversion factors to arrive at J. Round the answer to the correct number of significant figures (in this case, three because of the three significant figures in 225 Cal).

SOLUTION
$$225 \text{ Cal} \times \frac{1000 \text{ cal}}{1 \text{ Cal}} \times \frac{4.184 \text{ J}}{1 \text{ cal}} = 9.41 \times 10^5 \text{ J}$$

CHECK
Check your answer. Are the units correct? Does the answer make physical sense?

The units of the answer (J) are the desired units. The magnitude of the answer makes sense because the J is a smaller unit than the Cal; therefore, the quantity of energy in J should be greater than the quantity in Cal.

▶ **SKILLBUILDER 3.5** | **Conversion of Energy Units**
The complete combustion of a small wooden match produces approximately 512 cal of heat. How many kilojoules are produced?

▶ **SKILLBUILDER PLUS**   Convert 2.75×10^4 kJ to calories.

▶ **FOR MORE PRACTICE**   Example 3.16; Problems 51, 52, 53, 54, 55, 56, 57, 58.

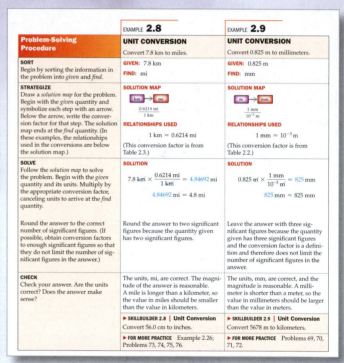

Problem-Solving Procedure	EXAMPLE **2.8** **UNIT CONVERSION** Convert 7.8 km to miles.	EXAMPLE **2.9** **UNIT CONVERSION** Convert 0.825 m to millimeters.
SORT Begin by sorting the information in the problem into *given* and *find*.	GIVEN: 7.8 km FIND: mi	GIVEN: 0.825 m FIND: mm
STRATEGIZE Draw a *solution map* for the problem. Begin with the *given* quantity and symbolize each step with an arrow. Below the arrow, write the conversion factor for that step. The solution map ends at the *find* quantity. (In these examples, the relationships used in the conversions are below the solution map.)	SOLUTION MAP km → mi $\frac{0.6214 \text{ mi}}{1 \text{ km}}$ RELATIONSHIPS USED 1 km = 0.6214 mi (This conversion factor is from Table 2.3.)	SOLUTION MAP m → mm $\frac{1 \text{ mm}}{10^{-3} \text{ m}}$ RELATIONSHIPS USED 1 mm = 10^{-3} m (This conversion factor is from Table 2.2.)
SOLVE Follow the *solution map* to solve the problem. Begin with the *given* quantity and its units. Multiply by the appropriate conversion factor, canceling units to arrive at the *find* quantity.	SOLUTION $7.8 \text{ km} \times \frac{0.6214 \text{ mi}}{1 \text{ km}} = 4.84692 \text{ mi}$ $4.84692 \text{ mi} = 4.8 \text{ mi}$	SOLUTION $0.825 \text{ m} \times \frac{1 \text{ mm}}{10^{-3} \text{ m}} = 825 \text{ mm}$ $825 \text{ mm} = 825 \text{ mm}$
Round the answer to the correct number of significant figures. (If possible, obtain conversion factors to enough significant figures so that they do not limit the number of significant figures in the answer.)	Round the answer to two significant figures because the quantity given has two significant figures.	Leave the answer with three significant figures because the quantity given has three significant figures and the conversion factor is a definition and therefore does not limit the number of significant figures in the answer.
CHECK Check your answer. Are the units correct? Does the answer make sense?	The units, mi, are correct. The magnitude of the answer is reasonable. A mile is longer than a kilometer, so the value in miles should be smaller than the value in kilometers.	The units, mm, are correct, and the magnitude is reasonable. A millimeter is shorter than a meter, so the value in millimeters should be larger than the value in meters.
	▶ **SKILLBUILDER 2.8**	**Unit Conversion** Convert 56.0 cm to inches. ▶ **FOR MORE PRACTICE** Example 2.26; Problems 73, 74, 75, 76.	▶ **SKILLBUILDER 2.9**	**Unit Conversion** Convert 5678 m to kilometers. ▶ **FOR MORE PRACTICE** Problems 69, 70, 71, 72.

NEW! and UPDATED! Interactive Worked Examples are digital versions of select worked examples from the text that make Nivaldo Tro's unique problem-solving strategies interactive. In these digital versions the author walks students through the problem-solving process, asking them to pause and answer questions along the way. Worked example videos are embedded in eText 2.0 and assignable in MasteringChemistry™.

Mass-to-Mass Conversions
Given: 58.5 g of CO_2
$$6\ CO_2(g) + 6\ H_2O(l) \xrightarrow{\text{sunlight}} 6\ O_2(g) + C_6H_{12}O_6(aq)$$
Find: g $C_6H_{12}O_6$

Solution Map   Mass A → Moles A → Moles B → Mass B

g CO_2 → mol CO_2 → mol $C_6H_{12}O_6$ → g $C_6H_{12}O_6$

$\frac{1 \text{ mol } CO_2}{44.01 \text{ g } CO_2}$

What is the correct form of the conversion factor between mol CO_2 and mol glucose? Use the balanced equation to obtain the conversion factor.

Students learn to think critically about information in the classroom and in everyday life.

NEW! Key Concept Videos combine artwork from the textbook with 2D and 3D animations to create a dynamic on-screen viewing experience and help students understand and apply important concepts throughout the text. Key Concept Videos are embedded in eText 2.0 and are assignable in MasteringChemistry™.

These short (3–5 minutes) videos combine animation and live-action clips of author Nivaldo Tro explaining the key concept of each chapter. Embedded questions in each video increase engagement and test student understanding. Follow-up questions are assignable in MasteringChemistry™.

UPDATED! Chapter-in-Review Exercises and Self-Assessment Quizzes have been revised using MasteringChemistry™ metadata to identify questions that students struggled with in previous editions. In addition to a full complement of end-of-chapter questions, each chapter features a 10–15 multiple-choice question quiz that help students assess their understanding of chapter content, building critical thinking skills and reinforcing key concepts.

Chapter 3 in Review

MasteringChemistry™ provides end-of-chapter exercises, feedback-enriched tutorial problems, animations, and interactive activities to encourage problem solving practice and deeper understanding of key concepts and topics.

Self-Assessment Quiz

Q1. Which substance is a pure compound?
(a) Gold
(b) Water
(c) Milk
(d) Fruit cake

Q2. Which property of trinitrotoluene (TNT) is most likely a chemical property?
(a) Yellow color
(b) Melting point is 80.1 °C
(c) Explosive
(d) None of the above

Q3. Which change is a chemical change?
(a) The condensation of dew on a cold night
(b) A forest fire
(c) The smoothening of rocks by ocean waves
(d) None of the above

Q4. Which process is endothermic?
(a) The burning of natural gas in a stove
(b) The metabolism of glucose by your body
(c) The melting of ice in a soft drink
(d) None of the above

Q5. A 35-g sample of potassium completely reacts with chlorine to form 67 g of potassium chloride. How many grams of chlorine must have reacted?
(a) 67 g
(b) 35 g
(c) 32 g
(d) 12 g

Q6. A runner burns 2.56×10^3 kJ during a five-mile run. How many nutritional Calories did the runner burn?
(a) 1.07×10^1 Cal
(b) 612 Cal
(c) 6.12×10^5 Cal
(d) 1.07×10^4 Cal

Q7. Convert the boiling point of water (100.00 °C) to K.
(a) −173.15 K
(b) 0 K
(c) 100.00 K
(d) 373.15 K

Q8. A European doctor reports that you have a fever of 39.2 °C. What is your fever in degrees Fahrenheit?
(a) 102.6 °F
(b) 128.26 °F
(c) 71.2 °F
(d) 4 °F

Q9. How much heat must be absorbed by 125 g of ethanol to change its temperature from 21.5 °C to 34.8 °C?
(a) 6.95 kJ
(b) 4.02×10^3 kJ
(c) 86.6 kJ
(d) 4.02 kJ

Q10. Substance A has a heat capacity that is much greater than that of substance B. If 10.0 g of substance A initially at 25.0 °C is brought into thermal contact with 10.0 g of B initially at 75.0 °C, what can you conclude about the final temperature of the two substances once the exchange of heat between the substances is complete?
(a) The final temperature will be between 25.0 °C and 50.0 °C.
(b) The final temperature will be between 50.0 °C and 75.0 °C.
(c) The final temperature will be 50.0 °C.
(d) You can conclude nothing about the final temperature without more information.

Answers: 1:b, 2:c, 3:b, 4:c, 5:c, 6:b, 7:d, 8:a, 9:d, 10:a

Multipart macroscopic and molecular images engage students in chemistry.

Multipart images allow students to see the relationship between the formulas they write down on paper (symbolic), the world they see around them (macroscopic), and the atoms and molecules that compose the world (molecular).

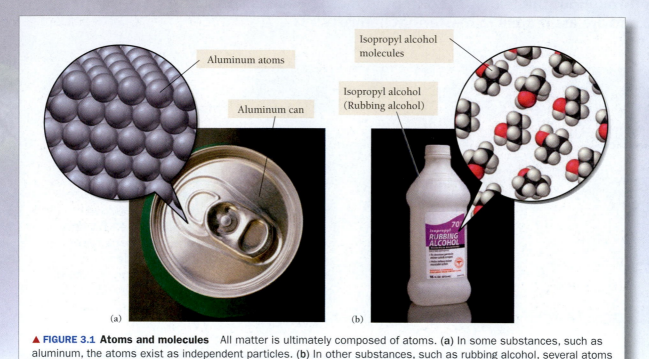

Aluminum atoms

Aluminum can

Isopropyl alcohol molecules

Isopropyl alcohol (Rubbing alcohol)

(a)

(b)

▲ **FIGURE 3.1 Atoms and molecules**   All matter is ultimately composed of atoms. **(a)** In some substances, such as aluminum, the atoms exist as independent particles. **(b)** In other substances, such as rubbing alcohol, several atoms bond together in well-defined structures called molecules.

Abundant molecular-level views show students the connection between everyday processes that are visible to the eye and the behavior of atoms and molecules.

▶ **FIGURE 3.4 Three states of matter**   Water exists as ice (solid), water (liquid), and steam (gas). In ice, the water molecules are closely spaced and, although they vibrate about a fixed point, they do not generally move relative to one another. In liquid water, the water molecules are also closely spaced but are free to move around and past each other. In steam, water molecules are separated by large distances and do not interact significantly with one another.

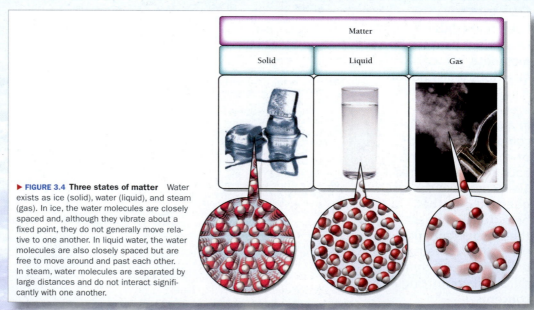

Matter

Solid

Liquid

Gas

A revised art program helps students make connections and see that chemistry is all around them.

Distillation

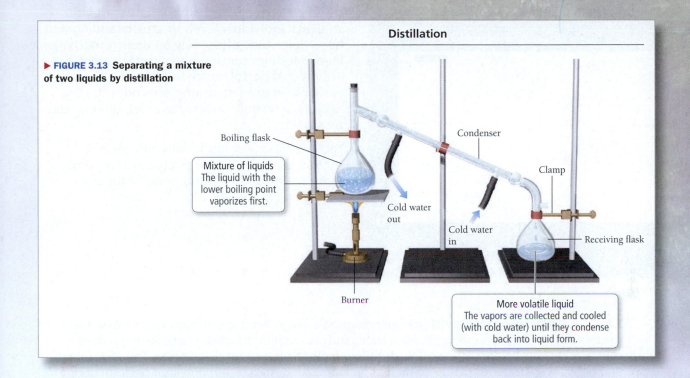

▶ FIGURE 3.13 **Separating a mixture of two liquids by distillation**

Boiling flask

Mixture of liquids
The liquid with the lower boiling point vaporizes first.

Condenser

Clamp

Cold water out

Cold water in

Receiving flask

Burner

More volatile liquid
The vapors are collected and cooled (with cold water) until they condense back into liquid form.

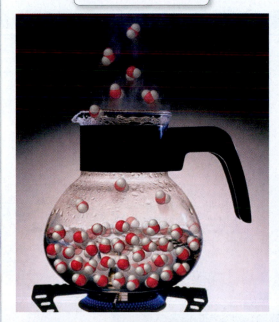

Water molecules are the same in water and steam.

▲ FIGURE 3.9 **A physical property** The boiling point of water is a physical property, and boiling is a physical change. When water boils, it turns into a gas, but the water molecules are the same in both the liquid water and the gaseous steam.

NEW and UPDATED! Illustrations include extensive labels and annotations to direct student attention to key elements in the art and promote understanding of the processes depicted. Numerous figures in the sixth edition have updated labels and annotations to focus readers on key concepts. Relevant information is placed where it is most needed and makes the art a vital study and review tool.

Dynamic Study Modules and the Chemistry Primer help students come to class prepared.

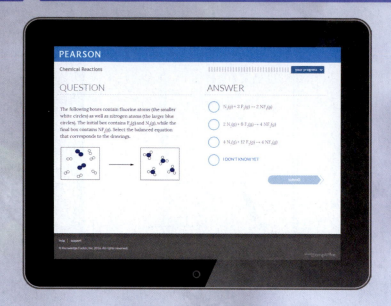

66 Dynamic Study Modules adapt to students' individual levels of understanding and help them study effectively on their own. Dynamic Study Modules continuously assess student activity and performance in real time. These are available as graded assignments prior to class and are accessible on smartphones, tablets, and computers.

Topics include key math skills and general chemistry concepts such as phases of matter, redox reactions, acids and bases, solutions, and chemical equilibrium.

The Chemistry Primer's pre-built diagnostic assignments get students up-to-speed at the beginning of the course, addressing topics such as math in the context of chemistry, basic chemical literacy, balancing chemical equations, mole theory, and stoichiometry. The Chemistry Primer scales to students' needs – remediation is only suggested to students that perform poorly on initial assessment, and involves Tutorials, Wrong-Answer Specific Feedback, Video Instruction, and Step-Wise Scaffolding to build student understanding.

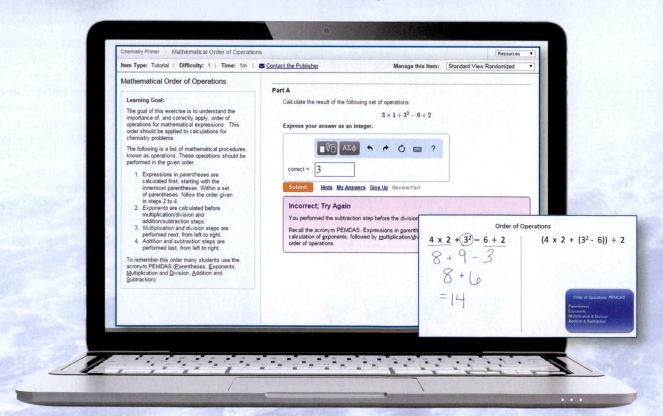

MasteringChemistry™ ensures student engagement before, during, and after class.

With questions specific to Tro's Introductory Chemistry, **Learning Catalytics™** generates class discussion, guides lecture, and promotes peer-to-peer learning with real-time analytics. Instructors can:

- **NEW!** Upload a full PowerPoint® deck for easy creation of slide questions
- Help students develop critical thinking skills
- Monitor responses to find out where students are struggling
- Adjust teaching strategy with real-time data
- Automatically group students for discussion, team-work, and peer-to-peer learning

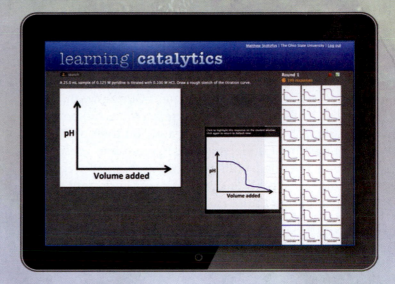

UPDATED! MasteringChemistry™ offers a wide variety of problems, ranging from multi-step tutorials with extensive hints and feedback to multiple-choice End-of-Chapter Problems and Test Bank questions.

To provide additional scaffolding for students moving from Tutorial Problems to End-of-Chapter Problems we created **NEW!** **Enhanced End-of-Chapter** Problems that now contain specific wrong-answer feedback.

Students can study anywhere with fully interactive and mobile eText 2.0 features.

NEWLY INTERACTIVE! Self-Assessment Quizzes and Conceptual Checkpoints allow students to interact with all Conceptual Checkpoints and Self-Assessment Quizzes within eText 2.0! With one click these activities are brought to life, allowing students to study on their own and test their understanding in real-time. These interactives help students extinguish misconceptions and deepen their understanding of important concepts and topics.

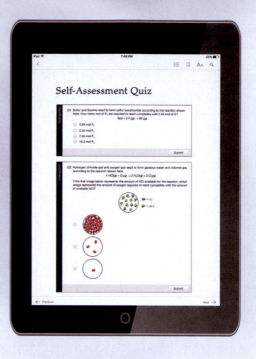

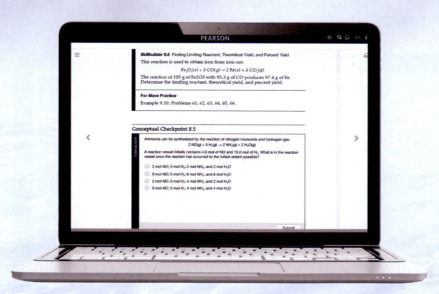

An icon in the text alerts students to interactive eText 2.0 features. The eText is fully optimized for use on mobile devices, allowing students to study anywhere.

In addition to **Conceptual Checkpoints** and **Self-Assessment Quizzes**, all **Key Concept Videos** and **Interactive Worked Examples** are available in the eText.

PEARSON
eText
2.0

INTRODUCTORY
CHEMISTRY
ESSENTIALS

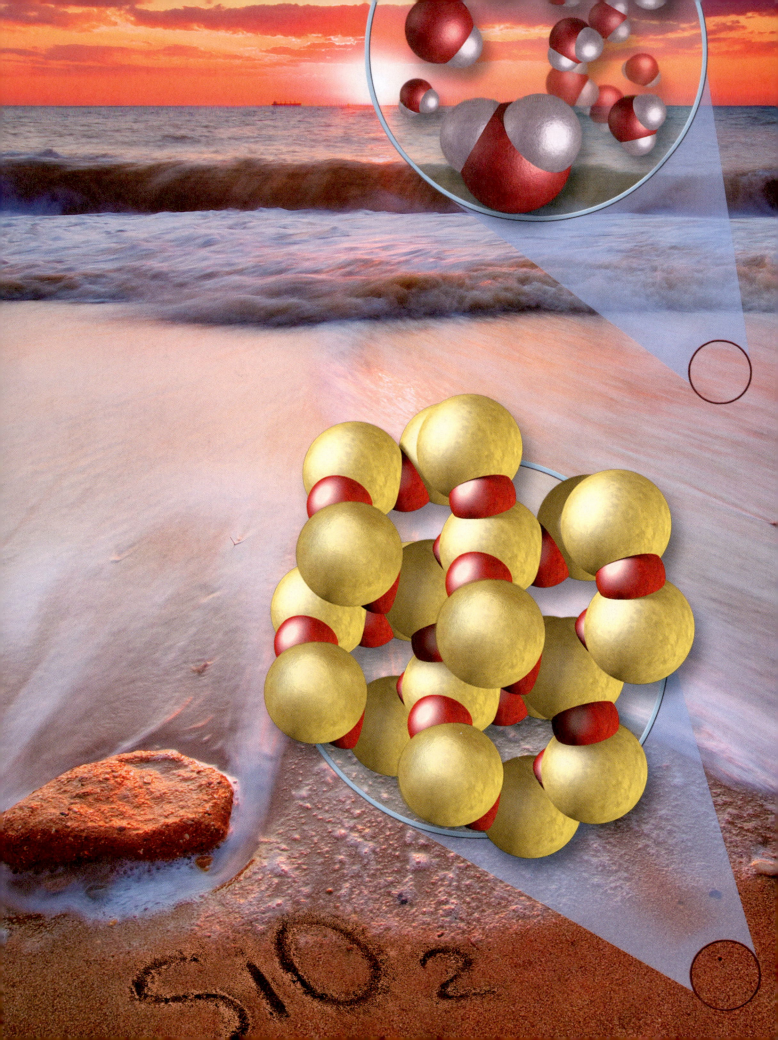

1 The Chemical World

"Imagination is more important than knowledge."
—Albert Einstein (1879–1955)

1.1 Sand and Water

▲ Richard Feynman (1918–1988), Nobel Prize–winning physicist and popular professor at California Institute of Technology.

I love the beach but hate sand. Sand gets everywhere and even comes home with you. Sand is annoying because sand particles are so small. They stick to your hands, to your feet, and to any food you might be trying to eat for lunch. But the smallness of sand particles pales in comparison to the smallness of the particles that compose them. Sand—like all other kinds of ordinary matter—is composed of atoms. Atoms are unimaginably small. A single sand grain contains more atoms than there are sand grains on the largest of beaches.

The idea that matter is composed of tiny particles is among the greatest discoveries of humankind. Nobel laureate Richard Feynman (1918–1988), in a lecture to first-year physics students at the California Institute of Technology, said that the most important idea in all human knowledge is that *all things are made of atoms*. Why is this idea so important? Because it establishes how we should go about understanding the properties of the things around us. If we want to understand how matter behaves, we must understand how the particles that compose that matter behave.

Atoms, and the molecules they compose, determine how matter behaves—if they were different, matter would be different. The nature of water molecules, for example, determines how water behaves. If water molecules were different—even in a small way—then water would be a different sort of substance. For example, we know that a water molecule is composed of two hydrogen atoms bonded to an oxygen atom with a shape that looks like this:

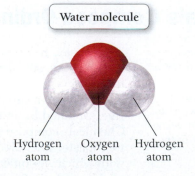

Water molecule

Hydrogen atom   Oxygen atom   Hydrogen atom

◀ A single grain of sand on a large beach contains more atoms than there are grains of sand on the entire beach.

3

How would water be different if the shape of the water molecule was different? What if the hydrogen atoms bonded to oxygen to form a linear molecule instead of a bent one?

Hypothetical linear water molecule

The answer to this question is not altogether simple. We don't know exactly how our hypothetical linear water would behave, but we do know it would be much different than ordinary water. For example, linear water would probably have a much lower boiling point than ordinary water. In fact, it may even be a gas (instead of a liquid) at room temperature. Imagine what our world would be like if water was a gas at room temperature. There would be no rivers, no lakes, no oceans, and probably no people (since liquid water is such an important part of what composes us).

There is a direct connection between the world of atoms and molecules and the world we experience every day (▼ FIGURE 1.1). Chemists explore this connection. They seek to understand it. A good, simple definition of **chemistry** is *the science that tries to understand how matter behaves by studying how atoms and molecules behave.*

▲ FIGURE 1.1 **Virtually everything around you is composed of chemicals.**

1.2 Chemicals Compose Ordinary Things

▶ Recognize that chemicals make up virtually everything we come into contact with in our world. *(Note: Most of the sections in the chapters in this book link to a Learning Objective (LO), which is listed at the beginning of the section.)*

We just saw how chemists are interested in substances such as sand and water. But are these substances chemicals? Yes. In fact, everything that we can hold or touch is made of chemicals. When most people think of chemicals, however, they may envision a can of paint thinner in their garage, or recall a headline about a river polluted by industrial waste. But chemicals compose ordinary things, too. Chemicals compose the air we breathe and the water we drink. They compose toothpaste, Tylenol®, and toilet paper. Chemicals make up virtually everything with which we come into contact. Chemistry explains the properties and behavior of chemicals, in the broadest sense, by helping us understand the molecules that compose them.

▲ Chemists are interested in knowing why ordinary things, such as water, are the way they are. When a chemist sees a pitcher of water, she thinks of the molecules that compose the liquid and how those molecules determine its properties.

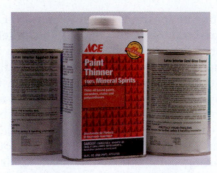 

▲ People often have a very narrow view of chemicals, thinking of them only as dangerous poisons or pollutants.

As you experience the world around you, molecules are interacting to create your reality. Imagine watching a sunset. Molecules are involved in every step. Molecules in the air interact with light from the sun, scattering away the blue and green light and leaving the red and orange light to create the color you see. Molecules in your eyes absorb that light and, as a result, are altered in a way that sends a signal to your brain. Molecules in your brain then interpret the signal to produce images and emotions. This whole process—mediated by molecules—creates the evocative experience of seeing a sunset.

Chemists are interested in why ordinary substances are the way they are. Why is water a liquid? Why is salt a solid? Why does soda fizz? Why is a sunset red? Throughout this book, you will learn the answers to these questions and many others. *You will learn the connections between the behavior of matter and the structure of the particles that compose it.*

1.3   The Scientific Method: How Chemists Think

▶ Identify and understand the key characteristics of the scientific method: observation, the formulation of hypotheses, the testing of hypotheses by experiment, and the formulation of laws and theories.

Chemists use the **scientific method**—a way of learning that emphasizes observation and experimentation—to understand the world. The scientific method stands in contrast to ancient Greek philosophies that emphasized *reason* as the way to understand the world. Although the scientific method is not a rigid procedure that automatically leads to a definitive answer, it does have key characteristics that distinguish it from other ways of acquiring knowledge. These key characteristics include observation, the formulation of hypotheses, the testing of hypotheses by experiment, and the formulation of laws and theories.

The first step in acquiring scientific knowledge (▼ FIGURE 1.2) is often the **observation** or measurement of some aspect of nature. Some observations are simple, requiring nothing more than the naked eye. Other observations rely

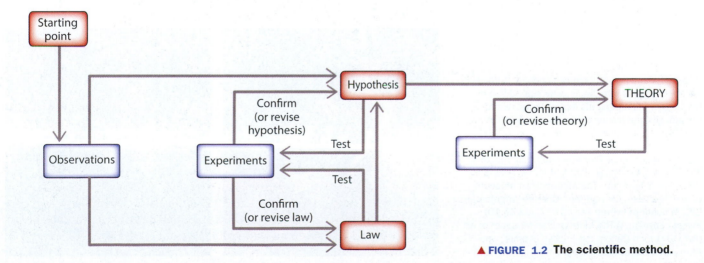

▲ FIGURE 1.2 The scientific method.

on the use of sensitive instrumentation. Occasionally, an important observation happens entirely by chance. Alexander Fleming (1881–1955), for example, discovered penicillin when he observed a bacteria-free circle around a certain mold that had accidentally grown on a culture plate. Regardless of how these observations occur, they usually involve the measurement or description of some aspect of the physical world. For example, Antoine Lavoisier (1743–1794), a French chemist who studied *combustion*, burned substances in closed containers. He carefully measured the mass of each container and its contents before and after burning the substance inside, noting that there was no change in the mass during combustion. Lavoisier made an *observation* about the physical world.

> Combustion means burning. The mass of an object is a measure of the quantity of matter within it.

Observations often lead scientists to formulate a **hypothesis**, a tentative interpretation or explanation of the observations. Lavoisier explained his observations on combustion by hypothesizing that combustion involved the combination of a substance with a component of air. A good hypothesis is *falsifiable*, which means that further testing has the potential to prove it wrong. Hypotheses are tested by **experiments**, highly controlled observations designed to validate or invalidate hypotheses. The results of an experiment may confirm a hypothesis or show the hypothesis to be mistaken in some way. In the latter case, the hypothesis may have to be modified, or even discarded and replaced by an alternative hypothesis. Either way, the new or revised hypothesis must also be tested through further experimentation.

Sometimes a number of similar observations lead to the development of a **scientific law**, a brief statement that summarizes past observations and predicts future ones. For example, based on his observations of combustion, Lavoisier developed the **law of conservation of mass**, which states, "In a chemical reaction matter is neither created nor destroyed." This statement grew out of Lavoisier's observations, and it predicted the outcome of similar experiments on *any* chemical reaction. Laws are also subject to experiments, which can prove them wrong or validate them.

> Scientific theories are also called *models*.

One or more well-established hypotheses may form the basis for a scientific **theory**. Theories provide a broader and deeper explanation for observations and laws. They are models of the way nature is, and they often predict behavior that extends well beyond the observations and laws on which they are founded. A good example of a theory is the **atomic theory** of John Dalton (1766–1844). Dalton explained the law of conservation of mass, as well as other laws and observations, by proposing that all matter was composed of small, indestructible particles called atoms. Dalton's theory was a model of the physical world—it went beyond the laws and observations of the time to explain these laws and observations.

▶ (Right) Painting of the French chemist Antoine Lavoisier and his wife, Marie, who helped him in his work by illustrating his experiments, recording results, and translating scientific articles from English. [*Source:* Jacques Louis David (French, 1748–1825). "Antoine-Laurent Lavoisier (1743–1794) and His Wife (Marie-Anne-Pierrette Paulze, 1758–1836)," 1788, oil on canvas, H. 102-1/4 in. W. 76-5/8 in. (259.7 × 194.6 cm). The Metropolitan Museum of Art, Purchase, Mr. and Mrs. Charles Wrightsman Gift, in honor of Everett Fahy, 1977. (1977.10) Image copyright © The Metropolitan Museum of Art.] (Far right) John Dalton, the English chemist who formulated the atomic theory.

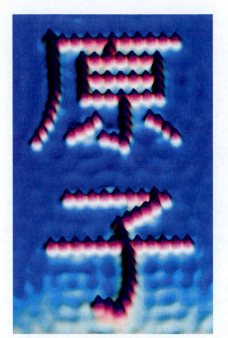

▲ **FIGURE 1.3 Are atoms real?** The atomic theory has 200 years of experimental evidence to support it, including recent images, such as this one, of atoms themselves. This image shows the Kanji (a system of Japanese writing using Chinese characters) for "atom" written with individual iron atoms on top of a copper surface.

Theories are also tested and validated by experiments. Notice that the scientific method begins with observation, and then laws, hypotheses, and theories are developed based on those observations. Experiments, which are carefully controlled observations, determine the validity of laws, hypotheses, or theories. If a law, hypothesis, or theory is inconsistent with the findings of an experiment, it must be revised and new experiments must be conducted to test the revisions. Over time, scientists eliminate poor theories, and good theories—those consistent with experiments—remain. Established theories with strong experimental support are the most powerful pieces of scientific knowledge. People unfamiliar with science sometimes say, "That is just a theory," as if theories were mere speculations. However, well-tested theories are as close to truth as we get in science. For example, the idea that all matter is made of atoms is "just a theory," but it is a theory with 200 years of experimental evidence to support it, including the recent imaging of atoms themselves (◄ FIGURE 1.3). Established theories should not be taken lightly—they are the pinnacle of scientific understanding.

The eText 2.0 icon indicates that this feature is embedded and interactive in the eText.

PEARSON eText 2.0

CONCEPTUAL ✔ CHECKPOINT 1.1

Which statement most resembles a scientific theory?

(a) When the pressure on a sample of oxygen gas is increased 10%, the volume of the gas decreases by 10%.

(b) The volume of a gas is inversely proportional to its pressure.

(c) A gas is composed of small particles in constant motion.

(d) A gas sample has a mass of 15.8 g and a volume of 10.5 L.

Note: The answers to all Conceptual Checkpoints appear at the end of the chapter.

EVERYDAY CHEMISTRY

Combustion and the Scientific Method

Early chemical theories attempted to explain common phenomena such as combustion. Why did things burn? What was happening to a substance when it burned? Could something that was burned be unburned? Early chemists burned different substances and made observations to try to answer these questions. They observed that substances stop burning when placed in a closed container. They found that many metals burn to form a white powder that they called a *calx* (now we know that these white powders are oxides of the metal) and that the metal could be recovered from the calx, or unburned, by combining the calx with charcoal and heating it.

Chemists in the first part of the eighteenth century formed a theory about combustion to explain these observations. In this theory, combustion involved a fundamental substance that they called *phlogiston*. This substance was present in anything that burned and was released during combustion. Flammable objects were flammable because they contained phlogiston. When things burned in a closed container, they didn't burn for very long because the space within the container became saturated with phlogiston. When things

burned in the open, they continued to burn until all of the phlogiston within them was gone. This theory also explained how metals that had burned could be unburned. Charcoal was a phlogiston-rich material—they knew this because it burned so well—and when it was combined with a calx, which was a metal that had been emptied of its phlogiston, it transferred some of its phlogiston into the calx, converting the calx back into the unburned form of the metal. The phlogiston theory was consistent with all of the observations of the time and was widely accepted as valid.

Like any theory, the phlogiston theory was tested continually by experiment. One set of experiments, conducted in the mid-eighteenth century by Louis-Bernard Guyton de Morveau (1737–1816), consisted of weighing metals before and after burning them. In every case the metals *gained* weight when they were burned. This observation is inconsistent with the phlogiston theory, which predicted that metals should *lose* weight because phlogiston was supposed to be lost during combustion. Clearly, the phlogiston theory needed modification.

continued on page 8

continued from page 7

The first modification was that phlogiston was a very light substance, so that it actually "buoyed up" the materials that contained it. Thus when phlogiston was released, the material became heavier. Such a modification seemed to fit the observations but also seemed far-fetched. Antoine Lavoisier developed a more likely explanation by devising a completely new theory of combustion. Lavoisier proposed that, when a substance burned, it actually took something *out* of the air, and when it unburned, it released something back into the air. Lavoisier said that burning objects *fixed* (attached or bonded) the air and that the *fixed* air was released during unburning. In a confirming experiment (▶ FIGURE 1.4), Lavoisier roasted a mixture of calx and charcoal with the aid of sunlight focused by a giant burning lens, and found that a huge volume of "fixed air" was released in the process. The scientific method worked. The phlogiston theory was proven wrong, and a new theory of combustion took its place—a theory that, with a few refinements, is still valid today.

B1.1 CAN YOU ANSWER THIS? *What is the difference between a law and a theory? How does the example of the phlogiston theory demonstrate this difference?*

▲ **FIGURE 1.4 Focusing on combustion** The great burning lens belonging to the Academy of Sciences. Lavoisier used a similar lens in 1777 to show that a mixture of *calx* (metal oxide) and charcoal released a large volume of *fixed air* when heated.

1.4 Analyzing and Interpreting Data

▶ Identify patterns in data and interpret graphs.

We just learned how early scientists such as Lavoisier and Dalton saw patterns in a series of related measurements. Sets of measurements constitute scientific *data*, and learning to analyze and interpret data is an important scientific skill.

Identifying Patterns in Data

Suppose you are an early chemist trying to understand the composition of water. You know that water is composed of the elements hydrogen and oxygen. You do several experiments in which you decompose different samples of water into hydrogen and oxygen, and you get the following results:

Sample	Mass of Water Sample	Mass of Hydrogen Formed	Mass of Oxygen Formed
A	20.0 g	2.2 g	17.8 g
B	50.0 g	5.6 g	44.4 g
C	100.0 g	11.1 g	88.9 g

Do you notice any patterns in this data? The first and easiest pattern to see is that the sum of the masses of oxygen and hydrogen always sums to the mass of the water sample. For example, for the first water sample, 2.2 g hydrogen + 17.8 g oxygen = 20.0 g water. The same is true for the other samples. Another pattern, which is a bit more difficult to see, is that the ratio of the masses of oxygen and hydrogen is the same for each sample.

Sample	Mass of Hydrogen Formed	Mass of Oxygen Formed	Mass Oxygen / Mass Hydrogen
A	2.2 g	17.8 g	8.1
B	5.6 g	44.4 g	7.9
C	11.1 g	88.9 g	8.01

The ratio is 8—the small variations are due to experimental error, which is common in all measurements and observations.

Atmospheric Carbon Dioxide

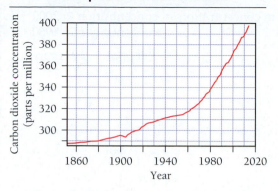

▲ FIGURE 1.5 Atmospheric carbon dioxide levels from 1860 to present.

Seeing patterns in data is a creative process that requires you to not just merely tabulate laboratory measurements, but to see relationships that may not always be obvious. The best scientists see patterns that others have missed. As you learn to interpret data in this course, be creative and try looking at data in new ways.

Interpreting Graphs

Data is often visualized using graphs or images, and scientists must constantly analyze and interpret graphs. For example, the graph in ◄ FIGURE 1.5 shows the concentration of carbon dioxide in Earth's atmosphere as a function of time. Carbon dioxide is a greenhouse gas that has been rising as result of the burning of fossil fuels (such as gasoline and coal). When you look at a graph such as this one, you should first examine the x and y axes and make sure you understand what each axis represents. You should also examine the numerical range of the axes. In Figure 1.5, the y axis does not begin at zero in order to better display the change that is occurring. How would this graph look different if the y axis began at zero instead of at 290? Notice also that, in this graph, the increase in carbon dioxide has not been constant over time. The rate of increase—represented by the slope of the line—has intensified since about 1960.

EXAMPLE **1.1**   **Interpreting Graphs**

Examine the graph in Figure 1.5 and answer each question.

(a) What was the concentration of carbon dioxide in 1960?
(b) What was the concentration in 2000?
(c) How much did the concentration increase between 1960 and 2000?
(d) What is the average rate of increase over this time?
(e) If the average rate of increase remains constant, estimate the carbon dioxide concentration in 2030.

SOLUTION

a) What was the concentration of carbon dioxide in 1960?

To determine the concentration of carbon dioxide in 1960, draw a vertical line at the year 1960. At the point where the vertical line intersects the carbon dioxide concentration curve, draw a horizontal line. The point where the horizontal line intercepts the y axis represents the concentration in 1960. So, the concentration in 1960 was 318 ppm.

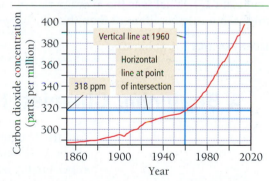

b) What was the concentration in 2000?

Apply the same procedure as in part a, but now shift the vertical line to the year 2000. The concentration in the year 2000 was 370 ppm.

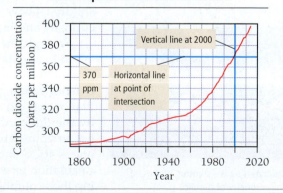

continued on page 10 ▶

continued from page 9

c) How much did the concentration increase between 1960 and 2000? The increase in the carbon dioxide concentration is the difference between the two concentrations. When calculating changes in quantities such as this, take the final quantity minus the initial quantity.	increase in concentration = concentration in 2000 − concentration in 1960 $= 390 \text{ ppm} - 318 \text{ ppm}$ $= 72 \text{ ppm}$
d) What is the average rate of increase over this time? The average rate of increase over this time is the change in the concentration divided by the number of years that passed. Determine the number of years that have passed by subtracting the initial year from the final year. Determine the average rate of increase by dividing the change in concentration (from part c) by the number of years that you just calculated.	number of years = final year − initial year $= 2000 - 1960$ $= 40 \text{ years}$ $\text{average rate} = \dfrac{\text{change in concentration}}{\text{number of years}}$ $= \dfrac{72 \text{ ppm}}{40 \text{ years}}$ $= \dfrac{1.8 \text{ ppm}}{\text{year}}$
e) If the average rate of increase remains constant, estimate the carbon dioxide concentration in 2030. Determine the increase in concentration between 2000 and 2030 by multiplying the number of years that pass in that time interval by the average rate of change (from part d). Lastly, determine the concentration in 2030 by adding the increase between 2000 and 2030 to the concentration in 2000.	$\text{increase} = 30 \text{ years} \times \dfrac{1.8 \text{ ppm}}{\text{year}}$ $= 54 \text{ ppm}$ concentration in 2030 = 370 ppm + 54 ppm $= 424 \text{ ppm}$

▶ **SKILLBUILDER 1.1** | What was the average rate of increase in carbon dioxide concentration between 1880 and 1920? Why might that rate be different than the rate between 1960 and 2000?

▶ **FOR MORE PRACTICE**   Problem 25.

1.5 A Beginning Chemist: How to Succeed

You are a beginning chemist. This may be your first chemistry course, but it may not be your last. To succeed as a beginning chemist, keep the following ideas in mind. First, chemistry requires curiosity and imagination. If you are content knowing that the sky is blue but don't care *why* it is blue, then you may have to rediscover your curiosity. I say "rediscover" because even children—or better, *especially* children—have this kind of curiosity. To succeed as a chemist, you must have the curiosity and imagination of a child—*you must want to know the why of things.*

Second, chemistry requires calculation. Throughout this course, you will be asked to calculate answers and quantify information. *Quantification* involves measurement as part of observation—it is one of the most important tools in science. Quantification allows you to go beyond merely saying that this object is hot and that one is cold or that this one is large and that one is small. It allows you to specify the difference precisely. For example, two samples of water may feel equally hot to your hand, but when you measure their temperatures, you may find that one is 40 °C and the other is 44 °C. Even small differences can be important in a calculation or experiment, so assigning numbers to observations and manipulating those numbers become very important in chemistry.

▲ To succeed as a scientist, you must have the curiosity of a child.

Lastly, chemistry requires commitment. To succeed in this course, you must commit to learning chemistry. Roald Hoffmann (1937–), winner of the 1981 Nobel Prize for chemistry, said,

> *I like the idea that human beings can do anything they want to. They need to be trained sometimes. They need a teacher to awaken the intelligence within them. But to be a chemist requires no special talent, I'm glad to say. Anyone can do it, with hard work.*

Professor Hoffmann is right. The key to success in this course is hard work—that requires commitment. You must do your work regularly and carefully. If you do, you will succeed, and you will be rewarded by seeing a whole new world—the world of molecules and atoms. This world exists beneath the surface of nearly everything you encounter. I welcome you to this world and consider it a privilege, together with your professor, to be your guide.

Chapter **1** in Review

MasteringChemistry™ provides end-of-chapter exercises, feedback-enriched tutorial problems, animations, and interactive activities to encourage problem solving practice and deeper understanding of key concepts and topics.

Self-Assessment Quiz

The eText 2.0 icon indicates that this feature is embedded and interactive in the eText.

Q1. Where can you find chemicals?
(a) In a hardware store
(b) In a chemical stockroom
(c) All around you and even inside of you
(d) All of the above

Q2. Which statement best defines chemistry?
(a) The science that studies solvents, drugs, and insecticides
(b) The science that studies the connections between the properties of matter and the particles that compose that matter
(c) The science that studies air and water pollution
(d) The science that seeks to understand processes that occur only in chemical laboratories

Q3. According to the scientific method, what is a law?
(a) A short statement that summarizes a large number of observations
(b) A fact that can never be refuted
(c) A model that gives insight into how nature is
(d) An initial guess with explanatory power

Q4. Which statement is an example of an observation?
(a) In a chemical reaction, matter is conserved.
(b) All matter is made of atoms.
(c) When a given sample of gasoline is burned in a closed container, the mass of the container and its contents does not change.
(d) Atoms bond to one another by sharing electrons.

Q5. The graph below shows the area of a circle as a function of its radius. What is the radius of a circle that has an area of 155 square inches?

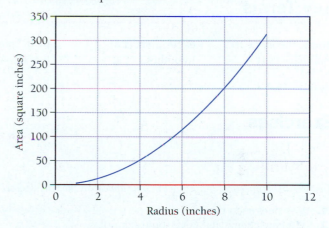

(a) 7.0 inches   (b) 6.5 inches
(c) 6.8 inches   (d) 6.2 inches

Q6. Which characteristic is necessary for success in understanding chemistry?
(a) Curiosity   (b) Calculation
(c) Commitment   (d) All of the above

Answers: 1: d, 2: b, 3: a, 4: c, 5: a, 6: d

Chemical Principles

Matter and Molecules

Chemists are interested in all matter, even ordinary matter such as water or air. You don't need to go to a chemical storeroom to find chemicals because they are all around you. Chemistry is the science that studies the connections between the properties of matter and the particles that compose that matter.

Relevance

Chemists want to understand matter for several reasons. First, chemists are simply curious—they want to know why. Why are some substances reactive and others not? Why are some substances gases, some liquids, and others solids? Chemists are also practical; they want to understand matter so that they can control it and produce substances that are useful to society and to humankind.

The Scientific Method

Chemists employ the scientific method, which makes use of observations, hypotheses, laws, theories, and experiments. Observations involve measuring or observing some aspect of nature. Hypotheses are tentative interpretations of observations. Laws summarize the results of a large number of observations, and theories are models that explain and give the underlying causes for observations and laws. Hypotheses, laws, and theories must be tested and validated by experiment. If they are not confirmed, they are revised and tested through further experimentation.

The scientific method is a way to understand the world. Since the inception of the scientific method, knowledge about the natural world has grown rapidly. The application of the scientific method has produced technologies that have raised living standards throughout the world with advances such as increased food production, rapid transportation, unparalleled access to information, and longer life spans.

Analyzing and Interpreting Data

A series of measurements are often referred to as data. Scientific data can be graphed to better see relationships between variables.

Virtually all scientists have to analyze and interpret the data they collect. This skill is an important part of understanding chemistry.

Success as a Beginning Chemist

To succeed as a beginning chemist, you must be curious and imaginative, be willing to do calculations, and be committed to learning the material.

To succeed as a beginning chemist, you must be curious and imaginative, be willing to do calculations, and be committed to learning the material.

Key Terms

atomic theory [1.3]
chemistry [1.1]
experiment [1.3]
hypothesis [1.3]
law of conservation
of mass [1.3]
observation [1.3]
scientific law [1.3]
scientific method [1.3]
theory [1.3]

Exercises

Questions

Answers to all questions numbered in blue appear in the Answers section at the back of the book.

1. Why does soda fizz?
2. What are chemicals? Give some examples.
3. What do chemists try to do? How do they understand the natural world?
4. What is meant by the statement, "Matter does what molecules do"? Give an example.
5. Define *chemistry*.
6. How is chemistry connected to everyday life? How is chemistry relevant outside the chemistry laboratory?
7. Explain the scientific method.
8. Cite an example from this chapter of the scientific method at work.
9. What is the difference between a law and a theory?
10. What is the difference between a hypothesis and a theory?
11. What is wrong with the statement, "It is just a theory"?
12. What is the law of conservation of mass, and who discovered it?
13. What is the atomic theory, and who formulated it?
14. What are three things you need to do to succeed in this course?

Problems

Note: The exercises in the Problems section are paired, and the answers to the odd-numbered exercises (numbered in blue) appear in the Answers section at the back of the book.

15. Classify each statement as an observation, a law, or a theory.
    (a) When a metal is burned in a closed container, the sum of the masses of the container and its contents does not change.
    (b) Matter is made of atoms.
    (c) Matter is conserved in chemical reactions.
    (d) When wood is burned in a closed container, its mass does not change.

16. Classify each statement as an observation, a law, or a theory.
    (a) The star closest to Earth is moving away from Earth at high speed.
    (b) A body in motion stays in motion unless acted upon by a force.
    (c) The universe began as a cosmic explosion called the Big Bang.
    (d) A stone dropped from an altitude of 450 m falls to the ground in 9.6 s.

17. A student prepares several samples of the same gas and measures their mass and volume. The results are tabulated as follows. Formulate a tentative law from the measurements.

Mass of Gas (in g)	Volume of Gas (in L)
22.5	1.60
35.8	2.55
70.2	5.00
98.5	7.01

18. A student measures the volume of a gas sample at several different temperatures. The results are tabulated as follows. Formulate a tentative law from the measurements.

Temperature of Gas (in K)	Volume of Gas (in L)
298	4.55
315	4.81
325	4.96
335	5.11

19. A chemist in an imaginary universe does an experiment that attempts to correlate the size of an atom with its chemical reactivity. The results are tabulated as follows.

Size of Atom	Chemical Reactivity
small	low
medium	intermediate
large	high

(a) Formulate a law from this data.
(b) Formulate a theory to explain the law.

20. A chemist decomposes several samples of water into hydrogen and oxygen and measures the mass of the hydrogen and the oxygen obtained. The results are tabulated as follows.

Sample Number	Grams of Hydrogen	Grams of Oxygen
1	1.5	12
2	2	16
3	2.5	20

(a) Summarize these observations in a short statement. Next, the chemist decomposes several samples of carbon dioxide into carbon and oxygen. The results are tabulated as follows:

Sample Number	Grams of Carbon	Grams of Oxygen
1	0.5	1.3
2	1.0	2.7
3	1.5	4.0

(b) Summarize these observations in a short statement.
(c) Formulate a law from the observations in (a) and (b).
(d) Formulate a theory that might explain your law in (c).

Questions for Group Work

Discuss these questions with the group and record your consensus answer.

21. The manufacturer of a particular brand of toothpaste claims that the brand contains "no chemicals." Using a few grammatically correct English sentences, describe what you think the company means by that statement. Would a scientist consider the manufacturer's statement to be correct? Why or why not?
22. Make a list (including up to ten items) of all the atoms or molecules group members can name off the top of their heads. Get at least one contribution from each group member.

23. In your own words, provide a brief definition for each of the following: observation, law, hypothesis, and theory.
24. How curious are you? How good are your quantitative skills? How hard are you willing to work to succeed in chemistry? Answer these questions individually on a scale of 1 (= not at all) to 5 (= very), then share your answers with your group. Report the group average for each question.

Data Interpretation and Analysis

25. The graph displays world population over time. Study the graph and answer the following questions.

(a) What was the world population in 1950?
(b) What was the world population in 2010?
(c) How much did the population increase between 1950 and 2010?
(d) What is the average rate of increase over this time?
(e) If the average rate of increase remains constant, estimate the world population in 2035.

World Population Versus Time

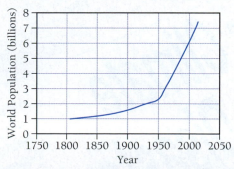

Source: http://www.worldometers.info/world-population/

Answers to Skillbuilder Exercises

Answers to Conceptual Checkpoints

1.1 (c) Answers a and d are observations. Answer b is a scientific law. Answer c is the only answer that proposes a *model* for what a gas is like.

2 Measurement and Problem Solving

"The important thing in science is not so much to obtain new facts as to discover new ways of thinking about them."

—Sir William Lawrence Bragg (1890–1971)

2.1 The Metric Mix-up: A $125 Million Unit Error

On December 11, 1998, NASA launched the Mars Climate Orbiter, which was to become the first weather satellite for a planet other than Earth. The Orbiter's mission was to monitor the atmosphere on Mars and to serve as a communications relay for the Mars Polar Lander, a probe that was to follow the Orbiter and land on the planet's surface three weeks later. Unfortunately, the mission ended in disaster. A unit mix-up caused the Orbiter to enter the Martian atmosphere at an altitude that was too low. Instead of settling into a stable orbit, the Orbiter likely disintegrated. The cost of the failed mission was estimated at $125 million.

Later investigations showed that the Orbiter had come within 57 km of the planet surface, which was too close. When a spacecraft enters a planet's atmosphere too close to the planet's surface, friction with the atmosphere can cause the spacecraft to burn up. The on-board computers that controlled the trajectory corrections were programmed in metric units (newton · second), but the ground engineers entered the trajectory corrections in English units (pound · second). The English and the metric units are not equivalent (1 pound · second = 4.45 newton · second). The corrections that the ground engineers entered were 4.45 times too small and did not alter the trajectory enough to keep the Orbiter at a sufficiently high altitude. In chemistry as in space exploration, *units* (see Section 2.5) are critical. If we get them wrong, the consequences can be disastrous.

A unit is a standard, agreed on quantity by which other quantities are measured.

2.2 Scientific Notation: Writing Large and Small Numbers

► Express very large and very small numbers using scientific notation.

Science constantly pushes the boundaries of the very large and the very small. We can, for example, now measure time periods as short as 0.000000000000001 second and distances as great as 14,000,000,000 light-years. Because the many zeros

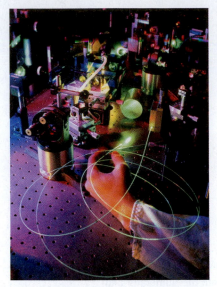

▲ Lasers such as this one can measure time periods as short as 1×10^{-15} s.

in these numbers are cumbersome to write, we use **scientific notation** to write them more compactly. In scientific notation, 0.000000000000001 is 1×10^{-15}, and 14,000,000,000 is 1.4×10^{10}. A number written in scientific notation consists of a **decimal part**, a number that is usually between 1 and 10, and an **exponential part**, 10 raised to an **exponent**, n.

$$\underset{\substack{\text{decimal} \\ \text{part}}}{1.2} \times \underset{\substack{\text{exponential} \\ \text{part}}}{10}^{-10} \leftarrow \text{exponent } (n)$$

A positive exponent (n) means 1 multiplied by 10 n times.

$$10^0 = 1$$
$$10^1 = 1 \times 10 = 10$$
$$10^2 = 1 \times 10 \times 10 = 100$$
$$10^3 = 1 \times 10 \times 10 \times 10 = 1000$$

A negative exponent ($-n$) means 1 divided by 10 n times.

$$10^{-1} = \frac{1}{10} = 0.1$$

$$10^{-2} = \frac{1}{10 \times 10} = 0.01$$

$$10^{-3} = \frac{1}{10 \times 10 \times 10} = 0.001$$

To convert a number to scientific notation, we move the decimal point (either to the left or to the right, as needed) to obtain a number between 1 and 10 and then multiply that number (the decimal part) by 10 raised to the power that reflects the movement of the decimal point. For example, to write 5983 in scientific notation, we move the decimal point to the left three places to get 5.983 (a number between 1 and 10) and then multiply the decimal part by 1000 to compensate for moving the decimal point.

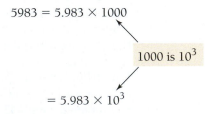

$$5983 = 5.983 \times 1000$$

$$1000 \text{ is } 10^3$$

$$= 5.983 \times 10^3$$

We can do this in one step by counting how many places we move the decimal point to obtain a number between 1 and 10 and then by writing the decimal part multiplied by 10 raised to the number of places we moved the decimal point.

$$5983 = 5.983 \times 10^3$$
$$\underset{3\,2\,1}{}$$

The Mathematics Review Appendix (p. MR-2) includes a review of mathematical operations for numbers written in scientific notation.

If the decimal point is moved to the left, as in the previous example, the exponent is positive. If the decimal is moved to the right, the exponent is negative.

$$0.00034 = 3.4 \times 10^{-4}$$
$$\underset{1\,2\,3\,4}{}$$

To express a number in scientific notation:

1. Move the decimal point to obtain a number between 1 and 10.
2. Write the result from Step 1 multiplied by 10 raised to the number of places you moved the decimal point.
   - The exponent is positive if you moved the decimal point to the left.
   - The exponent is negative if you moved the decimal point to the right.

> Remember, large numbers have positive exponents and small numbers have negative exponents.

EXAMPLE **2.1** | **Scientific Notation**

The 2016 U.S. population was estimated to be 323,000,000 people. Express this number in scientific notation.

To obtain a number between 1 and 10, move the decimal point to the left eight decimal places; the exponent is 8. Because you move the decimal point to the left, the sign of the exponent is positive.	**SOLUTION** 323,000,000 people = 3.23×10^8 people

▶ **SKILLBUILDER 2.1** | **Scientific Notation**

The total U.S. national debt in 2016 was approximately $18,416,000,000,000. Express this number in scientific notation.

Note: The answers to all Skillbuilders appear at the end of the chapter.

▶ **FOR MORE PRACTICE** Example 2.19; Problems 31, 32.

EXAMPLE **2.2** | **Scientific Notation**

The radius of a carbon atom is approximately 0.000000000070 m. Express this number in scientific notation.

To obtain a number between 1 and 10, move the decimal point to the right 11 decimal places; therefore, the exponent is 11. Because you moved the decimal point to the right, the sign of the exponent is negative.	**SOLUTION** 0.000000000070 m = 7.0×10^{-11} m

▶ **SKILLBUILDER 2.2** | **Scientific Notation**

Express the number 0.000038 in scientific notation.

▶ **FOR MORE PRACTICE** Problems 33, 34.

The eText 2.0 icon indicates that this feature is embedded and interactive in the eText.

CONCEPTUAL ✔ CHECKPOINT 2.1

The radius of a dust speck is 4.5×10^{-3} mm. What is the correct value of this number in decimal notation (i.e., express the number without using scientific notation)?

(a) 4500 mm   **(b)** 0.045 mm   **(c)** 0.0045 mm   **(d)** 0.00045 mm

Note: The answers to all Conceptual Checkpoints appear at the end of the chapter.

2.3 Significant Figures: Writing Numbers to Reflect Precision

▶ Report measured quantities to the right number of digits.
▶ Determine which digits in a number are significant.

Climate change has become a household term. Global climate affects agriculture, weather, and ocean levels. The report that global temperatures are increasing is based on the continuing work of scientists who analyze records from thousands of temperature-measuring stations around the world. To date, these scientists conclude that the average global temperature has risen by 0.7 °C since 1880.

Notice how the scientists reported their results. What if the scientists had included additional zeros in their results—for example, 0.70 °C or 0.700 °C—or if they had reported the number their computer displayed after averaging many measurements, something like 0.68759824 °C. Would these reported numbers convey the same information? Not really. Scientists adhere to a standard way of reporting measured quantities in which the number of reported digits reflects the precision in the measurement—more digits, more precision; fewer digits, less precision. Numbers are usually written so that the uncertainty is indicated by the last reported digit. For example, by reporting a temperature increase of 0.7 °C, the scientists mean 0.7 ± 0.1 °C (± means plus or minus). The temperature rise could be as much as 0.8 °C or as little as 0.6 °C, but it is not 1.0 °C. The degree of certainty in this particular measurement is critical, influencing political decisions that directly affect people's lives.

We report scientific numbers so that every digit is certain except the last, which we estimate. For example, if a reported measurement is:

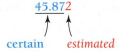

$$\underline{45.87}2$$

certain   *estimated*

We know that the first four digits are certain; the last digit is estimated. The number of digits reported depends on the precision of the measuring device. For example, consider the measurement shown here:

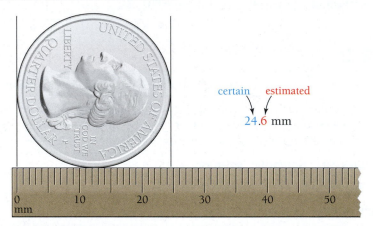

certain   estimated

24.6 mm

How should we report the width of the quarter as measured with this particular ruler? Since the ruler has markings every 1 mm, we can say for certain that the quarter is between 24 and 25 mm wide. We can then estimate the remaining width to the nearest 0.1 mm. We do this by mentally dividing the space between the 24 and 25 mm marks into 10 equal spaces and estimate the position of the line that indicates the width of the quarter. In this case, the line is a bit beyond the halfway point between the two markings on the ruler, so we might report 24.6 mm. Another person might report 24.7 mm or 24.5 mm—both of which are within the realm of the precision of the ruler. However, it would be incorrect to report 24.60 mm because reporting the measurement with four digits instead of three overstates the precision of the measuring device.

Suppose that we weigh an object on a balance with marks at every 1 g, and the pointer is between the 1-g mark and the 2-g mark (▶ FIGURE 2.1) but much closer to the 1-g mark. To record this measurement, we mentally divide the space between the 1- and 2-g marks into 10 equal spaces and estimate the position of the pointer. In this case, the pointer indicates about 1.2 g. We write the measurement as 1.2 g, indicating that we are sure of the "1" but have estimated the ".2."

If we measure an object using a balance with marks every tenth of a gram, we need to write the result with more digits. For example, suppose that on this more precise balance the pointer is between the 1.2-g mark and the 1.3-g mark (▶ FIGURE 2.2). We again divide the space between the two marks into 10 equal spaces and estimate the third digit. In the case of the nut shown in Figure 2.2, we report 1.26 g. Digital balances usually have readouts that report the mass to the correct number of digits.

1.2 g

1.26 g

Balance has marks every one gram.

Balance has marks every tenth of a gram.

▲ **FIGURE 2.1 Estimating tenths of a gram**   This balance has markings every 1 g, so we estimate to the tenths place. To estimate between markings, we mentally divide the space into 10 equal spaces and estimate the last digit. This reading is 1.2 g.

▲ **FIGURE 2.2 Estimating hundredths of a gram**   Because this scale has markings every 0.1 g, we estimate to the hundredths place. The correct reading is 1.26 g.

EXAMPLE **2.3**   **Reporting the Right Number of Digits**

The bathroom scale in ▼ **FIGURE 2.3** has markings at every 1 lb. Report the reading to the correct number of digits.

▲ **FIGURE 2.3 Reading a bathroom scale**

SOLUTION

Because the pointer is between the 147- and 148-lb markings, you mentally divide the space between the markings into 10 equal spaces and estimate the next digit. In this case, you should report the result as:

147.7 lb

What if you estimated a little differently and wrote 147.6 lb? In general, one unit of difference in the last digit is acceptable because the last digit is estimated and different people might estimate it slightly differently. However, if you wrote 147.2 lb, you would clearly be wrong.

▶ **SKILLBUILDER 2.3 | Reporting the Right Number of Digits**

You use a thermometer to measure the temperature of a backyard hot tub, and you obtain the reading shown in ▶ **FIGURE 2.4**. Record the temperature reading to the correct number of digits.

◀ **FIGURE 2.4 Reading a thermometer**

▶ **FOR MORE PRACTICE**   Example 2.20; Problems 41, 42.

Counting Significant Figures

The non–place-holding digits in a measurement are **significant figures** (or **significant digits**). As we have seen, these significant figures represent the precision of a measured quantity—the greater the number of significant figures, the greater the precision of the measurement. We can determine the number of significant figures in a written number fairly easily; however, if the number contains zeros, we must distinguish between the zeros that are significant and those that simply mark the decimal place. In the number 0.002, for example, the leading zeros mark the decimal place; they *do not* add to the precision of the measurement. In the number 0.00200, however, the trailing zeros *do* add to the precision of the measurement.

To determine the number of significant figures in a number, we follow these rules:

1. All nonzero digits are significant.

1.05   0.0110

2. Interior zeros (zeros between two numbers) are significant.

4.0208   50.1

3. Trailing zeros (zeros to the right of a nonzero number) that fall after a decimal point are significant.

5.10   3.00

> When a number is expressed in scientific notation, all trailing zeros are significant.

4. Trailing zeros that fall before a decimal point are significant.

50.00   1700.24

5. Leading zeros (zeros to the left of the first nonzero number) are not significant. They only serve to locate the decimal point.

For example, the number 0.0005 has only one significant digit.

> Some books put a decimal point after one or more trailing zeros if the zeros are to be considered significant. We avoid that practice in this book, but you should be aware of it.

6. Trailing zeros at the end of a number, but before an *implied* decimal point, are ambiguous and should be avoided by using scientific notation.

For example, it is unclear if the number 350 has two or three significant figures. We can avoid confusion by writing the number as 3.5×10^2 to indicate two significant figures or as 3.50×10^2 to indicate three.

Exact Numbers

Exact numbers have an unlimited number of significant figures. Exact numbers originate from three sources:

- Exact counting of discrete objects. For example, 10 pencils means 10.0000... pencils and 3 atoms means 3.00000...atoms.
- *Defined quantities*, such as the number of centimeters in 1 m. Because 100 cm is defined as 1 m,

100 cm = 1 m means 100.00000 . . . cm = 1.0000000 . . . m

Note that some conversion factors (see Section 2.6) are defined quantities while others are not.

- Integral numbers that are part of an equation. For example, in the equation, $radius = \dfrac{diameter}{2}$, the number 2 is exact and therefore has an unlimited number of significant figures.

EXAMPLE **2.4** **Determining the Number of Significant Figures in a Number**

How many significant figures are in each number?

(a) 0.0035   **(b)** 1.080   **(c)** 2371   **(d)** 2.97×10^5 **(e)** 1 dozen = 12   **(f)** 100.00   **(g)** 100,000

The 3 and the 5 are significant (rule 1). The leading zeros only mark the decimal place and are not significant (rule 5).	**SOLUTION** **(a)** 0.0035 two significant figures
The interior zero is significant (rule 2), and the trailing zero is significant (rule 3). The 1 and the 8 are also significant (rule 1).	**(b)** 1.080 four significant figures
All digits are significant (rule 1).	**(c)** 2371 four significant figures
All digits in the decimal part are significant (rule 1).	**(d)** 2.97×10^5 three significant figures
Defined numbers are exact and therefore have an unlimited number of significant figures.	**(e)** 1 dozen = 12 unlimited significant figures
The 1 is significant (rule 1), and the trailing zeros before the decimal point are significant (rule 4). The trailing zeros after the decimal point are also significant (rule 3).	**(f)** 100.00 five significant figures
This number is ambiguous. Write as 1×10^5 to indicate one significant figure or as 1.00000×10^5 to indicate six significant figures.	**(g)** 100,000 ambiguous

▶ **SKILLBUILDER 2.4 | Determining the Number of Significant Figures in a Number**

How many significant figures are in each number?

(a) 58.31   **(b)** 0.00250   **(c)** 2.7×10^3 **(d)** 1 cm = 0.01 m   **(e)** 0.500   **(f)** 2100

▶ **FOR MORE PRACTICE** Example 2.21; Problems 43, 44, 45, 46, 47, 48.

CHEMISTRY IN THE MEDIA

The COBE Satellite and Very Precise Measurements That Illuminate Our Cosmic Past

Since the earliest times, humans have wondered about the origins of our planet. Scientists have probed this question and developed theories for how the universe and the Earth began. The most accepted theory today about the origin of the universe is the Big Bang theory. According to this theory, the universe began in a tremendous expansion about 13.7 billion years ago and has been expanding ever since. A measurable prediction of this theory is the presence of a remnant "background radiation" from the expansion of the universe. That remnant radiation is characteristic of the current temperature of the universe. When the Big Bang occurred, the temperature of the universe was very hot and the associated radiation very bright. Today, 13.7 billion years later, the temperature of the universe is very cold and the background radiation is very faint.

In the early 1960s, Robert H. Dicke, P. J. E. Peebles, and their colleagues at Princeton University began to build a device to measure this background radiation and by doing so, they took a direct look into the cosmological past and

provided evidence for the Big Bang theory. At about the same time, quite by accident, Arno Penzias and Robert Wilson of Bell Telephone Laboratories measured excess radio noise on one of their communications satellites. As it turned out, this noise was the background radiation that the Princeton scientists were looking for. The two groups published papers together in 1965 reporting their findings along with the corresponding current temperature of the universe, about 3 degrees above absolute zero, or 3 K. We will define temperature measurement scales in Chapter 3. For now, know that 3 K is an extremely low temperature (454 degrees below zero on the Fahrenheit scale).

In 1989, NASA's Goddard Space Flight Center developed the Cosmic Background Explorer (COBE) satellite to measure the background radiation more precisely. The COBE satellite determined that the background radiation corresponded to a universe with a temperature of 2.735 K. (Notice the difference in significant figures from the previous measurement.) It went on to measure tiny fluctuations in the background radiation that amount to temperature differences of 1 part in 100,000. These fluctuations, though small, are an important prediction of the Big Bang theory. Scientists announced that the COBE satellite had produced the strongest evidence yet for the Big Bang theory of the creation of the universe. This is the way that science works. Measurement, and precision in measurement, are important to understanding the world—so important that we dedicate most of this chapter to the concept of measurement.

B2.1 CAN YOU ANSWER THIS? *How many significant figures are there in each of the preceding temperature measurements (3 K, 2.735 K)?*

▲ The COBE satellite, launched in 1989 to measure background radiation. Background radiation is a remnant of the Big Bang—the expansion that is believed to have formed the universe.

CONCEPTUAL ✔ CHECKPOINT 2.2

The Curiosity Rover on the surface of Mars recently measured a daily low temperature of −65.19 °C. What is the implied range of the actual temperature?

(a) between −65.190 °C and −65.199 °C   **(b)** between −65.18 °C and −65.20 °C
(c) between −65.1 °C and −65.2 °C   **(d)** exactly −65.19 °C

2.4 Significant Figures in Calculations

▶ Round numbers to the correct number of significant figures.
▶ Determine the correct number of significant figures in the results of multiplication and division calculations.
▶ Determine the correct number of significant figures in the results of addition and subtraction calculations.
▶ Determine the correct number of significant figures in the results of calculations involving both addition/subtraction and multiplication/division.

When we use measured quantities in calculations, the results of the calculation must reflect the precision of the measured quantities. We should not lose or gain precision during mathematical operations.

Multiplication and Division

In multiplication or division, the result carries the same number of significant figures as the factor with the fewest significant figures.

For example:

$$5.02 \times 89.665 \times 0.10 = 45.0118 = 45$$

(3 sig. figures)   (5 sig. figures)   (2 sig. figures)   (2 sig. figures)

We round the intermediate result (in blue) to two significant figures to reflect the least precisely known factor (0.10), which has two significant figures.

In division, we follow the same rule.

$$5.892 \div 6.10 = 0.96590 = 0.966$$

(4 sig. figures)   (3 sig. figures)   (3 sig. figures)

We round the intermediate result (in blue) to three significant figures to reflect the least precisely known factor (6.10), which has three significant figures.

Rounding

When we round to the correct number of significant figures:

We round down if the last (or leftmost) digit dropped is 4 or less;
we round up if the last (or leftmost) digit dropped is 5 or more.

For example, when we round each of these numbers to two significant figures:

2.33 rounds to 2.3
2.37 rounds to 2.4

2.34 rounds to 2.3
2.35 rounds to 2.4

We consider only the *last (or leftmost) digit being dropped* when we decide in which direction to round—we ignore all digits to the right of it. For example, to round 2.349 to two significant figures, only the 4 in the hundredths place (2.3**4**9) determines which direction to round—the 9 is irrelevant.

2.349 rounds to 2.3

For calculations involving multiple steps, we round only the final answer—we do not round the intermediate results. This prevents small rounding errors from affecting the final answer.

EXAMPLE **2.5** | **Significant Figures in Multiplication and Division**

Perform each calculation to the correct number of significant figures.

(a) $1.01 \times 0.12 \times 53.51 \div 96$
(b) $56.55 \times 0.920 \div 34.2585$

	SOLUTION
Round the intermediate result (in blue) to two significant figures to reflect the two significant figures in the least precisely known quantities (0.12 and 96).	(a) $1.01 \times 0.12 \times 53.51 \div 96 = 0.067556 = 0.068$
Round the intermediate result (in blue) to three significant figures to reflect the three significant figures in the least precisely known quantity (0.920).	(b) $56.55 \times 0.920 \div 34.2585 = 1.51863 = 1.52$

▶ **SKILLBUILDER 2.5** | **Significant Figures in Multiplication and Division**

Perform each calculation to the correct number of significant figures.

(a) $1.10 \times 0.512 \times 1.301 \times 0.005 \div 3.4$
(b) $4.562 \times 3.99870 \div 89.5$

▶ **FOR MORE PRACTICE** Examples 2.22, 2.23; Problems 57, 58, 59, 60.

Addition and Subtraction

In addition or subtraction, the result has the same number of decimal places as the quantity with the fewest decimal places.
For example:

$$
\begin{array}{r}
5.74 \\
0.823 \\
+\ 2.651 \\
\hline
9.214 = 9.21
\end{array}
$$

> It is sometimes helpful to draw a vertical line directly to the right of the number with the fewest decimal places. The line shows the number of decimal places that should be in the answer.

We round the intermediate answer (in blue) to two decimal places because the quantity with the fewest decimal places (5.74) has two decimal places.
For subtraction, we follow the same rule. For example:

$$
\begin{array}{r}
4.8 \\
-\ 3.965 \\
\hline
0.835 = 0.8
\end{array}
$$

We round the intermediate answer (in blue) to one decimal place because the quantity with the fewest decimal places (4.8) has one decimal place. Remember: *For multiplication and division, the quantity with the fewest **significant figures** determines* *the number of significant figures in the answer. For addition and subtraction, the quantity with the fewest* **decimal places** *determines the number of decimal places in the answer.* In multiplication and division we focus on significant figures, but in addition and subtraction we focus on decimal places. When a problem involves addition and subtraction, the answer may have a different number of significant figures than the initial quantities. For example:

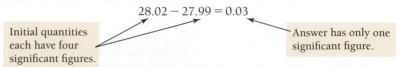

$$28.02 - 27.99 = 0.03$$

Initial quantities each have four significant figures.

Answer has only one significant figure.

The answer has only one significant figure, even though the initial quantities each had four significant figures.

EXAMPLE **2.6**   **Significant Figures in Addition and Subtraction**

Perform the calculations to the correct number of significant figures.

(a)   0.987
      +125.1
      −1.22

(b)   0.765
      −3.449
      −5.98

	SOLUTION

Round the intermediate answer (in blue) to one decimal place to reflect the quantity with the fewest decimal places (125.1). Notice that 125.1 is not the quantity with the fewest significant figures—it has four while the other quantities only have three—but because it has the fewest decimal places, it determines the number of decimal places in the answer.	(a) 0.987
	    +125.1
	     − 1.22
	    124.867 = 124.9

Round the intermediate answer (in blue) to two decimal places to reflect the quantity with the fewest decimal places (5.98).	(b) 0.765
	    − 3.449
	    − 5.98
	    −8.664 = −8.66

▶ **SKILLBUILDER 2.6** | **Significant Figures in Addition and Subtraction**

Perform the calculations to the correct number of significant figures.

(a)   2.18
      +5.621
      +1.5870
      −1.8

(b)   7.876
      −0.56
      +123.792

▶ **FOR MORE PRACTICE**   Example 2.24; Problems 61, 62, 63, 64.

Calculations Involving Both Multiplication/Division and Addition/Subtraction

In calculations involving both multiplication/division and addition/subtraction, we do the steps in parentheses first; determine the correct number of significant figures in the intermediate answer; then complete the remaining steps. For example:

$$3.489 \times (5.67 - 2.3)$$

We complete the subtraction step first.

$$5.67 - 2.3 = 3.37$$

We use the subtraction rule to determine that the intermediate answer (3.37) has only one significant decimal place. To avoid small errors, we do not round at this point; instead, we underline the least significant figure as a reminder.

$$= 3.489 \times 3.3\underline{7}$$

We then complete the multiplication step.

$$3.489 \times 3.3\underline{7} = 11.758 = 12$$

The multiplication rule indicates that the intermediate answer (11.758) rounds to two significant figures (12) because it is limited by the two significant figures in 3.3\underline{7}.

EXAMPLE 2.7

Significant Figures in Calculations Involving Both Multiplication/Division and Addition/Subtraction

Perform the calculations to the correct number of significant figures.

(a) $6.78 \times 5.903 \times (5.489 - 5.01)$
(b) $19.667 - (5.4 \times 0.916)$

Do the step in parentheses first. Use the subtraction rule to mark 0.479 to two decimal places because 5.01, the number in the parentheses with the least number of decimal places, has two.

Then perform the multiplication and round the answer to two significant figures because the number with the least number of significant figures has two.

SOLUTION

(a) $6.78 \times 5.903 \times (5.489 - 5.01)$
$= 6.78 \times 5.903 \times (0.479)$
$= 6.78 \times 5.903 \times 0.4\underline{79}$
$6.78 \times 5.903 \times 0.4\underline{79} = 19.1707$
$= 19$

Do the step in parentheses first. The number with the least number of significant figures within the parentheses (5.4) has two, so mark the answer to two significant figures.

Then perform the subtraction and round the answer to one decimal place because the number with the least number of decimal places has one.

(b) $19.667 - (5.4 \times 0.916)$
$= 19.667 - (4.9464)$
$= 19.667 - 4.\underline{9}464$
$19.667 - 4.\underline{9}464 = 14.7206$
$= 14.7$

▶ **SKILLBUILDER 2.7** | **Significant Figures in Calculations Involving Both Multiplication/Division and Addition/Subtraction**

Perform each calculation to the correct number of significant figures.

(a) $3.897 \times (782.3 - 451.88)$ **(b)** $(4.58 \div 1.239) - 0.578$

▶ **FOR MORE PRACTICE** Example 2.25; Problems 65, 66, 67, 68.

CONCEPTUAL CHECKPOINT 2.3

Which calculation would have its result reported to the *greater* number of significant figures?

(a) $3 + (15/12)$
(b) $(3 + 15)/12$

2.5 The Basic Units of Measurement

▶ Recognize and work with the SI base units of measurement, prefix multipliers, and derived units.

By themselves, numbers have little meaning. Read this sentence: When my son was 7 he walked 3, and when he was 4 he could throw his baseball 8 and tell us that his school was 5 away. The sentence is confusing. We don't know what the numbers mean because the **units** are missing. The meaning becomes clear, however, when we add the missing units to the numbers: When my son was 7 *months old* he walked 3 *steps*, and when he was 4 *years old* he could throw his baseball 8 *feet* and tell us that his school was 5 *minutes* away. Units make all the difference. In chemistry, units are critical. Never write a number by itself; always use its associated units—otherwise your work will be as confusing as the initial sentence.

The abbreviation *SI* comes from the French *le Système International*.

The two most common unit systems are the **English system**, used in the United States, and the **metric system**, used in most of the rest of the world. The English system uses units such as inches, yards, and pounds, while the metric system uses centimeters, meters, and kilograms. The most convenient system for science measurements is based on the metric system and is called the **International System** of units or **SI units**. SI units are a set of standard units agreed on by scientists throughout the world.

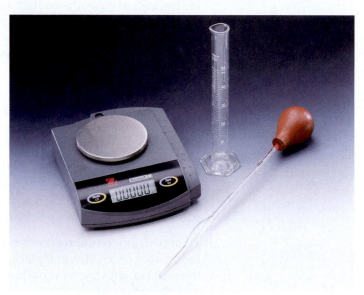

▲ Science uses instruments to make measurements. Every instrument is calibrated in a particular unit without which the measurements would be meaningless.

TABLE 2.1 Important SI Base Units		
Quantity	**Unit**	**Symbol**
length	meter	m
mass	kilogram	kg
time	second	s
temperature*	kelvin	K

*Temperature units are discussed in Chapter 3.

Because the mass of the block of metal used to define a kilogram has changed slightly over the years, scientists are currently working on alternate ways to define the kilogram.

The Base Units

Table 2.1 lists the common SI base units. They include the **meter (m)** as the base unit of length; the **kilogram (kg)** as the base unit of mass; and the **second (s)** as the base unit of time. Each of these base units is precisely defined. The meter is defined as the distance light travels in a certain period of time: $1/299{,}792{,}458$ s (▶ FIGURE 2.5). (The speed of light is 3.0×10^8 m/s.) The kilogram is defined as the mass of a block of metal kept at the International Bureau of Weights and Measures at Sèvres, France (▶ FIGURE 2.6). The second is defined using an atomic standard (▶ FIGURE 2.7).

Most people are familiar with the SI base unit of time, the second. However, if you live in the United States, you may be less familiar with the meter and the kilogram. The meter is slightly longer than a yard (a yard is 36 in., while a meter is 39.37 in.). A 100-yd football field measures only 91.4 m.

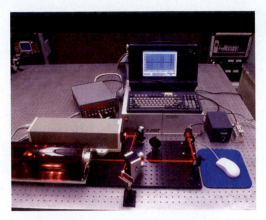

▲ **FIGURE 2.5 The base unit of length** The definition of a meter, established by international agreement in 1983, is the distance that light travels in a vacuum in 1/299,792,458 s.
Question: Why is such a precise standard necessary?

▲ **FIGURE 2.6 The base unit of mass** A duplicate of the international standard kilogram, called kilogram 20, is kept at the National Institute of Standards and Technology near Washington, D.C.

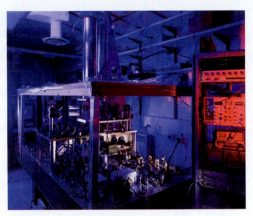

▲ **FIGURE 2.7 The base unit of time** The second is defined, using an atomic clock, as the duration of 9,192,631,770 periods of the radiation emitted from a certain transition in a cesium-133 atom.

A nickel (5 cents) has a mass of about 5 g.

The kilogram is a measure of mass, which is different from weight. The **mass** of an object is a measure of the quantity of matter within it, while the weight of an object is a measure of the gravitational pull on that matter. Consequently, weight depends on gravity, while mass does not. If you were to weigh yourself on Mars, for example, the lower gravity would pull you toward the scale less than Earth's gravity would, resulting in a lower weight. A 150-lb person on Earth weighs only 57 lb on Mars. However, the person's mass, the quantity of matter in his or her body, remains the same. A kilogram of mass is the equivalent of 2.205 lb of weight on Earth, so if we express mass in kilograms, a 150-lb person on Earth has a mass of approximately 68 kg. A second common unit of mass is the gram (g), defined as follows:

$$1000 \text{ g} = 10^3 \text{ g} = 1 \text{ kg}$$

Prefix Multipliers

The SI system employs **prefix multipliers** (Table 2.2, on the next page) with the base units. These multipliers change the value of the unit by powers of 10. For example, the kilometer (km) features the prefix *kilo-*, meaning 1000 or 10^3. Therefore:

$$1 \text{ km} = 1000 \text{ m} = 10^3 \text{ m}$$

Similarly, the millisecond (ms) has the prefix *milli-*, meaning 0.001 or 10^{-3}.

$$1 \text{ ms} = 0.001 \text{ s} = 10^{-3} \text{ s}$$

The prefix multipliers allow us to express a wide range of measurements in units that are similar in size to the quantity we are measuring. We choose the prefix multiplier that is most convenient for a particular measurement. For example, to measure the diameter of a coin, we might use centimeters or millimeters because coins have diameters between 1–3 cm (or 10–30 mm). A centimeter is a common metric unit and is about equivalent to the width of a pinky finger (2.54 cm = 1 in.). The millimeter could also conveniently express the diameter of a coin, but the kilometer would not work as well because, in that unit, a coin's diameter is 0.000010–0.000030 km. We pick a unit similar in size to (or smaller than) the quantity we are measuring. Consider expressing the length of a short chemical bond, about 1.2×10^{-10} m. Which prefix multiplier should we use? The most convenient one is probably the picometer (pico = 10^{-12}). Chemical bonds measure about 120 pm.

TABLE 2.2 SI Prefix Multipliers

Prefix	Symbol	Meaning	Multiplier	
tera-	T	trillion	1,000,000,000,000	(10^{12})
giga-	G	billion	1,000,000,000	(10^{9})
mega-	M	million	1,000,000	(10^{6})
kilo-	k	thousand	1,000	(10^{3})
hecto-	h	hundred	100	10^{2}
deca-	da	ten	10	10^{1}
deci-	d	tenth	0.1	(10^{-1})
centi-	c	hundredth	0.01	(10^{-2})
milli-	m	thousandth	0.001	(10^{-3})
micro-	μ	millionth	0.000001	(10^{-6})
nano-	n	billionth	0.000000001	(10^{-9})
pico-	p	trillionth	0.000000000001	(10^{-12})
femto-	f	quadrillionth	0.000000000000001	(10^{-15})

TABLE 2.3 Some Common Units and Their Equivalents

Length
1 kilometer (km) = 0.6214 mile (mi)
1 meter (m) = 39.37 inches (in.)
= 1.094 yards (yd)
1 foot (ft) = 30.48 centimeters (cm)
1 inch (in.) = 2.54 centimeters (cm)
(exact)

Mass
1 kilogram (kg) = 2.205 pounds (lb)
1 pound (lb) = 453.59 grams (g)
1 ounce (oz) = 28.35 grams (g)

Volume
1 liter (L) = 1000 milliliters (mL)
= 1000 cubic centimeters (cm³)
1 liter (L) = 1.057 quarts (qt)
1 U.S. gallon (gal) = 3.785 liters (L)

CONCEPTUAL CHECKPOINT 2.4

Which is the most convenient unit to use to express the dimensions of a polio virus, which is about 2.8×10^{-8} m in diameter?

(a) Mm (b) mm (c) μm (d) nm

PEARSON eText 2.0

Derived Units

A derived unit is formed from other units. For example, many units of **volume**, a measure of space, are derived units. Any unit of length, when cubed (raised to the third power), becomes a unit of volume. Therefore, cubic meters (m^3), cubic centimeters (cm^3), and cubic millimeters (mm^3) are all units of volume. A three-bedroom house has a volume of about 630 m^3, a can of soda pop has a volume of about 350 cm^3, and a rice grain has a volume of about 3 mm^3. We also use the **liter (L)** and milliliter (mL) to express volume (although these are not derived units). A gallon is equal to 3.785 L. A milliliter is equivalent to 1 cm^3. Table 2.3 lists some common units and their equivalents.

2.6 Problem Solving and Unit Conversion

▶ Convert between units.

Problem solving is one of the most important skills you will acquire in this course. Not only will this skill help you succeed in chemistry, but it will also help you learn how to think critically, which is important in every area of knowledge. When my daughter was a freshman in high school, she came to me for help on an algebra problem. The problem went something like this:

Sam and Sara live 11 miles apart. Sam leaves his house traveling at 6 miles per hour toward Sara's house. Sara leaves her house traveling at 3 miles per hour toward Sam's house. How much time elapses until Sam and Sara meet?

The Mathematics Review Appendix (p. MR-1) includes a review of how to solve algebraic problems for a variable.

Solving the problem requires setting up the equation $11 - 6t = 3t$. Although my daughter could solve this equation for t quite easily, getting to the equation from the problem statement was another matter—that process requires *critical thinking*, and that was the skill she needed to learn to successfully solve the problem. You can't succeed in chemistry—or in life, really—without developing critical thinking skills. Learning how to solve chemical problems helps you develop these kinds of skills.

Although no simple formula applies to every problem, you can learn problem-solving strategies and begin to develop some chemical intuition. You can think of many of the problems in this book as *unit conversion problems*, where you are given one or more quantities and asked to convert them into different units. Other problems require the use of *specific equations* to get to the information you are trying to find. In the sections that follow, you will find strategies to help you solve both of these types of problems. Of course, many problems contain both conversions and equations, requiring the combination of these strategies, and some problems may require an altogether different approach, but the basic tools you learn here can be applied to those problems as well.

Converting Between Units

Units are critical in calculations. Knowing how to work with and manipulate units in calculations is a crucial part of problem solving. In calculations, units help determine correctness. You should always include units in calculations, and you can think of many calculations as converting from one unit to another. You can multiply, divide, and cancel units like any other algebraic quantity.

Remember:

1. Always write every number with its associated unit. Never ignore units; they are critical.

2. Always include units in your calculations, dividing them and multiplying them as if they were algebraic quantities. Do not let units magically appear or disappear in calculations. Units must flow logically from beginning to end.

Consider converting 17.6 in. to centimeters. You know from Table 2.3 that 1 in. = 2.54 cm. To determine how many centimeters are in 17.6 in., perform the conversion:

$$17.6 \text{ in.} \times \frac{2.54 \text{ cm}}{1 \text{ in.}} = 44.7 \text{ cm}$$

The unit in. cancels and you are left with cm as your final unit. The quantity $\dfrac{2.54 \text{ cm}}{1 \text{ in.}}$ is a **conversion factor** between in. and cm—it is a quotient with cm on top and in. on bottom.

For most conversion problems, you are given a quantity in some units and asked to convert the quantity to another unit. These calculations take the form:

$$\text{information given} \times \text{conversion factor(s)} = \text{information sought}$$

$$\text{given unit} \times \frac{\text{desired unit}}{\text{given unit}} = \text{desired unit}$$

You can construct conversion factors from any two quantities known to be equivalent. In this example, 2.54 cm = 1 in., so construct the conversion factor by dividing both sides of the equality by 1 in. and canceling the units.

$$2.54 \text{ cm} = 1 \text{ in.}$$

$$\frac{2.54 \text{ cm}}{1 \text{ in.}} = \frac{1 \text{ in.}}{1 \text{ in.}}$$

$$\frac{2.54 \text{ cm}}{1 \text{ in.}} = 1$$

The quantity $\dfrac{2.54 \text{ cm}}{1 \text{ in.}}$ is equal to 1, and you can use it to convert between inches and centimeters.

> Using units as a guide to solving problems is called dimensional analysis.

What if you want to perform the conversion the other way, from centimeters to inches? If you try to use the same conversion factor, the units do not cancel correctly.

$$44.7 \text{ cm} \times \frac{2.54 \text{ cm}}{1 \text{ in.}} = \frac{114 \text{ cm}^2}{\text{in.}}$$

The units in the answer, as well as the value of the answer, are incorrect. The unit $\text{cm}^2/\text{in.}$ is not correct, and, based on your knowledge that centimeters are smaller than inches, you know that 44.7 cm cannot be equivalent to 114 in. In solving problems, always check if the final units are correct, and consider whether or not the magnitude of the answer makes sense. In this case, the mistake was in how the conversion factor was used. You must invert it.

$$44.7 \text{ cm} \times \frac{1 \text{ in.}}{2.54 \text{ cm}} = 17.6 \text{ in.}$$

You can invert conversion factors because they are equal to 1 and the inverse of 1 is 1.

$$\frac{1}{1} = 1$$

Therefore,

$$\frac{2.54 \text{ cm}}{1 \text{ in.}} = 1 = \frac{1 \text{ in.}}{2.54 \text{ cm}}$$

You can diagram conversions using a **solution map**. A solution map is a visual outline that shows the strategic route required to solve a problem. For unit conversion, the solution map focuses on units and how to convert from one unit to another. The solution map for converting from inches to centimeters is:

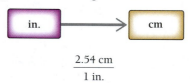

$$\frac{2.54 \text{ cm}}{1 \text{ in.}}$$

The solution map for converting from centimeters to inches is:

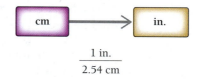

$$\frac{1 \text{ in.}}{2.54 \text{ cm}}$$

Each arrow in a solution map for a unit conversion has an associated conversion factor, with the units of the previous step in the denominator and the units of the following step in the numerator. For one-step problems such as these, the solution map is only moderately helpful, but for multistep problems, it becomes a powerful way to develop a problem-solving strategy. In the section that follows, you will learn how to incorporate solution maps into an overall problem-solving strategy.

PEARSON
eText
2.0

CONCEPTUAL ✓ CHECKPOINT 2.5

Which conversion factor should you use to convert 4 ft to inches (12 in. = 1 ft)?

(a) $\dfrac{12 \text{ in.}}{1 \text{ ft}}$ (b) $\dfrac{1 \text{ ft}}{12 \text{ in.}}$ (c) $\dfrac{1 \text{ in.}}{12 \text{ ft}}$ (d) $\dfrac{12 \text{ ft}}{1 \text{ in.}}$

General Problem-Solving Strategy

In this book, we use a standard problem-solving procedure that you can adapt to many of the problems encountered in chemistry and beyond. Solving any problem essentially requires that you assess the information given in the problem and devise a way to get to the requested information. In other words, you need to:

- Identify the starting point (the *given* information).
- Identify the endpoint (what you must *find*).
- Devise a way to get from the starting point to the endpoint using what is given as well as what you already know or can look up. You can use a *solution map* to diagram the steps required to get from the starting point to the endpoint.

In graphic form, this progression looks like this:

$$\textbf{Given} \longrightarrow \textbf{Solution Map} \longrightarrow \textbf{Find}$$

Beginning students often have trouble knowing how to start solving a chemistry problem. Although no problem-solving procedure is applicable to all problems, the following four-step procedure can be helpful in working through many of the numerical problems you will encounter in chemistry.

1. **Sort.** Begin by sorting the information in the problem. *Given* information is the basic data provided by the problem—often one or more numbers with their associated units. The given information is the starting point for the problem. *Find* indicates what the problem is asking you to find (the endpoint of the problem).

2. **Strategize.** This is usually the hardest part of solving a problem. In this process, you must create a solution map—the series of steps that will get you from the given information to the information you are trying to find. You have already seen solution maps for simple unit conversion problems. Each arrow in a solution map represents a computational step. On the left side of the arrow is the quantity (or quantities) you had before the step; on the right side of the arrow is the quantity (or quantities) you will have after the step; and below the arrow is the information you need to get from one to the other—the relationship between the quantities.

   Often such relationships will take the form of conversion factors or equations. These may be given in the problem, in which case you will have written them down under "Given" in Step 1. Usually, however, you will need other information—such as physical constants, formulas, or conversion factors—to help get you from what you are given to what you must find. You may recall this information from what you have learned, or you can look it up in the chapters or tables within the book.

   In some cases, you may get stuck at the strategize step. If you cannot figure out how to get from the given information to the information you are asked to find, you might try working backwards. For example, you may want to look at the units of the quantity you are trying to find and look for conversion factors to get to the units of the given quantity. You may even try a combination of strategies; work forward, backward, or some of both. If you persist, you will develop a strategy to solve the problem.

3. **Solve.** This is the most straightforward part of solving a problem. Once you set up the problem properly and devise a solution map, you follow the map to solve the problem. Carry out mathematical operations (paying attention to the rules for significant figures in calculations) and cancel units as needed.

4. **Check.** Beginning students often overlook this step. Experienced problem solvers always ask, Does this answer make physical sense? Are the units correct? Is the number of significant figures correct? When solving multistep problems, errors easily creep into the solution. You can catch most of these errors by simply checking the answer. For example, suppose you are calculating the number of atoms in a gold coin and end up with an answer of 1.1×10^{-6} atoms. Could the gold coin really be composed of one-millionth of one atom?

In Examples 2.8 and 2.9, you will find this problem-solving procedure applied to unit conversion problems. The left column summarizes the procedure, and the middle and right columns show two examples of applying the procedure. You will encounter this three-column format in selected examples throughout this text. It allows you to see how a particular procedure can be applied to two different problems. Work through one problem first (from top to bottom) and then examine how the other problem applies the same procedure. Recognizing the commonalities and differences between problems is a key part of problem solving.

Problem-Solving Procedure	EXAMPLE **2.8**	EXAMPLE **2.9**
	UNIT CONVERSION Convert 7.8 km to miles.	**UNIT CONVERSION** Convert 0.825 m to millimeters.
SORT Begin by sorting the information in the problem into *given* and *find*.	**GIVEN:** 7.8 km **FIND:** mi	**GIVEN:** 0.825 m **FIND:** mm
STRATEGIZE Draw a *solution map* for the problem. Begin with the *given* quantity and symbolize each step with an arrow. Below the arrow, write the conversion factor for that step. The solution map ends at the *find* quantity. (In these examples, the relationships used in the conversions are below the solution map.)	**SOLUTION MAP** km → mi $\dfrac{0.6214 \text{ mi}}{1 \text{ km}}$ **RELATIONSHIPS USED** 1 km = 0.6214 mi (This conversion factor is from Table 2.3.)	**SOLUTION MAP** m → mm $\dfrac{1 \text{ mm}}{10^{-3} \text{ m}}$ **RELATIONSHIPS USED** 1 mm = 10^{-3} m (This conversion factor is from Table 2.2.)
SOLVE Follow the *solution map* to solve the problem. Begin with the *given* quantity and its units. Multiply by the appropriate conversion factor, canceling units to arrive at the *find* quantity. Round the answer to the correct number of significant figures. (If possible, obtain conversion factors to enough significant figures so that they do not limit the number of significant figures in the answer.)	**SOLUTION** $7.8 \text{ km} \times \dfrac{0.6214 \text{ mi}}{1 \text{ km}} = 4.84692 \text{ mi}$ $4.84692 \text{ mi} = 4.8 \text{ mi}$ Round the answer to two significant figures because the quantity given has two significant figures.	**SOLUTION** $0.825 \text{ m} \times \dfrac{1 \text{ mm}}{10^{-3} \text{ m}} = 825 \text{ mm}$ $825 \text{ mm} = 825 \text{ mm}$ Leave the answer with three significant figures because the quantity given has three significant figures and the conversion factor is a definition and therefore does not limit the number of significant figures in the answer.
CHECK Check your answer. Are the units correct? Does the answer make sense?	The units, mi, are correct. The magnitude of the answer is reasonable. A mile is longer than a kilometer, so the value in miles should be smaller than the value in kilometers.	The units, mm, are correct, and the magnitude is reasonable. A millimeter is shorter than a meter, so the value in millimeters should be larger than the value in meters.
	▶ **SKILLBUILDER 2.8**	**Unit Conversion** Convert 56.0 cm to inches.	▶ **SKILLBUILDER 2.9**	**Unit Conversion** Convert 5678 m to kilometers.
	▶ **FOR MORE PRACTICE** Example 2.26; Problems 73, 74, 75, 76.	▶ **FOR MORE PRACTICE** Problems 69, 70, 71, 72.

CONCEPTUAL ✔ **CHECKPOINT** **2.6**

Which conversion factor should you use to convert a distance in meters to kilometers?

(a) $\dfrac{1\text{ m}}{10^3\text{ km}}$ (b) $\dfrac{10^3\text{ m}}{1\text{ km}}$ (c) $\dfrac{1\text{ km}}{10^3\text{ m}}$ (d) $\dfrac{10^3\text{ km}}{1\text{ m}}$

2.7 Solving Multistep Unit Conversion Problems

▶ Convert between units.

When solving multistep unit conversion problems, follow the preceding procedure, but add more steps to the solution map. Each step in the solution map should have a conversion factor, with the units of the previous step in the denominator and the units of the following step in the numerator. For example, suppose you want to convert 194 cm to ft. The solution map begins with cm, and you use the relationship 2.54 cm = 1 in. to convert to in. Then use the relationship 12 in. = 1 ft to convert to ft.

SOLUTION MAP

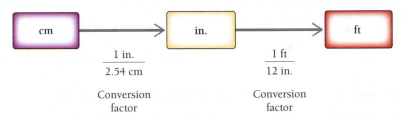

Once the solution map is complete, follow it to solve the problem.

SOLUTION

$$194\text{ cm} \times \frac{1\text{ in.}}{2.54\text{ cm}} \times \frac{1\text{ ft}}{12\text{ in.}} = 6.3648\text{ ft}$$

You then round to the correct number of significant figures—in this case, three (from 194 cm, which has three significant figures).

Because 1 foot is defined as 12 in., it does not limit significant figures.

$$6.3648\text{ ft} = 6.36\text{ ft}$$

Finally, check the answer. The units of the answer, feet, are the correct ones, and the magnitude seems about right. A foot is larger than a centimeter, so it is reasonable that the value in feet is smaller than the value in centimeters.

The eText 2.0 icon indicates that this feature is embedded and interactive in the eText.

Interactive Worked Example Video 2.10

EXAMPLE **2.10** **Solving Multistep Unit Conversion Problems**

A recipe for making creamy pasta sauce calls for 0.75 L of cream. Your measuring cup measures only in cups. How many cups of cream should you use? (4 cups = 1 quart)

SORT	
Begin by sorting the information in the problem into given and find.	**GIVEN:** 0.75 L **FIND:** cups

continued on page 34 ▶

continued from page 33

STRATEGIZE Draw a solution map for the problem. Begin with the *given* quantity and symbolize each step with an arrow. Below the arrow, write the conversion factor for that step. The solution map ends at the *find* quantity.	**SOLUTION MAP** 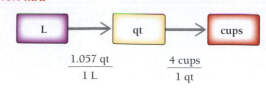 **RELATIONSHIPS USED** 1.057 qt = 1 L (from Table 2.3) 4 cups = 1 qt (given in problem statement)
SOLVE Follow the solution map to solve the problem. Begin with 0.75 L and multiply by the appropriate conversion factor, canceling units to arrive at qt. Then, use the second conversion factor to arrive at cups. Round the answer to the correct number of significant figures. In this case, you round the answer to two significant figures because the quantity given has two significant figures.	**SOLUTION** $$0.75 \; \cancel{L} \times \frac{1.057 \; \cancel{qt}}{1 \; \cancel{L}} \times \frac{4 \; \text{cups}}{1 \; \cancel{qt}} = 3.171 \; \text{cups}$$ $$3.171 \; \text{cups} = 3.2 \; \text{cups}$$
CHECK Check your answer. Are the units correct? Does the answer make physical sense?	The answer has the right units (cups) and seems reasonable. A cup is smaller than a liter, so the value in cups should be larger than the value in liters.

▶ **SKILLBUILDER 2.10** | **Solving Multistep Unit Conversion Problems**

A recipe calls for 1.2 cups of oil. How many liters of oil is this?

▶ **FOR MORE PRACTICE**   Problems 85, 86.

EXAMPLE **2.11**   **Solving Multistep Unit Conversion Problems**

One lap of a running track measures 255 m. To run 10.0 km, how many laps should you run?

SORT Begin by sorting the information in the problem into given and find. You are given a distance in km and asked to find the distance in laps. You are also given the quantity 255 m per lap, which is a conversion factor between m and laps.	**GIVEN:** 10.0 km 255 m = 1 lap **FIND:** number of laps
STRATEGIZE Build the solution map beginning with km and ending at laps. Focus on the units.	**SOLUTION MAP** 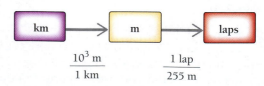 **RELATIONSHIPS USED** 1 km = 10³ m (from Table 2.2) 1 lap = 255 m (given in problem)

SOLVE	SOLUTION
Follow the solution map to solve the problem. Begin with 10.0 km and multiply by the appropriate conversion factor, canceling units to arrive at m. Then, use the second conversion factor to arrive at laps. Round the intermediate answer (in blue) to three significant figures because it is limited by the three significant figures in the given quantity, 10.0 km.	$$10.0 \ \cancel{km} \times \frac{10^3 \ \cancel{m}}{1 \ \cancel{km}} \times \frac{1 \ lap}{255 \ \cancel{m}} = 39.216 \ laps = 39.2 \ laps$$
CHECK Check your answer. Are the units correct? Does the answer make physical sense?	The units of the answer are correct, and the value of the answer makes sense. If a lap is 255 m, there are about 4 laps to each km (1000 m), so it seems reasonable that you would have to run about 40 laps to cover 10 km.

▶ **SKILLBUILDER 2.11** | **Solving Multistep Unit Conversion Problems**

A running track measures 1056 ft per lap. To run 15.0 km, how many laps should you run? (1 mi = 5280 ft)

▶ **SKILLBUILDER PLUS**

An island is 5.72 nautical mi from the coast. How far away is the island in meters? (1 nautical mi = 1.151 mi)

▶ **FOR MORE PRACTICE** Problems 83, 84.

2.8 Unit Conversion in Both the Numerator and Denominator

▶ Convert units in a quantity that has units in the numerator and the denominator.

Some unit conversion problems require converting the units in both the numerator and denominator of a fraction. For example, the Toyota Prius has an EPA estimated city gas mileage of 48.0 miles per gallon. In Europe, gasoline is sold in liters (L), and distances are measured in kilometers (km). How do we convert the Prius's mileage estimate from miles per gallon to kilometers per liter? The answer is to use two conversion factors: one from miles to kilometers and another from gallons to liters:

$$1 \ mi = 1.609 \ km$$
$$1 \ gal = 3.785 \ L$$

Begin with the quantity 48.0 mi/gal and write the conversion factors so that the units cancel correctly. First, convert the numerator to km and then the denominator to L:

SOLUTION MAP

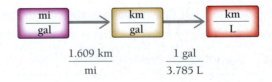

SOLUTION

$$48.0 \frac{\cancel{mi}}{\cancel{gal}} \times \frac{1.609 \ km}{\cancel{mi}} \times \frac{1 \ \cancel{gal}}{3.785 \ L} = 20.4 \ \frac{km}{L}$$

Notice that to convert the denominator from gal to L, you write the conversion factor with gal in the numerator and L in the denominator.

EXAMPLE **2.12** | **Solving Unit Conversions in the Numerator and Denominator**

A prescription medication requires 11.5 mg per kg of body weight. Convert this quantity to the number of grams required per pound of body weight and determine the correct dose (in g) for a 145-lb patient.

SORT	
Begin by sorting the information in the problem into given and find. You are given the dose of the drug in mg/kg and the weight of the patient in lb. You are asked to find the dose in g/lb and the dose in g for the 145-lb patient.	**GIVEN:** $11.5 \dfrac{mg}{kg}$ 145 lb **FIND:** $\dfrac{g}{lb}$; dose in g

STRATEGIZE	
The solution map has two parts. In the first part, convert from mg/kg to g/lb. In the second part, use the result from the first part to determine the correct dose for a 145-lb patient.	**SOLUTION MAP** 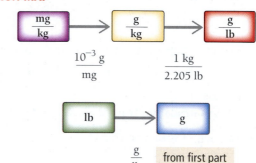 **RELATIONSHIPS USED** $1 \text{ mg} = 10^{-3} \text{ g}$ (from Table 2.2) $1 \text{ kg} = 2.205 \text{ lb}$ (from Table 2.3)

SOLVE	
Follow the solution map to solve the problem. For the first part, begin with 11.5 mg/kg and multiply by the two conversion factors to arrive at the dose in g/lb. Mark the answer to three significant figures to reflect the three significant figures in the least precisely known quantity.	**SOLUTION** $11.5 \dfrac{mg}{kg} \times \dfrac{10^{-3} \text{ g}}{mg} \times \dfrac{1 \text{ kg}}{2.205 \text{ lb}} = 0.00521\underline{5} \dfrac{g}{lb}$
For the second part, begin with 145 lb and use the dose obtained in the first part to convert to g. Then round the answer to the correct number of significant figures, which is three.	$145 \text{ lb} \times \dfrac{0.005215 \text{ g}}{lb} = 0.7561\underline{7} \text{ g} = 0.756 \text{ g}$

CHECK	
Check your answer. Are the units correct? Does the answer make physical sense?	The units of the answer are correct, and the value of the answer makes sense. Drug doses can vary over some range, but in many cases they are between 0 and 1 gram.

▶ **SKILLBUILDER 2.12 | Solving Unit Conversions in the Numerator and Denominator**

A car is driving at a velocity of 65 km/hr. What is the car's velocity in m/s?

▶ **FOR MORE PRACTICE** Example 2.27; Problems 95–98.

2.9 Units Raised to a Power

▶ Convert units raised to a power.

When converting quantities with units raised to a power, such as cubic centimeters (cm^3), you must raise the conversion factor to that power. For example, suppose you want to convert the size of a motorcycle engine reported as 1255 cm^3 to cubic inches. You know that:

$$2.54 \text{ cm} = 1 \text{ in.}$$

The unit cm³ is often abbreviated as cc.

Most tables of conversion factors do not include conversions between cubic units, but you can derive them from the conversion factors for the basic units. You cube both sides of the preceding equality to obtain the proper conversion factor.

$$(2.54 \text{ cm})^3 = (1 \text{ in.})^3$$
$$(2.54)^3 \text{ cm}^3 = 1^3 \text{ in.}^3$$
$$16.387 \text{ cm}^3 = 1 \text{ in.}^3$$

You can do the same thing in fractional form.

$$\frac{1 \text{ in.}}{2.54 \text{ cm}} = \frac{(1 \text{ in.})^3}{(2.54 \text{ cm})^3} = \frac{1 \text{ in.}^3}{16.387 \text{ cm}^3}$$

You then proceed with the conversion in the usual manner.

SOLUTION MAP

$$cm^3 \rightarrow in.^3$$

$$\frac{1 \text{ in.}^3}{16.387 \text{ cm}^3}$$

2.54 cm = 1 in. is an exact conversion factor. After cubing, you retain five significant figures so that the conversion factor does not limit the four significant figures of your original quantity (1255 cm³).

SOLUTION

$$1255 \text{ cm}^3 \times \frac{1 \text{ in.}^3}{16.387 \text{ cm}^3} = 76.5851 \text{ in.}^3 = 76.59 \text{ in.}^3$$

CHEMISTRY AND HEALTH

Drug Dosage

The unit of choice in specifying drug dosage is the milligram (mg). A bottle of aspirin, Tylenol®, or any other common drug lists the number of milligrams of the active ingredient contained in each tablet, as well as the number of tablets to take per dose. The following table shows the mass of the active ingredient per pill in several common pain relievers, all reported in milligrams. The remainder of each tablet is composed of inactive ingredients such as cellulose (or fiber) and starch.

The recommended adult dose for many of these pain relievers is one or two tablets every four to eight hours (depending on the specific pain reliever). Notice that the extra-strength version of each pain reliever just contains a higher dose of the same compound found in the regular-strength version. For the pain relievers listed, three regular-strength tablets are the equivalent of two extra-strength tablets (and probably cost less).

The dosages given in the table are fairly standard for each drug, regardless of the brand. On most drugstore shelves,

there are many different brands of regular-strength ibuprofen, some sold under the generic name and others sold under their brand names (such as Advil®). However, if you look closely at the labels, you will find that they all contain the same thing: 200 mg of the compound ibuprofen. There is no difference in the compound or in the amount of the compound. Yet these pain relievers will most likely all have different prices. Choose the least expensive. Why pay more for the same thing?

B2.2 CAN YOU ANSWER THIS? *Convert each of the doses in the table to ounces. Why are drug dosages not listed in ounces?*

Drug Mass per Pill for Common Pain Relievers	
Pain Reliever	**Mass of Active Ingredient per Pill**
aspirin	325 mg
aspirin, extra strength	500 mg
ibuprofen (Advil®)	200 mg
ibuprofen, extra strength	300 mg
acetaminophen (Tylenol®)	325 mg
acetaminophen, extra strength	500 mg

EXAMPLE **2.13** | **Converting Quantities Involving Units Raised to a Power**

A circle has an area of 2659 cm². What is its area in square meters?

SORT	
You are given an area in square centimeters and asked to convert the area to square meters.	**GIVEN:** 2659 cm² **FIND:** m²

STRATEGIZE	
Build a solution map beginning with cm² and ending with m². Remember that you must square the conversion factor.	**SOLUTION MAP** 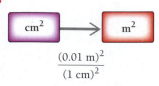 $$\frac{(0.01 \text{ m})^2}{(1 \text{ cm})^2}$$ **RELATIONSHIPS USED** 1 cm = 0.01 m (from Table 2.2)

SOLVE	
Follow the solution map to solve the problem. Square the conversion factor (both the units and the number) as you carry out the calculation. Round the answer to four significant figures to reflect the four significant figures in the given quantity. The conversion factor is exact and therefore does not limit the number of significant figures.	**SOLUTION** $$2659 \text{ cm}^2 \times \frac{(0.01 \text{ m})^2}{(1 \text{ cm})^2} = 2659 \cancel{\text{ cm}^2} \times \frac{10^{-4} \text{ m}^2}{1 \cancel{\text{ cm}^2}}$$ $$= 0.265900 \text{ m}^2$$ $$= 0.2659 \text{ m}^2$$

CHECK	
Check your answer. Are the units correct? Does the answer make physical sense?	The units of the answer are correct, and the magnitude makes physical sense. A square meter is much larger than a square centimeter, so the value in square meters should be much smaller than the value in square centimeters.

▶ **SKILLBUILDER 2.13** | **Converting Quantities Involving Units Raised to a Power**

An automobile engine has a displacement (a measure of the size of the engine) of 289.7 in.³ What is its displacement in cubic centimeters?

▶ **FOR MORE PRACTICE** Example 2.28; Problems 87, 88, 89, 90, 91, 92.

 Interactive Worked Example Video 2.14

EXAMPLE **2.14** | **Solving Multistep Conversion Problems Involving Units Raised to a Power**

The average annual per person crude oil consumption in the United States is 15,615 dm³. What is this value in cubic inches?

SORT	
You are given a volume in cubic decimeters and asked to convert it to cubic inches.	**GIVEN:** 15,615 dm³ **FIND:** in.³

STRATEGIZE	
Build a solution map beginning with dm³ and ending with in.³ You must cube each of the conversion factors because the quantities involve cubic units.	**SOLUTION MAP** 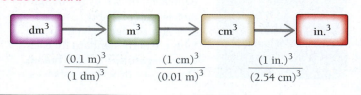 $$\frac{(0.1 \text{ m})^3}{(1 \text{ dm})^3} \qquad \frac{(1 \text{ cm})^3}{(0.01 \text{ m})^3} \qquad \frac{(1 \text{ in.})^3}{(2.54 \text{ cm})^3}$$

RELATIONSHIPS USED

1 dm = 0.1 m (from Table 2.2)

1 cm = 0.01 m (from Table 2.2)

2.54 cm = 1 in. (from Table 2.3)

SOLVE	**SOLUTION**
Follow the solution map to solve the problem. Begin with the given value in dm^3 and multiply by the string of conversion factors to arrive at $in.^3$ Be sure to cube each conversion factor as you carry out the calculation. Round the answer to five significant figures to reflect the five significant figures in the least precisely known quantity (15,615 dm^3). The conversion factors are all exact and therefore do not limit the number of significant figures.	$$15{,}615 \ dm^3 \times \frac{(0.1 \ m)^3}{(1 \ dm)^3} \times \frac{(1 \ cm)^3}{(0.01 \ m)^3} \times \frac{(1 \ in.)^3}{(2.54 \ cm)^3}$$ $$= 9.5289 \times 10^5 \ in.^3$$
CHECK Check your answer. Are the units correct? Does the answer make physical sense?	The units of the answer are correct, and the magnitude makes sense. A cubic inch is smaller than a cubic decimeter, so the value in cubic inches should be larger than the value in cubic decimeters.

▶ **SKILLBUILDER 2.14** | **Solving Multistep Problems Involving Units Raised to a Power**

How many cubic inches are there in 3.25 yd^3?

▶ **FOR MORE PRACTICE** Problems 93, 94.

PEARSON
eText
2.0

CONCEPTUAL ✔ CHECKPOINT 2.7

You know that there are 3 ft in a yard. How many cubic feet are there in a cubic yard?

(a) 3   **(b)** 6   **(c)** 9   **(d)** 27

2.10 Density

▶ Calculate the density of a substance.
▶ Use density as a conversion factor.

Why do some people pay over $3000 for a bicycle made of titanium? A steel frame would be just as strong for a fraction of the cost. The difference between the two bikes is their masses—the titanium bike is lighter. For a given volume of metal, titanium has less mass than steel. Titanium is said to be *less dense* than steel. The **density** of a substance is the ratio of its mass to its volume.

$$density = \frac{mass}{volume} \quad or \quad d = \frac{m}{V}$$

Density is a fundamental property of a substance and differs from one substance to another. The units of density are those of mass divided by those of volume, most conveniently expressed in grams per cubic centimeter (g/cm^3) or grams per milliliter (g/mL). Table 2.4 on the next page lists the densities of some common substances. Aluminum is among the least dense structural metals with a density of 2.70 g/cm^3, while platinum is among the densest with a density of 21.4 g/cm^3. Titanium has a density of 4.50 g/cm^3.

▲ Top-end bicycle frames are made of titanium because of titanium's low density and high relative strength. Titanium has a density of 4.50 g/cm^3, while iron, for example, has a density of 7.86 g/cm^3.

TABLE 2.4 Densities of Some Common Substances	
Substance	**Density (g/cm³)**
charcoal, oak	0.57
ethanol	0.789
ice	0.92
water	1.0
glass	2.6
aluminum	2.7
titanium	4.50
iron	7.86
copper	8.96
lead	11.4
gold	19.3
platinum	21.4

> Remember that cubic centimeters and milliliters are equivalent units.

Calculating Density

You can calculate the density of a substance by dividing the mass of a given amount of the substance by its volume. For example, a sample of liquid has a volume of 22.5 mL and a mass of 27.2 g. To find its density, use the equation $d = m/V$.

$$d = \frac{m}{V} = \frac{27.2 \text{ g}}{22.5 \text{ mL}} = 1.21 \text{ g/mL}$$

You can use a solution map to solve problems involving equations, but the solution map takes a slightly different form than one for a pure conversion problem. In a problem involving an equation, the solution map shows how the *equation* takes you from the *given* quantities to the *find* quantity. The solution map for this problem is:

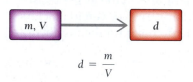

$$d = \frac{m}{V}$$

The solution map illustrates that the values of m and V, when substituted into the equation $d = \frac{m}{V}$, give the desired result, d.

EXAMPLE **2.15** **Calculating Density**

A jeweler offers to sell a ring to a woman and tells her that it is made of platinum. Noting that the ring feels a little light, the woman decides to perform a test to determine the ring's density. She places the ring on a balance and finds that it has a mass of 5.84 g. She also finds that the ring *displaces* 0.556 cm³ of water. Is the ring made of platinum? The density of platinum is 21.4 g/cm³. (The displacement of water is a common way to measure the volume of irregularly shaped objects. To say that an object *displaces* 0.556 cm³ of water means that when the object is submerged in a container of water filled to the brim, 0.556 cm³ overflows. Therefore, the volume of the object is 0.556 cm³.)

SORT You are given the mass and volume of the ring and asked to find the density.	**GIVEN:** $m = 5.84$ g $V = 0.556$ cm³ **FIND:** density in g/cm³
STRATEGIZE If the ring is platinum, its density should match that of platinum. Build a solution map that represents how you get from the given quantities (mass and volume) to the find quantity (density). Unlike in conversion problems, where you write a conversion factor beneath the arrow, here you write the equation for density beneath the arrow.	**SOLUTION MAP** 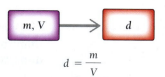 $d = \frac{m}{V}$ **RELATIONSHIPS USED** $d = \frac{m}{V}$ (equation for density)
SOLVE Follow the solution map. Substitute the given values into the density equation and calculate the density. Round the answer to three significant figures to reflect the three significant figures in the given quantities.	**SOLUTION** $d = \frac{m}{V} = \frac{5.84 \text{ g}}{0.556 \text{ cm}^3} = 10.5 \text{ g/cm}^3$ The density of the ring is much too low to be platinum; therefore, the ring is a fake.

CHECK	
Check your answer. Are the units correct? Does the answer make physical sense?	The units of the answer are correct, and it seems that the magnitude could be an actual density. As you can see from Table 2.4, the densities of liquids and solids range from below 1 g/cm³ to just over 20 g/cm³.

▶ **SKILLBUILDER 2.15** | **Calculating Density**

The woman takes the ring back to the jewelry shop, where she is met with endless apologies. The jeweler had accidentally made the ring out of silver rather than platinum. The jeweler gives her a new ring that she promises is platinum. This time when the customer checks the density, she finds the mass of the ring to be 9.67 g and its volume to be 0.452 cm³. Is this ring genuine?

▶ **FOR MORE PRACTICE**   Example 2.29; Problems 95, 96, 97, 98, 99, 100.

▲ A graduated cylinder is used to measure the volume of a liquid in the laboratory.

PEARSON
eText
2.0
Interactive
Worked Example
Video 2.16

Density as a Conversion Factor

You can use the density of a substance as a conversion factor between the mass of the substance and its volume. For example, suppose you need 68.4 g of a liquid with a density of 1.32 g/cm³ and want to measure the correct volume with a graduated cylinder (a piece of laboratory glassware used to measure volume). How much volume should you measure?

Start with the mass of the liquid and use the density as a conversion factor to convert mass to volume. However, you must use the inverted density expression 1 cm³/1.32 g because you want g, the unit you are converting from, to be on the bottom (in the denominator) and cm³, the unit you are converting to, on the top (in the numerator). The solution map takes this form:

SOLUTION MAP

$$ \boxed{g} \longrightarrow \boxed{cm^3} \longrightarrow \boxed{mL} $$

$$ \frac{1\ cm^3}{1.32\ g} \qquad \frac{1\ mL}{1\ cm^3} $$

SOLUTION

$$ 68.4\ \cancel{g} \times \frac{1\ \cancel{cm^3}}{1.32\ \cancel{g}} \times \frac{1\ mL}{1\ \cancel{cm^3}} = 51.8\ mL $$

You must measure 51.8 mL to obtain 68.4 g of the liquid.

EXAMPLE **2.16**   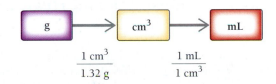 **Density as a Conversion Factor**

The gasoline in an automobile gas tank has a mass of 60.0 kg and a density of 0.752 g/cm³. What is its volume in cm³?

SORT	
You are given the mass in kilograms and asked to find the volume in cubic centimeters. Density is the conversion factor between mass and volume.	**GIVEN:** 60.0 kg density = 0.752 g/ cm³ **FIND:** volume in cm³

STRATEGIZE	
Build the solution map starting with kg and ending with cm³. Use the density (inverted) to convert from g to cm³.	**SOLUTION MAP** $\boxed{kg} \longrightarrow \boxed{g} \longrightarrow \boxed{cm^3}$ $\frac{1000\ g}{1\ kg} \qquad \frac{1\ cm^3}{0.752\ g}$

	RELATIONSHIPS USED
	0.752 g/cm^3 (given in problem)
	$1000 \text{ g} = 1 \text{ kg}$ (from Table 2.2)
SOLVE Follow the solution map to solve the problem. Round the answer to three significant figures to reflect the three significant figures in the given quantities.	**SOLUTION** $60.0 \text{ kg} \times \dfrac{1000 \text{ g}}{1 \text{ kg}} \times \dfrac{1 \text{ cm}^3}{0.752 \text{ g}} = 7.98 \times 10^4 \text{ cm}^3$
CHECK Check your answer. Are the units correct? Does the answer make physical sense?	The units of the answer are those of volume, so they are correct. The magnitude seems reasonable because the density is somewhat less than 1 g/cm^3; therefore, the volume of 60.0 kg should be somewhat more than $60.0 \times 10^3 \text{ cm}^3$.

▶ **SKILLBUILDER 2.16 | Density as a Conversion Factor**

A drop of acetone (nail polish remover) has a mass of 35 mg and a density of 0.788 g/cm^3. What is its volume in cubic centimeters?

▶ **SKILLBUILDER PLUS**

A steel cylinder has a volume of 246 cm^3 and a density of 7.93 g/cm^3. What is its mass in kilograms?

▶ **FOR MORE PRACTICE**   Example 2.30; Problems 101, 102.

CHEMISTRY AND HEALTH

Density, Cholesterol, and Heart Disease

Cholesterol is the fatty substance found in animal-derived foods such as beef, eggs, fish, poultry, and milk products. The body uses cholesterol for several purposes. However, excessive amounts of cholesterol in the blood—which can be caused by both genetic factors and diet—may result in the deposition of cholesterol in arterial walls, leading to a condition called atherosclerosis, or blocking of the arteries. These blockages are dangerous because they inhibit blood flow to important organs, causing heart attacks and strokes. The risk of stroke and heart attack increases with increasing blood cholesterol levels (Table 2.5). Cholesterol is carried in the bloodstream by a class of substances known as lipoproteins. Lipoproteins are often separated and classified according to their density.

The main carriers of blood cholesterol are low-density lipoproteins (LDLs). LDLs, also called bad cholesterol, have a density of 1.04 g/cm^3. They are bad because they tend to deposit cholesterol on arterial walls, increasing the risk of stroke and heart attack. Cholesterol is also carried by high-density lipoproteins (HDLs). HDLs, called good cholesterol, have a density of 1.13 g/cm^3. HDLs transport cholesterol to the liver for processing and excretion and therefore have a tendency to reduce cholesterol on arterial walls. Too low a level of HDLs (below 35 mg/100 mL) is considered a risk factor for heart disease. Exercise, along with a diet low in saturated fats, is believed to raise HDL levels in the blood while lowering LDL levels.

B2.3 CAN YOU ANSWER THIS?   *What mass of low-density lipoprotein is contained in a cylinder that is 1.25 cm long and 0.50 cm in diameter? (The volume of a cylinder, V, is given by $V = \pi r^2 \ell$, where r is the radius of the cylinder and ℓ is its length.)*

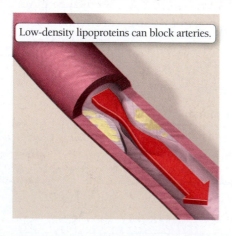

Low-density lipoproteins can block arteries.

TABLE 2.5   Risk of Stroke and Heart Attack versus Blood Cholesterol Level

Risk Level	Total Blood Cholesterol (mg/100 mL)	LDL (mg/100 mL)
low	< 200	< 130
borderline	200–239	130–159
high	> 240	> 160

2.11 Numerical Problem-Solving Strategies and the Solution Map

In this chapter, you have seen a few examples of how to solve numerical problems. In Section 2.6, you encountered a procedure to solve simple unit conversion problems. You then learned how to modify that procedure to work with multistep unit conversion problems and problems involving an equation. This section summarizes and generalizes these procedures and includes two additional examples. Just as was done in Section 2.6, the left column shows the general procedure for solving numerical problems, and the center and right columns illustrate the application of the procedure to the two examples.

Solving Numerical Problems	EXAMPLE **2.17** **UNIT CONVERSION** A chemist needs a 23.5-kg sample of ethanol for a large-scale reaction. What volume in liters of ethanol should the chemist use? The density of ethanol is 0.789 g/cm³.	EXAMPLE **2.18** **UNIT CONVERSION WITH EQUATION** A 55.9-kg person displaces 57.2 L of water when submerged in a water tank. What is the density of the person in grams per cubic centimeter?
1. SORT • Scan the problem for one or more numbers and their associated units. This number (or numbers) is (are) the starting point(s) of the calculation. Write them down as given. • Scan the problem to determine what you are asked to find. Sometimes the units of this quantity are implied; other times they are specified. Write down the quantity and/or units you are asked to find.	**GIVEN:** 23.5 kg ethanol density $= 0.789$ g/cm³ **FIND:** volume in L	**GIVEN:** $m = 55.9$ kg $V = 57.2$ L **FIND:** density in g/cm³
2. STRATEGIZE • For problems involving only conversions, focus on units. The solution map shows how to get from the units in the given quantity to the units in the quantity you are asked to find. • For problems involving equations, focus on the equation. The solution map shows how the equation takes you from the given quantity (or quantities) to the quantity you are asked to find. • Some problems may involve both unit conversions and equations, in which case the solution map employs both of the above points.	**SOLUTION MAP** 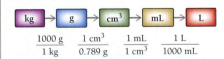 $$\frac{1000\ \text{g}}{1\ \text{kg}} \quad \frac{1\ \text{cm}^3}{0.789\ \text{g}} \quad \frac{1\ \text{mL}}{1\ \text{cm}^3} \quad \frac{1\ \text{L}}{1000\ \text{mL}}$$ **RELATIONSHIPS USED** 0.789 g/cm³ (given in problem) 1000 g $= 1$ kg (Table 2.2) 1000 mL $= 1$ L (Table 2.2) 1 mL $= 1$ cm³ (Table 2.3)	**SOLUTION MAP** $$d = \frac{m}{V}$$ **RELATIONSHIPS USED** $d = \dfrac{m}{V}$ (definition of density)

3. SOLVE

- For problems involving only conversions, begin with the given quantity and its units. Multiply by the appropriate conversion factor(s), canceling units, to arrive at the quantity you are asked to find.

- For problems involving equations, solve the equation to arrive at the quantity you are asked to find. (Use algebra to rearrange the equation so that the quantity you are asked to find is isolated on one side.) Gather each of the quantities that must go into the equation in the correct units. (Convert to the correct units using additional solution maps if necessary.) Finally, substitute the numerical values and their units into the equation and calculate the answer.

- Round the answer to the correct number of significant figures. Use the significant figure rules from Sections 2.3 and 2.4.

SOLUTION

$$23.5 \text{ kg} \times \frac{1000 \text{ g}}{1 \text{ kg}} \times \frac{1 \text{ cm}^3}{0.789 \text{ g}} \times$$

$$\frac{1 \text{ mL}}{1 \text{ cm}^3} \times \frac{1 \text{ L}}{1000 \text{ mL}} = 29.7845 \text{ L}$$

$$29.7845 \text{ L} = 29.8 \text{ L}$$

The equation is already solved for the find quantity. Convert mass from kilograms to grams.

$$m = 55.9 \text{ kg} \times \frac{1000 \text{ g}}{1 \text{ kg}}$$

$$= 5.59 \times 10^4 \text{ g}$$

Convert volume from liters to cubic centimeters.

$$V = 57.2 \text{ L} \times \frac{1000 \text{ mL}}{1 \text{ L}} \times \frac{1 \text{ cm}^3}{1 \text{ mL}}$$

$$= 57.2 \times 10^3 \text{ cm}^3$$

Calculate density.

$$d = \frac{m}{V} = \frac{55.9 \times 10^3 \text{ g}}{57.2 \times 10^3 \text{ cm}^3}$$

$$= 0.9772727 \frac{\text{g}}{\text{cm}^3}$$

$$= 0.977 \frac{\text{g}}{\text{cm}^3}$$

4. CHECK

- Does the magnitude of the answer make physical sense? Are the units correct?

The units are correct (L) and the magnitude is reasonable. Because the density is less than 1 g/cm³, the calculated volume (29.8 L) should be greater than the mass (23.5 kg).

The units are correct. Because the mass in kilograms and the volume in liters are very close to each other in magnitude, it makes sense that the density is close to 1 g/cm³.

▶ SKILLBUILDER 2.17 | Unit Conversion

A pure gold metal bar displaces 0.82 L of water. What is its mass in kilograms? (The density of gold is 19.3 g/cm³.)

▶ SKILLBUILDER 2.18 | Unit Conversion with Equation

A gold-colored pebble is found in a stream. Its mass is 23.2 mg, and its volume is 1.20 mm³. What is its density in grams per cubic centimeter? Is it gold? (The density of gold = 19.3 g/cm³.)

▶ FOR MORE PRACTICE Problems 103, 109, 110, 111, 112.

▶ FOR MORE PRACTICE Problems 103, 104, 105, 106.

Chapter 2 in Review

MasteringChemistry™ provides end-of-chapter exercises, feedback-enriched tutorial problems, animations, and interactive activities to encourage problem solving practice and deeper understanding of key concepts and topics.

Self-Assessment Quiz

PEARSON eText 2.0

The eText 2.0 icon indicates that this feature is embedded and interactive in the eText.

Q1. Express the number 0.000042 in scientific notation.
(a) 0.42×10^{-4}
(b) 4.2×10^{-5}
(c) 4.2×10^{-4}
(d) 4×10^{-5}

Q2. A graduated cylinder has markings every milliliter. Which measurement is accurately reported for this graduated cylinder?
(a) 21 mL
(b) 21.2 mL
(c) 21.23 mL
(d) 21.232 mL

Q3. How many significant figures are in the number 0.00620?
(a) 2
(b) 3
(c) 4
(d) 5

Q4. Round the number 89.04997 to three significant figures.
(a) 89.03
(b) 89.04
(c) 89.1
(d) 89.0

Q5. Perform this multiplication to the correct number of significant figures: $65.2 \times 0.0015 \times 12.02$
(a) 1.17
(b) 1.18
(c) 1.2
(d) 1.176

Q6. Perform this addition to the correct number of significant figures: $8.32 + 12.148 + 0.02$
(a) 20.488
(b) 20.49
(c) 20.5
(d) 21

Q7. Perform this calculation to the correct number of significant figures: $78.222 \times (12.02 - 11.52)$
(a) 39
(b) 39.1
(c) 39.11
(d) 39.111

Q8. Convert 76.8 cm to m.
(a) 0.0768 m
(b) 7.68 m
(c) 0.768 m
(d) 7.68×10^{-2} m

Q9. Convert 2855 mg to kg.
(a) 2.855×10^{-3} kg
(b) 2.855 kg
(c) 0.02855 kg
(d) 3.503×10^{-4} kg

Q10. A runner runs 4875 ft in 6.85 minutes. What is the runner's average speed in miles per hour?
(a) 1.34 mi/hr
(b) 0.0022 mi/hr
(c) 8.09 mi/hr
(d) 8.087 mi/hr

Q11. An automobile travels 97.2 km on 7.88 L of gasoline. What is the gas mileage for the automobile in miles per gallon?
(a) 2.02 mi/gal
(b) 7.67 mi/gal
(c) 0.034 mi/gal
(d) 29.0 mi/gal

Q12. Convert 876.9 in.^{3} to m^3.
(a) 0.01437 m^3
(b) 22.27 m^3
(c) 5.351×10^7 m^3
(d) 0.014 m^3

Q13. Convert 27 m/s to km/hr.
(a) 97 km/hr
(b) 7.5 km/hr
(c) 1.6 km/hr
(d) 0.027 km/hr

Q14. A cube measures 2.5 cm on each edge and has a mass of 66.9 g. Calculate the density of the material that composes the cube. (The volume of a cube is equal to the edge length cubed.)
(a) 10.7 g/cm^3
(b) 4.3 g/cm^3
(c) 0.234 g/cm^3
(d) 26.7 g/cm^3

Q15. What is the mass of 225 mL of a liquid that has a density of 0.880 g/mL?
(a) 198 g
(b) 0.0039 g
(c) 0.198 g
(d) 0.256 g

Q16. What is the edge length of a 155-g iron cube? (The density of iron is 7.86 g/cm^3, and the volume of a cube is equal to the edge length cubed.)
(a) 0.0197 cm
(b) 19.7 cm
(c) 1218 cm
(d) 2.70 cm

Answers: 1:b, 2:b, 3:b, 4:d, 5:c, 6:b, 7:a, 8:c, 9:a, 10:c, 11:d, 12:a, 13:a, 14:a, 15:b, 16:d

Chemical Principles

Relevance

Uncertainty

Scientists report measured quantities so that the number of digits reflects the certainty in the measurement. Write measured quantities so that every digit is certain except the last, which is estimated.

Measurement is a hallmark of science, and you must communicate the precision of a measurement so that others know how reliable the measurement is. When you write or manipulate measured quantities, you must show and retain the precision with which the original measurement was made.

Units

Measured quantities usually have units associated with them. The SI unit for length is the meter; for mass, the kilogram; and for time, the second. Prefix multipliers such as *kilo-* or *milli-* are often used in combination with these basic units. The SI units of volume are units of length raised to the third power; liters or milliliters are often used as well.

The units in a measured quantity communicate what the quantity actually is. Without an agreed-on system of units, scientists could not communicate their measurements. Units are also important in calculations, and the tracking of units throughout a calculation is essential.

Chemical Principles

Relevance

Density

The density of a substance is its mass divided by its volume, $d = m/V$, and is usually reported in units of grams per cubic centimeter or grams per milliliter. Density is a fundamental property of all substances and generally differs from one substance to another.

The density of substances is an important consideration in choosing materials for manufacturing and production. Airplanes, for example, are made of low-density materials, while bridges are made of higher-density materials. Density is important as a conversion factor between mass and volume and vice versa.

Chemical Skills

Examples

LO: Express very large and very small numbers using scientific notation (Section 2.2).

To express a number in scientific notation:

- Move the decimal point to obtain a number between 1 and 10.
- Write the decimal part multiplied by 10 raised to the number of places you moved the decimal point.
- The exponent is positive if you moved the decimal point to the left and negative if you moved the decimal point to the right.

EXAMPLE 2.19   Scientific Notation

Express the number 45,000,000 in scientific notation.

$$45,000,000$$

7 6 5 4 3 2 1

$$4.5 \times 10^7$$

LO: Report measured quantities to the right number of digits (Section 2.3).

Report measured quantities so that every digit is certain except the last, which is estimated.

EXAMPLE 2.20   Reporting Measured Quantities to the Right Number of Digits

Record the volume of liquid in the graduated cylinder to the correct number of digits. Laboratory glassware is calibrated (and should therefore be read) from the bottom of the meniscus, the curved surface at the top of a column of liquid (see figure).

Meniscus

Because the graduated cylinder has markings every 0.1 mL, you should record the measurement to the nearest 0.01 mL. In this case, that is 4.57 mL.

LO: Determine which digits in a number are significant (Section 2.3).

Always count the following as significant:

- nonzero digits
- interior zeros
- trailing zeros after a decimal point
- trailing zeros before a decimal point but after a nonzero number

Never count the following digits as significant:

- zeros to the left of the first nonzero number

The following digits are ambiguous, and you should avoid them by using scientific notation:

- zeros at the end of a number but before a decimal point

EXAMPLE 2.21 | **Counting Significant Digits**

How many significant figures are in the following numbers?

1.0050	five significant figures
0.00870	three significant figures
100.085	six significant digits
5400	It is not possible to tell in its current form.

The number must be written as 5.4×10^3, 5.40×10^3, or 5.400×10^3, depending on the number of significant figures intended.

LO: Round numbers to the correct number of significant figures (Section 2.4).

When rounding numbers to the correct number of significant figures, round down if the last digit dropped is 4 or less; round up if the last digit dropped is 5 or more.

EXAMPLE 2.22 | **Rounding**

Round 6.442 and 6.456 to two significant figures each.

6.442 rounds to 6.4
6.456 rounds to 6.5

LO: Determine the correct number of significant figures in the results of multiplication and division calculations (Section 2.4).

The result of a multiplication or division should carry the same number of significant figures as the factor with the least number of significant figures.

EXAMPLE 2.23 | **Significant Figures in Multiplication and Division**

Perform the calculation and report the answer to the correct number of significant figures.

$$8.54 \times 3.589 \div 4.2$$
$$= 7.2976$$
$$= 7.3$$

Round the final result to two significant figures to reflect the two significant figures in the factor with the least number of significant figures (4.2).

LO: Determine the correct number of significant figures in the results of addition and subtraction calculations (Section 2.4).

The result of an addition or subtraction should carry the same number of decimal places as the quantity carrying the least number of decimal places.

EXAMPLE 2.24 | **Significant Figures in Addition and Subtraction**

Perform the operation and report the answer to the correct number of significant figures.

$$
\begin{array}{r}
3.098 \\
+0.67 \\
-0.9452 \\
\hline
2.8228 = 2.82
\end{array}
$$

Round the final result to two decimal places to reflect the two decimal places in the quantity with the least number of decimal places (0.67).

LO: Determine the correct number of significant figures in the results of calculations involving both addition/subtraction and multiplication/division (Section 2.4).

In calculations involving both addition/subtraction and multiplication/division, do the steps in parentheses first, keeping track of how many significant figures are in the answer by underlining the least significant figure, then proceeding with the remaining steps. Do not round off until the very end.

EXAMPLE 2.25 | **Significant Figures in Calculations Involving Both Addition/Subtraction and Multiplication/Division**

Perform the operation and report the answer to the correct number of significant figures.

$$8.16 \times (5.4323 - 5.411)$$
$$= 8.16 \times 0.021\underline{3}$$
$$= 0.1738 = 0.17$$

LO: Convert between units (Sections 2.6, 2.7).

Solve unit conversion problems by following these steps.

1. **Sort** Write down the given quantity and its units and the quantity you are asked to find and its units.

2. **Strategize** Draw a solution map showing how to get from the given quantity to the quantity you are asked to find.

3. **Solve** Follow the solution map. Starting with the given quantity and its units, multiply by the appropriate conversion factor(s), canceling units, to arrive at the quantity to find in the desired units. Round the final answer to the correct number of significant figures.

4. **Check** Are the units correct? Does the answer make physical sense?

EXAMPLE 2.26 | **Unit Conversion**

Convert 108 ft to meters.

GIVEN: 108 ft

FIND: m

SOLUTION MAP

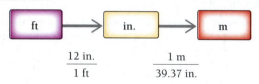

$$\frac{12 \text{ in.}}{1 \text{ ft}} \qquad \frac{1 \text{ m}}{39.37 \text{ in.}}$$

RELATIONSHIPS USED
1 m = 39.37 in. (Table 2.3)
1 ft = 12 in. (by definition)

SOLUTION

$$108 \text{ ft} \times \frac{12 \text{ in.}}{1 \text{ ft}} \times \frac{1 \text{ m}}{39.37 \text{ in.}}$$
$$= 32.918 \text{ m}$$
$$= 32.9 \text{ m}$$

The answer has the right units (meters), and it makes sense; because a meter is longer than a foot, the number of meters should be less than the number of feet.

LO: Convert units in a quantity that has units in the numerator and the denominator (Section 2.8).

1. **Sort** Write down the given quantity and its units and the quantity you are asked to find and its units.

2. **Strategize** Draw a solution map showing how to get from the given quantity to the quantity you are asked to find. Because the units are squared, you must square the conversion factor.

EXAMPLE 2.27 | **Unit Conversion in the Numerator and Denominator**

A drug company lists the dose of a drug as 0.012 g per pound of body weight. What is the dose of the drug in mg per kilogram of body weight?

GIVEN: $0.012 \dfrac{\text{g}}{\text{lb}}$

FIND: $\dfrac{\text{mg}}{\text{kg}}$

SOLUTION MAP

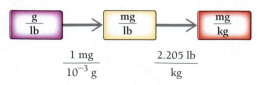

$$\frac{1 \text{ mg}}{10^{-3} \text{ g}} \qquad \frac{2.205 \text{ lb}}{\text{kg}}$$

3. **Solve** Follow the solution map. Starting with the given quantity and its units, multiply by the appropriate conversion factor(s), canceling units, to arrive at the quantity you are asked to find in the desired units. Don't forget to square the conversion factor for squared units.

4. **Check** Are the units correct? Does the answer make physical sense?

RELATIONSHIPS USED

1 mg = 10^{-3} g (Table 2.2)
1 kg = 2.205 lb (Table 2.3)

SOLUTION

$$0.012 \,\frac{\cancel{g}}{\cancel{lb}} \times \frac{mg}{10^{-3}\,\cancel{g}} \times \frac{2.205\,\cancel{lb}}{1\,kg} = 26.46 \,\frac{mg}{kg}$$

The units of the answer are correct, and the value of the answer makes sense. The dose in mg/kg should be larger than the dose in g/lb because the mg is a smaller unit than kg (so it takes more to make the same amount) and the kg is a larger unit than lb (so it takes more of the drug per kg than per lb).

LO: Convert units raised to a power (Section 2.9).

When working problems involving units raised to a power, raise the conversion factors to the same power.

How many square meters are in 1.0 km²?

GIVEN: 1.0 km²

FIND: m²

SOLUTION MAP

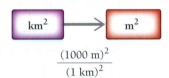

$$\frac{(1000\ m)^2}{(1\ km)^2}$$

1. **Sort** Begin by sorting the information in the problem into given and find. You are given the dose of the drug in g/lb. You are asked to find the dose in mg/kg.

2. **Strategize** Convert the numerator from g to mg; then convert the denominator from lb to kg.

RELATIONSHIPS USED

1 km = 1000 m (Table 2.2)

SOLUTION

3. **Solve** Follow the solution map to solve the problem. Begin with 0.012 g/lb and multiply by the two conversion factors to arrive at the dose in mg/kg. Round to two significant figures to reflect the two significant figures in the least precisely known quantity.

$$1.0\ km^2 \times \frac{(1000\ m)^2}{(1\ km)^2}$$

$$= 1.0\ \cancel{km^2} \times \frac{1 \times 10^6\ m^2}{1\ \cancel{km^2}}$$

$$= 1.0 \times 10^6\ m^2$$

4. **Check** Check your answer. Are the units correct? Does the answer make physical sense?

The units are correct. The answer makes physical sense. A square meter is much smaller than a square kilometer, so the number of square meters should be much larger than the number of square kilometers.

LO: Calculate the density of a substance (Section 2.10).

The density of an object or substance is its mass divided by its volume.

$$d = \frac{m}{V}$$

1. **Sort** Write down the given quantity and its units and the quantity you are asked to find and its units.

2. **Strategize** Draw a solution map showing how to get from the given quantity to the quantity you are asked to find. Use the definition of density as the equation that takes you from the mass and the volume to the density.

3. **Solve** Substitute the correct values into the equation for density.

4. **Check** Are the units correct? Does the answer make physical sense?

EXAMPLE **2.29**   **Calculating Density**

An object has a mass of 23.4 g and displaces 5.7 mL of water. Determine its density in grams per milliliter.

GIVEN:
$$m = 23.4 \text{ g}$$
$$V = 5.7 \text{ mL}$$

FIND: density in g/mL

SOLUTION MAP

$$d = \frac{m}{V}$$

RELATIONSHIPS USED

$d = \dfrac{m}{V}$ (definition of density)

SOLUTION

$$d = \frac{m}{V} = \frac{23.4 \text{ g}}{5.7 \text{ mL}} = 4.11 \text{ g/mL} = 4.1 \text{ g/mL}$$

The units (g/mL) are units of density. The answer is in the range of values for the densities of liquids and solids (see Table 2.4).

LO: Use density as a conversion factor (Section 2.10).

You can use density as a conversion factor from mass to volume or from volume to mass. To convert between volume and mass, use density directly. To convert between mass and volume, invert the density.

1. **Sort** Write down the given quantity and its units and the quantity you are asked to find and its units.

2. **Strategize** Draw a solution map showing how to get from the given quantity to the quantity you are asked to find. Use the inverse of the density to convert from g to mL.

3. **Solve** Begin with the given quantity and multiply by the appropriate conversion factors to arrive at the quantity you are asked to find. Round to the correct number of significant figures.

4. **Check** Are the units correct? Does the answer make physical sense?

EXAMPLE **2.30**   **Density as a Conversion Factor**

What is the volume in liters of 321 g of a liquid with a density of 0.84 g/mL?

GIVEN: 321 g

FIND: volume in L

SOLUTION MAP

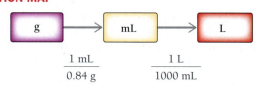

$$\frac{1 \text{ mL}}{0.84 \text{ g}} \qquad \frac{1 \text{ L}}{1000 \text{ mL}}$$

RELATIONSHIPS USED

0.84 g/mL (given in the problem)
1 L = 1000 mL (Table 2.2)

SOLUTION

$$321 \text{ g} \times \frac{1 \text{ mL}}{0.84 \text{ g}} \times \frac{1 \text{ L}}{1000 \text{ mL}}$$

$$= 0.382 \text{ L} = 0.38 \text{ L}$$

The answer is in the correct units. The magnitude seems right because the density is slightly less than 1; therefore, the volume (382 mL) should be slightly greater than the mass (321 g).

Key Terms

conversion factor [2.6]
decimal part [2.2]
density (*d*) [2.10]
English system [2.5]
exponent [2.2]

exponential part [2.2]
International System [2.5]
kilogram (kg) [2.5]
liter (L) [2.5]
mass [2.5]

meter (m) [2.5]
metric system [2.5]
prefix multipliers [2.5]
scientific notation [2.2]
second (s) [2.5]

SI units [2.5]
significant figures (digits) [2.3]
solution map [2.6]
units [2.5]
volume [2.5]

Exercises

Questions

Answers to all questions numbered in blue appear in the Answers section at the back of the book.

1. Why is it necessary to include units when reporting scientific measurements?
2. Why are the number of digits reported in scientific measurements important?
3. Why is scientific notation useful?
4. If a measured quantity is written correctly, which digits are certain? Which are uncertain?
5. When do zeros count as significant digits, and when don't they count?
6. How many significant digits are there in exact numbers? What kinds of numbers are exact?
7. What limits the number of significant digits in a calculation involving only multiplication and division?
8. What limits the number of significant digits in a calculation involving only addition and subtraction?
9. How do we determine significant figures in calculations involving both addition/subtraction and multiplication/division?
10. What are the rules for rounding numbers?
11. What are the basic SI units of length, mass, and time?
12. List the common units of volume.
13. Suppose you are trying to measure the diameter of a Frisbee. What unit and prefix multiplier should you use?
14. What is the difference between mass and weight?
15. Using a metric ruler, measure these objects to the correct number of significant figures.
    (a) quarter (diameter)   (b) dime (diameter)
    (c) notebook paper (width)   (d) this book (width)
16. Using a stopwatch, measure each time to the correct number of significant figures.
    (a) time between your heartbeats
    (b) time it takes you to do the next problem
    (c) time between your breaths

17. Explain why units are important in calculations.
18. How are units treated in a calculation?
19. What is a conversion factor?
20. Why does the fundamental value of a quantity not change when you multiply the quantity by a conversion factor?
21. Write the conversion factor that converts a measurement in inches to feet. How does the conversion factor change for converting a measurement in feet to inches?
22. Write conversion factors for each.
    (a) miles to kilometers
    (b) kilometers to miles
    (c) gallons to liters
    (d) liters to gallons
23. This book outlines a four-step problem-solving strategy. Describe each step and its significance.
    (a) Sort
    (b) Strategize
    (c) Solve
    (d) Check
24. Experienced problem solvers always consider both the value and units of their answer to a problem. Why?
25. Draw a solution map to convert a measurement in grams to pounds.
26. Draw a solution map to convert a measurement in milliliters to gallons.
27. Draw a solution map to convert a measurement in meters to feet.
28. Draw a solution map to convert a measurement in ounces to grams. (1 lb = 16 oz)
29. What is density? Explain why density can work as a conversion factor. Between what quantities does it convert?
30. Explain how you would calculate the density of a substance. Include a solution map in your explanation.

Problems

Note: The exercises in the Problems section are paired, and the answers to the odd-numbered exercises (numbered in blue) appear in the Answers section at the back of the book.

SCIENTIFIC NOTATION

31. Express each number in scientific notation.
    (a) 38,802,000 (population of California)
    (b) 1,419,000 (population of Hawaii)
    (c) 19,746,000 (population of New York)
    (d) 584,000 (population of Wyoming)

32. Express each number in scientific notation.
    (a) 7,376,000,000 (population of the world)
    (b) 1,404000,000 (population of China)
    (c) 11,258,000 (population of Cuba)
    (d) 4,677,000 (population of Ireland)

33. Express each number in scientific notation.
(a) 0.00000000007461 m (length of a hydrogen–hydrogen chemical bond)
(b) 0.0000158 mi (number of miles in an inch)
(c) 0.000000632 m (wavelength of red light)
(d) 0.000015 m (diameter of a human hair)

34. Express each number in scientific notation.
(a) 0.000000001 s (time it takes light to travel 1 ft)
(b) 0.143 s (time it takes light to travel around the world)
(c) 0.000000000001 s (time it takes a chemical bond to undergo one vibration)
(d) 0.000001 m (approximate size of a dust particle)

35. Express each number in decimal notation (i.e., express the number without using scientific notation).
(a) 6.022×10^{23} (number of carbon atoms in 12.01 g of carbon)
(b) 1.6×10^{-19} C (charge of a proton in coulombs)
(c) 2.99×10^{8} m/s (speed of light)
(d) 3.44×10^{2} m/s (speed of sound)

36. Express each number in decimal notation (i.e., express the number without using scientific notation).
(a) 450×10^{-9} m (wavelength of blue light)
(b) 13.7×10^{9} years (approximate age of the universe)
(c) 5×10^{9} years (approximate age of Earth)
(d) 5.0×10^{1} years (approximate age of this author)

37. Express each number in decimal notation (i.e., express the number without using scientific notation).
(a) 3.22×10^{7}
(b) 7.2×10^{-3}
(c) 1.18×10^{11}
(d) 9.43×10^{-6}

38. Express each number in decimal notation (i.e., express the number without using scientific notation).
(a) 1.30×10^{6}
(b) 1.1×10^{-4}
(c) 1.9×10^{2}
(d) 7.41×10^{-10}

39. Complete the table.

Decimal Notation	Scientific Notation
2,000,000,000	_____
_____	1.211×10^{9}
0.000874	_____
_____	3.2×10^{11}

40. Complete the table.

Decimal Notation	Scientific Notation
_____	4.2×10^{-3}
315,171,000	_____
_____	1.8×10^{-11}
1,232,000	_____

SIGNIFICANT FIGURES

41. Read each instrument to the correct number of significant figures. Laboratory glassware should always be read from the bottom of the *meniscus* (the curved surface at the top of the liquid column).

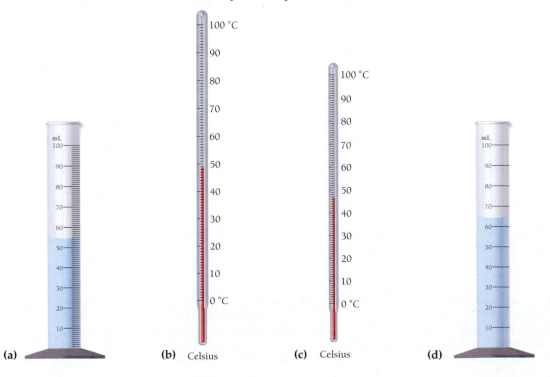

(a)　　　　(b) Celsius　　　(c) Celsius　　　(d)

42. Read each instrument to the correct number of significant figures. Laboratory glassware should always be read from the bottom of the meniscus (the curved surface at the top of the liquid column).

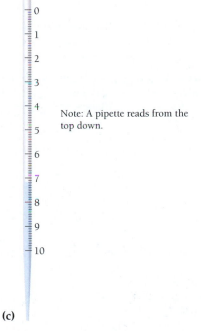

(a)

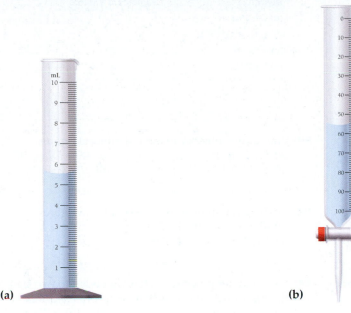

Note: A burette reads from the top down.

(b)

Note: A pipette reads from the top down.

(c)

Note: Digital balances normally display mass to the correct number of significant figures for that particular balance.

(d)

---

43. For each measured quantity, underline the zeros that are significant and draw an X through the zeros that are not.
(a) 0.005050 m
(b) 0.0000000000000060 s
(c) 220,103 kg
(d) 0.00108 in.

44. For each measured quantity, underline the zeros that are significant and draw an X through the zeros that are not.
(a) 0.00010320 s
(b) 1,322,600,324 kg
(c) 0.0001240 in.
(d) 0.02061 m

---

45. How many significant figures are in each measured quantity?
(a) 0.001125 m   (b) 0.1125 m
(c) 1.12500×10^4 m   (d) 11205 m

46. How many significant figures are in each measured quantity?
(a) 13001 kg   (b) 13111 kg
(c) 1.30×10^4 kg   (d) 0.00013 kg

47. Correct any entries in the table that are wrong.

Quantity	Significant Figures
(a) 895675 m	6
(b) 0.000869 kg	6
(c) 0.5672100 s	5
(d) 6.022×10^{23} atoms	4

48. Correct any entries in the table that are wrong.

Quantity	Significant Figures
(a) 24 days	2
(b) 5.6×10^{-12} s	3
(c) 3.14 m	3
(d) 0.00383 g	5

ROUNDING

49. Round each number to four significant figures.
(a) 255.98612   (b) 0.0004893222
(c) 2.900856×10^{-4} (d) 2,231,479

50. Round each number to three significant figures.
(a) 10,776.522   (b) 4.999902×10^6
(c) 1.3499999995   (d) 0.0000344988

51. Round each number to two significant figures.
(a) 2.34   (b) 2.35
(c) 2.349   (d) 2.359

52. Round each number to three significant figures.
(a) 65.74   (b) 65.749
(c) 65.75   (d) 65.750

53. Each number is supposed to be rounded to three significant figures. Correct the ones that are incorrectly rounded.
(a) 42.3492 to 42.4   (b) 56.9971 to 57.0
(c) 231.904 to 232   (d) 0.04555 to 0.046

54. Each number is supposed to be rounded to two significant figures. Correct the ones that are incorrectly rounded.
(a) 1.249×10^3 to 1.3×10^3 (b) 3.999×10^2 to 40
(c) 56.21 to 56.2   (d) 0.009964 to 0.010

55. Round the number on the left to the number of significant figures indicated by the example in the first row. (Use scientific notation as needed to avoid ambiguity.)

Number	Rounded to 4 Significant Figures	Rounded to 2 Significant Figures	Rounded to 1 Significant Figure
1.45815	1.458	1.5	1
8.32466			
84.57225			
132.5512			

56. Round the number on the left to the number of significant figures indicated by the example in the first row. (Use scientific notation as needed to avoid ambiguity.)

Number	Rounded to 4 Significant Figures	Rounded to 2 Significant Figures	Rounded to 1 Significant Figure
94.52118	94.52	95	9×10^1
105.4545			
0.455981			
0.009999991			

SIGNIFICANT FIGURES IN CALCULATIONS

57. Perform each calculation to the correct number of significant figures.
(a) $4.5 \times 0.03060 \times 0.391$
(b) $5.55 \div 8.97$
(c) $(7.890 \times 10^{12}) \div (6.7 \times 10^4)$
(d) $67.8 \times 9.8 \div 100.04$

58. Perform each calculation to the correct number of significant figures.
(a) $89.3 \times 77.0 \times 0.08$
(b) $(5.01 \times 10^5) \div (7.8 \times 10^2)$
(c) $4.005 \times 74 \times 0.007$
(d) $453 \div 2.031$

59. Correct any answers that have the incorrect number of significant figures.
(a) $34.00 \times 567 \div 4.564 = 4.2239 \times 10^3$
(b) $79.3 \div 0.004 \times 35.4 = 7 \times 10^5$
(c) $89.763 \div 22.4581 = 3.997$
(d) $(4.32 \times 10^{12}) \div (3.1 \times 10^{-4}) = 1.4 \times 10^{16}$

60. Correct any answers that have the incorrect number of significant figures.
a) $45.3254 \times 89.00205 = 4034.05$
b) $0.00740 \times 45.0901 = 0.334$
c) $49857 \div 904875 = 0.05510$
d) $0.009090 \times 6007.2 = 54.605$

61. Perform each calculation to the correct number of significant figures.
(a) $87.6 + 9.888 + 2.3 + 10.77$ (b) $43.7 - 2.341$
(c) $89.6 + 98.33 - 4.674$ (d) $6.99 - 5.772$

62. Perform each calculation to the correct number of significant figures.
(a) $1459.3 + 9.77 + 4.32$ (b) $0.004 + 0.09879$
(c) $432 + 7.3 - 28.523$ (d) $2.4 + 1.777$

63. Correct any answers that have the incorrect number of significant figures.
(a) $(3.8 \times 10^5) - (8.45 \times 10^5) = -4.7 \times 10^5$
(b) $0.00456 + 1.0936 = 1.10$
(c) $8475.45 - 34.899 = 8440.55$
(d) $908.87 - 905.34095 = 3.5291$

64. Correct any answers that have the incorrect number of significant figures.
(a) $78.9 + 890.43 - 23 = 9.5 \times 10^2$
(b) $9354 - 3489.56 + 34.3 = 5898.74$
(c) $0.00407 + 0.0943 = 0.0984$
(d) $0.00896 - 0.007 = 0.00196$

65. Perform each calculation to the correct number of significant figures.
(a) $(78.4 - 44.889) \div 0.0087$
(b) $(34.6784 \times 5.38) + 445.56$
(c) $(78.7 \times 10^5 \div 88.529) + 356.99$
(d) $(892 \div 986.7) + 5.44$

66. Perform each calculation to the correct number of significant figures.
(a) $(1.7 \times 10^6 \div 2.63 \times 10^5) + 7.33$
(b) $(568.99 - 232.1) \div 5.3$
(c) $(9443 + 45 - 9.9) \times 8.1 \times 10^6$
(d) $(3.14 \times 2.4367) - 2.34$

67. Correct any answers that have the incorrect number of significant figures.
(a) $(78.56 - 9.44) \times 45.6 = 3152$
(b) $(8.9 \times 10^5 \div 2.348 \times 10^2) + 121 = 3.9 \times 10^3$
(c) $(45.8 \div 3.2) - 12.3 = 2$
(d) $(4.5 \times 10^3 - 1.53 \times 10^3) \div 34.5 = 86$

68. Correct any answers that have the incorrect number of significant figures.
(a) $(908.4 - 3.4) \div 3.52 \times 10^4 = 0.026$
(b) $(1206.7 - 0.904) \times 89 = 1.07 \times 10^5$
(c) $(876.90 + 98.1) \div 56.998 = 17.11$
(d) $(4.55 \div 407859) + 1.00098 = 1.00210$

UNIT CONVERSION

69. Perform each conversion.
(a) 3.55 kg to grams (b) 8944 mm to meters
(c) 4598 mg to kilograms (d) 0.0187 L to milliliters

70. Perform each conversion.
(a) 155.5 cm to meters (b) 2491.6 g to kilograms
(c) 248 cm to millimeters (d) 6781 mL to liters

71. Perform each conversion.
(a) 5.88 dL to liters
(b) 3.41×10^{-5} g to micrograms
(c) 1.01×10^{-8} s to nanoseconds
(d) 2.19 pm to meters

72. Perform each conversion.
(a) 1.08 Mm to kilometers
(b) 4.88 fs to picoseconds
(c) 7.39×10^{11} m to gigameters
(d) 1.15×10^{-10} m to picometers

73. Perform each conversion.
(a) 22.5 in. to centimeters (b) 126 ft to meters
(c) 825 yd to kilometers (d) 2.4 in. to millimeters

74. Perform each conversion.
(a) 78.3 in. to centimeters (b) 445 yd to meters
(c) 336 ft to centimeters (d) 45.3 in. to millimeters

75. Perform each conversion.
(a) 40.0 cm to inches (b) 27.8 m to feet
(c) 10.0 km to miles (d) 3845 kg to pounds

76. Perform each conversion.
(a) 254 cm to inches (b) 89 mm to inches
(c) 7.5 L to quarts (d) 122 kg to pounds

77. Complete the table.

m	km	Mm	Gm	Tm
5.08×10^8 m	_____	508 Mm	_____	_____
_____	_____	27,976 Mm	_____	_____
_____	_____	_____	_____	1.77 Tm
_____	1.5×10^5 km	_____	_____	_____
_____	_____	_____	423 Gm	_____

78. Complete the table.

s	ms	μs	ns	ps
1.31×10^{-4} s	_____	131 μs	_____	_____
_____	_____	_____	_____	12.6 ps
_____	_____	_____	155 ns	_____
_____	1.99×10^{-3} ms	_____	_____	_____
_____	_____	8.66×10^{-5} μs	_____	_____

79. Convert 2.255×10^{10} g to each unit.
(a) kg (b) Mg
(c) mg (d) metric tons (1 metric ton = 1000 kg)

80. Convert 1.88×10^{-6} g to each unit.
(a) mg (b) cg
(c) ng (d) μg

81. A student loses 3.3 lb in one month. How many grams did he lose?

82. A student gains 1.9 lb in two weeks. How many grams did he gain?

83. A runner wants to run 10.0 km. She knows that her running pace is 7.5 mi/h. How many minutes must she run? *Hint:* Use 7.5 mi/h as a conversion factor between distance and time.

84. A cyclist rides at an average speed of 24 mi/h. If he wants to bike 195 km, how long (in hours) must he ride?

85. A recipe calls for 5.0 qt of milk. What is this quantity in cubic centimeters?

86. A gas can holds 2.0 gal of gasoline. What is this quantity in cubic centimeters?

UNITS RAISED TO A POWER

87. Fill in the blanks.
 (a) $1.0 \text{ km}^2 = $ _____ m^2
 (b) $1.0 \text{ cm}^3 = $ _____ m^3
 (c) $1.0 \text{ mm}^3 = $ _____ m^3

88. Fill in the blanks.
 (a) $1.0 \text{ ft}^2 = $ _____ in.^2
 (b) $1.0 \text{ yd}^2 = $ _____ ft^2
 (c) $1.0 \text{ m}^2 = $ _____ yd^2

89. The hydrogen atom has a volume of approximately $6.2 \times 10^{-31} \text{ m}^3$. What is this volume in each unit?
 (a) cubic picometers
 (b) cubic nanometers
 (c) cubic angstroms (1 angstrom = 10^{-10} m)

90. Earth has a surface area of 197 million square miles. What is its area in each unit?
 (a) square kilometers
 (b) square megameters
 (c) square decimeters

91. A house has an area of 215 m^2. What is its area in each unit?
 (a) km^2 (b) dm^2 (c) cm^2

92. A classroom has a volume of 285 m^3. What is its volume in each unit?
 (a) km^3 (b) dm^3 (c) cm^3

93. Total U.S. farmland occupies 954 million acres. How many square miles is this?
 (1 acre = 43,560 ft^2; 1 mi = 5280 ft)

94. The average U.S. farm occupies 435 acres. How many square miles is this?
 (1 acre = 43,560 ft^2; 1 mi = 5280 ft)

UNIT CONVERSION IN BOTH THE NUMERATOR AND DENOMINATOR

95. The speed limit on many U.S. highways is 65 mi/hr. Convert this speed into each alternative unit.
 (a) km/day   (b) ft/s   (c) m/s   (d) yd/min

96. A form of children's Tylenol is sold as a suspension (in which the drug is suspended in water) that contains 32 mg/mL. Convert this concentration into each alternative unit.
 (a) g/L   (b) g/dL   (c) kg/hL   (d) μg/μL

97. A prescription medication requires 7.55 mg per kg of body weight. Convert this quantity to the number of grams required per pound of body weight and determine the correct dose (in g) for a 175-lb patient.

98. A prescription medication requires 0.00225 g per lb of body weight. Convert this quantity to the number of mg required per kg of body weight and determine the correct dose (in mg) for a 105-kg patient.

DENSITY

99. A sample of an unknown metal has a mass of 35.4 g and a volume of 3.11 cm^3. Calculate its density and identify the metal by comparison to Table 2.4.

100. A new penny has a mass of 2.49 g and a volume of 0.349 cm^3. Is the penny pure copper?

101. Glycerol is a syrupy liquid often used in cosmetics and soaps. A 2.50-L sample of pure glycerol has a mass of 3.15×10^3 g. What is the density of glycerol in grams per cubic centimeter?

102. An aluminum engine block has a volume of 4.77 L and a mass of 12.88 kg. What is the density of the aluminum in grams per cubic centimeter?

103. A supposedly gold crown is tested to determine its density. It displaces 10.7 mL of water and has a mass of 206 g. Could the crown be made of gold?

104. A vase is said to be solid platinum. It displaces 18.65 mL of water and has a mass of 157 g. Could the vase be solid platinum?

105. Ethylene glycol (antifreeze) has a density of 1.11 g/cm^3.
   (a) What is the mass in grams of 387 mL of ethylene glycol?
   (b) What is the volume in liters of 3.46 kg of ethylene glycol?

106. Acetone (fingernail-polish remover) has a density of 0.7857 g/cm^3.
   (a) What is the mass in grams of 17.56 mL of acetone?
   (b) What is the volume in milliliters of 7.22 g of acetone?

Cumulative Problems

107. A thief uses a bag of sand to replace a gold statue that sits on a weight-sensitive, alarmed pedestal. The bag of sand and the statue have exactly the same volume, 1.75 L. (Assume that the mass of the bag is negligible.)
   (a) Calculate the mass of each object. (density of gold = 19.3 g/cm^3; density of sand = 3.00 g/cm^3)
   (b) Did the thief set off the alarm? Explain.

108. One of the particles in an atom is the proton. A proton has a radius of approximately 1.0×10^{-13} cm and a mass of 1.7×10^{-24} g. Determine the density of a proton. (*Hint:* Find the volume of the proton and then divide the mass by the volume to get the density.)

$$\text{(volume of a sphere} = \frac{4}{3}\pi r^3; \pi = 3.14)$$

109. A block of metal has a volume of 13.4 $in.^3$ and weighs 5.14 lb. What is its density in grams per cubic centimeter?

110. A log is either oak or pine. It displaces 2.7 gal of water and weighs 19.8 lb. Is the log oak or pine? (density of oak = 0.9 g/cm^3; density of pine = 0.4 g/cm^3)

111. The density of aluminum is 2.7 g/cm^3. What is its density in kilograms per cubic meter?

112. The density of platinum is 21.4 g/cm^3. What is its density in pounds per cubic inch?

113. A typical backyard swimming pool holds 150 yd^3 of water. What is the mass in pounds of the water?

114. An iceberg has a volume of 8975 ft^3. What is the mass in kilograms of the iceberg? The density of ice is 0.917 g/cm^3.

115. The mass of fuel in an airplane must be determined before takeoff. A jet contains 155,211 L of fuel after it has been filled with fuel. What is the mass of the fuel in kilograms if the fuel's density is 0.768 g/cm^3?

116. A backpacker carries 2.5 L of white gas as fuel for her stove. How many pounds does the fuel add to her load? Assume the density of white gas to be 0.79 g/cm^3.

117. Honda produces a hybrid electric car called the Honda Insight. The Insight has both a gasoline-powered engine and an electric motor and has an EPA gas mileage rating of 43 mi per gallon on the highway. What is the Insight's rating in kilometers per liter?

118. You rent a car in Germany with a gas mileage rating of 12.8 km/L. What is its rating in miles per gallon?

119. A car has a mileage rating of 38 mi per gallon of gasoline. How many miles can the car travel on 76.5 L of gasoline?

120. A hybrid SUV consumes fuel at a rate of 12.8 km/L. How many miles can the car travel on 22.5 gal of gasoline?

121. Consider these observations on two blocks of different unknown metals:

	Volume
Block A	125 cm^3
Block B	145 cm^3

If block A has a greater mass than block B, what can be said of the relative densities of the two metals? (Assume that both blocks are solid.)

122. Consider these observations on two blocks of different unknown metals:

	Volume
Block A	125 cm^3
Block B	105 cm^3

If block A has a greater mass than block B, what can be said of the relative densities of the two metals? (Assume that both blocks are solid.)

123. You measure the masses and volumes of two cylinders. The mass of cylinder 1 is 1.35 times the mass of cylinder 2. The volume of cylinder 1 is 0.792 times the volume of cylinder 2. If the density of cylinder 1 is 3.85 g/cm³, what is the density of cylinder 2?

124. A bag contains a mixture of copper and lead BBs. The average density of the BBs is 9.87 g/cm³. Assuming that the copper and lead are pure, determine the relative amounts of each kind of BB.

125. An aluminum sphere has a mass of 25.8 g. Find the radius of the sphere. (The density of aluminum is 2.7 g/cm³, and the volume of a sphere is given by the equation $V = \frac{4}{3}\pi r^3$.)

126. A copper cube has a mass of 87.2 g. Find the edge length of the cube. (The density of copper is 8.96 g/cm³, and the volume of a cube is equal to the edge length cubed.)

Highlight Problems

127. Recall from Section 2.1 that NASA lost the Mars Climate Orbiter because one group of engineers used metric units in their calculations while another group used English units. Consequently, the Orbiter descended too far into the Martian atmosphere and burned up. Suppose that the Orbiter was to have established orbit at 155 km and that one group of engineers specified this distance as 1.55×10^5 m. Suppose further that a second group of engineers programmed the Orbiter to go to 1.55×10^5 ft. What was the difference in kilometers between the two altitudes? How low did the probe go?

◀ The $94 million Mars Climate Orbiter was lost in the Martian atmosphere in 1999 because two groups of engineers failed to communicate with each other about the units in their calculations.

128. A NASA satellite showed that in 2012 the ozone hole over Antarctica had a maximum surface area of 21.2 million km². The largest ozone hole on record occurred in 2006 and had a surface area of 29.6 million km². Calculate the difference in diameter (in meters) between the ozone hole in 2012 and in 2006.

129. In 1999, scientists discovered a new class of black holes with masses 100 to 10,000 times the mass of our sun but occupying less space than our moon. Suppose that one of these black holes has a mass of 1×10^3 suns and a radius equal to one-half the radius of our moon. What is its density in grams per cubic centimeter? The mass of the sun is 2.0×10^{30} kg, and the radius of the moon is 2.16×10^3 mi. (Volume of a sphere $= \frac{4}{3}\pi r^3$.)

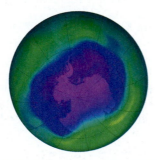

◀ A layer of ozone gas (a form of oxygen) in the upper atmosphere protects Earth from harmful ultraviolet radiation in sunlight. Human-made chemicals react with the ozone and deplete it, especially over the Antarctic at certain times of the year (the so-called ozone hole). The region of low-ozone concentration in 2006 (represented here by the dark purple color) was the largest on record.

130. A titanium bicycle frame contains the same amount of titanium as a titanium cube measuring 6.8 cm on a side. Use the density of titanium to calculate the mass in kilograms of titanium in the frame. What would be the mass of a similar frame composed of iron?

▶ A titanium bicycle frame contains the same amount of titanium as a titanium cube measuring 6.8 cm on a side.

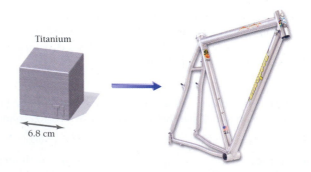

Titanium

6.8 cm

Questions for Group Work

Discuss these questions with the group and record your consensus answer.

131. Look up the thickness of a human hair.
   (a) Convert it to an SI standard unit (if it isn't already).
   (b) Write it in scientific notation.
   (c) Write it without scientific notation (you may need some zeros!).

(d) Write it with an appropriate prefix on a base unit.
Now complete the same exercises with the distance from the Earth to the sun.

132. The following statements are all true.
   (a) Jessica's house is 5 km from the grocery store.
   (b) Jessica's house is 4.73 km from the grocery store.
   (c) Jessica's house is 4.73297 km from the grocery store.
   How can they all be true? What does the number of digits communicate? What sort of device would Jessica need to make the claim in each statement?

Data Interpretation and Analysis

134. The density of a substance can change with temperature. ▶ **FIGURE A** displays the density of water from $-150\ °C$ to $100\ °C$. Examine the graph and answer the questions that follow.
   (a) Water undergoes a large change in density at $0\ °C$ as it freezes to form ice. Calculate the percent change in density that occurs at $0\ °C$.
   (b) Calculate the volume (in cm^3) of 54 g of water at $1\ °C$ and the volume of the same mass of ice at $-1\ °C$. What is the change in volume?
   (c) Antarctica contains 26.5 million cubic kilometers of ice. Assume the temperature of the ice is $-20\ °C$. If all of this ice were heated to $1\ °C$ and melted to form water, what volume of liquid water would form?

133. Convert the height of each group member from feet and inches into meters. Once you determine your heights in meters, calculate the sum of all your heights. Use appropriate rules for significant figures at each step.

(d) A 1.00-L sample of water is heated from $1\ °C$ to $100\ °C$, What is the volume of the water after it is heated?

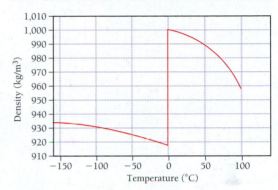

▲ **FIGURE A** Density of Water as a Function of Temperature at 1.00 atm

Answers to Skillbuilder Exercises

Skillbuilder 2.1	$\$1.8416 \times 10^{13}$
Skillbuilder 2.2	3.8×10^{-5}
Skillbuilder 2.3	103.4 °F
Skillbuilder 2.4	**(a)** four significant figures
	(b) three significant figures
	(c) two significant figures
	(d) unlimited significant figures
	(e) three significant figures
	(f) ambiguous
Skillbuilder 2.5	**(a)** 0.001 or 1×10^{-3}
	(b) 0.204
Skillbuilder 2.6	**(a)** 7.6
	(b) 131.11
Skillbuilder 2.7	**(a)** 1288
	(b) 3.12

Skillbuilder 2.8	22.0 in.
Skillbuilder 2.9	5.678 km
Skillbuilder 2.10	0.28 L
Skillbuilder 2.11	46.6 laps
Skillbuilder Plus, p. 35	1.06×10^4 m
Skillbuilder 2.12	18 m/s
Skillbuilder 2.13	$4747\ cm^3$
Skillbuilder 2.14	1.52×10^5 in.^{3}
Skillbuilder 2.15	Yes, the density is $21.4\ g/cm^3$ and matches that of platinum.
Skillbuilder 2.16	$4.4 \times 10^{-2}\ cm^3$
Skillbuilder Plus, p. 42	1.95 kg
Skillbuilder 2.17	16 kg
Skillbuilder 2.18	$d = 19.3\ g/cm^3$; yes, the density is consistent with that of gold.

Answers to Conceptual Checkpoints

2.1 (c) Multiplying by 10^{-3} is equivalent to moving the decimal point three places to the left.

2.2 (b) The last digit is considered to be uncertain by ± 1.

2.3 (b) The result of the calculation in **(a)** would be reported as 4; the result of the calculation in **(b)** would be reported as 1.5.

2.4 (d) The diameter would be expressed as 28 nm.

2.5 (a) The unit of ft should be in the denominator, and the conversion factor in **(c)** is incorrect (because 1 in. $\neq$ 12 ft).

2.6 (c) Kilometers must appear in the numerator and meters in the denominator, and the conversion factor in **(d)** is incorrect (10^3 km $\neq$ 1 m).

2.7 (d) (3 ft) $\times$ (3 ft) $\times$ (3 ft) = 27 ft^3

3

Matter and Energy

"Thus, the task is, not so much to see what no one has yet seen; but to think what nobody has yet thought, about that which everybody sees."

—Erwin Schrödinger (1887–1961)

3.1 In Your Room

Look around the room you are in—what do you see? You might see your desk, your bed, or a glass of water. Maybe you have a window and can see trees, grass, or mountains. You can certainly see this book and possibly the table it sits on. What are these things made of? They are all made of *matter*, which we will define more carefully shortly. For now, know that all you see is matter—your desk, your bed, the glass of water, the trees, the mountains, and this book. Some of what you don't see is matter as well. For example, you are constantly breathing air, which is also matter, into and out of your lungs. You feel the matter in air when you feel wind on your skin. Virtually everything is made of matter.

Think about the differences between different kinds of matter. Air is different from water, and water is different from wood. One of our first tasks as we learn about matter is to identify the similarities and differences among different kinds of matter. How are sugar and salt similar? How are air and water different? Why are they different? Why is a mixture of sugar and water similar to a mixture of salt and water but different from a mixture of sand and water? As students of chemistry, we are particularly interested in the similarities and differences between various kinds of matter and how these reflect the similarities and differences between their component atoms and molecules. We strive to understand the connection between the macroscopic world and the molecular one.

◀ Everything that you can see in this room is made of matter. As students of chemistry, we are interested in how the differences between different kinds of matter are related to the differences between the molecules and atoms that compose the matter. The molecular structures shown here are water molecules on the left and carbon atoms in graphite on the right.

3.2 What Is Matter?

▶ Define matter, atoms, and molecules.

We define **matter** as anything that occupies space and has mass. Some types of matter—such as steel, water, wood, and plastic—are readily visible to our eyes. Other types of matter—such as air or microscopic dust—are not visible to our naked eyes. Matter may sometimes appear smooth and continuous, but actually it is not. Matter is ultimately composed of **atoms**, submicroscopic particles that are the fundamental building blocks of matter (▼ FIGURE 3.1a). In many cases, these atoms are bonded together to form **molecules**, two or more atoms joined to one another in specific geometric arrangements (▼ FIGURE 3.1b). Advances in microscopy have allowed us to image the atoms (▼ FIGURE 3.2) and molecules (▼ FIGURE 3.3) that compose matter, sometimes with stunning clarity.

Aluminum atoms

Aluminum can

Isopropyl alcohol molecules

Isopropyl alcohol (Rubbing alcohol)

(a) (b)

▲ **FIGURE 3.1 Atoms and molecules** All matter is ultimately composed of atoms. **(a)** In some substances, such as aluminum, the atoms exist as independent particles. **(b)** In other substances, such as rubbing alcohol, several atoms bond together in well-defined structures called molecules.

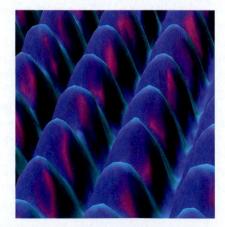

▲ **FIGURE 3.2 Scanning tunneling microscope image of nickel atoms** A scanning tunneling microscope (STM) creates an image by scanning a surface with a tip of atomic dimensions. It can distinguish individual atoms, seen as blue bumps in this image.

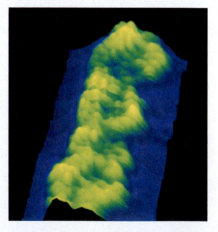

▲ **FIGURE 3.3 Scanning tunneling microscope image of a DNA molecule** DNA is the hereditary material that encodes the operating instructions for most cells in living organisms. In this image, the DNA molecule is yellow, and the double-stranded structure of DNA is discernible.

3.3 Classifying Matter According to Its State: Solid, Liquid, and Gas

▶ Classify matter as solid, liquid, or gas.

PEARSON eText 2.0

Key Concept Video
Classifying Matter

The eText 2.0 icon indicates that this feature is embedded and interactive in the eText.

The common **states of matter** are **solid**, **liquid**, and **gas** (▼ FIGURE 3.4). In solid matter, atoms or molecules pack close to each other in fixed locations. Although neighboring atoms or molecules in a solid may vibrate or oscillate, they do not move around each other, giving solids their familiar fixed volume and rigid shape.

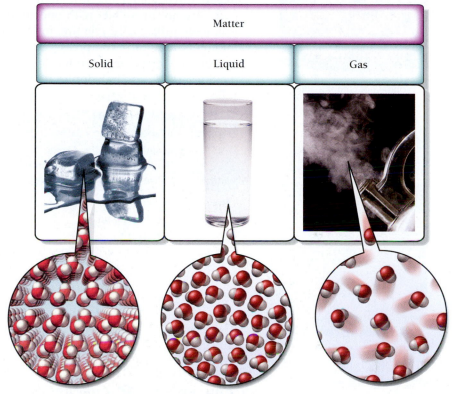

▶ **FIGURE 3.4 Three states of matter** Water exists as ice (solid), water (liquid), and steam (gas). In ice, the water molecules are closely spaced and, although they vibrate about a fixed point, they do not generally move relative to one another. In liquid water, the water molecules are also closely spaced but are free to move around and past each other. In steam, water molecules are separated by large distances and do not interact significantly with one another.

Well-ordered, three-dimensional structure

(a) Crystalline solid

No long-range order

(b) Amorphous solid

▲ **FIGURE 3.5 Types of solid matter** (a) In a crystalline solid, atoms or molecules occupy specific positions to create a well-ordered, three-dimensional structure. (b) In an amorphous solid, atoms do not have any long-range order.

Ice, diamond, quartz, and iron are examples of solid matter. Solid matter may be **crystalline**, in which case its atoms or molecules arrange in geometric patterns with long-range, repeating order (◀ FIGURE 3.5a), or it may be **amorphous**, in which case its atoms or molecules do not have long-range order (◀ FIGURE 3.5b). Examples of *crystalline* solids include salt (▼ FIGURE 3.6) and diamond; the well-ordered, geometric shapes of salt and diamond crystals

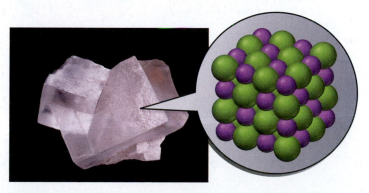

▲ **FIGURE 3.6 Salt: a crystalline solid** Sodium chloride is an example of a crystalline solid. The well-ordered, cubic shape of salt crystals is due to the well-ordered, cubic arrangement of its atoms.

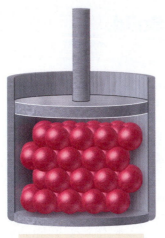

Solid—not compressible

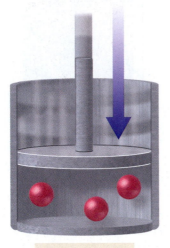

Gas—compressible

▲ FIGURE 3.7 **Gases are compressible** Because the atoms or molecules that compose gases are not in contact with one another, we can compress gases.

PEARSON
eText
2.0

The eText 2.0 icon indicates that this feature is embedded and interactive in the eText.

reflect the well-ordered geometric arrangement of their atoms. Examples of *amorphous* solids include glass, rubber, and plastic.

In liquid matter, atoms or molecules are close to each other (about as close as molecules in a solid), but they are free to move around and by each other. Like solids, liquids have a fixed volume because their atoms or molecules are in close contact. Unlike solids, however, liquids assume the shape of their container because the atoms or molecules are free to move relative to one another. Water, gasoline, alcohol, and mercury are examples of liquid matter.

In gaseous matter, atoms or molecules are separated by large distances and are free to move relative to one another. Because the atoms or molecules that compose gases are not in contact with one another, gases are **compressible** (◄ FIGURE 3.7). When we inflate a bicycle tire, for example, we push more atoms and molecules into the same space, compressing them and making the tire harder. Gases always assume the shape and volume of their containers. Oxygen, helium, and carbon dioxide are all good examples of gases. Table 3.1 summarizes the properties of solids, liquids, and gases.

TABLE 3.1	Properties of Solids, Liquids, and Gases				
State	**Atomic/ Molecular Motion**	**Atomic/ Molecular Spacing**	**Shape**	**Volume**	**Compressibility**
Solid	Oscillation/ vibration about fixed point	Close together	Definite	Definite	Incompressible
Liquid	Free to move relative to one another	Close together	Indefinite	Definite	Incompressible
Gas	Free to move relative to one another	Far apart	Indefinite	Indefinite	Compressible

CONCEPTUAL ✔ CHECKPOINT **3.1**

Which image best represents matter in the gas state?

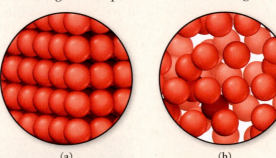
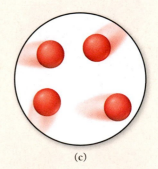

(a)   (b)   (c)

Note: You can find the answers to all Conceptual Checkpoints at the end of the chapter.

3.4 Classifying Matter According to Its Composition: Elements, Compounds, and Mixtures

▶ Classify matter as element, compound, or mixture.

In addition to classifying matter according to its state, we can classify it according to its composition (▶ FIGURE 3.8). Matter may be either a **pure substance**, composed of only one type of atom or molecule, or a **mixture**, composed of two or more different types of atoms or molecules combined in variable proportions.

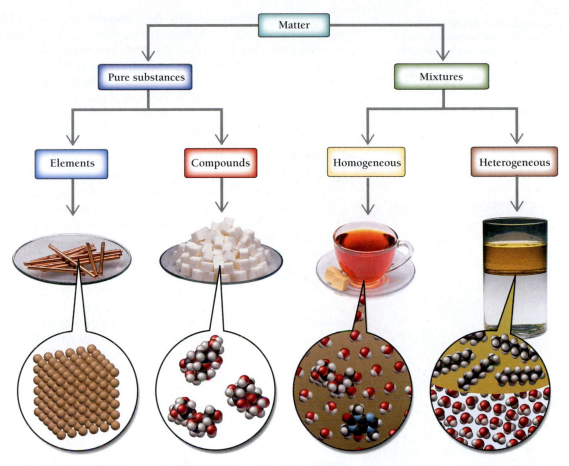

▲ **FIGURE 3.8 Classification of matter**   Matter may be a pure substance or a mixture. A pure substance may be either an element (such as copper) or a compound (such as sugar), and a mixture may be either homogeneous (such as sweetened tea) or heterogeneous (such as gasoline and water).

Pure substances are composed of only one type of atom or molecule. Helium and water are both pure substances. The atoms that compose helium are all helium atoms, and the molecules that compose water are all water molecules—no other atoms or molecules are mixed in.

Pure substances can themselves be divided into two types: elements and compounds. Copper is an example of an **element**, a substance that cannot be broken down into simpler substances. The graphite in pencils is also an element—carbon. No chemical transformation can decompose graphite into simpler substances; it is pure carbon. All known elements are listed in the periodic table in the inside front cover of this book and in alphabetical order on the inside back cover of this book.

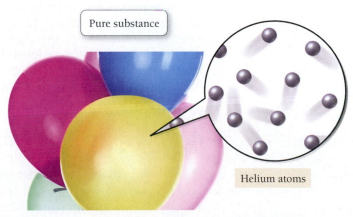

▲ Helium is a pure substance composed only of helium atoms.

A compound is composed of different atoms that are chemically united (bonded). A mixture is composed of different substances that are not chemically united, but simply mixed together.

A pure substance can also be a **compound**, a substance composed of two or more elements in fixed definite proportions. Compounds are more common than pure elements because most elements are chemically reactive and combine with other elements to form compounds. Water, table salt, and sugar are examples of compounds; they can all be decomposed into simpler substances. If you heat sugar in a pan over a flame, you decompose it into several substances including carbon (an element) and gaseous water (a different compound). The black substance left on your pan after burning is the carbon; the water escapes into the air as steam.

The majority of matter that we encounter is in the form of mixtures. Apple juice, a flame, salad dressing, and soil are all examples of mixtures; they each contain several substances with proportions that vary from one sample to another. Other common mixtures include air, seawater, and brass. Air is a mixture composed primarily of nitrogen and oxygen gas; seawater is a mixture composed primarily of salt and water; and brass is a mixture composed of copper and zinc. Each of these mixtures can have different proportions of its constituent components. For example, metallurgists vary the relative amounts of copper and zinc in brass to tailor the metal's properties to its intended use—the higher the zinc content relative to the copper content, the more brittle the brass.

Pure substance

Water molecules

▲ Water is a pure substance composed only of water molecules.

Air and seawater are mixtures.

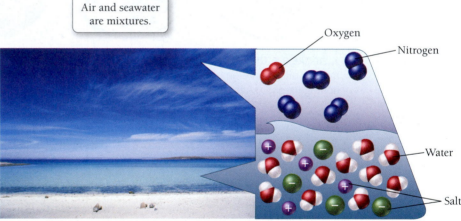

▲ Air and seawater are examples of mixtures. Air contains primarily nitrogen and oxygen. Seawater contains primarily salt and water.

We can also classify mixtures according to how uniformly the substances within them mix. In a **heterogeneous mixture**, such as oil and water, the composition varies from one region to another. In a **homogeneous mixture**, such as salt water or sweetened tea, the composition is the same throughout. Homogeneous mixtures have uniform compositions because the atoms or molecules that compose them mix uniformly. Remember that the properties of matter are determined by the atoms or molecules that compose it.

To summarize, as shown in Figure 3.8 (on p. 65):

- Matter may be a pure substance, or it may be a mixture.
- A pure substance may be either an element or a compound.
- A mixture may be either homogeneous or heterogeneous.
- Mixtures may be composed of two or more elements, two or more compounds, or a combination of both.

EXAMPLE **3.1**   **Classifying Matter**

Classify each type of matter as a pure substance or a mixture. If it is a pure substance, classify it as an element or a compound; if it is a mixture, classify it as homogeneous or heterogeneous.

(a) a lead weight   **(b)** seawater
(c) distilled water   **(d)** Italian salad dressing

SOLUTION

Begin by examining the alphabetical listing of pure elements inside the back cover of this text. If the substance appears in that table, it is a pure substance and an element. If it is not in the table but is a pure substance, then it is a compound. If the substance is not a pure substance, then it is a mixture. Think about your everyday experience with each mixture to determine if it is homogeneous or heterogeneous.

(a) Lead is listed in the table of elements. It is a pure substance and an element.

(b) Seawater is composed of several substances, including salt and water; it is a mixture. It has a uniform composition, so it is a homogeneous mixture.

(c) Distilled water is not listed in the table of elements, but it is a pure substance (water); therefore, it is a compound.

(d) Italian salad dressing contains a number of substances and is therefore a mixture. It usually separates into at least two distinct regions with different composition and is therefore a heterogeneous mixture.

▶ **SKILLBUILDER 3.1 | Classifying Matter**

Classify each type of matter as a pure substance or a mixture. If it is a pure substance, classify it as an element or a compound. If it is a mixture, classify it as homogeneous or heterogeneous.

(a) mercury in a thermometer
(b) exhaled air
(c) chicken noodle soup
(d) sugar

▶ **FOR MORE PRACTICE**   Example 3.12; Problems 31, 32, 33, 34, 35, 36.

Note: The answers to all Skillbuilder exercises appear at the end of the chapter.

CONCEPTUAL ✔ **CHECKPOINT 3.2**

Which of the substances depicted here is a pure substance?

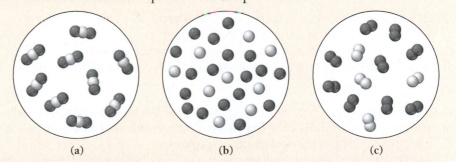

(a)   (b)   (c)

3.5 Differences in Matter: Physical and Chemical Properties

▶ Distinguish between physical and chemical properties.

The characteristics that distinguish one substance from another are called **properties**. Different substances have unique properties that characterize them and distinguish them from other substances. For example, we can distinguish water from alcohol based on their different smells, or we can distinguish gold from silver based on their different colors.

In chemistry, we categorize properties into two different types: physical and chemical. A **physical property** is one that a substance displays without changing

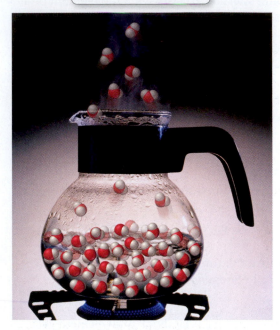

Water molecules are the same in water and steam.

▲ **FIGURE 3.9 A physical property** The boiling point of water is a physical property, and boiling is a physical change. When water boils, it turns into a gas, but the water molecules are the same in both the liquid water and the gaseous steam.

its composition. A **chemical property** is one that a substance displays only through changing its composition. For example, the characteristic odor of gasoline is a physical property—gasoline does not change its composition when it exhibits its odor. In contrast, the flammability of gasoline is a chemical property—gasoline does change its composition when it burns.

The atomic or molecular composition of a substance does not change when the substance displays its physical properties. For example, the boiling point of water—a physical property—is 100 °C. When water boils, it changes from a liquid to a gas, but the gas is still water (◀ FIGURE 3.9). On the contrary, the susceptibility of iron to rust is a chemical property—iron must change into iron(III) oxide to display this property (▼ FIGURE 3.10). Physical properties include odor, taste, color, appearance, melting point, boiling point, and density. Chemical properties include corrosiveness, flammability, acidity, and toxicity.

Iron(III) oxide or rust

Iron atoms

▲ **FIGURE 3.10 A chemical property** The susceptibility of iron to rusting is a chemical property, and rusting is a chemical change. When iron rusts, it turns from iron to iron(III) oxide.

EXAMPLE **3.2** | **Physical and Chemical Properties**

Classify each property as physical or chemical.

(a) the tendency of copper to turn green when exposed to air
(b) the tendency of automobile paint to dull over time
(c) the tendency of gasoline to evaporate quickly when spilled
(d) the low mass (for a given volume) of aluminum relative to other metals

SOLUTION

(a) Copper turns green because it reacts with gases in air to form compounds; this is a chemical property.
(b) Automobile paint dulls over time because it can fade (decompose) due to sunlight or it can react with oxygen in air. In either case, this is a chemical property.
(c) Gasoline evaporates quickly because it has a low boiling point; this is a physical property.
(d) Aluminum's low mass (for a given volume) relative to other metals is due to its low density; this is a physical property.

▶ **SKILLBUILDER 3.2** | **Physical and Chemical Properties**

Classify each property as physical or chemical.

(a) the explosiveness of hydrogen gas
(b) the bronze color of copper
(c) the shiny appearance of silver
(d) the ability of dry ice to sublime (change from solid directly to vapor)

▶ **FOR MORE PRACTICE** Example 3.13; Problems 37, 38, 39, 40.

3.6 Changes in Matter: Physical and Chemical Changes

▶ Distinguish between physical and chemical changes.

Every day, we witness changes in matter: Ice melts, iron rusts, and fruit ripens. What happens to the atoms and molecules that make up these substances during these changes? The answer depends on the kind of change. In a **physical change**, matter changes its appearance but not its composition. For example, when ice melts, it looks different—water looks different from ice—but its composition is the same. Solid ice and liquid water are both composed of water molecules, so melting is a physical change. Similarly, when glass shatters, it looks different, but its composition remains the same—it is still glass. Again, this is a physical change. In a **chemical change**, however, matter *does* change its composition. For example, copper turns green upon continued exposure to air because it reacts with gases in air to form new compounds. This is a chemical change. Matter undergoes a chemical change when it undergoes a **chemical reaction**. In a chemical reaction, the substances present before the chemical change are called **reactants**, and the substances present after the change are called **products**:

$$\text{Reactants} \xrightarrow[\text{Change}]{\text{Chemical}} \text{Products}$$

We will cover chemical reactions in much more detail in Chapter 7.

State changes—transformations from one state of matter (such as solid or liquid) to another—are always physical changes.

The differences between physical and chemical changes are not always apparent. Only chemical examination of the substances before and after the change can verify whether the change is physical or chemical. For many cases, however, we can identify chemical and physical changes based on what we know about the changes. Changes in state, such as melting or boiling, or changes that involve merely appearance, such as those produced by cutting or crushing, are always physical changes. Changes involving chemical reactions—often evidenced by heat exchange or color changes—are always chemical changes.

The main difference between chemical and physical changes is related to the changes at the molecular and atomic level. In physical changes, the atoms that compose the matter *do not* change their fundamental associations, even though the matter may change its appearance. In chemical changes, atoms do change their fundamental associations, resulting in matter with a new identity. *A physical change results in a different form of the same substance, while a chemical change results in a completely new substance.*

Consider physical and chemical changes in liquid butane, the substance used to fuel butane lighters. In many lighters, you can see the liquid butane through the plastic case of the lighter. If you push the fuel button on the lighter without turning the flint, some of the liquid butane *vaporizes* (changes from liquid to gas). You cannot see the gaseous butane, but if you listen carefully you can usually hear hissing as it leaks out (◀ FIGURE 3.11). Since the liquid butane and the gaseous butane are both composed of butane molecules, the change is physical. On the other hand, if you push the button *and* turn the flint to create a spark, a chemical change occurs. The butane molecules react with oxygen molecules in air to form new molecules, carbon dioxide, and water (◀ FIGURE 3.12). The change is chemical because the molecular composition changes upon burning.

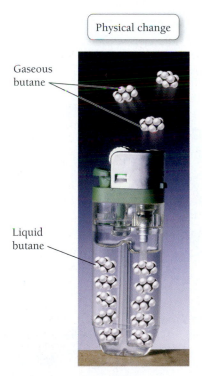

Physical change

Gaseous butane

Liquid butane

▲ FIGURE 3.11 Vaporization: a physical change   If you push the button on a lighter without turning the flint, some of the liquid butane vaporizes to gaseous butane. Since the liquid butane and the gaseous butane are both composed of butane molecules, this is a physical change.

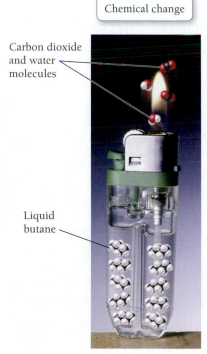

Chemical change

Carbon dioxide and water molecules

Liquid butane

▲ FIGURE 3.12 Burning: a chemical change   If you push the button *and* turn the flint to create a spark, you produce a flame. The butane molecules react with oxygen molecules in air to form new molecules, carbon dioxide, and water. This is a chemical change.

EXAMPLE **3.3** | **Physical and Chemical Changes**

Classify each change as physical or chemical.

(a) the rusting of iron
(b) the evaporation of fingernail-polish remover (acetone) from the skin
(c) the burning of coal
(d) the fading of a carpet upon repeated exposure to sunlight

SOLUTION

(a) Iron rusts because it reacts with oxygen in air to form iron(III) oxide; therefore, this is a chemical change.
(b) When fingernail-polish remover (acetone) evaporates, it changes from liquid to gas, but it remains acetone; therefore, this is a physical change.
(c) Coal burns because it reacts with oxygen in air to form carbon dioxide; this is a chemical change.
(d) A carpet fades on repeated exposure to sunlight because the molecules that give the carpet its color are decomposed by sunlight; this is a chemical change.

▶ **SKILLBUILDER 3.3** | **Physical and Chemical Changes**

Classify each change as physical or chemical.

(a) copper metal forming a blue solution when it is dropped into colorless nitric acid
(b) a train flattening a penny placed on a railroad track
(c) ice melting into liquid water
(d) a match igniting a firework

▶ **FOR MORE PRACTICE** Example 3.14; Problems 41, 42, 43, 44.

CONCEPTUAL ✔ **CHECKPOINT 3.3**

In this figure, liquid water is vaporizing into steam.

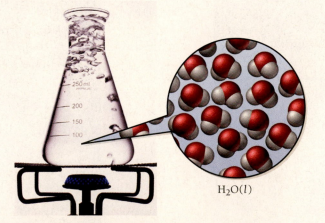

$H_2O(l)$

Which diagram best represents the molecules in the steam?

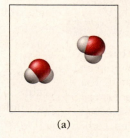
(a)

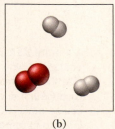
(b)

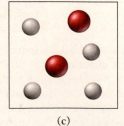
(c)

Distillation

▶ **FIGURE 3.13 Separating a mixture of two liquids by distillation**

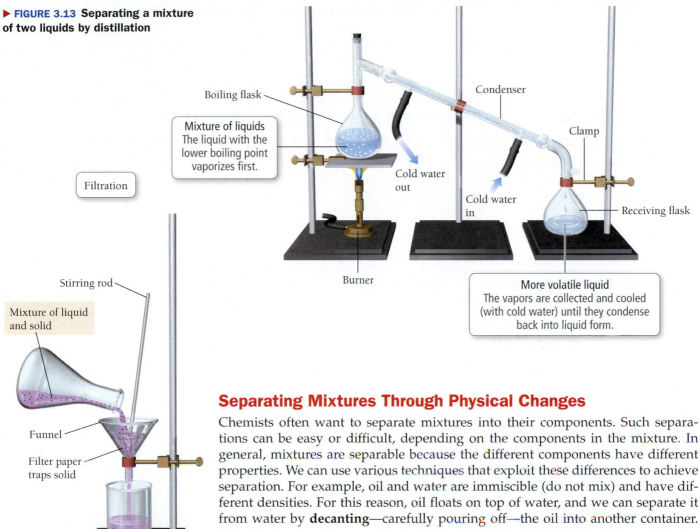

Boiling flask

Mixture of liquids
The liquid with the lower boiling point vaporizes first.

Condenser

Clamp

Filtration

Cold water out

Cold water in

Receiving flask

Burner

More volatile liquid
The vapors are collected and cooled (with cold water) until they condense back into liquid form.

Stirring rod

Mixture of liquid and solid

Funnel

Filter paper traps solid

Liquid component of mixture

▲ **FIGURE 3.14 Separating a solid from a liquid by filtration**

Separating Mixtures Through Physical Changes

Chemists often want to separate mixtures into their components. Such separations can be easy or difficult, depending on the components in the mixture. In general, mixtures are separable because the different components have different properties. We can use various techniques that exploit these differences to achieve separation. For example, oil and water are immiscible (do not mix) and have different densities. For this reason, oil floats on top of water, and we can separate it from water by **decanting**—carefully pouring off—the oil into another container. We can separate mixtures of miscible liquids by **distillation**, in which we heat the mixture to boil off the more **volatile**—the more easily vaporizable—liquid. We then recondense the volatile liquid in a condenser and collect it in a separate flask (▲ FIGURE 3.13). If a mixture is composed of a solid and a liquid, we can separate the two by **filtration**, in which we pour the mixture through filter paper usually held in a funnel (◀ FIGURE 3.14).

3.7 Conservation of Mass: There Is No New Matter

▶ Apply the law of conservation of mass.

This law is a slight oversimplification. In nuclear reactions, covered in Chapter 17, significant changes in mass can occur. In chemical reactions, however, the changes are so minute that they can be ignored.

We examine the quantitative relationships in chemical reactions in Chapter 8.

As we have seen, our planet, our air, and even our own bodies are composed of matter. Physical and chemical changes do not destroy matter, nor do they create new matter. Recall from Chapter 1 that Antoine Lavoisier, by studying combustion, established the law of conservation of mass, which states:

Matter is neither created nor destroyed in a chemical reaction.

During physical and chemical changes, the total amount of matter remains constant even though it may not initially appear that it has. When we burn butane in a lighter, for example, the butane slowly disappears. Where does it go? It combines with oxygen to form carbon dioxide and water that travel into the surrounding air. The mass of the carbon dioxide and water that forms, however, exactly equals the mass of the butane and oxygen that combined.

Suppose that we burn 58 g of butane in a lighter. It will react with 208 g of oxygen to form 176 g of carbon dioxide and 90 g of water.

$$\underbrace{\text{Butane} + \text{Oxygen}}_{\text{58 g} + \text{208 g}} \longrightarrow \underbrace{\text{Carbon Dioxide} + \text{Water}}_{\text{176 g} + \text{90 g}}$$

$$\underbrace{}_{266\ g} \qquad \underbrace{}_{266\ g}$$

The sum of the masses of the butane and oxygen, 266 g, is equal to the sum of the masses of the carbon dioxide and water, which is also 266 g. In this chemical reaction, as in all chemical reactions, matter is conserved.

EXAMPLE 3.4 Conservation of Mass

A chemist forms 16.6 g of potassium iodide by combining 3.9 g of potassium with 12.7 g of iodine. Show that these results are consistent with the law of conservation of mass.

SOLUTION

The sum of the masses of the potassium and iodine is:

$$3.9\ g + 12.7\ g = 16.6\ g$$

The sum of the masses of potassium and iodine equals the mass of the product, potassium iodide. The results are consistent with the law of conservation of mass.

▶ **SKILLBUILDER 3.4 | Conservation of Mass**

Suppose 12 g of natural gas combines with 48 g of oxygen in a flame. The chemical change produces 33 g of carbon dioxide. How many grams of water form?

▶ **FOR MORE PRACTICE** Example 3.15; Problems 45, 46, 47, 48, 49, 50.

CONCEPTUAL ✔ CHECKPOINT 3.4

Consider a drop of water that is put into a flask, sealed with a cap, and heated until the droplet vaporizes. Is the mass of the container and water different after heating?

3.8 Energy

▶ Recognize the different forms of energy.
▶ Identify and convert between energy units.

Matter is one of the two major components of our universe. The other major component is **energy**, *the capacity to do work*. **Work** is defined as the result of a force acting on a distance. For example, if you push this book across your desk, you have done work. You may at first think that in chemistry we are only concerned with matter, but the behavior of matter is driven in large part by energy, so understanding energy is critical to understanding chemistry. Like matter, energy is conserved.

The **law of conservation of energy** states that *energy is neither created nor destroyed*. The total amount of energy is constant; energy can be changed from one form to another or transferred from one object to another, but it cannot be created out of nothing, and it does not vanish into nothing.

Virtually all samples of matter have energy. The total energy of a sample of matter is the sum of its **kinetic energy**, the energy associated with its motion, and its **potential energy**, the energy associated with its position or composition. For example, a moving billiard ball contains *kinetic energy* because it is *moving* at some speed across the billiard table. Water behind a dam contains *potential energy* because it is held at a high *position* in the Earth's gravitational field by the dam. When the water flows through the dam from a higher position to a lower position, it can turn a turbine and produce electrical

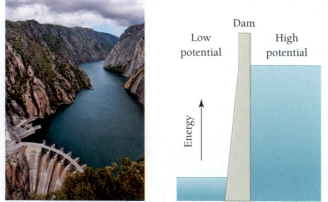

▲ Water behind a dam contains potential energy.

CHEMISTRY IN THE ENVIRONMENT

Getting Energy out of Nothing?

The law of conservation of energy has significant implications for energy use. The best we can do with energy is break even (and even that is not really possible); we can't continually draw energy from a device without putting energy into it. A device that supposedly produces energy without the need for energy input is sometimes called a *perpetual motion machine* (▶ FIGURE 3.15). According to the law of conservation of energy, such a machine cannot exist. Occasionally, the media report or speculate on the discovery of a system that appears to produce more energy than it consumes. For example, I once heard a radio talk show on the subject of energy and gasoline costs. The reporter suggested that we simply design an electric car that recharges itself while being driven. The battery in the electric car would charge during operation in the same way that the battery in a conventional car recharges, except the electric car would run with energy from the battery. Although people have dreamed of machines such as this for decades, such ideas violate the law of conservation of energy because they produce energy without any energy input. In the case of the perpetually moving electric car, the fault lies in the idea that driving the electric car can recharge the battery—it can't.

The battery in a conventional car recharges because energy from gasoline combustion is converted into electrical energy that then charges the battery. The electric car needs energy to move forward, and the battery eventually

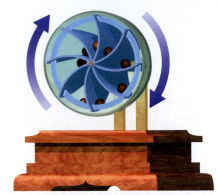

◀ **FIGURE 3.15**
A proposed perpetual motion machine The rolling balls supposedly keep the wheel perpetually spinning.
Question: Can you explain why this would not work?

discharges as it provides that energy. Hybrid cars (electric and gasoline-powered) such as the Toyota Prius can capture some limited energy from braking and use that energy to recharge the battery. However, they could never run indefinitely without the addition of fuel. Our society has a continual need for energy, and as our current energy resources dwindle, new energy sources are required. Unfortunately, those sources must also follow the law of conservation of energy—energy must be conserved.

B3.1 CAN YOU ANSWER THIS? *A friend asks you to invest in a new flashlight he invented that never needs batteries. What questions should you ask before writing a check?*

---

energy. **Electrical energy** is the energy associated with the flow of electrical charge. **Thermal energy** is the energy associated with the random motions of atoms and molecules in matter. The hotter an object, the more thermal energy it contains.

Chemical systems contain **chemical energy**, a form of potential energy associated with the positions of the particles that compose the chemical system. For example, the molecules that compose gasoline contain a substantial amount of chemical energy. They are a bit like the water behind a dam. Burning the gasoline is analogous to releasing the water from the dam. The chemical energy present in the gasoline is released upon burning. When we drive a car, we use that chemical energy to move the car forward. When we heat a home, we use chemical energy stored in natural gas to produce heat and warm the air in the house.

Units of Energy

We commonly use several different energy units. The SI unit of energy is the joule (J), named after the English scientist James Joule (1818–1889), who demonstrated that energy could be converted from one type to another as long as the total energy was conserved. A second unit of energy is the **calorie (cal)**, the amount of energy required to raise the temperature of 1 g of water by 1 °C. A calorie is a larger unit than a joule: 1 cal = 4.184 J. A related energy unit is the nutritional or *capital C* **Calorie (Cal)**, equivalent to 1000 *little c* calories. Electricity bills usually come in yet another energy unit, the **kilowatt-hour (kWh)**. The average cost of residential electricity in the United States is about $0.12 per kilowatt-hour. Table 3.2 lists various energy units and their conversion factors. Table 3.3 shows the amount of energy required for various processes in each of these units.

TABLE 3.2 Energy Conversion Factors

1 calorie (cal)	= 4.184 joules (J)
1 Calorie (Cal)	= 1000 calories (cal)
1 kilowatt-hour (kWh)	= 3.60×10^6 joules (J)

TABLE 3.3 Energy Use in Various Units

Unit	Energy Required to Raise Temperature of 1 g of Water by 1 °C	Energy Required to Light 100-W Bulb for 1 Hour	Total Energy Used by Average U.S. Citizen in 1 Day
joule (J)	4.18	3.6×10^5	9.0×10^8
calorie (cal)	1.00	8.60×10^4	2.2×10^8
Calorie (Cal)	0.00100	86.0	2.2×10^5
kilowatt-hour (kWh)	1.16×10^{-6}	0.100	2.50×10^2

PEARSON
eText
2.0

Interactive
Worked Example
Video 3.5

EXAMPLE **3.5** | **Conversion of Energy Units**

A candy bar contains 225 Cal of nutritional energy. How many joules does it contain?

SORT

Begin by sorting the information in the problem. Here you are *given* energy in Calories and asked to *find* energy in joules.

GIVEN: 225 Cal

FIND: J

STRATEGIZE

Draw a solution map. Begin with Cal, convert to cal, and then convert to J.

SOLUTION MAP

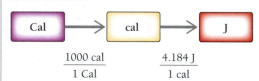

$$\frac{1000\ cal}{1\ Cal} \qquad \frac{4.184\ J}{1\ cal}$$

RELATIONSHIPS USED

1000 calories = 1 Cal (Table 3.2)
4.184 J = 1 cal (Table 3.2)

SOLVE

Follow the solution map to solve the problem. Begin with 225 Cal and multiply by the appropriate conversion factors to arrive at J. Round the answer to the correct number of significant figures (in this case, three because of the three significant figures in 225 Cal).

SOLUTION

$$225\ \cancel{Cal} \times \frac{1000\ \cancel{cal}}{1\ \cancel{Cal}} \times \frac{4.184\ J}{1\ \cancel{cal}} = 9.41 \times 10^5\ J$$

CHECK

Check your answer. Are the units correct? Does the answer make physical sense?

The units of the answer (J) are the desired units. The magnitude of the answer makes sense because the J is a smaller unit than the Cal; therefore, the quantity of energy in J should be greater than the quantity in Cal.

▶ **SKILLBUILDER 3.5** | **Conversion of Energy Units**

The complete combustion of a small wooden match produces approximately 512 cal of heat. How many kilojoules are produced?

▶ **SKILLBUILDER PLUS** Convert 2.75×10^4 kJ to calories.

▶ **FOR MORE PRACTICE** Example 3.16; Problems 51, 52, 53, 54, 55, 56, 57, 58.

PEARSON
eText
2.0

CONCEPTUAL ✔ CHECKPOINT **3.5**

Suppose a salesperson wants to make an appliance seem as efficient as possible. In which units does the yearly energy consumption of the appliance have the lowest numerical value and therefore seem most efficient?

(a) J (b) cal (c) Cal (d) kWh

3.9 Energy and Chemical and Physical Change

▶ Distinguish between exothermic and endothermic reactions.

> When discussing energy transfer, we often define the object of our study (such as a flask in which a chemical reaction is occurring) as the *system*. The system then exchanges energy with its *surroundings*. In other words, we view energy changes as an exchange of energy between the system and the surroundings.

The physical and chemical changes that we discussed in Section 3.6 are usually accompanied by energy changes. For example, when water evaporates from skin (a physical change), the water molecules absorb energy, cooling the skin. When we burn natural gas on a stove (a chemical change), energy is released, heating the food we are cooking.

The release of energy during a chemical reaction is analogous to the release of energy that occurs when a weight falls to the ground. When you lift a weight, you raise its potential energy; when you drop it, you release the potential energy (◀ FIGURE 3.16). *Systems with high potential energy—like the raised weight—have a tendency to change in a way that lowers their potential energy.* For this reason, objects or systems with high potential energy tend to be *unstable*. A weight lifted several meters from the ground is unstable because it contains a significant amount of localized potential energy. Unless restrained, the weight will fall, lowering its potential energy.

Some chemical substances are like a raised weight. For example, the molecules that compose TNT (trinitrotoluene) have a relatively high potential energy—energy is concentrated in them just as energy is concentrated in the raised weight. TNT molecules therefore tend to undergo rapid chemical changes that lower their potential energy, which is why TNT is explosive. Chemical reactions that *release* energy, like the explosion of TNT, are **exothermic**.

Some chemical reactions behave in just the opposite way—they *absorb* energy from their surroundings as they occur. Such reactions are **endothermic**. The reaction that occurs in a chemical cold pack is a good example of an endothermic reaction. When you break the barrier separating the reactants in the chemical cold pack, the substances mix, react, and absorb heat from the surroundings. The surroundings—possibly including your bruised ankle—get colder.

We can represent the energy changes that occur during a chemical reaction with an energy diagram, as shown in Figure 3.17. In an exothermic reaction (▼ FIGURE 3.17a), the reactants have greater energy than the products, and energy is released as the reaction occurs. In an endothermic reaction (▼ FIGURE 3.17b), the products have more energy than the reactants, and energy is absorbed as the reaction occurs.

If a particular reaction or process is exothermic, then the reverse process must be endothermic. For example, the evaporation of water from skin is endothermic (and therefore cools you off), but the condensation of water onto skin is exothermic (which is why steam burns can be so painful).

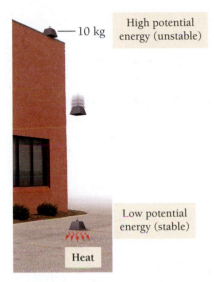

▲ FIGURE 3.16 **Potential energy of raised weight** A weight lifted off the ground has a high potential and will tend to fall toward the ground to lower its potential energy.

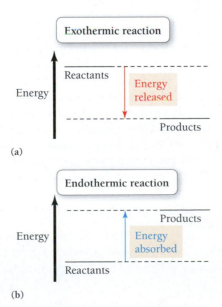

◀ FIGURE 3.17 **Exothermic and endothermic reactions** (a) In an exothermic reaction, energy is released. (b) In an endothermic reaction, energy is absorbed.

EXAMPLE **3.6** | **Exothermic and Endothermic Processes**

Classify each change as exothermic or endothermic.

(a) wood burning in a fire
(b) ice melting

SOLUTION

(a) When wood burns, it emits heat into the surroundings. Therefore, the process is exothermic.
(b) When ice melts, it absorbs heat from the surroundings. For example, when ice melts in a glass of water, it cools the water as the melting ice absorbs heat from the water. Therefore, the process is endothermic.

▶ **SKILLBUILDER 3.6** | **Exothermic and Endothermic Processes**

Classify each change as exothermic or endothermic.

(a) water freezing into ice
(b) natural gas burning

▶ **FOR MORE PRACTICE** Problems 61, 62, 63, 64.

3.10 Temperature: Random Motion of Molecules and Atoms

▶ Convert between Fahrenheit, Celsius, and Kelvin temperature scales.

The atoms and molecules that compose matter are in constant random motion—they contain *thermal energy*. The **temperature** of a substance is a measure of its thermal energy. The hotter an object, the greater the random motion of the atoms and molecules that compose it, and the higher its temperature. We must be careful to not confuse *temperature* with *heat*. **Heat**, which has units of energy, is the *transfer* or *exchange* of thermal energy caused by a temperature difference. For example, when a cold ice cube is dropped into a warm cup of water, energy is transferred as heat from the water to the ice, resulting in the cooling of the water. Temperature, by contrast, is a *measure* of the thermal energy of matter (not the exchange of thermal energy).

The most familiar temperature scale in the United States is the **Fahrenheit** (**°F**) **scale**. On the Fahrenheit scale, water freezes at 32 °F and boils at 212 °F. Room temperature is approximately 72 °F. The Fahrenheit scale was initially set up by assigning 0 °F to the freezing point of a concentrated saltwater solution and 96 °F to normal body temperature (although body temperature is now known to be 98.6 °F).

The scale scientists use is the **Celsius** (**°C**) **scale**. On this scale, water freezes at 0 °C and boils at 100 °C. Room temperature is approximately 22 °C.

The Fahrenheit and Celsius scales differ in both the size of their respective degrees and the temperature each calls "zero" (▶ **FIGURE 3.18**). Both the Fahrenheit and Celsius scales contain negative temperatures. A third temperature scale, called the **Kelvin (K) scale**, avoids negative temperatures by assigning 0 K to the coldest temperature possible, absolute zero. Absolute zero (-273.15 °C or -459.7 °F) is the temperature at which molecular motion virtually stops. There is no lower temperature. The kelvin degree, or kelvin (K), is the same size as the Celsius degree; the only difference is the temperature that each scale designates as zero.

We can convert between these temperature scales using the following formulas.

The degree symbol is used with the Celsius and Fahrenheit scales but not with the Kelvin scale.

$$K = °C + 273.15$$

$$°C = \frac{°F - 32}{1.8}$$

▶ **FIGURE 3.18 Comparison of the Fahrenheit, Celsius, and Kelvin temperature scales**   The Fahrenheit degree is five-ninths the size of a Celsius degree. The Celsius degree and the kelvin degree are the same size.

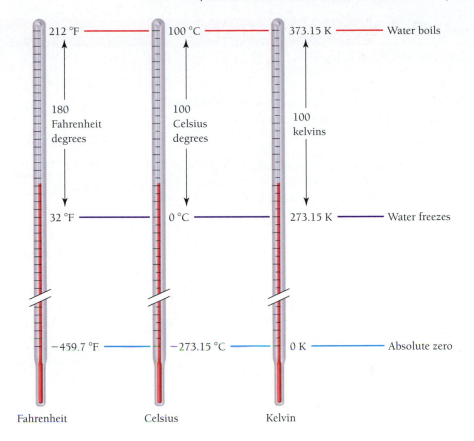

For example, suppose we want to convert 212 K to Celsius. Following the procedure for solving numerical problems (Section 2.6), we first sort the information in the problem statement:

GIVEN: 212 K

FIND: °C

We then strategize by building a solution map.

SOLUTION MAP

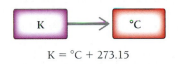

$$K = °C + 273.15$$

In a solution map involving a formula, the formula establishes the relationship between the variables. However, the formula under the arrow is not necessarily solved for the correct variable until later, as is the case here.

RELATIONSHIPS USED

$K = °C + 273.15$ (This equation relates the *given* quantity (K) to the *find* quantity (°C) and is given in this section.)

SOLUTION

Finally, we follow the solution map to solve the problem. The equation below the arrow shows the relationship between K and °C, but it is not solved for the correct variable. Before using the equation, we must solve it for °C.

$$K = °C + 273.15$$
$$°C = K - 273.15$$

We can now substitute the given value for K and calculate the answer to the correct number of significant figures.

$$°C = 212 - 273.15$$
$$= -61 \, °C$$

EXAMPLE **3.7** | **Converting between Celsius and Kelvin Temperature Scales**

Convert −25 °C to kelvins.

SORT	
You are given a temperature in degrees Celsius and asked to find the value of the temperature in kelvins.	**GIVEN:** −25 °C **FIND:** K

STRATEGIZE	
Draw a solution map. Use the equation that relates the temperature in kelvins to the temperature in Celsius to convert from the given quantity to the quantity you want to find.	**SOLUTION MAP**  $$K = °C + 273.15$$ **RELATIONSHIPS USED** $K = °C + 273.15$ (presented in this section)

SOLVE	
Follow the solution map by substituting the correct value for °C and calculating the answer to the correct number of significant figures.	**SOLUTION** $$K = °C + 273.15$$ $$K = -25 \,°C + 273.15 = 248 \text{ K}$$ The significant figures are limited to the ones place because the given temperature (−25 °C) is only given to the ones place. Remember that for addition and subtraction, the number of *decimal places* in the least precisely known quantity determines the number of *decimal places* in the result.

CHECK	
Check your answer. Are the units correct? Does the answer make physical sense?	The units (K) are correct. The answer makes sense because the value in kelvins should be a more positive number than the value in degrees Celsius.

▶ **SKILLBUILDER 3.7** | **Converting between Celsius and Kelvin Temperature Scales**

Convert 358 K to Celsius.

▶ **FOR MORE PRACTICE** Example 3.17; Problems 65c, 66d.

EXAMPLE **3.8** | **Converting between Fahrenheit and Celsius Temperature Scales**

Convert 55 °F to Celsius.

SORT	
You are given a temperature in degrees Fahrenheit and asked to find the value of the temperature in degrees Celsius.	**GIVEN:** 55 °F **FIND:** °C

STRATEGIZE	
Draw the solution map. Use the equation that shows the relationship between the *given* quantity (°F) and the *find* quantity (°C).	**SOLUTION MAP** 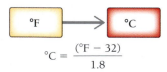 $$°C = \frac{(°F - 32)}{1.8}$$ **RELATIONSHIPS USED** $$°C = \frac{(°F - 32)}{1.8} \text{ (presented in this section)}$$

SOLVE	
Substitute the given value into the equation and calculate the answer to the correct number of significant figures.	**SOLUTION** $$°C = \frac{(°F - 32)}{1.8}$$ $$°C = \frac{(55 - 32)}{1.8} = 12.778 \,°C = 13 \,°C$$

CHECK	The units (°C) are correct. The value of the answer (13 °C) is smaller than the value in degrees Fahrenheit. For positive temperatures, the value of a temperature in degrees Celsius will always be smaller than the value in degrees Fahrenheit because the Fahrenheit degree is smaller than the Celsius degree and the Fahrenheit scale is offset by 32 degrees (see Figure 3.18).
Check your answer. Are the units correct? Does the answer make physical sense?	

▶ **SKILLBUILDER 3.8 | Converting between Fahrenheit and Celsius Temperature Scales**

Convert 139 °C to Fahrenheit.

▶ **FOR MORE PRACTICE** Example 3.18; Problems 65a, 66a, c.

EXAMPLE **3.9**   **Converting between Fahrenheit and Kelvin Temperature Scales**

Convert 310 K to Fahrenheit.

SORT	**GIVEN:** 310 K
You are given a temperature in kelvins and asked to find the value of the temperature in degrees Fahrenheit.	**FIND:** °F

STRATEGIZE	**SOLUTION MAP**
Build the solution map, which requires two steps: one to convert kelvins to degrees Celsius and one to convert degrees Celsius to degrees Fahrenheit.	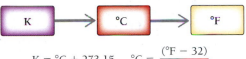

$$K = °C + 273.15 \qquad °C = \frac{(°F - 32)}{1.8}$$

RELATIONSHIPS USED

$K = °C + 273.15$ (presented in this section)

$°C = \dfrac{(°F - 32)}{1.8}$ (presented in this section) |

SOLVE	**SOLUTION**
Solve the first equation for °C and substitute the *given* quantity in K to convert it to °C.	$K = °C + 273.15$
Solve the second equation for °F. Substitute the value of the temperature in °C (from the previous step) to convert it to °F and round the answer to the correct number of significant figures.	$°C = K - 273.15$
$°C = 310 - 273.15 = 36.85 \,°C$
$°C = \dfrac{(°F - 32)}{1.8}$
$1.8\,(°C) = (°F - 32)$
$°F = 1.8\,(°C) + 32$
$°F = 1.8\,(36.85) + 32 = 98.33 \,°F = 98 \,°F$

CHECK	The units (°F) are correct. The magnitude of the answer is a bit trickier to judge. In this temperature range, a temperature in Fahrenheit should indeed be smaller than a temperature in kelvins. However, because the Fahrenheit degree is smaller, temperatures in Fahrenheit become larger than temperatures in kelvins above 575 °F.
Check your answer. Are the units correct? Does the answer make physical sense?	

▶ **SKILLBUILDER 3.9 | Converting between Fahrenheit and Kelvin Temperature Scales**

Convert −321 °F to kelvins.

▶ **FOR MORE PRACTICE** Problems 65b, d, 66b.

PEARSON eText 2.0

CONCEPTUAL ✓ CHECKPOINT 3.6

Which temperature is identical on both the Celsius and the Fahrenheit scales?

(a) 100°

(b) 32°

(c) 0°

(d) −40°

3.11 Temperature Changes: Heat Capacity

▶ Relate energy, temperature change, and heat capacity.

All substances change temperature when they are heated, but how much they change for a given amount of heat varies significantly from one substance to another. For example, if you put a steel skillet on a flame, its temperature rises rapidly. However, if you put some water in the skillet, the temperature increases more slowly. Why? One reason is that when you add water, the same amount of heat energy must warm more matter, so the temperature rise is slower. The second, and more interesting reason, is that water is more resistant to temperature change than steel because water has a higher *specific heat capacity*. The **specific heat capacity** (or **specific heat**) of a substance is the quantity of heat (usually in joules) required to change the temperature of 1 g of the substance by 1 °C. Specific heat capacity has units of joules per gram per degree Celsius (J/g °C). Table 3.4 lists the values of the specific heat capacity for several substances.

Notice that water has the highest specific heat capacity on the list—changing the temperature of water requires a lot of heat. If you have traveled from an inland geographical region to a coastal one and have felt a drop in temperature, you have experienced the effects of water's high specific heat capacity. On a summer day in California, for example, the temperature difference between Sacramento (an inland city) and San Francisco (a coastal city) can be 30 °F (17 °C); San Francisco enjoys a cool 68 °F (20 °C), while Sacramento bakes at near 100 °F (37 °C), yet the intensity of sunlight falling on these two cities

TABLE 3.4 Specific Heat Capacities of Some Common Substances

Substance	Specific Heat Capacity (J/g °C)
Lead	0.128
Gold	0.128
Silver	0.235
Copper	0.385
Iron	0.449
Aluminum	0.903
Ethanol	2.42
Water	4.184

▲ San Francisco enjoys cool weather even in summer months because of the high heat capacity of the surrounding ocean.

is the same. Why the large temperature difference? The difference is due to the presence of the Pacific Ocean, which practically surrounds San Francisco. On the one hand, water, with its high heat capacity, absorbs much of the sun's heat without undergoing a large increase in temperature, keeping San Francisco cool. The land surrounding Sacramento, on the other hand, with its lower heat capacity, cannot absorb a lot of heat without a large increase in temperature—it has a lower *capacity* to absorb heat without a large temperature increase.

Similarly, only two U.S. states have never recorded a temperature above 100 °F (37 °C). One of them is obvious: Alaska. It is too far north to get that hot. The other one, however, may come as a surprise. It is Hawaii. The water that surrounds America's only island state moderates the temperature, preventing Hawaii from ever getting too hot.

EVERYDAY CHEMISTRY

Coolers, Camping, and the Heat Capacity of Water

Have you ever loaded a cooler with ice and then added room-temperature drinks? If you have, you know that the ice quickly melts. In contrast, if you load your cooler with chilled drinks, the ice lasts for hours. Why the difference? The answer is related to the high heat capacity of the water within the drinks. As you just learned, water must absorb a lot of heat to raise its temperature, and it must also release a lot of heat to lower its temperature. When the warm drinks are placed into the ice, they release heat, which then melts the ice. The chilled drinks, however, are already cold, so they do not release much heat. It is always better to load your cooler with chilled drinks—that way, the ice will last the rest of the day.

B3.2 CAN YOU ANSWER THIS? *Suppose you are cold-weather camping and decide to heat some objects to bring into your sleeping bag for added warmth. You place a large water jug and a rock of equal mass close to the fire. Over time, both the rock and the water jug warm to about 38 °C (100 °F). If you could bring only one into your sleeping bag, which one would keep you warmer? Why?*

▲ The ice in a cooler loaded with cold drinks lasts much longer than the ice in a cooler loaded with warm drinks.

QUESTION: Can you explain why?

CONCEPTUAL ✔ CHECKPOINT 3.7

If you want to heat a metal plate to as high a temperature as possible for a given energy input, what metal should you use? (Assume all the plates have the same mass.)

(a) copper

(b) iron

(c) aluminum

(d) it would make no difference

3.12 Energy and Heat Capacity Calculations

▶ Perform calculations involving transfer of heat and changes in temperature.

When a substance absorbs heat (which we represent with the symbol q), its temperature change (which we represent as ΔT) is in direct proportion to the amount of heat absorbed.

$$\xrightarrow[q]{} \boxed{\text{System}}$$
$$\Delta T$$

In other words, the more heat absorbed, the greater the temperature change. We can use the specific heat capacity of the substance to *quantify* the relationship between the amount of heat added to a given amount of the substance and the corresponding temperature increase. The equation that relates these quantities is:

$$\text{heat} = \text{mass} \times \text{specific heat capacity} \times \text{temperature change}$$
$$q = m \quad \times \quad C \quad \times \quad \Delta T$$

where q is the amount of heat in joules, m is the mass of the substance in grams, C is the specific heat capacity in joules per gram per degree Celsius, and ΔT is the temperature change in Celsius. The symbol Δ means *the change in*, so ΔT means *the change in temperature*. For example, suppose we are making a cup of tea and want to know how much heat energy will warm 235 g of water (about 8 oz) from 25 °C to 100.0 °C (boiling).

We begin by sorting the information in the problem statement.

GIVEN: 235 g water (m)
25 °C initial temperature (T_i)
100.0 °C final temperature (T_f)

FIND: amount of heat needed (q)

Then we strategize by building a solution map.

SOLUTION MAP

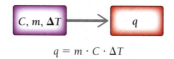

$$q = m \cdot C \cdot \Delta T$$

In addition to m and ΔT, the equation requires C, the specific heat capacity of water. The next step is to gather all of the required quantities for the equation (C, m, and ΔT) in the correct units. These are:

$$C = 4.18 \, \text{J/g} \,°\text{C}$$
$$m = 235 \, \text{g}$$

The other quantity we require is ΔT. The change in temperature is the difference between the final temperature (T_f) and the initial temperature (T_i).

$$\Delta T = T_f - T_i$$
$$= 100.0 \,°\text{C} - 25 \,°\text{C} = 75 \,°\text{C}$$

SOLUTION
Finally, we solve the problem. We substitute the correct values into the equation and calculate the answer to the correct number of significant figures.

$$q = m \cdot C \cdot \Delta T$$

$$= 235 \, \cancel{\text{g}} \times 4.18 \frac{\text{J}}{\cancel{\text{g}} \,\cancel{°\text{C}}} \times 75 \,\cancel{°\text{C}}$$

$$= 7.367 \times 10^4 \, \text{J} = 7.4 \times 10^4 \, \text{J}$$

It is critical that we substitute each of the correct variables into the equation in the correct units and cancel units as we calculate the answer. If, during this process, we learn that one of our variables is not in the correct units, we convert it to the correct units using the skills we learned in Chapter 2. Notice that the sign of q is positive (+) if the substance is increasing in temperature (heat entering the substance) and negative (−) if the substance is decreasing in temperature (heat leaving the substance).

ΔT in °C is equal to ΔT in K but is not equal to ΔT in °F.

Interactive
Worked Example
Video 3.10

EXAMPLE **3.10** | **Relating Heat Energy to Temperature Changes**

Gallium is a solid metal at room temperature but melts at 29.9 °C. If you hold gallium in your hand, it melts from your body heat. How much heat must 2.5 g of gallium absorb from your hand to raise the temperature of the gallium from 25.0 °C to 29.9 °C? The specific heat capacity of gallium is 0.372 J/g °C.

SORT

You are given the mass of gallium, its initial and final temperatures, and its specific heat capacity, and are asked to find the amount of heat absorbed by the gallium.

GIVEN: 2.5 g gallium (m)

$T_i = 25.0$ °C

$T_f = 29.9$ °C

$C = 0.372$ J/g °C

FIND: q

STRATEGIZE

The equation that relates the *given* and *find* quantities is the specific heat capacity equation. The solution map indicates that this equation takes you from the *given* quantities to the quantity you are asked to *find*.

SOLUTION MAP

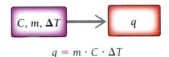

$$q = m \cdot C \cdot \Delta T$$

RELATIONSHIPS USED

$q = m \cdot C \cdot \Delta T$ (presented in this section)

SOLVE

Before solving the problem, you must gather the necessary quantities—C, m, and ΔT—in the correct units.

SOLUTION

$C = 0.372$ J/g °C

$m = 2.5$ g

$$\Delta T = T_f - T_i$$

$$= 29.9 \text{ °C} - 25.0 \text{ °C}$$

$$= 4.9 \text{ °C}$$

Substitute C, m, and ΔT into the equation, canceling units, and calculate the answer to the correct number of significant figures.

$$q = m \cdot C \cdot \Delta T$$

$$= 2.5 \text{ g} \times 0.372 \frac{1}{\text{g °C}} \times 4.9 \text{ °C} = 4.557 \text{ J} = 4.6 \text{ J*}$$

CHECK

Check your answer. Are the units correct? Does the answer make physical sense?

The units (J) are correct. The magnitude of the answer makes sense because it takes almost 1 J to heat the 2.5 g sample of the metal by 1 °C; therefore, it should take about 5 J to heat the sample by 5 °C.

▶ **SKILLBUILDER 3.10** | **Relating Heat Energy to Temperature Changes**

You find a 1979 copper penny (pre-1982 pennies are nearly pure copper) in the snow and pick it up. How much heat does the penny absorb as it warms from the temperature of the snow, −5.0 °C, to the temperature of your body, 37.0 °C? Assume the penny is pure copper and has a mass of 3.10 g. You can find the heat capacity of copper in Table 3.4.

▶ **SKILLBUILDER PLUS**   The temperature of a lead fishing weight rises from 26 °C to 38 °C as it absorbs 11.3 J of heat. What is the mass of the fishing weight in grams?

▶ **FOR MORE PRACTICE**   Example 3.19; Problems 75, 76, 77, 78.

* This is the amount of heat required to raise the temperature to the melting point. Actually, melting the gallium requires additional heat.

EXAMPLE **3.11** | **Relating Heat Capacity to Temperature Changes**

A chemistry student finds a shiny rock that she suspects is gold. She weighs the rock on a balance and determines that its mass is 14.3 g. She then finds that the temperature of the rock rises from 25 °C to 52 °C upon absorption of 174 J of heat. Find the heat capacity of the rock and determine whether the value is consistent with the heat capacity of gold (which is listed in Table 3.4).

SORT	**GIVEN:** 14.3 g
You are given the mass of the "gold" rock, the amount of heat absorbed, and the initial and final temperatures. You are asked to find the heat capacity of the rock.	174 J of heat absorbed $T_i = 25\,°C$ $T_f = 52\,°C$ **FIND:** C

STRATEGIZE	**SOLUTION MAP**
The solution map shows how the heat capacity equation relates the *given* and *find* quantities.	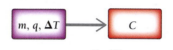 $q = m \cdot C \cdot \Delta T$ **RELATIONSHIPS USED** $q = m \cdot C \cdot \Delta T$ (presented in this section)

SOLVE	**SOLUTION**
First, gather the necessary quantities—*m*, *q*, and ΔT—in the correct units.	$m = 14.3\ \text{g}$ $q = 174\ \text{J}$ $\Delta T = 52\,°C - 25\,°C = 27\,°C$
Then solve the equation for *C* and substitute the correct variables into the equation. Finally, calculate the answer to the right number of significant figures.	$q = m \cdot C \cdot \Delta T$ $C = \dfrac{q}{m \cdot \Delta T}$ $C = \dfrac{174\ \text{J}}{14.3\ \text{g} \times 27\,°C}$ $= 0.4507\ \dfrac{\text{J}}{\text{g}\,°C} = 0.45\ \dfrac{\text{J}}{\text{g}\,°C}$ By comparing the calculated value of the specific heat capacity (0.45 J/g °C) with the specific heat capacity of gold from Table 3.4 (0.128 J/g °C), you conclude that the rock is not pure gold.

CHECK	
Check your answer. Are the units correct? Does the answer make physical sense?	The units of the answer are those of specific heat capacity, so they are correct. The magnitude of the answer falls in the range of specific heat capacities given in Table 3.4. A value of heat capacity that falls far outside this range would immediately be suspect.

▶ **SKILLBUILDER 3.11** | **Relating Heat Capacity to Temperature Changes**

A 328-g sample of water absorbs 5.78×10^3 J of heat. Calculate the change in temperature for the water. If the water is initially at 25.0 °C, what is its final temperature?

▶ **FOR MORE PRACTICE**   Problems 85, 86, 87, 88.

CONCEPTUAL ✔ **CHECKPOINT 3.8**

The heat capacity of substance A is twice that of substance B. If samples of equal mass of the two substances absorb the same amount of heat, which substance undergoes the larger change in temperature?

Chapter 3 in Review

MasteringChemistry™ provides end-of-chapter exercises, feedback-enriched tutorial problems, animations, and interactive activities to encourage problem solving practice and deeper understanding of key concepts and topics.

Self-Assessment Quiz

PEARSON
eText
2.0

Q1. Which substance is a pure compound?
(a) Gold
(b) Water
(c) Milk
(d) Fruit cake

Q2. Which property of trinitrotoluene (TNT) is most likely a chemical property?
(a) Yellow color
(b) Melting point is 80.1 °C
(c) Explosive
(d) None of the above

Q3. Which change is a chemical change?
(a) The condensation of dew on a cold night
(b) A forest fire
(c) The smoothening of rocks by ocean waves
(d) None of the above

Q4. Which process is endothermic?
(a) The burning of natural gas in a stove
(b) The metabolism of glucose by your body
(c) The melting of ice in a soft drink
(d) None of the above

Q5. A 35-g sample of potassium completely reacts with chlorine to form 67 g of potassium chloride. How many grams of chlorine must have reacted?
(a) 67 g
(b) 35 g
(c) 32 g
(d) 12 g

Q6. A runner burns 2.56×10^3 kJ during a five-mile run. How many nutritional Calories did the runner burn?
(a) 1.07×10^1 Cal
(b) 612 Cal
(c) 6.12×10^5 Cal
(d) 1.07×10^4 Cal

Q7. Convert the boiling point of water (100.00 °C) to K.
(a) −173.15 K
(b) 0 K
(c) 100.00 K
(d) 373.15 K

Q8. A European doctor reports that you have a fever of 39.2 °C. What is your fever in degrees Fahrenheit?
(a) 102.6 °F
(b) 128.26 °F
(c) 71.2 °F
(d) 4 °F

Q9. How much heat must be absorbed by 125 g of ethanol to change its temperature from 21.5 °C to 34.8 °C?
(a) 6.95 kJ
(b) 4.02×10^3 kJ
(c) 86.6 kJ
(d) 4.02 kJ

Q10. Substance A has a heat capacity that is much greater than that of substance B. If 10.0 g of substance A initially at 25.0 °C is brought into thermal contact with 10.0 g of B initially at 75.0 °C, what can you conclude about the final temperature of the two substances once the exchange of heat between the substances is complete?
(a) The final temperature will be between 25.0 °C and 50.0 °C.
(b) The final temperature will be between 50.0 °C and 75.0 °C.
(c) The final temperature will be 50.0 °C.
(d) You can conclude nothing about the final temperature without more information.

Answers: 1:b, 2:c, 3:b, 4:c, 5:c, 6:b, 7:d, 8:a, 9:d, 10:a

Chemical Principles

Matter
Matter is anything that occupies space and has mass. It is composed of atoms, which are often bonded together as molecules. Matter can exist as a solid, a liquid, or a gas. Solid matter can be either amorphous or crystalline.

Classification of Matter
We can classify matter according to its composition. Pure matter is composed of only one type of substance; that substance may be an element (a substance that cannot be decomposed into simpler substances), or it may be a compound (a substance composed of two or more elements in fixed definite proportions). Mixtures are composed of two or more different substances, the proportions of which may vary from one sample to the next. Mixtures can be either homogeneous, having the same composition throughout, or heterogeneous, having a composition that varies from region to region.

Relevance

Everything is made of matter—you, me, the chair you sit on, and the air we breathe. The physical universe basically contains only two things: matter and energy. We begin our study of chemistry by defining and classifying these two building blocks of the universe.

Since ancient times, humans have tried to understand matter and harness it for their purposes. The earliest humans shaped matter into tools and used the transformation of matter—especially fire—to keep warm and to cook food. To manipulate matter, we must understand it. Fundamental to this understanding is the connection between the properties of matter and the molecules and atoms that compose it.

Properties and Changes of Matter

We can divide the properties of matter into two types: physical and chemical. The physical properties of matter do not involve a change in composition. The chemical properties of matter involve a change in composition. We can divide changes in matter into physical and chemical. In a physical change, the appearance of matter may change, but its composition does not. In a chemical change, the composition of matter changes.

The physical and chemical properties of matter make the world around us the way it is. For example, a physical property of water is its boiling point at sea level—100 °C. The physical properties of water—and all matter—are determined by the atoms and molecules that compose it. If water molecules were different—even slightly different—water would boil at a different temperature. Imagine a world where water boiled at room temperature.

Conservation of Mass

Whether the changes in matter are chemical or physical, matter is always conserved. In a chemical change, the masses of the matter undergoing the chemical change must equal the sum of the masses of matter resulting from the chemical change.

The conservation of matter is relevant to, for example, pollution. We often think that humans create pollution, but, actually, we are powerless to create anything. Matter cannot be created. So, pollution is simply misplaced matter—matter that we have put into places where it does not belong.

Energy

Besides matter, energy is the other major component of our universe. Like matter, energy is conserved—it can be neither created nor destroyed. Energy exists in various different types, and these can be converted from one to another. Some common units of energy are the joule (J), the calorie (cal), the nutritional Calorie (Cal), and the kilowatt-hour (kWh). Chemical reactions that emit energy are exothermic; those that absorb energy are endothermic.

Our society's energy sources will not last forever because as we burn fossil fuels—our primary energy source—we convert chemical energy, stored in molecules, to kinetic and thermal energy. The kinetic and thermal energy is not readily available to be used again. Consequently, our energy resources are dwindling, and the conservation of energy implies that we will not be able simply to create new energy—it must come from somewhere. All of the chemical reactions that we use for energy are exothermic.

Temperature

The temperature of matter is related to the random motions of the molecules and atoms that compose it—the greater the motion, the higher the temperature. To measure temperature we use three scales: Fahrenheit (°F), Celsius (°C), and Kelvin (K).

The temperature of matter and its measurement are relevant to many everyday phenomena. Humans are understandably interested in the weather, and air temperature is a fundamental part of weather. We use body temperature as one measure of human health and global temperature as one measure of the planet's health.

Heat Capacity

The temperature change that a sample of matter undergoes upon absorption of a given amount of heat relates to the heat capacity of the substance composing the matter. Water has one of the highest heat capacities, meaning that it is most resistant to rapid temperature changes.

The heat capacity of water explains why it is cooler in coastal areas, which are near large bodies of high-heat-capacity water, than in inland areas, which are surrounded by low-heat-capacity land. It also explains why it takes longer to cool a refrigerator filled with liquids than an empty one.

Chemical Skills

Examples

LO: Classify matter as element, compound, or mixture (Section 3.4).

Begin by examining the alphabetical listing of elements in the back of this book. If the substance is listed in that table, it is a pure substance and an element.

If the substance is not listed in that table, refer to your everyday experience with the substance to determine whether it is a pure substance. If it is a pure substance not listed in the table, then it is a compound.

If it is not a pure substance, then it is a mixture. Refer to your everyday experience with the mixture to determine whether it has uniform composition (homogeneous) or non-uniform composition (heterogeneous).

EXAMPLE **3.12** | **Classifying Matter**

Classify each type of matter as a pure substance or a mixture. If it is a pure substance, classify it as an element or a compound. If it is a mixture, classify it as homogeneous or heterogeneous.

(a) pure silver
(b) swimming-pool water
(c) dry ice (solid carbon dioxide)
(d) blueberry muffin

SOLUTION
(a) Pure element; silver appears in the element table.
(b) Homogeneous mixture; pool water contains at least water and chlorine, and it is uniform throughout.
(c) Compound; dry ice is a pure substance (carbon dioxide), but it is not listed in the table.
(d) Heterogeneous mixture; a blueberry muffin is a mixture of several things and has nonuniform composition.

LO: Distinguish between physical and chemical properties (Section 3.5).

To distinguish between physical and chemical properties, consider whether the substance changes composition while displaying the property. If it *does not* change composition, the property is physical; if it *does*, the property is chemical.

EXAMPLE 3.13 Distinguishing between Physical and Chemical Properties

Classify each property as physical or chemical.

(a) the tendency for platinum jewelry to scratch easily
(b) the ability of sulfuric acid to burn the skin
(c) the ability of hydrogen peroxide to bleach hair
(d) the density of lead relative to other metals

SOLUTION

(b) Physical; scratched platinum is still platinum.
(c) Chemical; the acid chemically reacts with the skin to produce the burn.
(d) Chemical; the hydrogen peroxide chemically reacts with hair to change the hair.
(e) Physical; you can determine the density of lead by measuring the volume and mass of a lead sample.

LO: Distinguish between physical and chemical changes (Section 3.6).

To distinguish between physical and chemical changes, consider whether the substance changes composition during the change. If it *does not* change composition, the change is physical; if it *does*, the change is chemical.

EXAMPLE 3.14 Distinguishing between Physical and Chemical Changes

Classify each change as physical or chemical.

(a) the explosion of gunpowder in the barrel of a gun
(b) the melting of gold in a furnace
(c) the bubbling that occurs when you mix baking soda and vinegar
(d) the bubbling that occurs when water boils

SOLUTION

(a) Chemical; the gunpowder reacts with oxygen during the explosion.
(b) Physical; the liquid gold is still gold.
(c) Chemical; the bubbling is a result of a chemical reaction between the two substances to form new substances, one of which is carbon dioxide released as bubbles.
(d) Physical; the bubbling is due to liquid water turning into gaseous water, but it is still water.

LO: Apply the law of conservation of mass (Section 3.7).

The sum of the masses of the substances involved in a chemical change must be the same before and after the change.

EXAMPLE 3.15 Applying the Law of Conservation of Mass

An automobile runs for 10 minutes and burns 47 g of gasoline. The gasoline combines with oxygen from air and forms 132 g of carbon dioxide and 34 g of water. How much oxygen is consumed in the process?

SOLUTION

The total mass after the chemical change is:

$$132 \text{ g} + 34 \text{ g} = 166 \text{ g}$$

The total mass before the change must also be 166 g.

$$47 \text{ g} + \text{oxygen} = 166 \text{ g}$$

So, the mass of oxygen consumed is the total mass (166 g) minus the mass of gasoline (47 g).

$$\text{grams of oxygen} = 166 \text{ g} - 47 \text{ g} = 119 \text{ g}$$

LO: Identify and convert among energy units (Section 3.8).

Solve unit conversion problems using the problem-solving strategies outlined in Section 2.6.

SORT

You are given an amount of energy in kilowatt-hours and asked to find the amount in calories.

STRATEGIZE

Draw a solution map. Begin with kilowatt-hours and determine the conversion factors to get to calories.

SOLVE

Follow the solution map to solve the problem. Begin with the *given* quantity and multiply by the conversion factors to arrive at calories. Round the answer to the correct number of significant figures.

CHECK

Are the units correct? Does the answer make physical sense?

EXAMPLE **3.16**	**Converting Energy Units**

Convert 1.7×10^3 kWh (the amount of energy the average U.S. citizen uses in one week) into calories.

GIVEN: 1.7×10^3 kWh

FIND: cal

SOLUTION MAP

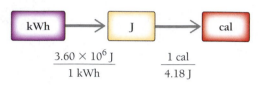

$$\frac{3.60 \times 10^6 \text{ J}}{1 \text{ kWh}} \qquad \frac{1 \text{ cal}}{4.18 \text{ J}}$$

RELATIONSHIPS USED

$$1 \text{ kWh} = 3.60 \times 10^6 \text{ J (Table 3.2)}$$
$$1 \text{ cal} = 4.18 \text{ J} \qquad \text{(Table 3.2)}$$

SOLUTION

$$1.7 \times 10^3 \text{ kWh} \times \frac{3.60 \times 10^6 \text{ J}}{1 \text{ kWh}} \times \frac{1 \text{ cal}}{4.18 \text{ J}}$$
$$= 1.464 \times 10^9 \text{ cal}$$
$$1.464 \times 10^9 \text{ cal} = 1.5 \times 10^9 \text{ cal}$$

The unit of the answer, cal, is correct. The magnitude of the answer makes sense because cal is a smaller unit than kWh; therefore, the value in cal should be larger than the value in kWh.

LO: Convert between Fahrenheit, Celsius, and Kelvin temperature scales (Section 3.10).

Solve temperature conversion problems using the problem-solving procedure in Sections 2.6 and 2.10. Take the steps appropriate for equations.

SORT

You are given the temperature in kelvins and asked to convert it to degrees Celsius.

STRATEGIZE

Draw a solution map. Use the equation that relates the *given* quantity to the *find* quantity.

SOLVE

Solve the equation for the *find* quantity (°C) and substitute the temperature in K into the equation. Calculate the answer to the correct number of significant figures.

CHECK

Are the units correct? Does the answer make physical sense?

	Converting between Celsius and Kelvin
EXAMPLE **3.17**	**Temperature Scales**

Convert 257 K to Celsius.

GIVEN: 257 K

FIND: °C

SOLUTION MAP

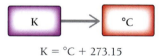

$$K = {}^{\circ}C + 273.15$$

RELATIONSHIPS USED

$$K = {}^{\circ}C + 273.15 \text{ (Section 3.10)}$$

SOLUTION

$$K = {}^{\circ}C + 273.15$$
$${}^{\circ}C = K - 273.15$$
$${}^{\circ}C = 257 - 273.15 = -16\,{}^{\circ}C$$

The answer has the correct unit, and its magnitude seems correct (see Figure 3.18).

LO: Convert between Fahrenheit, Celsius, and Kelvin temperature scales (Section 3.10).

Solve temperature conversion problems using the problem-solving procedure in Sections 2.6 and 2.10. Take the steps appropriate for equations.

SORT
You are given the temperature in degrees Celsius and asked to convert it to degrees Fahrenheit.

STRATEGIZE
Draw a solution map. Use the equation that relates the *given* quantity to the *find* quantity.

SOLVE
Solve the equation for the *find* quantity (°F) and substitute the temperature in °C into the equation. Calculate the answer to the correct number of significant figures.

CHECK
Are the units correct? Does the answer make physical sense?

EXAMPLE 3.18 | Converting between Fahrenheit and Celsius Temperature Scales

Convert 62.0 °C to Fahrenheit.

GIVEN: 62.0 °C

FIND: °F

SOLUTION MAP

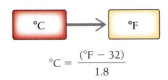

$$°C = \frac{(°F - 32)}{1.8}$$

RELATIONSHIPS USED

$$°C = \frac{(°F - 32)}{1.8}$$

SOLUTION

$$°C = \frac{(°F - 32)}{1.8}$$

$$1.8\,(°C) = °F - 32$$

$$°F = 1.8\,(°C) + 32$$

$$°F = 1.8\,(62.0) + 32 = 143.60\ °F = 144\ °F$$

The answer is in the correct units, and its magnitude seems correct (see Figure 3.18).

LO: Perform calculations involving transfer of heat and changes in temperature (Section 3.12).

Solve heat capacity problems using the problem-solving procedure in Sections 2.6 and 2.10. Take the steps appropriate for equations.

SORT
You are given the volume of water and the amount of heat absorbed. You are asked to find the change in temperature.

STRATEGIZE
The solution map shows how the heat-capacity equation relates the *given* and *find* quantities.

SOLVE
First, gather the necessary quantities—m, q, and ΔT—in the correct units. You must convert the value for q from kJ to J.

You also must convert the value for m from milliliters to grams; use the density of water, 1.0 g/mL, to convert milliliters to grams.

Look up the heat capacity for water in Table 3.4.

EXAMPLE 3.19 | Relating Energy, Temperature Change, and Heat Capacity

What is the temperature change in 355 mL of water upon absorption of 34 kJ of heat?

GIVEN: 355 mL water
34 kJ of heat

FIND: ΔT

SOLUTION MAP

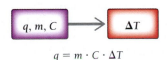

$$q = m \cdot C \cdot \Delta T$$

RELATIONSHIPS USED

$$q = m \cdot C \cdot \Delta T$$

SOLUTION

$$q = 34\ \text{kJ} \times \frac{1000\ \text{J}}{1\ \text{kJ}} = 3.4 \times 10^4\ \text{J}$$

$$m = 355\ \text{mL} \times \frac{1.0\ \text{g}}{1\ \text{mL}} = 355\ \text{g}$$

$$C = 4.18\ \text{J/g}\ °C$$

$$q = m \cdot C \cdot \Delta T$$

Then solve the equation for ΔT and substitute the correct variables into the equation. Finally, calculate the answer to the right number of significant figures.

$$\Delta T = \frac{q}{mC}$$

$$\Delta T = \frac{3.4 \times 10^4 \text{ J}}{355 \text{ g} \times 4.18 \text{ J/g} \, ^\circ\text{C}}$$

$$= 22.91 \, ^\circ\text{C} = 23 \, ^\circ\text{C}$$

CHECK

Check your answer. Are the units correct? Does the answer make physical sense?

The answer has the correct units, and the magnitude seems correct. If the magnitude of the answer were a huge number—3×10^6, for example—you would need to go back and look for a mistake. Above 100 °C, water boils, so such a large answer would be unlikely.

Key Terms

amorphous [3.3]
atom [3.2]
calorie (cal) [3.8]
Calorie (Cal) [3.8]
Celsius (°C) scale [3.10]
chemical change [3.6]
chemical energy [3.8]
chemical property [3.5]
chemical reaction [3.6]
compound [3.4]
compressible [3.3]
crystalline [3.3]
decanting [3.6]

distillation [3.6]
electrical energy [3.8]
element [3.4]
endothermic [3.9]
energy [3.8]
exothermic [3.9]
Fahrenheit (°F) scale [3.10]
filtration [3.6]
gas [3.3]
heat [3.10]
heterogeneous mixture [3.4]
homogeneous mixture [3.4]
Kelvin (K) scale [3.10]

kilowatt-hour (kWh) [3.8]
kinetic energy [3.8]
law of conservation of energy [3.8]
liquid [3.3]
matter [3.2]
mixture [3.4]
molecule [3.2]
physical change [3.6]
physical property [3.5]
potential energy [3.8]
product [3.6]
property [3.5]

pure substance [3.4]
reactants [3.6]
solid [3.3]
specific heat capacity (specific heat) [3.11]
state of matter [3.3]
temperature [3.10]
thermal energy [3.8]
volatile [3.6]
work [3.8]

Exercises

Questions

Answers to all odd-numbered questions (numbered in blue) appear in the Answers section at the back of the book.

1. Define matter and list some examples.
2. What is matter composed of?
3. What are the three states of matter?
4. What are the properties of a solid?
5. What is the difference between a crystalline solid and an amorphous solid?
6. What are the properties of a liquid?
7. What are the properties of a gas?
8. Why are gases compressible?
9. What is a mixture?
10. What is the difference between a homogeneous mixture and a heterogeneous mixture?
11. What is a pure substance?
12. What is an element? A compound?
13. What is the difference between a mixture and a compound?
14. What is the definition of a physical property? What is the definition of a chemical property?
15. What is the difference between a physical change and a chemical change?

16. What is the law of conservation of mass?
17. What is the definition of energy?
18. What is the law of conservation of energy?
19. Explain the difference between kinetic energy and potential energy.
20. What is chemical energy? List some examples of common substances that contain chemical energy.
21. List three common units for energy.
22. What is an exothermic reaction? Which has greater energy in an exothermic reaction, the reactants or the products?
23. What is an endothermic reaction? Which has greater energy in an endothermic reaction, the reactants or the products?
24. List three common units for measuring temperature.
25. Explain the difference between heat and temperature.
26. How do the three temperature scales differ?
27. What is heat capacity?
28. Why are coastal geographic regions normally cooler in the summer than inland geographic regions?

29. The following equation can be used to convert Fahrenheit temperature to Celsius temperature.

$$°C = \frac{(°F - 32)}{1.8}$$

Use algebra to change the equation to convert Celsius temperature to Fahrenheit temperature.

30. The following equation can be used to convert Celsius temperature to Kelvin temperature.

$$K = °C + 273.15$$

Use algebra to change the equation to convert Kelvin temperature to Celsius temperature.

Problems

Note: The exercises in the Problems section are paired, and the answers to the odd-numbered exercises (numbered in blue) appear in the Answers section at the back of the book.

CLASSIFYING MATTER

31. Classify each pure substance as an element or a compound.
 (a) aluminum
 (b) sulfur
 (c) methane
 (d) acetone

32. Classify each pure substance as an element or a compound.
 (a) carbon
 (b) baking soda (sodium bicarbonate)
 (c) nickel
 (d) gold

33. Classify each mixture as homogeneous or heterogeneous.
 (a) coffee
 (b) chocolate sundae
 (c) apple juice
 (d) gasoline

34. Classify each mixture as homogeneous or heterogeneous.
 (a) baby oil
 (b) chocolate chip cookie
 (c) water and gasoline
 (d) wine

35. Classify each substance as a pure substance or a mixture. If it is a pure substance, classify it as an element or a compound. If it is a mixture, classify it as homogeneous or heterogeneous.
 (a) helium gas
 (b) clean air
 (c) rocky road ice cream
 (d) concrete

36. Classify each substance as a pure substance or a mixture. If it is a pure substance, classify it as an element or a compound. If it is a mixture, classify it as homogeneous or heterogeneous.
 (a) urine
 (b) pure water
 (c) Snickers™ bar
 (d) soil

PHYSICAL AND CHEMICAL PROPERTIES AND PHYSICAL AND CHEMICAL CHANGES

37. Classify each property as physical or chemical.
 (a) the tendency of silver to tarnish
 (b) the shine of chrome
 (c) the color of gold
 (d) the flammability of propane gas

38. Classify each property as physical or chemical.
 (a) the boiling point of ethyl alcohol
 (b) the temperature at which dry ice sublimes (turns from a solid into a gas)
 (c) the flammability of ethyl alcohol
 (d) the smell of perfume

39. Which of the following properties of ethylene (a ripening agent for bananas) are physical properties, and which are chemical?
 • colorless
 • odorless
 • flammable
 • gas at room temperature
 • 1 L has a mass of 1.260 g under standard conditions
 • mixes with acetone
 • polymerizes to form polyethylene

40. Which of the following properties of ozone (a pollutant in the lower atmosphere but part of a protective shield against UV light in the upper atmosphere) are physical, and which are chemical?
 • bluish color
 • pungent odor
 • very reactive
 • decomposes on exposure to ultraviolet light
 • gas at room temperature

41. Classify each change as physical or chemical.
  (a) A balloon filled with hydrogen gas explodes upon contact with a spark.
  (b) The liquid propane in a barbecue evaporates away because someone left the valve open.
  (c) The liquid propane in a barbecue ignites upon contact with a spark.
  (d) Copper metal turns green on exposure to air and water.

42. Classify each change as physical or chemical.
  (a) Sugar dissolves in hot water.
  (b) Sugar burns in a pot.
  (c) A metal surface becomes dull because of continued abrasion.
  (d) A metal surface becomes dull on exposure to air.

43. A block of aluminum is **(a)** ground into aluminum powder and then **(b)** ignited. It then emits flames and smoke. Classify **(a)** and **(b)** as chemical or physical changes.

44. Several pieces of graphite from a mechanical pencil are **(a)** broken into tiny pieces. Then the pile of graphite is **(b)** ignited with a hot flame. Classify **(a)** and **(b)** as chemical or physical changes.

THE CONSERVATION OF MASS

45. An automobile gasoline tank holds 42 kg of gasoline. When the gasoline burns, 168 kg of oxygen are consumed and carbon dioxide and water are produced. What total combined mass of carbon dioxide and water is produced?

46. In the explosion of a hydrogen-filled balloon, 0.50 g of hydrogen reacts with 4.0 g of oxygen. How many grams of water vapor are formed? (Water vapor is the only product.)

47. Are these data sets on chemical changes consistent with the law of conservation of mass?
  (a) A 7.5-g sample of hydrogen gas completely reacts with 60.0 g of oxygen gas to form 67.5 g of water.
  (b) A 60.5-g sample of gasoline completely reacts with 243 g of oxygen to form 206 g of carbon dioxide and 88 g of water.

48. Are these data sets on chemical changes consistent with the law of conservation of mass?
  (a) A 12.8-g sample of sodium completely reacts with 19.6 g of chlorine to form 32.4 g of sodium chloride.
  (b) An 8-g sample of natural gas completely reacts with 32 g of oxygen gas to form 17 g of carbon dioxide and 16 g of water.

49. In a butane lighter, 9.7 g of butane combine with 34.7 g of oxygen to form 29.3 g carbon dioxide and how many grams of water?

50. A 56-g sample of iron reacts with 24 g of oxygen to form how many grams of iron oxide?

CONVERSION OF ENERGY UNITS

51. Perform each conversion.
  (a) 588 cal to joules
  (b) 17.4 J to Calories
  (c) 134 kJ to Calories
  (d) 56.2 Cal to joules

52. Perform each conversion.
  (a) 45.6 J to calories
  (b) 355 cal to joules
  (c) 43.8 kJ to calories
  (d) 215 cal to kilojoules

53. Perform each conversion.
  (a) 25 kWh to joules
  (b) 249 cal to Calories
  (c) 113 cal to kilowatt-hours
  (d) 44 kJ to calories

54. Perform each conversion.
  (a) 345 Cal to kilowatt-hours
  (b) 23 J to calories
  (c) 5.7×10^3 J to kilojoules
  (d) 326 kJ to joules

55. Complete the table:

J	cal	Cal	kWh
225 J	_____	5.38×10^{-2} Cal	_____
_____	8.21×10^5 cal	_____	_____
_____	_____	_____	295 kWh
_____	_____	155 Cal	_____

56. Complete the table:

J	cal	Cal	kWh
7.88×10^6 J	1.88×10^6 cal	_____	_____
_____	_____	1154 Cal	_____
_____	88.4 cal	_____	_____
_____	_____	_____	125 kWh

57. An energy bill indicates that a customer used 1027 kWh in July. How many joules did the customer use?

58. A television uses 32 kWh of energy per year. How many joules does it use?

59. An adult eats food whose nutritional energy totals approximately 2.2×10^3 Cal per day. The adult burns 2.0×10^3 Cal per day. How much excess nutritional energy, in kilojoules, does the adult consume per day? If 1 lb of fat is stored by the body for each 14.6×10^3 kJ of excess nutritional energy consumed, how long will it take this person to gain 1 lb?

60. How many joules of nutritional energy are in a bag of chips with a label that lists 245 Cal? If 1 lb of fat is stored by the body for each 14.6×10^3 kJ of excess nutritional energy consumed, how many bags of chips contain enough nutritional energy to result in 1 lb of body fat?

ENERGY AND CHEMICAL AND PHYSICAL CHANGE

61. A common type of handwarmer contains iron powder that reacts with oxygen to form an oxide of iron. As soon as the handwarmer is exposed to air, the reaction begins and heat is emitted. Is the reaction between the iron and oxygen exothermic or endothermic? Draw an energy diagram showing the relative energies of the reactants and products in the reaction.

62. In a chemical cold pack, two substances are kept separate by a divider. When the divider is broken, the substances mix and absorb heat from the surroundings. The chemical cold pack feels cold. Is the reaction exothermic or endothermic? Draw an energy diagram showing the relative energies of the reactants and products in the reaction.

63. Classify each process as exothermic or endothermic.
(a) gasoline burning in a car
(b) isopropyl alcohol evaporating from skin
(c) water condensing as dew during the night

64. Classify each process as exothermic or endothermic.
(a) dry ice subliming (changing from a solid directly to a gas)
(b) the wax in a candle burning
(c) a match burning

CONVERTING BETWEEN TEMPERATURE SCALES

65. Perform each temperature conversion.
(a) 212 °F to Celsius (temperature of boiling water)
(b) 77 K to Fahrenheit (temperature of liquid nitrogen)
(c) 25 °C to kelvins (room temperature)
(d) 98.6 °F to kelvins (body temperature)

66. Perform each temperature conversion.
(a) 102 °F to Celsius
(b) 0 K to Fahrenheit
(c) −48 °C to Fahrenheit
(d) 273 K to Celsius

67. The coldest temperature ever measured in the United States was −80 °F on January 23, 1971, in Prospect Creek, Alaska. Convert that temperature to degrees Celsius and Kelvin. (Assume that −80 °F is accurate to two significant figures.)

68. The warmest temperature ever measured in the United States was 134 °F on July 10, 1913, in Death Valley, California. Convert that temperature to degrees Celsius and Kelvin.

69. Vodka does not freeze in the freezer because it contains a high percentage of ethanol. The freezing point of pure ethanol is −114 °C. Convert that temperature to degrees Fahrenheit and Kelvin.

70. Liquid helium boils at 4.2 K. Convert this temperature to degrees Fahrenheit and Celsius.

71. The temperature in the South Pole during the Antarctic winter is so cold that planes cannot land or take off, effectively leaving the inhabitants of the South Pole isolated for the winter. The average daily temperature at the South Pole in July is −59.7 °C. Convert this temperature to degrees Fahrenheit.

72. The coldest temperature ever recorded in Iowa was −47 °F on February 3, 1998. Convert this temperature to kelvins and degrees Celsius.

73. Complete the table.

Kelvin	Fahrenheit	Celsius
0.0 K	_____	−273.0 °C
_____	82.5 °F	_____
_____	_____	8.5 °C

74. Complete the table.

Kelvin	Fahrenheit	Celsius
273.0 K	_____	0.0 °C
_____	−40.0 °F	_____
385 K	_____	_____

ENERGY, HEAT CAPACITY, AND TEMPERATURE CHANGES

75. Calculate the amount of heat required to raise the temperature of a 65-g sample of water from 32 °C to 65 °C.

76. Calculate the amount of heat required to raise the temperature of a 22-g sample of water from 7 °C to 18 °C.

77. Calculate the amount of heat required to heat a 45-kg sample of ethanol from 11.0 °C to 19.0 °C.

78. Calculate the amount of heat required to heat a 3.5-kg gold bar from 21 °C to 67 °C.

79. If 89 J of heat are added to a pure gold coin with a mass of 12 g, what is its temperature change?

80. If 57 J of heat are added to an aluminum can with a mass of 17.1 g, what is its temperature change?

81. An iron nail with a mass of 12 g absorbs 15 J of heat. If the nail was initially at 28 °C, what is its final temperature?

82. A 45-kg sample of water absorbs 345 kJ of heat. If the water was initially at 22.1 °C, what is its final temperature?

83. Calculate the temperature change that occurs when 248 cal of heat are added to 24 g of water.

84. A lead fishing weight with a mass of 57 g absorbs 146 cal of heat. If its initial temperature is 47 °C, what is its final temperature?

85. An unknown metal with a mass of 28 g absorbs 58 J of heat. Its temperature rises from 31.1 °C to 39.9 °C. Calculate the heat capacity of the metal and identify it, referring to Table 3.4.

86. When 2.8 J of heat are added to 5.6 g of an unknown metal, its temperature rises by 3.9 °C. Are these data consistent with the metal being gold?

87. When 56 J of heat are added to 11 g of a liquid, its temperature rises from 10.4 °C to 12.7 °C. What is the heat capacity of the liquid?

88. When 47.5 J of heat are added to 13.2 g of a liquid, its temperature rises by 1.72 °C. What is the heat capacity of the liquid?

89. Two identical coolers are packed for a picnic. Each cooler is packed with eighteen 12-oz soft drinks and 3 lb of ice. However, the drinks that went into cooler A were refrigerated for several hours before they were packed in the cooler, while the drinks that went into cooler B were at room temperature. When the two coolers are opened three hours later, most of the ice in cooler A is still ice, while nearly all of the ice in cooler B has melted. Explain.

90. A 100-g block of iron metal and 100 g of water are each warmed to 75 °C and placed into two identical insulated containers. Two hours later, the two containers are opened and the temperature of each substance is measured. The iron metal has cooled to 38 °C while the water has cooled only to 69 °C. Explain.

91. How much energy (in J) is lost when a sample of iron with a mass of 25.7 g cools from 75.0 °C to 22.0 °C?

92. A sample of aluminum with mass of 53.2 g is initially at 155 °C. What is the temperature of the aluminum after it loses 2.87×10^3 J?

Cumulative Problems

93. 245 mL of water with an initial temperature of 32 °C absorbs 17 kJ of heat. Find the final temperature of the water. (density of water = 1.0 g/mL)

94. 32 mL of ethanol with an initial temperature of 11 °C absorbs 562 J of heat. Find the final temperature of the ethanol. (density of ethanol = 0.789 g/mL)

95. A pure gold ring with a volume of 1.57 cm^3 is initially at 11.4 °C. When it is put on, it warms to final temperature of 29.5 °C. How much heat (in J) does the ring absorb? (density of gold = 19.3 g/cm^3)

96. A block of aluminum with a volume of 98.5 cm^3 absorbs 67.4 J of heat. If its initial temperature is 32.5 °C, what is its final temperature? (density of aluminum = 2.70 g/cm^3)

97. How much heat in kilojoules is required to heat 56 L of water from an initial temperature of 85 °F to a final temperature of 212 °F? (The density of water is 1.00 g/mL.)

98. How much heat in joules is required to heat a 43-g sample of aluminum from an initial temperature of 72 °F to a final temperature of 145 °F? (The density of water is 1.00 g/mL.)

99. What is the temperature change (ΔT) in Celsius when 29.5 L of water absorbs 2.3 kWh of heat?

100. If 1.45 L of water has an initial temperature of 25.0 °C, what is its final temperature after absorption of 9.4×10^{22} kWh of heat?

101. A water heater contains 55 gal of water. How many kilo-watt-hours of energy are necessary to heat the water in the water heater by 25 °C? (*Hint:* $\Delta T = 25$ °C)

102. A room contains 48 kg of air. How many kilowatt-hours of energy are necessary to heat the air in the room from an initial temperature of 7 °C to a final temperature of 28 °C? The heat capacity of air is 1.03 J/g °C.

103. A backpacker wants to carry enough fuel to heat 2.5 kg of water from 25 °C to 100.0 °C. If the fuel she carries produces 36 kJ of heat per gram when it burns, how much fuel should she carry? (For the sake of simplicity, assume that the transfer of heat is 100% efficient.)

104. A cook wants to heat 1.35 kg of water from 32.0 °C to 100.0 °C. If he uses the combustion of natural gas (which is exothermic) to heat the water, how much natural gas will he need to burn? Natural gas produces 49.3 kJ of heat per gram. (For the sake of simplicity, assume that the transfer of heat is 100% efficient.)

105. Evaporating sweat cools the body because evaporation is endothermic and absorbs 2.44 kJ per gram of water evaporated. Estimate the mass of water that must evaporate from the skin to cool a body by 0.50 °C, if the mass of the body is 95 kg and its heat capacity is 4.0 J/g °C. (Assume that the heat transfer is 100% efficient.)

106. When ice melts, it absorbs 0.33 kJ per gram. How much ice is required to cool a 12.0-oz drink from 75 °F to 35 °F, if the heat capacity of the drink is 4.18 J/g °C? (Assume that the heat transfer is 100% efficient.)

107. A 15.7-g aluminum block is warmed to 53.2 °C and plunged into an insulated beaker containing 32.5 g of water initially at 24.5 °C. The aluminum and the water are allowed to come to thermal equilibrium. Assuming that no heat is lost, what is the final temperature of the water and aluminum?

108. A 25.0-mL sample of ethanol (density = 0.789 g/mL) initially at 7.0 °C is mixed with 35.0 mL of water (density = 1.0 g/mL) initially at 25.3 °C in an insulated beaker. Assuming that no heat is lost, what is the final temperature of the mixture?

109. The wattage of an appliance indicates its average power consumption in watts (W), where 1 W = 1 J/s. What is the difference in the number of kJ of energy consumed per month between a refrigeration unit that consumes 625 W and one that consumes 855 W? If electricity costs $0.15 per kWh, what is the monthly cost difference to operate the two refrigerators? (Assume 30.0 days in one month and 24.0 hours per day.)

110. A portable electric water heater transfers 255 watts (W) of power to 5.5 L of water, where 1 W = 1 J/s. How much time (in minutes) does it take for the water heater to heat the 5.5 L of water from 25 °C to 42 °C? (Assume that the water has a density of 1.0 g/mL.)

111. What temperature is the same whether it is expressed on the Celsius or Fahrenheit scale?

112. What temperature on the Celsius scale is equal to twice its value when expressed on the Fahrenheit scale?

Highlight Problems

113. Classify each as a pure substance or a mixture.

(a)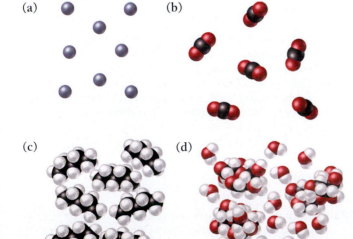

(b)

(c)

(d)

114. Classify each as a pure substance or a mixture. If it is a pure substance, classify it as an element or a compound. If it is a mixture, classify it as homogeneous or heterogeneous.

(a)

(b)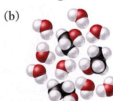

(c)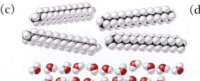

(d)

115. This molecular drawing shows images of acetone molecules before and after a change. Was the change chemical or physical?

116. This molecular drawing shows images of methane molecules and oxygen molecules before and after a change. Was the change chemical or physical?

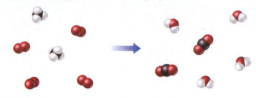

117. A major event affecting global climate is the El Niño/La Niña cycle. In this cycle, equatorial Pacific Ocean waters warm by several degrees Celsius above normal (El Niño) and then cool by several degrees Celsius below normal (La Niña). This cycle affects weather not only in North and South America, but also in places as far away as Africa. Why does a seemingly small change in ocean temperature have such a large impact on weather?

March 1, 2016

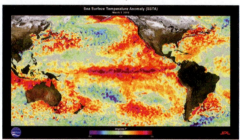

◀ Temperature anomaly plot of the world's oceans. The large red-orange section in the middle of the map indicates the El Niño effect, a warming of the Pacific Ocean along the equator.

118. Global warming refers to the rise in average global temperature due to the increased concentration of certain gases, called greenhouse gases, in our atmosphere. Earth's oceans, because of their high heat capacity, absorb heat and therefore act to slow down global warming. How much heat would be required to warm Earth's oceans by 1.0 °C? Assume that the volume of water in Earth's oceans is 137×10^7 km^3 and that the density of seawater is 1.03 g/cm^3. Also assume that the heat capacity of seawater is the same as that of water.

▲ Earth's oceans moderate temperatures by absorbing heat during warm periods.

119. Examine the data for the maximum and minimum average temperatures of San Francisco and Sacramento in the summer and in the winter.

San Francisco (Coastal City)

January		August	
High	Low	High	Low
57.4 °F	43.8 °F	64.4 °F	54.5 °F

Sacramento (Inland City)

January		August	
High	Low	High	Low
53.2 °F	37.7 °F	91.5 °F	57.7 °F

(a) Notice the difference between the August high in San Francisco and Sacramento. Why is it much hotter in the summer in Sacramento?

(b) Notice the difference between the January low in San Francisco and Sacramento. How might the heat capacity of the ocean contribute to this difference?

Questions for Group Work

Discuss the following questions with your group. Record the answer that your group agrees on.

120. Using white and black circles to represent different kinds of atoms, make a drawing that accurately represents each of the following: a solid element, a liquid compound, a homogeneous mixture, and a heterogeneous mixture.

121. Make a drawing (clearly showing *before* and *after*) depicting your liquid compound from Question 120 undergoing a physical change. Make a drawing depicting your solid element undergoing a chemical change.

122. A friend asks you to invest in a new motorbike that he invented that never needs gasoline and does not use batteries. What questions should you ask before investing?

123. In a grammatically correct sentence or two (and in your own words), describe what heat and temperature have in common and how they are different. Make sure you are using the words correctly with their precise scientific meanings.

Data Interpretation and Analysis

124. The graph at right shows U.S. energy consumption by source from 1980 to 2040 (based on projections). The consumption is measured in quadrillion BTUs or quads (1 quad $= 1.055 \times 10^{18}$ J).

   (a) What were the three largest sources of U.S. energy in 2013 in descending order? What total percent of U.S. energy do these three sources provide?

   (b) What percent of total U.S. energy is provided by renewables in 2013?

   (c) Which two sources of U.S. energy decline as a percentage of total energy use between 1989 and 2040 (based on projections)?

   (d) How much U.S. energy (in joules) was produced by nuclear power in 1990?

U.S. Energy Consumption by Source

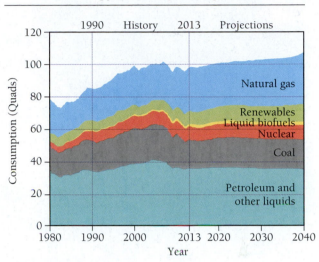

Answers to Skillbuilder Exercises

Skillbuilder 3.1 .... **(a)** pure substance, element
                                    (b) mixture, homogeneous
                                    (c) mixture, heterogeneous
                                    (d) pure substance, compound

Skillbuilder 3.2 .... **(a)** chemical
                                    (b) physical
                                    (c) physical
                                    (d) physical

Skillbuilder 3.3 .... **(a)** chemical
                                    (b) physical
                                    (c) physical
                                    (d) chemical

Skillbuilder 3.4 .... 27 g
Skillbuilder 3.5 .... 2.14 kJ
Skillbuilder Plus, p. 74 6.57×10^6 cal
Skillbuilder 3.6 .... **(a)** exothermic
                                    (b) exothermic
Skillbuilder 3.7 .... 85 °C
Skillbuilder 3.8 .... 282 °F
Skillbuilder 3.9 .... 77 K
Skillbuilder 3.10 .... 50.1 J
Skillbuilder Plus, p. 83   7.4 g
Skillbuilder 3.11 .... $\Delta T = 4.21\,°C; T_f = 29.2\,°C$

Answers to Conceptual Checkpoints

3.1 (c) The particles are far apart and moving relative to one another.

3.2 (a) The substance is composed of only one type of particle (even though each particle is composed of two different type of atoms), so it is a pure substance.

3.3 (a) Vaporization is a physical change, so the water molecules are the same before and after the boiling.

3.4 No In the vaporization, the liquid water becomes gaseous, but its mass does not change. Like chemical changes, physical changes also follow the law of conservation of mass.

3.5 (d) kWh is the largest of the four units listed, so the numerical value of the yearly energy consumption is lowest if expressed in kWh.

3.6 (d) You can confirm this by substituting each of the Fahrenheit temperatures into the equation in Section 3.10 and solving for the Celsius temperature.

3.7 (a) Because copper has the lowest specific heat capacity of the three metals, it experiences the greatest temperature change for a given energy input.

3.8 Substance B will undergo a greater change in temperature because it has the lower heat capacity. A substance with a lower heat capacity is less resistant to temperature changes.

4 Atoms and Elements

"Nothing exists except atoms and empty space; everything else is opinion."
—Democritus (460–370 B.C.)

4.1 Experiencing Atoms at Tiburon

My wife and I recently enjoyed a visit to the northern California seaside town of Tiburon. Tiburon sits next to San Francisco Bay and enjoys views of the water, the city of San Francisco, and the surrounding mountains. As we walked along a waterside path, I could feel the wind as it blew over the bay. I could hear the water splashing on the shore, and I could smell the sea air. What was the cause of these sensations? The answer is simple—atoms.

Because all matter is made of atoms, atoms are at the foundation of our sensations. The atom is the fundamental building block of everything you hear, feel, see, and experience. When you feel wind on your skin, you are feeling atoms. When you hear sounds, you are in a sense hearing atoms. When you touch a shoreside rock, you are touching atoms, and when you smell sea air, you are smelling atoms. You eat atoms, you breathe atoms, and you excrete atoms. Atoms are the building blocks of matter; they are the basic units from which nature builds. They are all around us and compose everything, including our own bodies.

Atoms are incredibly small. A single pebble from the shoreline contains more atoms than you could ever count. The number of atoms in a single pebble far exceeds the number of pebbles on the bottom of San Francisco Bay. To get an idea of how small atoms are, imagine this: if every atom within a small pebble were the size of the pebble itself, the pebble would be larger than Mount Everest (▶ **FIGURE 4.1** on the next page). Atoms are small—yet they compose everything.

The key to connecting the submicroscopic world with the macroscopic world is the atom. Atoms compose matter; the properties of atoms determine the properties of matter. An **atom** is the smallest identifiable unit of an element. Recall from Section 3.4 that an *element* is a substance that cannot be broken down into simpler substances. There are about 91 different elements in nature and consequently about 91 different

> As we learned in Chapter 3, many atoms exist not as free particles but as groups of atoms bound together to form molecules. Nevertheless, all matter is ultimately made of atoms.

> The exact number of naturally occurring elements is controversial because some elements previously considered only synthetic may actually occur in nature in very small quantities.

◀ Seaside rocks are typically composed of silicates, compounds of silicon and oxygen atoms. Seaside air, like all air, contains nitrogen and oxygen molecules, and it often also contains substances called amines. The amine shown here is triethylamine, which is emitted by decaying fish. Triethylamine is one of the compounds responsible for the fishy smell of the seaside.

▲ **FIGURE 4.1 The size of the atom** If every atom within a pebble were the size of the pebble itself, then the pebble would be larger than Mount Everest.

kinds of atoms. In addition, scientists have succeeded in making over 20 synthetic elements (not found in nature). In this chapter, we examine atoms: what they are made of, how they differ from one another, and how they are structured. We also examine the elements that atoms compose and some of the properties of those elements.

4.2 Indivisible: The Atomic Theory

▶ Recognize that all matter is composed of atoms.

▲ Diogenes and Democritus, as imagined by a medieval artist. Democritus is the first person on record to have postulated that matter was composed of atoms.

▶ **FIGURE 4.2 Writing with atoms** Scientists at IBM used a special microscope, called a scanning tunneling microscope (STM), to move xenon atoms to form the letters I, B, and M. The cone shape of these atoms is due to the peculiarities of the instrumentation. Atoms are, in general, spherical in shape.

If we look at matter, even under a microscope, it is not obvious that matter is composed of tiny particles. In fact, it appears to be just the opposite. If we divide a sample of matter into smaller and smaller pieces, it seems that we could divide it forever. From our perspective, matter seems continuous. The first people recorded as thinking otherwise were Leucippus (fifth century B.C.E., exact dates unknown) and Democritus (460–370 B.C.E.). These Greek philosophers theorized that matter was ultimately composed of small, indivisible particles. Democritus suggested that if you divided matter into smaller and smaller pieces, you would eventually end up with tiny, indestructible particles called *atomos*, or "atoms." The word *atomos* means "indivisible."

The ideas of Leucippus and Democritus were not widely accepted. It was not until 1808—over 2000 years later—that John Dalton formalized a theory of atoms that gained broad acceptance. Dalton's atomic theory has three parts:

1. Each element is composed of tiny indestructible particles called atoms.
2. All atoms of a given element have the same mass and other properties that distinguish them from the atoms of other elements.
3. Atoms combine in simple, whole-number ratios to form compounds.

Today, the evidence for the atomic theory is overwhelming. Recent advances in microscopy have allowed scientists not only to image individual atoms but also to pick them up and move them (▼ FIGURE 4.2). Matter is indeed composed of atoms.

Xenon atoms

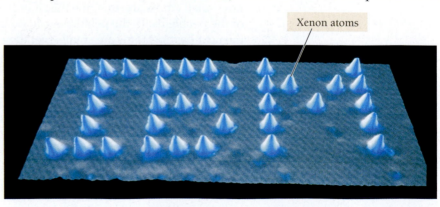

4.3   The Nuclear Atom

▶ Explain how the experiments of Thomson and Rutherford led to the development of the nuclear theory of the atom.

Electric charge is more fully defined in Section 4.4. For now, think of it as an inherent property of electrons that causes them to interact with other charged particles.

By the end of the nineteenth century, scientists were convinced that matter was composed of atoms, the permanent, indestructible building blocks from which all substances are constructed. However, an English physicist named J. J. Thomson (1856–1940) complicated the picture by discovering an even smaller and more fundamental particle called the **electron**. Thomson discovered that electrons are negatively charged, that they are much smaller and lighter than atoms, and that they are uniformly present in many different kinds of substances. His experiments proved that the indestructible building block called the atom could apparently be "chipped."

The discovery of negatively charged particles within atoms raised the question of a balancing positive charge. Atoms were known to be charge-neutral, so it was believed that they must contain positive charge that balanced the negative charge of electrons. But how do the positive and negative charges within the atom fit together? Are atoms just a jumble of even more fundamental particles? Are they solid spheres, or do they have some internal structure? Thomson proposed that the negatively charged electrons were small particles held within a positively charged sphere. This model, the most popular of the time, became known as the plum-pudding model (plum pudding is an English dessert) (◀ FIGURE 4.3). The picture suggested by Thomson was—to those of us not familiar with plum pudding—like a blueberry muffin, where the blueberries are the electrons and the muffin is the positively charged sphere.

In 1909, Ernest Rutherford (1871–1937), who had worked under Thomson and adhered to his plum-pudding model, performed an experiment in an attempt to confirm it. His experiment instead proved the plum-pudding model wrong. In his experiment, Rutherford directed tiny, positively charged particles—called alpha particles—at an ultrathin sheet of gold foil (▼ FIGURE 4.4). Alpha particles are about 7000 times more massive than electrons and carry a positive charge. These particles were to act as probes of the gold atoms' structure. If the gold atoms were indeed like blueberry muffins or plum pudding—with their mass

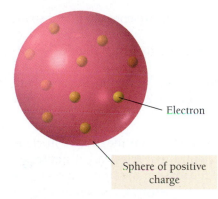

▲ FIGURE 4.3 **Plum-pudding model of the atom**   In the model suggested by J. J. Thomson, negatively charged electrons (yellow) are held in a sphere of positive charge (red).

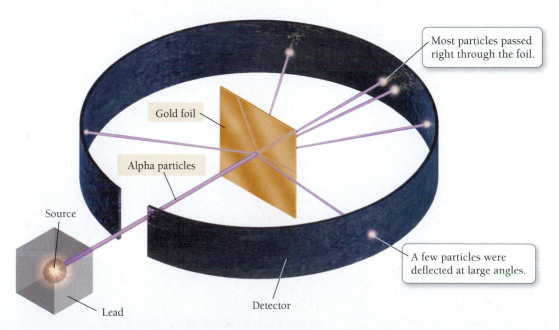

▲ FIGURE 4.4 **Rutherford's gold foil experiment**   Rutherford directed tiny particles called alpha particles at a thin sheet of gold foil. Most of the particles passed directly through the foil. A few, however, were deflected—some of them at sharp angles.

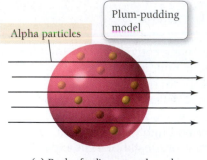

(a) Rutherford's expected result

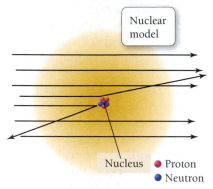

(b) Rutherford's actual result

▲ **FIGURE 4.5 Discovery of the atomic nucleus** **(a)** If the plum-pudding model were correct, the alpha particles would pass right through the gold foil with minimal deflection. **(b)** A small number of alpha particles were deflected or bounced back. The only way to explain the deflections was to suggest that most of the mass and all of the positive charge of an atom must be concentrated in a space much smaller than the size of the atom itself—the nucleus. The nucleus itself is composed of positively charged particles (protons) and neutral particles (neutrons).

and charge spread throughout the entire volume of the atom—these speeding probes should pass right through the gold foil with minimum deflection. Rutherford's results were not as he expected. A majority of the particles did pass directly through the foil, but some particles were deflected, and some (1 in 20,000) even bounced back. The results puzzled Rutherford, who found them "about as credible as if you had fired a 15-inch shell at a piece of tissue paper and it came back and hit you." What must the structure of the atom be in order to explain this odd behavior?

Rutherford created a new model to explain his results (◀ **FIGURE 4.5**). He concluded that matter must not be as uniform as it appears. It contains large regions of empty space dotted with small regions of very dense matter. In order to explain the deflections he observed, the mass and positive charge of an atom must all be concentrated in a space much smaller than the size of the atom itself. Based on this idea, he developed the **nuclear theory of the atom**, which has three basic parts:

1. Most of the atom's mass and all of its positive charge are contained in a small core called the **nucleus**.
2. Most of the volume of the atom is empty space through which the tiny, negatively charged electrons are dispersed.
3. There are as many negatively charged electrons outside the nucleus as there are positively charged particles (*protons*) inside the nucleus, so that the atom is electrically neutral.

Later work by Rutherford and others demonstrated that the atom's nucleus contains both positively charged **protons** and neutral particles called **neutrons**. The dense nucleus makes up more than 99.9% of the mass of the atom but occupies only a small fraction of its volume. The electrons are distributed through a much larger region but don't have much mass (▼ **FIGURE 4.6**). For now, you can think of these electrons as akin to the water droplets that make up a cloud—they are dispersed throughout a large volume but weigh almost nothing.

Rutherford's nuclear theory is still valid today. The revolutionary part of this theory is the idea that matter—at its core—is much less uniform than it appears. If the nucleus of the atom were the size of this dot ·, the average electron would be about 10 m away. Yet the dot would contain almost the entire mass of the atom. What if matter were composed of atomic nuclei piled on top of each other like

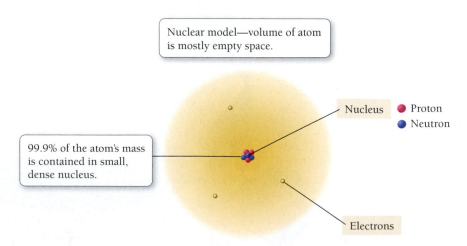

▲ **FIGURE 4.6 The nuclear atom** In this model, the atom's mass is concentrated in a nucleus that contains protons and neutrons. The number of electrons outside the nucleus is equal to the number of protons inside the nucleus. In this image, the nucleus is greatly enlarged and the electrons are portrayed as particles.

marbles? Such matter would be incredibly dense; a single grain of sand composed of solid atomic nuclei would have a mass of 5 million kg (or a weight of about 10 million lb). Astronomers believe that black holes and neutron stars are composed of this kind of incredibly dense matter.

4.4 The Properties of Protons, Neutrons, and Electrons

▶ Describe the respective properties and charges of electrons, neutrons, and protons.

Key Concept Video
Subatomic Particles and Isotope Symbols

◀ If a proton had the mass of a baseball, an electron would have the mass of a rice grain. The proton is nearly 2000 times as massive as an electron.

Protons and neutrons have similar masses. In SI units, the mass of the proton is 1.67262×10^{-27} kg, and the mass of the neutron is a close 1.67493×10^{-27} kg. A more common unit to express these masses, however, is the **atomic mass unit (amu)**, defined as one-twelfth of the mass of a carbon atom containing six protons and six neutrons. A proton has a mass of 1.0073 amu, and a neutron has a mass of 1.0087 amu. Electrons, by contrast, have an almost negligible mass of 0.00091×10^{-27} kg, or approximately 0.00055 amu.

The proton and the electron both have electrical **charge**. The proton's charge is 1+ and the electron's charge is 1−. The charges of the proton and the electron are equal in magnitude but opposite in sign, so that when paired, the two particles' charges exactly cancel. The neutron has no charge.

What is electrical charge? Electrical charge is a fundamental property of protons and electrons, just as mass is a fundamental property of matter. Most matter is charge-neutral because protons and electrons occur together and their charges cancel. However, you may have experienced excess electrical charge when brushing your hair on a dry day. The brushing action results in the accumulation of electrical charge on the hair strands, which then repel each other, causing your hair to stand on end.

EVERYDAY CHEMISTRY

Solid Matter?

If matter really is mostly empty space as Rutherford suggested, then why does it appear so solid? Why can you tap your knuckles on the table and feel a solid thump? Matter appears solid because the variation in the density is on such a small scale that our eyes can't see it. Imagine a jungle gym 100 stories high and the size of a football field. It is mostly empty space. Yet if you were to view it from an airplane, it would appear as a solid mass. Matter is similar. When you tap your knuckles on the table, it is much like one giant jungle gym (your finger) crashing into another (the table). Even though they are both primarily empty space, one does not fall into the other.

B4.1 CAN YOU ANSWER THIS? *Use the jungle gym analogy to explain why most of Rutherford's alpha particles went right through the gold foil and why a few bounced back. Remember that his gold foil was extremely thin.*

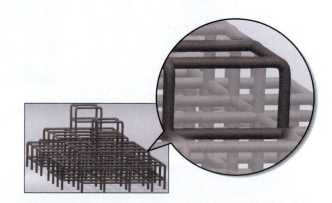

▲ Matter appears solid and uniform because the variation in density is on a scale too small for our eyes to see. Just as this scaffolding appears solid at a distance, so matter appears solid to us.

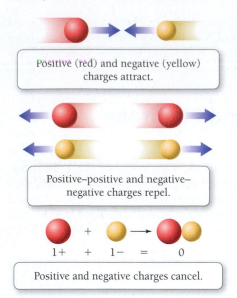

Positive (red) and negative (yellow) charges attract.

Positive–positive and negative–negative charges repel.

1+ + 1− = 0

Positive and negative charges cancel.

▲ **FIGURE 4.7 The properties of electrical charge**

We can summarize the nature of electrical charge as follows (◄ FIGURE 4.7).

- Electrical charge is a fundamental property of protons and electrons.
- Positive and negative electrical charges attract each other.
- Positive–positive and negative–negative charges repel each other.
- Positive and negative charges cancel each other so that a proton and an electron, when paired, are charge-neutral.

Note that matter is usually charge-neutral due to the canceling effect of protons and electrons. When matter does acquire charge imbalances, these imbalances usually equalize quickly, often in dramatic ways. For example, the shock you receive when touching a doorknob during dry weather is the equalization of a charge imbalance that developed as you walked across the carpet. Lightning is an equalization of charge imbalances that develop during electrical storms.

If you had a sample of matter—even a tiny sample, such as a sand grain—that was composed of only protons or only electrons, the forces around that matter would be extraordinary, and the matter would be unstable. Fortunately, matter is not that way—protons and electrons exist together, canceling each other's charge and making matter charge-neutral. Table 4.1 summarizes the properties of protons, neutrons, and electrons.

TABLE 4.1	Subatomic Particles		
	Mass (kg)	Mass (amu)	Charge
proton	1.67262×10^{-27}	1.0073	1+
neutron	1.67493×10^{-27}	1.0087	0
electron	0.00091×10^{-27}	0.00055	1−

Negative charge buildup

Charge equalization

Positive charge buildup

▶ Matter is normally charge-neutral, having equal numbers of positive and negative charges that exactly cancel. When the charge balance of matter is disturbed, as in an electrical storm, it quickly rebalances, often in dramatic ways such as lightning.

PEARSON
eText
2.0

CONCEPTUAL ✔ CHECKPOINT 4.1

An atom composed of which of these particles would have a mass of approximately 12 amu and be charge-neutral?

(a) 6 protons and 6 electrons

(b) 3 protons, 3 neutrons, and 6 electrons

(c) 6 protons, 6 neutrons, and 6 electrons

(d) 12 neutrons and 12 electrons

4.5 Elements: Defined by Their Numbers of Protons

▶ Determine an element's atomic symbol and atomic number using the periodic table.

We have seen that atoms are composed of protons, neutrons, and electrons. However, it is the number of protons in the nucleus of an atom that identifies it as a particular element. For example, atoms with 2 protons in their nucleus are helium atoms, atoms with 13 protons in their nucleus are aluminum atoms, and atoms with 92 protons in their nucleus are uranium atoms. The number of protons in an atom's nucleus defines the element (▼ FIGURE 4.8). Every aluminum atom has 13 protons in its nucleus; if it had a different number of protons, it would be a different element. The number of protons in the nucleus of an atom is its **atomic number** and is represented with the symbol Z.

The periodic table of the elements (▼ FIGURE 4.9) lists all known elements according to their atomic numbers. Each element is represented by a unique

Helium nucleus — 2 protons

Aluminum nucleus — 13 protons

▲ **FIGURE 4.8 The number of protons in the nucleus defines the element**

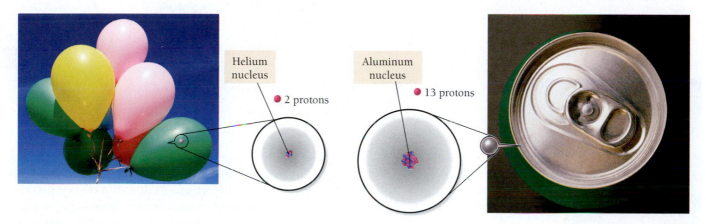

▲ **FIGURE 4.9 The periodic table of the elements**

chemical symbol, a one- or two-letter abbreviation for the element that appears directly below its atomic number on the periodic table. The chemical symbol for helium is He; for aluminum, Al; and for uranium, U. The chemical symbol and the atomic number always go together. If the atomic number is 13, the chemical symbol must be Al. If the atomic number is 92, the chemical symbol must be U. This is just another way of saying that the number of protons defines the element.

Most chemical symbols are based on the English name of the element. For example, the symbol for carbon is C; for silicon, Si; and for bromine, Br. Some elements, however, have symbols based on their Latin names. For example, the symbol for potassium is K, from the Latin *kalium*, and the symbol for sodium is Na, from the Latin *natrium*. Additional elements with symbols based on their Greek or Latin names include:

lead	Pb	*plumbum*
mercury	Hg	*hydrargyrum*
iron	Fe	*ferrum*
silver	Ag	*argentum*
tin	Sn	*stannum*
copper	Cu	*cuprum*

Early scientists often gave newly discovered elements names that reflected their properties. For example, *argon* originates from the Greek word *argos*, meaning "inactive," referring to argon's chemical inertness (it does not react with other elements). *Bromine* originates from the Greek word *bromos*, meaning "stench," referring to bromine's strong odor. Scientists named other elements after countries. For example, polonium was named after Poland, francium after France, and americium after the United States of America. Still other elements were named after scientists. Curium was named after Marie Curie, and mendelevium after Dmitri Mendeleev. You can find every element's name, symbol, and atomic number in the periodic table (inside front cover) and in an alphabetical listing (inside back cover) in this book.

▲ The name *bromine* originates from the Greek word *bromos*, meaning "stench." Bromine vapor, seen as the red-brown gas in this photograph, has a strong odor.

▶ Curium is named after Marie Curie (1867–1934), a chemist who helped discover radioactivity and also discovered two new elements. Curie won two Nobel Prizes for her work.

Curium
96
Cm
(247)

EXAMPLE **4.1** | **Atomic Number, Atomic Symbol, and Element Name**

List the atomic symbol and atomic number for each element.

(a) silicon
(b) potassium
(c) gold
(d) antimony

SOLUTION

As you become familiar with the periodic table, you will be able to quickly locate elements on it. At first you may find it easier to locate them in the alphabetical listing on the inside back cover of this book, but you should become familiar with their positions in the periodic table.

Element	Symbol	Atomic Number
silicon	Si	14
potassium	K	19
gold	Au	79
antimony	Sb	51

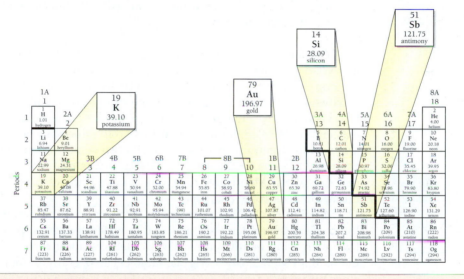

▶ **SKILLBUILDER 4.1** | **Atomic Number, Atomic Symbol, and Element Name**

Find the name and atomic number for each element.
(a) Na
(b) Ni
(c) P
(d) Ta

▶ **FOR MORE PRACTICE**   Problems 41, 42, 45, 46, 47, 48, 49, 50.

4.6   Looking for Patterns: The Periodic Law and the Periodic Table

▶ Use the periodic table to classify elements by group.

The organization of the periodic table is credited primarily to the work of Russian chemist Dmitri Mendeleev (1834–1907); however, German chemist Julius Lothar Meyer (1830–1895) had independently suggested a similar organization.

▲ Dmitri Mendeleev, a Russian chemistry professor who arranged early versions of the periodic table.

In Mendeleev's time, chemists had discovered about 65 different elements along with their relative masses, chemical activity, and some of their physical properties. However, there was no systematic way of organizing them.

> The properties (colors) of these elements form a repeating pattern.

1	2	3	4	5	6	7	8	9	10	11	12	13	14	15	16	17	18	19	20
H	He	Li	Be	B	C	N	O	F	Ne	Na	Mg	Al	Si	P	S	Cl	Ar	K	Ca

▲ **FIGURE 4.10 Recurring properties**   These elements are listed in order of increasing atomic number (Mendeleev used relative mass, which is similar). The color of each element represents its properties. Notice that the properties (colors) of these elements form a repeating pattern.

In 1869, Mendeleev noticed that certain groups of elements had similar properties. He found that if he listed the elements in order of increasing relative mass, those similar properties recurred in a regular pattern (▲ **FIGURE 4.10**). Mendeleev summarized these observations in the **periodic law**:

> When the elements are arranged in order of increasing relative mass, certain sets of properties recur periodically.

| *Periodic* means "recurring regularly." |

Mendeleev organized all the known elements in a table in which relative mass increased from left to right and elements with similar properties were aligned in the same vertical columns (◀ **FIGURE 4.11**). Because many elements had not yet been discovered, Mendeleev's table contained some gaps, which allowed him to predict the existence of yet-undiscovered elements. For example, Mendeleev predicted the existence of an element he called *eka-silicon,* which fell below silicon on the table and between gallium and arsenic. In 1886, eka-silicon was discovered by German chemist Clemens Winkler (1838–1904) and was found to have almost exactly the properties that Mendeleev had anticipated. Winkler named the element germanium, after his home country.

Mendeleev's original listing has evolved into the modern **periodic table**. In the modern table, elements are listed in order of increasing atomic number rather than increasing relative mass. The modern periodic table also contains more elements than Mendeleev's original table because many more have been discovered since his time.

> Elements with similar properties align in vertical columns.

1							2
H							He
3	4	5	6	7	8	9	10
Li	Be	B	C	N	O	F	Ne
11	12	13	14	15	16	17	18
Na	Mg	Al	Si	P	S	Cl	Ar
19	20						
K	Ca						

▲ **FIGURE 4.11 Making a periodic table**   If we place the elements from Figure 4.10 in a table, we can arrange them in rows so that similar properties align in the same vertical columns. This is similar to Mendeleev's first periodic table.

Mendeleev's periodic law was based on observation. Like all scientific laws, the periodic law summarized many observations but did not give the underlying reason for the observation—only theories do that. For now, we accept the periodic law as it is, but in Chapter 9 we will examine a powerful theory that explains the law and gives the underlying reasons for it.

We can broadly classify the elements in the periodic table as metals, nonmetals, and metalloids (▶ **FIGURE 4.12**). **Metals** occupy the left side of the periodic table and have similar properties: They are good conductors of heat and electricity; they can be pounded into flat sheets (malleability); they can be drawn into wires (ductility); they are often shiny; and they tend to lose electrons when they undergo chemical changes. Good examples of metals are iron, magnesium, chromium, and sodium.

Nonmetals occupy the upper right side of the periodic table. The dividing line between metals and nonmetals is the zigzag diagonal line running from boron to astatine in Figure 4.12. Nonmetals have more varied properties—some are solids at room temperature, others are gases—but as a whole, they tend to be poor conductors of heat and electricity, and they all tend to gain electrons when they undergo chemical changes. Good examples of nonmetals are oxygen, nitrogen, chlorine, and iodine.

Most of the elements that lie along the zigzag diagonal line dividing metals and nonmetals are **metalloids**, or semimetals, and display mixed properties.

1A 1																	8A 18
1 **1** **H**	2A 2											3A 13	4A 14	5A 15	6A 16	7A 17	2 **He**
2 **3** **Li**	**4** **Be**											**5** **B**	6 **C**	7 **N**	8 **O**	9 **F**	10 **Ne**
3 **11** **Na**	**12** **Mg**	3B 3	4B 4	5B 5	6B 6	7B 7	8B 8	9	10	1B 11	2B 12	**13** **Al**	14 **Si**	15 **P**	16 **S**	17 **Cl**	18 **Ar**
4 **19** **K**	**20** **Ca**	**21** **Sc**	**22** **Ti**	**23** **V**	**24** **Cr**	**25** **Mn**	**26** **Fe**	**27** **Co**	**28** **Ni**	**29** **Cu**	**30** **Zn**	**31** **Ga**	32 **Ge**	33 **As**	34 **Se**	35 **Br**	36 **Kr**
5 **37** **Rb**	**38** **Sr**	**39** **Y**	**40** **Zr**	**41** **Nb**	**42** **Mo**	**43** **Tc**	**44** **Ru**	**45** **Rh**	**46** **Pd**	**47** **Ag**	**48** **Cd**	**49** **In**	**50** **Sn**	51 **Sb**	52 **Te**	53 **I**	54 **Xe**
6 **55** **Cs**	**56** **Ba**	**57** **La**	**72** **Hf**	**73** **Ta**	**74** **W**	**75** **Re**	**76** **Os**	**77** **Ir**	**78** **Pt**	**79** **Au**	**80** **Hg**	**81** **Tl**	**82** **Pb**	**83** **Bi**	84 **Po**	85 **At**	86 **Rn**
7 **87** **Fr**	**88** **Ra**	**89** **Ac**	**104** **Rf**	**105** **Db**	**106** **Sg**	**107** **Bh**	**108** **Hs**	**109** **Mt**	**110** **Ds**	**111** **Rg**	**112** **Cn**	**113** **Nh**	**114** **Fl**	**115** **Mc**	**116** **Lv**	**117** **Ts**	**118** **Og**

Metals
Nonmetals
Metalloids

Lanthanides	58 **Ce**	59 **Pr**	60 **Nd**	61 **Pm**	62 **Sm**	63 **Eu**	64 **Gd**	65 **Tb**	66 **Dy**	67 **Ho**	68 **Er**	69 **Tm**	70 **Yb**	71 **Lu**
Actinides	90 **Th**	91 **Pa**	92 **U**	93 **Np**	94 **Pu**	95 **Am**	96 **Cm**	97 **Bk**	98 **Cf**	99 **Es**	100 **Fm**	101 **Md**	102 **No**	103 **Lr**

▲ **FIGURE 4.12 Metals, nonmetals, and metalloids** The elements in the periodic table can be broadly classified as metals, nonmetals, or metalloids.

We also call metalloids **semiconductors** because of their intermediate electrical conductivity, which can be changed and controlled. This property makes semiconductors useful in the manufacture of the electronic devices that are central to computers, cell phones, and many other technological gadgets. Silicon, arsenic, and germanium are good examples of metalloids.

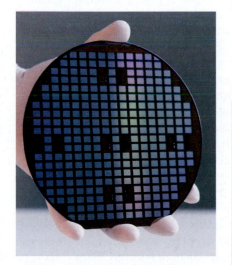

▲ Silicon is a metalloid used extensively in the computer and electronics industries.

EXAMPLE **4.2** **Classifying Elements as Metals, Nonmetals, or Metalloids**

Classify each element as a metal, nonmetal, or metalloid.

(a) Ba   (b) I   (c) O   (d) Te

SOLUTION

(a) Barium is on the left side of the periodic table; it is a metal.
(b) Iodine is on the right side of the periodic table; it is a nonmetal.
(c) Oxygen is on the right side of the periodic table; it is a nonmetal.
(d) Tellurium is in the middle-right section of the periodic table, along the line that divides the metals from the nonmetals; it is a metalloid.

▶ **SKILLBUILDER 4.2 | Classifying Elements as Metals, Nonmetals, or Metalloids**

Classify each element as a metal, nonmetal, or metalloid.
(a) S   (b) Cl   (c) Ti   (d) Sb

▶ **FOR MORE PRACTICE** Problems 51, 52, 53, 54.

We can also broadly divide the periodic table into **main-group elements**, whose properties tend to be more predictable based on their position in the periodic table, and **transition elements** or **transition metals**, whose properties are less

▶ FIGURE 4.13 **Main-group and transition elements** We broadly divide the periodic table into main-group elements, whose properties we can generally predict based on their position, and transition elements, whose properties tend to be less predictable based on their position.

easily predictable based simply on their position in the periodic table (▲ FIGURE 4.13). Main-group elements are in columns labeled with a number and the letter A. Transition elements are in columns labeled with a number and the letter B. A competing numbering system does not use letters but only the numbers 1–18. We show both numbering systems in the periodic table in the inside front cover of this book.

PEARSON
eText
2.0

CONCEPTUAL ✔ CHECKPOINT 4.2

Which element is a main-group metal?

(a) O   **(b)** Ag   **(c)** P   **(d)** Pb

The noble gases are inert (unreactive) compared to other elements. However, some noble gases, especially the heavier ones, form a limited number of compounds with other elements under special conditions.

Each column within the periodic table is a **family** or **group** of elements. The elements within a family of main-group elements usually have similar properties, and some have a group name. For example, the Group 8A elements, the **noble gases**, are chemically inert gases. The most familiar noble gas is probably helium, used to fill balloons. Helium, like the other noble gases, is chemically stable—it won't combine with other elements to form compounds—and is therefore safe to put into balloons. Other noble gases include neon, often used in neon signs; argon, which makes up a small percentage of our atmosphere; krypton; and xenon. The Group 1A elements, the **alkali metals**, are all very reactive metals. A marble-sized piece of sodium can explode when dropped into water. Other alkali metals include lithium, potassium, and rubidium. The Group 2A elements, the **alkaline earth metals**, are also fairly reactive, although not quite as reactive as the alkali metals. Calcium, for example, reacts fairly vigorously when dropped into water but does not explode as readily as sodium. Other alkaline earth metals are magnesium, a common low-density structural metal; strontium; and barium. The Group 7A elements, the **halogens**, are very reactive nonmetals. Chlorine, a greenish-yellow gas with a pungent odor, is probably the most familiar halogen. Because of its reactivity, people often use chlorine as a sterilizing and disinfecting agent (it reacts with and kills bacteria and other microscopic organisms). Other halogens include bromine, a red-brown liquid that readily evaporates into a gas; iodine, a purple solid; and fluorine, a pale yellow gas.

Noble gases

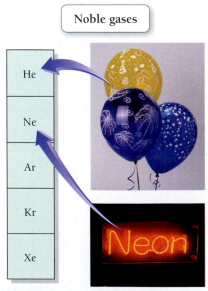

▲ The noble gases include helium (used in balloons), neon (used in neon signs), argon, krypton, and xenon.

Alkali metals

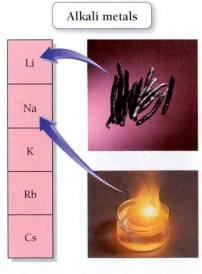

▲ The alkali metals include lithium, sodium (shown in the second photo reacting with water), potassium, rubidium, and cesium.

▶ The periodic table with Groups 1A, 2A, 7A, and 8A highlighted.

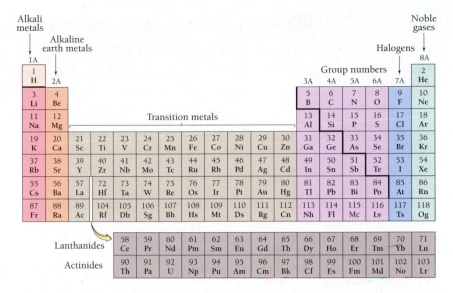

EXAMPLE **4.3**   **Groups and Families of Elements**

To which group or family of elements does each element belong?

(a) Mg   **(b)** N   **(c)** K   **(d)** Br

SOLUTION

(a) Mg is in Group 2A; it is an alkaline earth metal.
(b) N is in Group 5A.
(c) K is in Group 1A; it is an alkali metal.
(d) Br is in Group 7A; it is a halogen.

▶ **SKILLBUILDER 4.3** | **Groups and Families of Elements**

To which group or family of elements does each element belong?

(a) Li   **(b)** B   **(c)** I   **(d)** Ar

▶ **FOR MORE PRACTICE**   Problems 57, 58, 59, 60, 61, 62, 63, 64.

Alkaline earth metals

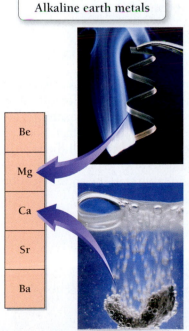

◀ The alkaline earth metals include beryllium, magnesium (shown burning in the first photo), calcium (shown reacting with water in the second photo), strontium, and barium.

Halogens

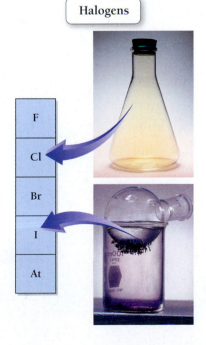

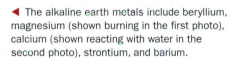

▶ The halogens include fluorine, chlorine, bromine, iodine, and astatine.

CONCEPTUAL ✔ CHECKPOINT 4.3

Which statement is NEVER true?

(a) An element can be both a transition element and a metal.

(b) An element can be both a transition element and a metalloid.

(c) An element can be both a metalloid and a halogen.

(d) An element can be both a main-group element and a halogen.

4.7 Ions: Losing and Gaining Electrons

▶ Determine ion charge from the number of protons and electrons.
▶ Determine the number of protons and electrons in an ion.

In chemical reactions, atoms often lose or gain electrons to form charged particles called **ions**. For example, a neutral lithium (Li) atom contains 3 protons and 3 electrons; however, in reactions, a lithium atom loses 1 electron (e^-) to form a Li^+ ion.

$$Li \longrightarrow Li^+ + e^-$$

> The charge of an ion is indicated in the upper right corner of the symbol.

The Li^+ *ion* contains 3 protons but only 2 electrons, resulting in a net charge of 1+. We usually write ion charges with the magnitude of the charge first followed by the sign of the charge. For example, we write a positive two charge as 2+ and a negative two charge as 2−. The charge of an ion depends on how many electrons were gained or lost and is given by the formula:

$$\text{ion charge} = \text{number of protons} - \text{number of electrons}$$
$$= \#p^+ - \#e^-$$

where p^+ stands for *proton* and e^- stands for *electron*.

For the Li^+ ion with 3 protons and 2 electrons, the charge is:

$$\text{ion charge} = 3 - 2 = 1+$$

A neutral fluorine (F) atom contains 9 protons and 9 electrons; however, in chemical reactions a fluorine atom gains 1 electron to form F^- ions:

$$F + e^- \longrightarrow F^-$$

The F^- *ion* contains 9 protons and 10 electrons, resulting in a 1− charge.

$$\text{ion charge} = 9 - 10$$
$$= 1-$$

Positively charged ions, such as Li^+, are **cations**, and negatively charged ions, such as F^-, are **anions**. Ions, on the one hand, behave very differently than the atoms from which they are formed. Neutral sodium atoms, for example, are extremely reactive, interacting violently with most things they contact. Sodium cations (Na^+), on the other hand, are relatively inert—we eat them all the time in sodium chloride (table salt). In nature, cations and anions always occur together so that, again, matter is charge-neutral. For example, in table salt, the sodium cation occurs together with the chloride anion (Cl^-).

CONCEPTUAL ✔ CHECKPOINT 4.4

What is the symbol for the ion that forms when oxygen gains two electrons?

$$O + 2e^- \longrightarrow \underline{\quad\quad}$$

EXAMPLE 4.4

Determining Ion Charge from Numbers of Protons and Electrons

Determine the charge of each ion.

(a) a magnesium ion with 10 electrons
(b) a sulfur ion with 18 electrons
(c) an iron ion with 23 electrons

SOLUTION

To determine the charge of each ion, use the ion charge equation.

$$\text{ion charge} = \#p^+ - \#e^-$$

You are given the number of electrons in the problem. You can obtain the number of protons from the element's atomic number in the periodic table.

(a) Magnesium's atomic number is 12.

$$\text{ion charge} = 12 - 10 = 2+ \ (Mg^{2+})$$

(b) Sulfur's atomic number is 16.

$$\text{ion charge} = 16 - 18 = 2- \ (S^{2-})$$

(c) Iron's atomic number is 26.

$$\text{ion charge} = 26 - 23 = 3+ \ (Fe^{3+})$$

▶ **SKILLBUILDER 4.4 | Determining Ion Charge from Numbers of Protons and Electrons**

Determine the charge of each ion.

(a) a nickel ion with 26 electrons
(b) a bromine ion with 36 electrons
(c) a phosphorus ion with 18 electrons

▶ **FOR MORE PRACTICE**   Example 4.10; Problems 75, 76.

EXAMPLE 4.5

Determining the Number of Protons and Electrons in an Ion

Determine the number of protons and electrons in the Ca^{2+} ion.

The periodic table indicates that the atomic number for calcium is 20, so calcium has 20 protons. You can find the number of electrons using the ion charge equation.

SOLUTION

$$\text{ion charge} = \#p^+ - \#e^-$$
$$2+ = 20 - \#e^-$$
$$\#e^- = 20 - 2 = 18$$

The number of electrons is 18. The Ca^{2+} ion has 20 protons and 18 electrons.

▶ **SKILLBUILDER 4.5 | Determining the Number of Protons and Electrons in an Ion**

Determine the number of protons and electrons in the S^{2-} ion.

▶ **FOR MORE PRACTICE**   Example 4.11; Problems 77, 78.

Ions and the Periodic Table

For many main-group elements, we can use the periodic table to predict how many electrons tend to be lost or gained when an atom of that particular element ionizes. The number associated with the letter A above each *main-group* column in the periodic table—1 through 8—gives the number of *valence electrons* for the elements in

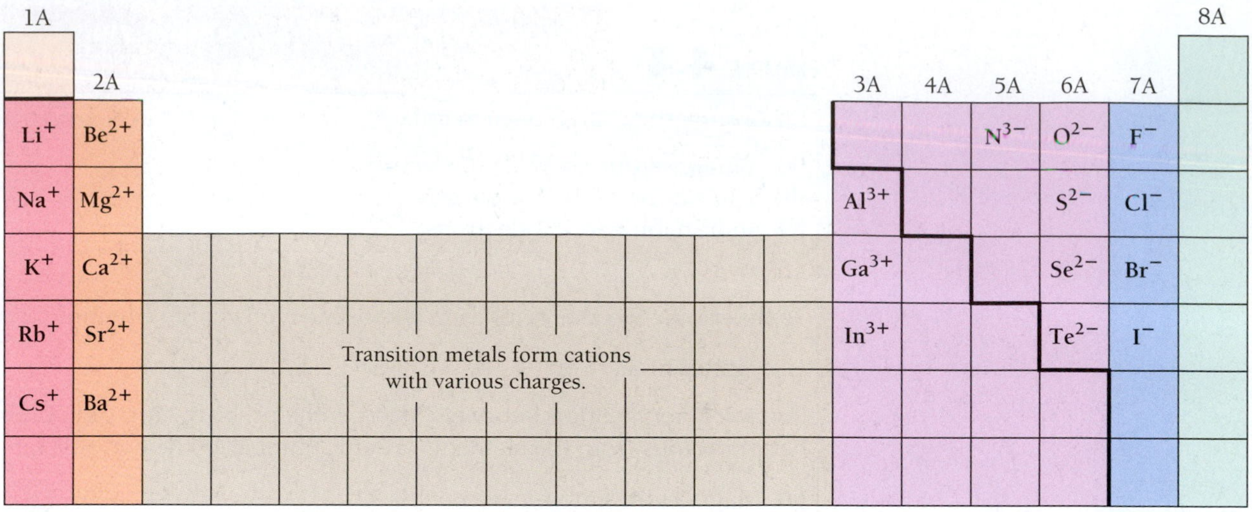

▲ FIGURE 4.14 **Elements that form predictable ions**

that column. We will discuss the concept of valence electrons more fully in Chapter 9; for now, you can think of valence electrons as the outermost electrons in an atom. Because oxygen is in column 6A, we can deduce that it has 6 valence electrons; because magnesium is in column 2A, it has 2 valence electrons, and so on. An important exception to this rule is helium—it is in column 8A but has only 2 valence electrons. Valence electrons are particularly important because, as we shall see in Chapter 10, these electrons are the ones that are most important in chemical bonding.

We can predict the charge acquired by a particular element when it ionizes from its position in the periodic table relative to the noble gases.

Main-group elements tend to form ions that have the same number of valence electrons as the nearest noble gas.

For example, the closest noble gas to oxygen is neon. When oxygen ionizes, it *acquires* two additional electrons for a total of 8 valence electrons—the same number as neon. When determining the closest noble gas, we can move either forward or backward on the periodic table. For example, the closest noble gas to magnesium is also neon, even though neon (atomic number 10) falls before magnesium (atomic number 12) in the periodic table. Magnesium *loses* its 2 valence electrons to attain the same number of valence electrons as neon.

In accordance with this principle, the alkali metals (Group 1A) tend to lose 1 electron and form 1+ ions, while the alkaline earth metals (Group 2A) tend to lose 2 electrons and form 2+ ions. The halogens (Group 7A) tend to gain 1 electron and form 1- ions. The groups in the periodic table that form predictable ions are shown in ▲ FIGURE 4.14. Become familiar with these groups and the ions they form. In Chapter 9, we will examine a theory that more fully explains why these groups form ions as they do.

EXAMPLE 4.6 Charge of Ions from Position in Periodic Table

Based on their position in the periodic table, what ions do barium and iodine tend to form?

SOLUTION

Because barium is in Group 2A, it tends to form a cation with a 2+ charge (Ba^{2+}). Because iodine is in Group 7A, it tends to form an anion with a 1- charge (I^-).

▶ SKILLBUILDER 4.6 | **Charge of Ions from Position in Periodic Table**

Based on their position in the periodic table, what ions do potassium and selenium tend to form?

▶ FOR MORE PRACTICE Problems 81, 82.

CONCEPTUAL ✓ CHECKPOINT 4.5

Which pair of ions has the same total number of electrons?

(a) Na^+ and Mg^{2+} (b) F^- and Cl^-

(c) O^- and O^{2-} (d) Ga^{3+} and Fe^{3+}

4.8   Isotopes: When the Number of Neutrons Varies

▶ Determine atomic numbers, mass numbers, and isotope symbols for an isotope.
▶ Determine number of protons and neutrons from isotope symbols.

Recent studies have shown that for some elements, the relative amounts of each different isotope vary depending on the history of the sample. However, these variations are usually small and beyond the scope of this book.

Percent means "per hundred." 90.48% means that 90.48 atoms out of 100 are the isotope with 10 neutrons.

All atoms of a given element have the same number of protons; however, they do not necessarily have the same number of neutrons. Because neutrons and protons have nearly the same mass (approximately 1 amu), and the number of neutrons in the atoms of a given element can vary, all atoms of a given element *do not* have the same mass (contrary to what John Dalton originally proposed in his atomic theory). For example, all neon atoms in nature contain 10 protons, but they may have 10, 11, or 12 neutrons (▼ FIGURE 4.15). All three types of neon atoms exist, and each has a slightly different mass. Atoms with the same number of protons but different numbers of neutrons are **isotopes**. Some elements, such as beryllium (Be) and aluminum (Al), have only one naturally occurring isotope, while other elements, such as neon (Ne) and chlorine (Cl), have two or more.

For a given element, the relative amounts of each different isotope in a naturally occurring sample of that element are always the same. For example, in any natural sample of neon atoms, 90.48% of the atoms are the isotope with 10 neutrons, 0.27% are the isotope with 11 neutrons, and 9.25% are the isotope with 12 neutrons as summarized in Table 4.2. This means that in a sample of 10,000 neon atoms, 9048 have 10 neutrons, 27 have 11 neutrons, and 925 have 12 neutrons. These percentages are the **percent natural abundance** of the isotopes. The preceding numbers are for neon only; each element has its own unique percent natural abundance of isotopes.

TABLE 4.2   Neon Isotopes

Symbol	Number of Protons	Number of Neutrons	A (Mass Number)	Percent Natural Abundance
Ne-20 or $^{20}_{10}$Ne	10	10	20	90.48%
Ne-21 or $^{21}_{10}$Ne	10	11	21	0.27%
Ne-22 or $^{22}_{10}$Ne	10	12	22	9.25%

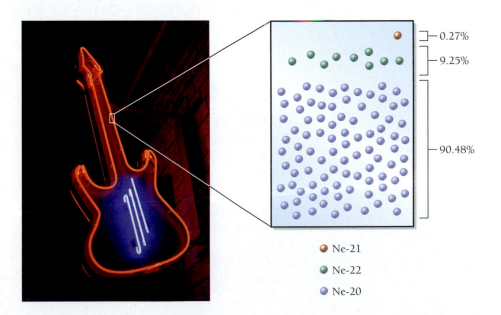

▶ **FIGURE 4.15 Isotopes of neon** Naturally occurring neon contains three different isotopes: Ne-20 (with 10 neutrons), Ne-21 (with 11 neutrons), and Ne-22 (with 12 neutrons).

The sum of the number of neutrons and protons in an atom is its **mass number** and is given the symbol **A**.

$$A = \text{number of protons} + \text{number of neutrons}$$

For neon, which has 10 protons, the mass numbers of the three different naturally occurring isotopes are 20, 21, and 22, corresponding to 10, 11, and 12 neutrons, respectively.

We often symbolize isotopes in the following way:

where X is the chemical symbol, A is the mass number, and Z is the atomic number.
For example, the symbols for the neon isotopes are:

$$^{20}_{10}\text{Ne} \quad ^{21}_{10}\text{Ne} \quad ^{22}_{10}\text{Ne}$$

Notice that the chemical symbol, Ne, and the atomic number, 10, are redundant. If the atomic number is 10, the symbol must be Ne, and vice versa. The mass numbers, however, are different, reflecting the different number of neutrons in each isotope.

A second common notation for isotopes is the chemical symbol (or chemical name) followed by a hyphen and the mass number of the isotope:

In this notation, the neon isotopes are:

Ne-20	neon-20
Ne-21	neon-21
Ne-22	neon-22

Notice that all isotopes of a given element have the same number of protons (otherwise they would be a different element). Notice also that the mass number is the *sum* of the number of protons and the number of neutrons. The number of neutrons in an isotope is the difference between the mass number and the atomic number.

> In general, mass number increases with increasing atomic number.

PEARSON
eText
2.0

CONCEPTUAL ✓ CHECKPOINT **4.6**

Carbon has two naturally occurring isotopes: $^{12}_{6}\text{C}$ and $^{13}_{6}\text{C}$. Using circles to represent protons and squares to represent neutrons, draw the nucleus of each isotope.

EXAMPLE **4.7** | **Atomic Numbers, Mass Numbers, and Isotope Symbols**

What are the atomic number (Z), mass number (A), and symbols of the carbon isotope that has 7 neutrons?

SOLUTION

You can determine that the atomic number (Z) of carbon is 6 (from the periodic table). This means that carbon atoms have 6 protons. The mass number (A) for the isotope with 7 neutrons is the sum of the number of protons and the number of neutrons.

$$A = 6 + 7 = 13$$

So, $Z = 6$, $A = 13$, and the symbols for the isotope are C-13 and $^{13}_{6}\text{C}$.

▶ **SKILLBUILDER 4.7** | **Atomic Numbers, Mass Numbers, and Isotope Symbols**

What are the atomic number, mass number, and symbols for the chlorine isotope with 18 neutrons?

▶ **FOR MORE PRACTICE**   Example 4.12; Problems 87, 89, 91, 92.

Interactive
Worked Example
Video 4.8

EXAMPLE **4.8**

Numbers of Protons and Neutrons from Isotope Symbols

How many protons and neutrons are in the chromium isotope $^{52}_{24}Cr$?	
The number of protons is equal to Z (lower left number).	**SOLUTION**
	$\#p^+ = Z = 24$
The number of neutrons is equal to A (upper left number) − Z (lower left number).	$\#n = A - Z$ $= 52 - 24$ $= 28$

▶ **SKILLBUILDER 4.8** | **Numbers of Protons and Neutrons from Isotope Symbols**

How many protons and neutrons are in the potassium isotope $^{39}_{19}K$?

▶ **FOR MORE PRACTICE**   Example 4.13; Problems 93, 94.

CONCEPTUAL ✔ **CHECKPOINT 4.7**

If an atom with a mass number of 27 has 14 neutrons, it is an isotope of which element?

(a) silicon

(b) aluminum

(c) cobalt

(d) niobium

CONCEPTUAL ✔ **CHECKPOINT 4.8**

Throughout this book, we represent atoms as spheres. For example, we represent a carbon atom by a black sphere as shown here. In light of the nuclear theory of the atom, when represented this way, would C-12 and C-13 look different? Why or why not?

Carbon

4.9 Atomic Mass: The Average Mass of an Element's Atoms

▶ Calculate atomic mass from percent natural abundances and isotopic masses.

An important part of Dalton's atomic theory was that all atoms of a given element have the same mass. But as we just learned, the atoms of a given element may have different masses (because of isotopes). So Dalton was not completely correct. We can, however, calculate an average mass—called the **atomic mass**—for each element. You can find the atomic mass of each element in the periodic table directly beneath the element's symbol; it represents the average mass of the atoms that compose that element. For example, the periodic table lists the atomic mass of chlorine as 35.45 amu. Naturally occurring chlorine consists of 75.77% chlorine-35 (mass 34.97 amu) and 24.23% chlorine-37 (mass 36.97 amu). Its atomic mass is:

Some books use the term *average atomic mass* or *atomic weight* instead of simply *atomic mass*.

$$\text{atomic mass} = (0.7577 \times 34.97 \text{ amu}) + (0.2423 \times 36.97 \text{ amu})$$
$$= 35.45 \text{ amu}$$

CHEMISTRY IN THE ENVIRONMENT

Radioactive Isotopes at Hanford, Washington

The nuclei of the isotopes of a given element are not all equally stable. For example, naturally occurring lead is composed primarily of Pb-206, Pb-207, and Pb-208. Other isotopes of lead also exist, but their nuclei are unstable. Scientists can make some of these other isotopes, such as Pb-185, in the laboratory. However, within seconds Pb-185 atoms emit a few energetic subatomic particles from their nuclei and change into different isotopes of different elements (which are themselves unstable). These emitted subatomic particles are called **nuclear radiation**, and we call the isotopes that emit them **radioactive**. Nuclear radiation, always associated with unstable nuclei, can be harmful to humans and other living organisms because the emitted energetic particles interact with and damage biological molecules. Some isotopes, such as Pb-185, emit significant amounts of radiation only for a very short time. Others remain radioactive for a long time—in some cases millions or even billions of years.

The nuclear power and nuclear weapons industries produce by-products containing unstable isotopes of several different elements. Many of these isotopes emit nuclear radiation for a long time, and their disposal is an environmental problem. For example, in Hanford, Washington, which for 50 years produced fuel for nuclear weapons,

177 underground storage tanks contain 55 million gallons of highly radioactive nuclear waste. Certain radioactive isotopes within that waste will produce nuclear radiation for the foreseeable future. Unfortunately, some of the underground storage tanks in Hanford are aging, and leaks have allowed some of the waste to seep into the environment. While the danger from short-term external exposure to this waste is minimal, ingestion of the waste through contamination of drinking water or food supplies would pose significant health risks. Consequently, Hanford is now the site of the largest environmental cleanup project in U.S. history, involving 11,000 workers. The U.S. government expects the project to last for decades, and current costs are about $3 billion per year.

Radioactive isotopes are not always harmful, however, and many have beneficial uses. For example, physicians give technetium-99 (Tc-99) to patients to diagnose disease. The radiation emitted by Tc-99 helps doctors image internal organs or detect infection.

B4.2 CAN YOU ANSWER THIS? *Give the number of neutrons in each of the following isotopes: Pb-206, Pb-207, Pb-208, Pb-185, Tc-99.*

◀ Storage tanks at Hanford, Washington, contain 55 million gallons of high-level nuclear waste. Each tank pictured here holds 1 million gallons.

Notice that the atomic mass of chlorine is closer to 35 than 37 because naturally occurring chlorine contains more chlorine-35 atoms than chlorine-37 atoms. Notice also that when we use percentages in these calculations, we must always convert them to their decimal value. To convert a percentage to its decimal value, we divide by 100. For example:

$$75.77\% = 75.77/100 = 0.7577$$
$$24.23\% = 24.23/100 = 0.2423$$

In general, we calculate atomic mass according to the following equation:

$$\text{atomic mass} = (\text{fraction of isotope 1} \times \text{mass of isotope 1}) +$$
$$(\text{fraction of isotope 2} \times \text{mass of isotope 2}) +$$
$$(\text{fraction of isotope 3} \times \text{mass of isotope 3}) + \ldots$$

where the fractions of each isotope are the percent natural abundances converted to their decimal values. Atomic mass is useful because it allows us to assign a characteristic mass to each element, and, as we will see in Chapter 6, it allows us to quantify the number of atoms in a sample of that element.

 Interactive Worked Example Video 4.9

EXAMPLE **4.9** **Calculating Atomic Mass**

Gallium has two naturally occurring isotopes: Ga-69 with mass 68.9256 amu and a natural abundance of 60.11%, and Ga-71 with mass 70.9247 amu and a natural abundance of 39.89%. Calculate the atomic mass of gallium.

Remember to convert the percent natural abundances into decimal form by dividing by 100.	**SOLUTION** $$\text{fraction Ga-69} = \frac{60.11}{100} = 0.6011$$ $$\text{fraction Ga-71} = \frac{39.89}{100} = 0.3989$$
Use the fractional abundances and the atomic masses of the isotopes to calculate the atomic mass according to the atomic mass definition.	$$\text{atomic mass} = (0.6011 \times 68.9256 \text{ amu}) + (0.3989 \times 70.9247 \text{ amu})$$ $$= 41.4312 \text{ amu} + 28.2919 \text{ amu}$$ $$= 69.7231 = 69.72 \text{ amu}$$

▶ **SKILLBUILDER 4.9 | Calculating Atomic Mass**

Magnesium has three naturally occurring isotopes with masses of 23.99, 24.99, and 25.98 amu and natural abundances of 78.99%, 10.00%, and 11.01%. Calculate the atomic mass of magnesium.

▶ **FOR MORE PRACTICE** Example 4.14; Problems 97, 98.

CONCEPTUAL ✔ **CHECKPOINT 4.9**

A fictitious element is composed of isotopes A and B with masses of 61.9887 and 64.9846 amu, respectively. The atomic mass of the element is 64.52. What can you conclude about the natural abundances of the two isotopes?

(a) The natural abundance of isotope A must be greater than the natural abundance of isotope B.

(b) The natural abundance of isotope B must be greater than the natural abundance of isotope A.

(c) The natural abundances of both isotopes must be about equal.

(d) Nothing can be concluded about the natural abundances of the two isotopes from the given information.

Chapter 4 in Review

MasteringChemistry™ provides end-of-chapter exercises, feedback-enriched tutorial problems, animations, and interactive activities to encourage problem solving practice and deeper understanding of key concepts and topics.

Self-Assessment Quiz

Q1. Which statement is not part of Dalton's atomic theory?
(a) Each element is composed of indestructible particles called atoms.
(b) All atoms of a given element have the same mass and other properties.
(c) Atoms are themselves composed of protons, neutrons, and electrons.
(d) Atoms combine in simple whole-number ratios to form compounds.

Q2. Which statement best summarizes the nuclear model of the atom that emerged from Rutherford's gold foil experiment?
(a) The atom is composed of a dense core that contains most of its mass and all of its positive charge, while low-mass negatively charged particles compose most of its volume.
(b) The atom is composed of a sphere of positive charge with many negatively charged particles within the sphere.
(c) Most of the mass of the atom is evenly distributed throughout its volume.
(d) All of the particles that compose an atom have exactly the same mass.

Q3. An ion composed of which of these particles would have a mass of approximately 16 amu and a charge of 2−?
(a) 8 protons and 8 electrons
(b) 8 protons, 8 neutrons, and 10 electrons
(c) 8 protons, 8 neutrons, and 8 electrons
(d) 8 protons, 8 neutrons, and 6 electrons

Q4. Which element is a main-group metal with an even atomic number?
(a) K　　(b) Ca　　(c) Cr　　(d) Se

Q5. Which element is a halogen?
(a) Ne　　(b) O　　(c) Ca　　(d) I

Q6. Which pair of elements has the most similar properties?
(a) Sr and Ba　　(b) S and Ar
(c) H and He　　(d) K and Se

Q7. Which element is a row 4 noble gas?
(a) Ne　　(b) Br　　(c) Zr　　(d) Kr

Q8. How many electrons does the predictable (most common) ion of fluorine contain?
(a) 1　　(b) 4　　(c) 9　　(d) 10

Q9. How many neutrons does the Fe-56 isotope contain?
(a) 26　　(b) 30　　(c) 56　　(d) 112

Q10. Determine the number of protons, neutrons, and electrons in $^{32}_{16}S^{2-}$.
(a) 16 protons; 32 neutrons; and 18 electrons
(b) 16 protons; 16 neutrons; and 18 electrons
(c) 32 protons; 16 neutrons; and 2 electrons
(d) 16 protons; 48 neutrons; and 16 electrons

Q11. What is the charge of the Cr ion that contains 21 electrons?
(a) 2−　　(b) 3−　　(c) 2+　　(d) 3+

Q12. An element has four naturally occurring isotopes; the table lists the mass and natural abundance of each isotope. Find the atomic mass of the element.

Isotope	Mass (amu)	Natural Abundance
A	203.9730	1.4%
B	205.9744	24.1%
C	206.9758	22.1%
D	207.9766	52.4%

(a) 207.2 amu
(b) 2.072×10^4 amu
(c) 206.2 amu
(d) 206.5 amu

Answers: 1:c, 2:a, 3:b, 4:b, 5:d, 6:a, 7:d, 8:d, 9:b, 10:b, 11:d, 12:a

Chemical Principles

The Atomic Theory

Democritus and Leucippus, ancient Greek philosophers, were the first to assert that matter is ultimately composed of small, indestructible particles. It was not until 2000 years later, however, that John Dalton introduced a formal atomic theory stating that matter is composed of atoms; atoms of a given element have unique properties that distinguish them from atoms of other elements; and atoms combine in simple, whole-number ratios to form compounds.

Relevance

The concept of atoms is important because it explains the physical world. Humans and everything we see are made of atoms. To understand the physical world, we must begin by understanding atoms. Atoms are the key concept—they determine the properties of matter.

Discovery of the Atom's Nucleus

Rutherford's gold foil experiment probed atomic structure, and his results led to the nuclear model of the atom, which, with minor modifications to accommodate neutrons, is still valid today. According to this model, the atom is composed of protons and neutrons—which compose most of the atom's mass and are grouped together in a dense nucleus—and electrons—which compose most of the atom's volume. Protons and neutrons have similar masses (1 amu), while electrons have a much smaller mass (0.00055 amu).

We can understand why this is relevant by asking, what if it were otherwise? What if matter were *not* mostly empty space? While we cannot know for certain, it seems probable that such matter would not form the diversity of substances required for life—and then, of course, we would not be around to ask the question.

Charge

Protons and electrons both have electrical charge; the charge of the proton is 1+, and the charge of the electron is 1−. The neutron has no charge. When protons and electrons combine in atoms, their charges cancel.

Electrical charge is relevant to much of our modern world. Many of the machines and computers we depend on are powered by electricity, which is the movement of electrical charge.

The Periodic Table

The periodic table tabulates all known elements in order of increasing atomic number. The periodic table is arranged so that similar elements are grouped in columns. Columns of elements in the periodic table have similar properties and are called groups or families. Elements on the left side of the periodic table are metals and tend to lose electrons in their chemical changes. Elements on the upper right side of the periodic table are nonmetals and tend to gain electrons in their chemical changes. Elements between the two are metalloids.

The periodic table helps us organize the elements in ways that allow us to predict their properties. Helium, for example, is not toxic in small amounts because it is an inert gas—it does not react with anything. The gases in the column below it on the periodic table are also inert gases and form a family or group of elements called the noble gases. By tabulating the elements and grouping similar ones together, we begin to understand their properties.

Atomic Number

The characteristic that defines an element is the number of protons in the nuclei of its atoms; this number is the atomic number (Z).

Elements are the fundamental building blocks from which all compounds are made.

Ions

When an atom gains or loses electrons, it becomes an ion. Positively charged ions are cations, and negatively charged ions are anions. Cations and anions occur together so that matter is ordinarily charge-neutral.

Ions occur in many common compounds, such as sodium chloride.

Isotopes

While all atoms of a given element have the same number of protons, they do not necessarily have the same number of neutrons. Atoms of the same element with different numbers of neutrons are isotopes. Isotopes are characterized by their mass number (A), the sum of the number of protons and the number of neutrons in the nucleus.

Each naturally occurring sample of an element has the same percent natural abundance of each isotope. We can use these percentages, together with the mass of each isotope, to calculate the atomic mass of the element, a weighted average of the masses of the individual isotopes.

Isotopes are relevant because they influence atomic masses. To understand these masses, we must understand the presence and abundance of isotopes. In nuclear processes—processes in which the nuclei of atoms actually change—the presence of different isotopes becomes even more important.

Some isotopes are not stable—they lose subatomic particles and transform into other elements. The emission of subatomic particles by unstable nuclei is called radioactive decay. In many situations, such as in diagnosing and treating certain diseases, nuclear radiation is extremely useful. In other situations, such as in the disposal of radioactive waste, it can pose environmental problems.

Chemical Skills	*Examples*

LO: Determine ion charge from the number of protons and electrons (Section 4.7).

- Refer to the periodic table or the alphabetical list of elements to find the atomic number of the element; this number is equal to the number of protons.
- Use the ion charge equation to calculate charge.

$$\text{Ion charge} = \#p^+ - \#e^-$$

EXAMPLE **4.10** **Determining Ion Charge from Numbers of Protons and Electrons**

Determine the charge of a selenium ion with 36 electrons.

SOLUTION

Selenium is atomic number 34; therefore, it has 34 protons.

$$\text{ion charge} = 34 - 36 = 2-$$

LO: Determine the number of protons and electrons in an ion (Section 4.7).

- Refer to the periodic table or the alphabetical list of elements to find the atomic number of the element; this number is equal to the number of protons.
- Use the ion charge equation and substitute in the known values.

$$\text{Ion charge} = \#p^+ - \#e^-$$

- Solve the equation for the number of electrons.

EXAMPLE **4.11** **Determining the Number of Protons and Electrons in an Ion**

Find the number of protons and electrons in the O^{2-} ion.

SOLUTION

The atomic number of O is 8; therefore, it has 8 protons.

$$\text{ion charge} = \#p^+ - \#e^-$$
$$2- = 8 - \#e^-$$
$$\#e^- = 8 + 2 = 10$$

The ion has 8 protons and 10 electrons.

LO: Determine atomic numbers, mass numbers, and isotope symbols for an isotope (Section 4.8).

- Refer to the periodic table or the alphabetical list of elements to find the atomic number of the element.
- The mass number (A) is equal to the atomic number plus the number of neutrons.
- Write the symbol for the isotope by writing the symbol for the element with the mass number in the upper left corner and the atomic number in the lower left corner.
- The other symbol for the isotope is simply the chemical symbol followed by a hyphen and the mass number.

EXAMPLE **4.12** **Determining Atomic Numbers, Mass Numbers, and Isotope Symbols for an Isotope**

What are the atomic number (Z), mass number (A), and symbols for the iron isotope with 30 neutrons?

SOLUTION

The atomic number of iron is 26.

$$A = 26 + 30 = 56$$

The mass number is 56.

$$^{56}_{26}Fe$$
$$Fe\text{-}56$$

LO: Determine number of protons and neutrons from isotope symbols (Section 4.8).

- The number of protons is equal to Z (lower left number).
- The number of neutrons is equal to

$$A \text{ (upper left number)} - Z \text{ (lower left number)}$$

EXAMPLE **4.13** **Determining Number of Protons and Neutrons from Isotope Symbols**

How many protons and neutrons are in $^{62}_{28}Ni$?

SOLUTION

28 protons

$$\#n = 62 - 28 = 34 \text{ neutrons}$$

LO: Calculate atomic mass from percent natural abundances and isotopic masses (Section 4.9).

- Convert the natural abundances from percent to decimal values by dividing by 100.
- Find the atomic mass by multiplying the fractions of each isotope by its respective mass and summing them.
- Round to the correct number of significant figures.
- Check your work.

EXAMPLE **4.14** | **Calculating Atomic Mass from Percent Natural Abundances and Isotopic Masses**

Copper has two naturally occurring isotopes: Cu-63 with mass 62.9395 amu and a natural abundance of 69.17%, and Cu-65 with mass 64.9278 amu and a natural abundance of 30.83%. Calculate the atomic mass of copper.

SOLUTION

$$\text{fraction Cu-63} = \frac{69.17}{100} = 0.6917$$

$$\text{fraction Cu-65} = \frac{30.83}{100} = 0.3083$$

$$\text{atomic mass} = (0.6917 \times 62.9395 \text{ amu}) + (0.3083 \times 64.9278)$$

$$= 43.5353 \text{ amu} + 20.0172 \text{ amu}$$

$$= 63.5525 \text{ amu}$$

$$= 63.55 \text{ amu}$$

Key Terms

alkali metals [4.6]	chemical symbol [4.5]	metalloids [4.6]	percent natural
alkaline earth metals [4.6]	electron [4.3]	metals [4.6]	abundance [4.8]
anion [4.7]	family (of elements) [4.6]	neutron [4.3]	periodic law [4.6]
atom [4.1]	group (of elements) [4.6]	noble gases [4.6]	periodic table [4.6]
atomic mass [4.9]	halogens [4.6]	nonmetals [4.6]	proton [4.3]
atomic mass unit (amu) [4.4]	ion [4.7]	nuclear radiation [4.9]	radioactive [4.9]
atomic number (Z) [4.5]	isotope [4.8]	nuclear theory of the	semiconductor [4.6]
cation [4.7]	main-group elements [4.6]	atom [4.3]	transition elements [4.6]
charge [4.4]	mass number (A) [4.8]	nucleus (of an atom) [4.3]	transition metals [4.6]

Exercises

Questions

1. What did Democritus contribute to our modern understanding of matter?
2. What are three main ideas in Dalton's atomic theory?
3. Describe Rutherford's gold foil experiment and the results of that experiment. How did these results refute the plum-pudding model of the atom?
4. What are the main ideas in the nuclear theory of the atom?
5. List the three subatomic particles and their properties.
6. What is electrical charge?
7. Is matter usually charge-neutral? How would matter be different if it were not charge-neutral?
8. What does the atomic number of an element specify?
9. What is a chemical symbol?
10. List some examples of how elements were named.
11. What was Dmitri Mendeleev's main contribution to our modern understanding of chemistry?
12. What is the main idea in the periodic law?
13. How is the periodic table organized?
14. What are the properties of metals? Where are metals found on the periodic table?
15. What are the properties of nonmetals? Where are nonmetals found on the periodic table?
16. Where on the periodic table are metalloids found?
17. What is a family or group of elements?
18. Locate each group of elements on the periodic table and list its group number.
    (a) alkali metals
    (b) alkaline earth metals
    (c) halogens
    (d) noble gases
19. What is an ion?
20. What is an anion? What is a cation?

21. Locate each group on the periodic table and list the charge of the ions it tends to form.
    (a) Group 1A
    (b) Group 2A
    (c) Group 3A
    (d) Group 6A
    (e) Group 7A

22. What are isotopes?
23. What is the percent natural abundance of isotopes?
24. What is the mass number of an isotope?
25. What notations are commonly used to specify isotopes? What do each of the numbers in these symbols mean?
26. What is the atomic mass of an element?

Problems

ATOMIC AND NUCLEAR THEORY

27. Which statements are *inconsistent* with Dalton's atomic theory as it was originally stated? Explain your answers.
    (a) All carbon atoms are identical.
    (b) Helium atoms can be split into two hydrogen atoms.
    (c) An oxygen atom combines with 1.5 hydrogen atoms to form water molecules.
    (d) Two oxygen atoms combine with a carbon atom to form carbon dioxide molecules.

28. Which statements are *consistent* with Dalton's atomic theory as it was originally stated? Explain your answers.
    (a) Calcium and titanium atoms have the same mass.
    (b) Neon and argon atoms are the same.
    (c) All cobalt atoms are identical.
    (d) Sodium and chlorine atoms combine in a 1:1 ratio to form sodium chloride.

29. Which statements are *inconsistent* with Rutherford's nuclear theory as it was originally stated? Explain your answers.
    (a) Helium atoms have two protons in the nucleus and two electrons outside the nucleus.
    (b) Most of the volume of hydrogen atoms is due to the nucleus.
    (c) Aluminum atoms have 13 protons in the nucleus and 22 electrons outside the nucleus.
    (d) The majority of the mass of nitrogen atoms is due to their 7 electrons.

30. Which statements are *consistent* with Rutherford's nuclear theory as it was originally stated? Explain your answers.
    (a) Atomic nuclei are small compared to the size of atoms.
    (b) The volume of an atom is mostly empty space.
    (c) Neutral potassium atoms contain more protons than electrons.
    (d) Neutral potassium atoms contain more neutrons than protons.

31. If atoms are mostly empty space and atoms compose all ordinary matter, why does solid matter seem to have no space within it?

32. Rutherford's experiment indicated that matter was not as uniform as it appears. What part of his experimental results implied this idea? Explain.

PROTONS, NEUTRONS, AND ELECTRONS

33. Which statement about electrons is true?
    (a) Electrons attract one another.
    (b) Electrons are repelled by protons.
    (c) Some electrons have a charge of 1− and some have no charge.
    (d) Electrons are much lighter than neutrons.

34. Which statement about electrons is false?
    (a) Most atoms have more electrons than protons.
    (b) Electrons have a charge of 1−.
    (c) If an atom has an equal number of protons and electrons, it will be charge-neutral.
    (d) Electrons experience an attraction to protons.

35. Which statement about protons is true?
    (a) Protons have twice the mass of neutrons.
    (b) Protons have the same magnitude of charge as electrons but are opposite in sign.
    (c) Most atoms have more protons than electrons.
    (d) Protons have a charge of 1−.

36. Which statement about protons is false?
    (a) Protons have about the same mass as neutrons.
    (b) Protons have about the same mass as electrons.
    (c) All atoms have protons.
    (d) Protons have the same magnitude of charge as neutrons but are opposite in sign.

37. How many electrons would it take to equal the mass of a proton?

38. A helium nucleus has two protons and two neutrons. How many electrons would it take to equal the mass of a helium nucleus?

39. What mass of electrons is required to neutralize the charge of 1.0 g of protons?

40. What mass of protons is required to neutralize the charge of 1.0 g of electrons?

ELEMENTS, SYMBOLS, AND NAMES

41. Find the atomic number (Z) for each element.
(a) Fr
(b) Kr
(c) Pa
(d) Ge
(e) Al

42. Find the atomic number (Z) for each element.
(a) Si
(b) W
(c) Ni
(d) Rn
(e) Sr

43. How many protons are in the nucleus of an atom of each element?
(a) Ar
(b) Sn
(c) Xe
(d) O
(e) Tl

44. How many protons are in the nucleus of an atom of each element?
(a) Ti
(b) Li
(c) U
(d) Br
(e) F

45. List the symbol and atomic number of each element.
(a) carbon
(b) nitrogen
(c) sodium
(d) potassium
(e) copper

46. List the symbol and atomic number of each element.
(a) boron
(b) neon
(c) silver
(d) mercury
(e) curium

47. List the name and the atomic number of each element.
(a) Mn
(b) Ag
(c) Au
(d) Pb
(e) S

48. List the name and the atomic number of each element.
(a) Y
(b) N
(c) Ne
(d) K
(e) Mo

49. Fill in the blanks to complete the table.

Element Name	Element Symbol	Atomic Number
____	Au	79
Tin	____	____
____	As	____
Copper	____	29
____	Fe	____
____	____	80

50. Fill in the blanks to complete the table.

Element Name	Element Symbol	Atomic Number
____	Al	13
Iodine	____	____
____	Sb	____
Sodium	____	____
____	Rn	86
____	____	82

THE PERIODIC TABLE

51. Classify each element as a metal, nonmetal, or metalloid.
(a) Sr
(b) Mg
(c) F
(d) N
(e) As

52. Classify each element as a metal, nonmetal, or metalloid.
(a) Na
(b) Ge
(c) Si
(d) Br
(e) Ag

53. Which elements would you expect to lose electrons in chemical changes?
(a) potassium
(b) sulfur
(c) fluorine
(d) barium
(e) copper

54. Which elements would you expect to gain electrons in chemical changes?
(a) nitrogen
(b) iodine
(c) tungsten
(d) strontium
(e) gold

55. Which elements are main-group elements?
(a) Te
(b) K
(c) V
(d) Re
(e) Ag

56. Which elements are *not* main-group elements?
(a) Al
(b) Br
(c) Mo
(d) Cs
(e) Pb

57. Which elements are alkaline earth metals?
(a) sodium
(b) aluminum
(c) calcium
(d) barium
(e) lithium

58. Which elements are alkaline earth metals?
(a) rubidium
(b) tungsten
(c) magnesium
(d) cesium
(e) beryllium

59. Which elements are alkali metals?
(a) barium
(b) sodium
(c) gold
(d) tin
(e) rubidium

60. Which elements are alkali metals?
(a) scandium
(b) iron
(c) potassium
(d) lithium
(e) cobalt

61. Classify each element as a halogen, a noble gas, or neither.
(a) Cl
(b) Kr
(c) F
(d) Ga
(e) He

62. Classify each element as a halogen, a noble gas, or neither.
(a) Ne
(b) Br
(c) S
(d) Xe
(e) I

63. To what group number does each element belong?
(a) oxygen
(b) aluminum
(c) silicon
(d) tin
(e) phosphorus

64. To what group number does each element belong?
(a) germanium
(b) nitrogen
(c) sulfur
(d) carbon
(e) boron

65. Which element do you expect to be most like sulfur? Why?
(a) nitrogen
(b) oxygen
(c) fluorine
(d) lithium
(e) potassium

66. Which element do you expect to be most like magnesium? Why?
(a) potassium
(b) silver
(c) bromine
(d) calcium
(e) lead

67. Which pair of elements do you expect to be most similar? Why?
(a) Si and P
(b) Cl and F
(c) Na and Mg
(d) Mo and Sn
(e) N and Ni

68. Which pair of elements do you expect to be most similar? Why?
(a) Ti and Ga
(b) N and O
(c) Li and Na
(d) Ar and Br
(e) Ge and Ga

69. Which element is a main-group nonmetal?
(a) K
(b) Fe
(c) Sn
(d) S

70. Which element is a row 5 transition element?
(a) Sr
(b) Pd
(c) P
(d) V

71. Fill in the blanks to complete the table.

Chemical Symbol	Group Number	Group Name	Metal or Nonmetal
K	___	___	metal
Br	___	halogens	___
Sr	___	___	___
He	8A	___	___
Ar	___	___	___

72. Fill in the blanks to complete the table.

Chemical Symbol	Group Number	Group Name	Metal or Nonmetal
Cl	7A	___	___
Ca	___	___	metal
Xe	___	___	nonmetal
Na	___	alkali metal	___
F	___	___	___

IONS

73. Complete each ionization equation.
(a) $Na \longrightarrow Na^+ + $ ___
(b) $O + 2e^- \longrightarrow$ ___
(c) $Ca \longrightarrow Ca^{2+} + $ ___
(d) $Cl + e^- \longrightarrow$ ___

74. Complete each ionization equation.
(a) $Mg \longrightarrow$ ___ $+ 2e^-$
(b) $Ba \longrightarrow Ba^{2+} + $ ___
(c) $I + e^- \longrightarrow$ ___
(d) $Al \longrightarrow$ ___ $+ 3e^-$

75. Determine the charge of each ion.
(a) oxygen ion with 10 electrons
(b) aluminum ion with 10 electrons
(c) titanium ion with 18 electrons
(d) iodine ion with 54 electrons

76. Determine the charge of each ion.
(a) tungsten ion with 68 electrons
(b) tellurium ion with 54 electrons
(c) nitrogen ion with 10 electrons
(d) barium ion with 54 electrons

77. Determine the number of protons and electrons in each ion.
(a) Na^+
(b) Ba^{2+}
(c) O^{2-}
(d) Co^{3+}

78. Determine the number of protons and electrons in each ion.
(a) Al^{3+}
(b) S^{2-}
(c) I^-
(d) Ag^+

79. Determine whether each statement is true or false. If false, correct it.
(a) The Ti^{2+} ion contains 22 protons and 24 electrons.
(b) The I^- ion contains 53 protons and 54 electrons.
(c) The Mg^{2+} ion contains 14 protons and 12 electrons.
(d) The O^{2-} ion contains 8 protons and 10 electrons.

80. Determine whether each statement is true or false. If false, correct it.
(a) The Fe^{2+} ion contains 29 protons and 26 electrons.
(b) The Cs^+ ion contains 55 protons and 56 electrons.
(c) The Se^{2-} ion contains 32 protons and 34 electrons.
(d) The Li^+ ion contains 3 protons and 2 electrons.

81. Predict the ion formed by each element.
(a) Rb
(b) K
(c) Al
(d) O

82. Predict the ion formed by each element.
(a) F
(b) N
(c) Mg
(d) Na

83. Predict how many electrons each element will most likely gain or lose.
(a) Ga
(b) Li
(c) Br
(d) S

84. Predict how many electrons each element will most likely gain or lose.
(a) I
(b) Ba
(c) Cs
(d) Se

85. Fill in the blanks to complete the table.

Symbol	Ion Commonly Formed	Number of Electrons in Ion	Number of Protons in Ion
Te	——	54	——
In	——		49
Sr	Sr^{2+}	——	——
——	Mg^{2+}	——	12
Cl	——	——	

86. Fill in the table.

Symbol	Ion Commonly Formed	Number of Electrons in Ion	Number of Protons in Ion
F	——	——	9
——	Be^{2+}	2	——
Br	——	36	——
Al	——	——	13
O	——	——	——

ISOTOPES

87. Determine the atomic number and mass number for each isotope.
  (a) the hydrogen isotope with 2 neutrons
  (b) the chromium isotope with 28 neutrons
  (c) the calcium isotope with 22 neutrons
  (d) the tantalum isotope with 109 neutrons

88. How many neutrons are in an atom with each atomic number and mass number?
  (a) $Z = 28$, $A = 59$
  (b) $Z = 92$, $A = 235$
  (c) $Z = 21$, $A = 46$
  (d) $Z = 18$, $A = 42$

89. Write isotopic symbols in the form $_Z^A X$ for each isotope.
  (a) the oxygen isotope with 8 neutrons
  (b) the fluorine isotope with 10 neutrons
  (c) the sodium isotope with 12 neutrons
  (d) the aluminum isotope with 14 neutrons

90. Write isotopic symbols in the form X-A (for example, C-13) for each isotope.
  (a) the iodine isotope with 74 neutrons
  (b) the phosphorus isotope with 16 neutrons
  (c) the uranium isotope with 234 neutrons
  (d) the argon isotope with 22 neutrons

91. Write the symbol for each isotope in the form $_Z^A X$.
  (a) cobalt-60
  (b) neon-22
  (c) iodine-131
  (d) plutonium-244

92. Write the symbol for each isotope in the form $_Z^A X$.
  (a) U-235
  (b) V-52
  (c) P-32
  (d) Xe-144

93. Determine the number of protons and neutrons in each isotope.
  (a) $_{11}^{23}Na$
  (b) $_{88}^{266}Ra$
  (c) $_{82}^{208}Pb$
  (d) $_{7}^{14}N$

94. Determine the number of protons and neutrons in each isotope.
  (a) $_{15}^{33}P$
  (b) $_{19}^{40}K$
  (c) $_{86}^{222}Rn$
  (d) $_{43}^{99}Tc$

95. Carbon-14, present within living organisms and substances derived from living organisms, is often used to establish the age of fossils and artifacts. Determine the number of protons and neutrons in a carbon-14 isotope and write its symbol in the form $_Z^A X$.

96. Plutonium-239 is used in nuclear bombs. Determine the number of protons and neutrons in plutonium-239 and write its symbol in the form $_Z^A X$.

ATOMIC MASS

97. Rubidium has two naturally occurring isotopes: Rb-85 with mass 84.9118 amu and a natural abundance of 72.17%, and Rb-87 with mass 86.9092 amu and a natural abundance of 27.83%. Calculate the atomic mass of rubidium.

98. Silicon has three naturally occurring isotopes: Si-28 with mass 27.9769 amu and a natural abundance of 92.21%, Si-29 with mass 28.9765 amu and a natural abundance of 4.69%, and Si-30 with mass 29.9737 amu and a natural abundance of 3.10%. Calculate the atomic mass of silicon.

99. Bromine has two naturally occurring isotopes (Br-79 and Br-81) and an atomic mass of 79.904 amu.
  (a) If the natural abundance of Br-79 is 50.69%, what is the natural abundance of Br-81?
  (b) If the mass of Br-81 is 80.9163 amu, what is the mass of Br-79?

100. Silver has two naturally occurring isotopes (Ag-107 and Ag-109).
  (a) Use the periodic table to find the atomic mass of silver.
  (b) If the natural abundance of Ag-107 is 51.84%, what is the natural abundance of Ag-109?
  (c) If the mass of Ag-107 is 106.905 amu, what is the mass of Ag-109?

101. An element has two naturally occurring isotopes. Isotope 1 has a mass of 120.9038 amu and a relative abundance of 57.4%, and isotope 2 has a mass of 122.9042 amu and a relative abundance of 42.6%. Find the atomic mass of this element and, referring to the periodic table, identify it.

102. Copper has two naturally occurring isotopes. Cu-63 has a mass of 62.939 amu and relative abundance of 69.17%. Use the atomic weight of copper to determine the mass of the other copper isotope.

Cumulative Problems

103. Electrical charge is sometimes reported in coulombs (C). On this scale, 1 electron has a charge of -1.6×10^{-19} C. Suppose your body acquires -125 mC (millicoulombs) of charge on a dry day. How many excess electrons has it acquired? (*Hint*: Use the charge of an electron in coulombs as a conversion factor between charge and number of electrons.)

104. How many excess protons are in a positively charged object with a charge of $+398$ mC (millicoulombs)? The charge of 1 proton is $+1.6 \times 10^{-19}$ C. (*Hint*: Use the charge of the proton in coulombs as a conversion factor between charge and number of protons.)

105. The hydrogen atom contains 1 proton and 1 electron. The radius of the proton is approximately 1.0 fm (femtometer), and the radius of the hydrogen atom is approximately 53 pm (picometers). Calculate the volume of the nucleus and the volume of the atom for hydrogen. What percentage of the hydrogen atom's volume does the nucleus occupy? (*Hint*: Convert both given radii to m, and then calculate their volumes using the formula for the volume of a sphere, which is $V = \frac{4}{3}\pi r^3$.)

106. Carbon-12 contains 6 protons and 6 neutrons. The radius of the nucleus is approximately 2.7 fm, and the radius of the atom is approximately 70 pm. Calculate the volume of the nucleus and the volume of the atom. What percentage of the carbon atom's volume does the nucleus occupy? (*Hint*: Convert both given radii to m, and then calculate their volumes using the formula for the volume of a sphere, which is $V = \frac{4}{3}\pi r^3$.)

107. Prepare a table like Table 4.2 for the four different isotopes of Sr that have the natural abundances and masses listed here.

Sr-84	0.56%	83.9134 amu
Sr-86	9.86%	85.9093 amu
Sr-87	7.00%	86.9089 amu
Sr-88	82.58%	87.9056 amu

Use your table and the listed atomic masses to calculate the atomic mass of strontium.

108. Determine the number of protons and neutrons in each isotope of chromium and use the listed natural abundances and masses to calculate its atomic mass.

Cr-50	4.345%	49.9460 amu
Cr-52	83.79%	51.9405 amu
Cr-53	9.50%	52.9407 amu
Cr-54	2.365%	53.9389 amu

109. Fill in the blanks to complete the table.

Symbol	Z	A	Number of Protons	Number of Electrons	Number of Neutrons	Charge
Zn^{2+}	___	___	___		34	2+
___	25	55	___	22	___	___
___	___	___	15	15	16	___
O^{2-}	___	16	___	___	___	2−
___	___	___	16	18	18	___

110. Fill in the blanks to complete the table.

Symbol	Z	A	Number of Protons	Number of Electrons	Number of Neutrons	Charge
Mg^{2+}	___	25	___		13	2+
___	22	48	___	18	___	___
___	16	___	___	___	16	2−
Ga^{3+}	___	71	___	___	___	___
___	___	___	82	80	125	___

111. Europium has two naturally occurring isotopes: Eu-151 with a mass of 150.9198 amu and a natural abundance of 47.8%, and Eu-153. Use the atomic mass of europium to find the mass and natural abundance of Eu-153.

112. Rhenium has two naturally occurring isotopes: Re-185 with a natural abundance of 37.40% and Re-187 with a natural abundance of 62.60%. The sum of the masses of the two isotopes is 371.9087 amu. Find the masses of the individual isotopes.

113. Chapter 1 describes the difference between observations, laws, and theories. Cite two examples of theories from Chapter 4 and explain why they are theories.

114. Chapter 1 describes the difference between observations, laws, and theories. Cite one example of a law from Chapter 4 and explain why it is a law.

115. The atomic mass of fluorine is 19.00 amu, and all fluorine atoms in a naturally occurring sample of fluorine have this mass. The atomic mass of chlorine is 35.45 amu, but no chlorine atoms in a naturally occurring sample of chlorine have this mass. Provide an explanation for the difference.

116. The atomic mass of germanium is 72.61 amu. Is it likely that any individual germanium atoms have a mass of 72.61 amu?

117. Copper has only two naturally occurring isotopes, Cu-63 and Cu-65. The mass of Cu-63 is 62.9396 amu, and the mass of Cu-65 is 64.9278 amu. Use the atomic mass of copper to determine the relative abundance of each isotope in a naturally occurring sample. (*Hint*: The relative abundances of the two isotopes sum to 100%.)

118. Gallium has only two naturally occurring isotopes, Ga-69 and Ga-71. The mass of Ga-69 is 68.9256 amu, and the mass of Ga-71 is 70.9247 amu. Use the atomic mass of gallium to determine the relative abundance of each isotope in a naturally occurring sample.

Highlight Problems

119. The figure shown here is a representation of 50 atoms of a fictitious element with the symbol Nt and atomic number 120. Nt has three isotopes represented by the following colors: Nt-304 (red), Nt-305 (blue), and Nt-306 (green).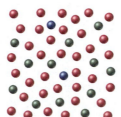
  (a) Assuming that the figure is statistically representative of naturally occurring Nt, what is the percent natural abundance of each Nt isotope?
  (b) Use the listed masses of each isotope to calculate the atomic mass of Nt. Then draw a box for the element similar to the boxes for each element shown in the periodic table in the inside front cover of this book. Make sure your box includes the atomic number, symbol, and atomic mass. (Assume that the percentages from part a are correct to four significant figures.)

Nt-304	303.956 amu
Nt-305	304.962 amu
Nt-306	305.978 amu

120. Neutron stars are believed to be composed of solid nuclear matter, primarily neutrons.
  (a) If the radius of a neutron is 1.0×10^{-13} cm, calculate its density in g/cm^3.

  (volume of a sphere $= \frac{4}{3}\pi r^3$)

  (b) Assuming that a neutron star has the same density as a neutron, calculate the mass in kilograms of a small piece of a neutron star the size of a spherical pebble with a radius of 0.10 mm.

Questions for Group Work

Discuss these questions with the group and record your consensus answer.

121. Complete the following table.

Particle	Mass (amu)	Charge	In the nucleus? (yes/no)	# in ^{32}S atom	# in $^{79}Br^-$ ion
Proton					
Neutron					
Electron					

122. Make a sketch of an oxygen atom. Include the correct number of protons, electrons, and neutrons for the most abundant isotope. Use the following symbols: proton = •, neutron = o, electron = •.

123. The table at right includes data similar to that used by Mendeleev when he made the periodic table. Write on a small card the symbol, atomic mass, and a stable compound formed by each element. Arrange your cards in order of increasing atomic mass. Do you observe any repeating patterns? Describe any patterns you observe. (*Hint*: There is one missing element somewhere in the pattern.)

Element	Atomic Mass	Stable Compound	Element	Atomic Mass	Stable Compound
Be	9	$BeCl_2$	O	16	H_2O
S	32	H_2S	Ga	69.7	GaH_3
F	19	F_2	As	75	AsF_3
Ca	40	$CaCl_2$	C	12	CH_4
Li	7	$LiCl$	K	39	KCl
Si	28	SiH_4	Mg	24.3	$MgCl_2$
Cl	35.4	Cl_2	Se	79	H_2Se
B	10.8	BH_3	Al	27	AlH_3
Ge	72.6	GeH_4	Br	80	Br_2
N	14	NF_3	Na	23	$NaCl$

124. Arrange the cards from Question 123 so that mass increases from left to right and elements with similar properties are above and below each other. Copy the periodic table you have invented onto a piece of paper. There is one element missing. Predict its mass and a stable compound it might form.

Data Interpretation and Analysis

125. The graph at the right shows the atomic radius for the 19 elements in the periodic table.

    (a) Describe the trend in atomic radius in going from H (atomic number 1) to K (atomic number 19).

    (b) Find the three elements represented with blue dots on a periodic table. What do their placements in the table have in common?

    (c) Find the three elements represented with red dots on a periodic table. What do their placements in the table have in common?

    (d) Based on the graph, what is the radius of C?

Atomic radius of elements 1-19

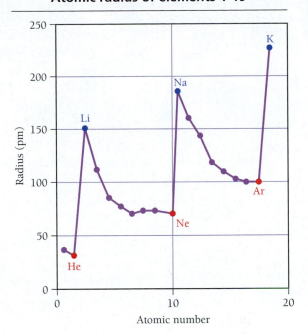

Answers to Skillbuilder Exercises

Skillbuilder 4.1 .... **(a)** sodium, 11
    (b) nickel, 28
    (c) phosphorus, 15
    (d) tantalum, 73

Skillbuilder 4.2 .... **(a)** nonmetal
    (b) nonmetal
    (c) metal
    (d) metalloid

Skillbuilder 4.3 .... **(a)** alkali metal, Group 1A
    (b) Group 3A

    (c) halogen, Group 7A
    (d) noble gas, Group 8A

Skillbuilder 4.4 .... **(a)** 2+
    (b) 1−
    (c) 3−

Skillbuilder 4.5 .... 16 protons, 18 electrons
Skillbuilder 4.6 .... K^+ and Se^{2-}
Skillbuilder 4.7 .... Z = 17, A = 35, Cl-35, and $^{35}_{17}Cl$
Skillbuilder 4.8 .... 19 protons, 20 neutrons
Skillbuilder 4.9 .... 24.31 amu

Answers to Conceptual Checkpoints

4.1 (c) The mass in amu is approximately equal to the number of protons plus the number of neutrons. In order to be charge-neutral, the number of protons must equal the number of electrons.

4.2 (d) Lead is a metal (see Figure 4.12) and a main-group element (see Figure 4.13).

4.3 (b) All of the metalloids are main-group elements (see Figures 4.12 and 4.13).

4.4 O^{2-}

4.5 (a) Both of these ions have 10 electrons.

4.6

C-12 nucleus

C-13 nucleus

4.7 (b) This atom must have (27 − 14) = 13 protons; the element with an atomic number of 13 is Al.

4.8 The isotopes C-12 and C-13 would not look different in this representation of atoms because the only difference between the two isotopes is that C-13 has an extra neutron in the nucleus. The illustration represents the whole atom and does not attempt to illustrate its nucleus. Because the nucleus of an atom is miniscule compared to the size of the atom itself, the extra neutron would not affect the size of the atom.

4.9 (b) The natural abundance of isotope B must be greater than the natural abundance of isotope A because the atomic mass is closer to the mass of isotope B than to the mass of isotope A.

Molecules and Compounds

$C_{12}H_{22}O_{11}$

H_2O
Sucrose

O_2
Caffeine

H_3C

Sucrose

Sucrose

Sucrose

TUESDAY 23RD

MASTERING HOMEWORK DUE
CHEMISTRY

CALL

5 Molecules and Compounds

Almost all aspects of life are engineered at the molecular level, and without understanding molecules, we can only have a very sketchy understanding of life itself.

—Francis Harry Compton Crick (1916–2004)

5.1 Sugar and Salt

Sodium, a shiny metal (◀ FIGURE 5.1) that dulls almost instantly upon exposure to air, is extremely reactive and poisonous. If you were to consume any appreciable amount of elemental sodium, you would need immediate medical help. Chlorine, a pale yellow gas (▼ FIGURE 5.2), is equally reactive and poisonous. Yet the compound formed from these two elements, sodium chloride, is the relatively harmless flavor enhancer that we call table salt (▼ FIGURE 5.3). When elements combine to form compounds, their properties completely change.

▲ FIGURE 5.1 **Elemental sodium** Sodium is an extremely reactive metal that dulls almost instantly upon exposure to air.

▶ FIGURE 5.2 **Elemental chlorine** Chlorine is a yellow gas with a pungent odor. It is highly reactive and poisonous.

▲ FIGURE 5.3 **Sodium chloride**   The compound formed by sodium and chlorine is table salt.

◀ Ordinary table sugar is a compound called sucrose. A sucrose molecule, such as the one shown here, contains carbon, hydrogen, and oxygen atoms. The properties of sucrose are, however, very different from those of carbon (also shown in the form of graphite), hydrogen, and oxygen. The properties of a compound are, in general, different from the properties of the elements that compose it.

Consider also ordinary sugar. Sugar is a compound composed of carbon, hydrogen, and oxygen. Each of these elements has its own unique properties. Carbon is most familiar to us as the graphite found in pencils or as the diamonds in jewelry. Hydrogen is an extremely flammable gas used as a fuel for rocket engines, and oxygen is one of the gases that compose air. When these three elements combine to form sugar, however, a sweet, white, crystalline solid results.

In Chapter 4, you learned how protons, neutrons, and electrons combine to form different elements, each with its own properties and its own chemistry, each different from the other. In this chapter, you will learn how these elements combine with each other to form different compounds, each with its own properties and its own chemistry, each different from all the others and different from the elements that compose it. This is the great wonder of nature: how from such simplicity—protons, neutrons, and electrons—we get such great complexity. It is exactly this complexity that makes life possible. Life could not exist with just 91 different elements if they did not combine to form compounds. It takes compounds in all of their diversity to make living organisms.

5.2 Compounds Display Constant Composition

▶ Restate and apply the law of constant composition.

Although some of the substances you encounter in everyday life are elements, most are not—they are compounds. Free atoms are rare in nature. As you learned in Chapter 3, a compound is different from a mixture of elements. In a compound, the elements combine in fixed, definite proportions, whereas in a mixture, they can have any proportions whatsoever. Consider the difference between a mixture of hydrogen and oxygen gas (▼ FIGURE 5.4) and the compound water (▼ FIGURE 5.5). A mixture of hydrogen and oxygen gas can contain any proportions of hydrogen and oxygen. Water, on the other hand, is composed of water molecules that consist of two hydrogen atoms bonded to one oxygen atom. Consequently, water has a definite proportion of hydrogen to oxygen.

The ratio of hydrogen to oxygen in water is fixed.

The ratio of hydrogen to oxygen in a mixture is variable.

Hydrogen molecule

Water molecule

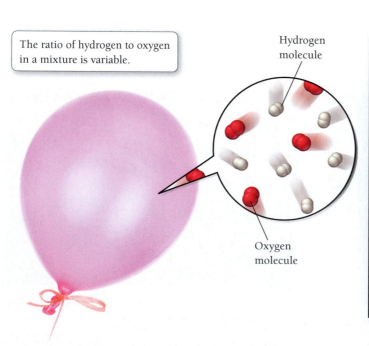

Oxygen molecule

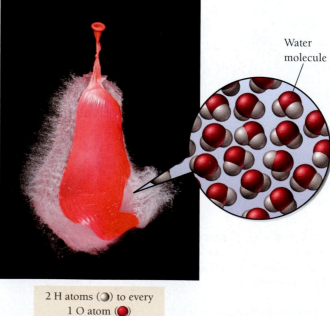

2 H atoms (◖) to every 1 O atom (⬤)

▲ FIGURE 5.4 **A mixture**   This balloon is filled with a mixture of hydrogen and oxygen gas. The relative amounts of hydrogen and oxygen are variable. We could easily add either more hydrogen or more oxygen to the balloon.

▲ FIGURE 5.5 **A chemical compound**   This balloon is filled with water, composed of molecules that have a fixed ratio of hydrogen to oxygen.

The first chemist to formally state the idea that elements combine in fixed proportions to form compounds was Joseph Proust (1754–1826) in the **law of constant composition**, which states:

All samples of a given compound have the same proportions of their constituent elements.

For example, if we decompose an 18.0 g sample of water, we get 16.0 g of oxygen and 2.0 g of hydrogen, or an oxygen-to-hydrogen mass ratio of

Even though atoms combine in whole-number ratios, their mass ratios are not necessarily whole numbers.

$$\text{mass ratio} = \frac{16.0 \text{ g O}}{2.0 \text{ g H}} = 8.0 \quad \text{or} \quad 8.0\text{:}1$$

This is true of any sample of pure water, no matter what its origin. The law of constant composition applies not only to water but to every compound. If we decompose a 17.0 g sample of ammonia, a compound composed of nitrogen and hydrogen, we get 14.0 g of nitrogen and 3.0 g of hydrogen, or a nitrogen-to-hydrogen mass ratio of:

$$\text{mass ratio} = \frac{14.0 \text{ g N}}{3.0 \text{ g H}} = 4.7 \quad \text{or} \quad 4.7\text{:}1$$

Again, this ratio is the same for every sample of ammonia—the composition of each compound is constant.

EXAMPLE **5.1**   **Constant Composition of Compounds**

Two samples of carbon dioxide, obtained from different sources, are decomposed into their constituent elements. One sample produces 4.8 g of oxygen and 1.8 g of carbon, and the other sample produces 17.1 g of oxygen and 6.4 g of carbon. Show that these results are consistent with the law of constant composition.

Calculate the mass ratio of one element to the other by dividing the larger mass by the smaller one.	**SOLUTION**
For the first sample:	$\dfrac{\text{mass oxygen}}{\text{mass carbon}} = \dfrac{4.8 \text{ g}}{1.8 \text{ g}} = 2.7$
For the second sample:	$\dfrac{\text{mass oxygen}}{\text{mass carbon}} = \dfrac{17.1 \text{ g}}{6.4 \text{ g}} = 2.7$

Because the ratios are the same for the two samples, these results are consistent with the law of constant composition.

▶ **SKILLBUILDER 5.1** | **Constant Composition of Compounds**

Two samples of carbon monoxide, obtained from different sources, are decomposed into their constituent elements. One sample produces 4.3 g of oxygen and 3.2 g of carbon, and the other sample produces 7.5 g of oxygen and 5.6 g of carbon. Are these results consistent with the law of constant composition?

▶ **FOR MORE PRACTICE**   Example 5.16; Problems 25, 26.

CONCEPTUAL ✔ **CHECKPOINT 5.1**

A compound composed of two elements A and B has a ratio of $\dfrac{\text{mass A}}{\text{mass B}} = 3.0$. Decomposition of the compound produces 9.0 g of element A. What mass of element B is produced?

(a) 27.0 g B   (b) 9.0 g B   (c) 3.0 g B   (d) 1.0 g B

5.3 Chemical Formulas: How to Represent Compounds

▶ Write chemical formulas.

▶ Determine the total number of each type of atom in a chemical formula.

We represent a compound with a **chemical formula**, which indicates the elements present in the compound and the relative number of atoms of each element. For example, H_2O is the chemical formula for water; it indicates that water consists of hydrogen and oxygen atoms in a 2:1 ratio. (Note that the ratio in a chemical

> Compounds have constant composition with respect to mass (as you learned in the previous section) because they are composed of atoms in fixed ratios.

formula is a ratio of atoms, not a ratio of masses.) The formula contains the symbol for each element, accompanied by a subscript indicating the number of atoms of that element. By convention, a subscript of 1 is omitted.

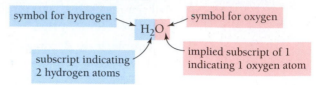

Other common chemical formulas include NaCl for table salt, indicating sodium and chlorine atoms in a 1:1 ratio; CO_2 for carbon dioxide, indicating carbon and oxygen atoms in a 1:2 ratio; and $C_{12}H_{22}O_{11}$ for table sugar (sucrose), indicating carbon, hydrogen, and oxygen atoms in a 12:22:11 ratio. The subscripts in a chemical formula are part of the compound's definition—if the subscripts change, the formula no longer specifies the same compound. For example, CO is the chemical formula for carbon monoxide, an air pollutant with adverse health effects on humans. When inhaled, carbon monoxide interferes with the blood's ability to carry oxygen, which can be fatal. CO is the primary substance responsible for the deaths of people who inhale too much automobile exhaust. If we change the subscript of the O in CO from 1 to 2, however, we get the formula for a totally different compound. CO_2 is the chemical formula for carbon dioxide, the relatively harmless product of combustion and human respiration. We breathe small amounts of CO_2 all the time with no harmful effects. So, remember that:

CO CO_2

The subscripts in a chemical formula represent the relative numbers of each type of atom in a chemical compound; they never change for a given compound.

Chemical formulas normally list the most metallic elements first. Therefore, the formula for table salt is NaCl, not ClNa. In compounds that do not include a metal, we list the more metal-like element first. Recall from Chapter 4 that metals occupy the left side of the periodic table and nonmetals the upper right side. Among nonmetals, those to the left in the periodic table are more metal-like than those to the right and are normally listed first. Therefore, we write CO_2 and NO, not O_2C and ON. Within a single column in the periodic table, elements toward the bottom are more metal-like than elements toward the top. So, we write SO_2, not O_2S. Table 5.1 lists the specific order for listing nonmetal elements in a chemical formula.

> There are a few historical exceptions in which the most metallic element is not listed first, such as the hydroxide ion, which we write as OH^-.

TABLE 5.1   Order of Listing Nonmetal Elements in a Chemical Formula

C	P	N	H	S	I	Br	Cl	O	F

Elements on the left are generally listed before elements on the right.

EXAMPLE 5.2   Writing Chemical Formulas

Write a chemical formula for each compound.

(a) the compound containing two aluminum atoms to every three oxygen atoms
(b) the compound containing three oxygen atoms to every sulfur atom
(c) the compound containing four chlorine atoms to every carbon atom

Aluminum is the metal, so list it first.	**SOLUTION** (a) Al_2O_3
Sulfur is below oxygen on the periodic table and occurs before oxygen in Table 5.1, so list it first.	(b) SO_3
Carbon is to the left of chlorine on the periodic table and occurs before chlorine in Table 5.1, so list it first.	(c) CCl_4

▶ **SKILLBUILDER 5.2** | **Writing Chemical Formulas**

Write a chemical formula for each compound.

(a) the compound containing two silver atoms to every sulfur atom
(b) the compound containing two nitrogen atoms to every oxygen atom
(c) the compound containing two oxygen atoms to every titanium atom

▶ **FOR MORE PRACTICE**   Example 5.17; Problems 31, 32, 33, 34.

Polyatomic Ions in Chemical Formulas

Some chemical formulas contain groups of atoms that act as a unit. When more than one group of the same kind is present, we set their formula off in parentheses with a subscript to indicate the number of units of that group. Many of these groups of atoms have a charge associated with them and are called **polyatomic ions**. For example, NO_3^- is a polyatomic ion with a $1-$ charge. We describe polyatomic ions in more detail in Section 5.5.

To determine the total number of each type of atom in a compound containing a group within parentheses, we multiply the subscript outside the parentheses by the subscript for each atom inside the parentheses. For example, $Mg(NO_3)_2$ indicates a compound containing one magnesium atom (present as the Mg^{2+} ion) and two NO_3^- groups.

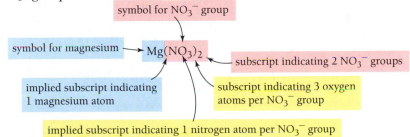

Therefore, the formula $Mg(NO_3)_2$ has the following numbers of each type of atom.

Mg: 1 Mg
N: $1 \times 2 = 2$ N (implied 1 inside parentheses times 2 outside parentheses)
O: $3 \times 2 = 6$ O (3 inside parentheses times 2 outside parentheses)

EXAMPLE **5.3** | **Total Number of Each Type of Atom in a Chemical Formula**

Determine the number of each type of atom in $Mg_3(PO_4)_2$.

SOLUTION

Mg: There are three Mg atoms (present as Mg^{2+} ions), as indicated by the subscript 3.

P: There are two P atoms. We determine this by multiplying the subscript outside the parentheses (2) by the subscript for P inside the parentheses, which is 1 (implied).

O: There are eight O atoms. We determine this by multiplying the subscript outside the parentheses (2) by the subscript for O inside the parentheses (4).

▶ **SKILLBUILDER 5.3** | **Total Number of Each Type of Atom in a Chemical Formula**

Determine the number of each type of atom in K_2SO_4.

▶ **SKILLBUILDER PLUS**

Determine the number of each type of atom in $Al_2(SO_4)_3$.

▶ **FOR MORE PRACTICE**   Example 5.18; Problems 35, 36, 37, 38.

CONCEPTUAL ✔ **CHECKPOINT 5.2**

Which formula represents the greatest total number of atoms?

(a) $Al(C_2H_3O_2)_3$ (b) $Al_2(Cr_2O_7)_3$ (c) $Pb(HSO_4)_4$

(d) $Pb_3(PO_4)_4$ (e) $(NH_4)_3 PO_4$

Types of Chemical Formulas

We categorize chemical formulas into three types: empirical, molecular, and structural. An **empirical formula** is the simplest whole-number ratio of atoms of each element in a compound. A **molecular formula** is the *actual* number of atoms of each element in a molecule of the compound. For example, the molecular formula for hydrogen peroxide is H_2O_2, and its empirical formula is HO. The molecular formula is always a whole-number multiple of the empirical formula. For many compounds, the molecular and empirical formulas are the same. For example, the empirical and molecular formula for water is H_2O because water molecules contain two hydrogen atoms and one oxygen atom; no simpler whole-number ratio can express the number of hydrogen atoms relative to oxygen atoms.

A **structural formula** uses lines to represent chemical bonds and shows how the atoms in a molecule are connected to each other. The structural formula for hydrogen peroxide is H—O—O—H. We can also use **molecular models**—three-dimensional representations of molecules—to represent compounds. In this book, we use two types of molecular models: ball-and-stick and space-filling. In **ball-and-stick models**, we represent atoms as balls and chemical bonds as sticks. The balls and sticks are connected to represent the molecule's shape. The balls are color coded, and we assign each element a color as shown in the margin.

In **space-filling models**, atoms fill the space between each other to more closely represent our best idea for how a molecule might appear if we could scale it to a visible size. Consider the following ways to represent a molecule of methane, the main component of natural gas:

○ Hydrogen

● Carbon

● Nitrogen

● Oxygen

● Fluorine

● Phosphorus

● Sulfur

● Chlorine

CH_4

$$H-\overset{\overset{\displaystyle H}{|}}{\underset{\underset{\displaystyle H}{|}}{C}}-H$$

Molecular formula   Structural formula   Ball-and-stick model   Space-filling model

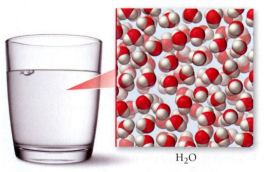

Macroscopic   Molecular

H_2O

Symbolic

The molecular formula of methane indicates that methane has one carbon atom and four hydrogen atoms. The structural formula shows how the atoms are connected: each hydrogen atom is bonded to the central carbon atom. The ball-and-stick model and the space-filling model illustrate the *geometry* of the molecule: how the atoms are arranged in three dimensions.

Throughout this book, you have seen and will continue to see images that show the connection between the *macroscopic world* (what we see), the *atomic and molecular world* (the particles that compose matter), and the *symbolic way* that chemists represent the atomic and molecular world. For example, at left is a representation of water using this kind of image.

The main goal of these images is to help you visualize the main theme of this book: *the connection between the world around us and the world of atoms and molecules.*

CONCEPTUAL ✔ **CHECKPOINT 5.3**

Write a formula for the compound represented by this space-filling model.

5.4   A Molecular View of Elements and Compounds

▶ Classify elements as atomic or molecular.

▶ Classify compounds as ionic or molecular.

Recall from Chapter 3 that we can categorize pure substances as either elements or compounds. We can further subcategorize elements and compounds according to the basic units that compose them (▼ FIGURE 5.6). Pure substances may be either elements or compounds. Elements may be either atomic or molecular. Compounds may be either molecular or ionic.

Atomic Elements

Atomic elements have single atoms as their basic units. Most elements fall into this category. For example, helium is composed of helium atoms, copper is composed of copper atoms, and mercury of mercury atoms (▼ FIGURE 5.7).

Molecular Elements

Molecular elements do not normally exist in nature with single atoms as their basic units. Instead, these elements exist as *diatomic molecules*—two atoms of that element bonded together—as their basic units. For example, hydrogen is composed of H_2 molecules, oxygen is composed of O_2 molecules, and chlorine of Cl_2 molecules (▼ FIGURE 5.8). Table 5.2 and ▶ FIGURE 5.9, on the next page, list elements that exist as diatomic molecules.

A few molecular elements, such as S_8 and P_4, are composed of molecules containing several atoms.

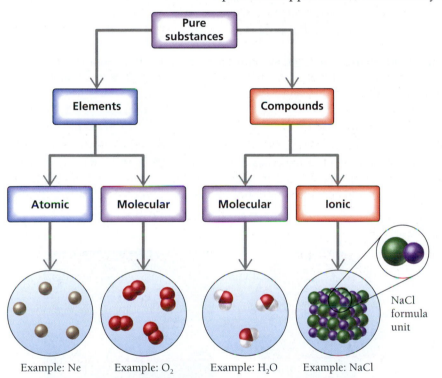

Example: Ne   Example: O_2 Example: H_2O Example: NaCl

NaCl formula unit

▲ FIGURE 5.6 **A molecular view of elements and compounds**

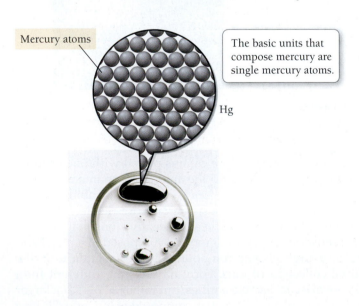

Mercury atoms

The basic units that compose mercury are single mercury atoms.

Hg

▲ FIGURE 5.7 **An atomic element**

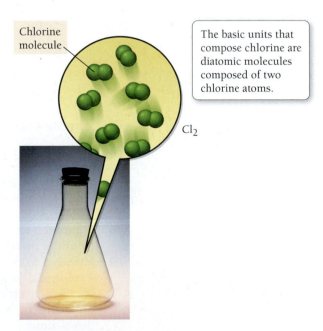

Chlorine molecule

The basic units that compose chlorine are diatomic molecules composed of two chlorine atoms.

Cl_2

▲ FIGURE 5.8 **A molecular element**

TABLE 5.2 Elements That Occur as Diatomic Molecules

Name of Element	Formula of Basic Unit
hydrogen	H_2
nitrogen	N_2
oxygen	O_2
fluorine	F_2
chlorine	Cl_2
bromine	Br_2
iodine	I_2

Main group — Main group

Elements that exist as diatomic molecules

Transition metals

Lanthanides: Ce 58, Pr 59, Nd 60, Pm 61, Sm 62, Eu 63, Gd 64, Tb 65, Dy 66, Ho 67, Er 68, Tm 69, Yb 70, Lu 71

Actinides: Th 90, Pa 91, U 92, Np 93, Pu 94, Am 95, Cm 96, Bk 97, Cf 98, Es 99, Fm 100, Md 101, No 102, Lr 103

▲ **FIGURE 5.9 Elements that form diatomic molecules** Elements that normally exist as diatomic molecules are highlighted in yellow on this periodic table. Note that they are all nonmetals and include four of the halogens.

Molecular Compounds

Molecular compounds are composed of two or more nonmetals. The basic units of molecular compounds are molecules composed of the constituent atoms. For example, water is composed of H_2O molecules, dry ice is composed of CO_2 molecules (▼ **FIGURE 5.10**), and acetone (finger nail–polish remover) of C_3H_6O molecules.

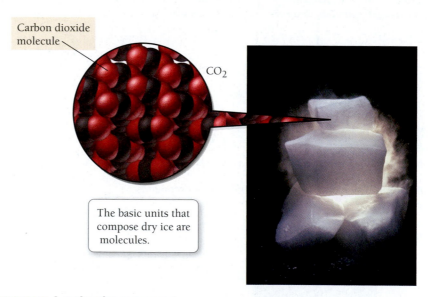

Carbon dioxide molecule

CO_2

The basic units that compose dry ice are molecules.

▲ **FIGURE 5.10 A molecular compound**

Ionic Compounds

Ionic compounds are composed of one or more cations paired with one or more anions. In most cases, the cations are metals and the anions are nonmetals. When a metal, which has a tendency to lose electrons (see Section 4.6), combines with a nonmetal, which has a tendency to gain electrons, one or more electrons transfer from the metal to the nonmetal, creating positive and negative ions that are then attracted to each other. We can assume that a compound composed of a metal and a nonmetal is ionic. The basic unit of ionic compounds is the **formula unit**, the smallest electrically neutral collection of ions. Formula units are different from molecules in that they do not exist as discrete entities, but rather as part of a larger three-dimensional array. For example, salt (NaCl) is composed of Na^+ and Cl^- ions

Sodium chloride formula unit

The basic units that compose table salt are NaCl formula units.

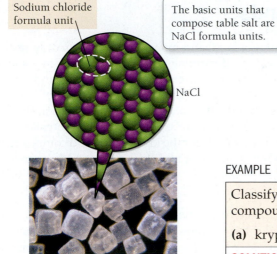

NaCl

▲ **FIGURE 5.11 An ionic compound** Unlike molecular compounds, ionic compounds do not contain individual molecules but rather sodium and chloride ions in an alternating three-dimensional array.

in a 1:1 ratio. In table salt, Na^+ and Cl^- ions exist in an alternating three-dimensional array (◄ **FIGURE 5.11**). However, any one Na^+ ion does not pair with one specific Cl^- ion. Sometimes chemists refer to formula units as molecules, but this is not strictly correct since ionic compounds do not contain distinct molecules.

EXAMPLE 5.4 **Classifying Substances as Atomic Elements, Molecular Elements, Molecular Compounds, or Ionic Compounds**

Classify each substance as an atomic element, molecular element, molecular compound, or ionic compound.

(a) krypton   **(b)** $CoCl_2$ **(c)** nitrogen   **(d)** SO_2 **(e)** KNO_3

SOLUTION

(a) Krypton is an element that is not listed as diatomic in Table 5.2; therefore, it is an atomic element.

(b) $CoCl_2$ is a compound composed of a metal (left side of periodic table) and nonmetal (right side of the periodic table); therefore, it is an ionic compound.

(c) Nitrogen is an element that is listed as diatomic in Table 5.2; therefore, it is a molecular element.

(d) SO_2 is a compound composed of two nonmetals; therefore, it is a molecular compound.

(e) KNO_3 is a compound composed of a metal and two nonmetals; therefore, it is an ionic compound.

▶ **SKILLBUILDER 5.4** | **Classifying Substances as Atomic Elements, Molecular Elements, Molecular Compounds, or Ionic Compounds**

Classify each substance as an atomic element, molecular element, molecular compound, or ionic compound.

(a) chlorine   **(b)** NO   **(c)** Au   **(d)** Na_2O **(e)** $CrCl_3$

▶ **FOR MORE PRACTICE** Example 5.19, Example 5.20; Problems 43, 44, 45, 46.

PEARSON eText 2.0

CONCEPTUAL ✅ **CHECKPOINT 5.4**

Which image represents a molecular compound?

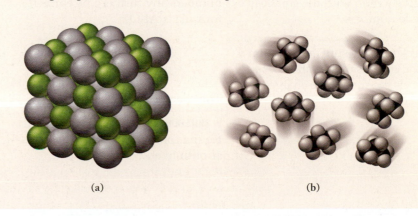

(a)   (b)

5.5 Writing Formulas for Ionic Compounds

▶ Write formulas for ionic compounds.

Revisit Section 4.7 and Figure 4.14 to review the elements that form ions with a predictable charge.

Because ionic compounds must be charge-neutral and many elements form only one type of ion with a predictable charge, we can determine the formulas for many ionic compounds based on their constituent elements. For example, on the one hand, the formula for the ionic compound composed of sodium and chlorine must be NaCl and not anything else because in ionic compounds Na always forms 1+ cations and Cl always forms 1− anions. In order for the compound to be charge-neutral, it must contain one Na^+ cation to every Cl^- anion. The formula for the ionic compound composed of magnesium and chlorine, on the other hand, must be $MgCl_2$ because Mg always forms 2+ cations and Cl always forms 1− anions. In order for the compound to be charge-neutral, it must contain one Mg^{2+} cation to every two Cl^- anions. In general:

• Ionic compounds always contain positive and negative ions.
• In the chemical formula, the sum of the charges of the positive ions (cations) must always exactly cancel the sum of the charges of the negative ions (anions).

Writing Formulas for Ionic Compounds Containing Only Monoatomic Ions

To write the formula for an ionic compound containing monoatomic ions (no polyatomic ions), follow the procedure in the left column of the following examples. The center and right columns provide two examples of how to apply the procedure.

PEARSON
eText
2.0
Interactive
Worked Example
Video 5.5

Writing Formulas for Ionic Compounds	EXAMPLE 5.5 Write a formula for the ionic compound that forms from aluminum and oxygen.	EXAMPLE 5.6 Write a formula for the ionic compound that forms from magnesium and oxygen.
1. Write the symbol for the metal and its charge followed by the symbol of the nonmetal and its charge. For many elements, you can determine these charges from their group number in the periodic table (refer to Figure 4.14).	SOLUTION Al^{3+} O^{2-}	SOLUTION Mg^{2+} O^{2-}
2. Use the magnitude of the charge on each ion (without the sign) as the subscript for the other ion.	Al^{3+} O^{2-} Al_2O_3	Mg^{2+} O^{2-} Mg_2O_2
3. If possible, reduce the subscripts to give a ratio with the smallest whole numbers.	In this case, you cannot reduce the numbers any further; the correct formula is Al_2O_3.	To reduce the subscripts, divide both subscripts by 2. $Mg_2O_2 \div 2 = MgO$
4. Check to make sure that the sum of the charges of the cations exactly cancels the sum of the charges of the anions.	Cations: $2(3+) = 6+$ Anions: $3(2-) = 6-$ The charges cancel.	Cations: 2+ Anions: 2− The charges cancel.
	▶ SKILLBUILDER 5.5	Write a formula for the compound that forms from strontium and chlorine.	▶ SKILLBUILDER 5.6	Write a formula for the compound that forms from aluminum and nitrogen.
		▶ FOR MORE PRACTICE Problems 53, 54, 57.

Writing Formulas for Ionic Compounds Containing Polyatomic Ions

As noted previously, some ionic compounds contain polyatomic ions (ions that are themselves composed of a group of atoms with an overall charge). Table 5.3 lists the most common polyatomic ions. You need to be able to recognize polyatomic ions in a chemical formula, so it is good idea to become familiar with Table 5.3. To write a formula for ionic compounds containing polyatomic ions, use the formula and charge of the polyatomic ion as demonstrated in Example 5.7.

TABLE 5.3 Some Common Polyatomic Ions

Name	Formula	Name	Formula
acetate	$C_2H_3O_2^-$	hypochlorite	ClO^-
carbonate	CO_3^{2-}	chlorite	ClO_2^-
hydrogen carbonate (or bicarbonate)	HCO_3^-	chlorate	ClO_3^-
hydroxide	OH^-	perchlorate	ClO_4^-
nitrate	NO_3^-	permanganate	MnO_4^-
nitrite	NO_2^-	sulfate	SO_4^{2-}
chromate	CrO_4^{2-}	sulfite	SO_3^{2-}
dichromate	$Cr_2O_7^{2-}$	hydrogen sulfite (or bisulfite)	HSO_3^-
phosphate	PO_4^{3-}	hydrogen sulfate (or bisulfate)	HSO_4^-
hydrogen phosphate	HPO_4^{2-}	peroxide	O_2^{2-}
ammonium	NH_4^+	cyanide	CN^-

EXAMPLE **5.7**

Writing Formulas for Ionic Compounds Containing Polyatomic Ions

Write a formula for the compound that forms from calcium and nitrate ions.

1. Write the symbol for the metal ion followed by the symbol for the polyatomic ion and their charges. You can deduce these charges from the group numbers in the periodic table. For polyatomic ions, look up the charges and names in Table 5.3.	**SOLUTION** Ca^{2+} NO_3^-
2. Use the magnitude of the charge on each ion as the subscript for the other ion.	$Ca_1(NO_3)_2$
3. Check to see if the subscripts can be reduced to simpler whole numbers. You should drop subscripts of 1 because they are implied.	In this case, you cannot further reduce the subscripts, but you can drop the subscript of 1. $Ca(NO_3)_2$
4. Confirm that the sum of the charges of the cations exactly cancels the sum of the charges of the anions.	Cations Anions 2+ $2(1-) = 2-$

▶ **SKILLBUILDER 5.7** | Write the formula for the compound that forms between aluminum and phosphate ions.

▶ **SKILLBUILDER PLUS** | Write the formula for the compound that forms between sodium and sulfite ions.

▶ **FOR MORE PRACTICE**   Example 5.21; Problems 55, 56, 58.

CONCEPTUAL ✓ **CHECKPOINT 5.5**

Some metals can form ions of different charges in different compounds. Deduce the charge of the Cr ion in the compound $Cr(NO_3)_3$.

(a) 1+ **(b)** 2+ **(c)** 3+ **(d)** 1−

5.6 Nomenclature: Naming Compounds

▶ Distinguish between common and systematic names for compounds.

Because there are so many different compounds, chemists have developed systematic ways to name them. These naming rules can help you examine a compound's formula and determine its name, or vice versa. Many compounds also have a common name. For example, H_2O has the common name *water* and the systematic name *dihydrogen monoxide*. A common name is like a nickname for a compound, used by those who are familiar with it. Since water is such a familiar compound, everyone uses its common name and not its systematic name. In the sections that follow, you'll learn how to systematically name simple ionic and molecular compounds. Keep in mind, however, that some compounds also have common names that are often used instead of the systematic name. Common names can be learned only through familiarity.

5.7 Naming Ionic Compounds

▶ Name binary ionic compounds containing a metal that forms only one type of ion.
▶ Name binary ionic compounds containing a metal that forms more than one type of ion.
▶ Name ionic compounds containing a polyatomic ion.

The first step in naming an ionic compound is identifying it as one. Remember, any time we have a metal and one or more nonmetals together in a chemical formula, we can assume the compound is ionic. Ionic compounds are categorized into two types (▼ FIGURE 5.12) depending on the metal in the compound. The first type (sometimes called Type I) contains a metal with an invariant charge—one that does not vary from one compound to another. Sodium, for instance, has a 1+ charge in all of its compounds. ▼ FIGURE 5.13 lists examples of metals whose charge is invariant from one compound to another. The charge of most of these metals can be inferred from their group number in the periodic table (see Figure 4.14).

The second type of ionic compound (sometimes called Type II) contains a metal with a charge that can differ in different compounds. In other words, the

PEARSON eText 2.0

Key Concept Video
Naming Ionic Compounds

Metals Whose Charge Is Invariant from One Compound to Another

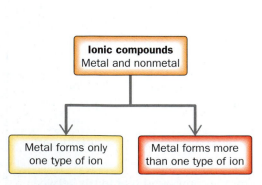

▲ **FIGURE 5.12 Classification of ionic compounds** We classify ionic compounds into two types, depending on the metal in the compound.

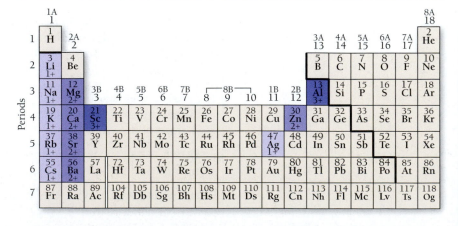

▲ **FIGURE 5.13 Metals with invariant charges** The metals highlighted in this periodic table always form an ion with the same charge in all of their compounds.

TABLE 5.4   Some Metals That Form More Than One Type of Ion and Their Common Charges (*This list is not exhaustive but meant to show examples.*)

Metal	Symbol Ion	Name	Older Name*
chromium	Cr^{2+}	chromium(II)	chromous
	Cr^{3+}	chromium(III)	chromic
iron	Fe^{2+}	iron(II)	ferrous
	Fe^{3+}	iron(III)	ferric
cobalt	Co^{2+}	cobalt(II)	cobaltous
	Co^{3+}	cobalt(III)	cobaltic
copper	Cu^{+}	copper(I)	cuprous
	Cu^{2+}	copper(II)	cupric
tin	Sn^{2+}	tin(II)	stannous
	Sn^{4+}	tin(IV)	stannic
mercury	Hg_2^{2+}	mercury(I)	mercurous
	Hg^{2+}	mercury(II)	mercuric
lead	Pb^{2+}	lead(II)	plumbous
	Pb^{4+}	lead(IV)	plumbic

* An older naming system substitutes the names found in this column for the name of the metal and its charge. Under this system, chromium(II) oxide is named chromous oxide. We do *not* use this older system in this text.

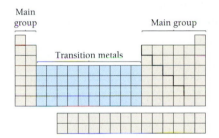

▲ **FIGURE 5.14 The transition metals** The metals that form more than one type of ion are usually (but not always) transition metals.

metal in this second type of ionic compound can form more than one kind of cation (depending on the compound). Iron, for instance, has a 2+ charge in some of its compounds and a 3+ charge in others. The best way to remember these is by elimination. For our purposes, you can assume that any metal *not* highlighted in Figure 5.13 is of this second type. Table 5.4 lists some examples of metals of this type. Metals that form cations whose charges can vary in different compounds are usually (but not always) found in the **transition metals** section of the periodic table (◄ **FIGURE 5.14**). The exceptions are Zn and Ag, which are transition metals, but form cations with the same charge in all of their compounds (as you can see from Figure 5.13). Two other exceptions are Pb and Sn, which are *not* transition metals but still form cations whose charges can vary in different compounds.

Naming Binary Ionic Compounds Containing a Metal That Forms Only One Type of Cation

Binary compounds contain only two different elements. The names for binary ionic compounds containing a metal that forms only one type of ion have the form:

> | name of cation (metal) | base name of anion (nonmetal) + *-ide* |

Because the charge of the metal is always the same for these types of compounds, we do not need to specify it in the compound's name. For example, the name for NaCl consists of the name of the cation, *sodium*, followed by the base name of the anion, *chlor*, with the ending *-ide*. The full name is *sodium chloride*.

> NaCl   sodium chloride

The name for $CaBr_2$ consists of the name of the cation, *calcium*, followed by the base name of the anion, *brom*, with the ending *-ide*. The full name is *calcium bromide*.

> $CaBr_2$ calcium bromide

Table 5.5 (on the next page) contains the base names for various nonmetals and their most common charges in ionic compounds.

The name of the cation in ionic compounds is the same as the name of the metal.

TABLE 5.5 Some Common Anions

Nonmetal	Symbol for Ion	Base Name	Anion Name
fluorine	F^-	fluor-	fluoride
chlorine	Cl^-	chlor-	chloride
bromine	Br^-	brom-	bromide
iodine	I^-	iod-	iodide
oxygen	O^{2-}	ox-	oxide
sulfur	S^{2-}	sulf-	sulfide
nitrogen	N^{3-}	nitr-	nitride

EXAMPLE **5.8**

Naming Ionic Compounds Containing a Metal That Forms Only One Type of Cation

Name the compound MgF_2.

SOLUTION

The cation is magnesium. The anion is fluorine, which becomes *fluoride*. Its correct name is *magnesium fluoride*.

▶ **SKILLBUILDER 5.8 | Naming Ionic Compounds Containing a Metal That Forms Only One Type of Ion**

Name the compound KBr.

▶ **SKILLBUILDER PLUS**   Name the compound Zn_3N_2.

▶ **FOR MORE PRACTICE**   Example 5.22; Problems 59, 60.

Naming Binary Ionic Compounds Containing a Metal That Forms More Than One Type of Cation

Because the charge of the metal cation in these types of compounds is not always the same, we must specify the charge in the metal's name. We specify the charge with a roman numeral (in parentheses) following the name of the metal. For example, we distinguish between Cu^+ and Cu^{2+} by writing a (I) to indicate the 1+ ion or a (II) to indicate the 2+ ion:

$$Cu^+ \quad copper(I)$$

$$Cu^{2+} \quad copper(II)$$

The full names for these types of compounds have the form:

name of cation (metal)	(charge of cation (metal) in roman numerals in parentheses)	base name of anion (nonmetal) + -ide

We can determine the charge of the metal from the chemical formula of the compound—remember that the sum of all the charges must be zero. For example, the charge of iron in $FeCl_3$ must be 3+ in order for the compound to be charge-neutral with the three Cl^- anions. The name for $FeCl_3$ is therefore the name of the cation, *iron*, followed by the charge of the cation in parentheses *(III)*, followed by the base name of the anion, *chlor*, with the ending *-ide*. The full name is *iron(III) chloride*.

$$FeCl_3 \quad iron(III)\ chloride$$

Likewise, the name for CrO consists of the name of the cation, *chromium*, followed by the charge of the cation in parentheses *(II)*, followed by the base name of the anion, *ox-*, with the ending *-ide*. The full name is *chromium(II) oxide*.

CrO   chromium(II) oxide

The charge of chromium must be 2+ in order for the compound to be charge-neutral with one O^{2-} anion.

EXAMPLE **5.9** **Naming Ionic Compounds Containing a Metal That Forms More Than One Type of Cation**

Name the compound $PbCl_4$.

SOLUTION

The name for $PbCl_4$ consists of the name of the cation, *lead*, followed by the charge of the cation in parentheses *(IV)*, followed by the base name of the anion, *chlor-*, with the ending *-ide*. The full name is *lead(IV) chloride*. We know the charge on Pb is 4+ because the charge on Cl is 1−. Since there are 4 Cl^- anions, the Pb cation must be Pb^{4+}.

$PbCl_4$ lead(IV) chloride

▶ **SKILLBUILDER 5.9** | **Naming Ionic Compounds Containing a Metal That Forms More Than One Type of Cation**

Name the compound PbO.

▶ **FOR MORE PRACTICE**   Example 5.23; Problems 61, 62.

CONCEPTUAL ✔ CHECKPOINT 5.6

Explain why CaO is NOT named calcium(II) oxide.

Naming Ionic Compounds Containing a Polyatomic Ion

We name ionic compounds containing polyatomic ions using the same procedure we applied to other ionic compounds, except that we use the name of the polyatomic ion whenever it occurs (see Table 5.3). For example, we name KNO_3 using its cation, K^+, *potassium*, and its polyatomic anion, NO_3^-, *nitrate*. The full name is *potassium nitrate*.

KNO$_3$ potassium nitrate

We name $Fe(OH)_2$ according to its cation, *iron*, its charge *(II)*, and its polyatomic ion, *hydroxide*. Its full name is *iron(II) hydroxide*.

Fe(OH)$_2$ iron(II) hydroxide

If the compound contains both a polyatomic cation and a polyatomic anion, we use the names of both polyatomic ions. For example, NH_4NO_3 is *ammonium nitrate*.

NH$_4$NO$_3$ ammonium nitrate

Most polyatomic ions are **oxyanions**, anions containing oxygen. Notice in Table 5.3 that when a series of oxyanions contain different numbers of oxygen atoms, we name them systematically according to the number of oxygen atoms in the ion. If there are two ions in the series, we give the one with more oxygen atoms the ending *-ate* and we give the one with fewer oxygen atoms the ending *-ite*. For example, NO_3^- is *nitrate* and NO_2^- is *nitrite*.

NO$_3^-$ nitrate
NO$_2^-$ nitrite

If there are more than two ions in the series, then we use the prefixes *hypo-*, meaning "less than," and *per-*, meaning "more than." So we call ClO^- *hypochlorite*, meaning "less oxygen than chlorite," and we call ClO_4^- *perchlorate*, meaning "more oxygen than chlorate."

ClO^-	hypochlorite
ClO_2^-	chlorite
ClO_3^-	chlorate
ClO_4^-	perchlorate

EXAMPLE **5.10** | **Naming Ionic Compounds Containing a Polyatomic Ion**

Name the compound K_2CrO_4.

SOLUTION

The name for K_2CrO_4 consists of the name of the cation, *potassium*, followed by the name of the polyatomic ion, *chromate*.

$$K_2CrO_4 \quad \text{potassium chromate}$$

▶ **SKILLBUILDER 5.10** | **Naming Ionic Compounds Containing a Polyatomic Ion**
Name the compound $Mn(NO_3)_2$.

▶ **FOR MORE PRACTICE** Example 5.24; Problems 65, 66.

PEARSON
eText
2.0

CONCEPTUAL ✔ **CHECKPOINT 5.7**

We just saw that the anion ClO_3^- is named chlorate. What is the name of the anion IO_3^-?

EVERYDAY CHEMISTRY

Polyatomic Ions

A glance at the labels of household products reveals the importance of polyatomic ions in everyday compounds. For example, the active ingredient in household bleach is sodium hypochlorite, which acts to decompose color-causing molecules in clothes (bleaching action) and to kill bacteria (disinfection). A box of baking soda contains sodium bicarbonate (sodium hydrogen carbonate), which acts as an antacid when consumed in small quantities and as a source of carbon dioxide gas in baking. The pockets of carbon dioxide gas make baked goods fluffy rather than flat.

Calcium carbonate is the active ingredient in many antacids such as Tums™ and Alka-Mints™. It neutralizes stomach acids, relieving the symptoms of indigestion and heartburn. Too much calcium carbonate, however, can cause constipation, so Tums should not be overused. Sodium nitrite is a common food additive used to preserve packaged meats such as ham, hot dogs, and bologna. Sodium nitrite inhibits the growth of bacteria, especially those that cause botulism, an often fatal type of food poisoning.

▲ Compounds containing polyatomic ions are present in many consumer products.

▲ The active ingredient in bleach is sodium hypochlorite.

B5.1 CAN YOU ANSWER THIS? *Write a formula for each of these compounds that contain polyatomic ions: sodium hypochlorite, sodium bicarbonate, calcium carbonate, sodium nitrite.*

5.8 Naming Molecular Compounds

▶ Name molecular compounds.

Key Concept Video
Naming Molecular Compounds

The first step in naming a molecular compound is identifying it as one. Remember, nearly all molecular compounds form from two or more nonmetals. In this section, we discuss how to name binary (two-element) molecular compounds. Their names have the form:

When writing the name of a molecular compound, as when writing the formula, the first element is the more metal-like one (see Table 5.1). The prefixes given to each element indicate the number of atoms present.

mono- 1	*hexa-* 6
di- 2	*hepta-* 7
tri- 3	*octa-* 8
tetra- 4	*nona-* 9
penta- 5	*deca-* 10

If there is only one atom of the *first element* in the formula, the prefix *mono-* is normally omitted. For example, the name for CO_2 begins with *carbon*, without a prefix because *mono-* is omitted for the first element, followed by the prefix *di-*, to indicate two oxygen atoms, followed by the base name of the second element, *ox*, with the ending *-ide*.

carbon di- ox -ide

The full name is *carbon dioxide*.

CO_2 carbon dioxide

> When the prefix ends with a vowel and the base name starts with a vowel, we sometimes drop the first vowel, especially in the case of mono oxide, which becomes monoxide.

The name for the compound N_2O, also called laughing gas, begins with the first element, *nitrogen*, with the prefix *di-*, to indicate that there are two of them, followed by the base name of the second element, *ox*, prefixed by *mono-*, to indicate one, and the suffix *-ide*. Because *mono-* ends with a vowel and *oxide* begins with one, we drop an *o* and combine the two as *monoxide*. The entire name is *dinitrogen monoxide*.

N_2O dinitrogen monoxide

EXAMPLE 5.11 | Naming Molecular Compounds

Name each compound.

(a) CCl_4 **(b)** BCl_3 **(c)** SF_6

SOLUTION

(a) The name of the compound is the name of the first element, *carbon*, followed by the base name of the second element, *chlor*, prefixed by *tetra-* to indicate four, and the suffix *-ide*.

CCl_4 carbon tetrachloride

(b) The name of the compound is the name of the first element, *boron*, followed by the base name of the second element, *chlor*, prefixed by *tri-* to indicate three, and the suffix *-ide*.

BCl_3 boron trichloride

(c) The name of the compound is the name of the first element, *sulfur*, followed by the base name of the second element, *fluor*, prefixed by *hexa-* to indicate six, and the suffix *-ide*. The entire name is *sulfur hexafluoride*.

SF_6 sulfur hexafluoride

▶ **SKILLBUILDER 5.11 | Naming Molecular Compounds**

Name the compound N_2O_4.

▶ **FOR MORE PRACTICE** Example 5.25; Problems 71, 72.

CONCEPTUAL ✔ **CHECKPOINT 5.8**

The compound NCl_3 is named nitrogen trichloride, while $AlCl_3$ is simply aluminum chloride. Why the difference?

5.9 Naming Acids

▶ Name binary acids.
▶ Name oxyacids containing an oxyanion ending in *-ate*.
▶ Name oxyacids containing an oxyanion ending in *-ite*.

HCl(*g*) refers to HCl molecules in the gas state.

Acids are molecular compounds that produce H^+ ions when dissolved in water. They are composed of hydrogen, which we usually write first in their formula, and one or more nonmetals, which we usually write second. Acids are characterized by their sour taste and their ability to dissolve some metals. For example, HCl(*aq*) is an acid—the (*aq*) indicates that the compound is "aqueous" or "dissolved in water." HCl(*aq*) has a characteristically sour taste. Since HCl(*aq*) is present in stomach fluids, its sour taste becomes painfully obvious during vomiting. HCl(*aq*) also dissolves some metals. If we drop a strip of zinc into a beaker of HCl(*aq*), it will slowly disappear as the acid converts the zinc metal into dissolved Zn^{2+} cations.

Acids are present in many foods, such as lemons and limes, and they are used in some household products such as toilet bowl cleaners and Lime-A-Way. In this section, we only learn how to name them, but in Chapter 14 we will learn more about the properties of acids. We categorize acids into two groups: **binary acids**, those containing only hydrogen and a nonmetal, and oxyacids, those containing hydrogen, a nonmetal, and oxygen (▼ **FIGURE 5.15**).

▶ **FIGURE 5.15 Classification of acids** We classify acids into two types, depending on the number of elements in the acid. If the acid contains only two elements, it is a binary acid. If it contains oxygen, it is an oxyacid.

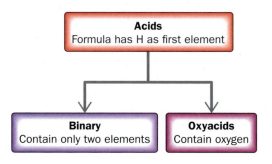

Naming Binary Acids

Binary acids are composed of hydrogen and a nonmetal. The names for binary acids have the following form:

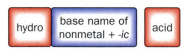

For example, HCl(*aq*) is hydro*chlor*ic acid and HBr(*aq*) is hydro*brom*ic acid.

| HCl(*aq*) | hydrochloric acid | HBr(*aq*) | hydrobromic acid |

EXAMPLE **5.12**	**Naming Binary Acids**
Give the name of H_2S(*aq*).	
The base name of S is *sulfur*, so the name is *hydrosulfuric acid*.	**SOLUTION** H_2S(*aq*) hydrosulfuric acid

▶ **SKILLBUILDER 5.12** | **Naming Binary Acids**
Name HF(*aq*).

▶ **FOR MORE PRACTICE** Example 5.26; Problems 77b, 78d.

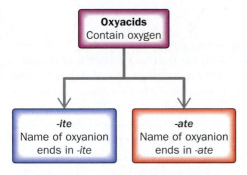

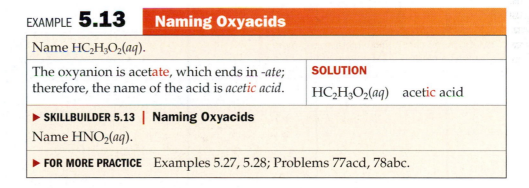

▲ **FIGURE 5.16 Classification of oxyacids**
Oxyacids are classified into two types, depending on the endings of the oxyanions that they contain.

You can use the saying, "*ic* I *ate* an acid" to remember the association of *-ic* with *-ate*.

Naming Oxyacids

Oxyacids are acids that contain oxyanions, which are listed in the table of polyatomic ions (Table 5.3). For example, $HNO_3(aq)$ contains the nitrate (NO_3^-) ion, $H_2SO_3(aq)$ contains the sulfite (SO_3^{2-}) ion, and $H_2SO_4(aq)$ contains the sulfate (SO_4^{2-}) ion. All of these acids are a combination of one or more H^+ ions with an oxyanion. The number of H^+ ions depends on the charge of the oxyanion, so that the formula is always charge-neutral. The names of oxyacids depend on the ending of the oxyanion (◀ **FIGURE 5.16**).

The names of acids containing oxyanions ending with *-ite* take this form:

The names of acids containing oxyanions ending with *-ate* take this form:

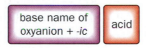

So H_2SO_3 is *sulfurous* acid (oxyanion is sulfite), and HNO_3 is *nitric acid* (oxyanion is nitrate).

$H_2SO_3(aq)$ sulfurous acid $HNO_3(aq)$ nitric acid

Table 5.6 lists some common oxyacids and their oxyanions.

EXAMPLE **5.13** | **Naming Oxyacids**

Name $HC_2H_3O_2(aq)$.

The oxyanion is acetate, which ends in *-ate*; therefore, the name of the acid is *acetic acid*.

SOLUTION

$HC_2H_3O_2(aq)$ acetic acid

▶ **SKILLBUILDER 5.13 | Naming Oxyacids**
Name $HNO_2(aq)$.

▶ **FOR MORE PRACTICE**   Examples 5.27, 5.28; Problems 77acd, 78abc.

TABLE 5.6   Names of Some Common Oxyacids and Their Oxyanions

Acid Formula	Acid Name	Oxyanion Name	Oxyanion Formula
HNO_2	nitrous acid	nitrite	NO_2^-
HNO_3	nitric acid	nitrate	NO_3^-
H_2SO_3	sulfurous acid	sulfite	SO_3^{2-}
H_2SO_4	sulfuric acid	sulfate	SO_4^{2-}
$HClO_2$	chlorous acid	chlorite	ClO_2^-
$HClO_3$	chloric acid	chlorate	ClO_3^-
$HC_2H_3O_2$	acetic acid	acetate	$C_2H_3O_2^-$
H_2CO_3	carbonic acid	carbonate	CO_3^{2-}

5.10 Nomenclature Summary

▶ Recognize and name chemical compounds.

> Acids are technically a subclass of molecular compounds; that is, they are molecular compounds that form H^+ ions when dissolved in water.

Naming compounds requires several steps. The flowchart in ▼ FIGURE 5.17 summarizes the different categories of compounds that we have covered in the chapter and how to identify and name them. The first step is to decide whether the compound is ionic, molecular, or an acid. We can recognize ionic compounds by the presence of a metal and a nonmetal, molecular compounds by two or more nonmetals, and acids by the presence of hydrogen (written first) and one or more nonmetals.

Ionic Compounds

For an ionic compound, we must next decide whether the metal forms only one type of ion or more than one type of ion. Group 1A (alkali) metals, Group 2A (alkaline earth) metals, and aluminum always form only one type of ion (Figure 5.13). Most of the transition metals (except Zn, Sc, and Ag) form more than one type of ion. Once we have identified the type of ionic compound, we name it according to the scheme in the chart. If the ionic compound contains a polyatomic ion—something we must learn to recognize by familiarity—we insert the name of the polyatomic ion in place of the metal (positive polyatomic ion) or the nonmetal (negative polyatomic ion).

Molecular Compounds

We have learned how to name only one type of molecular compound, the binary (two-element) compound. If we identify a compound as molecular, we name it according to the scheme in Figure 5.17.

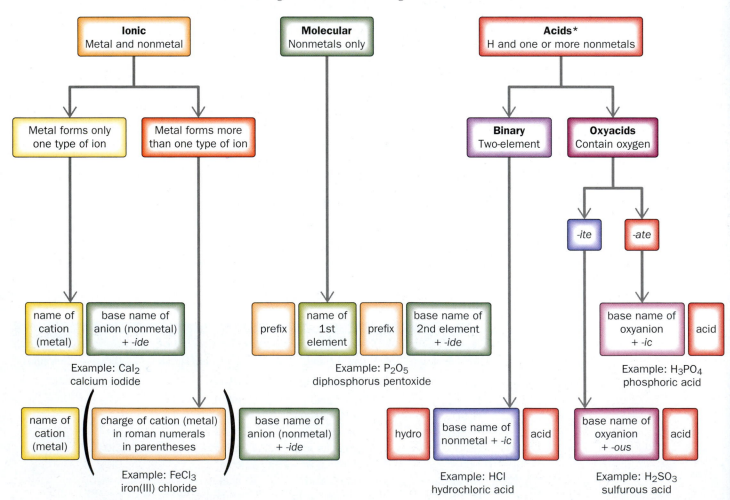

▲ **FIGURE 5.17 Nomenclature flowchart**

Acids

To name an acid, we must first decide whether it is a binary (two-element) acid or an oxyacid (an acid containing oxygen). We name binary acids according to the scheme in Figure 5.17. We must further subdivide oxyacids based on the name of their corresponding oxyanion. If the oxyanion ends in *-ite*, we use one scheme; if it ends with *-ate*, we use the other.

EXAMPLE 5.14   Nomenclature Using Figure 5.17

Name each compound: CO, CaF_2, HF(*aq*), $Fe(NO_3)_3$, $HClO_4$(*aq*), H_2SO_3(*aq*).

SOLUTION

The table illustrates how to use Figure 5.17 to arrive at a name for each compound.

Formula	Flowchart Path	Name
CO	molecular	carbon monoxide
CaF_2	ionic ⟶ one type of ion ⟶	calcium fluoride
HF(*aq*)	acid ⟶ binary ⟶	hydrofluoric acid
$Fe(NO_3)_3$	ionic ⟶ more than one type of ion ⟶	iron(III) nitrate
$HClO_4$(*aq*)	acid ⟶ oxyacid ⟶ *-ate* ⟶	perchloric acid
H_2SO_3(*aq*)	acid ⟶ oxyacid ⟶ *-ite* ⟶	sulfurous acid

▶ **FOR MORE PRACTICE**   Problems 93, 94.

5.11 Formula Mass: The Mass of a Molecule or Formula Unit

▶ Calculate formula mass.

The terms *molecular mass* and *molecular weight*, which are also commonly used, have the same meaning as formula mass.

In Chapter 4, we discussed atoms and elements and defined the average mass of the atoms that compose an element as the *atomic mass* for that element. Similarly, in this chapter, which introduces molecules and compounds, we designate the average mass of the molecules (or formula units) that compose a compound as the **formula mass**.

For any compound, the *formula mass* is the sum of the atomic masses of all the atoms in its chemical formula:

$$\text{formula mass} = \left(\begin{array}{c} \text{\# atoms of 1st} \\ \text{element in} \\ \text{chemical formula} \end{array} \times \begin{array}{c} \text{atomic mass} \\ \text{of} \\ \text{1st element} \end{array} \right) + \left(\begin{array}{c} \text{\# atoms of 2nd} \\ \text{element in} \\ \text{chemical formula} \end{array} \times \begin{array}{c} \text{atomic mass} \\ \text{of} \\ \text{2nd element} \end{array} \right) + \ldots$$

Like atomic mass for atoms, formula mass characterizes the average mass of a molecule or formula unit. For example, the formula mass of water, H_2O, is:

$$\text{formula mass} = 2(1.01 \text{ amu}) + 16.00 \text{ amu}$$
$$= 18.02 \text{ amu}$$

and that of sodium chloride, NaCl, is:

$$\text{formula mass} = 22.99 \text{ amu} + 35.45 \text{ amu}$$
$$= 58.44 \text{ amu}$$

In addition to giving a characteristic mass to the molecules or formula units of a compound, formula mass—as we will discuss in Chapter 6—allows us to quantify the number of molecules or formula units in a sample of a given mass.

EXAMPLE **5.15** | **Calculating Formula Mass**

Calculate the formula mass of carbon tetrachloride, CCl_4.

SOLUTION

To find the formula mass, sum the atomic masses of each atom in the chemical formula.

$$\text{formula mass} = 1 \times (\text{atomic mass C}) + 4 \times (\text{atomic mass Cl})$$
$$= 12.01 \text{ amu} + 4(35.45 \text{ amu})$$
$$= 12.01 \text{ amu} + 141.80 \text{ amu}$$
$$= 153.8 \text{ amu}$$

▶ **SKILLBUILDER 5.15** | **Calculating Formula Masses**

Calculate the formula mass of dinitrogen monoxide, N_2O, also called laughing gas.

▶ **FOR MORE PRACTICE** Example 5.29; Problems 83, 84.

CONCEPTUAL ✓ CHECKPOINT 5.9

Which substance has the greatest formula mass?

(a) O_2 (b) O_3 (c) H_2O (d) H_2O_2

Chapter 5 in Review

MasteringChemistry™ provides end-of-chapter exercises, feedback-enriched tutorial problems, animations, and interactive activities to encourage problem solving practice and deeper understanding of key concepts and topics.

Self-Assessment Quiz

Q1. Carbon tetrachloride has a chlorine-to-carbon mass ratio of 11.8:1. If a sample of carbon tetrachloride contains 35 g of chlorine, what mass of carbon does it contain?
(a) 0.34 g C
(b) 1.0 g C
(c) 3.0 g C
(d) 11.8 g C

Q2. Write a chemical formula for a compound that contains two chlorine atoms to every one oxygen atom.
(a) Cl_2O
(b) ClO_2
(c) $2ClO$
(d) $Cl(O_2)_2$

Q3. How many oxygen atoms are in the chemical formula $Fe_2(SO_4)_3$?
(a) 2
(b) 3
(c) 4
(d) 12

Q4. Which element is a molecular element?
(a) copper
(b) iodine
(c) krypton
(d) potassium

Q5. Which compound is ionic?
(a) BrF_5
(b) HNO_3
(c) $MgSO_4$
(d) NI_3

Q6. Write a formula for the compound that forms between Sr and Br.
(a) SrBr
(b) Sr_2Br
(c) $SrBr_2$
(d) Sr_2Br_2

Q7. Write a formula for the compound that forms between sodium and chlorite ions.
(a) $NaClO_2$
(b) Na_2ClO_2
(c) $Na(ClO)_2$
(d) $NaClO_3$

Q8. Name the compound Li_3N.
(a) trilithium mononitride
(b) trilithium nitride
(c) lithium(I) nitride
(d) lithium nitride

Q9. Name the compound $CrCl_3$.
(a) monochromium trichloride
(b) chromium trichloride
(c) chromium chloride
(d) chromium(III) chloride

Q10. Name the compound $BaSO_4$.
(a) barium sulfate
(b) barium(II) sulfate
(c) barium monosulfur tetraoxygen
(d) barium tetrasulfate

Q11. Name the compound PF_5.
(a) monophosphorus pentafluoride
(b) phosphorus pentafluoride
(c) phosphorus fluoride
(d) phosphorus(III) fluoride

Q12. What is the formula for manganese(III) oxide?
(a) MnO
(b) Mn_3O
(c) Mn_2O_3
(d) MnO_3

Q13. Name the acid $H_3PO_4(aq)$.
(a) hydrogen phosphate
(b) phosphoric acid
(c) phosphorus acid
(d) hydrophosphic acid

Q14. What is the formula for hydrobromic acid?
(a) HBr
(b) HBrO
(c) $HBrO_2$
(d) $HBrO_3$

Q15. Determine the formula mass of CCl_2F_2.
(a) 66.46 amu
(b) 108.9 amu
(c) 132.92 amu
(d) 120.91 amu

Answers: 1:c, 2:a, 3:d, 4:b, 5:c, 6:c, 7:a, 8:d, 9:d, 10:a, 11:b, 12:c, 13:b, 14:a, 15:d

Chemical Principles

Relevance

Compounds

Matter is ultimately composed of atoms, and those atoms are often combined in compounds. The most important characteristic of a compound is its constant composition. The elements that make up a particular compound are in fixed, definite proportions in all samples of the compound.

Most of the matter you encounter is in the form of compounds. Water, salt, and carbon dioxide are all examples of common simple compounds. More complex compounds include caffeine, aspirin, acetone, and testosterone.

Chemical Formulas

Chemical formulas represent compounds. Formulas indicate the elements present in the compound and the relative number of atoms of each. These formulas represent the basic units that make up a compound. Pure substances can be categorized according to the basic units that compose them. Elements can be composed of atoms or molecules. Compounds can be molecular, in which case their basic units are molecules, or ionic, in which case their basic units are formula units (composed of cations and anions). We can write the formulas for many ionic compounds simply by knowing the elements in the compound.

To understand compounds, you must understand their composition, which is represented by a chemical formula. The connection between the molecular world and the macroscopic world hinges on the particles that compose matter. Since most matter is in the form of compounds, the properties of most matter depend on the molecules or ions that compose it. Molecular matter does what its molecules do; ionic matter does what its ions do. The world you see and experience is governed by what these particles are doing.

Chemical Nomenclature

We can write the names of simple ionic compounds, molecular compounds, and acids, by examining their chemical formulas. The nomenclature flowchart (Figure 5.17) shows the basic procedure for determining these names.

Because there are so many compounds, there must be a systematic way to name them. By learning these few simple rules, you will be able to name thousands of different compounds. The next time you look at the label on a consumer product, try to identify as many of the compounds as you can by examining their names.

Formula Mass

The formula mass of a compound is the sum of the atomic masses of all the atoms in the chemical formula for the compound. Like atomic mass for elements, formula mass characterizes the average mass of a molecule or formula unit.

Besides being the characteristic mass of a molecule or formula unit, formula mass is important in many calculations involving the composition of compounds and quantities in chemical reactions.

Chemical Skills

Examples

LO: Restate and apply the law of constant composition (Section 5.2).

The law of constant composition states that all samples of a given compound have the same ratio of their constituent elements.

To determine whether experimental data are consistent with the law of constant composition, calculate the ratios of the masses of each element in all samples. When calculating these ratios, it is most convenient to put the larger number in the numerator (top) and the smaller one in the denominator (bottom); that way, the ratio is greater than 1. If the ratios are the same, then the data are consistent with the law of constant composition.

EXAMPLE 5.16 Constant Composition of Compounds

Two samples said to be carbon disulfide (CS_2) are decomposed into their constituent elements. One sample produces 8.08 g S and 1.51 g C, while the other produces 31.3 g S and 3.85 g C. Are these results consistent with the law of constant composition?

SOLUTION

Sample 1

$$\frac{\text{mass S}}{\text{mass C}} = \frac{8.08 \text{ g}}{1.51 \text{ g}} = 5.35$$

Sample 2

$$\frac{\text{mass S}}{\text{mass C}} = \frac{31.3 \text{ g}}{3.85 \text{ g}} = 8.13$$

These results are not consistent with the law of constant composition, so the information that the two samples are the same substance must therefore be in error.

LO: Write chemical formulas (Section 5.3).

Chemical formulas indicate the elements present in a compound and the relative number of atoms of each. When writing formulas, put the more metallic element first.

EXAMPLE 5.17 Writing Chemical Formulas

Write a chemical formula for the compound containing one nitrogen atom for every two oxygen atoms.

SOLUTION

NO_2

LO: Determine the total number of each type of atom in a chemical formula (Section 5.3).

The numbers of atoms not enclosed in parentheses are given directly by their subscript.

Find the numbers of atoms within parentheses by multiplying their subscript within the parentheses by their subscript outside the parentheses.

EXAMPLE 5.18 Determining the Total Number of Each Type of Atom in a Chemical Formula

Determine the number of each type of atom in $Pb(ClO_3)_2$.

SOLUTION

one Pb atom

two Cl atoms

six O atoms

LO: Classify elements as atomic or molecular (Section 5.4).

Most elements exist as atomic elements; their basic units in nature are individual atoms. However, several elements (H_2, N_2, O_2, F_2, Cl_2, Br_2, and I_2) exist as molecular elements; their basic units in nature are diatomic molecules.

EXAMPLE 5.19 Classifying Elements as Atomic or Molecular

Classify each element as atomic or molecular: sodium, iodine, and nitrogen.

SOLUTION

sodium: atomic

iodine: molecular (I_2)

nitrogen: molecular (N_2)

LO: Classify compounds as ionic or molecular (Section 5.4).

Compounds containing a metal and a nonmetal are ionic. If the metal is a transition metal, it is likely to form more than one type of ion (see exceptions in Figure 5.13). If the metal is not a transition metal, it is likely to form only one type of ion (see exceptions in Table 5.4).

Compounds composed of only nonmetals are molecular.

EXAMPLE 5.20 Classifying Compounds as Ionic or Molecular

Classify each compound as ionic or molecular. If they are ionic, determine whether the metal forms only one type of ion or more than one type of ion.

$$FeCl_3, K_2SO_4, CCl_4$$

SOLUTION

$FeCl_3$: ionic, metal forms more than one type of ion

K_2SO_4: ionic, metal forms only one type of ion

CCl_4: molecular

LO: Write formulas for ionic compounds (Section 5.5).

1. Write the symbol for the metal ion followed by the symbol for the nonmetal ion (or polyatomic ion) and their charges. These charges can be deduced from the group numbers in the periodic table. (In the case of polyatomic ions, the charges come from Table 5.3.)

2. Use the magnitude of the charge on each ion as the subscript for the other ion.

3. Check to see if you can reduce the subscripts to simpler whole numbers. Drop subscripts of 1; they are implied.

4. Confirm that the sum of the charges of the cations exactly cancels the sum of the charges of the anions.

EXAMPLE 5.21 Writing Formulas for Ionic Compounds

Write a formula for the compound that forms from lithium and sulfate ions.

SOLUTION

$$Li^+ \quad SO_4^{2-}$$

$$Li_2(SO_4)$$

In this case, the subscripts cannot be further reduced.

$$Li_2SO_4$$

Cations	Anions
$2(1+) = 2+$	$2-$

LO: Name binary ionic compounds containing a metal that forms only one type of ion (Section 5.7).

The name of the metal is unchanged. The name of the nonmetal is its base name with the ending *-ide*.

EXAMPLE 5.22 Naming Binary Ionic Compounds Containing a Metal That Forms Only One Type of Ion

Name the compound Al_2O_3.

SOLUTION

aluminum oxide

LO: Name binary ionic compounds containing a metal that forms more than one type of ion (Section 5.7).

Because the names of these compounds include the charge of the metal ion, first determine that charge by calculating the total charge of the nonmetal ions.

The total charge of the metal ions must equal the total charge of the nonmetal ions, but have the opposite sign.

The name of the compound is the name of the metal ion, followed by the charge of the metal ion, followed by the base name of the nonmetal + *-ide*.

EXAMPLE 5.23 Naming Binary Ionic Compounds Containing a Metal That Forms More Than One Type of Ion

Name the compound Fe_2S_3.

SOLUTION

3 sulfide ions $\times$ $(2-) = 6-$

2 iron ions $\times$ (*ion charge*) $= 6+$

ion charge $= 3+$

charge of each iron ion $= 3+$
iron(III) sulfide

LO: Name compounds containing a polyatomic ion (Section 5.7).

Name ionic compounds containing a polyatomic ion in the normal way, except substitute the name of the polyatomic ion (from Table 5.3) in place of the nonmetal.

Because the metal in this example forms more than one type of ion, you need to determine the charge on the metal ion. The charge of the metal ion must be equal in magnitude to the sum of the charges of the polyatomic ions but opposite in sign.

The name of the compound is the name of the metal ion, followed by the charge of the metal ion, followed by the name of the polyatomic ion.

EXAMPLE **5.24** **Naming Compounds Containing a Polyatomic Ion**

Name the compound $Co(ClO_4)_2$.

SOLUTION

2 perchlorate ions $\times$ (1−) = 2−

charge of cobalt ion = 2+

cobalt(II) perchlorate

LO: Name molecular compounds (Section 5.8).

The name consists of a prefix indicating the number of atoms of the first element, followed by the name of the first element, and a prefix for the number of atoms of the second element, followed by the base name of the second element plus the suffix -*ide*. The prefix -*mono* is normally dropped on the first element.

EXAMPLE **5.25** **Naming Molecular Compounds**

Name the compound NO_2.

SOLUTION

nitrogen dioxide

LO: Name binary acids (Section 5.9).

The name begins with *hydro-*, followed by the base name of the nonmetal, plus the suffix -*ic*, and the word *acid*.

EXAMPLE **5.26** **Naming Binary Acids**

Name the acid $HI(aq)$.

SOLUTION

hydroiodic acid

LO: Name oxyacids containing an oxyanion ending in –*ate* (Section 5.9).

The name is the base name of the oxyanion + -*ic*, followed by the word *acid* (sulfate violates the rule somewhat; in strict terms, the base name would be *sulf*).

EXAMPLE **5.27** **Naming Oxyacids Containing an Oxyanion Ending in -*ate***

Name the acid $H_2SO_4(aq)$.

SOLUTION

The oxyanion is sulfate. The name of the acid is *sulfuric acid*.

LO: Name oxyacids containing an oxyanion ending in –*ite* (Section 5.9).

The name is the base name of the oxyanion + -*ous*, followed by the word *acid*.

EXAMPLE **5.28** **Naming Oxyacids Containing an Oxyanion Ending in -*ite***

Name the acid $HClO_2(aq)$.

SOLUTION

The oxyanion is chlorite. The name of the acid is *chlorous acid*.

LO: Calculate formula mass (Section 5.11).

The formula mass is the sum of the atomic masses of all the atoms in the chemical formula. In determining the number of each type of atom, multiply subscripts inside parentheses by subscripts outside parentheses.

EXAMPLE **5.29** **Calculating Formula Mass**

Calculate the formula mass of $Mg(NO_3)_2$.

SOLUTION

$$formula\ mass = 24.31 + 2(14.01) + 6(16.00)$$
$$= 148.33\ amu$$

Key Terms

acid [5.9]	empirical formula [5.3]	molecular compound [5.4]	polyatomic ion [5.3]
atomic element [5.4]	formula mass [5.11]	molecular element [5.4]	space-filling model [5.3]
ball-and-stick model [5.3]	formula unit [5.4]	molecular formula [5.3]	structural formula [5.3]
binary acid [5.9]	ionic compound [5.4]	molecular model [5.3]	transition metals [5.7]
binary compound [5.7]	law of constant	oxyacid [5.9]	
chemical formula [5.3]	composition [5.2]	oxyanion [5.7]	

Exercises

Questions

1. Do the properties of an element change when it combines with another element to form a compound? Explain.
2. How might the world be different if elements did not combine to form compounds?
3. What is the law of constant composition? Who discovered it?
4. What is a chemical formula? List some examples.
5. In a chemical formula, which element is listed first?
6. In a chemical formula, how do you calculate the number of atoms of an element within parentheses? Provide an example.
7. Explain the difference between a molecular formula and an empirical formula.
8. What is a structural formula? What is the difference between a structural formula and a molecular model?
9. What is the difference between a molecular element and an atomic element? List the elements that occur as diatomic molecules.
10. What is the difference between an ionic compound and a molecular compound?
11. What is the difference between a common name for a compound and a systematic name?
12. List the metals that form only one type of ion (that is, metals whose charge is invariant from one compound to another). What are the group numbers of these metals?
13. Identify the block in the periodic table of metals that tend to form more than one type of ion.
14. What is the basic form for the names of ionic compounds containing a metal that forms only one type of ion?
15. What is the basic form for the names of ionic compounds containing a metal that forms more than one type of ion?
16. Why are roman numerals needed in the names of ionic compounds containing a metal that forms more than one type of ion?
17. How are compounds containing a polyatomic ion named?
18. Which polyatomic ions have a 2− charge? Which polyatomic ions have a 3− charge?
19. What is the basic form for the names of molecular compounds?
20. How many atoms does each prefix specify? *mono-, di-, tri-, tetra-, penta-, hexa-*.
21. What is the basic form for the names of binary acids?
22. What is the basic form for the name of oxyacids whose oxyanions end with *-ate*?
23. What is the basic form for the name of oxyacids whose oxyanions end with *-ite*?
24. What is the formula mass of a compound?

Problems

CONSTANT COMPOSITION OF COMPOUNDS

25. Two samples of sodium chloride are decomposed into their constituent elements. One sample produces 4.65 g of sodium and 7.16 g of chlorine, and the other sample produces 7.45 g of sodium and 11.5 g of chlorine. Are these results consistent with the law of constant composition? Explain your answer.

26. Two samples of carbon tetrachloride are decomposed into their constituent elements. One sample produces 32.4 g of carbon and 373 g of chlorine, and the other sample produces 12.3 g of carbon and 112 g of chlorine. Are these results consistent with the law of constant composition? Explain your answer.

27. Upon decomposition, one sample of magnesium fluoride produced 1.65 kg of magnesium and 2.57 kg of fluorine. A second sample produced 1.32 kg of magnesium. How much fluorine (in grams) did the second sample produce? Remember that, according to the law of constant composition, the ratio of the masses of the two elements must be the same in both samples.

28. The mass ratio of sodium to fluorine in sodium fluoride is 1.21:1. A sample of sodium fluoride produces 34.5 g of sodium upon decomposition. How much fluorine (in grams) forms? *Hint:* the ratio $\dfrac{\text{mass sodium}}{\text{mass fluorine}} = 1.21$.

29. Use the law of constant composition to complete the table summarizing the amounts of nitrogen and oxygen produced upon the decomposition of several samples of dinitrogen monoxide. Remember that, according to the law of constant composition, the ratio of the masses of the two elements $\left(\dfrac{\text{mass nitrogen}}{\text{mass oxygen}}\right)$ must be the same in all samples.

	Mass N_2O	Mass N	Mass O
Sample A	2.85 g	1.82 g	1.03 g
Sample B	4.55 g	_____	_____
Sample C	_____	_____	1.35 g
Sample D	_____	1.11 g	_____

30. Use the law of constant composition to complete the table summarizing the amounts of iron and chlorine produced upon the decomposition of several samples of iron(III) chloride. Remember that, according to the law of constant composition, the ratio of the masses of the two elements $\left(\dfrac{\text{mass chlorine}}{\text{mass iron}}\right)$ must be the same in all samples.

	Mass $FeCl_3$	Mass Fe	Mass Cl
Sample A	3.785 g	1.302 g	2.483 g
Sample B	2.175 g	_____	_____
Sample C	_____	2.012 g	_____
Sample D	_____	_____	2.329 g

CHEMICAL FORMULAS

31. Write a chemical formula for the compound containing one nitrogen atom for every three iodine atoms.

32. Write a chemical formula for the compound containing one carbon atom for every four bromine atoms.

33. Write chemical formulas for compounds containing:
(a) three iron atoms for every four oxygen atoms
(b) one phosphorus atom for every three chlorine atoms
(c) one phosphorus atom for every five chlorine atoms
(d) two silver atoms for every oxygen atom

34. Write chemical formulas for compounds containing:
(a) one calcium atom for every two iodine atoms
(b) two nitrogen atoms for every four oxygen atoms
(c) one silicon atom for every two oxygen atoms
(d) one zinc atom for every two chlorine atoms

35. How many oxygen atoms are in each chemical formula?
(a) H_3PO_4 (b) Na_2HPO_4
(c) $Ca(HCO_3)_2$ (d) $Ba(C_2H_3O_2)_2$

36. How many hydrogen atoms are in each of the formulas in Question 35?

37. Determine the number of each type of atom in each formula.
(a) $MgCl_2$ (b) $NaNO_3$
(c) $Ca(NO_2)_2$ (d) $Sr(OH)_2$

38. Determine the number of each type of atom in each formula.
(a) NH_4Cl (b) $Mg_3(PO_4)_2$
(c) $NaCN$ (d) $Ba(HCO_3)_2$

39. Complete the table.

Formula	Number of $C_2H_3O_2^-$ Units	Number of Carbon Atoms	Number of Hydrogen Atoms	Number of Oxygen Atoms	Number of Metal Atoms
$Mg(C_2H_3O_2)_2$	____	____	____	____	____
$NaC_2H_3O_2$	____	____	____	____	____
$Cr_2(C_2H_3O_2)_4$	____	____	____	____	____

40. Complete the table.

Formula	Number of SO_4^{2-} Units	Number of Sulfur Atoms	Number of Oxygen Atoms	Number of Metal Atoms
$CaSO_4$	____	____	____	____
$Al_2(SO_4)_3$	____	____	____	____
K_2SO_4	____	____	____	____

41. Give the empirical formula that corresponds to each molecular formula.
(a) C_2H_6 (b) N_2O_4 (c) $C_4H_6O_2$ (d) NH_3

42. Give the empirical formula that corresponds to each molecular formula.
(a) C_2H_2 (b) CO_2 (c) $C_6H_{12}O_6$ (d) B_2H_6

MOLECULAR VIEW OF ELEMENTS AND COMPOUNDS

43. Classify each element as atomic or molecular.
(a) chlorine (b) argon
(c) cobalt (d) hydrogen

44. Which elements have molecules as their basic units?
(a) helium (b) oxygen
(c) iron (d) bromine

45. Classify each compound as ionic or molecular.
(a) CS_2 (b) CuO (c) KI (d) PCl_3

46. Classify each compound as ionic or molecular.
(a) PtO_2 (b) CF_2Cl_2 (c) CO (d) SO_3

47. Match the substances on the left with the basic units that compose them on the right. Remember that atomic elements are composed of atoms, molecular elements are composed of diatomic molecules, molecular compounds are composed of molecules, and ionic compounds are composed of formula units.

helium	molecules
CCl_4	formula units
K_2SO_4	diatomic molecules
bromine	single atoms

48. Match the substances on the left with the basic units that compose them on the right. Remember that atomic elements are composed of atoms, molecular elements are composed of diatomic molecules, molecular compounds are composed of molecules, and ionic compounds are composed of formula units.

NI_3	molecules
copper metal	single atoms
$SrCl_2$	diatomic molecules
nitrogen	formula units

49. What are the basic units—single atoms, molecules, or formula units—that compose each substance?
(a) $BaBr_2$ (b) Ne (c) I_2 (d) CO

50. What are the basic units—single atoms, molecules, or formula units—that compose each substance?
(a) Rb_2O (b) N_2 (c) $Fe(NO_3)_2$ (d) N_2F_4

51. Classify each compound as ionic or molecular. If it is ionic, determine whether the metal forms only one type of ion or more than one type of ion.
  (a) KCl   (b) CBr_4 (c) NO_2 (d) $Sn(SO_4)_2$

52. Classify each compound as ionic or molecular. If it is ionic, determine whether the metal forms only one type of ion or more than one type of ion.
  (a) $CoCl_2$ (b) CF_4 (c) $BaSO_4$ (d) NO

WRITING FORMULAS FOR IONIC COMPOUNDS

53. Write a formula for the ionic compound that forms from each pair of elements.
  (a) sodium and sulfur
  (b) strontium and oxygen
  (c) aluminum and sulfur
  (d) magnesium and chlorine

54. Write a formula for the ionic compound that forms from each pair of elements.
  (a) aluminum and oxygen
  (b) beryllium and iodine
  (c) calcium and sulfur
  (d) calcium and iodine

55. Write a formula for the compound that forms from potassium and
  (a) acetate   (b) chromate
  (c) phosphate   (d) cyanide

56. Write a formula for the compound that forms from calcium and
  (a) hydroxide   (b) carbonate
  (c) phosphate   (d) hydrogen phosphate

57. Write formulas for the compounds formed from the element on the left and each of the elements on the right.
  (a) Li   N, O, F
  (b) Ba   N, O, F
  (c) Al   N, O, F

58. Write formulas for the compounds formed from the element on the left and each polyatomic ion on the right.
  (a) Rb NO_3^-, SO_4^{2-}, PO_4^{3-}
  (b) Sr NO_3^-, SO_4^{2-}, PO_4^{3-}
  (c) In NO_3^-, SO_4^{2-}, PO_4^{3-}
  (Assume In charge is 3+.)

NAMING IONIC COMPOUNDS

59. Name each ionic compound. In each of these compounds, the metal forms only one type of ion.
  (a) CsCl   (b) $SrBr_2$ (c) K_2O (d) LiF

60. Name each ionic compound. In each of these compounds, the metal forms only one type of ion.
  (a) LiI   (b) MgS   (c) BaF_2 (d) NaF

61. Name each ionic compound. In each of these compounds, the metal forms more than one type of ion.
  (a) $CrCl_2$ (b) $CrCl_3$ (c) SnO_2 (d) PbI_2

62. Name each ionic compound. In each of these compounds, the metal forms more than one type of ion.
  (a) $HgBr_2$ (b) Fe_2O_3 (c) CuI_2 (d) $SnCl_4$

63. Determine whether the metal in each ionic compound forms only one type of ion or more than one type of ion and name the compound accordingly.
  (a) Cr_2O_3 (b) NaI   (c) $CaBr_2$ (d) SnO

64. Determine whether the metal in each ionic compound forms only one type of ion or more than one type of ion and name the compound accordingly.
  (a) FeI_3 (b) $PbCl_4$ (c) SrI_2 (d) BaO

65. Name each ionic compound containing a polyatomic ion.
  (a) $Ba(NO_3)_2$ (b) $Pb(C_2H_3O_2)_2$
  (c) NH_4I (d) $KClO_3$
  (e) $CoSO_4$ (f) $NaClO_4$

66. Name each ionic compound containing a polyatomic ion.
  (a) $Ba(OH)_2$ (b) $Fe(OH)_3$
  (c) $Cu(NO_2)_2$ (d) $PbSO_4$
  (e) KClO   (f) $Mg(C_2H_3O_2)_2$

67. Name each polyatomic ion.
  (a) BrO^- (b) BrO_2^- (c) BrO_3^- (d) BrO_4^-

68. Name each polyatomic ion.
  (a) IO^- (b) IO_2^- (c) IO_3^- (d) IO_4^-

69. Write a formula for each ionic compound.
  (a) copper(II) bromide
  (b) silver nitrate
  (c) potassium hydroxide
  (d) sodium sulfate
  (e) potassium hydrogen sulfate
  (f) sodium hydrogen carbonate

70. Write a formula for each ionic compound.
  (a) copper(I) chlorate
  (b) potassium permanganate
  (c) lead(II) chromate
  (d) calcium fluoride
  (e) iron(II) phosphate
  (f) lithium hydrogen sulfite

NAMING MOLECULAR COMPOUNDS

71. Name each molecular compound.
  (a) SO_2 (b) NI_3 (c) BrF_5
  (d) NO (e) N_4Se_4

72. Name each molecular compound.
  (a) XeF_4 (b) PI_3 (c) SO_3
  (d) $SiCl_4$ (e) I_2O_5

73. Write a formula for each molecular compound.
  (a) carbon monoxide
  (b) disulfur tetrafluoride
  (c) dichlorine monoxide
  (d) phosphorus pentafluoride
  (e) boron tribromide
  (f) diphosphorus pentasulfide

74. Write a formula for each molecular compound.
  (a) chlorine monoxide
  (b) xenon tetroxide
  (c) xenon hexafluoride
  (d) carbon tetrabromide
  (e) diboron tetrachloride
  (f) tetraphosphorus triselenide

75. Determine whether the name shown for each molecular compound is correct. If not, provide the compound's correct name.
  (a) PBr_5 phosphorus(V) pentabromide
  (b) P_2O_3 phosphorus trioxide
  (c) SF_4 monosulfur hexafluoride
  (d) NF_3 nitrogen trifluoride

76. Determine whether the name shown for each molecular compound is correct. If not, provide the compound's correct name.
  (a) NCl_3 nitrogen chloride
  (b) CI_4 carbon(IV) iodide
  (c) CO carbon oxide
  (d) SCl_4 sulfur tetrachloride

NAMING ACIDS

77. Determine whether each acid is a binary acid or an oxyacid and name each acid. If the acid is an oxyacid, provide the name of the oxyanion.
  (a) $HNO_2(aq)$ (b) $HI(aq)$
  (c) $H_2SO_4(aq)$ (d) $HNO_3(aq)$

78. Determine whether each acid is a binary acid or an oxyacid and name each acid. If the acid is an oxyacid, provide the name of the oxyanion.
  (a) $H_2CO_3(aq)$ (b) $HC_2H_3O_2(aq)$
  (c) $H_3PO_4(aq)$ (d) $HCl(aq)$

79. Name each acid.
  (a) $HClO$ (b) $HClO_2$ (c) $HClO_3$ (d) $HClO_4$

80. Name each acid. (*Hint:* The names of the oxyanions are analogous to the names of the oxyanions of chlorine.)
  (a) $HBrO_3$ (b) HIO_3

81. Write a formula for each acid.
  (a) phosphoric acid   (b) hydrobromic acid
  (c) sulfurous acid

82. Write a formula for each acid.
  (a) hydrofluoric acid   (b) hydrocyanic acid
  (c) chlorous acid

FORMULA MASS

83. Calculate the formula mass for each compound.
  (a) HNO_3 (b) $CaBr_2$ (c) CCl_4 (d) $Sr(NO_3)_2$

84. Calculate the formula mass for each compound.
  (a) CS_2 (b) $C_6H_{12}O_6$ (c) $Fe(NO_3)_3$ (d) C_7H_{16}

85. Arrange the compounds in order of decreasing formula mass.
  $$Ag_2O, PtO_2, Al(NO_3)_3, PBr_3$$

86. Arrange the compounds in order of decreasing formula mass.
  $$WO_2, Rb_2SO_4, Pb(C_2H_3O_2)_2, RbI$$

Cumulative Problems

87. Write a molecular formula for each molecular model. (White = hydrogen; red = oxygen; black = carbon; blue = nitrogen; yellow = sulfur)

(a)   (b)   (c)

88. Write a molecular formula for each molecular model. (White = hydrogen; red = oxygen; black = carbon; blue = nitrogen; yellow = sulfur)

(a)   (b)   (c)

89. How many chlorine atoms are in each set?
(a) three carbon tetrachloride molecules
(b) two calcium chloride formula units
(c) four phosphorus trichloride molecules
(d) seven sodium chloride formula units

90. How many oxygen atoms are in each set?
(a) four dinitrogen monoxide molecules
(b) two calcium carbonate formula units
(c) three sulfur dioxide molecules
(d) five perchlorate ions

91. Specify the number of hydrogen atoms (white) represented in each set of molecular models:

(a)　　　　　(b)　　　　　(c)

92. Specify the number of oxygen atoms (red) represented in each set of molecular models:

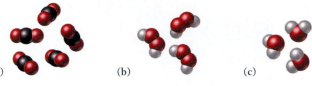

(a)　　　　　(b)　　　　　(c)

93. Complete the table:

Formula	Type of Compound (Ionic, Molecular, Acid)	Name
N_2H_4	molecular	_____
_____	_____	potassium chloride
$H_2CrO_4(aq)$	_____	_____
_____	_____	cobalt(III) cyanide

94. Complete the table:

Formula	Type of Compound (Ionic, Molecular, Acid)	Name
$K_2Cr_2O_7$	ionic	_____
$HBr(aq)$	_____	hydrobromic acid
_____	_____	dinitrogen pentoxide
PbO_2	_____	_____

95. Is each name correct for the given formula? If not, provide the correct name.
(a) $Ca(NO_2)_2$　　　calcium nitrate
(b) K_2O　　　　　dipotassium monoxide
(c) PCl_3　　　　phosphorus chloride
(d) $PbCO_3$　　　lead(II) carbonate
(e) KIO_2　　　　potassium hypoiodite

96. Is each name correct for the given formula? If not, provide the correct name.
(a) $HNO_3(aq)$　　hydrogen nitrate
(b) $NaClO$　　　sodium hypochlorite
(c) CaI_2　　　　calcium diiodide
(d) $SnCrO_4$　　tin chromate
(e) $NaBrO_3$　　sodium bromite

97. For each compound, list the correct formula and calculate the formula mass.
(a) tin(IV) sulfate
(b) nitrous acid
(c) sodium bicarbonate
(d) phosphorus pentafluoride

98. For each compound, list the correct formula and calculate the formula mass.
(a) barium bromide
(b) dinitrogen trioxide
(c) copper(I) sulfate
(d) hydrobromic acid

99. Name each compound and calculate its formula mass.
(a) PtO_2
(b) N_2O_5
(c) $Al(ClO_3)_3$
(d) PBr_5

100. Name each compound and calculate its formula mass.
(a) $Al_2(SO_4)_3$
(b) P_2O_3
(c) $HClO(aq)$
(d) $Cr(C_2H_3O_2)_3$

101. A compound contains only carbon and hydrogen and has a formula mass of 28.06 amu. What is its molecular formula?

102. A compound contains only nitrogen and oxygen and has a formula mass of 44.02 amu. What is its molecular formula?

103. Carbon has two naturally occurring isotopes: carbon-12 (mass = 12.00 amu) and carbon-13 (mass = 13.00 amu). Chlorine also has two naturally occurring isotopes: chlorine-35 (mass = 34.97 amu) and chlorine-37 (mass = 36.97 amu). How many CCl_4 molecules of different masses can exist? Determine the mass (in amu) of each of them.

104. Nitrogen has two naturally occurring isotopes: nitrogen-14 (mass = 14.00 amu) and nitrogen-15 (mass = 15.00 amu). Bromine also has two naturally occurring isotopes: bromine-79 (mass = 78.92 amu) and bromine-81 (mass = 80.92 amu). How many types of NBr_3 molecules of different masses can exist? Determine the mass (in amu) of each of them.

Highlight Problems

105. Examine each substance and the corresponding molecular view and classify it as an atomic element, a molecular element, a molecular compound, or an ionic compound.

(a)

(b)

(c)

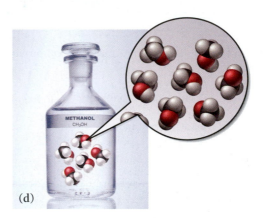

(d)

106. Molecules can be as small as two atoms or as large as thousands of atoms. In 1962, Max F. Perutz and John C. Kendrew were awarded the Nobel Prize for their discovery of the structure of hemoglobin, a very large molecule that transports oxygen from the lungs to cells through the bloodstream. The chemical formula of hemoglobin is $C_{2952}H_{4664}O_{832}N_{812}S_8Fe_4$. Calculate the formula mass of hemoglobin.

▲ Max Perutz and John C. Kendrew won a Nobel Prize in 1962 for determining the structure of hemoglobin by X-ray diffraction.

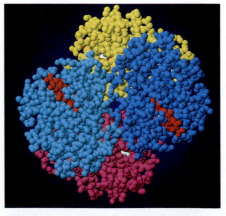

▲ Computer-generated model of hemoglobin.

107. Examine each consumer product label. Write chemical formulas for as many of the compounds as possible based on what you have learned in this chapter.

(a)

Active Ingredient:
Sodium Hypochlorite 6.0%
Other Ingredients:........ 94.0%
Total:.................. 100.0%
(Yields 5.7% available chlorine)

KEEP OUT OF REACH OF CHILDREN

DANGER: CORROSIVE.

FIRST AID: IF IN EYES: Hold eye open and rinse slowly and gently with water for 15–20 minutes. Remove contact lenses, if present, after the first 5 minutes, then continue rinsing eye. **IF ON SKIN OR CLOTHING:** Take off contaminated clothing. Rinse skin immediately with plenty of water for 15–20 minutes. **IN EITHER CASE, CALL A POISON CONTROL CENTER OR DOCTOR IMMEDIATELY FOR TREATMENT ADVICE.** See back panel for additional precautionary labeling.

Kills Methicillin Resistant Staphylococcus aureus (MRSA)
*Staphylococcus aureus, Streptococcus pyogenes, Salmonella enterica, Escherichia coli O157:H7 and Influenza A2

3 QT (96 FL OZ) 2.83 L

(b)

Drug Facts

Active ingredients (in each 5 mL teaspoon)	Purposes
Aluminum hydroxide (equivalent to dried gel, USP) 400 mg	Antacid
Magnesium hydroxide 400 mg	Antacid
Simethicone 40 mg	Antigas

Use relieves: ■ heartburn ■ acid indigestion ■ sour stomach ■ upset stomach due to these symptoms ■ pressure and bloating commonly referred to as gas

Warnings
Ask a doctor before use if you have
■ kidney disease ■ a magnesium-restricted diet

Ask a doctor or pharmacist if you are taking a prescription drug. Antacids may interact with certain prescription drugs.

Stop use and ask a doctor if symptoms last more than 2 weeks.

Keep out of reach of children.

Directions ■ shake well ■ adults/children 12 years and older: take 2-4 teaspoonfuls between meals, at bedtime, or as directed by a doctor ■ do not take more than 12 teaspoonfuls in a 24-hour period, or use the maximum dosage for more than 2 weeks ■ children under 12 years: ask a doctor

Other information ■ each teaspoon contains: magnesium 171 mg ■ do not use if breakaway band on plastic cap is broken or missing ■ does not meet USP requirements for preservative effectiveness ■ do not freeze

Inactive ingredients butylparaben, carboxymethylcellulose sodium, flavors, hypromellose, microcrystalline cellulose, propylparaben, purified water, sodium saccharin, sorbitol

Questions or comments?
1-800-469-5268 (English) or
1-888-466-8746 (Spanish)

Johnson&Johnson • MERCK   7644154

Consumer Pharmaceuticals Co.
FORT WASHINGTON, PA 19034 USA   © 2009 JJMCPC

(c)

NOTICE! PROTECTIVE INNER SEAL BENEATH CAP. IF MISSING OR DAMAGED, DO NOT USE CONTENTS.

Drug Facts

Active ingredients (in each tablet)	Purpose
Calcium carbonate 1000 mg	Antacid
Simethicone 60 mg	Antigas

Uses for the relief of
• acid indigestion
• heartburn
• sour stomach
• upset stomach associated with these symptoms
• bloating and pressure commonly referred to as gas

Warning
Do not take more than 8 tablets in a 24-hour period or use the maximum dosage for more than 2 weeks except under the advice and supervision of a physician

Ask a doctor before use if you have
• kidney stones   • a calcium-restricted diet

Ask a doctor or pharmacist before use if your are presently taking a prescription drug. Antacids may interact with certain prescription drugs.

When using this product
• at maximum dose, constipation may occur

Stop use and ask a doctor if
• symptoms last more than two weeks

Keep out of reach of children.

Directions

(d)

Nutrition Facts

Serving Size 1/8 tsp (0.6g)
Servings Per Container about 472

Amount Per Serving

Calories 0

	% Daily Value*
Total Fat 0g	0%
Sodium 65mg	3%
Total Carb. 0g	0%
Protein 0g	
Calcium 2%	

Not a significant source of calories from fat, saturated fat, trans fat, cholesterol, dietary fiber, sugars, vitamin A, vitamin C and iron.

*Percent Daily Values are based on a 2,000 calorie diet.

Ingredients: Cornstarch, Sodium Bicarbonate, Sodium Aluminum Sulfate, Monocalcium Phosphate.

CLABBER GIRL CORPORATION TERRE HAUTE, IN 47808

davisbakingpowder.com

MADE IN USA

Questions for Group Work

Discuss these questions with the group and record your consensus answer.

108. Write the correct formula for each species: carbon monoxide, carbon dioxide, the carbonate ion. List as many similarities and differences between these three species as you can. Try to get at least one contribution from each group member.

109. What questions do you need to ask about a substance in order to determine whether it is (1) an atomic element, (2) a molecular element, (3) a molecular compound, or (4) an ionic compound? Write a detailed set of instructions describing how to determine the classification of a substance based on the answers to your questions.

110. Name each compound: $Ca(NO_3)_2$, N_2O, NaS, $CrCl_3$. For each compound, include a detailed step-by-step description of the process you used to determine the name. (*Tip:* Each group member could name one compound and present it to the whole group.)

111. Calculate the formula mass for each compound in Group Work Question 110.

Data Interpretation and Analysis

112. Climate scientists have become increasingly concerned that rising levels of carbon dioxide in the atmosphere (produced by the burning of fossil fuels) will affect the global climate in harmful ways such as increased temperatures, rising sea levels, and coastal flooding. The graph at right shows the concentration of carbon dioxide in the atmosphere from the mid-1800s to the present time. Study the graph and answer the questions that follow.

(a) What were the carbon dioxide concentrations in 1950 and 2000? How much did the carbon dioxide concentration increase during these 50 years?

(b) What was the average yearly increase between 1950 and 2000?

(c) Beginning from the carbon dioxide concentration in 2010 (390 ppm), and assuming the average yearly increase you calculated in part b, what will the carbon dioxide concentration be in 2050?

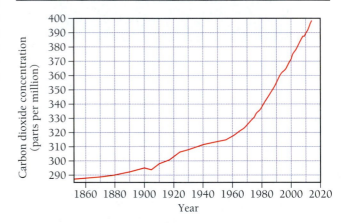

Atmospheric Carbon Dioxide

Answers to Skillbuilder Exercises

Skillbuilder 5.1 .... Yes, because in both cases

$$\frac{\text{Mass O}}{\text{Mass C}} = 1.3$$

Skillbuilder 5.2 ....
(a) Ag_2S
(b) N_2O
(c) TiO_2

Skillbuilder 5.3 .... two K atoms, one S atom, four O atoms

Skillbuilder Plus, p. 137 .... two Al atoms, three S atoms, twelve O atoms

Skillbuilder 5.4 ....
(a) molecular element
(b) molecular compound
(c) atomic element
(d) ionic compound
(e) ionic compound

Skillbuilder 5.5 .... $SrCl_2$
Skillbuilder 5.6 .... AlN
Skillbuilder 5.7 .... $AlPO_4$
Skillbuilder Plus, p. 143 .... Na_2SO_3
Skillbuilder 5.8 .... potassium bromide
Skillbuilder Plus, p. 146 .... zinc nitride
Skillbuilder 5.9 .... lead(II) oxide
Skillbuilder 5.10 .... manganese(II) nitrate
Skillbuilder 5.11 .... dinitrogen tetroxide
Skillbuilder 5.12 .... hydrofluoric acid
Skillbuilder 5.13 .... nitrous acid
Skillbuilder 5.15 .... 44.02 amu

Answers to Conceptual Checkpoints

5.1 (c) The ratio of A/B is 3.0, and A is 9.0 g, so B must be 3.0 g.

5.2 (b) This formula represents 2 Al atoms + 3(2 Cr atoms +7 O atoms) = 29 atoms.

5.3 H_2O_2

5.4 (b) The figure represents a molecular compound because the compound exists as individual molecules. Figure (a) represents an ionic compound with formula units in a lattice structure.

5.5 (c) Because the nitrate ion has a charge of 1−, the three nitrate ions together have a charge of 3−. Because the compound must be charge-neutral, Cr must have a charge of 3+.

5.6 Because calcium forms only one type of ion (Ca^{2+}); therefore, the charge of the ion is not included in the name (it is always the same, 2+).

5.7 Iodate

5.8 This question addresses one of the most common errors in nomenclature: the failure to correctly categorize the compound. NCl_3 is a molecular compound (two or more nonmetals) and therefore requires prefixes to indicate the number of each type of atom. $AlCl_3$ is an ionic compound (metal and nonmetal) and therefore requires no such prefixes.

5.9 (b)

Moonlight Diner
Specials

Breaksfast Special
Two eggs, Three strips
of bacon, Sausage link,
Toast , NaCl $5.95

Lunch Special

1500 milligrams a day

Omelettes
Served with side of NaCl
Denver Omelette
BP Delight
Atherosclerosis
Sodium Chloride
Hypertension

142/93 mmHg

Symptoms:

Chest pain
Confusion
Ear noise or buzzing
Irregular heartbeat
Nosebleed
Tiredness

On the Healthy Side
120/80 mmHg

6 Chemical Composition

In science, you don't ask why, you ask how much.

—Erwin Chargaff (1905–2002)

6.1 How Much Sodium?

Sodium is an important dietary mineral that we eat in food, primarily as sodium chloride (table salt). Sodium helps regulate body fluids, and eating too much of it can lead to high blood pressure. High blood pressure, in turn, increases the risk of stroke and heart attack. Consequently, people with high blood pressure should limit their sodium intake. The FDA recommends that a person consume less than 2.4 g (2400 mg) of sodium per day. However, sodium is usually consumed as sodium chloride, so the mass of sodium that we eat is not the same as the mass of sodium chloride that we eat. How many grams of sodium chloride can we consume and still stay below the FDA recommendation for sodium?

To answer this question, we need to know the *chemical composition* of sodium chloride. From Chapter 5, we are familiar with its formula, NaCl, which indicates that there is one sodium ion to every chloride ion. However, because the masses of sodium and chlorine are different, the relationship between the mass of sodium and the mass of sodium chloride is not clear from the chemical formula alone. In this chapter, we learn how to use the information in a chemical formula, together with atomic and formula masses, to calculate the amount of a constituent element in a given amount of a compound (or vice versa).

Chemical composition is important not just for assessing dietary sodium intake but for addressing many other questions as well. A company that mines iron, for example, wants to know how much iron it can extract from a given amount of iron ore; an organization interested in developing hydrogen as a potential fuel would want to know how much hydrogen it can extract from a given amount of water. Many environmental issues also require knowledge of chemical composition. An estimate of the threat of

▲ The mining of iron requires knowing how much iron is in a given amount of iron ore.

◄ Ordinary table salt is a compound called sodium chloride. The sodium within sodium chloride is linked to high blood pressure. In this chapter, we learn how to determine how much sodium is in a given amount of sodium chloride.

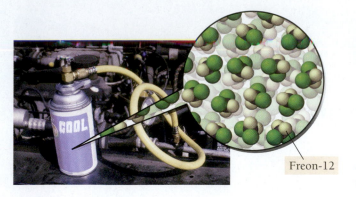

ozone depletion requires knowing how much chlorine is in a given amount of a particular chlorofluorocarbon such as freon-12. To determine these kinds of quantities, we must understand the relationships inherent in a chemical formula and the relationship between numbers of atoms or molecules and their masses. In this chapter, we examine these relationships.

Freon-12

◀ Estimating the threat of ozone depletion requires knowing the amount of chlorine in a given amount of a chlorofluorocarbon.

6.2   Counting Nails by the Pound

▶ Recognize that we use the mass of atoms to count them because they are too small and numerous to count individually.

Some hardware stores sell nails by the pound, which is easier than selling them by the nail because customers often need hundreds of nails and counting them takes too long. However, a customer may still want to know the number of nails contained in a given weight of nails. This problem is similar to asking how many atoms are in a given mass of an element. With atoms, we *must* use their mass as a way to count them because atoms are too small and too numerous to count individually. Even if you could see atoms and counted them 24 hours a day for as long as you lived, you would barely begin to count the number of atoms in something as small as a grain of sand. However, just as the hardware store customer wants to know the number of nails in a given weight, we want to know the number of atoms in a given mass. How do we do that?

Suppose the hardware store customer buys 2.60 lb of medium-sized nails and a dozen nails weigh 0.150 lb. How many nails did the customer buy? This calculation requires two conversions: one between pounds and dozens and another between dozens and number of nails. The conversion factor for the first part is the weight per dozen nails.

3.4 lb nails

$$0.150 \text{ lb nails} = 1 \text{ doz nails}$$

The conversion factor for the second part is the number of nails in one dozen.

$$1 \text{ doz nails} = 12 \text{ nails}$$

The solution map for the problem is:

| lb nails → doz nails → number of nails |

$$\frac{1 \text{ doz nails}}{0.150 \text{ lb nails}} \qquad \frac{12 \text{ nails}}{1 \text{ doz nails}}$$

Beginning with 2.60 lb and using the solution map as a guide, we convert from lb to number of nails.

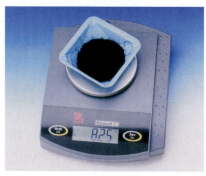

8.25 grams carbon

$$2.60 \text{ lb nails} \times \frac{1 \text{ doz nails}}{0.150 \text{ lb nails}} \times \frac{12 \text{ nails}}{1 \text{ doz nails}} = 208 \text{ nails}$$

▲ Asking how many nails are in a given weight of nails is similar to asking how many atoms are in a given mass of an element. In both cases, we count the objects by weighing them.

The customer who bought 2.60 lb of nails has 208 nails. She counted the nails by weighing them. If the customer purchased a different size of nail, the first conversion factor—relating pounds to dozens—would change, but the second conversion factor would not. One dozen corresponds to 12 nails, regardless of their size.

PEARSON
eText
2.0

CONCEPTUAL ✓ CHECKPOINT **6.1**

A certain type of nail weighs 0.50 lb per dozen. How many nails are contained in 3.5 lb of these nails?

(a) 84   (b) 21   (c) 0.58   (d) 12

6.3 Counting Atoms by the Gram

▶ Convert between moles and number of atoms.

▶ Convert between grams and moles.

▶ Convert between grams and number of atoms.

1 mol of copper atoms

▲ Twenty-two *copper* pennies contain approximately one mole of copper atoms. Pennies were mostly copper until 1982, at which point the U.S. Mint started making them out of zinc with a copper coating (because copper became too valuable).

Determining the number of atoms in a sample with a certain mass is similar to determining the number of nails in a sample with a certain weight. With nails, we used a dozen as a convenient number in our conversions, but a dozen is too small to use with atoms. We need a larger number because atoms are so small. The chemist's "dozen" is called the **mole (mol)** and has a value of 6.022×10^{23}.

$$1 \text{ mol} = 6.022 \times 10^{23}$$

This is **Avogadro's number**, named after Amadeo Avogadro (1776–1856).

The first thing to understand about the mole is that it can specify Avogadro's number of anything. *One mole of anything is 6.022×10^{23} units of that thing.* For example, 1 mol of marbles corresponds to 6.022×10^{23} marbles, and 1 mol of sand grains corresponds to 6.022×10^{23} sand grains. One mole of atoms, ions, or molecules generally makes up objects of reasonable size. For example, 22 copper pennies contain approximately 1 mol of copper (Cu) atoms, and a couple of large helium balloons contain approximately 1 mol of helium (He) atoms.

The second thing to understand about the mole is how it gets its specific value. *The numerical value of the mole is defined as being equal to the number of atoms in exactly 12 g of pure carbon-12.*

This definition of the mole establishes a relationship between mass (grams of carbon) and number of atoms (Avogadro's number). This relationship, as we will see shortly, allows us to count atoms by weighing them.

Converting between Moles and Number of Atoms

Converting between moles and number of atoms is similar to converting between dozens and number of nails. To convert between moles of atoms and number of atoms, we use these conversion factors:

$$\frac{1 \text{ mol}}{6.022 \times 10^{23} \text{ atoms}} \quad \text{or} \quad \frac{6.022 \times 10^{23} \text{ atoms}}{1 \text{ mol}}$$

For example, suppose we want to convert 3.5 mol of helium to a number of helium atoms. We set up the problem in the standard way.

GIVEN: 3.5 mol He

FIND: He atoms

RELATIONSHIPS USED 1 mol He = 6.022×10^{23} He atoms

SOLUTION MAP We draw a solution map showing the conversion from moles of He to He atoms.

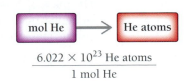

$$\frac{6.022 \times 10^{23} \text{ He atoms}}{1 \text{ mol He}}$$

1 mol of helium atoms

▲ Two large helium balloons contain approximately one mole of helium atoms.

SOLUTION

Beginning with 3.5 mol He, we use the conversion factor to get to He atoms.

$$3.5 \text{ mol He} \times \frac{6.022 \times 10^{23} \text{ He atoms}}{1 \text{ mol He}} = 2.1 \times 10^{24} \text{ He atoms}$$

EXAMPLE **6.1** | **Converting between Moles and Number of Atoms**

A silver ring contains 1.1×10^{22} silver atoms. How many moles of silver are in the ring?

SORT You are given the number of silver atoms and asked to find the number of moles.	**GIVEN:** 1.1×10^{22} Ag atoms **FIND:** mol Ag
STRATEGIZE Draw a solution map, beginning with silver atoms and ending at moles. The conversion factor is Avogadro's number.	**SOLUTION MAP** Ag atoms ⟶ mol Ag $$\frac{1 \text{ mol Ag}}{6.022 \times 10^{23} \text{ Ag atoms}}$$ **RELATIONSHIPS USED** 1 mol Ag = 6.022×10^{23} Ag atoms (Avogadro's number)
SOLVE Follow the solution map to solve the problem. Beginning with 1.1×10^{22} Ag atoms, use the conversion factor to determine the moles of Ag.	**SOLUTION** $$1.1 \times 10^{22} \text{ Ag atoms} \times \frac{1 \text{ mol Ag}}{6.022 \times 10^{23} \text{ Ag atoms}}$$ $$= 1.8 \times 10^{-2} \text{ mol Ag}$$
CHECK Are the units correct? Does the answer make physical sense?	The units, mol Ag, are the desired units. The magnitude of the answer is orders of magnitude smaller than the given quantity because it takes many atoms to make a mole, so you expect the answer to be orders of magnitude smaller than the given quantity.

▶ **SKILLBUILDER 6.1 | Converting between Moles and Number of Atoms**

How many gold atoms are in a pure gold ring containing 8.83×10^{-2} mol Au?

▶ **FOR MORE PRACTICE** Example 6.14; Problems 17, 18, 19, 20.

Converting between Grams and Moles of an Element

We just explained how to convert between moles and number of atoms, which is like converting between dozens and number of nails. We need one more conversion factor to convert from the mass of a sample to the number of atoms in the sample. For nails, we used the weight of one dozen nails; for atoms, we use the mass of 1 mol of atoms.

The mass of 1 mol of atoms of an element is its **molar mass**. The value of an element's molar mass in grams per mole is numerically equal to the element's atomic mass in atomic mass units.

Recall that Avogadro's number, the number of atoms in a mole, is defined as the number of atoms in exactly 12 g of carbon-12. The atomic mass unit is defined as one-twelfth of the mass of a carbon-12 atom, so it follows that the molar mass of any element—the mass of 1 mol of atoms in grams of that element—is equal to the atomic mass of that element expressed in atomic mass units. For example, copper has an atomic mass of 63.55 amu; therefore, 1 mol of copper atoms has a mass of 63.55 g, and the molar mass of copper is 63.55 g/mol. Just as the weight of 1 doz nails changes for different types of nails, so the mass of 1 mol of atoms changes for different elements: 1 mol of sulfur atoms (sulfur atoms are lighter than copper atoms) has a mass of 32.06 g; 1 mol of carbon atoms (lighter than sulfur) has a mass of 12.01 g; and 1 mol of lithium atoms (lighter yet) has a mass of 6.94 g.

$$32.06 \text{ g sulfur} = 1 \text{ mol sulfur} = 6.022 \times 10^{23} \text{ S atoms}$$

$$12.01 \text{ g carbon} = 1 \text{ mol carbon} = 6.022 \times 10^{23} \text{ C atoms}$$

$$6.94 \text{ g lithium} = 1 \text{ mol lithium} = 6.022 \times 10^{23} \text{ Li atoms}$$

The lighter the atom, the less mass in 1 mol of that atom (▼ FIGURE 6.1).

The molar mass of any element is a conversion factor between grams of that element and moles of that element. For carbon:

$$12.01 \text{ g C} = 1 \text{ mol C} \quad \text{or} \quad \frac{12.01 \text{ g C}}{1 \text{ mol C}} \quad \text{or} \quad \frac{1 \text{ mol C}}{12.01 \text{ g C}}$$

1 dozen large nails

1 dozen small nails

(a)

1 mole S (32.06 g)

1 mole C (12.01 g)

(b)

▶ FIGURE 6.1 The mass of 1 mol (a) Each of these pictures shows the same number of nails: 12. As you can see, 12 large nails have more weight and occupy more space than 12 small nails. The same is true for atoms. (b) Each of these samples has the same number of atoms: 6.022×10^{23}. Because sulfur atoms are more massive and larger than carbon atoms, 1 mol of S atoms is heavier and occupies more space than 1 mol of C atoms.

> A 0.58-g diamond is about a three-carat diamond.

Suppose we want to calculate the number of moles of carbon in a 0.58-g diamond (pure carbon).

We first sort the information in the problem.

GIVEN: 0.58 g C

FIND: mol C

SOLUTION MAP We then strategize by drawing a solution map showing the conversion from grams of C to moles of C. The conversion factor is the molar mass of carbon.

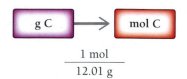

$$\frac{1 \text{ mol}}{12.01 \text{ g}}$$

RELATIONSHIPS USED
12.01 g C = 1 mol C (molar mass of carbon, from periodic table)

SOLUTION
Finally, we solve the problem by following the solution map.

$$0.58 \text{ g C} \times \frac{1 \text{ mol C}}{12.01 \text{ g C}} = 4.8 \times 10^{-2} \text{ mol C}$$

EXAMPLE **6.2** **The Mole Concept—Converting between Grams and Moles**

Calculate the number of moles of sulfur in 57.8 g of sulfur.

SORT	
Begin by sorting the information in the problem. You are given the mass of sulfur and asked to find the number of moles.	**GIVEN:** 57.8 g S **FIND:** mol S

STRATEGIZE	
Draw a solution map showing the conversion from g S to mol S. The conversion factor is the molar mass of sulfur.	**SOLUTION MAP** 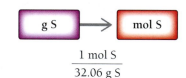 $$\frac{1 \text{ mol S}}{32.06 \text{ g S}}$$ **RELATIONSHIPS USED** 32.06 g S = 1 mol S (molar mass of sulfur, from periodic table)

SOLVE	
Follow the solution map to solve the problem. Begin with 57.8 g S and use the conversion factor to determine mol S.	**SOLUTION** $$57.8 \text{ g S} \times \frac{1 \text{ mol C}}{32.06 \text{ g S}} = 1.80 \text{ mol S}$$

CHECK	
Check your answer. Are the units correct? Does the answer make physical sense?	The units (mol S) are correct. The magnitude of the answer makes sense because 1 mol of S has a mass of 32.06 g; therefore, 57.8 g of S should be close to 2 mols.

▶ **SKILLBUILDER 6.2 | The Mole Concept—Converting between Grams and Moles**

Calculate the number of grams of sulfur in 2.78 mol of sulfur.

▶ **FOR MORE PRACTICE** Example 6.15; Problems 25, 26, 27, 28, 29, 30.

Converting between Grams of an Element and Number of Atoms

Suppose we want to know the number of carbon *atoms* in the 0.58-g diamond. We first convert from grams to moles and then from moles to number of atoms. The solution map is:

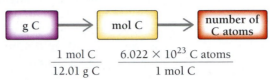

Notice the similarity between this solution map and the one we used for nails:

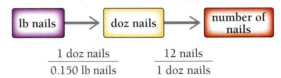

Beginning with 0.58 g carbon and using the solution map as a guide, we convert to the number of carbon atoms.

$$0.58 \ \cancel{\text{g C}} \times \frac{1 \ \cancel{\text{mol C}}}{12.01 \ \cancel{\text{g C}}} \times \frac{6.022 \times 10^{23} \ \text{C atoms}}{1 \ \cancel{\text{mol C}}} = 2.9 \times 10^{22} \ \text{C atoms}$$

EXAMPLE **6.3** | **The Mole Concept—Converting between Grams and Number of Atoms**

How many aluminum atoms are in an aluminum can with a mass of 16.2 g?

SORT	
You are given the mass of aluminum and asked to find the number of aluminum atoms.	**GIVEN:** 16.2 g Al **FIND:** Al atoms

STRATEGIZE

The solution map has two steps. In the first step, convert from g Al to mol Al. In the second step, convert from mol Al to the number of Al atoms. The required conversion factors are the molar mass of aluminum and the number of atoms in a mole.

SOLUTION MAP

g Al → mol Al → number of Al atoms

$$\frac{1 \ \text{mol Al}}{26.98 \ \text{g Al}} \quad \frac{6.022 \times 10^{23} \ \text{Al atoms}}{1 \ \text{mol Al}}$$

RELATIONSHIPS USED

26.98 g Al = 1 mol Al (molar mass of aluminum, from periodic table)

6.022×10^{23} = 1 mol (Avogadro's number)

SOLVE

Follow the solution map to solve the problem, beginning with 16.2 g Al and multiplying by the appropriate conversion factors to arrive at Al atoms.

SOLUTION

$$16.2 \ \cancel{\text{g Al}} \times \frac{1 \ \cancel{\text{mol Al}}}{26.98 \ \cancel{\text{g Al}}} \times \frac{6.022 \times 10^{23} \ \text{Al atoms}}{1 \ \cancel{\text{mol Al}}} = 3.62 \times 10^{23} \ \text{Al atoms}$$

CHECK

Are the units correct? Does the answer make physical sense?

The units, Al atoms, are correct. The answer makes sense because the number of atoms in any macroscopic-sized sample of matter is very large.

▶ **SKILLBUILDER 6.3** | **The Mole Concept—Converting between Grams and Number of Atoms**

Calculate the mass of 1.23×10^{24} helium atoms.

▶ **FOR MORE PRACTICE**   Example 6.16; Problems 35, 36, 37, 38, 39, 40, 41, 42.

PEARSON eText 2.0   Interactive Worked Example Video 6.3

Before we move on, notice that numbers with large exponents, such as 6.022×10^{23}, are almost unimaginably large. Twenty-two copper pennies contain 6.022×10^{23} or 1 mol of copper atoms; 6.022×10^{23} pennies would cover Earth's entire surface to a depth of 300 m. Even objects that are small by everyday standards occupy a huge space when we have a mole of them. For example, one crystal of granulated sugar has a mass of less than 1 mg and a diameter of less than 0.1 mm, yet 1 mol of sugar crystals would cover the state of Texas to a depth of several feet. For every increase of 1 in the exponent of a number, the number increases by 10. So a number with an exponent of 23 is incredibly large. A mole has to be a large number because atoms are so small.

CONCEPTUAL ✔ CHECKPOINT 6.2

Which statement is *always* true for samples of atomic elements, regardless of the type of element present in the samples?

(a) If two samples of different elements contain the same number of atoms, they contain the same number of moles.

(b) If two samples of different elements have the same mass, they contain the same number of moles.

(c) If two samples of different elements have the same mass, they contain the same number of atoms.

CONCEPTUAL ✔ CHECKPOINT 6.3

Without doing any calculations, determine which sample contains the most atoms.

(a) one gram of cobalt

(b) one gram of carbon

(c) one gram of lead

6.4 Counting Molecules by the Gram

▶ Convert between grams and moles of a compound.

▶ Convert between mass of a compound and number of molecules.

The calculations we just performed for atoms can also be applied to molecules for covalent compounds or formula units for ionic compounds. We first convert between the mass of a compound and moles of the compound, and then we calculate the number of molecules (or formula units) from moles.

Converting between Grams and Moles of a Compound

Remember, ionic compounds do not contain individual molecules. The smallest electrically neutral collection of ions is a formula unit.

For elements, the molar mass is the mass of 1 mol of atoms of that element. For compounds, the molar mass is the mass of 1 mol of molecules or formula units of that compound. The molar mass of a compound in grams per mole is numerically equal to the formula mass of the compound in atomic mass units. For example, the formula mass of CO_2 is:

Remember, the formula mass for a compound is the sum of the atomic masses of all the atoms in a chemical formula.

$$\text{formula mass} = 1(\text{atomic mass of C}) + 2(\text{atomic mass of O})$$
$$= 1(12.01 \text{ amu}) + 2(16.00 \text{ amu})$$
$$= 44.01 \text{ amu}$$

The molar mass of CO_2 is therefore:

$$\text{molar mass} = 44.01 \text{ g/mol}$$

Just as the molar mass of an element serves as a conversion factor between grams and moles of that element, the molar mass of a compound serves as a conversion factor between grams and moles of that compound. For example, suppose we want to find the number of moles in a 22.5-g sample of dry ice (solid CO_2). We begin by sorting the information.

GIVEN: 22.5 g CO_2

FIND: mol CO_2

SOLUTION MAP
We then strategize by drawing a solution map that shows how the molar mass converts grams of the compound to moles of the compound.

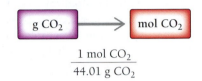

$$\frac{1 \text{ mol } CO_2}{44.01 \text{ g } CO_2}$$

RELATIONSHIPS USED
44.01 g CO_2 = 1 mol CO_2 (molar mass of CO_2)

SOLUTION
Finally, we solve the problem.

$$22.5 \text{ g} \times \frac{1 \text{ mol } CO_2}{44.01 \text{ g}} = 0.511 \text{ mol } CO_2$$

EXAMPLE **6.4** **The Mole Concept—Converting between Grams and Moles for Compounds**

Calculate the mass (in grams) of 1.75 mol of water.	
SORT You are given moles of water and asked to find the mass.	**GIVEN:** 1.75 mol H_2O **FIND:** g H_2O
STRATEGIZE Draw a solution map showing the conversion from mol H_2O to g H_2O. The conversion factor is the molar mass of water, which you can determine by summing the atomic masses of all the atoms in the chemical formula.	**SOLUTION MAP** mol H_2O $\longrightarrow$ g H_2O $\dfrac{18.02 \text{ g } H_2O}{1 \text{ mol } H_2O}$ **RELATIONSHIPS USED** H_2O molar mass = 2(atomic mass H) + 1(atomic mass O) = 2(1.01) + 1(16.00) = 18.02 g/mol
SOLVE Follow the solution map to solve the problem. Begin with 1.75 mol of water and use the molar mass to convert to grams of water.	**SOLUTION** $1.75 \text{ mol } H_2O \times \dfrac{18.02 \text{ g } H_2O}{\text{mol } H_2O} = 31.5 \text{ g } H_2O$
CHECK Check your answer. Are the units correct? Does the answer make physical sense?	The units (g H_2O) are the desired units. The magnitude of the answer makes sense because 1 mol of water has a mass of 18.02 g; therefore, 1.75 mol should have a mass that is slightly less than 36 g.

▶ **SKILLBUILDER 6.4 | The Mole Concept—Converting between Grams and Moles**

Calculate the number of moles of NO_2 in 1.18 g of NO_2.

▶ **FOR MORE PRACTICE** Problems 47, 48, 49, 50.

Converting between Grams of a Compound and Number of Molecules

Suppose that we want to find the *number of* CO_2 *molecules* in a sample of dry ice (solid CO_2) with a mass of 22.5 g. The solution map for the problem is:

$$\boxed{\text{g } CO_2} \longrightarrow \boxed{\text{mol } CO_2} \longrightarrow \boxed{\text{CO}_2 \text{ molecules}}$$

$$\frac{1 \text{ mol } CO_2}{44.01 \text{ g } CO_2} \qquad \frac{6.022 \times 10^{23} \text{ CO}_2 \text{ molecules}}{1 \text{ mol } CO_2}$$

Notice that the first part of the solution map is identical to calculating the number of moles of CO_2 in 22.5 g of dry ice. The second part of the solution map shows the conversion from moles to number of molecules. Following the solution map, we calculate:

$$22.5 \text{ g } \cancel{CO_2} \times \frac{1 \text{ mol } \cancel{CO_2}}{44.01 \text{ g } \cancel{CO_2}} \times \frac{6.022 \times 10^{23} \text{ CO}_2 \text{ molecules}}{\cancel{\text{mol } CO_2}}$$

$$= 3.08 \times 10^{23} \text{ CO}_2 \text{ molecules}$$

EXAMPLE **6.5** | **The Mole Concept—Converting between Mass of a Compound and Number of Molecules**

What is the mass of 4.78×10^{24} NO_2 molecules?

SORT	**GIVEN:** 4.78×10^{24} NO_2 molecules
You are given the number of NO_2 molecules and asked to find the mass.	**FIND:** g NO_2
STRATEGIZE	**SOLUTION MAP**
The solution map has two steps. In the first step, convert from molecules of NO_2 to moles of NO_2. In the second step, convert from moles of NO_2 to mass of NO_2. The required conversion factors are the molar mass of NO_2 and the number of molecules in a mole.	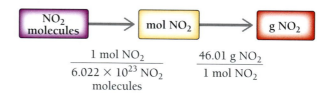 $$\frac{1 \text{ mol } NO_2}{6.022 \times 10^{23} \text{ NO}_2 \text{ molecules}} \qquad \frac{46.01 \text{ g } NO_2}{1 \text{ mol } NO_2}$$ **RELATIONSHIPS USED** 6.022×10^{23} molecules = 1 mol (Avogadro's number) NO_2 molar mass = 1(atomic mass N) + 2(atomic mass O) $$= 14.01 + 2(16.00)$$ $$= 46.01 \text{ g/mol}$$
SOLVE	**SOLUTION**
Using the solution map as a guide, begin with molecules of NO_2 and multiply by the appropriate conversion factors to arrive at g NO_2.	$$4.78 \times 10^{24} \cancel{\text{NO}_2 \text{ molecules}} \times \frac{1 \text{ mol } \cancel{NO_2}}{6.022 \times 10^{23} \cancel{\text{NO}_2 \text{ molecules}}}$$ $$\times \frac{46.01 \text{ g } NO_2}{1 \cancel{\text{ mol } NO_2}} = 365 \text{ g } NO_2$$
CHECK	
Check your answer. Are the units correct? Does the answer make physical sense?	The units, g NO_2, are correct. Because the number of NO_2 molecules is more than one mole, the answer should be more than one molar mass (more than 46.01 g), which it is; therefore, the magnitude of the answer is reasonable.

▶ **SKILLBUILDER 6.5** | **The Mole Concept—Converting between Mass and Number of Molecules**

How many H_2O molecules are in a sample of water with a mass of 3.64 g?

▶ **FOR MORE PRACTICE** Problems 51, 52, 53, 54.

CONCEPTUAL ✓ CHECKPOINT **6.4**

Compound A has a molar mass of 100 g/mol, and Compound B has a molar mass of 200 g/mol. If you have samples of equal mass of both compounds, which sample contains the greater number of molecules?

6.5 Chemical Formulas as Conversion Factors

▶ Convert between moles of a compound and moles of a constituent element.

▶ Convert between grams of a compound and grams of a constituent element.

3 leaves : 1 clover

▲ We know that each clover has three leaves. We can express that as a ratio: 3 leaves : 1 clover.

We are almost ready to address the sodium problem posed in Section 6.1. To determine how much of a particular element (such as sodium) is in a given amount of a particular compound (such as sodium chloride), we must understand the numerical relationships inherent in a chemical formula. We can understand these relationships with a straightforward analogy: Asking how much sodium is in a given amount of sodium chloride is similar to asking how many leaves are on a given number of clovers. For example, suppose we want to know the number of leaves on 14 clovers. We need a conversion factor between leaves and clovers. For clovers, the conversion factor comes from our everyday knowledge about them—we know that each clover has three leaves. We can express that relationship as a ratio between clovers and leaves.

3 leaves : 1 clover

Like other conversion factors, this ratio gives the relationship between leaves and clovers. With this ratio, we can write a conversion factor to determine the number of leaves in 14 clovers. The solution map is:

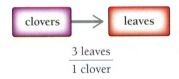

$$\frac{3 \text{ leaves}}{1 \text{ clover}}$$

We solve the problem by beginning with clovers and converting to leaves.

$$14 \text{ clovers} \times \frac{3 \text{ leaves}}{1 \text{ clover}} = 42 \text{ leaves}$$

Similarly, a chemical formula gives us ratios between elements and molecules for a particular compound. For example, the formula for carbon dioxide (CO_2) indicates that there are two O atoms per CO_2 molecule. We write this as:

2 O atoms : 1 CO_2 molecule

Just as 3 leaves : 1 clover can also be written as 3 dozen leaves : 1 dozen clovers, for molecules we can write:

2 doz O atoms : 1 doz CO_2 molecules

However, for atoms and molecules, we normally work in moles.

2 mol O : 1 mol CO_2

Chemical formulas are discussed in Chapter 5.

With conversion factors such as these—which come directly from the chemical formula—we can determine the amounts of the constituent elements present in a given amount of a compound.

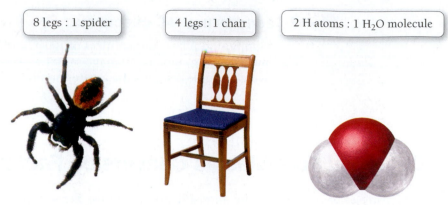

| 8 legs : 1 spider | 4 legs : 1 chair | 2 H atoms : 1 H_2O molecule |

▲ Each of these shows a ratio.

Converting between Moles of a Compound and Moles of a Constituent Element

Suppose we want to know the number of moles of O in 18 mol of CO_2. Our solution map is:

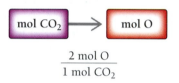

$$\boxed{\text{mol } CO_2} \longrightarrow \boxed{\text{mol O}}$$

$$\frac{2 \text{ mol O}}{1 \text{ mol } CO_2}$$

We can then calculate the moles of O.

$$18 \text{ mol } CO_2 \times \frac{2 \text{ mol O}}{1 \text{ mol } CO_2} = 36 \text{ mol O}$$

EXAMPLE **6.6**

Chemical Formulas as Conversion Factors—Converting between Moles of a Compound and Moles of a Constituent Element

Determine the number of moles of O in 1.7 mol of $CaCO_3$.

SORT	
You are given the number of moles of $CaCO_3$ and asked to find the number of moles of O.	**GIVEN:** 1.7 mol $CaCO_3$ **FIND:** mol O

STRATEGIZE	
The solution map begins with moles of calcium carbonate and ends with moles of oxygen. Determine the conversion factor from the chemical formula, which indicates three O atoms for every $CaCO_3$ unit.	**SOLUTION MAP** $\boxed{\text{mol } CaCO_3} \longrightarrow \boxed{\text{mol O}}$ $\dfrac{3 \text{ mol O}}{1 \text{ mol } CaCO_3}$ **RELATIONSHIPS USED** 3 mol O : 1 mol $CaCO_3$ (from chemical formula)

SOLVE	
Follow the solution map to solve the problem. The subscripts in a chemical formula are exact, so they never limit significant figures.	**SOLUTION** $1.7 \text{ mol } CaCO_3 \times \dfrac{3 \text{ mol O}}{1 \text{ mol } CaCO_3} = 5.1 \text{ mol O}$

CHECK	
Check your answer. Are the units correct? Does the answer make physical sense?	The units (mol O) are correct. The magnitude is reasonable as the number of moles of oxygen should be larger than the number of moles of $CaCO_3$ (because each $CaCO_3$ unit contains 3 O atoms).

▶ **SKILLBUILDER 6.6 | Chemical Formulas as Conversion Factors—Converting between Moles of a Compound and Moles of a Constituent Element**

Determine the number of moles of O in 1.4 mol of H_2SO_4.

▶ **FOR MORE PRACTICE**   Example 6.17; Problems 63, 64.

CONCEPTUAL ✓ CHECKPOINT 6.5

How many moles of H are present in 12 moles of CH_4?

Converting between Grams of a Compound and Grams of a Constituent Element

Now, we have the tools we need to solve our sodium problem from the beginning of the chapter. Suppose we want to know the mass of sodium in 15 g of NaCl. The chemical formula gives us the relationship between moles of Na and moles of NaCl:

$$1 \text{ mol Na} : 1 \text{ mol NaCl}$$

To use this relationship, we need *mol* NaCl, but we have *g* NaCl. We can use the *molar mass* of NaCl to convert from g NaCl to mol NaCl. Then we use the conversion factor from the chemical formula to convert to mol Na. Finally, we use the molar mass of Na to convert to g Na. The solution map is:

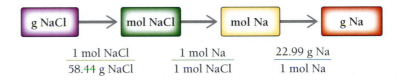

Notice that we must convert from g NaCl to mol NaCl *before* we can use the chemical formula as a conversion factor.

> *The chemical formula gives us a relationship between moles of substances, not between grams.*

We follow the solution map to solve the problem.

$$15 \text{ g NaCl} \times \frac{1 \text{ mol NaCl}}{58.44 \text{ g NaCl}} \times \frac{1 \text{ mol Na}}{1 \text{ mol NaCl}} \times \frac{22.99 \text{ g Na}}{1 \text{ mol Na}} = 5.9 \text{ g Na}$$

The general form for solving problems where you are asked to find the mass of an element present in a given mass of a compound is:

mass compound ⟶ **moles** compound ⟶ **moles** element ⟶ **mass** element

Use the atomic or molar mass to convert between mass and moles, and use the relationships inherent in the chemical formula to convert between moles and moles (▼ **FIGURE 6.2**).

▶ **FIGURE 6.2 Mole relationships from a chemical formula** The relationships inherent in a chemical formula allow us to convert between moles of the compound and moles of a constituent element (or vice versa).

1 mol CCl_4 : 4 mol Cl

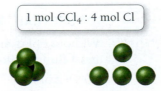

PEARSON eText 2.0
Interactive
Worked Example
Video 6.7

EXAMPLE **6.7**

Chemical Formulas as Conversion Factors—Converting between Grams of a Compound and Grams of a Constituent Element

Carvone ($C_{10}H_{14}O$) is the main component of spearmint oil. It has a pleasant aroma and mint flavor. Carvone is added to chewing gum, liqueurs, soaps, and perfumes. Calculate the mass of carbon in 55.4 g of carvone.

SORT	GIVEN: 55.4 g $C_{10}H_{14}O$
You are given the mass of carvone and asked to find the mass of one of its constituent elements.	FIND: g C

STRATEGIZE

Base the solution map on:

grams $\longrightarrow$ mole $\longrightarrow$ mole $\longrightarrow$ grams

SOLUTION MAP

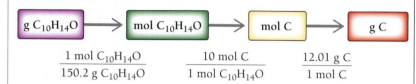

$$\frac{1 \text{ mol } C_{10}H_{14}O}{150.2 \text{ g } C_{10}H_{14}O} \qquad \frac{10 \text{ mol } C}{1 \text{ mol } C_{10}H_{14}O} \qquad \frac{12.01 \text{ g } C}{1 \text{ mol } C}$$

You need three conversion factors.
The first is the molar mass of carvone.

RELATIONSHIPS USED

$$\begin{aligned} \text{molar mass carvone} &= 10(12.01) + 14(1.01) + 1(16.00) \\ &= 120.1 + 14.14 + 16.00 \\ &= 150.2 \text{ g/mol} \end{aligned}$$

The second conversion factor is the relationship between moles of carbon and moles of carvone from the molecular formula.

10 mol C : 1 mol $C_{10}H_{14}O$ (from chemical formula)

The third conversion factor is the molar mass of carbon.

1 mol C = 12.01 g C (molar mass C, from periodic table)

SOLVE

Follow the solution map to solve the problem, beginning with g $C_{10}H_{14}O$ and multiplying by the appropriate conversion factors to arrive at g C.

SOLUTION

$$55.4 \text{ g } \cancel{C_{10}H_{14}O} \times \frac{1 \text{ mol } \cancel{C_{10}H_{14}O}}{150.2 \text{ g } \cancel{C_{10}H_{14}O}}$$

$$\times \frac{10 \text{ mol } \cancel{C}}{1 \text{ mol } \cancel{C_{10}H_{14}O}} \times \frac{12.01 \text{ g C}}{1 \text{ mol } \cancel{C}} = 44.3 \text{ g C}$$

CHECK

Check your answer. Are the units correct? Does the answer make physical sense?

The units, g C, are correct. The magnitude of the answer is reasonable because the mass of carbon with the compound must be less than the mass of the compound itself. If you had arrived at a mass of carbon that was greater than the mass of the compound, you would know that you had made a mistake; the mass of a constituent element can never be greater than the mass of the compound itself.

▶ **SKILLBUILDER 6.7** | **Chemical Formulas as Conversion Factors—Converting between Grams of a Compound and Grams of a Constituent Element**

Determine the mass of oxygen in a 5.8-g sample of sodium bicarbonate ($NaHCO_3$).

▶ **SKILLBUILDER PLUS** Determine the mass of oxygen in a 7.20-g sample of $Al_2(SO_4)_3$.

▶ **FOR MORE PRACTICE** Example 6.18; Problems 67, 68, 69, 70.

PEARSON eText 2.0

CONCEPTUAL ✔ **CHECKPOINT 6.6**

Without doing any detailed calculations, determine which sample contains the most fluorine atoms.

(a) 25 g of HF   (b) 1.5 mol of CH_3F (c) 1.0 mol of F_2

6.6 Mass Percent Composition of Compounds

▶ Use mass percent composition as a conversion factor.

Another way to express how much of an element is in a given compound is to use the element's mass percent composition for that compound. The **mass percent composition** or simply **mass percent** of an element is the element's percentage of the total mass of the compound. For example, the mass percent composition of sodium in sodium chloride is 39%. This information tells us that a 100-g sample of sodium chloride contains 39 g of sodium. We can determine the mass percent composition for a compound from experimental data using the formula:

$$\text{mass percent of element } X = \frac{\text{mass of } X \text{ in a sample of the compound}}{\text{mass of the sample of the compound}} \times 100\%$$

Suppose a 0.358-g sample of chromium reacts with oxygen to form 0.523 g of the metal oxide. Then the mass percent of chromium is:

$$\text{mass percent Cr} = \frac{\text{mass Cr}}{\text{mass metal oxide}} \times 100\%$$

$$= \frac{0.358\,\text{g}}{0.523\,\text{g}} \times 100\% = 68.5\%$$

We can use mass percent composition as a conversion factor between grams of a constituent element and grams of the compound. For example, we just saw that the mass percent composition of sodium in sodium chloride is 39%. This can be written as:

$$39 \text{ g sodium} : 100 \text{ g sodium chloride}$$

or in fractional form:

$$\frac{39 \text{ g Na}}{100 \text{ g NaCl}} \quad \text{or} \quad \frac{100 \text{ g NaCl}}{39 \text{ g Na}}$$

These fractions are conversion factors between g Na and g NaCl, as shown in Example 6.8.

EXAMPLE **6.8** | **Using Mass Percent Composition as a Conversion Factor**

The FDA recommends that adults consume less than 2.4 g of sodium per day. How many grams of sodium chloride can you consume and still be within the FDA guidelines? Sodium chloride is 39% sodium by mass.

SORT You are given the mass of sodium and the mass percent of sodium in sodium chloride. When mass percent is given, write it as a fraction. *Percent means per hundred*, so 39% sodium indicates that there are 39 g Na per 100 g NaCl. You are asked to find the mass of sodium chloride that contains the given mass of sodium.	**GIVEN:** $2.4\,\text{g Na}$ $\dfrac{39\,\text{g Na}}{100\,\text{g NaCl}}$ **FIND:** g NaCl
STRATEGIZE Draw a solution map that starts with the mass of sodium and uses the mass percent as a conversion factor to get to the mass of sodium chloride.	**SOLUTION MAP** 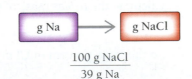 $\dfrac{100 \text{ g NaCl}}{39 \text{ g Na}}$ **RELATIONSHIPS USED** 39 g Na : 100 g NaCl (given in the problem)

continued on page 184 ▶

continued from page 183

SOLVE	SOLUTION
Follow the solution map to solve the problem, beginning with grams Na and ending with grams of NaCl. The amount of salt you can consume and still be within the FDA guideline is 6.2 g NaCl.	$$2.4 \text{ g Na} \times \frac{100 \text{ g NaCl}}{39 \text{ g Na}} = 6.2 \text{ g NaCl}$$

CHECK	SOLUTION
Check your answer. Are the units correct? Does the answer make physical sense?	The units, g NaCl, are correct. The answer makes physical sense because the mass of NaCl should be *larger* than the mass of Na. The mass of a compound containing a given mass of a particular element is always larger than the mass of the element itself. 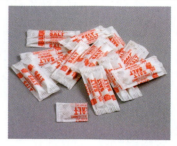 ▲ Twelve and a half packets of salt contain 6.2 g NaCl.

▶ **SKILLBUILDER 6.8 | Using Mass Percent Composition as a Conversion Factor**

If a woman consumes 22 g of sodium chloride, how much sodium does she consume? Sodium chloride is 39% sodium by mass.

▶ **FOR MORE PRACTICE** Example 6.19; Problems 75, 76, 77, 78.

6.7 Mass Percent Composition from a Chemical Formula

▶ Determine mass percent composition from a chemical formula.

In the previous section, we demonstrated how to calculate mass percent composition from experimental data and how to use mass percent composition as a conversion factor. We can also calculate the mass percent of any element in a compound from the chemical formula for the compound. Based on the chemical formula, the mass percent of element X in a compound is:

$$\text{mass percent of element } X = \frac{\text{mass of element } X \text{ in 1 mol of compound}}{\text{mass of 1 mol of compound}} \times 100\%$$

Suppose, for example, that we want to calculate the mass percent composition of Cl in the chlorofluorocarbon CCl_2F_2. The mass percent of Cl is given by:

CCl_2F_2

$$\text{mass percent Cl} = \frac{2 \times \text{molar mass Cl}}{\text{molar mass } CCl_2F_2} \times 100\%$$

We must multiply the molar mass of Cl by 2 because the chemical formula has a subscript of 2 for Cl, meaning that 1 mol of CCl_2F_2 contains 2 mol of Cl atoms. We calculate the molar mass of CCl_2F_2 as follows:

$$\text{molar mass} = 1(12.01) + 2(35.45) + 2(19.00) = 120.91 \text{ g/mol}$$

So the mass percent of Cl in CCl_2F_2 is:

$$\text{mass percent Cl} = \frac{2 \times \text{molar mass Cl}}{\text{molar mass } CCl_2F_2} \times 100\% = \frac{2 \times 35.45 \text{ g}}{120.91 \text{ g}} \times 100\%$$

$$= 58.64\%$$

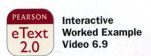

PEARSON eText 2.0 Interactive Worked Example Video 6.9

EXAMPLE 6.9 Mass Percent Composition

Calculate the mass percent of Cl in freon-114 ($C_2Cl_4F_2$).

SORT

You are given the molecular formula of freon-114 and asked to find the mass percent of Cl.

GIVEN: $C_2Cl_4F_2$

FIND: Mass % Cl

STRATEGIZE

You can use the information in the chemical formula to substitute into the mass percent equation and obtain the mass percent Cl.

SOLUTION MAP

chemical formula → mass % Cl

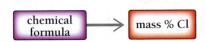

$$\text{mass \% Cl} = \frac{4 \times \text{molar mass Cl}}{\text{molar mass } C_2Cl_4F_2} \times 100\%$$

RELATIONSHIPS USED

mass percent of element X =

$$\frac{\text{mass of element } X \text{ in 1 mol of compound}}{\text{mass of 1 mol of compound}} \times 100\%$$

(mass percent equation, introduced in this section)

SOLVE

Calculate the molar mass of freon-114 and substitute the values into the equation to find mass percent Cl.

SOLUTION

$$4 \times \text{molar mass Cl} = 4(35.45 \text{ g}) = 141.8 \text{ g}$$

$$\begin{aligned}\text{molar mass } C_2Cl_4F_2 &= 2(12.01) + 4(35.45) + 2(19.00) \\ &= 24.02 + 141.8 + 38.00 \\ &= \frac{203.8 \text{ g}}{\text{mol}}\end{aligned}$$

$$\begin{aligned}\text{mass \% Cl} &= \frac{4 \times \text{molar mass Cl}}{\text{molar mass } C_2Cl_4F_2} \times 100\% \\ &= \frac{141.8 \text{ g}}{203.8 \text{ g}} \times 100\% \\ &= 69.58\%\end{aligned}$$

CHECK

Check your answer. Are the units correct? Does the answer make physical sense?

The units (%) are correct. The answer makes physical sense. Mass percent composition should never exceed 100%. If your answer is greater than 100%, you have made an error.

▶ **SKILLBUILDER 6.9 | Mass Percent Composition**

Acetic acid ($HC_2H_3O_2$) is the active ingredient in vinegar. Calculate the mass percent composition of O in acetic acid.

▶ **FOR MORE PRACTICE** Example 6.20; Problems 79, 80, 81, 82, 85, 86.

PEARSON eText 2.0

CONCEPTUAL CHECKPOINT 6.7

Which compound has the highest mass percent of O? (You should not have to perform any detailed calculations to answer this question.)

(a) CrO

(b) CrO_2

(c) Cr_2O_3

CHEMISTRY AND HEALTH

Fluoridation of Drinking Water

In the early 1900s, scientists discovered that people whose drinking water naturally contained fluoride (F^-) ions had fewer cavities than people whose water did not. At appropriate levels, fluoride strengthens tooth enamel, which prevents tooth decay. In an effort to improve public health, fluoride has been artificially added to drinking water supplies since 1945. In the United States today, about 62% of the population drinks artificially fluoridated drinking water. The American Dental Association and public health agencies estimate that water fluoridation reduces tooth decay by 40 to 65%.

The fluoridation of public drinking water, however, is often controversial. Some opponents argue that fluoride is available from other sources—such as toothpaste, mouthwash, drops, and pills—and therefore should not be added to drinking water. Anyone who wants fluoride can get it from these optional sources, they argue, and the government should not impose fluoride on the population. Other opponents argue that the risks associated with fluoridation are too great. Indeed, too much fluoride can cause teeth to become brown and spotted, a condition known as dental fluorosis. Extremely high levels can lead to skeletal fluorosis, a condition in which the bones become brittle and arthritic.

The scientific consensus is that, like many minerals, fluoride shows some health benefits at certain levels—about 1–4 mg/day for adults—but can have detrimental effects at higher levels. Consequently, most major cities fluoridate their drinking water at a level of about 0.7 mg/L. Most adults drink between 1 and 2 L of water per day, so they receive beneficial amounts of fluoride from the water. Bottled water does not normally contain fluoride. Fluoridated bottled water can sometimes be found in the infant section of supermarkets.

B6.1 CAN YOU ANSWER THIS? *Fluoride is often added to water as sodium fluoride (NaF). What is the mass percent composition of F^- in NaF? How many grams of NaF must be added to 1500 L of water to fluoridate it at a level of 0.7 mg F^-/L?*

6.8 Calculating Empirical Formulas for Compounds

▶ Determine an empirical formula from experimental data.

▶ Calculate an empirical formula from reaction data.

In Section 6.7, we learned how to calculate mass percent composition from a chemical formula. But can we go the other way? Can we calculate a chemical formula from mass percent composition? This is important because laboratory analyses of compounds do not often give chemical formulas directly; rather, they give the relative masses of each element present in a compound. For example, if we decompose water into hydrogen and oxygen in the laboratory, we could measure the masses of hydrogen and oxygen produced. Can we determine the chemical formula for water from this kind of data?

▶ We just learned how to go from the chemical formula of a compound to its mass percent composition. Can we also go the other way?

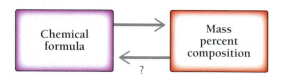

The answer is a qualified yes. We can determine a chemical formula, but it is the **empirical formula**, not the molecular formula. As we saw in Section 5.3, an empirical formula gives the smallest whole-number ratio of each type of atom in a compound, not the specific number of each type of atom in a molecule. Recall that the **molecular formula** is always a whole-number multiple of the empirical formula: Molecular formula = empirical × n, where $n = 1, 2, 3 \ldots$.

A chemical formula represents a ratio of atoms or moles of atoms, not a ratio of masses.

For example, the molecular formula for hydrogen peroxide is H_2O_2, and its empirical formula is HO.

$$HO \times 2 \longrightarrow H_2O_2$$

Calculating an Empirical Formula from Experimental Data

Suppose we decompose a sample of water in the laboratory and find that it produces 3.0 g of hydrogen and 24 g of oxygen. How do we determine an empirical formula from these data?

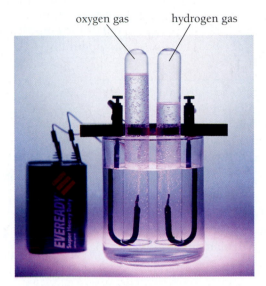

oxygen gas   hydrogen gas

▲ Water can be decomposed by an electric current into hydrogen and oxygen. How can we find the empirical formula for water from the masses of its component elements?

We know that an empirical formula represents a ratio of atoms or a ratio of moles of atoms, but it *does not* represent a ratio of masses. So the first thing we must do is convert our data from grams to moles. How many moles of each element formed during the decomposition? To convert to moles, we divide each mass by the molar mass of that element.

$$\text{mol H} = 3.0 \text{ g H} \times \frac{1 \text{ mol H}}{1.01 \text{ g H}} = 3.0 \text{ mol H}$$

$$\text{mol O} = 24 \text{ g O} \times \frac{1 \text{ mol O}}{16.00 \text{ g O}} = 1.5 \text{ mol O}$$

From these data, we know there are 3 mol of H for every 1.5 mol of O. We can now write a pseudoformula for water:

$$H_3O_{1.5}$$

To get whole-number subscripts in our formula, we divide all the subscripts by the smallest one, in this case 1.5.

$$H_{\frac{3}{1.5}}O_{\frac{1.5}{1.5}} = H_2O$$

Our empirical formula for water, which in this case also happens to be the molecular formula, is H_2O. The following procedure can be used to obtain the empirical formula of any compound from experimental data. The left column outlines the procedure, and the center and right columns contain two examples of how to apply the procedure.

PEARSON
eText 2.0

Interactive
Worked Example
Video 6.11

Obtaining an Empirical Formula from Experimental Data

EXAMPLE 6.10

You decompose a compound containing nitrogen and oxygen in the laboratory and produce 24.5 g of nitrogen and 70.0 g of oxygen. Calculate the empirical formula of the compound.

EXAMPLE 6.11

A laboratory analysis of aspirin determines the following mass percent composition:

C 60.00%
H 4.48%
O 35.53%

Find the empirical formula.

---

1. Write down (or calculate) the masses of each element present in a sample of the compound. If you are given mass percent composition, assume a 100-g sample and calculate the masses of each element from the given percentages.

GIVEN: 24.5 g N
70.0 g O

FIND: empirical formula

GIVEN: In a 100-g sample:

60.00 g C
4.48 g H
35.53 g O

FIND: empirical formula

---

2. Convert each of the masses in Step 1 to moles by using the appropriate molar mass for each element as a conversion factor.

SOLUTION

$$24.5 \, \text{g N} \times \frac{1 \, \text{mol N}}{14.01 \, \text{g N}}$$
$$= 1.75 \, \text{mol N}$$

$$70.0 \, \text{g O} \times \frac{1 \, \text{mol O}}{16.00 \, \text{g O}}$$
$$= 4.38 \, \text{mol O}$$

SOLUTION

$$60.00 \, \text{g C} \times \frac{1 \, \text{mol C}}{12.01 \, \text{g C}}$$
$$= 4.996 \, \text{mol C}$$

$$4.48 \, \text{g H} \times \frac{1 \, \text{mol H}}{1.01 \, \text{g H}}$$
$$= 4.44 \, \text{mol H}$$

$$35.53 \, \text{g O} \times \frac{1 \, \text{mol O}}{16.00 \, \text{g O}}$$
$$= 2.221 \, \text{mol O}$$

---

3. Write down a pseudoformula for the compound, using the moles of each element (from Step 2) as subscripts.

$N_{1.75}O_{4.38}$

$C_{4.996}H_{4.44}O_{2.221}$

---

4. Divide all the subscripts in the formula by the smallest subscript.

$N_{\frac{1.75}{1.75}} O_{\frac{4.38}{1.75}} \longrightarrow N_1O_{2.5}$

$C_{\frac{4.996}{2.221}} H_{\frac{4.44}{2.221}} O_{\frac{2.221}{2.221}} \longrightarrow C_{2.25}H_2O_1$

---

5. If the subscripts are not whole numbers, multiply all the subscripts by a small whole number (see the following table) to arrive at whole-number subscripts.

Fractional Subscript	Multiply by This Number to Get Whole-Number Subscripts
_.10	10
_.20	5
_.25	4
_.33	3
_.50	2
_.66	3
_.75	4

$N_1O_{2.5} \times 2 \longrightarrow N_2O_5$

The correct empirical formula is N_2O_5.

$C_{2.25}H_2O_1 \times 4 \longrightarrow C_9H_8O_4$

The correct empirical formula is $C_9H_8O_4$.

▶ **SKILLBUILDER 6.10** | A sample of a compound is decomposed in the laboratory and produces 165 g of carbon, 27.8 g of hydrogen, and 220.2 g O. Calculate the empirical formula of the compound.

▶ **FOR MORE PRACTICE**
Problems 87, 88, 89, 90.

▶ **SKILLBUILDER 6.11** | Ibuprofen, an aspirin substitute, has the mass percent composition: C 75.69%; H 8.80%; O 15.51%. Calculate the empirical formula of ibuprofen.

▶ **FOR MORE PRACTICE**
Example 6.21; Problems 91, 92, 93, 94.

EXAMPLE **6.12** | **Calculating an Empirical Formula from Reaction Data**

A 3.24-g sample of titanium reacts with oxygen to form 5.40 g of the metal oxide. What is the empirical formula of the metal oxide?

You are given the mass of titanium and the mass of the metal oxide that forms. You are asked to find the empirical formula. You need to recognize this problem as one requiring a special procedure and apply that procedure, which is outlined below.	**GIVEN:** 3.24 g Ti 　　　　5.40 g metal oxide **FIND:** empirical formula
1. Write down (or calculate) the masses of each element present in a sample of the compound. 　In this case, you are given the mass of the initial Ti sample and the mass of its oxide after the sample reacts with oxygen. The mass of oxygen is the difference between the mass of the oxide and the mass of titanium.	**SOLUTION** 　3.24 g Ti 　mass O = mass oxide − mass titanium 　　　　　 = 5.40 g − 3.24 g 　　　　　 = 2.16 g O
2. Convert each of the masses in Step 1 to moles by using the appropriate molar mass for each element as a conversion factor.	$3.24 \text{ g Ti} \times \dfrac{1 \text{ mol Ti}}{47.88 \text{ g Ti}} = 0.0677 \text{ mol Ti}$ $2.16 \text{ g O} \times \dfrac{1 \text{ mol O}}{16.00 \text{ g O}} = 0.135 \text{ mol O}$
3. Write down a pseudoformula for the compound, using the moles of each element obtained in Step 2 as subscripts.	$Ti_{0.0677}O_{0.135}$
4. Divide all the subscripts in the formula by the smallest subscript.	$Ti_{\frac{0.0677}{0.0677}} O_{\frac{0.135}{0.0677}} \longrightarrow TiO_2$
5. If the subscripts are not whole numbers, multiply all the subscripts by a small whole number to arrive at whole-number subscripts.	As the subscripts are already whole numbers, this last step is unnecessary. The correct empirical formula is TiO_2.

▶ **SKILLBUILDER 6.12** | **Calculating an Empirical Formula from Reaction Data**

A 1.56-g sample of copper reacts with oxygen to form 1.95 g of the metal oxide. What is the formula of the metal oxide?

▶ **FOR MORE PRACTICE** Problems 95, 96, 97, 98.

6.9 Calculating Molecular Formulas for Compounds

▶ Calculate a molecular formula from an empirical formula and molar mass.

We can determine the *molecular* formula of a compound from the empirical formula if we also know the molar mass of the compound. Recall from Section 6.8 that the molecular formula is always a whole-number multiple of the empirical formula.

$$\text{molecular formula} = \text{empirical formula} \times n, \text{ where } n = 1, 2, 3 \ldots$$

Suppose we want to find the molecular formula for fructose (a sugar found in fruit) from its empirical formula, CH_2O, and its molar mass, 180.2 g/mol. We know that the molecular formula is a whole-number multiple of CH_2O.

$$\text{molecular formula} = CH_2O \times n$$

We also know that the molar mass is a whole-number multiple of the **empirical formula molar mass**, the sum of the masses of all the atoms in the empirical formula.

$$\text{molar mass} = \text{empirical formula molar mass} \times n$$

▲ Fructose, a sugar found in fruit.

For a particular compound, the value of n in both cases is the same. Therefore, we can find n by calculating the ratio of the molar mass to the empirical formula molar mass.

$$n = \frac{\text{molar mass}}{\text{empirical formula molar mass}}$$

For fructose, the empirical formula molar mass is:

empirical formula molar mass = $1(12.01) + 2(1.01) + 16.00 = 30.03 \text{ g/mol}$

Therefore, n is:

$$n = \frac{180.2 \text{ g/mol}}{30.03 \text{ g/mol}} = 6$$

We can then use this value of n to find the molecular formula.

molecular formula = $CH_2O \times 6 = C_6H_{12}O_6$

EXAMPLE **6.13** | **Calculating Molecular Formula from Empirical Formula and Molar Mass**

Naphthalene is a compound containing carbon and hydrogen that is used in mothballs. Its empirical formula is C_5H_4 and its molar mass is 128.16 g/mol. What is its molecular formula?

SORT

You are given the empirical formula and the molar mass of a compound and asked to find its molecular formula.

GIVEN: empirical formula = C_5H_4
molar mass = 128.16 g/mol

FIND: molecular formula

STRATEGIZE

In the first step, use the molar mass (which is given) and the empirical formula molar mass (which you can calculate based on the empirical formula) to determine n (the integer by which you must multiply the empirical formula to determine the molecular formula).

In the second step, multiply the subscripts in the empirical formula by n to arrive at the molecular formula.

SOLUTION MAP

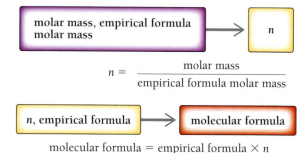

$$n = \frac{\text{molar mass}}{\text{empirical formula molar mass}}$$

molecular formula = empirical formula $\times$ n

SOLVE

First find the empirical formula molar mass. Next follow the solution map. Find n by dividing the molar mass by the empirical formula molar mass (which you just calculated). Multiply the empirical formula by n to determine the molecular formula.

SOLUTION

empirical formula molar mass = $5(12.01) + 4(1.01)$
$= 64.09 \text{ g/mol}$

$$n = \frac{\text{molar mass}}{\text{empirical formula mass}} = \frac{128.16 \text{ g/mol}}{64.09 \text{ g/mol}} = 2$$

molecular formula = $C_5H_4 \times 2 = C_{10}H_8$

CHECK

Check your answer. Does the answer make physical sense?

The answer makes physical sense because it is a whole-number multiple of the empirical formula. Any answer containing fractional subscripts would be an error.

▶ **SKILLBUILDER 6.13 | Calculating Molecular Formula from Empirical Formula and Molar Mass**

Butane is a compound containing carbon and hydrogen used as a fuel in butane lighters. Its empirical formula is C_2H_5, and its molar mass is 58.12 g/mol. Find its molecular formula.

▶ **SKILLBUILDER PLUS** A compound with the following mass percent composition has a molar mass of 60.10 g/mol. Find its molecular formula.

C 39.97%   H 13.41%   N 46.62%

▶ **FOR MORE PRACTICE** Example 6.22; Problems 99, 100, 101, 102.

Chapter **6** in Review

MasteringChemistry™ provides end-of-chapter exercises, feedback-enriched tutorial problems, animations, and interactive activities to encourage problem solving practice and deeper understanding of key concepts and topics.

Self-Assessment Quiz

PEARSON
eText
2.0

Q1. How many atoms are there in 5.8 mol helium?
(a) 23.2 atoms
(b) 9.6×10^{-24} atoms
(c) 5.8×10^{23} atoms
(d) 3.5×10^{24} atoms

Q2. A sample of pure silver has a mass of 155 g. How many moles of silver are in the sample?
(a) 1.44 mol
(b) 1.67×10^4 mol
(c) 0.696 mol
(d) 155 mol

Q3. How many carbon atoms are there in a 12.5-kg sample of carbon?
(a) 6.27×10^{20} atoms
(b) 9.04×10^{28} atoms
(c) 6.27×10^{26} atoms
(d) 1.73×10^{-21} atoms

Q4. Which sample contains the greatest number of atoms?
(a) 15 g Ne
(b) 15 g Ar
(c) 15 g Kr
(d) None of the above (all contain the same number of atoms).

Q5. What is the average mass (in grams) of a single carbon dioxide molecule?
(a) 3.8×10^{-26} g
(b) 7.31×10^{-23} g
(c) 2.65×10^{25} g
(d) 44.01 g

Q6. How many moles of O are in 1.6 mol of $Ca(NO_3)_2$?
(a) 1.6 mol O
(b) 3.2 mol O
(c) 4.8 mol O
(d) 9.6 mol O

Q7. How many grams of Cl are in 25.8 g CF_2Cl_2?
(a) 7.56 g
(b) 3.78 g
(c) 15.1 g
(d) 0.427 g

Q8. Which sample contains the greatest number of F atoms?
(a) 2.0 mol HF
(b) 1.5 mol F_2
(c) 1.0 mol CF_4
(d) 0.5 mol CH_2F_2

Q9. The compound A_2X is 35.8% A by mass. What mass of the compound contains 55.1 g A?
(a) 308 g
(b) 154 g
(c) 19.7 g
(d) 35.8 g

Q10. Which compound has the highest mass percent C?
(a) CO
(b) CO_2
(c) H_2CO_3
(d) H_2CO

Q11. What is the mass percent N in $C_2H_8N_2$?
(a) 23.3% N
(b) 16.6% N
(c) 215% N
(d) 46.6% N

Q12. A compound is 52.14% C, 13.13% H, and 34.73% O by mass. What is the empirical formula of the compound?
(a) C_4HO_3
(b) C_2H_6O
(c) $C_2H_8O_3$
(d) C_3HO_6

Q13. A compound has the empirical formula CH_2O and a formula mass of 120.10 amu. What is the molecular formula of the compound?
(a) CH_2O
(b) $C_2H_4O_2$
(c) $C_3H_6O_3$
(d) $C_4H_8O_4$

Q14. A compound is decomposed in the laboratory and produces 1.40 g N and 0.20 g H. What is the empirical formula of the compound?
(a) NH
(b) N_2H
(c) NH_2
(d) N_7H

Answers: 1:d, 2:a, 3:c, 4:a, 5:b, 6:d, 7:c, 8:c, 9:b, 10:a, 11:d, 12:b, 13:d, 14:c

Chemical Principles

Relevance

The Mole Concept

The mole is a specific number (6.022×10^{23}) that allows us to count atoms or molecules by weighing them. One mole of any element has a mass equivalent to its atomic mass in grams, and a mole of any compound has a mass equivalent to its formula mass in grams. The mass of 1 mol of an element or compound is its molar mass.

The mole concept allows us to determine the number of atoms or molecules in a sample from its mass. Just as a hardware store customer wants to know the number of nails in a certain weight of nails, we want to know the number of atoms in a certain mass of atoms. Because atoms are too small to count, we use their mass.

Chemical Formulas and Chemical Composition

Chemical formulas indicate the relative number of each kind of element in a compound. These numbers are based on atoms or moles. By using molar masses, we can use the information in a chemical formula to determine the relative masses of each kind of element in a compound. We can then relate the mass of a sample of a compound to the masses of the elements contained in the compound.

The chemical composition of compounds is important because it lets us determine how much of a particular element is contained within a particular compound. For example, to assess the threat to the Earth's ozone layer from chlorofluorocarbons (CFCs), we need to know how much chlorine is in a particular CFC.

Empirical and Molecular Formulas from Laboratory Data

The relative masses of the elements within a compound allow us to determine the empirical formula of the compound. If we know the molar mass of the compound, we can also determine its molecular formula.

The first thing we want to know about an unknown compound is its chemical formula because the formula reveals the compound's composition. Chemists often arrive at formulas by analyzing compounds in the laboratory—either by decomposing them or by synthesizing them—to determine the relative masses of the elements they contain.

Chemical Skills

Examples

LO: Convert between moles and number of atoms (Section 6.3).

EXAMPLE **6.14** | **Converting between Moles and Number of Atoms**

Calculate the number of atoms in 4.8 mol of copper.

GIVEN: 4.8 mol Cu

FIND: Cu atoms

SORT
You are given moles of copper and asked to find the number of copper atoms.

SOLUTION MAP

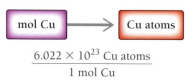

$$\frac{6.022 \times 10^{23} \text{ Cu atoms}}{1 \text{ mol Cu}}$$

STRATEGIZE
To convert between moles and number of atoms, use Avogadro's number, 6.022×10^{23} atoms = 1 mol, as a conversion factor.

RELATIONSHIPS USED
1 mol Cu = 6.022×10^{23} Cu atoms (Avogadro's number, from inside back cover)

SOLUTION

SOLVE
Follow the solution map to solve the problem.

$$4.8 \text{ mol Cu} \times \frac{6.022 \times 10^{23} \text{ Cu atoms}}{1 \text{ mol Cu}} =$$

$$2.9 \times 10^{24} \text{ Cu atoms}$$

CHECK
Check your answer. Are the units correct? Does the answer make physical sense?

The units, Cu atoms, are correct. The answer makes physical sense because the number is very large, as you would expect for nearly 5 mol of atoms.

LO: Convert between grams and moles (Section 6.3).

SORT
You are given the number of moles of aluminum and asked to find the mass of aluminum in grams.

STRATEGIZE
Use the molar mass of aluminum to convert between moles and grams.

SOLVE
Follow the solution map to solve the problem.

CHECK
Check your answer. Are the units correct? Does the answer make physical sense?

EXAMPLE **6.15** | **Converting between Grams and Moles**

Calculate the mass of aluminum (in grams) of 6.73 mol of aluminum.

GIVEN: 6.73 mol Al

FIND: g Al

SOLUTION MAP

$$\boxed{\text{mol Al}} \longrightarrow \boxed{\text{g Al}}$$

$$\frac{26.98 \text{ g Al}}{1 \text{ mol Al}}$$

RELATIONSHIPS USED
26.98 g Al = 1 mol Al (molar mass of Al from periodic table)

SOLUTION

$$6.73 \text{ mol Al} \times \frac{26.98 \text{ g Al}}{1 \text{ mol Al}} = 182 \text{ g Al}$$

The units, g Al, are correct. The answer makes physical sense because each mole has a mass of about 27 g; therefore, nearly 7 mol should have a mass of nearly 190 g.

LO: Convert between grams and number of atoms or molecules (Section 6.3).

SORT
You are given the mass of a zinc sample and asked to find the number of Zn atoms that it contains.

STRATEGIZE
Use the molar mass of the element to convert from grams to moles, and then use Avogadro's number to convert moles to number of atoms.

SOLVE
Follow the solution map to solve the problem.

CHECK
Check your answer. Are the units correct? Does the answer make physical sense?

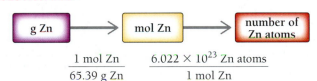

EXAMPLE **6.16** | **Converting between Grams and Number of Atoms or Molecules**

Determine the number of atoms in a 48.3-g sample of zinc.

GIVEN: 48.3 g Zn

FIND: Zn atoms

SOLUTION MAP

$$\boxed{\text{g Zn}} \longrightarrow \boxed{\text{mol Zn}} \longrightarrow \boxed{\begin{array}{c}\text{number of}\\ \text{Zn atoms}\end{array}}$$

$$\frac{1 \text{ mol Zn}}{65.39 \text{ g Zn}} \qquad \frac{6.022 \times 10^{23} \text{ Zn atoms}}{1 \text{ mol Zn}}$$

RELATIONSHIPS USED
65.39 g Zn = 1 mol Zn (molar mass of Zn from periodic table)

1 mol = 6.022×10^{23} atoms (Avogadro's number, from inside back cover)

SOLUTION

$$48.3 \text{ g Zn} \times \frac{1 \text{ mol Zn}}{65.39 \text{ g Zn}} \times \frac{6.022 \times 10^{23} \text{ Zn atoms}}{1 \text{ mol Zn}}$$

$$= 4.45 \times 10^{23} \text{ Zn atoms}$$

The units, Zn atoms, are correct. The answer makes physical sense because the number of atoms in any macroscopic-sized sample should be very large.

LO: Convert between moles of a compound and moles of a constituent element (Section 6.5).

EXAMPLE **6.17** | **Converting between Moles of a Compound and Moles of a Constituent Element**

Determine the number of moles of oxygen in 7.20 mol of H_2SO_4.

GIVEN: 7.20 mol H_2SO_4

FIND: mol O

SOLUTION MAP

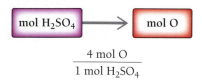

$$\frac{4\ mol\ O}{1\ mol\ H_2SO_4}$$

RELATIONSHIPS USED

4 mol O : 1 mol H_2SO_4

SORT

You are given the number of moles of sulfuric acid and asked to find the number of moles of oxygen.

STRATEGIZE

To convert between moles of a compound and moles of a constituent element, use the chemical formula of the compound to determine a ratio between the moles of the element and the moles of the compound.

SOLVE

Follow the solution map to solve the problem.

SOLUTION

$$7.20\ \text{mol}\ H_2SO_4 \times \frac{4\ mol\ O}{1\ \text{mol}\ H_2SO_4} = 28.8\ mol\ O$$

CHECK

Check your answer. Are the units correct? Does the answer make physical sense?

The units, mol O, are correct. The answer makes physical sense because the number of moles of an element in a compound is equal to or greater than the number of moles of the compound itself.

LO: Convert between grams of a compound and grams of a constituent element (Section 6.5).

EXAMPLE **6.18** | **Converting between Grams of a Compound and Grams of a Constituent Element**

Find the grams of iron in 79.2 g of Fe_2O_3.

GIVEN: 79.2 g Fe_2O_3

FIND: g Fe

SOLUTION MAP

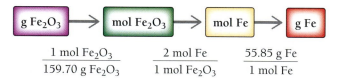

$$\frac{1\ mol\ Fe_2O_3}{159.70\ g\ Fe_2O_3} \qquad \frac{2\ mol\ Fe}{1\ mol\ Fe_2O_3} \qquad \frac{55.85\ g\ Fe}{1\ mol\ Fe}$$

RELATIONSHIPS USED

molar mass Fe_2O_3

= 2(55.85) + 3(16.00)

= 159.70 g/mol

2 mol Fe : 1 mol Fe_2O_3 (from given chemical formula)

SORT

You are given the mass of iron(III) oxide and asked to find the mass of iron contained within it.

STRATEGIZE

Use the molar mass of the compound to convert from grams of the compound to moles of the compound. Then use the chemical formula to obtain a conversion factor to convert from moles of the compound to moles of the constituent element. Finally, use the molar mass of the constituent element to convert from moles of the element to grams of the element.

SOLVE

Follow the solution map to solve the problem.

SOLUTION

$$79.2\ g\ Fe_2O_3 \times \frac{1\ mol\ Fe_2O_3}{159.70\ g\ Fe_2O_3} \times \frac{2\ mol\ Fe}{1\ mol\ Fe_2O_3} \times$$

$$\frac{55.85\ g\ Fe}{1\ mol\ Fe} = 55.4\ g\ Fe$$

CHECK

Check your answer. Are the units correct? Does the answer make physical sense?

The units, g Fe, are correct. The answer makes physical sense because the mass of a constituent element within a compound should be less than the mass of the compound itself.

LO: Use mass percent composition as a conversion factor (Section 6.6).

EXAMPLE 6.19 Using Mass Percent Composition as a Conversion Factor

Determine the mass of titanium in 57.2 g of titanium(IV) oxide. The mass percent of titanium in titanium(IV) oxide is 59.9%.

SORT
You are given the mass of titanium(IV) oxide and the mass percent titanium in the oxide. You are asked to find the mass of titanium in the sample.

GIVEN: 57.2 g TiO$_2$

$$\frac{59.9 \text{ g Ti}}{100 \text{ g TiO}_2}$$

FIND: g Ti

STRATEGIZE
Use the percent composition as a conversion factor between grams of titanium(IV) oxide and grams of titanium.

SOLUTION MAP

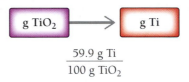

$$\frac{59.9 \text{ g Ti}}{100 \text{ g TiO}_2}$$

RELATIONSHIPS USED
59.9 g Ti : 100 g TiO$_2$

SOLVE
Follow the solution map to solve the problem.

SOLUTION

$$57.2 \text{ g } \cancel{\text{TiO}_2} \times \frac{59.9 \text{ g Ti}}{100 \text{ g } \cancel{\text{TiO}_2}} = 34.3 \text{ g Ti}$$

CHECK
Check your answer. Are the units correct? Does the answer make physical sense?

The units, g Ti, are correct. The answer makes physical sense because the mass of an element within a compound should be less than the mass of the compound itself.

LO: Determine mass percent composition from a chemical formula (Section 6.7).

EXAMPLE 6.20 Determining Mass Percent Composition from a Chemical Formula

Calculate the mass percent composition of potassium in potassium oxide (K$_2$O).

SORT
You are given the formula of potassium oxide and asked to determine the mass percent of potassium within it.

GIVEN: K$_2$O

FIND: mass % K

STRATEGIZE
The solution map shows how you can substitute the information derived from the chemical formula into the mass percent equation to yield the mass percent of the element.

SOLUTION MAP

$$\text{mass \% K} = \frac{2 \times \text{molar mass K}}{\text{molar mass K}_2\text{O}} \times 100\%$$

RELATIONSHIPS USED

mass percent of element X

$$= \frac{\text{mass of element } X \text{ in 1 mol of compound}}{\text{mass of 1 mol of compound}} \times 100\%$$

(mass percent equation, from Section 6.6)

SOLVE
Calculate the molar mass of potassium oxide and follow the solution map to solve the problem.

SOLUTION

molar mass K$_2$O = 2(39.10) + 16.00
= 94.20 g/mol

$$\text{mass \% K} = \frac{2(39.10 \text{ g K})}{94.20 \text{ g K}_2\text{O}} \times 100\% = 83.01\% \text{ K}$$

CHECK
Check your answer. Are the units correct? Does the answer make physical sense?

The units, % K, are correct. The answer makes physical sense because it should be below 100%.

LO: Determine an empirical formula from experimental data (Section 6.8).

You need to recognize this problem as one requiring a special procedure. Follow these steps to solve the problem.

EXAMPLE **6.21** **Determining an Empirical Formula from Experimental Data**

A laboratory analysis of vanillin, the favoring agent in vanilla, determined the mass percent composition: C, 63.15%; H, 5.30%; O, 31.55%. Determine the empirical formula of vanillin.

GIVEN: 63.15% C, 5.30% H, and 31.55% O

FIND: empirical formula

SOLUTION
In a 100-g sample:

    63.15 g C
    5.30 g H
    31.55 g O

1. Write down (or calculate) the masses of each element present in a sample of the compound. If you are given mass percent composition, assume a 100-g sample and calculate the masses of each element from the given percentages.

2. Convert each of the masses in Step 1 to moles by using the appropriate molar mass for each element as a conversion factor.

$$63.15 \text{ g C} \times \frac{1 \text{ mol C}}{12.01 \text{ g C}} = 5.258 \text{ mol C}$$

$$5.30 \text{ g H} \times \frac{1 \text{ mol H}}{1.01 \text{ g H}} = 5.25 \text{ mol H}$$

$$31.55 \text{ g O} \times \frac{1 \text{ mol O}}{16.01 \text{ g O}} = 1.972 \text{ mol O}$$

3. Write down a pseudoformula for the compound using the moles of each element (from Step 2) as subscripts.

$$C_{5.258}H_{5.25}O_{1.972}$$

4. Divide all the subscripts in the formula by the smallest subscript.

$$C_{\frac{5.258}{1.972}} H_{\frac{5.25}{1.972}} O_{\frac{1.972}{1.972}} \longrightarrow C_{2.67}H_{2.66}O_1$$

5. If the subscripts are not whole numbers, multiply all the subscripts by a small whole number to arrive at whole-number subscripts.

$$C_{2.67}H_{2.66}O_1 \times 3 \longrightarrow C_8H_8O_3$$

The correct empirical formula is $C_8H_8O_3$.

LO: Calculate a molecular formula from an empirical formula and molar mass (Section 6.9).

EXAMPLE **6.22** **Calculating a Molecular Formula from an Empirical Formula and Molar Mass**

Acetylene, a gas used in welding torches, has the empirical formula CH and a molar mass of 26.04 g/mol. Find its molecular formula.

GIVEN: empirical formula = CH

          molar mass = 26.04 g/mol

FIND: molecular formula

SORT
You are given the empirical formula and molar mass of acetylene and asked to find the molecular formula.

STRATEGIZE
In the first step, use the molar mass (which is given) and the empirical formula molar mass (which you can calculate based on the empirical formula) to determine n (the integer by which you must multiply the empirical formula to arrive at the molecular formula).

In the second step, multiply the coefficients in the empirical formula by n to arrive at the molecular formula.

SOLUTION MAP

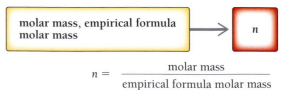

$$n = \frac{\text{molar mass}}{\text{empirical formula molar mass}}$$

$$\text{molecular formula} = \text{empirical formula} \times n$$

SOLVE

Follow the solution map to solve the problem. Calculate the empirical formula molar mass, which is the sum of the masses of all the atoms in the empirical formula.

Next, find n, the ratio of the molar mass to empirical mass.

Finally, multiply the empirical formula by n to get the molecular formula.

CHECK

Check your answer. Does the answer make physical sense?

SOLUTION

empirical formula molar mass

$$= 12.01 + 1.01$$

$$= 13.02 \text{ g/mol}$$

$$n = \frac{\text{molar mass}}{\text{empirical formula molar mass}}$$

$$= \frac{26.04 \text{ g/mol}}{13.02 \text{ g/mol}} = 2$$

molecular formula $= \text{CH} \times 2 \longrightarrow \text{C}_2\text{H}_2$

The answer makes physical sense because the formula subscripts are all whole numbers. Any answer with non-whole numbers is incorrrect.

Key Terms

Avogadro's number [6.3]
empirical formula [6.8]

empirical formula molar
mass [6.9]

mass percent (composition) [6.6]
molar mass [6.3]

mole (mol) [6.3]
molecular formula [6.8]

Exercises

Questions

1. Why is chemical composition important?
2. How can you efficiently determine the number of atoms in a sample of an element? Why is counting them not an option?
3. How many atoms are in 1 mol of atoms?
4. How many molecules are in 1 mol of molecules?
5. What is the mass of 1 mol of atoms for an element?
6. What is the mass of 1 mol of molecules for a compound?
7. What is the mass of 1 mol of atoms of each element?
   (a) P   (b) Pt   (c) C   (d) Cr
8. What is the mass of 1 mol of molecules of each compound?
   (a) CO_2 (b) CH_2Cl_2 (c) $\text{C}_{12}\text{H}_{22}\text{O}_{11}$ (d) SO_2
9. The subscripts in a chemical formula give relationships between moles of the constituent elements and moles of the compound. Explain why these subscripts *do not* give relationships between grams of the constituent elements and grams of the compound.
10. Write the conversion factors between moles of each constituent element and moles of the compound for $\text{C}_{12}\text{H}_{22}\text{O}_{11}$.

11. You can use mass percent composition as a conversion factor between grams of a constituent element and grams of the compound. Write the conversion factor (including units) inherent in each mass percent composition.
    (a) Water is 11.19% hydrogen by mass.
    (b) Fructose, also known as fruit sugar, is 53.29% oxygen by mass.
    (c) Octane, a component of gasoline, is 84.12% carbon by mass.
    (d) Ethanol, the alcohol in alcoholic beverages, is 52.14% carbon by mass.
12. What is the mathematical formula for calculating mass percent composition from a chemical formula?
13. How are the empirical formula and the molecular formula of a compound related?
14. Why is it important to be able to calculate an empirical formula from experimental data?
15. What is the empirical formula mass of a compound?
16. How are the molar mass and empirical formula mass for a compound related?

Problems

THE MOLE CONCEPT

17. How many mercury atoms are in 5.8 mol of mercury?

18. How many moles of gold atoms do 3.45×10^{24} gold atoms constitute?

19. How many atoms are in each elemental sample?
    (a) 3.4 mol Cu   (b) 9.7×10^{-3} mol C
    (c) 22.9 mol Hg   (d) 0.215 mol Na

20. How many moles of atoms are in each elemental sample?
    (a) 4.6×10^{24} Pb atoms   (b) 2.87×10^{22} He atoms
    (c) 7.91×10^{23} K atoms   (d) 4.41×10^{21} Ca atoms

21. Complete the table.

Element	Moles	Number of Atoms
Ne	0.552	____
Ar	____	3.25×10^{24}
Xe	1.78	____
He	____	1.08×10^{20}

22. Complete the table.

Element	Moles	Number of Atoms
Cr	____	9.61×10^{23}
Fe	1.52×10^{-5}	____
Ti	0.0365	____
Hg	____	1.09×10^{23}

23. Consider these definitions.

1 doz = 12   1 gross = 144
1 ream = 500   1 mol = 6.022×10^{23}

Suppose you have 872 sheets of paper. How many _____ of paper sheets do you have?
(a) dozens   **(b)** gross   **(c)** reams   **(d)** moles

24. A pure copper penny contains approximately 3.0×10^{22} copper atoms. Use the definitions in the previous problem to determine how many _____ of copper atoms are in a penny.
(a) dozens   **(b)** gross   **(c)** reams   **(d)** moles

25. How many moles of tin atoms are in a pure tin cup with a mass of 38.1 g?

26. A lead fishing weight contains 0.12 mol of lead atoms. What is its mass?

27. A pure gold coin contains 0.145 mol of gold. What is its mass?

28. A helium balloon contains 0.46 g of helium. How many moles of helium does it contain?

29. How many moles of atoms are in each elemental sample?
(a) 1.34 g Zn   **(b)** 24.9 g Ar
(c) 72.5 g Ta   **(d)** 0.0223 g Li

30. What is the mass in grams of each elemental sample?
(a) 6.64 mol W   **(b)** 0.581 mol Ba
(c) 68.1 mol Xe   **(d)** 1.57 mol S

31. Complete the table.

Element	Moles	Mass
Ne	____	22.5 g
Ar	0.117	____
Xe	____	1.00 kg
He	1.44×10^{-4}	____

32. Complete the table.

Element	Moles	Mass
Cr	0.00442	____
Fe	____	73.5 mg
Ti	1.009×10^{-3}	____
Hg	____	1.78 kg

33. A pure silver ring contains 0.0134 mmol (millimol) Ag. How many silver atoms does it contain?

34. A pure gold ring contains 0.0102 mmol (millimol) Au. How many gold atoms does it contain?

35. How many aluminum atoms are in 3.78 g of aluminum?

36. What is the mass of 4.91×10^{21} platinum atoms?

37. How many atoms are in each elemental sample?
(a) 16.9 g Sr
(b) 26.1 g Fe
(c) 8.55 g Bi
(d) 38.2 g P

38. Calculate the mass in grams of each elemental sample.
(a) 1.32×10^{20} uranium atoms
(b) 2.55×10^{22} zinc atoms
(c) 4.11×10^{23} lead atoms
(d) 6.59×10^{24} silicon atoms

39. How many carbon atoms are in a diamond (pure carbon) with a mass of 38 mg?

40. How many helium atoms are in a helium blimp containing 495 kg of helium?

41. How many titanium atoms are in a pure titanium bicycle frame with a mass of 1.28 kg?

42. How many copper atoms are in a pure copper statue with a mass of 133 kg?

43. Complete the table.

Element	Mass	Moles	Number of Atoms
Na	38.5 mg	____	____
C	____	1.12	____
V	____	____	214
Hg	1.44 kg	____	____

44. Complete the table.

Element	Mass	Moles	Number of Atoms
Pt	____	0.0449	____
Fe	____	____	1.14×10^{25}
Ti	23.8 mg	____	____
Hg	____	2.05	____

45. Which sample contains the greatest number of atoms?
(a) 27.2 g Cr   **(b)** 55.1 g Ti   **(c)** 205 g Pb

46. Which sample contains the greatest number of atoms?
(a) 10.0 g He   **(b)** 25.0 g Ne   **(c)** 115 g Xe

47. Determine the number of moles of molecules (or formula units) in each sample.
(a) 38.2 g sodium chloride
(b) 36.5 g nitrogen monoxide
(c) 4.25 kg carbon dioxide
(d) 2.71 mg carbon tetrachloride

48. Determine the mass of each sample.
(a) 1.32 mol carbon tetrafluoride
(b) 0.555 mol magnesium fluoride
(c) 1.29 mmol carbon disulfide
(d) 1.89 kmol sulfur trioxide

49. Complete the table.

Compound	Mass	Moles	Number of Molecules
H_2O	112 kg	____	____
N_2O	6.33 g	____	____
SO_2	____	2.44	____
CH_2Cl_2	____	0.0643	____

50. Complete the table.

Compound	Mass	Moles	Number of Molecules
CO_2	____	0.0153	____
CO	____	0.0150	____
BrI	23.8 mg	____	____
CF_2Cl_2	1.02 kg	____	____

51. A mothball, composed of naphthalene ($C_{10}H_8$), has a mass of 1.32 g. How many naphthalene molecules does it contain?

52. Calculate the mass in grams of a single water molecule.

53. How many molecules are in each sample?
(a) 3.5 g H_2O
(b) 56.1 g N_2
(c) 89 g CCl_4
(d) 19 g $C_6H_{12}O_6$

54. Calculate the mass in grams of each sample.
(a) 5.94×10^{20} H_2O_2 molecules
(b) 2.8×10^{22} SO_2 molecules
(c) 4.5×10^{25} O_3 molecules
(d) 9.85×10^{19} CH_4 molecules

55. A sugar crystal contains approximately 1.8×10^{17} sucrose ($C_{12}H_{22}O_{11}$) molecules. What is its mass in milligrams?

56. A salt crystal has a mass of 0.12 mg. How many NaCl formula units does it contain?

57. How much money, in dollars, does 1 mol of pennies represent? If this amount of money were evenly distributed to the entire world's population (about 7.1 billion people), how much would each person get? Would each person be a millionaire? Billionaire? Trillionaire?

58. A typical dust particle has a diameter of about 10.0 mm. If 1.0 mol of dust particles were laid end to end along the equator, how many times would they encircle the planet? The circumference of the Earth at the equator is 40,076 km.

CHEMICAL FORMULAS AS CONVERSION FACTORS

59. Determine the number of moles of Cl in 2.7 mol $CaCl_2$.

60. How many moles of O are in 12.4 mol $Fe(NO_3)_3$?

61. Which sample contains the greatest number of moles of O?
(a) 2.3 mol H_2O **(b)** 1.2 mol H_2O_2
(c) 0.9 mol $NaNO_3$ **(d)** 0.5 mol $Ca(NO_3)_2$

62. Which sample contains the greatest number of moles of Cl?
(a) 3.8 mol HCl   **(b)** 1.7 mol CH_2Cl_2
(c) 4.2 mol $NaClO_3$ **(d)** 2.2 mol $Mg(ClO_4)_2$

63. Determine the number of moles of C in each sample.
  (a) 2.5 mol CH_4 **(b)** 0.115 mol C_2H_6
  (c) 5.67 mol C_4H_{10} **(d)** 25.1 mol C_8H_{18}

64. Determine the number of moles of H in each sample.
  (a) 4.67 mol H_2O **(b)** 8.39 mol NH_3
  (c) 0.117 mol N_2H_4 **(d)** 35.8 mol $C_{10}H_{22}$

65. For each set of molecular models, write a relationship between moles of hydrogen and moles of molecules. Then determine the total number of hydrogen atoms present. (H—white; O—red; C—black; N—blue)

(a)   (b)   (c)

66. For each set of molecular models, write a relationship between moles of oxygen and moles of molecules. Then determine the total number of oxygen atoms present. (H— white; O—red; C—black; S—yellow)

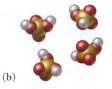

(a)   (b)   (c)

67. How many grams of Cl are in 38.0 g of each sample of chlorofluorocarbons (CFCs)?
  (a) CF_2Cl_2
  (b) $CFCl_3$
  (c) $C_2F_3Cl_3$
  (d) CF_3Cl

68. Calculate the number of grams of sodium in 1.00 g of each sodium-containing food additive.
  (a) NaCl (table salt)
  (b) Na_3PO_4 (sodium phosphate)
  (c) $NaC_7H_5O_2$ (sodium benzoate)
  (d) $Na_2C_6H_6O_7$ (sodium hydrogen citrate)

69. Iron is found in Earth's crust as several different iron compounds. Calculate the mass (in kg) of each compound that contains 1.0×10^3 kg of iron.
  (a) Fe_2O_3 (hematite)
  (b) Fe_3O_4 (magnetite)
  (c) $FeCO_3$ (siderite)

70. Lead is found in Earth's crust as several lead compounds. Calculate the mass (in kg) of each compound that contains 1.0×10^3 kg of lead.
  (a) PbS (galena)
  (b) $PbCO_3$ (cerussite)
  (c) $PbSO_4$ (anglesite)

MASS PERCENT COMPOSITION

71. A 2.45-g sample of strontium completely reacts with oxygen to form 2.89 g of strontium oxide. Use this data to calculate the mass percent composition of strontium in strontium oxide.

72. A 3.53-g sample of aluminum completely reacts with oxygen to form 6.67 g of aluminum oxide. Use this data to calculate the mass percent composition of aluminum in aluminum oxide.

73. A 1.912-g sample of calcium chloride is decomposed into its constituent elements and found to contain 0.690 g Ca and 1.222 g Cl. Calculate the mass percent composition of Ca and Cl in calcium chloride.

74. A 0.45-g sample of aspirin is decomposed into its constituent elements and found to contain 0.27 g C, 0.020 g H, and 0.16 g O. Calculate the mass percent composition of C, H, and O in aspirin.

75. Copper(II) fluoride contains 37.42% F by mass. Use this percentage to calculate the mass of fluorine in grams contained in 28.5 g of copper(II) fluoride.

76. Silver chloride, used in silver plating, contains 75.27% Ag. Calculate the mass of silver chloride in grams required to make 4.8 g of silver plating.

77. In small amounts, the fluoride ion (often consumed as NaF) prevents tooth decay. According to the American Dental Association, an adult female should consume 3.0 mg of fluorine per day. Calculate the amount of sodium fluoride (45.24% F) that a woman should consume to get the recommended amount of fluorine.

78. The iodide ion, usually consumed as potassium iodide, is a dietary mineral essential to good nutrition. In countries where potassium iodide is added to salt, iodine deficiency or goiter has been almost completely eliminated. The recommended daily allowance (RDA) for iodine is 150 μg/day. How much potassium iodide (76.45% I) should you consume to meet the RDA?

MASS PERCENT COMPOSITION FROM CHEMICAL FORMULA

79. Calculate the mass percent composition of nitrogen in each compound.
(a) N_2O (b) NO (c) NO_2 (d) N_2O_5

80. Calculate the mass percent composition of carbon in each compound.
(a) C_2H_2 (b) C_3H_6 (c) C_2H_6 (d) C_2H_6O

81. Calculate the mass percent composition of each element in each compound.
(a) $C_2H_4O_2$ (b) CH_2O_2 (c) C_3H_9N (d) $C_4H_{12}N_2$

82. Calculate the mass percent composition of each element in each compound.
(a) $FeCl_3$ (b) TiO_2 (c) H_3PO_4 (d) HNO_3

83. Calculate the mass percent composition of O in each compound.
(a) calcium nitrate (b) iron(II) sulfate
(c) carbon dioxide

84. Calculate the mass percent composition of Cl in each compound.
(a) carbon tetrachloride (b) calcium hypochlorite
(c) perchloric acid

85. Various iron ores have different amounts of iron per kilogram of ore. Calculate the mass percent composition of iron for each iron ore: Fe_2O_3 (hematite), Fe_3O_4 (magnetite), $FeCO_3$ (siderite). Which ore has the highest iron content?

86. Plants need nitrogen to grow, so many fertilizers consist of nitrogen-containing compounds. Calculate the mass percent composition of nitrogen in each fertilizer: NH_3, $CO(NH_2)_2$, NH_4NO_3, $(NH_4)_2SO_4$. Which fertilizer has the highest nitrogen content?

CALCULATING EMPIRICAL FORMULAS

87. A compound containing nitrogen and oxygen is decomposed in the laboratory and produces 1.78 g of nitrogen and 4.05 g of oxygen. Calculate the empirical formula of the compound.

88. A compound containing selenium and fluorine is decomposed in the laboratory and produces 2.231 g of selenium and 3.221 g of fluorine. Calculate the empirical formula of the compound.

89. Samples of several compounds are decomposed, and the masses of their constituent elements are measured. Calculate the empirical formula for each compound.
(a) 1.245 g Ni, 5.381 g I
(b) 1.443 g Se, 5.841 g Br
(c) 2.128 g Be, 7.557 g S, 15.107 g O

90. Samples of several compounds are decomposed, and the masses of their constituent elements are measured. Calculate the empirical formula for each compound.
(a) 2.677 g Ba, 3.115 g Br
(b) 1.651 g Ag, 0.1224 g O
(c) 0.672 g Co, 0.569 g As, 0.486 g O

91. The rotten smell of a decaying animal carcass is partially due to a nitrogen-containing compound called putrescine. Elemental analysis of putrescine indicates that it consists of 54.50% C, 13.73% H, and 31.77% N. Calculate the empirical formula of putrescine. (*Hint*: Begin by assuming a 100-g sample and determining the mass of each element in that 100-g sample.)

92. Citric acid, the compound responsible for the sour taste of lemons, has the elemental composition: C, 37.51%; H, 4.20%; O, 58.29%. Calculate the empirical formula of citric acid. (*Hint*: Begin by assuming a 100-g sample and determining the mass of each element in that 100-g sample.)

93. These compounds are found in many natural favors and scents. Calculate the empirical formula for each compound.
(a) ethyl butyrate (pineapple oil): C, 62.04%; H, 10.41%; O, 27.55%
(b) methyl butyrate (apple flavor): C, 58.80%; H, 9.87%; O, 31.33%
(c) benzyl acetate (oil of jasmine): C, 71.98%; H, 6.71%; O, 21.31%

94. Calculate the empirical formula for each over-the-counter pain reliever
(a) acetaminophen (Tylenol): C, 63.56%; H, 6.00%; N, 9.27%; O, 21.17%
(b) naproxen (Aleve): C, 73.03%; H, 6.13%; O, 20.84%

95. A 1.45-g sample of phosphorus burns in air and forms 2.57 g of a phosphorus oxide. Calculate the empirical formula of the oxide. (*Hint*: Determine the mass of oxygen in the 2.57 g of phosphorus oxide by determining the difference in mass before and after the phosphorus burns in air.)

96. A 2.241-g sample of nickel reacts with oxygen to form 2.852 g of the metal oxide. Calculate the empirical formula of the oxide.

97. A 0.77-mg sample of nitrogen reacts with chlorine to form 6.61 mg of the chloride. What is the empirical formula of the nitrogen chloride?

98. A 45.2-mg sample of phosphorus reacts with selenium to form 131.6 mg of the selenide. What is the empirical formula of the phosphorus selenide?

CALCULATING MOLECULAR FORMULAS

99. A compound containing carbon and hydrogen has a molar mass of 56.11 g/mol and an empirical formula of CH_2. Determine its molecular formula.

100. A compound containing phosphorus and oxygen has a molar mass of 219.9 g/mol and an empirical formula of P_2O_3. Determine its molecular formula.

101. The molar masses and empirical formulas of several compounds containing carbon and chlorine are listed here. Find the molecular formula of each compound.
(a) 284.77 g/mol, CCl   (b) 131.39 g/mol, C_2HCl_3
(c) 181.44 g/mol, C_2HCl

102. The molar masses and empirical formulas of several compounds containing carbon and nitrogen are listed here. Find the molecular formula of each compound.
(a) 163.26 g/mol, $C_{11}H_{17}N$ (b) 186.24 g/mol, C_6H_7N
(c) 312.29 g/mol, C_3H_2N

Cumulative Problems

103. A pure copper cube has an edge length of 1.42 cm. How many copper atoms does it contain? (volume of a cube = (edge length)^3; density of copper = 8.96 g/cm^3) *Hint*: Start by calculating the volume of the copper cube.

104. A pure silver sphere has a radius of 0.886 cm. How many silver atoms does it contain? (volume of a sphere = $\frac{4}{3}\pi r^3$; density of silver = 10.5 g/cm^3) *Hint*: Start by calculating the volume of the silver sphere.

105. A drop of water has a volume of approximately 0.05 mL. How many water molecules does it contain? (density of water = 1.0 g/cm^3)

106. Fingernail–polish remover is primarily acetone (C_3H_6O). How many acetone molecules are in a bottle of acetone with a volume of 325 mL? (density of acetone = 0.788 g/cm^3)

107. Complete the table.

Substance	Mass	Moles	Number of Particles (atoms or molecules)
Ar	___	4.5×10^{-4}	___
NO_2	___	___	1.09×10^{20}
K	22.4 mg	___	___
C_8H_{18}	3.76 kg	___	___

108. Complete the table.

Substance	Mass	Moles	Number of Particles (atoms or molecules)
$C_6H_{12}O_6$	15.8 g	___	___
Pb	___	___	9.04×10^{21}
CF_4	22.5 mg	___	___
C	___	0.0388	___

109. Determine the chemical formula of each compound and refer to the formula to calculate the mass percent composition of each constituent element.
(a) copper(II) iodide
(b) sodium nitrate
(c) lead(II) sulfate
(d) calcium fluoride

110. Determine the chemical formula of each compound and refer to the formula to calculate the mass percent composition of each constituent element.
(a) nitrogen triiodide
(b) xenon tetrafluoride
(c) phosphorus trichloride
(d) carbon monoxide

111. The rock in a particular iron ore deposit contains 78% Fe_2O_3 by mass. How many kilograms of the rock must a mining company process to obtain 1.0×10^3 kg of iron?

112. The rock in a lead ore deposit contains 84% PbS by mass. How many kilograms of the rock must a mining company process to obtain 1.0 kg of Pb?

113. A leak in the air conditioning system of an office building releases 12 kg of CHF_2Cl per month. If the leak continues, how many kilograms of Cl are emitted into the atmosphere each year?

114. A leak in the air conditioning system of an older car releases 55 g of CF_2Cl_2 per month. How much Cl is emitted into the atmosphere each year by this car?

115. Hydrogen is a possible future fuel. However, elemental hydrogen is rare, so it must be obtained from a hydrogen-containing compound such as water. If hydrogen were obtained from water, how much hydrogen, in grams, could be obtained from 1.0 L of water? (density of water = $1.0 \, \text{g/cm}^3$)

116. Hydrogen, a possible future fuel mentioned in Problem 115, can also be obtained from ethanol. Ethanol can be made from the fermentation of crops such as corn. How much hydrogen, in grams, could be obtained from 1.0 kg of ethanol (C_2H_5OH)?

117. Complete the table of compounds that contain only carbon and hydrogen.

Formula	Molar Mass	% C (by mass)	% H (by mass)
C_2H_4	___	___	___
___	58.12	82.66%	___
C_4H_8	___	___	___
___	44.09	___	18.29%

118. Complete the table of compounds that contain only chromium and oxygen.

Formula	Name	Molar Mass	% Cr (by mass)	% O (by mass)
___	Chromium(III) oxide	___	___	___
___	___	84.00	61.90%	___
___	___	100.00	___	48.00%

119. Butanedione, a component of butter and body odor, has a cheesy smell. Elemental analysis of butanedione gave the mass percent composition: C, 55.80%; H, 7.03%; O, 37.17%. The molar mass of butanedione is 86.09 g/mol. Determine the molecular formula of butanedione.

120. Caffeine, a stimulant found in coffee and soda, has the mass percent composition: C, 49.48%; H, 5.19%; N, 28.85%; O, 16.48%. The molar mass of caffeine is 194.19 g/mol. Find the molecular formula of caffeine.

121. Nicotine, a stimulant found in tobacco, has the mass percent composition: C, 74.03%; H, 8.70%; N, 17.27%. The molar mass of nicotine is 162.26 g/mol. Find the molecular formula of nicotine.

122. Estradiol is a female sexual hormone that causes maturation and maintenance of the female reproductive system. Elemental analysis of estradiol gave the mass percent composition: C, 79.37%; H, 8.88%; O, 11.75%. The molar mass of estradiol is 272.37 g/mol. Find the molecular formula of estradiol.

123. A sample contains both KBr and KI in unknown quantities. If the sample has a total mass of 5.00 g and contains 1.51 g K, what are the percentages of KBr and KI in the sample by mass?

124. A sample contains both CO_2 and Ne in unknown quantities. If the sample contains a combined total of 1.75 mol and has a total mass of 65.3 g, what are the percentages of CO_2 and Ne in the sample by mole?

125. Ethanethiol (C_2H_6S) is a compound with a disagreeable odor that is used to impart an odor to natural gas. When ethanethiol is burned, the sulfur reacts with oxygen to form SO_2. What mass of SO_2 forms upon the complete combustion of 28.7 g of ethanethiol?

126. Methanethiol (CH_4S) has a disagreeable odor and is often a component of bad breath. When methanethiol is burned, the sulfur reacts with oxygen to form SO_2. What mass of SO_2 forms upon the complete combustion of 1.89 g of methanethiol?

127. An iron ore contains 38% Fe_2O_3 by mass. What is the maximum mass of iron that can be recovered from 10.0 kg of this ore?

128. Seawater contains approximately 3.5% NaCl by mass and has a density of 1.02 g/mL. What volume of seawater contains 1.0 g of sodium?

Highlight Problems

129. You can use the concepts in this chapter to obtain an estimate of the number of atoms in the universe. These steps will guide you through this calculation.
  (a) Begin by calculating the number of atoms in the sun. Assume that the sun is pure hydrogen with a density of 1.4 g/cm^3. The radius of the sun is 7×10^8 m, and the volume of a sphere is $V = \frac{4}{3}\pi r^3$.
  (b) The sun is an average-sized star, and stars are believed to compose most of the mass of the visible universe (planets are so small they can be ignored), so we can estimate the number of atoms in a galaxy by assuming that every star in the galaxy has the same number of atoms as our sun. The Milky Way galaxy is believed to contain 1×10^{11} stars. Use your answer from part a to calculate the number of atoms in the Milky Way galaxy.

(c) Astronomers estimate that the universe contains approximately 1×10^{11} galaxies. If each of these galaxies contains the same number of atoms as the Milky Way galaxy, what is the total number of atoms in the universe?

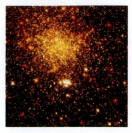 

▲ Our sun is one of the 100 billion stars in the Milky Way galaxy. The universe is estimated to contain about 100 billion galaxies.

130. Because of increasing evidence of damage to the ozone layer, chlorofluorocarbon (CFC) production was banned in 1996. However, about 100 million auto air conditioners still use CFC-12 (CF_2Cl_2). These air conditioners are recharged from stockpiled supplies of CFC-12. If each of the 100 million automobiles contains 1.1 kg of CFC-12 and leaks 25% of its CFC-12 into the atmosphere per year, how much Cl in kilograms do auto air conditioners add to the atmosphere each year? (Assume two significant figures in your calculations.)

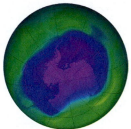

▲ The ozone hole over Antarctica on September 24, 2009. The dark blue and purple areas over the South Pole represent depressed ozone concentrations.

131. In 1996, the media reported that possible evidence of life on Mars was found on a meteorite called Allan Hills 84001 (AH 84001). The meteorite was discovered in Antarctica in 1984 and is believed to have originated on Mars. Elemental analysis of substances within its crevices revealed carbon-containing compounds that normally derive only from living organisms. Suppose that one of those compounds had a molar mass of 202.23 g/mol and the mass percent composition: C, 95.02%; H, 4.98%. What is the molecular formula for the carbon-containing compound?

▲ The Allan Hills 84001 meteorite. Elemental analysis of the substances within the crevices of this meteorite revealed carbon-containing compounds that normally originate from living organisms.

Questions for Group Work

Discuss these questions with the group and record your consensus answer.

132. Using grammatically correct English sentences, describe the relationship between the number 6.022×10^{23} and the gram.

133. Imagine that all our balances displayed mass in units of slugs instead of grams. Would we still want to use the number 6.022×10^{23} for Avogadro's number? Why or why not? If not, what number would we want to use instead? (A slug is equal to 14,954 g.)

134. Amylose is a "polysaccharide" that plants use to store energy. It is made of repeating subunits of $C_6H_{10}O_5$. If a particular amylose molecule has 2537 of these subunits, what is its molecular formula? What is its molar mass? What is the empirical formula for amylose?

Data Interpretation and Analysis

135. Public water systems often add fluoride to drinking water because, in the proper amounts, fluoride improves dental health and prevents cavities. Too much fluoride, however, can cause fluorosis, which stains teeth. In 2015, the U.S. Public Health Service (PHS) revised its 1962 recommendations for the amount of fluoride in public water systems. The 1962 recommendations depended on the average temperature for the region in question as shown here.

1962 Fluoride Recommendations

Annual Average of Maximum Daily Air Temperatures (°C)	Recommended Fluoride Concentration (mg/L)			Maximum Allowable Fluoride Concentration (mg/L)
	Lower	Optimum	Upper	
10.0–12.0	0.9	1.2	1.7	2.4
12.1–14.6	0.8	1.1	1.5	2.2
14.7–17.7	0.8	1.0	1.3	2.0
17.8–21.4	0.7	0.9	1.2	1.8
21.5–26.3	0.7	0.8	1.0	1.6
26.4–31.5	0.6	0.7	0.8	1.4

The new recommendation is simply for municipalities to fluoridate public water systems at a level of 0.7 mg/L. Notice that this level is at the lower end of previous recommendations. The recommended level was lowered because U.S. citizens are now getting fluoride from other sources, including toothpaste and mouthwash. The recommended level balances the need for fluoride to improve dental health with the risk of developing fluorosis from too much fluoride. Examine the data in the table and answer the following questions:

(a) Determine the percent change in optimum recommended fluoride concentration for a water system with annual average maximum daily temperatures of 17.8–21.4 °C. *Hint*: the percent change is given by

$$\% \text{ change} = \frac{\text{final value} - \text{initial value}}{\text{initial value}} \times 100\%.$$

(b) Sodium fluoride (NaF) and sodium fluorosilicate (Na_2SiF_6) are commercially available in 100.0-lb bags. Calculate the mass in kg of fluoride in a 100.0-lb bag for each of these compounds. If the compounds cost about the same per 1000.0 lb, which compound would be the better choice from an economic point of view?

(c) The National Institutes of Health (NIH) recommends a fluoride intake of 3.1 mg/day for adult females and 3.8 mg/day for adult males. If drinking water contains 0.7 mg/L, how much water should a person consume daily to meet the NIH recommendation for women? For men?

Answers to Skillbuilder Exercises

Skillbuilder 6.1	5.32×10^{22} Au atoms
Skillbuilder 6.2	89.1 g S
Skillbuilder 6.3	8.17 g He
Skillbuilder 6.4	2.56×10^{-2} mol NO_2
Skillbuilder 6.5	1.22×10^{23} H_2O molecules
Skillbuilder 6.6	5.6 mol O
Skillbuilder 6.7	3.3 g O
Skillbuilder Plus, p. 182	4.04 g O
Skillbuilder 6.8	8.6 g Na
Skillbuilder 6.9	53.28% O
Skillbuilder 6.10	CH_2O
Skillbuilder 6.11	$C_{13}H_{18}O_2$
Skillbuilder 6.12	CuO
Skillbuilder 6.13	C_4H_{10}
Skillbuilder Plus, p. 190	$C_2H_8N_2$

Answers to Conceptual Checkpoints

6.1 (a) $3.5 \text{ lb} \times \dfrac{1 \text{ doz}}{0.50 \text{ lb}} \times \dfrac{12 \text{ nails}}{\text{doz}} = 84 \text{ nails}$

6.2 (a) The mole is a counting unit; it represents a definite number (Avogadro's number, 6.022×10^{23}). Therefore, a given number of atoms always represents a precise number of moles, regardless of what atom is involved. Atoms of different elements have different masses. So if samples of different elements have the same mass, they *cannot* contain the same number of atoms or moles.

6.3 (b) Because carbon has lower molar mass than cobalt or lead, a 1-g sample of carbon contains more atoms than 1 g of cobalt or lead.

6.4 Sample A would have the greater number of molecules. Sample A has a lower molar mass than sample B, so a given mass of sample A has more moles and therefore more molecules than the same mass of sample B.

6.5 48 mol of H

6.6 (c) 1.0 mol of F_2 contains 2.0 mol of F atoms. Each of the other two options contains less than 2 mol of F atoms.

6.7 (b) This compound has the highest ratio of oxygen atoms to chromium atoms and therefore has the greatest mass percent of oxygen.

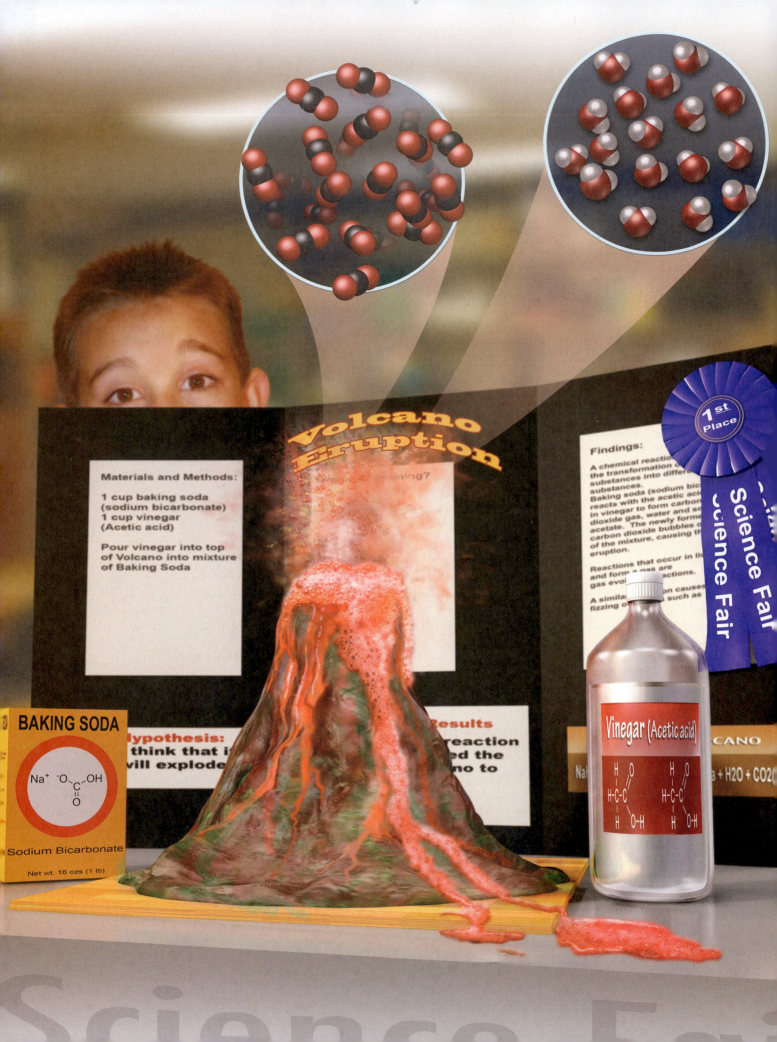

7

Chemical Reactions

Chemistry . . . is one of the broadest branches of science if for no other reason than, when we think about it, everything is chemistry.

—Luciano Caglioti (1933–)

7.1   Grade School Volcanoes, Automobiles, and Laundry Detergents

Did you make a clay volcano in grade school that erupted when filled with vinegar and baking soda? Have you pushed the gas pedal of a car and felt the acceleration as the car moved forward? Have you wondered why laundry detergents work better than hand soap to clean your clothes? Each of these processes involves a *chemical reaction*—the transformation of one or more substances into different substances.

In the classic grade school volcano, baking soda (which is sodium bicarbonate) reacts with acetic acid in vinegar to form carbon dioxide gas, water, and sodium acetate. The newly formed carbon dioxide bubbles out of the mixture, causing the eruption. Reactions that occur in liquids and form gases are *gas evolution reactions*. A similar reaction causes the fizzing of antacids such as Alka-Seltzer™.

When you drive a car, hydrocarbons such as octane (in gasoline) react with oxygen from the air to form carbon dioxide gas and water (▼ FIGURE 7.1).

▶ **FIGURE 7.1 A combustion reaction** In an automobile engine, hydrocarbons such as octane (C_8H_{18}) from gasoline combine with oxygen from the air and react to form carbon dioxide and water.

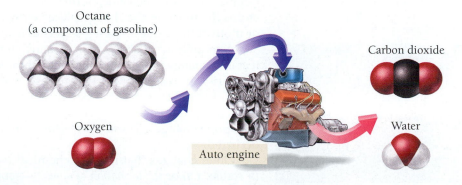

Octane (a component of gasoline)

Oxygen

Auto engine

Carbon dioxide

Water

◀ Schoolchildren sometimes make clay volcanoes that erupt by combining vinegar and baking soda, which react to produce the bubbling and splattering.

Soap in pure water	Soap in hard water

▲ FIGURE 7.2 **Soap and water** Soap forms suds with pure water (left) but reacts with the ions in hard water (right) to form a gray residue that adheres to clothes.

This reaction produces heat, which expands the gases in the car's cylinders, accelerating it forward. Reactions such as this one—in which a substance reacts with oxygen, emitting heat and forming one or more oxygen-containing compounds—are **combustion reactions**. Combustion reactions are a subcategory of *oxidation–reduction reactions*, in which electrons are transferred from one substance to another. The formation of rust and the dulling of automobile paint are other examples of oxidation–reduction reactions.

Laundry detergent works better than hand soap to wash clothes because it contains substances that soften hard water. Hard water contains dissolved calcium (Ca^{2+}) and magnesium (Mg^{2+}) ions. These ions react with soap to form a gray, slimy substance called *curd* or *soap scum* (◀ FIGURE 7.2). If you have ever tried to do your laundry with hand soap, you may have noticed gray soap scum residue on your clothes.

Laundry detergents prevent curd formation by removing Ca^{2+} and Mg^{2+} from water. Why? Because they contain substances that react with Ca^{2+} and Mg^{2+}. For example, many laundry detergents contain the carbonate (CO_3^{2-}) ion. Carbonate ions react with calcium and magnesium ions in the hard water to form solid calcium carbonate ($CaCO_3$) and solid magnesium carbonate ($MgCO_3$). These solids simply settle to the bottom of the laundry mixture, resulting in the removal of the ions from the water. In other words, laundry detergents contain substances that react with the ions in hard water to immobilize them. Reactions such as these—that form solid substances in water—are *precipitation reactions*. Precipitation reactions are also used to remove dissolved toxic metals in industrial wastes.

Chemical reactions take place all around us and even inside us. They are involved in many of the products we use daily and in many of our experiences. Chemical reactions can be relatively simple, like the combination of hydrogen and oxygen to form water, or they can be complex, like the synthesis of a protein molecule from thousands of simpler molecules. In some cases, such as the neutralization reaction that occurs in a swimming pool when acid is added to adjust the water's acidity level, chemical reactions are not noticeable to the naked eye. In other cases, such as the combustion reaction that produces a pillar of smoke and fire under a rocket during liftoff, chemical reactions are very obvious. In all cases, however, chemical reactions produce changes in the arrangements of the molecules and atoms that compose matter. Often, these molecular changes cause macroscopic changes that we can experience directly.

7.2 Evidence of a Chemical Reaction

▶ Identify evidence of a chemical reaction.

Color change

▲ A child's temperature-sensitive spoon changes color upon warming due to a reaction induced by the higher temperature.

If we could see the atoms and molecules that compose matter, we could easily identify a chemical reaction. Do the atoms combine with other atoms to form compounds? Do new molecules form? Do the original molecules decompose? Do the atoms in one molecule change places with atoms in another? If the answer to one or more of these questions is yes, a chemical reaction has occurred.

Although we can't see atoms, many chemical reactions do produce easily detectable changes as they occur. For example, when the color-causing molecules in a brightly colored shirt decompose with repeated exposure to sunlight, the color of the shirt fades. Similarly, when the molecules embedded in the plastic of a child's temperature-sensitive spoon transform upon warming, the color of the spoon changes. These *color changes* are evidence that a chemical reaction has occurred.

Other changes that identify chemical reactions include the *formation of a solid* (▶ FIGURE 7.3) or *the formation of a gas* (▶ FIGURE 7.4). Dropping Alka-Seltzer tablets into water or combining baking soda and vinegar (as in our opening example of the grade school volcano) are both good examples of chemical reactions that produce a gas—the gas is visible as bubbles in the liquid.

Solid formation

Gas formation

▲ **FIGURE 7.3 A precipitation reaction** The formation of a solid in a previously clear solution is evidence of a chemical reaction.

▲ **FIGURE 7.4 A gas evolution reaction** The formation of a gas is evidence of a chemical reaction.

Recall from Section 3.9 that a reaction that emits heat is an *exothermic* reaction and one that absorbs heat is an *endothermic* reaction.

Heat absorption and *emission*, as well as *light emission*, are also evidence of reactions. For example, a natural gas flame produces heat and light. A chemical cold pack becomes cold when the plastic barrier separating two substances is broken. Both of these changes suggest that a chemical reaction is occurring.

Heat absorption

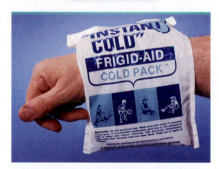

▲ A change in temperature due to absorption or emission of heat is evidence of a chemical reaction. This chemical cold pack becomes cold when the barrier separating two substances is broken and the substances combine.

▲ **FIGURE 7.5 Boiling: A physical change** When water boils, bubbles are formed and a gas is evolved. However, no chemical change has occurred because the gas, like the liquid water, is also composed of water molecules.

While these changes provide evidence of a chemical reaction, they are not *definitive* evidence. Only chemical analysis showing that the initial substances have changed into other substances conclusively proves that a chemical reaction has occurred. We can be fooled. For example, when water boils, bubbles form, but no chemical reaction has occurred. Boiling water forms gaseous steam, but both water and steam are composed of water molecules—no chemical change has occurred (◄ **FIGURE 7.5**). On the other hand, chemical reactions may occur without any obvious signs, yet chemical analysis may show that a reaction has indeed occurred. The changes occurring at the atomic and molecular level determine whether a chemical reaction has taken place.

In summary, each of the following provides evidence of a chemical reaction:

• *a color change*

• the *emission of light*

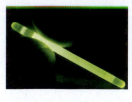

• the *formation of a solid* in a previously clear (unclouded) solution

• the *emission* or *absorption of heat*

• the *formation of a gas* when we add a substance to a solution

EXAMPLE **7.1** | **Evidence of a Chemical Reaction**

Which changes involve a chemical reaction? Explain your answers.

(a) ice melting upon warming
(b) an electric current passing through water, resulting in the formation of hydrogen and oxygen gas that appears as bubbles rising in the water
(c) iron rusting
(d) bubbles forming when a soda can is opened

SOLUTION

(a) not a chemical reaction; melting ice forms water, but both the ice and water are composed of water molecules.
(b) chemical reaction; water decomposes into hydrogen and oxygen, as evidenced by the bubbling.
(c) chemical reaction; iron changes into iron oxide, changing color in the process.
(d) not a chemical reaction; even though there is bubbling, it is just carbon dioxide coming out of the liquid.

▶ SKILLBUILDER 7.1 | **Evidence of a Chemical Reaction**

Which changes involve a chemical reaction? Explain your answers.

(a) butane burning in a butane lighter
(b) butane evaporating out of a butane lighter
(c) wood burning
(d) dry ice subliming

▶ **FOR MORE PRACTICE** Example 7.16; Problems 25, 26, 27, 28, 29, 30.

CONCEPTUAL ✔ CHECKPOINT 7.1

These images portray molecular views of various substances before and after a change. Determine whether a chemical reaction has occurred in each case.

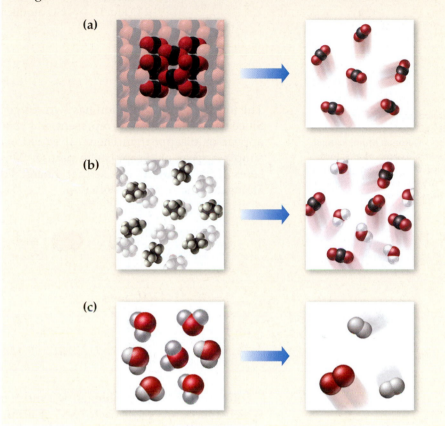

(a)

(b)

(c)

7.3 The Chemical Equation

▶ Identify balanced chemical equations.

Key Concept Video
Writing and Balancing Chemical Equations

TABLE 7.1 Abbreviations Indicating the States of Reactants and Products in Chemical Equations

Abbreviation	State
(g)	gas
(l)	liquid
(s)	solid
(aq)	aqueous (water solution)*

*The (aq) designation stands for *aqueous*, which indicates that a substance is dissolved in water. When a substance dissolves in water, the mixture is called a *solution*. We discuss solutions in greater detail in Section 7.5.

Recall from Section 3.6 that we represent chemical reactions with *chemical equations*. For example, the reaction occurring in a natural-gas flame, such as the flame on a kitchen stove, is methane (CH_4) reacting with oxygen (O_2) to form carbon dioxide (CO_2) and water (H_2O). We represent this reaction with the equation:

$$\underset{\text{reactants}}{CH_4 + O_2} \longrightarrow \underset{\text{products}}{CO_2 + H_2O}$$

The substances on the left side of the equation are the *reactants*, and the substances on the right side are the *products*. We often specify the state of each reactant or product in parentheses next to the formula. If we add states to our equation, it becomes:

$$CH_4(g) + O_2(g) \longrightarrow CO_2(g) + H_2O(g)$$

The (g) indicates that these substances are gases in the reaction. Table 7.1 summarizes the common states of reactants and products and the symbols used in chemical reactions.

Let's look more closely at the equation for the burning of natural gas. How many oxygen atoms are on each side of the equation?

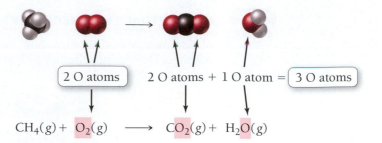

$$CH_4(g) + O_2(g) \longrightarrow CO_2(g) + H_2O(g)$$

The left side of the equation has two oxygen atoms, and the right side has three. Since chemical equations represent real chemical reactions, atoms cannot simply appear or disappear in chemical equations because, as we know, atoms don't simply appear or disappear in nature. We must account for the atoms on both sides of the equation. Notice that the left side of the equation has four hydrogen atoms and the right side only two.

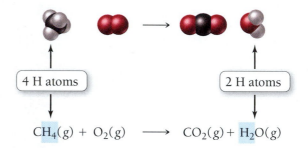

$$CH_4(g) + O_2(g) \longrightarrow CO_2(g) + H_2O(g)$$

To correct these problems, we must create a **balanced equation**, one in which the numbers of each type of atom on both sides of the equation are equal. To balance an equation, we insert coefficients—not subscripts—in front of the chemical formulas as needed to make the number of each type of atom in the reactants equal to the number of each type of atom in the products. New atoms do not form during a reaction, nor do atoms vanish—matter must be conserved.

When we balance chemical equations by inserting coefficients in front of the formulas of the reactants and products, it changes the number of molecules in the equation, but it does not change the *kinds* of molecules. To balance the preceding equation, for example, we put the coefficient 2 before O_2 in the reactants, and the coefficient 2 before H_2O in the products:

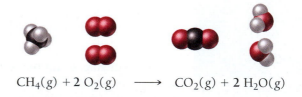

$$CH_4(g) + 2\,O_2(g) \longrightarrow CO_2(g) + 2\,H_2O(g)$$

The equation is now balanced because the numbers of each type of atom on both sides of the equation are equal. We can verify this by summing the number of each type of atom.

We determine the number of a particular type of atom within a chemical formula in an equation by multiplying the subscript for the atom by the coefficient for the chemical formula. If there is no coefficient or subscript, a 1 is implied. The balanced equation for the combustion of natural gas is:

$$CH_4(g) + 2\,O_2(g) \longrightarrow CO_2(g) + 2\,H_2O(g)$$

Reactants	Products
1 C atom (1 × $\underline{C}H_4$)	1 C atom (1 × $\underline{C}O_2$)
4 H atoms (1 × $C\underline{H}_4$)	4 H atoms (2 × $\underline{H}_2O$)
4 O atoms (2 × $\underline{O}_2$)	4 O atoms (1 × $C\underline{O}_2$ + 2 × $H_2\underline{O}$)

The numbers of each type of atom on both sides of the equation are equal—the equation is balanced.

$$CH_4(g) + 2\,O_2(g) \longrightarrow CO_2(g) + 2\,H_2O(g)$$

▶ A balanced chemical equation represents a chemical reaction. In this equation, methane molecules combine with oxygen to form carbon dioxide and water. The photo shows a stove burning methane and the balanced equation for the reaction above it.

PEARSON
eText
2.0

CONCEPTUAL ✔ **CHECKPOINT** 7.2

In photosynthesis, plants make the sugar glucose, $C_6H_{12}O_6$, from carbon dioxide and water. The equation for the reaction is:

$$6\,CO_2 + 6\,H_2O \longrightarrow C_6H_{12}O_6 + x\,O_2$$

In order for this equation to be balanced, the coefficient x must be

(a) 3   **(b)** 6   **(c)** 9   **(d)** 12

7.4 How to Write Balanced Chemical Equations

▶ Write balanced chemical equations.

The following procedure details the steps for writing balanced chemical equations. As in other procedures in the book, we show the steps in the left column and examples of applying each step in the center and right columns. Remember, change only the *coefficients* to balance a chemical equation; *never change the subscripts because changing the subscripts changes the kinds of molecules, not the number of molecules.*

PEARSON eText 2.0 Interactive Worked Example Video 7.2

Writing Balanced Chemical Equations	EXAMPLE **7.2** Write a balanced equation for the reaction between solid silicon dioxide and solid carbon to produce solid silicon monocarbide and carbon monoxide gas.	EXAMPLE **7.3** Write a balanced equation for the combustion reaction between liquid octane (C_8H_{18}), a component of gasoline, and gaseous oxygen to form gaseous carbon dioxide and gaseous water.
1. Write the unbalanced equation by writing chemical formulas for each of the reactants and products. Review Chapter 5 for nomenclature rules. (If the unbalanced equation is provided in the problem, skip this step and go to Step 2.)	**SOLUTION** $$SiO_2(s) + C(s) \longrightarrow SiC(s) + CO(g)$$	**SOLUTION** $$C_8H_{18}(l) + O_2(g) \longrightarrow CO_2(g) + H_2O(g)$$
2. If an element occurs in only one compound on both sides of the equation, balance it first. If there is more than one such element, balance metals before nonmetals.	**BEGIN WITH SI** $$SiO_2(s) + C(s) \longrightarrow SiC(s) + CO(g)$$ **1 Si atom $\longrightarrow$ 1 Si atom** Si is already balanced. **BALANCE O NEXT** $$SiO_2(s) + C(s) \longrightarrow SiC(s) + CO(g)$$ **2 O atoms $\longrightarrow$ 1 O atom** To balance O, put a 2 before CO(g). $$SiO_2(s) + C(s) \longrightarrow SiC(s) + 2\,CO(g)$$ **2 O atoms $\longrightarrow$ 2 O atoms**	**BEGIN WITH C** $$C_8H_{18}(l) + O_2(g) \longrightarrow CO_2(g) + H_2O(g)$$ **8 C atoms $\longrightarrow$ 1 C atom** To balance C, put an 8 before $CO_2(g)$. $$C_8H_{18}(l) + O_2(g) \longrightarrow 8\,CO_2(g) + H_2O(g)$$ **8 C atoms $\longrightarrow$ 8 C atoms** **BALANCE H NEXT** $$C_8H_{18}(l) + O_2(g) \longrightarrow 8\,CO_2(g) + H_2O(g)$$ **18 H atoms $\longrightarrow$ 2 H atoms** To balance H, put a 9 before $H_2O(g)$. $$C_8H_{18}(l) + O_2(g) \longrightarrow 8\,CO_2(g) + 9\,H_2O(g)$$ **18 H atoms $\longrightarrow$ 18 H atoms**
3. If an element occurs as a free element (not as part of a compound) on either side of the chemical equation, balance it last. Always balance free elements by adjusting the coefficient *on the free element*.	**BALANCE C** $$SiO_2(s) + C(s) \longrightarrow SiC(s) + 2\,CO(g)$$ **1 C atom $\longrightarrow$ 1 C + 2 C = 3 C atoms** To balance C, put a 3 before C(s). $$SiO_2(s) + 3\,C(s) \longrightarrow SiC(s) + 2\,CO(g)$$ **3 C atoms $\longrightarrow$ 1 C + 2 C = 3 C atoms**	**BALANCE O** $$C_8H_{18}(l) + O_2(g) \longrightarrow 8\,CO_2(g) + 9\,H_2O(g)$$ **2 O atoms $\longrightarrow$ 16 O + 9 O = 25 O atoms** To balance O, put a $\frac{25}{2}$ before $O_2(g)$. $$C_8H_{18}(l) + \tfrac{25}{2}\,O_2(g) \longrightarrow 8\,CO_2(g) + 9\,H_2O(g)$$ **25 O atoms $\longrightarrow$ 16 O + 9 O = 25 O atoms**
4. If the balanced equation contains coefficient fractions, change these into whole numbers by multiplying the entire equation by the appropriate factor.	This step is not necessary in this example. Proceed to Step 5.	$$[C_8H_{18}(l) + \tfrac{25}{2}\,O_2(g) \longrightarrow 8\,CO_2(g) + 9\,H_2O(g)] \times 2$$ $$2\,C_8H_{18}(l) + 25\,O_2(g) \longrightarrow 16\,CO_2(g) + 18\,H_2O(g)$$

5. Check to make certain the equation is balanced by summing the total number of each type of atom on both sides of the equation.

$$SiO_2(s) + 3\,C(s) \longrightarrow SiC(s) + 2\,CO(g)$$

Reactants		Products
1 Si atom	$\longrightarrow$	1 Si atom
2 O atoms	$\longrightarrow$	2 O atoms
3 C atoms	$\longrightarrow$	3 C atoms

The equation is balanced.

▶ **SKILLBUILDER 7.2** | Write a balanced equation for the reaction between solid chromium(III) oxide and solid carbon to produce solid chromium and carbon dioxide gas.

$$2\,C_8H_{18}(l) + 25\,O_2(g) \longrightarrow$$
$$16\,CO_2(g) + 18\,H_2O(g)$$

Reactants		Products
16 C atoms	$\longrightarrow$	16 C atoms
36 H atoms	$\longrightarrow$	36 H atoms
50 O atoms	$\longrightarrow$	50 O atoms

The equation is balanced.

▶ **SKILLBUILDER 7.3** | Write a balanced equation for the combustion reaction of gaseous C_4H_{10} and gaseous oxygen to form gaseous carbon dioxide and gaseous water.

▶ **FOR MORE PRACTICE**   Example 7.17; Problems 35, 36, 37, 38.

EXAMPLE **7.4**   **Balancing Chemical Equations**

Write a balanced equation for the reaction of solid aluminum with aqueous sulfuric acid to form aqueous aluminum sulfate and hydrogen gas.

Use your knowledge of chemical nomenclature from Chapter 5 to write a skeletal equation containing formulas for each of the reactants and products. The formulas for each compound MUST BE CORRECT before you begin to balance the equation.	**SOLUTION** $$Al(s) + H_2SO_4(aq) \longrightarrow Al_2(SO_4)_3(aq) + H_2(g)$$
Since both aluminum and hydrogen occur as free elements, balance those last. Sulfur and oxygen occur in only one compound on each side of the equation, so balance these first. Sulfur and oxygen are also part of a polyatomic ion that stays intact on both sides of the equation. *Balance polyatomic ions such as these as a unit.* There are 3 SO_4^{2-} ions on the right side of the equation, so put a 3 in front of H_2SO_4.	$$Al(s) + 3\,H_2SO_4(aq) \longrightarrow Al_2(SO_4)_3(aq) + H_2(g)$$
Balance Al next. Since there are 2 Al atoms on the right side of the equation, place a 2 in front of Al on the left side of the equation.	$$2\,Al(s) + 3\,H_2SO_4(aq) \longrightarrow Al_2(SO_4)_3(aq) + H_2(g)$$
Balance H next. Since there are 6 H atoms on the left side, place a 3 in front of $H_2(g)$ on the right side.	$$2\,Al(s) + 3\,H_2SO_4(aq) \longrightarrow Al_2(SO_4)_3(aq) + 3\,H_2(g)$$
Finally, sum the number of atoms on each side to make sure that the equation is balanced.	$$2\,Al(s) + 3\,H_2SO_4(aq) \longrightarrow Al_2(SO_4)_3(aq) + 3\,H_2(g)$$

Reactants		Products
2 Al atoms	$\longrightarrow$	2 Al atoms
6 H atoms	$\longrightarrow$	6 H atoms
3 S atoms	$\longrightarrow$	3 S atoms
12 O atoms	$\longrightarrow$	12 O atoms

▶ **SKILLBUILDER 7.4** | **Balancing Chemical Equations**

Write a balanced equation for the reaction of aqueous lead(II) acetate with aqueous potassium iodide to form solid lead(II) iodide and aqueous potassium acetate.

▶ **FOR MORE PRACTICE**   Problems 39, 40, 41, 42, 43, 44.

EXAMPLE **7.5**   **Balancing Chemical Equations**

Balance the chemical equation.

$$Fe(s) + HCl(aq) \longrightarrow FeCl_3(aq) + H_2(g)$$

Since Cl occurs in only one compound on each side of the equation, balance it first. One Cl atom is on the left side of the equation, and 3 Cl atoms are on the right side. To balance Cl, place a 3 in front of HCl.	**SOLUTION** $Fe(s) + 3\,HCl(aq) \longrightarrow FeCl_3(aq) + H_2(g)$
Since H and Fe occur as free elements, balance them last. There is 1 Fe atom on the left side of the equation and 1 Fe atom on the right, so Fe is balanced. There are 3 H atoms on the left and 2 H atoms on the right. Balance H by placing a $\frac{3}{2}$ in front of H_2. (That way you don't alter other elements that are already balanced.)	$Fe(s) + 3\,HCl(aq) \longrightarrow FeCl_3(aq) + \frac{3}{2}H_2(g)$
The equation now contains a coefficient fraction; clear it by multiplying the entire equation (both sides) by 2.	$[Fe(s) + 3\,HCl(aq) \longrightarrow FeCl_3(aq) + \frac{3}{2}H_2(g)] \times 2$ $2\,Fe(s) + 6\,HCl(aq) \longrightarrow 2\,FeCl_3(aq) + 3\,H_2(g)$
Finally, sum the number of atoms on each side to check that the equation is balanced.	$2\,Fe(s) + 6\,HCl(aq) \longrightarrow 2\,FeCl_3(aq) + 3\,H_2(g)$

Reactants		Products
2 Fe atoms	$\longrightarrow$	2 Fe atoms
6 Cl atoms	$\longrightarrow$	6 Cl atoms
6 H atoms	$\longrightarrow$	6 H atoms

▶ **SKILLBUILDER 7.5** | **Balancing Chemical Equations**

Balance the chemical equation.

$$HCl(g) + O_2(g) \longrightarrow H_2O(l) + Cl_2(g)$$

▶ **FOR MORE PRACTICE**   Problems 45, 46, 47, 48, 49, 50.

CONCEPTUAL ✔ **CHECKPOINT 7.3**

Which quantity must always be the same on both sides of a balanced chemical equation?

(a) the number of each type of atom

(b) the number of each type of molecule

(c) the sum of all of the coefficients

7.5 Aqueous Solutions and Solubility: Compounds Dissolved in Water

▶ Determine whether a compound is soluble.

In the previous section, we balanced chemical equations that represent chemical reactions. We now turn to investigating several types of reactions.

Aqueous Solutions

Since many of these reactions occur in water, we must first understand *aqueous solutions*. Reactions occurring in aqueous solutions are among the most common and important. An **aqueous solution** is a homogeneous mixture of a substance with water. For example, a sodium chloride (NaCl) solution (also called a saline solution) is composed of sodium chloride dissolved in water. Sodium chloride

Strong electrolyte solution

Battery

Battery

No ions to conduct electricity

Na^+ and Cl^- ions conduct electricity

Pure water

NaCl solution

(a)

(b)

▲ **FIGURE 7.6 Ions as conductors**   (a) Pure water does not conduct electricity. **(b)** Ions in a sodium chloride solution conduct electricity, causing the bulb to light. Solutions such as NaCl are called strong electrolyte solutions.

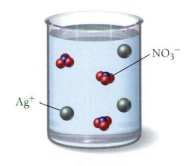

A sodium chloride solution contains independent Na^+ and Cl^- ions.

A silver nitrate solution contains independent Ag^+ and NO_3^- ions.

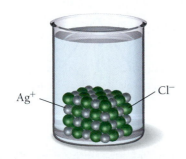

When silver chloride is added to water, it remains as solid AgCl—it does not dissolve into independent ions.

solutions are common both in the oceans and in living cells. You can form a sodium chloride solution yourself by adding table salt to water. As you stir the salt into the water, it seems to disappear. However, you know the salt is still there because if you taste the water, it has a salty flavor. How does sodium chloride dissolve in water?

When ionic compounds such as NaCl dissolve in water, they usually dissociate into their component ions. A sodium chloride solution, represented as NaCl(*aq*), does not contain any NaCl units; only dissolved Na^+ ions and Cl^- ions are present.

We know that NaCl is present as independent sodium and chloride ions in solution because sodium chloride solutions conduct electricity, which requires the presence of freely moving charged particles. Substances (such as NaCl) that completely dissociate into ions in solution are *strong electrolytes*, and the resultant solutions are **strong electrolyte solutions** (▲ **FIGURE 7.6**). Similarly, a silver nitrate solution, represented as AgNO$_3$(*aq*), does not contain any AgNO$_3$ units, but only dissolved Ag^+ ions and NO_3^- ions. It, too, is a strong electrolyte solution. When compounds containing polyatomic ions such as NO_3^- dissolve, the polyatomic ions dissolve as intact units.

Not all ionic compounds, however, dissolve in water. AgCl, for example, does not. If we add AgCl to water, it remains as solid AgCl and appears as a white solid at the bottom of the beaker.

Solubility

A compound is **soluble** in a particular liquid if it dissolves in that liquid; a compound is **insoluble** if it does not dissolve in the liquid. NaCl, for example, is soluble in water. If we mix solid sodium chloride into water, it dissolves and forms a strong electrolyte solution. AgCl, on the other hand, is insoluble in water. If we mix solid silver chloride into water, it remains as a solid within the liquid water.

There is no easy way to predict whether a particular compound will be soluble or insoluble in water. For ionic compounds, however, empirical rules have been deduced from observations of many compounds. These **solubility rules** are summarized in Table 7.2 and ▼ FIGURE 7.7. For example, the solubility rules indicate that compounds containing the lithium ion are *soluble*. That means that compounds such as LiBr, LiNO$_3$, Li$_2$SO$_4$, LiOH, and Li$_2$CO$_3$ all dissolve in water to form strong electrolyte solutions. If a compound contains Li$^+$, it is soluble. Similarly, the solubility rules state that compounds containing the NO$_3^-$ ion are soluble. Compounds such as AgNO$_3$, Pb(NO$_3$)$_2$, NaNO$_3$, Ca(NO$_3$)$_2$, and Sr(NO$_3$)$_2$ all dissolve in water to form strong electrolyte solutions.

The solubility rules also state that, with some exceptions, compounds containing the CO$_3^{2-}$ ion are *insoluble*. Compounds such as CuCO$_3$, CaCO$_3$, SrCO$_3$, and FeCO$_3$ do not dissolve in water. Note that the solubility rules have many exceptions. For example, compounds containing CO$_3^{2-}$ are *soluble when paired with Li$^+$, Na$^+$, K$^+$, or NH$_4^+$*. Thus Li$_2$CO$_3$, Na$_2$CO$_3$, K$_2$CO$_3$, and (NH$_4$)$_2$CO$_3$ are all soluble.

The solubility rules apply only to the solubility of the compounds in water.

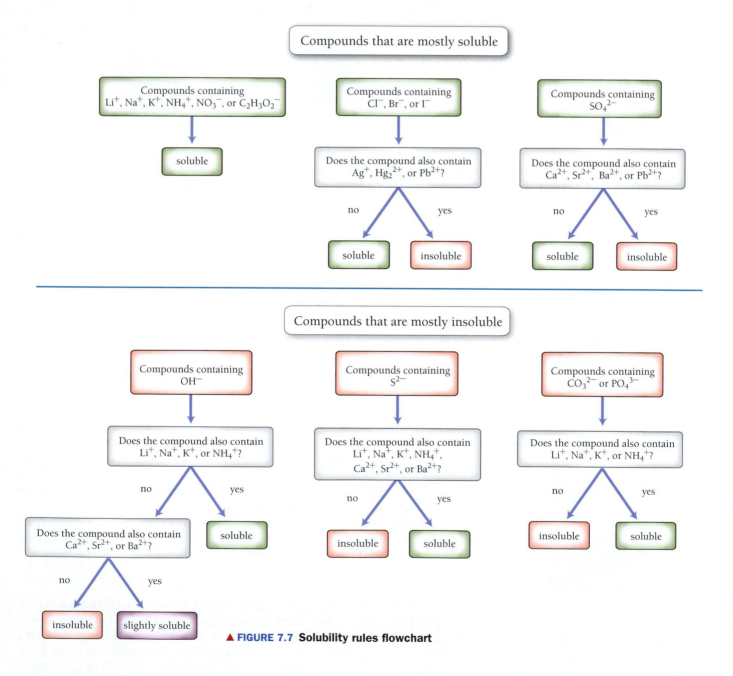

▲ FIGURE 7.7 Solubility rules flowchart

TABLE 7.2 Solubility Rules

Compounds Containing the Following Ions Are Mostly Soluble	Exceptions
Li^+, Na^+, K^+, NH_4^+	None
NO_3^-, $C_2H_3O_2^-$	None
Cl^-, Br^-, I^-	When any of these ions pair with Ag^+, Hg_2^{2+}, or Pb^{2+}, the compound is insoluble.
SO_4^{2-}	When SO_4^{2-} pairs with Ca^{2+}, Sr^{2+}, Ba^{2+}, or Pb^{2+}, the compound is insoluble.

Compounds Containing the Following Ions Are Mostly Insoluble	Exceptions
OH^-, S^{2-}	When either of these ions pairs with Li^+, Na^+, K^+, or NH_4^+, the compound is soluble. When S^{2-} pairs with Ca^{2+}, Sr^{2+}, or Ba^{2+}, the compound is soluble. When OH^- pairs with Ca^{2+}, Sr^{2+}, or Ba^{2+}, the compound is slightly soluble.*
CO_3^{2-}, PO_4^{3-}	When either of these ions pairs with Li^+, Na^+, K^+, or NH_4^+, the compound is soluble.

*For many purposes these can be considered insoluble.

EXAMPLE **7.6** **Determining Whether a Compound is Soluble**

Is each compound soluble or insoluble?

(a) AgBr   (b) $CaCl_2$ (c) $Pb(NO_3)_2$ (d) $PbSO_4$

SOLUTION

(a) Insoluble; compounds containing Br^- are normally soluble, but Ag^+ is an exception.
(b) Soluble; compounds containing Cl^- are normally soluble, and Ca^{2+} is not an exception.
(c) Soluble; compounds containing NO_3^- are always soluble.
(d) Insoluble; compounds containing SO_4^{2-} are normally soluble, but Pb^{2+} is an exception.

▶ **SKILLBUILDER 7.6 | Determining Whether a Compound Is Soluble**

Is each compound soluble or insoluble?

(a) CuS   (b) $FeSO_4$ (c) $PbCO_3$ (d) NH_4Cl

▶ **FOR MORE PRACTICE** Example 7.18; Problems 57, 58, 59, 60, 61, 62.

PEARSON eText 2.0

CONCEPTUAL ✓ CHECKPOINT 7.4

Which image best depicts a mixture of $BaCl_2$ and water?

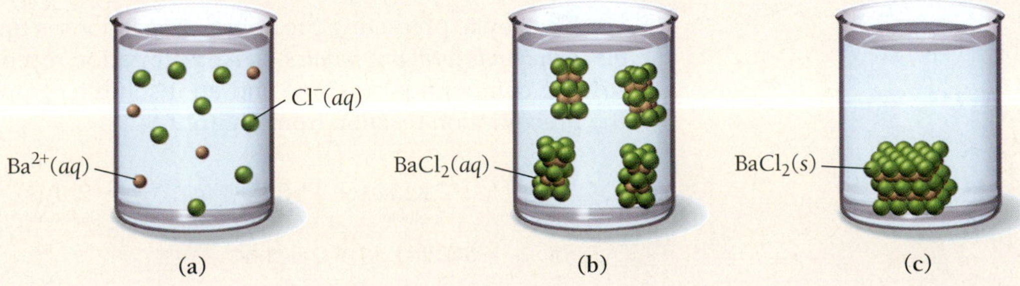

$Cl^-(aq)$
$Ba^{2+}(aq)$
(a)

$BaCl_2(aq)$
(b)

$BaCl_2(s)$
(c)

7.6 Precipitation Reactions: Reactions in Aqueous Solution That Form a Solid

▶ Predict and write equations for precipitation reactions.

Recall from Section 7.1 that sodium carbonate in laundry detergent reacts with dissolved Mg^{2+} and Ca^{2+} ions to form solids that precipitate (come out of) solution. This reaction is an example of a **precipitation reaction**—a reaction that forms a solid, called a **precipitate**, when two aqueous solutions are mixed.

Precipitation reactions are common in chemistry. Potassium iodide and lead nitrate, for example, both form colorless, strong electrolyte solutions when dissolved in water (see the solubility rules in Section 7.5). When the two solutions are combined, however, a brilliant yellow precipitate forms (▼ FIGURE 7.8). We describe this precipitation reaction with the chemical equation:

$$2\,KI(aq) + Pb(NO_3)_2(aq) \longrightarrow PbI_2(s) + 2\,KNO_3(aq)$$

Precipitation reactions do not always occur when two aqueous solutions mix. For example, when we combine solutions of $KI(aq)$ and $NaCl(aq)$, nothing happens (▼ FIGURE 7.9).

$$KI(aq) + NaCl(aq) \longrightarrow NO\ REACTION$$

$$2\,KI(aq) + Pb(NO_3)_2(aq) \longrightarrow PbI_2(s) + 2\,KNO_3(aq)$$

▲ FIGURE 7.8 **Precipitation** When we mix a potassium iodide solution with a lead(II) nitrate solution, a brilliant yellow precipitate of $PbI_2(s)$ forms.

$$KI(aq) + NaCl(aq) \longrightarrow NO\ REACTION$$

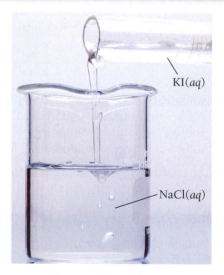

▲ FIGURE 7.9 **No reaction** When we mix a potassium iodide solution with a sodium chloride solution, no reaction occurs.

The key to predicting precipitation reactions is understanding that *only insoluble compounds form precipitates*. In a precipitation reaction, two solutions containing soluble compounds combine and an insoluble compound precipitates. Consider the precipitation reaction from Figure 7.8:

$$\underset{\text{soluble}}{2\,KI(aq)} + \underset{\text{soluble}}{Pb(NO_3)_2(aq)} \longrightarrow \underset{\text{insoluble}}{PbI_2(s)} + \underset{\text{soluble}}{2\,KNO_3(aq)}$$

KI and $Pb(NO_3)_2$ are both soluble, but the precipitate, PbI_2, is *insoluble*. Before mixing, $KI(aq)$ and $Pb(NO_3)_2(aq)$ are each dissociated in their respective solutions.

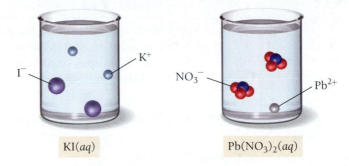

KI(*aq*)   Pb(NO$_3$)$_2$(*aq*)

The instant that the solutions are mixed, all four ions are present.

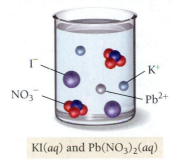

KI(*aq*) and Pb(NO$_3$)$_2$(*aq*)

However, new compounds—potentially insoluble ones—are now possible. Specifically, the cation from one compound can pair with the anion from the other compound to form new (and potentially insoluble) products:

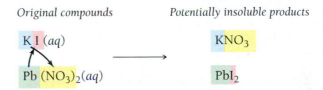

Original compounds   *Potentially insoluble products*

K I (*aq*)   KNO$_3$

Pb (NO$_3$)$_2$(*aq*)   PbI$_2$

If, on the one hand, the *potentially insoluble* products are both *soluble*, no reaction occurs. If, on the other hand, one or both of the potentially insoluble products are *indeed insoluble*, a precipitation reaction occurs. In this case, KNO_3 is soluble, but PbI_2 is insoluble. Consequently, PbI_2 precipitates.

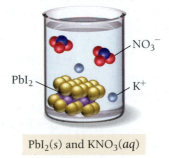

PbI$_2$(*s*) and KNO$_3$(*aq*)

To predict whether a precipitation reaction occurs when two solutions are mixed and to write an equation for the reaction, we follow the steps in the procedure that accompanies Examples 7.7 and 7.8. As usual, the steps are shown in the left column, and two examples of applying the procedure are shown in the center and right columns.

PEARSON eText 2.0 — Interactive Worked Example Video 7.7

Writing Equations for Precipitation Reactions	EXAMPLE **7.7** Write an equation for the precipitation reaction that occurs (if any) when solutions of sodium carbonate and copper(II) chloride are mixed.	EXAMPLE **7.8** Write an equation for the precipitation reaction that occurs (if any) when solutions of lithium nitrate and sodium sulfate are mixed.
1. Write the formulas of the two compounds being mixed as reactants in a chemical equation.	**SOLUTION** $Na_2CO_3(aq) + CuCl_2(aq) \longrightarrow$	**SOLUTION** $LiNO_3(aq) + Na_2SO_4(aq) \longrightarrow$
2. Below the equation, write the formulas of the potentially insoluble products that could form from the reactants. Obtain these by combining the cation from one reactant with the anion from the other. Make sure to write correct (charge-neutral) formulas for these ionic compounds as described in Section 5.5.	$Na_2CO_3(aq) + CuCl_2(aq) \longrightarrow$ **Potentially insoluble products** NaCl $CuCO_3$	$LiNO_3(aq) + Na_2SO_4(aq) \longrightarrow$ **Potentially insoluble products** $NaNO_3$ Li_2SO_4
3. Use the solubility rules from Section 7.5 to determine whether any of the potential new products are insoluble.	NaCl is *soluble* (compounds containing Cl^- are usually soluble, and Na^+ is not an exception). $CuCO_3$ is *insoluble* (compounds containing CO_3^{2-} are usually insoluble, and Cu^{2+} is not an exception).	$NaNO_3$ is *soluble* (compounds containing NO_3^- are soluble, and Na^+ is not an exception). Li_2SO_4 is *soluble* (compounds containing SO_4^{2-} are soluble, and Li^+ is not an exception).
4. If all of the potentially insoluble products are soluble, there will be no precipitate. Write NO REACTION next to the arrow.	Because this example has an insoluble product, you proceed to the next step.	$LiNO_3(aq) + Na_2SO_4(aq) \longrightarrow$ NO REACTION
5. If one or both of the potentially insoluble products are insoluble, write their formula(s) as the product(s) of the reaction, using (s) to indicate solid. Follow any soluble products with (aq) to indicate aqueous.	$Na_2CO_3(aq) + CuCl_2(aq) \longrightarrow$ $CuCO_3(s) + NaCl(aq)$	
6. Balance the equation. Remember to adjust only coefficients, not subscripts.	$Na_2CO_3(aq) + CuCl_2(aq) \longrightarrow$ $CuCO_3(s) + 2\,NaCl(aq)$	
	▶ **SKILLBUILDER 7.7** | Write an equation for the precipitation reaction that occurs (if any) when solutions of potassium hydroxide and nickel(II) bromide are mixed.	▶ **SKILLBUILDER 7.8** | Write an equation for the precipitation reaction that occurs (if any) when solutions of ammonium chloride and iron(III) nitrate are mixed. ▶ **FOR MORE PRACTICE** Example 7.19; Problems 63, 64, 65, 66.

EXAMPLE **7.9** | **Predicting and Writing Equations for Precipitation Reactions**

Write an equation for the precipitation reaction (if any) that occurs when solutions of lead(II) acetate and sodium sulfate are mixed. If no reaction occurs, write *NO REACTION*.

1. Write the formulas of the two compounds being mixed as reactants in a chemical equation.	**SOLUTION** $Pb(C_2H_3O_2)_2(aq) + Na_2SO_4(aq) \longrightarrow$
2. Below the equation, write the formulas of the potentially insoluble products that could form from the reactants. Determine these by combining the cation from one reactant with the anion from the other reactant. Make sure to adjust the subscripts so that all formulas are charge-neutral.	$Pb(C_2H_3O_2)_2(aq) + Na_2SO_4(aq) \longrightarrow$ **Potentially insoluble products** $NaC_2H_3O_2 \qquad PbSO_4$
3. Use the solubility rules from Section 7.5 to determine whether any of the potentially insoluble products are insoluble.	$NaC_2H_3O_2$ is *soluble* (compounds containing Na^+ are always soluble). $PbSO_4$ is *insoluble* (compounds containing SO_4^{2-} are normally soluble, but Pb^{2+} is an exception).
4. If all of the potentially insoluble products are soluble, there will be no precipitate. Write *NO REACTION* next to the arrow.	This reaction has an insoluble product so you proceed to the next step.
5. If one or both of the potentially insoluble products are insoluble, write their formula(s) as the product(s) of the reaction, using (s) to indicate solid. Follow any soluble products with (aq) to indicate aqueous.	$Pb(C_2H_3O_2)_2(aq) + Na_2SO_4(aq) \longrightarrow PbSO_4(s) + NaC_2H_3O_2(aq)$
6. Balance the equation.	$Pb(C_2H_3O_2)_2(aq) + Na_2SO_4(aq) \longrightarrow PbSO_4(s) + 2\,NaC_2H_3O_2(aq)$

▶ **SKILLBUILDER 7.9 | Predicting and Writing Equations for Precipitation Reactions**

Write an equation for the precipitation reaction (if any) that occurs when solutions of potassium sulfate and strontium nitrate are mixed. If no reaction occurs, write *NO REACTION*.

▶ **FOR MORE PRACTICE** Problems 67, 68.

PEARSON
eText
2.0

CONCEPTUAL ✔ CHECKPOINT 7.5

Which reaction results in the formation of a precipitate?
(a) $NaNO_3(aq) + CaS(aq)$
(b) $MgSO_4(aq) + CaS(aq)$
(c) $NaNO_3(aq) + MgSO_4(aq)$

7.7 Writing Chemical Equations for Reactions in Solution: Molecular, Complete Ionic, and Net Ionic Equations

▶ Write molecular, complete ionic, and net ionic equations.

Consider the following equation for a precipitation reaction:

$$AgNO_3(aq) + NaCl(aq) \longrightarrow AgCl(s) + NaNO_3(aq)$$

This equation is a **molecular equation**, an equation showing the complete neutral formulas for every compound in the reaction. We can also write equations for reactions occurring in aqueous solution to show that aqueous ionic compounds normally dissociate in solution. For example, we can write the previous equation as:

$$Ag^+(aq) + NO_3^-(aq) + Na^+(aq) + Cl^-(aq) \longrightarrow AgCl(s) + Na^+(aq) + NO_3^-(aq)$$

Equations such as this one, which show the reactants and products as they are actually present in solution, are **complete ionic equations**.

Notice that in the complete ionic equation, some of the ions in solution appear unchanged on both sides of the equation. These ions are called **spectator ions** because they do not participate in the reaction.

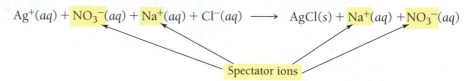

To simplify the equation and to more clearly show what is happening, spectator ions can be omitted.

$$Ag^+(aq) + Cl^-(aq) \longrightarrow AgCl(s)$$

Equations such as this one, which show only the *species* that actually participate in the reaction, are **net ionic equations**.

As another example, consider the reaction between HCl(*aq*) and NaOH(*aq*).

$$HCl(aq) + NaOH(aq) \longrightarrow H_2O(l) + NaCl(aq)$$

HCl, NaOH, and NaCl exist in solution as independent ions. The *complete ionic equation* for this reaction is:

$$H^+(aq) + Cl^-(aq) + Na^+(aq) + OH^-(aq) \longrightarrow H_2O(l) + Na^+(aq) + Cl^-(aq)$$

To write the *net ionic equation*, we remove the spectator ions, those that are unchanged on both sides of the equation.

$$H^+(aq) + \cancel{Cl^-(aq)} + \cancel{Na^+(aq)} + OH^-(aq) \longrightarrow H_2O(l) + \cancel{Na^+(aq)} + \cancel{Cl^-(aq)}$$

Spectator ions

The net ionic equation is $H^+(aq) + OH^-(aq) \longrightarrow H_2O(l)$.

To summarize:

- A molecular equation is a chemical equation showing the complete, neutral formulas for every compound in a reaction.
- A complete ionic equation is a chemical equation showing all of the species as they are actually present in solution.
- A net ionic equation is an equation showing only the species that actually participate in the reaction.

CONCEPTUAL ✓ CHECKPOINT 7.6

Which chemical equation is a net ionic equation?

(a) $K_2SO_4(aq) + BaCl_2(aq) \longrightarrow BaSO_4(s) + 2\ KCl(aq)$

(b) $2\ K^+(aq) + SO_4^{2-}(aq) + Ba^{2+}(aq) + 2\ Cl^-(aq) \longrightarrow$

$$BaSO_4(s) + 2\ K^+(aq) + 2\ Cl^-(aq)$$

(c) $Ba^{2+}(aq) + SO_4^{2-}(aq) \longrightarrow BaSO_4(s)$

EXAMPLE **7.10** **Writing Complete Ionic and Net Ionic Equations**

Consider this precipitation reaction occurring in aqueous solution.

$$Pb(NO_3)_2(aq) + 2\,LiCl(aq) \longrightarrow PbCl_2(s) + 2\,LiNO_3(aq)$$

Write a complete ionic equation and a net ionic equation for the reaction.

	SOLUTION
Write the complete ionic equation by separating aqueous ionic compounds into their constituent ions. The $PbCl_2(s)$ remains as one unit because it does not dissociate in solution (it is insoluble).	**Complete ionic equation** $Pb^{2+}(aq) + 2\,NO_3^-(aq) + 2\,Li^+(aq) + 2\,Cl^-(aq) \longrightarrow PbCl_2(s) + 2\,Li^+(aq) + 2\,NO_3^-(aq)$
Write the net ionic equation by eliminating the spectator ions, those that do not change during the reaction.	**Net ionic equation** $Pb^{2+}(aq) + 2\,Cl^-(aq) \longrightarrow PbCl_2(s)$

▶ **SKILLBUILDER 7.10** | **Writing Complete Ionic and Net Ionic Equations**

Write a complete ionic equation and a net ionic equation for this reaction occurring in aqueous solution.

$$2\,HBr(aq) + Ca(OH)_2(aq) \longrightarrow 2\,H_2O(l) + CaBr_2(aq)$$

▶ **FOR MORE PRACTICE** Example 7.20; Problems 69, 70, 71, 72.

7.8 Acid–Base and Gas Evolution Reactions

- ▶ Identify and write equations for acid–base reactions.
- ▶ Identify and write equations for gas evolution reactions.

Two other kinds of reactions that occur in solution are **acid–base reactions**—reactions that form water upon mixing of an acid and a base—and **gas evolution reactions**—reactions that evolve a gas. Like precipitation reactions, these reactions occur when the cation of one reactant combines with the anion of another. As we will see in Section 7.9, many gas evolution reactions also happen to be acid–base reactions.

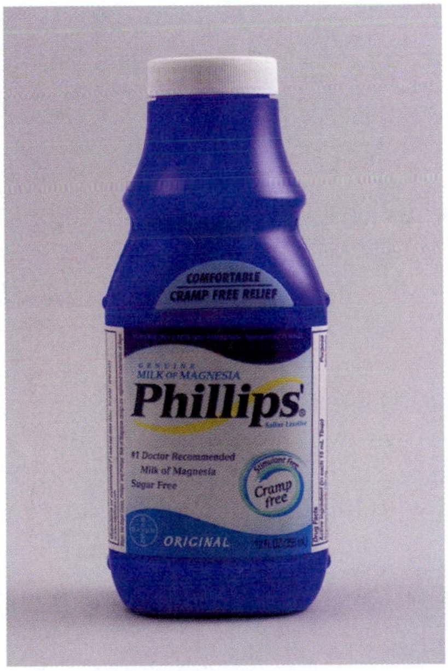

▲ Milk of magnesia is basic and tastes bitter.

Even though coffee is acidic overall, it contains some naturally occurring bases (such as caffeine) that give it a bitter taste.

Acid–Base (Neutralization) Reactions

As we saw in Chapter 5, an acid is a compound characterized by its sour taste, its ability to dissolve some metals, and its tendency to form H^+ ions in solution. A base is a compound characterized by its bitter taste, its slippery feel, and its tendency to form OH^- ions in solution. Table 7.3 lists some common acids and bases. Acids and bases are also found in many everyday substances. Foods such as lemons, limes, and vinegar contain acids. Soap, coffee, and milk of magnesia all contain bases.

When an acid and a base are mixed, the $H^+(aq)$ from the acid combines with the $OH^-(aq)$ from the base to form $H_2O(l)$. Consider the reaction between hydrochloric acid and sodium hydroxide mentioned earlier.

$$\underset{\text{Acid}}{HCl(aq)} + \underset{\text{Base}}{NaOH(aq)} \longrightarrow \underset{\text{Water}}{H_2O(l)} + \underset{\text{Salt}}{NaCl(aq)}$$

TABLE 7.3 Some Common Acids and Bases

Acid	Formula	Base	Formula
hydrochloric acid	HCl	sodium hydroxide	$NaOH$
hydrobromic acid	HBr	lithium hydroxide	$LiOH$
nitric acid	HNO_3	potassium hydroxide	KOH
sulfuric acid	H_2SO_4	calcium hydroxide	$Ca(OH)_2$
perchloric acid	$HClO_4$	barium hydroxide	$Ba(OH)_2$
acetic acid	$HC_2H_3O_2$		

▶ Common foods and everyday substances such as oranges, lemons, vinegar, and vitamin C contain acids.

Acid–base reactions (also called **neutralization reactions**) generally form water and an ionic compound—called a **salt**—that usually remains dissolved in the solution. The net ionic equation for many acid–base reactions is:

$$H^+(aq) + OH^-(aq) \longrightarrow H_2O(l)$$

Another acid–base reaction is the reaction that occurs between sulfuric acid and potassium hydroxide.

$$\underset{\text{Acid}}{H_2SO_4(aq)} + \underset{\text{Base}}{2\ KOH} \longrightarrow \underset{\text{Water}}{2\ H_2O(l)} + \underset{\text{Salt}}{K_2SO_4(aq)}$$

Notice the pattern of acid and base reacting to form water and a salt.

$$\text{Acid} + \text{Base} \longrightarrow \text{Water} + \text{Salt} \quad \text{(acid–base reactions)}$$

When writing equations for acid–base reactions, write the formula of the salt using the procedure for writing formulas of ionic compounds presented in Section 5.5.

EXAMPLE **7.11** | **Writing Equations for Acid–Base Reactions**

Write a molecular and a net ionic equation for the reaction between aqueous HNO_3 and aqueous $Ca(OH)_2$.

You must recognize these substances as an acid and a base. Write the skeletal reaction following the general pattern of acid plus base produces water plus salt.	**SOLUTION** $\underset{\text{Acid}}{HNO_3(aq)} + \underset{\text{Base}}{Ca(OH)_2(aq)} \longrightarrow \underset{\text{Water}}{H_2O(l)} + \underset{\text{Salt}}{Ca(NO_3)_2(aq)}$
Next, balance the equation.	$2\ HNO_3(aq) + Ca(OH)_2(aq) \longrightarrow 2\ H_2O(l) + Ca(NO_3)_2(aq)$
Write the net ionic equation by eliminating the ions that remain the same on both sides of the equation.	$2\ H^+(aq) + 2\ OH^-(aq) \longrightarrow 2\ H_2O(l)$ or simply $H^+(aq) + OH^-(aq) \longrightarrow H_2O(l)$

▶ **SKILLBUILDER 7.11** | **Writing Equations for Acid–Base Reactions**

Write a molecular and a net ionic equation for the reaction that occurs between aqueous H_2SO_4 and aqueous KOH.

▶ **FOR MORE PRACTICE**   Example 7.21; Problems 77, 78, 79, 80.

Gas Evolution Reactions

Some aqueous reactions form a gas as a product. These reactions, as we learned in Section 7.1, are gas evolution reactions. Some gas evolution reactions form a gaseous product directly when the cation of one reactant reacts with the anion of the other. For example, when sulfuric acid reacts with lithium sulfide, dihydrogen monosulfide gas forms:

$$H_2SO_4(aq) + Li_2S(aq) \longrightarrow \underset{\text{Gas}}{H_2S(g)} + Li_2SO_4(aq)$$

Many gas evolution reactions such as this one are also acid–base reactions. In Chapter 14 we learn how ions such as HCO_3^- act as bases in aqueous solution.

Other gas evolution reactions form an intermediate product that then decomposes into a gas. For example, when aqueous hydrochloric acid is mixed with aqueous sodium bicarbonate, the following reaction occurs:

$$HCl(aq) + NaHCO_3(aq) \longrightarrow H_2CO_3(aq) + NaCl(aq) \longrightarrow \underset{\text{Gas}}{H_2O(l) + CO_2(g)} + NaCl(aq)$$

Gas evolution reaction

$HC_2H_3O_2(aq) + NaHCO_3(aq) \longrightarrow$
$H_2O(l) + CO_2(g) + NaC_2H_3O_2(aq)$

▲ In this gas evolution reaction, vinegar (a dilute solution of acetic acid) and baking soda (sodium bicarbonate) produce carbon dioxide.

The intermediate product, H_2CO_3, is not stable and decomposes to form H_2O and gaseous CO_2. This reaction is almost identical to the reaction in the grade school volcano of Section 7.1, which involves the mixing of acetic acid and sodium bicarbonate:

$$HC_2H_3O_2(aq) + NaHCO_3(aq) \longrightarrow H_2CO_3(aq) + NaC_2H_3O_2(aq) \longrightarrow$$
$$H_2O(l) + CO_2(g) + NaC_2H_3O_2(aq)$$

The bubbling of the classroom volcano is caused by the newly formed carbon dioxide gas.

Other important gas evolution reactions form either H_2SO_3 or NH_4OH as intermediate products:

$$HCl(aq) + NaHSO_3(aq) \longrightarrow H_2SO_3(aq) + NaCl(aq) \longrightarrow$$
$$H_2O(l) + SO_2(g) + NaCl(aq)$$

$$NH_4Cl(aq) + NaOH(aq) \longrightarrow NH_4OH(aq) + NaCl(aq) \longrightarrow$$
$$H_2O(l) + NH_3(g) + NaCl(aq)$$

Table 7.4 lists the main types of compounds that form gases in aqueous reactions, as well as the gases that they form.

TABLE 7.4 Types of Compounds That Undergo Gas Evolution Reactions

Reactant Type	Intermediate Product	Gas Evolved	Example
sulfides	none	H_2S	$2\ HCl(aq) + K_2S(aq) \longrightarrow H_2S(g) + 2\ KCl(aq)$
carbonates and bicarbonates	H_2CO_3	CO_2	$2\ HCl(aq) + K_2CO_3(aq) \longrightarrow H_2O(l) + CO_2(g) + 2\ KCl(aq)$
sulfites and bisulfites	H_2SO_3	SO_2	$2\ HCl(aq) + K_2SO_3(aq) \longrightarrow H_2O(l) + SO_2(g) + 2\ KCl(aq)$
ammonium	NH_4OH	NH_3	$NH_4Cl(aq) + KOH(aq) \longrightarrow H_2O(l) + NH_3(g) + KCl(aq)$

EXAMPLE **7.12** **Writing Equations for Gas Evolution Reactions**

Write a molecular equation for the gas evolution reaction that occurs when you mix aqueous nitric acid and aqueous sodium carbonate.

Begin by writing a skeletal equation that includes the reactants and products that form when the cation of each reactant combines with the anion of the other.	**SOLUTION** $$HNO_3(aq) + Na_2CO_3(aq) \longrightarrow H_2CO_3(aq) + NaNO_3(aq)$$
You must recognize that $H_2CO_3(aq)$ decomposes into $H_2O(l)$ and $CO_2(g)$ and write the corresponding equation.	$$HNO_3(aq) + Na_2CO_3(aq) \longrightarrow H_2O(l) + CO_2(g) + NaNO_3(aq)$$
Finally, balance the equation.	$$2\ HNO_3(aq) + Na_2CO_3(aq) \longrightarrow H_2O(l) + CO_2(g) + 2\ NaNO_3(aq)$$

▶ **SKILLBUILDER 7.12 | Writing Equations for Gas Evolution Reactions**

Write a molecular equation for the gas evolution reaction that occurs when you mix aqueous hydrobromic acid and aqueous potassium sulfite.

▶ **SKILLBUILDER PLUS** | Write a net ionic equation for the previous reaction.

▶ **FOR MORE PRACTICE** Example 7.22; Problems 81, 82.

CHEMISTRY AND HEALTH

Neutralizing Excess Stomach Acid

Heartburn is a painful burning sensation in the esophagus (the tube that joins the throat and stomach). The pain is casued by stomach acid, which helps to break down food during digestion. Sometimes—especially after a large meal—some of that stomach acid can work its way up into the esophagus, causing the pain. You can relieve mild heartburn simply by swallowing repeatedly. Saliva contains the bicarbonate ion (HCO_3^-), which acts as a base to neutralize the acid. You can treat more severe heartburn with antacids, over-the-counter medicines that work by reacting with and neutralizing stomach acid. Antacids employ different bases as neutralizing agents. Tums™, for example, contains $CaCO_3$; milk of magnesia contains $Mg(OH)_2$; and Mylanta™ contains $Al(OH)_3$. They all, however, have the same effect of neutralizing stomach acid and relieving heartburn.

B7.1 CAN YOU ANSWER THIS? *Assume that stomach acid is HCl and write equations showing how each of these antacids neutralizes stomach acid.*

▲ The base in an antacid neutralizes excess stomach acid, relieving heartburn and acid stomach.

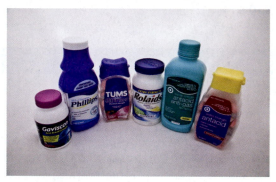

▲ Antacids contain bases such as $Mg(OH)_2$, $Al(OH)_3$, and $NaHCO_3$.

7.9 Oxidation–Reduction Reactions

▶ Identify redox reactions.

▶ Identify and write equations for combustion reactions.

We will cover oxidation–reduction reactions in more detail in Chapter 16.

Reactions involving the transfer of electrons are **oxidation–reduction reactions** or **redox reactions**. Redox reactions are responsible for the rusting of iron, the bleaching of hair, and the production of electricity in batteries. Many redox reactions involve the reaction of a substance with oxygen.

$$2\,H_2(g) + O_2(g) \longrightarrow 2\,H_2O(g)$$
(reaction that powers the space shuttle)

$$4\,Fe(s) + 3\,O_2(g) \longrightarrow 2\,Fe_2O_3(s)$$
(rusting of iron)

$$CH_4(g) + 2\,O_2(g) \longrightarrow CO_2(g) + 2\,H_2O(g)$$
(combustion of natural gas)

However, redox reactions don't always have to involve oxygen. Consider, for example, the reaction between sodium and chlorine to form table salt (NaCl).

$$2\,Na(s) + Cl_2(g) \longrightarrow 2\,NaCl(s)$$

The reaction between sodium and oxygen also forms other oxides besides Na_2O.

This reaction is similar to the reaction between sodium and oxygen, which can form sodium oxide.

$$4\,Na(s) + O_2(g) \longrightarrow 2\,Na_2O(s)$$

What do these two reactions have in common? In both cases, sodium (a metal with a tendency to lose electrons) reacts with a nonmetal (that has a tendency to gain electrons). In both cases, sodium atoms lose electrons to nonmetal atoms. A fundamental definition of oxidation is *the loss of electrons*, and a fundamental definition of reduction is *the gain of electrons*.

Helpful mnemonics:
OIL RIG—Oxidation Is Loss; Reduction Is Gain.
LEO GER—Lose Electrons Oxidation; Gain Electrons Reduction.

Notice that oxidation and reduction must occur together. If one substance loses electrons (oxidation), then another substance must gain electrons (reduction). For now, you simply need to be able to identify redox reactions. In Chapter 16 we will examine them more thoroughly.

Redox reactions are those in which:

A reaction can be classified as a redox reaction if it meets any one of these requirements.

- A substance reacts with elemental oxygen.
- A metal reacts with a nonmetal.
- More generally, one substance transfers electrons to another substance.

EXAMPLE **7.13** | **Identifying Redox Reactions**

Which of these are redox reactions?

(a) $2\,Mg(s) + O_2(g) \longrightarrow 2\,MgO(s)$
(b) $2\,HBr(aq) + Ca(OH)_2(aq) \longrightarrow 2\,H_2O(l) + CaBr_2(aq)$
(c) $Ca(s) + Cl_2(g) \longrightarrow CaCl_2(s)$
(d) $Zn(s) + Fe^{2+}(aq) \longrightarrow Zn^{2+}(aq) + Fe(s)$

SOLUTION

(a) Redox reaction; Mg reacts with elemental oxygen.
(b) Not a redox reaction; it is an acid–base reaction.
(c) Redox reaction; a metal reacts with a nonmetal.
(d) Redox reaction; Zn transfers two electrons to Fe^{2+}.

▶ SKILLBUILDER 7.13 | **Identifying Redox Reactions**

Which of these are redox reactions?

(a) $2\,Li(s) + Cl_2(g) \longrightarrow 2\,LiCl(s)$
(b) $2\,Al(s) + 3\,Sn^{2+}(aq) \longrightarrow 2\,Al^{3+}(aq) + 3\,Sn(s)$
(c) $Pb(NO_3)_2(aq) + 2\,LiCl(aq) \longrightarrow PbCl_2(s) + 2\,LiNO_3(aq)$
(d) $C(s) + O_2(g) \longrightarrow CO_2(g)$

▶ FOR MORE PRACTICE Example 7.23; Problems 83, 84.

The water formed in combustion reactions may be gaseous (*g*) or liquid (*l*) depending on the reaction conditions.

Combustion reactions are a type of redox reaction. They are important because most of our society's energy is derived from combustion reactions. Combustion reactions are characterized by the reaction of a substance with O_2 to form one or more oxygen-containing compounds, often including water. Combustion reactions are exothermic (they emit heat). For example, as we saw in Section 7.3, natural gas (CH_4) reacts with oxygen to form carbon dioxide and water.

$$CH_4(g) + 2\,O_2(g) \longrightarrow CO_2(g) + 2\,H_2O(g)$$

As mentioned in Section 7.1, combustion reactions power automobiles. For example, octane, a component of gasoline, reacts with oxygen to form carbon dioxide and water.

$$2\,C_8H_{18}(l) + 25\,O_2(g) \longrightarrow 16\,CO_2(g) + 18\,H_2O(g)$$

Combustion

▲ Combustion of octane occurs in the cylinders of an automobile engine.

Ethanol, the alcohol in alcoholic beverages, also reacts with oxygen in a combustion reaction to form carbon dioxide and water.

$$C_2H_5OH(l) + 3\,O_2(g) \longrightarrow 2\,CO_2(g) + 3\,H_2O(g)$$

Compounds containing carbon and hydrogen—or carbon, hydrogen, and oxygen—always form carbon dioxide and water upon combustion. Other combustion reactions include the reaction of carbon with oxygen to form carbon dioxide:

$$C(s) + O_2(g) \longrightarrow CO_2(g)$$

and the reaction of hydrogen with oxygen to form water:

$$2\,H_2(g) + O_2(g) \longrightarrow 2\,H_2O(g)$$

EXAMPLE **7.14**	**Writing Combustion Reactions**
Write a balanced equation for the combustion of liquid methyl alcohol (CH_3OH).	
Write a skeletal equation showing the reaction of CH_3OH with O_2 to form CO_2 and H_2O.	**SOLUTION** $CH_3OH(l) + O_2(g) \longrightarrow CO_2(g) + H_2O(g)$
Balance the skeletal equation using the rules in Section 7.4.	$2\,CH_3OH(l) + 3\,O_2(g) \longrightarrow 2\,CO_2(g) + 4\,H_2O(g)$

▶ **SKILLBUILDER 7.14 | Writing Combustion Reactions**
Write a balanced equation for the combustion of liquid pentane (C_5H_{12}), a component of gasoline.

▶ **SKILLBUILDER PLUS |** Write a balanced equation for the combustion of liquid propanol (C_3H_7OH).

▶ **FOR MORE PRACTICE** Example 7.24; Problems 85, 86.

7.10 Classifying Chemical Reactions

▶ Classify chemical reactions.

Throughout this chapter, we have examined different types of chemical reactions. We have seen examples of precipitation reactions, acid–base reactions, gas evolution reactions, oxidation–reduction reactions, and combustion reactions. We can organize these different types of reactions with the following flowchart:

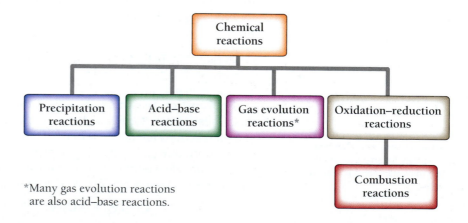

*Many gas evolution reactions are also acid–base reactions.

This classification scheme focuses on the type of chemistry or phenomenon that is occurring during the reaction (such as the formation of a precipitate or the transfer of electrons). However, an alternative way to classify chemical reactions is by what atoms or groups of atoms do during the reaction.

Classifying Chemical Reactions by What Atoms Do

In an alternative way of classifying reactions, we focus on the pattern of the reaction by classifying it into one of the following four categories. In this classification scheme, the letters (A, B, C, D) represent atoms or groups of atoms.

Type of Reaction	Generic Equation
synthesis or combination	$A + B \longrightarrow AB$
decomposition	$AB \longrightarrow A + B$
single-displacement	$A + BC \longrightarrow AC + B$
double-displacement	$AB + CD \longrightarrow AD + CB$

Synthesis or Combination Reactions

In a **synthesis** or **combination reaction**, simple substances combine to form more complex substances. The simpler substances may be elements, such as sodium and chlorine combining to form sodium chloride.

$$2\,Na(s) + Cl_2(g) \longrightarrow 2\,NaCl(s)$$

The simpler substances may also be compounds, such as calcium oxide and carbon dioxide combining to form calcium carbonate.

$$CaO(s) + CO_2(g) \longrightarrow CaCO_3(s)$$

In either case, a synthesis reaction follows the general equation:

$$A + B \longrightarrow AB$$

$$2\,Na(s) + Cl_2(g) \longrightarrow 2\,NaCl(s)$$

▲ In a synthesis reaction, two simpler substances combine to make a more complex substance. When sodium metal and chlorine gas combine, a chemical reaction occurs that forms sodium chloride.

Note that the first two of these reactions are also redox reactions.

Other examples of synthesis reactions include:

$$2\,H_2(g) + O_2(g) \longrightarrow 2\,H_2O(l)$$

$$2\,Mg(s) + O_2(g) \longrightarrow 2\,MgO(s)$$

$$SO_3(g) + H_2O(l) \longrightarrow H_2SO_4(aq)$$

Decomposition Reactions

In a **decomposition reaction**, a complex substance decomposes to form simpler substances. The simpler substances may be elements, such as the hydrogen and oxygen gases that form upon the decomposition of water when electrical current passes through it.

$$2\,H_2O(l) \xrightarrow{\text{electrical current}} 2\,H_2(g) + O_2(g)$$

The simpler substances may also be compounds, such as the calcium oxide and carbon dioxide that form upon heating calcium carbonate.

$$CaCO_3(s) \xrightarrow{\text{heat}} CaO(s) + CO_2(g)$$

In either case, a decomposition reaction follows the general equation:

$$AB \longrightarrow A + B$$

Other examples of decomposition reactions include:

$$2\,HgO(s) \xrightarrow{\text{heat}} 2\,Hg(l) + O_2(g)$$

$$2\,KClO_3(s) \xrightarrow{\text{heat}} 2\,KCl(s) + 3\,O_2(g)$$

$$CH_3I(g) \xrightarrow{\text{light}} CH_3(g) + I(g)$$

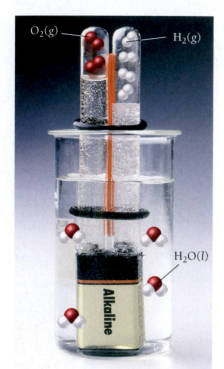

$O_2(g)$ ── ── $H_2(g)$

$H_2O(l)$

Alkaline

$$2\,H_2O(l) \longrightarrow 2\,H_2(g) + O_2(g)$$

▲ When electrical current is passed through water, the water undergoes a decomposition reaction to form hydrogen gas and oxygen gas.

Notice that these decomposition reactions require energy in the form of heat, electrical current, or light to make them happen. This is because compounds are normally stable and energy is required to decompose them. A number of decomposition reactions require *ultraviolet* or *UV light,* which is light in the ultraviolet region of the spectrum. UV light carries more energy than visible light and can therefore initiate the decomposition of many compounds. (We will discuss light in more detail in Chapter 9.)

Displacement Reactions

In a **displacement** or **single-displacement reaction**, one element displaces another in a compound. For example, when we add metallic zinc to a solution of copper(II) chloride, the zinc replaces the copper.

$$Zn(s) + CuCl_2(aq) \longrightarrow ZnCl_2(aq) + Cu(s)$$

A displacement reaction follows the general equation:

$$A + BC \longrightarrow AC + B$$

Other examples of displacement reactions include:

$$Mg(s) + 2\,HCl(aq) \longrightarrow MgCl_2(aq) + H_2(g)$$

$$2\,Na(s) + 2\,H_2O(l) \longrightarrow 2\,NaOH(aq) + H_2(g)$$

The last reaction can be identified more easily as a displacement reaction if we write water as $HOH(l)$.

$$2\,Na(s) + 2\,HOH(l) \longrightarrow 2\,NaOH(aq) + H_2(g)$$

▶ In a single-displacement reaction, one element displaces another in a compound. When zinc metal is immersed in a copper(II) chloride solution, the zinc atoms displace the copper ions in solution and the copper ions coat onto the zinc metal.

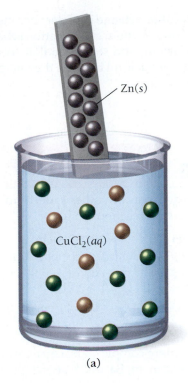

Zn(s)

CuCl₂(aq)

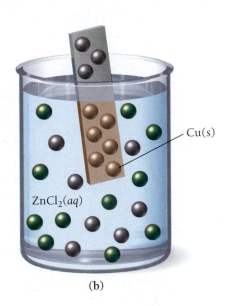

Cu(s)

ZnCl₂(aq)

(a)   (b)

Double-Displacement Reactions

In a **double-displacement reaction**, two elements or groups of elements in two different compounds exchange places to form two new compounds. For example, in aqueous solution, the silver in silver nitrate changes places with the sodium in sodium chloride and solid silver chloride and aqueous sodium nitrate form.

This double-displacement reaction is also a precipitation reaction.

$$AgNO_3(aq) + NaCl(aq) \longrightarrow AgCl(s) + NaNO_3(aq)$$

A double-displacement reaction follows the general form:

$$AB + CD \longrightarrow AD + CB$$

Other examples of double-displacement reactions include:

These double-displacement reactions are also acid–base reactions.

$$HCl(aq) + NaOH(aq) \longrightarrow H_2O(l) + NaCl(aq)$$

$$2\,HCl(aq) + Na_2CO_3(aq) \longrightarrow H_2CO_3(aq) + 2\,NaCl(aq)$$

As we learned in Section 7.8, $H_2CO_3(aq)$ is not stable and decomposes to form $H_2O(l) + CO_2(g)$, so the overall equation is:

This double-displacement reaction is also a gas evolution reaction and an acid–base reaction.

$$2\,HCl(aq) + Na_2CO_3(aq) \longrightarrow H_2O(l) + CO_2(g) + 2\,NaCl(aq)$$

Classification Flowchart

A flowchart for this classification scheme of chemical reactions is as follows:

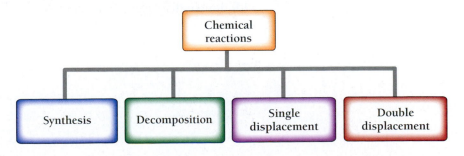

Chemical reactions

Synthesis — Decomposition — Single displacement — Double displacement

No single classification scheme is perfect because all chemical reactions are unique in some sense. However, both classification schemes—one that focuses on the type of chemistry occurring and the other that focuses on what atoms or groups of atoms are doing—are helpful because they help us see differences and similarities among chemical reactions.

EXAMPLE **7.15** | **Classifying Chemical Reactions According to What Atoms Do**

Classify each reaction as a synthesis, decomposition, single-displacement, or double-displacement reaction.

(a) $Na_2O(s) + H_2O(l) \longrightarrow 2\,NaOH(aq)$
(b) $Ba(NO_3)_2(aq) + K_2SO_4(aq) \longrightarrow BaSO_4(s) + 2\,KNO_3(aq)$
(c) $2\,Al(s) + Fe_2O_3(s) \longrightarrow Al_2O_3(s) + 2\,Fe(l)$
(d) $2\,H_2O_2(aq) \longrightarrow 2\,H_2O(l) + O_2(g)$
(e) $Ca(s) + Cl_2(g) \longrightarrow CaCl_2(s)$

SOLUTION

(a) Synthesis; a more complex substance forms from two simpler ones.
(b) Double-displacment; Ba and K switch places to form two new compounds.
(c) Single-displacement; Al displaces Fe in Fe_2O_3.
(d) Decomposition; a complex substance decomposes into simpler ones.
(e) Synthesis; a more complex substance forms from two simpler ones.

▶ SKILLBUILDER 7.15 | **Classifying Chemical Reactions According to What Atoms Do**

Classify each reaction as a synthesis, decomposition, single-displacement, or double-displacement reaction.

(a) $2\,Al(s) + 2\,H_3PO_4(aq) \longrightarrow 2\,AlPO_4(aq) + 3\,H_2(g)$
(b) $CuSO_4(aq) + 2\,KOH(aq) \longrightarrow Cu(OH)_2(s) + K_2SO_4(aq)$
(c) $2\,K(s) + Br_2(l) \longrightarrow 2\,KBr(s)$
(d) $CuCl_2(aq) \xrightarrow[\text{electrical current}]{} Cu(s) + Cl_2(g)$

▶ FOR MORE PRACTICE Example 7.25; Problems 89, 90, 91, 92.

PEARSON eText 2.0

CONCEPTUAL ✔ CHECKPOINT 7.7

Both precipitation reactions and acid–base reactions can also be classified as:

(a) synthesis reactions

(b) decomposition reactions

(c) single-displacement reactions

(d) double-displacement reactions

Chapter **7** in Review

MasteringChemistry™ provides end-of-chapter exercises, feedback-enriched tutorial problems, animations, and interactive activities to encourage problem solving practice and deeper understanding of key concepts and topics.

Self-Assessment Quiz

Q1. Which process is a chemical reaction?
(a) Gasoline evaporating from a gasoline tank
(b) Iron rusting when left outdoors
(c) Dew condensing on grass during the night
(d) Water boiling on a stove top

Q2. How many oxygen atoms are on the reactant side of this chemical equation?

$$K_2CO_3(aq) + Pb(NO_3)_2(aq) \longrightarrow 2\,KNO_3(aq) + PbCO_3(s)$$

(a) 3
(b) 6
(c) 9
(d) 12

Q3. What is the coefficient for hydrogen in the balanced equation for the reaction of solid iron(III) oxide with gaseous hydrogen to form solid iron and liquid water?
(a) 2
(b) 3
(c) 4
(d) 6

Q4. Determine the correct set of coefficients to balance the chemical equation.

$$\_C_6H_6(l) + \_O_2(g) \longrightarrow \_CO_2(g) + \_H_2O(g)$$

(a) 2, 15, 12, 6
(b) 1, 15, 6, 3
(c) 2, 15, 12, 12
(d) 1, 7, 6, 3

Q5. Which compound is soluble in water?
(a) $Fe(OH)_2$
(b) CuS
(c) $AgCl$
(d) $CuCl_2$

Q6. Name the precipitate that forms (if any) when aqueous solutions of barium nitrate and potassium sulfate are mixed.
(a) $BaK(s)$
(b) $NO_3SO_4(s)$
(c) $BaSO_4(s)$
(d) $KNO_3(s)$

Q7. Which set of reactants forms a solid precipitate when mixed?
(a) $NaNO_3(aq)$ and $KCl(aq)$
(b) $KOH(aq)$ and $Na_2CO_3(aq)$
(c) $CuCl_2(aq)$ and $NaC_2H_3O_2(aq)$
(d) $Na_2CO_3(aq)$ and $CaCl_2(aq)$

Q8. What is the net ionic equation for the reaction between $Pb(C_2H_3O_2)_2(aq)$ and $KBr(aq)$?
(a) $Pb(C_2H_3O_2)_2(aq) + 2\,KBr(aq) \longrightarrow$
$$PbBr_2(s) + 2\,KC_2H_3O_2(aq)$$
(b) $Pb^{2+}(aq) + 2\,Br^-(aq) \longrightarrow PbBr_2(s)$
(c) $K^+(aq) + C_2H_3O_2^-(aq) \longrightarrow KC_2H_3O_2(s)$
(d) $Pb(C_2H_3O_2)_2(aq) + 2\,KBr(aq) \longrightarrow$
$$2\,KC_2H_3O_2(s) + PbBr_2(aq)$$

Q9. Complete the equation:
$$HBr(aq) + NaOH\,(aq) \longrightarrow \underline{\hspace{3cm}}$$
(a) $NaH(s) + BrOH(aq)$
(b) $NaBr(s) + NaOH(aq)$
(c) $H_2O(l) + NaBr(aq)$
(d) No reaction occurs.

Q10. Complete the equation:
$$HNO_3(aq) + KHCO_3(aq) \longrightarrow \underline{\hspace{3cm}}$$
(a) $H_2O(l) + CO_2(g) + KNO_3(aq)$
(b) $KNO_3(s) + H_2CO_3(aq)$
(c) $HK(aq) + NO_3HCO_3$
(d) No reaction occurs.

Q11. What are the products of the balanced equation for the combustion of C_4H_9OH?
(a) $C_4H_9(s) + NaOH(aq)$
(b) $4\,CO_2(g) + 5\,H_2O(l)$
(c) $4\,O_2(g) + 5\,H_2O(l)$
(d) $2\,C_2H_4(g) + H_2O(l)$

Q12. Precipitation reactions are best classified as which type of reaction?
(a) Single displacement
(b) Double displacement
(c) Decomposition
(d) None of the above

Answers: 1:b, 2:c, 3:b, 4:a, 5:d, 6:c, 7:d, 8:b, 9:c, 10:a, 11:b, 12:b

Chemical Principles	*Relevance*

Chemical Reactions

In a chemical reaction, one or more substances—either elements or compounds—change into a different substance.

Chemical reactions are central to many processes, including transportation, energy generation, manufacturing of household products, vision, and life itself.

Evidence of a Chemical Reaction

The only absolute evidence for a chemical reaction is chemical analysis showing that one or more substances have changed into another substance. However, at least one of the following observations is often evidence of a chemical reaction: a color change; the formation of a solid or precipitate; the formation of a gas; the emission of light; and the emission or absorption of heat.

We can often perceive the changes that accompany chemical reactions. In fact, we often employ chemical reactions for the changes they produce. For example, we use the heat emitted by the combustion of fossil fuels to warm our homes, drive our cars, and generate electricity.

Chemical Equations

Chemical equations represent chemical reactions. They include formulas for the reactants (the substances present before the reaction) and for the products (the new substances formed by the reaction). Chemical equations must be balanced to reflect the conservation of matter in nature; atoms do not spontaneously appear or disappear.

Chemical equations allow us to represent and understand chemical reactions. For example, the equations for the combustion reactions of fossil fuels let us see that carbon dioxide, a gas that contributes to global warming, is one of the products of these reactions.

Aqueous Solutions and Solubility

Aqueous solutions are mixtures of a substance dissolved in water. If a substance dissolves in water, it is soluble; otherwise, it is insoluble.

Aqueous solutions are common. Oceans, lakes, and most of the fluids in our bodies are aqueous solutions.

Some Specific Types of Reactions

Precipitation reaction: A solid or precipitate forms upon mixing two aqueous solutions.

Acid–base reaction: Water form(s) upon mixing an acid and base.

Gas evolution reaction: A gas forms upon mixing two aqueous solutions.

Redox reaction: Electrons transfer from one substance to another.

Combustion reaction: A substance reacts with oxygen, emitting heat, and forming an oxygen-containing compound and, in many cases, water.

Many of the specific types of reactions discussed in this chapter occur in aqueous solutions and are therefore important to living organisms. Acid–base reactions, for example, constantly occur in the blood of living organisms to maintain constant blood acidity levels. In humans, a small change in blood acidity levels would result in death, so our bodies carry out chemical reactions to prevent this. Combustion reactions are important because they are the main energy source for our society.

Classifying Chemical Reactions

We can classify many chemical reactions into one of the following four categories according to what atoms or groups of atoms do:

- synthesis: $(A + B \longrightarrow AB)$
- decomposition: $(AB \longrightarrow A + B)$
- single-displacement: $(A + BC \longrightarrow AC + B)$
- double-displacement: $(AB + CD \longrightarrow AD + CB)$

We classify chemical reactions to better understand them and to recognize similarities and differences among reactions.

Chemical Skills

Examples

LO: Identify evidence of a chemical reaction (Section 7.2).

EXAMPLE **7.16** **Identifying a Chemical Reaction**

Which of these are chemical reactions?

(a) Copper turns green on exposure to air.
(b) When sodium bicarbonate is combined with hydrochloric acid, bubbling is observed.
(c) Liquid water freezes to form solid ice.
(d) A pure copper penny forms bubbles of a dark brown gas when dropped into nitric acid. The nitric acid solution turns blue.

To identify a chemical reaction, determine whether one or more of the initial substances changed into a different substance. If so, a chemical reaction occurred. One or more of the following often accompanies a chemical reaction: a color change; the formation of a solid or precipitate; the formation of a gas; the emission of light; and the emission or absorption of heat.

SOLUTION

(a) Chemical reaction, as evidenced by the color change.
(b) Chemical reaction, as evidenced by the evolution of a gas.
(c) Not a chemical reaction; solid ice is still water.
(d) Chemical reaction, as evidenced by the evolution of a gas and by a color change.

LO: Write balanced chemical equations (Sections 7.3, 7.4).

To write balanced chemical equations, follow these steps.

1. Write a skeletal equation by writing chemical formulas for each of the reactants and products. (If a skeletal equation is provided, proceed to Step 2.)

2. If an element occurs in only one compound on both sides of the equation, balance that element first. If there is more than one such element, and the equation contains both metals and nonmetals, balance metals before nonmetals.

3. If an element occurs as a free element on either side of the chemical equation, balance that element last.

4. If the balanced equation contains coefficient fractions, clear these by multiplying the entire equation by the appropriate factor.

5. Check to make certain the equation is balanced by summing the total number of each type of atom on both sides of the equation.

Reminders

• Change only the *coefficients* to balance a chemical equation, *never the subscripts*. Changing the subscripts would change the compounds themselves.

• If the equation contains polyatomic ions that stay intact on both sides of the equation, balance the polyatomic ions as a group.

EXAMPLE **7.17** **Writing Balanced Chemical Equations**

Write a balanced chemical equation for the reaction of solid vanadium(V) oxide with hydrogen gas to form solid vanadium(III) oxide and liquid water.

$$V_2O_5(s) + H_2(g) \longrightarrow V_2O_3(s) + H_2O(l)$$

SOLUTION

A skeletal equation is given. Proceed to Step 2.

Vanadium occurs in only one compound on both sides of the equation. However, it is balanced, so you can proceed and balance oxygen by placing a 2 in front of H_2O on the right side.

$$V_2O_5(s) + H_2(g) \longrightarrow V_2O_3(s) + 2\,H_2O(l)$$

Hydrogen occurs as a free element; balance it last by placing a 2 in front of H_2 on the left side.

$$V_2O_5(s) + 2\,H_2(g) \longrightarrow V_2O_3(s) + 2\,H_2O(l)$$

The equation does not contain coefficient fractions. Proceed to Step 5.

Check the equation.

$$V_2O_5(s) + 2\,H_2(g) \longrightarrow V_2O_3(s) + 2\,H_2O(l)$$

Reactants		Products
2 V atoms	$\longrightarrow$	2 V atoms
5 O atoms	$\longrightarrow$	5 O atoms
4 H atoms	$\longrightarrow$	4 H atoms

LO: Determine whether a compound is soluble (Section 7.5).

To determine whether or not a compound is soluble, refer to the solubility rules in Table 7.2. It is simplest to begin by looking for those ions that always form soluble compounds (Li^+, Na^+, K^+, NH_4^+, NO_3^-, and $C_2H_3O_2^-$). If a compound contains one of those, it is soluble. If it does not, determine if the anion is mostly soluble (Cl^-, Br^-, I^-, or SO_4^{2-}) or mostly insoluble (OH^-, S^{2-}, CO_3^{2-}, or PO_4^{3-}). Look at the cation as well to determine whether it is one of the exceptions.

EXAMPLE **7.18** | **Determining Whether a Compound is Soluble**

Is each compound soluble or insoluble?

(a) $CuCO_3$
(b) $BaSO_4$
(c) $Fe(NO_3)_3$

SOLUTION

(a) Insoluble; compounds containing CO_3^{2-} are insoluble, and Cu^{2+} is not an exception.
(b) Insoluble; compounds containing SO_4^{2-} are usually soluble, but Ba^{2+} is an exception.
(c) Soluble; all compounds containing NO_3^- are soluble.

LO: Predict and write equations for precipitation reactions (Section 7.6).

To predict whether a precipitation reaction occurs when two solutions are mixed and to write an equation for the reaction, follow these steps.

1. Write the formulas of the two compounds being mixed as reactants in a chemical equation.

2. Below the equation, write the formulas of the potentially insoluble products that could form from the reactants. Determine these by combining the cation from one reactant with the anion from the other. Make sure to adjust the subscripts so that all formulas are charge-neutral.

3. Use the solubility rules to determine whether any of the potentially insoluble products are indeed insoluble.

4. If all of the potentially insoluble products are soluble, there will be no precipitate. Write *NO REACTION* next to the arrow.

5. If one or both of the potentially insoluble products are insoluble, write their formula(s) as the product(s) of the reaction using (s) to indicate *solid*. Write any soluble products with (aq) to indicate *aqueous*.

6. Balance the equation.

EXAMPLE **7.19** | **Predicting Precipitation Reactions**

Write an equation for the precipitation reaction that occurs, if any, when solutions of sodium phosphate and cobalt(II) chloride are mixed.

SOLUTION

$$Na_3PO_4(aq) + CoCl_2(aq) \longrightarrow$$

Potentially Insoluble Products:

$$NaCl \qquad Co_3(PO_4)_2$$

NaCl is soluble.

$Co_3(PO_4)_2$ is insoluble.

Reaction contains an insoluble product; proceed to Step 5.

$$Na_3PO_4(aq) + CoCl_2(aq) \longrightarrow Co_3(PO_4)_2(s) + NaCl(aq)$$

$$2\,Na_3PO_4(aq) + 3\,CoCl_2(aq) \longrightarrow Co_3(PO_4)_2(s) + 6\,NaCl(aq)$$

LO: Write molecular, complete ionic, and net ionic equations (Section 7.7).

To write a molecular equation, include the complete, neutral formulas for every compound in the reaction.

To write a complete ionic equation from a molecular equation, separate all aqueous ionic compounds into independent ions. Do not separate solid, liquid, or gaseous compounds.

To write a net ionic equation from a complete ionic equation, eliminate all species that do not change (spectator ions) in the course of the reaction.

EXAMPLE **7.20** | **Writing Complete Ionic and Net Ionic Equations**

Write a complete ionic and a net ionic equation for the reaction.

$$2\,NH_4Cl(aq) + Hg_2(NO_3)_2(aq) \longrightarrow Hg_2Cl_2(s) + 2\,NH_4NO_3(aq)$$

SOLUTION

Complete ionic equation:

$$2\,NH_4^+(aq) + 2\,Cl^-(aq) + Hg_2^{2+}(aq) + 2\,NO_3^-(aq) \longrightarrow Hg_2Cl_2(s) + 2\,NH_4^+(aq) + 2\,NO_3^-(aq)$$

Net ionic equation:

$$2\,Cl^-(aq) + Hg_2^{2+}(aq) \longrightarrow Hg_2Cl_2(s)$$

LO: Identify and write equations for acid–base reactions (Section 7.8).

When you see an acid and a base (see Table 7.3) as reactants in an equation, write a reaction in which the acid and the base react to form water and a salt.

EXAMPLE 7.21 Writing Equations for Acid–Base Reactions

Write an equation for the reaction that occurs when aqueous hydroiodic acid is mixed with aqueous barium hydroxide.

SOLUTION

$$2\,HI(aq) + Ba(OH)_2(aq) \longrightarrow 2\,H_2O(l) + BaI_2(aq)$$
Acid · Base · Water · Salt

LO: Identify and write equations for gas evolution reactions (Section 7.8).

Refer to Table 7.4 to identify gas evolution reactions.

EXAMPLE 7.22 Writing Equations for Gas Evolution Reactions

Write an equation for the reaction that occurs when aqueous hydrobromic acid is mixed with aqueous potassium bisulfite.

SOLUTION

$$HBr(aq) + KHSO_3(aq) \longrightarrow H_2SO_3(aq) + KBr(aq) \longrightarrow$$
$$H_2O(l) + SO_2(g) + KBr(aq)$$

LO: Identify redox reactions (Section 7.9).

Redox reactions are those in which any of the following occurs:

- A substance reacts with elemental oxygen.
- A metal reacts with a nonmetal.
- One substance transfers electrons to another substance.

EXAMPLE 7.23 Identifying Redox Reactions

Which of these reactions is a redox reaction?

(a) $4\,Fe(s) + 3\,O_2(g) \longrightarrow 2\,Fe_2O_3(s)$
(b) $CaO(s) + CO_2(g) \longrightarrow CaCO_3(s)$
(c) $AgNO_3(aq) + NaCl(aq) \longrightarrow AgCl(s) + NaNO_3(aq)$

SOLUTION

Only **(a)** is a redox reaction.

LO: Identify and write equations for combustion reactions (Section 7.9).

In a combustion reaction, a substance reacts with O_2 to form one or more oxygen-containing compounds and, in many cases, water.

EXAMPLE 7.24 Writing Equations for Combustion Reactions

Write a balanced equation for the combustion of gaseous ethane (C_2H_6), a minority component of natural gas.

SOLUTION

The skeletal equation is:

$$C_2H_6(g) + O_2(g) \longrightarrow CO_2(g) + H_2O(g)$$

The balanced equation is:

$$2\,C_2H_6(g) + 7\,O_2(g) \longrightarrow 4\,CO_2(g) + 6\,H_2O(g)$$

LO: Classify chemical reactions (Section 7.10).

You can classify chemical reactions by inspection. The four major categories are:

Synthesis or combination

$$A + B \longrightarrow AB$$

Decomposition

$$AB \longrightarrow A + B$$

Single-displacement

$$A + BC \longrightarrow AC + B$$

Double-displacement

$$AB + CD \longrightarrow AD + CB$$

EXAMPLE 7.25 Classifying Chemical Reactions

Classify each chemical reaction as a synthesis, decomposition, single-displacement, or double-displacement reaction.

(a) $2\,K(s) + Br_2(g) \longrightarrow 2\,KBr(s)$
(b) $Fe(s) + 2\,AgNO_3(aq) \longrightarrow Fe(NO_3)_2(aq) + 2\,Ag(s)$
(c) $CaSO_3(s) \longrightarrow CaO(s) + SO_2(g)$
(d) $CaCl_2(aq) + Li_2SO_4(aq) \longrightarrow CaSO_4(s) + 2\,LiCl(aq)$

SOLUTION

(a) Synthesis; KBr, a more complex substance, is formed from simpler substances.
(b) Single-displacement; Fe displaces Ag in $AgNO_3$.
(c) Decomposition; $CaSO_3$ decomposes into simpler substances.
(d) Double-displacement; Ca and Li switch places to form new compounds.

Key Terms

acid–base reaction [7.8]
aqueous solution [7.5]
balanced equation [7.3]
combination reaction [7.10]
combustion reaction [7.9]
complete ionic equation [7.7]
decomposition reaction [7.10]

displacement reaction [7.10]
double-displacement
   reaction [7.10]
gas evolution reaction [7.8]
insoluble [7.5]
molecular equation [7.7]
net ionic equation [7.7]

neutralization reaction
   [7.8]
oxidation–reduction (redox)
   reaction [7.9]
precipitate [7.6]
precipitation reaction [7.6]
salt [7.8]

single-displacement reaction
   [7.10]
solubility rules [7.5]
soluble [7.5]
spectator ion [7.7]
strong electrolyte solution [7.5]
synthesis reaction [7.10]

Exercises

Questions

1. What is a chemical reaction? List some examples.
2. If you could observe atoms and molecules with the naked eye, what would you look for as conclusive evidence of a chemical reaction?
3. What are the main indications that a chemical reaction has occurred?
4. What is a chemical equation? Provide an example and identify the reactants and products.
5. What does each abbreviation, often used in chemical equations, represent?
   (a) (g) **(b)** (l) **(c)** (s) **(d)** (aq)
6. To balance a chemical equation, adjust the _____ as necessary to make the numbers of each type of atom on both sides of the equation equal. Never adjust the _____ to balance a chemical equation.
7. Is the chemical equation balanced? Why or why not?

$$2 Ag_2O(s) + C(s) \longrightarrow CO_2(g) + 4 Ag(s)$$

8. What is an aqueous solution? List two examples.
9. What does it mean if a compound is referred to as soluble? insoluble?
10. Explain what happens to an ionic substance when it dissolves in water.
11. Do polyatomic ions dissociate when they dissolve in water, or do they remain intact?

12. What is a strong electrolyte solution?
13. What are the solubility rules, and how are they useful?
14. What is a precipitation reaction? Provide an example and identify the precipitate.
15. Is the precipitate in a precipitation reaction always a compound that is soluble or insoluble? Explain.
16. Describe the differences between a molecular equation, a complete ionic equation, and net ionic equation. Give an example of each to illustrate the differences.
17. What is an acid–base reaction? List an example and identify the acid and the base.
18. What are the distinguishing properties of acids and bases?
19. What is a gas evolution reaction? Give an example.
20. What is a redox reaction? Give an example.
21. What is a combustion reaction? Give an example.
22. What are two different ways to classify chemical reactions presented in Section 7.10? Explain the differences between the two methods.
23. Explain the difference between a synthesis reaction and a decomposition reaction and provide an example of each.
24. Explain the difference between a single-displacement reaction and a double-displacement reaction and provide an example of each.

Problems

EVIDENCE OF CHEMICAL REACTIONS

25. Which observation is consistent with a chemical reaction occurring? Why?
    (a) Solid copper deposits on a piece of aluminum foil when the foil is placed in a blue copper nitrate solution. The blue color of the solution fades.
    (b) Liquid ethyl alcohol turns into a solid when placed in a low-temperature freezer.
    (c) A white precipitate forms when solutions of barium nitrate and sodium sulfate are mixed.
    (d) A mixture of sugar and water bubbles when yeasts are added. After several days, the sugar is gone and ethyl alcohol is found in the water.

26. Which observation is consistent with a chemical reaction occurring? Why?
    (a) Propane forms a flame and emits heat as it burns.
    (b) Acetone feels cold as it evaporates from the skin.
    (c) Bubbling occurs when potassium carbonate and hydrochloric acid solutions are mixed.
    (d) Heat is felt when a warm object is placed in your hand.

27. Vinegar forms bubbles when it is poured onto the calcium deposits on a faucet, and some of the calcium dissolves. Has a chemical reaction occurred? Explain your answer.

28. When a chemical drain opener is added to a clogged sink, bubbles form and the water in the sink gets warmer. Has a chemical reaction occurred? Explain your answer.

29. When a commercial hair bleaching mixture is applied to brown hair, the hair turns blond. Has a chemical reaction occurred? Explain your answer.

30. When water is boiled in a pot, it bubbles. Has a chemical reaction occurred? Explain your answer.

WRITING AND BALANCING CHEMICAL EQUATIONS

31. For each chemical equation (which may or may not be balanced), list the number of each type of atom on each side of the equation, and determine if the equation is balanced.
  (a) $Pb(NO_3)_2(aq) + 2\,NaCl(aq) \longrightarrow$
  $$PbCl_2(s) + 2\,NaNO_3(aq)$$
  (b) $C_3H_8(g) + O_2(g) \longrightarrow 3\,CO_2(g) + 4\,H_2O(g)$

32. For each chemical equation (which may or may not be balanced), list the number of each type of atom on each side of the equation, and determine if the equation is balanced.
  (a) $MgS(aq) + 2\,CuCl_2(aq) \longrightarrow 2\,CuS(s) + MgCl_2(aq)$
  (b) $2\,C_6H_{14}(l) + 19\,O_2(g) \longrightarrow 12\,CO_2(g) + 14\,H_2O(g)$

33. Consider the unbalanced chemical equation.

$$H_2O(l) \xrightarrow[\text{electrical current}]{} H_2(g) + O_2(g)$$

A chemistry student tries to balance the equation by placing the subscript 2 after the oxygen atom in H_2O. Explain why this is not correct. What is the correct balanced equation?

34. Consider the unbalanced chemical equation.

$$Al(s) + Cl_2(g) \longrightarrow AlCl_3(s)$$

A student tries to balance the equation by changing the subscript 2 on Cl to a 3. Explain why this is not correct. What is the correct balanced equation?

35. Write a balanced chemical equation for each chemical reaction.
  (a) Solid lead(II) sulfide reacts with aqueous hydrochloric acid to form solid lead(II) chloride and dihydrogen sulfide gas.
  (b) Gaseous carbon monoxide reacts with hydrogen gas to form gaseous methane (CH_4) and liquid water.
  (c) Solid iron(III) oxide reacts with hydrogen gas to form solid iron and liquid water.
  (d) Gaseous ammonia (NH_3) reacts with gaseous oxygen to form gaseous nitrogen monoxide and gaseous water.

36. Write a balanced chemical equation for each chemical reaction.
  (a) Solid copper reacts with solid sulfur to form solid copper(I) sulfide.
  (b) Sulfur dioxide gas reacts with oxygen gas to form sulfur trioxide gas.
  (c) Aqueous hydrochloric acid reacts with solid manganese(IV) oxide to form aqueous manganese(II) chloride, liquid water, and chlorine gas.
  (d) Liquid benzene (C_6H_6) reacts with gaseous oxygen to form carbon dioxide and liquid water.

37. Write a balanced chemical equation for each chemical reaction.
  (a) Solid magnesium reacts with aqueous copper(I) nitrate to form aqueous magnesium nitrate and solid copper.
  (b) Gaseous dinitrogen pentoxide decomposes to form nitrogen dioxide and oxygen gas.
  (c) Solid calcium reacts with aqueous nitric acid to form aqueous calcium nitrate and hydrogen gas.
  (d) Liquid methanol (CH_3OH) reacts with oxygen gas to form gaseous carbon dioxide and gaseous water.

38. Write a balanced chemical equation for each chemical reaction.
  (a) Gaseous acetylene (C_2H_2) reacts with oxygen gas to form gaseous carbon dioxide and gaseous water.
  (b) Chlorine gas reacts with aqueous potassium iodide to form solid iodine and aqueous potassium chloride.
  (c) Solid lithium oxide reacts with liquid water to form aqueous lithium hydroxide.
  (d) Gaseous carbon monoxide reacts with oxygen gas to form carbon dioxide gas.

39. When solid sodium is added to liquid water, it reacts with the water to produce hydrogen gas and aqueous sodium hydroxide. Write a balanced chemical equation for this reaction.

40. When iron rusts, solid iron reacts with gaseous oxygen to form solid iron(III) oxide. Write a balanced chemical equation for this reaction.

41. Sulfuric acid in acid rain forms when gaseous sulfur dioxide pollutant reacts with gaseous oxygen and liquid water to form aqueous sulfuric acid. Write a balanced chemical equation for this reaction.

42. Nitric acid in acid rain forms when gaseous nitrogen dioxide pollutant reacts with gaseous oxygen and liquid water to form aqueous nitric acid. Write a balanced chemical equation for this reaction.

43. Write a balanced chemical equation for the reaction of solid vanadium(V) oxide with hydrogen gas to form solid vanadium(III) oxide and liquid water.

44. Write a balanced chemical equation for the reaction of gaseous nitrogen dioxide with hydrogen gas to form gaseous ammonia and liquid water.

45. Write a balanced chemical equation for the fermentation of sugar ($C_{12}H_{22}O_{11}$) by yeasts in which the aqueous sugar reacts with water to form aqueous ethyl alcohol (C_2H_5OH) and carbon dioxide gas.

46. Write a balanced chemical equation for the photosynthesis reaction in which gaseous carbon dioxide and liquid water react in the presence of chlorophyll to produce aqueous glucose ($C_6H_{12}O_6$) and oxygen gas.

47. Balance each chemical equation.
(a) $Na_2S(aq) + Cu(NO_3)_2(aq) \longrightarrow NaNO_3(aq) + CuS(s)$
(b) $HCl(aq) + O_2(g) \longrightarrow H_2O(l) + Cl_2(g)$
(c) $H_2(g) + O_2(g) \longrightarrow H_2O(l)$
(d) $FeS(s) + HCl(aq) \longrightarrow FeCl_2(aq) + H_2S(g)$

48. Balance each chemical equation.
(a) $N_2H_4(l) \longrightarrow NH_3(g) + N_2(g)$
(b) $H_2(g) + N_2(g) \longrightarrow NH_3(g)$
(c) $Cu_2O(s) + C(s) \longrightarrow Cu(s) + CO(g)$
(d) $H_2(g) + Cl_2(g) \longrightarrow HCl(g)$

49. Balance each chemical equation.
(a) $BaO_2(s) + H_2SO_4(aq) \longrightarrow BaSO_4(s) + H_2O_2(aq)$
(b) $Co(NO_3)_3(aq) + (NH_4)_2S(aq) \longrightarrow$
$Co_2S_3(s) + NH_4NO_3(aq)$
(c) $Li_2O(s) + H_2O(l) \longrightarrow LiOH(aq)$
(d) $Hg_2(C_2H_3O_2)_2(aq) + KCl(aq) \longrightarrow$
$Hg_2Cl_2(s) + KC_2H_3O_2(aq)$

50. Balance each chemical equation.
(a) $MnO_2(s) + HCl(aq) \longrightarrow$
$Cl_2(g) + MnCl_2(aq) + H_2O(l)$
(b) $Co_2(g) + CaSiO_3(s) + H_2O(l) \longrightarrow$
$SiO_2(s) + Ca(HCO_3)_2(aq)$
(c) $Fe(s) + S(l) \longrightarrow Fe_2S_3(s)$
(d) $NO_2(g) + H_2O(l) \longrightarrow HNO_3(aq) + NO(g)$

51. Is each chemical equation correctly balanced? If not, correct it.
(a) $Rb(s) + H_2O(l) \longrightarrow RbOH(aq) + H_2(g)$
(b) $2 N_2H_4(g) + N_2O_4(g) \longrightarrow 3 N_2(g) + 4 H_2O(g)$
(c) $NiS(s) + O_2(g) \longrightarrow NiO(s) + SO_2(g)$
(d) $PbO(s) + 2 NH_3(g) \longrightarrow Pb(s) + N_2(g) + H_2O(l)$

52. Is each chemical equation correctly balanced? If not, correct it.
(a) $SiO_2(s) + 4 HF(aq) \longrightarrow SiF_4(g) + 2 H_2O(l)$
(b) $2 Cr(s) + 3 O_2(g) \longrightarrow Cr_2O_3(s)$
(c) $Al_2S_3(s) + H_2O(l) \longrightarrow 2 Al(OH)_3(s) + 3 H_2S(g)$
(d) $Fe_2O_3(s) + CO(g) \longrightarrow 2 Fe(s) + CO_2(g)$

53. Human cells obtain energy from a reaction called cellular respiration. Balance the skeletal equation for cellular respiration.

$$C_6H_{12}O_6(aq) + O_2(g) \longrightarrow CO_2(g) + H_2O(l)$$

54. Propane camping stoves produce heat by the combustion of gaseous propane (C_3H_8). Balance the skeletal equation for the combustion of propane.

$$C_3H_8(g) + O_2(g) \longrightarrow CO_2(g) + H_2O(g)$$

55. Catalytic converters work to remove nitrogen oxides and carbon monoxide from exhaust. Balance the skeletal equation for one of the reactions that occurs in a catalytic converter.

$$NO(g) + CO(g) \longrightarrow N_2(g) + CO_2(g)$$

56. Billions of pounds of urea are produced annually for use as fertilizer. Balance the skeletal equation for the synthesis of urea.

$$NH_3(g) + CO_2(g) \longrightarrow CO(NH_2)_2(s) + H_2O(l)$$

SOLUBILITY

57. Is each compound soluble or insoluble? For the soluble compounds, identify the ions present in solution.
(a) $NaC_2H_3O_2$ (b) $Sn(NO_3)_2$
(c) AgI (d) $Na_3(PO_4)$

58. Is each compound soluble or insoluble? For the soluble compounds, identify the ions present in solution.
(a) $(NH_4)_2S$ (b) $CuCO_3$
(c) ZnS (d) $Pb(C_2H_3O_2)_2$

59. Pair each cation on the left with an anion on the right that will form an *insoluble* compound with it and write a formula for the insoluble compound. Use each anion only once.

Ag^+ SO_4^{2-}
Ba^{2+} Cl^-
Cu^{2+} CO_3^{2-}
Fe^{3+} S^{2-}

60. Pair each cation on the left with an anion on the right that will form a *soluble* compound with it and write a formula for the soluble compound. Use each anion only once.

Na^+ NO_3^-
Sr^{2+} SO_4^{2-}
Co^{2+} S^{2-}
Pb^{2+} CO_3^{2-}

61. Move any misplaced compounds to the correct column.

Soluble	Insoluble
K_2S	K_2SO_4
$PbSO_4$	Hg_2I_2
BaS	$Cu_3(PO_4)_2$
$PbCl_2$	MgS
Hg_2Cl_2	$CaSO_4$
NH_4Cl	SrS
Na_2CO_3	Li_2S

62. Move any misplaced compounds to the correct column.

Soluble	Insoluble
LiOH	$CaCl_2$
Na_2CO_3	$Cu(OH)_2$
AgCl	$Ca(C_2H_3O_2)_2$
K_3PO_4	$SrSO_4$
CuI_2	Hg_2Br_2
$Pb(NO_3)_2$	$PbBr_2$
$CoCO_3$	PbI_2

PRECIPITATION REACTIONS

63. Complete and balance each equation. If no reaction occurs, write *NO REACTION*.
(a) $KI(aq) + BaS(aq) \longrightarrow$
(b) $K_2SO_4(aq) + BaBr_2(aq) \longrightarrow$
(c) $NaCl(aq) + Hg_2(C_2H_3O_2)_2(aq) \longrightarrow$
(d) $NaC_2H_3O_2(aq) + Pb(NO_3)_2(aq) \longrightarrow$

64. Complete and balance each equation. If no reaction occurs, write *NO REACTION*.
(a) $NaOH(aq) + FeBr_3(aq) \longrightarrow$
(b) $BaCl_2(aq) + AgNO_3(aq) \longrightarrow$
(c) $Na_2CO_3(aq) + CoCl_2(aq) \longrightarrow$
(d) $K_2S(aq) + BaCl_2(aq) \longrightarrow$

65. Write a molecular equation for the precipitation reaction that occurs (if any) when each pair of solutions is mixed. If no reaction occurs, write *NO REACTION*.
(a) sodium carbonate and lead(II) nitrate
(b) potassium sulfate and lead(II) acetate
(c) copper(II) nitrate and barium sulfide
(d) calcium nitrate and sodium iodide

66. Write a molecular equation for the precipitation reaction that occurs (if any) when each pair of solutions is mixed. If no reaction occurs, write *NO REACTION*.
(a) potassium chloride and lead(II) acetate
(b) lithium sulfate and strontium chloride
(c) potassium bromide and calcium sulfide
(d) chromium(III) nitrate and potassium phosphate

67. Correct any incorrect equations. If no reaction occurs, write *NO REACTION*.
(a) $Ba(NO_3)_2(aq) + (NH_4)_2SO_4(aq) \longrightarrow$
$BaSO_4(s) + 2NH_4NO_3(aq)$
(b) $BaS(aq) + 2KCl(aq) \longrightarrow BaCl_2(s) + K_2S(aq)$
(c) $2KI(aq) + Pb(NO_3)_2(aq) \longrightarrow PbI_2(s) + 2KNO_3(aq)$
(d) $Pb(NO_3)_2(aq) + 2LiCl(aq) \longrightarrow$
$2LiNO_3(s) + PbCl_2(aq)$

68. Correct any incorrect equations. If no reaction occurs, write *NO REACTION*.
(a) $AgNO_3(aq) + NaCl(aq) \longrightarrow NaCl(s) + AgNO_3(aq)$
(b) $K_2SO_4(aq) + Co(NO_3)_2(aq) \longrightarrow$
$CoSO_4(s) + 2KNO_3(aq)$
(c) $Cu(NO_3)_2(aq) + (NH_4)_2S(aq) \longrightarrow$
$CuS(s) + 2NH_4NO_3(aq)$
(d) $Hg_2(NO_3)_2(aq) + 2LiCl(aq) \longrightarrow$
$Hg_2Cl_2(s) + 2LiNO_3(aq)$

IONIC AND NET IONIC EQUATIONS

69. Identify the spectator ions in the complete ionic equation.

$2K^+(aq) + S^{2-}(aq) + Pb^{2+}(aq) + 2NO_3^-(aq) \longrightarrow$
$PbS(s) + 2K^+(aq) + 2NO_3^-(aq)$

70. Identify the spectator ions in the complete ionic equation.

$Ba^{2+}(aq) + 2I^-(aq) + 2Na^+(aq) + SO_4^{2-}(aq) \longrightarrow$
$BaSO_4(s) + 2I^-(aq) + 2Na^+(aq)$

71. Write balanced complete ionic and net ionic equations for each reaction.
(a) $AgNO_3(aq) + KCl(aq) \longrightarrow AgCl(s) + KNO_3(aq)$
(b) $CaS(aq) + CuCl_2(aq) \longrightarrow CuS(s) + CaCl_2(aq)$
(c) $NaOH(aq) + HNO_3(aq) \longrightarrow H_2O(l) + NaNO_3(aq)$
(d) $2K_3PO_4(aq) + 3NiCl_2(aq) \longrightarrow$
$Ni_3(PO_4)_2(s) + 6KCl(aq)$

72. Write balanced complete ionic and net ionic equations for each reaction.
(a) $HI(aq) + KOH(aq) \longrightarrow H_2O(l) + KI(aq)$
(b) $Na_2SO_4(aq) + CaI_2(aq) \longrightarrow CaSO_4(s) + 2NaI(aq)$
(c) $2HC_2H_3O_2(aq) + Na_2CO_3(aq) \longrightarrow$
$H_2O(l) + CO_2(g) + 2NaC_2H_3O_2(aq)$
(d) $NH_4Cl(aq) + NaOH(aq) \longrightarrow$
$H_2O(l) + NH_3(g) + NaCl(aq)$

73. Mercury(I) ions (Hg_2^{2+}) can be removed from solution by precipitation with Cl^-. Suppose a solution contains aqueous $Hg_2(NO_3)_2$. Write complete ionic and net ionic equations to show the reaction of aqueous $Hg_2(NO_3)_2$ with aqueous sodium chloride to form solid Hg_2Cl_2 and aqueous sodium nitrate.

74. Lead ions can be removed from solution by precipitation with sulfate ions. Suppose a solution contains lead(II) nitrate. Write a complete ionic and net ionic equation to show the reaction of aqueous lead(II) nitrate with aqueous potassium sulfate to form solid lead(II) sulfate and aqueous potassium nitrate.

75. Write complete ionic and net ionic equations for each of the reactions in Problem 67.

76. Write complete ionic and net ionic equations for each of the reactions in Problem 68.

ACID–BASE AND GAS EVOLUTION REACTIONS

77. When a hydrochloric acid solution is combined with a potassium hydroxide solution, an acid–base reaction occurs. Write a balanced molecular equation and a net ionic equation for this reaction.

78. A beaker of nitric acid is neutralized with calcium hydroxide. Write a balanced molecular equation and a net ionic equation for this reaction.

79. Complete and balance each acid–base reaction.
(a) $HCl(aq) + Ba(OH)_2(aq) \longrightarrow$
(b) $H_2SO_4(aq) + KOH(aq) \longrightarrow$
(c) $HClO_4(aq) + NaOH(aq) \longrightarrow$

80. Complete and balance each acid–base reaction.
(a) $HC_2H_3O_2(aq) + Ca(OH)_2(aq) \longrightarrow$
(b) $HBr(aq) + LiOH(aq) \longrightarrow$
(c) $H_2SO_4(aq) + Ba(OH)_2(aq) \longrightarrow$

81. Complete and balance each gas evolution reaction.
(a) $HBr(aq) + NaHCO_3(aq) \longrightarrow$
(b) $NH_4I(aq) + KOH(aq) \longrightarrow$
(c) $HNO_3(aq) + K_2SO_3(aq) \longrightarrow$
(d) $HI(aq) + Li_2S(aq) \longrightarrow$

82. Complete and balance each gas evolution reaction.
(a) $HClO_4(aq) + K_2CO_3(aq) \longrightarrow$
(b) $HC_2H_3O_2(aq) + LiHSO_3(aq) \longrightarrow$
(c) $(NH_4)_2SO_4(aq) + Ca(OH)_2(aq) \longrightarrow$
(d) $HCl(aq) + ZnS(s) \longrightarrow$

OXIDATION–REDUCTION AND COMBUSTION

83. Which reactions are redox reactions?
(a) $Ba(NO_3)_2(aq) + K_2SO_4(aq) \longrightarrow$
$$BaSO_4(s) + 2\,KNO_3(aq)$$
(b) $Ca(s) + Cl_2(g) \longrightarrow CaCl_2(s)$
(c) $HCl(aq) + NaOH(aq) \longrightarrow H_2O(l) + NaCl(aq)$
(d) $Zn(s) + Fe^{2+}(aq) \longrightarrow Zn^{2+}(aq) + Fe(s)$

84. Which reactions are redox reactions?
(a) $Al(s) + 3\,Ag^+(aq) \longrightarrow Al^{3+}(aq) + 3\,Ag(s)$
(b) $4\,K(s) + O_2(g) \longrightarrow 2\,K_2O(s)$
(c) $SO_3(g) + H_2O(l) \longrightarrow H_2SO_4(aq)$
(d) $Mg(s) + Br_2(l) \longrightarrow MgBr_2(s)$

85. Complete and balance each combustion reaction.
(a) $C_2H_6(g) + O_2(g) \longrightarrow$
(b) $Ca(s) + O_2(g) \longrightarrow$
(c) $C_3H_8O(l) + O_2(g) \longrightarrow$
(d) $C_4H_{10}S(l) + O_2(g) \longrightarrow$

86. Complete and balance each combustion reaction.
(a) $S(s) + O_2(g) \longrightarrow$
(b) $C_7H_{16}(l) + O_2(g) \longrightarrow$
(c) $C_4H_{10}O(l) + O_2(g) \longrightarrow$
(d) $CS_2(l) + O_2(g) \longrightarrow$

87. Write a balanced chemical equation for the synthesis reaction of $Br_2(g)$ with each metal.
(a) $Ag(s)$ (b) $K(s)$ (c) $Al(s)$ (d) $Ca(s)$

88. Write a balanced chemical equation for the synthesis reaction of $Cl_2(g)$ with each metal.
(a) $Zn(s)$ (b) $Ga(s)$ (c) $Rb(s)$ (d) $Mg(s)$

CLASSIFYING CHEMICAL REACTIONS BY WHAT ATOMS DO

89. Classify each chemical reaction as a synthesis, decomposition, single-displacement, or double-displacement reaction.
(a) $K_2S(aq) + Co(NO_3)_2(aq) \longrightarrow 2\,KNO_3(aq) + CoS(s)$
(b) $3\,H_2(g) + N_2(g) \longrightarrow 2\,NH_3(g)$
(c) $Zn(s) + CoCl_2(aq) \longrightarrow ZnCl_2(aq) + Co(s)$
(d) $CH_3Br(g) \xrightarrow{\text{UV light}} CH_3(g) + Br(g)$

90. Classify each chemical reaction as a synthesis, decomposition, single-displacement, or double-displacement reaction.
(a) $CaSO_4(g) \xrightarrow{\text{heat}} CaO(s) + SO_3(g)$
(b) $2\,Na(s) + O_2(g) \longrightarrow Na_2O_2(s)$
(c) $Pb(s) + 2\,AgNO_3(aq) \longrightarrow Pb(NO_3)_2(aq) + 2\,Ag(s)$
(d) $HI(aq) + NaOH(aq) \longrightarrow H_2O(l) + NaI(aq)$

91. NO is a pollutant emitted by motor vehicles. It is formed by the reaction:
(a) $N_2(g) + O_2(g) \longrightarrow 2\,NO(g)$
Once in the atmosphere, NO (through a series of reactions) adds one oxygen atom to form NO_2. NO_2 then interacts with UV light according to the reaction:
(b) $NO_2(g) \xrightarrow{\text{UV light}} NO(g) + O(g)$

These freshly formed oxygen atoms then react with O_2 in the air to form ozone (O_3), a main component of smog:
(c) $O(g) + O_2(g) \longrightarrow O_3(g)$
Classify each of the preceding reactions (a, b, c) as a synthesis, decomposition, single-displacement, or double-displacement reaction.

92. A main source of sulfur oxide pollutants are smelters where sulfide ores are converted into metals. The first step in this process is the reaction of the sulfide ore with oxygen in reactions such as:
(a) $2\,PbS(s) + 3\,O_2(g) \xrightarrow{\text{UV light}} 2\,PbO(s) + 2\,SO_2(g)$
Sulfur dioxide can then react with oxygen in air to form sulfur trioxide:
(b) $2\,SO_2(g) + O_2(g) \longrightarrow 2\,SO_3(g)$
Sulfur trioxide can then react with water from rain to form sulfuric acid that falls as acid rain:
(c) $SO_3(g) + H_2O(l) \longrightarrow H_2SO_4(aq)$
Classify each of the preceding reactions (a, b, c) as a synthesis, decomposition, single-displacement, or double-displacement reaction.

Cumulative Problems

93. Predict the products of each reaction and write balanced complete ionic and net ionic equations for each. If no reaction occurs, write *NO REACTION*.
(a) $NaI(aq) + Hg_2(NO_3)_2(aq) \longrightarrow$
(b) $HClO_4(aq) + Ba(OH)_2(aq) \longrightarrow$
(c) $Li_2CO_3(aq) + NaCl(aq) \longrightarrow$
(d) $HCl(aq) + Li_2CO_3(aq) \longrightarrow$

94. Predict the products of each reaction and write balanced complete ionic and net ionic equations for each. If no reaction occurs, write *NO REACTION*.
(a) $LiCl(aq) + AgNO_3(aq) \longrightarrow$
(b) $H_2SO_4(aq) + Li_2SO_3(aq) \longrightarrow$
(c) $HC_2H_3O_2(aq) + Ca(OH)_2(aq) \longrightarrow$
(d) $HCl(aq) + KBr(aq) \longrightarrow$

95. Predict the products of each reaction and write balanced complete ionic and net ionic equations for each. If no reaction occurs, write *NO REACTION*.
(a) $BaS(aq) + NH_4Cl(aq) \longrightarrow$
(b) $NaC_2H_3O_2(aq) + KCl(aq) \longrightarrow$
(c) $KHSO_3(aq) + HNO_3(aq) \longrightarrow$
(d) $MnCl_3(aq) + K_3PO_4(aq) \longrightarrow$

96. Predict the products of each reaction and write balanced complete ionic and net ionic equations for each. If no reaction occurs, write *NO REACTION*.
(a) $H_2SO_4(aq) + HNO_3(aq) \longrightarrow$
(b) $NaOH(aq) + LiOH(aq) \longrightarrow$
(c) $Cr(NO_3)_3(aq) + LiOH(aq) \longrightarrow$
(d) $HCl(aq) + Hg_2(NO_3)_2(aq) \longrightarrow$

97. Predict the type of reaction (if any) that occurs between each pair of substances. Write balanced molecular equations for each. If no reaction occurs, write *NO REACTION*.
(a) aqueous potassium hydroxide and aqueous acetic acid
(b) aqueous hydrobromic acid and aqueous potassium carbonate
(c) gaseous hydrogen and gaseous oxygen
(d) aqueous ammonium chloride and aqueous lead(II) nitrate

98. Predict the type of reaction (if any) that occurs between each pair of substances. Write balanced molecular equations for each. If no reaction occurs, write *NO REACTION*.
(a) aqueous hydrochloric acid and aqueous copper(II) nitrate
(b) liquid pentanol ($C_5H_{12}O$) and gaseous oxygen
(c) aqueous ammonium chloride and aqueous calcium hydroxide
(d) aqueous strontium sulfide and aqueous copper(II) sulfate

99. Classify each reaction in as many ways as possible.
(a) $2\,Al(s) + 3\,Cu(NO_3)_2(aq) \longrightarrow$
$\qquad\qquad 2\,Al(NO_3)_3(aq) + 3\,Cu(s)$
(b) $HBr(aq) + KHSO_3(aq) \longrightarrow$
$\qquad\qquad H_2O(l) + SO_2(g) + NaBr(aq)$
(c) $2\,HI(aq) + Na_2S(aq) \longrightarrow H_2S(g) + 2\,NaI(aq)$
(d) $K_2CO_3(aq) + FeBr_2(aq) \longrightarrow FeCO_3(s) + 2\,KBr(aq)$

100. Classify each reaction in as many ways as possible.
(a) $NaCl(aq) + AgNO_3(aq) \longrightarrow$
$\qquad\qquad AgCl(s) + NaNO_3(aq)$
(b) $2\,Rb(s) + Br_2(g) \longrightarrow 2\,RbBr(s)$
(c) $Zn(s) + NiBr_2(aq) \longrightarrow Ni(s) + ZnBr_2(aq)$
(d) $Ca(s) + 2\,H_2O(l) \longrightarrow Ca(OH)_2(aq) + H_2(g)$

101. Hard water often contains dissolved Ca^{2+} and Mg^{2+} ions. One way to soften water is to add phosphates. The phosphate ion forms insoluble precipitates with calcium and magnesium ions, removing them from solution. Suppose that a solution contains aqueous calcium chloride and aqueous magnesium nitrate. Write molecular, complete ionic, and net ionic equations showing how the addition of sodium phosphate precipitates the calcium and magnesium ions.

102. Lakes that have been acidified by acid rain (HNO_3 and H_2SO_4) can be neutralized by a process called *liming*, in which limestone ($CaCO_3$) is added to the acidified water. Write ionic and net ionic equations to show how limestone reacts with HNO_3 and H_2SO_4 to neutralize them. How would you be able to tell if the neutralization process was working?

103. What solution can you add to each cation mixture to precipitate one cation while keeping the other cation in solution? Write a net ionic equation for the precipitation reaction that occurs.
(a) $Fe^{2+}(aq)$ and $Pb^{2+}(aq)$ (b) $K^+(aq)$ and $Ca^{2+}(aq)$
(c) $Ag^+(aq)$ and $Ba^{2+}(aq)$ (d) $Cu^{2+}(aq)$ and $Hg_2^{2+}(aq)$

104. What solution can you add to each cation mixture to precipitate one cation while keeping the other cation in solution? Write a net ionic equation for the precipitation reaction that occurs.
(a) $Sr^{2+}(aq)$ and $Hg_2^{2+}(aq)$ (b) $NH_4^+(aq)$ and $Ca^{2+}(aq)$
(c) $Ba^{2+}(aq)$ and $Mg^{2+}(aq)$ (d) $Ag^+(aq)$ and $Zn^{2+}(aq)$

105. A solution contains an unknown amount of dissolved calcium. Addition of 0.112 mol of K_3PO_4 causes complete precipitation of all of the calcium. How many moles of calcium were dissolved in the solution? What mass of calcium was dissolved in the solution?

106. A solution contains an unknown amount of dissolved magnesium. Addition of 0.0877 mol of Na_2CO_3 causes complete precipitation of all of the magnesium. What mass of magnesium was dissolved in the solution?

107. A solution contains 0.133 g of dissolved lead. How many moles of sodium chloride must be added to the solution to completely precipitate all of the dissolved lead? What mass of sodium chloride must be added?

108. A solution contains 1.77 g of dissolved silver. How many moles of potassium chloride must be added to the solution to completely precipitate all of the silver? What mass of potassium chloride must be added?

Highlight Problems

109. Shown here are molecular views of two different possible mechanisms by which an automobile airbag might function. One of these mechanisms involves a chemical reaction and the other does not. By looking at the molecular views, can you tell which mechanism operates via a chemical reaction?

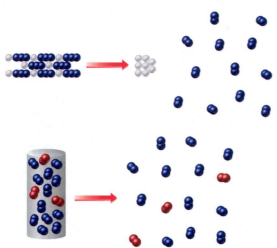

▲ When an airbag is detonated, the bag inflates. These figures show two possible ways in which the inflation may happen.

110. Precipitation reactions often produce brilliant colors. Look at the photographs of each precipitation reaction and write molecular, complete ionic, and net ionic equations for each one.

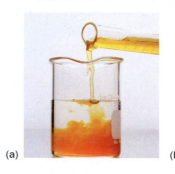

(a)

(b)

(c)

▲ (a) The precipitation reaction that occurs when aqueous iron(III) nitrate is added to aqueous sodium hydroxide. (b) The precipitation reaction that occurs when aqueous cobalt(II) chloride is added to aqueous potassium hydroxide. (c) The precipitation reaction that occurs when aqueous $AgNO_3$ is added to aqueous sodium iodide.

Questions for Group Work

Discuss these questions with the group and record your consensus answer.

111. Use group members to represent atoms or ions and act out a specific balanced chemical reaction.

112. Memorize the solubility rules. Without referring back to the rules, have each group member list two ionic compounds that are expected to be soluble and two that are expected to be insoluble. Include at least one exception. Check the work of the other members of your group.

113. Define and give an example of each of the following classes of reactions: precipitation, acid–base, gas evolution, redox (noncombustion), and combustion. Have each group member define one type and provide an example, and then present his or her reaction to the group.

Data Interpretation and Analysis

114. Water samples often contain dissolved ions such as Ca^{2+} and Fe^{2+}. The presence of these ions can often be detected by adding a precipitation agent—a substance that causes one of the dissolved ions to precipitate. For example, if sodium chloride is added to a water sample containing dissolved Ag^+, a white precipitate forms. If no precipitate forms, then the water sample does not contain dissolved Ag^+. Use the solubility rules from Section 7.5 and the data provided to determine the ions present in the water samples.

(a) Water Sample A may contain Ag^+, Ca^{2+}, and Cu^{2+}. Use the tabulated data to determine the ions present in Sample A.

Substance Added to Water Sample A	Observation
NaCl	No precipitate forms
Na₂SO₄	Precipitate forms
Na₂CO₃ (after filtering off precipitate from previous step)	Precipitate forms

(b) Water Sample B may contain Hg_2^{2+}, Ba^{2+}, and Fe^{2+}. Use the tabulated data to determine the ions present in Sample B.

Substance Added to Water Sample B	Observation
KCl	Precipitate forms
K₂SO₄ (after filtering off precipitate from previous step)	No precipitate forms
K₂CO₃	Precipitate forms

Answers to Skillbuilder Exercises

Skillbuilder 7.1
  (a) Chemical reaction; heat and light are emitted.
  (b) Not a chemical reaction; gaseous and liquid butane are both butane.
  (c) Chemical reaction; heat and light are emitted.
  (d) Not a chemical reaction; solid dry ice is made of carbon dioxide, which sublimes (evaporates) as carbon dioxide gas.

Skillbuilder 7.2
  $2 Cr_2O_3(s) + 3 C(s) \longrightarrow 4 Cr(s) + 3 CO_2(g)$

Skillbuilder 7.3
  $2 C_4H_{10}(g) + 13 O_2(g) \longrightarrow 8 CO_2(g) + 10 H_2O(g)$

Skillbuilder 7.4
  $Pb(C_2H_3O_2)_2(aq) + 2 KI(aq) \longrightarrow PbI_2(s) + 2 KC_2H_3O_2(aq)$

Skillbuilder 7.5
  $4 HCl(g) + O_2(g) \longrightarrow 2 H_2O(l) + 2 Cl_2(g)$

Skillbuilder 7.6
  (a) insoluble
  (b) soluble
  (c) insoluble
  (d) soluble

Skillbuilder 7.7
  $2 KOH(aq) + NiBr_2(aq) \longrightarrow Ni(OH)_2(s) + 2 KBr(aq)$

Skillbuilder 7.8
  $NH_4Cl(aq) + Fe(NO_3)_3(aq) \longrightarrow$ NO REACTION

Skillbuilder 7.9
  $K_2SO_4(aq) + Sr(NO_3)_2(aq) \longrightarrow SrSO_4(s) + 2 KNO_3(aq)$

Skillbuilder 7.10
Complete ionic equation:
  $$2 H^+(aq) + 2 Br^-(aq) + Ca^{2+}(aq) + 2 OH^-(aq) \longrightarrow$$
  $$2 H_2O(l) + Ca^{2+}(aq) + 2 Br^-(aq)$$
Net ionic equation:
  $2 H^+(aq) + 2 OH^-(aq) \longrightarrow 2 H_2O(l)$, or simply
  $H^+(aq) + OH^-(aq) \longrightarrow H_2O(l)$

Skillbuilder 7.11
Molecular equation:
  $H_2SO_4(aq) + 2 KOH(aq) \longrightarrow 2 H_2O(l) + K_2SO_4(aq)$
Net ionic equation:
  $H^+(aq) + OH^-(aq) \longrightarrow H_2O(l)$

Skillbuilder 7.12
  $2 HBr(aq) + K_2SO_3(aq) \longrightarrow H_2O(l) + SO_2(g) + 2 KBr(aq)$

Skillbuilder Plus, p. 227
  $2 H^+(aq) + SO_3^{2-}(aq) \longrightarrow H_2O(l) + SO_2(g)$

Skillbuilder 7.13
(a), (b), and (d) are all redox reactions; (c) is a precipitation reaction.

Skillbuilder 7.14
  $C_5H_{12}(l) + 8 O_2(g) \longrightarrow 5 CO_2(g) + 6 H_2O(g)$

Skillbuilder Plus, p. 230
  $2 C_3H_7OH(l) + 9 O_2(g) \longrightarrow 6 CO_2(g) + 8 H_2O(g)$

Skillbuilder 7.15
(a) single-displacement (b) double-displacement (c) synthesis (d) decomposition

Answers to Conceptual Checkpoints

7.1 (a) No reaction occurred. The molecules are the same before and after the change.
  (b) A reaction occurred; the molecules have changed.
  (c) A reaction occurred; the molecules have changed.
7.2 (b) There are 18 oxygen atoms on the left side of the equation, so the same number is needed on the right: $6 + 6(2) = 18$.
7.3 (a) The number of each type of atom must be the same on both sides of a balanced chemical equation. Since molecules change during a chemical reaction, their number is not the same on both sides (b), nor is the sum of all of the coefficients the same (c).
7.4 (a) Since chlorides are usually soluble and Ba^{2+} is not an exception, $BaCl_2$ is soluble and will dissolve in water. When it dissolves, it dissociates into its component ions, as shown in (a).

7.5 (b) Both of the possible products, MgS and $CaSO_4$, are insoluble. The possible products of the other reactions—Na_2S, $Ca(NO_3)_2$, Na_2SO_4, and $Mg(NO_3)_2$—are all soluble.
7.6 (c) The net ionic equation shows only the species that actually participate in the reaction.
7.7 (d) In a precipitation reaction, cations and anions "exchange partners" to produce at least one product. In an acid–base reaction, H^+ and OH^- combine to form water, and their partners pair off to a salt.

8 Quantities in Chemical Reactions

Man masters nature not by force but by understanding. That is why science has succeeded where magic failed: because it has looked for no spell to cast.

—Jacob Bronowski (1908–1974)

8.1 Climate Change: Too Much Carbon Dioxide

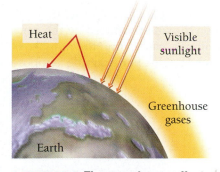

Outgoing heat is trapped by atmospheric greenhouse gases.

Heat

Visible sunlight

Greenhouse gases

Earth

▲ **FIGURE 8.1 The greenhouse effect** Greenhouse gases act like glass in a greenhouse, allowing visible-light energy to enter the atmosphere but preventing heat energy from escaping.

Average global temperatures depend on the balance between incoming sunlight, which warms Earth, and outgoing heat lost to space, which cools it. Certain gases in Earth's atmosphere, called **greenhouse gases**, affect that balance by acting like glass in a greenhouse. They allow sunlight into the atmosphere to warm Earth but prevent heat from escaping (◀ FIGURE 8.1). Without greenhouse gases, more heat would escape, and Earth's average temperature would be about 60°F colder. Caribbean tourists would freeze at an icy 21°F (−6°C), instead of baking at a tropical 81°F (27°C). On the other hand, if the concentration of greenhouse gases in the atmosphere were to increase, Earth's average temperature would rise.

In recent decades, scientists have become concerned because the atmospheric concentration of carbon dioxide (CO_2)—Earth's most significant greenhouse gas in terms of its contribution to climate—is rising. This rise in CO_2 concentration enhances the atmosphere's ability to hold heat and therefore leads to **climate change**, seen most clearly as an increase in Earth's average temperature. Since 1880, atmospheric CO_2 levels have risen by 38%, and Earth's average temperature has increased by 0.8 °C (about 1.4 °F) (▶ FIGURE 8.2, on the next page).

The primary cause of rising atmospheric CO_2 concentration is the burning of fossil fuels. Fossil fuels—natural gas, petroleum, and coal—provide approximately 90% of our society's energy. Combustion of fossil fuels, however, produces CO_2. As an example, consider the combustion of octane (C_8H_{18}), a component of gasoline:

$$2\,C_8H_{18}(l) + 25\,O_2(g) \longrightarrow 16\,CO_2(g) + 18\,H_2O(g)$$

The balanced chemical equation shows that 16 mol of CO_2 are produced for every 2 mol of octane burned. Because we know the world's annual fossil fuel consumption, we can estimate the world's annual CO_2 production. A simple calculation

◀ The combustion of fossil fuels such as octane (shown here) produces water and carbon dioxide as products. Carbon dioxide is a greenhouse gas that most climate scientists believe is responsible for climate change.

▶ **FIGURE 8.2 Climate change**
Yearly temperature differences from the 120-year average temperature. Earth's average temperature has increased by about 0.8 °C since 1880.

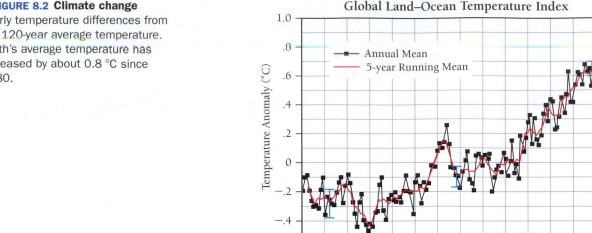

Global Land–Ocean Temperature Index

shows that the world's annual CO_2 production—from fossil fuel combustion—matches the measured annual atmospheric CO_2 increase. This implies that fossil fuel combustion is indeed responsible for increased atmospheric CO_2 levels.

The numerical relationship between chemical quantities in a balanced chemical equation is called reaction **stoichiometry**. Stoichiometry allows us to predict the amounts of products that form in a chemical reaction based on the amounts of reactants. Stoichiometry also allows us to predict how much of the reactants is necessary to form a given amount of product or how much of one reactant is required to completely react with another reactant. These calculations are central to chemistry, allowing chemists to plan and carry out chemical reactions to obtain products in desired quantities.

8.2 Making Pancakes: Relationships between Ingredients

▶ Recognize the numerical relationship between chemical quantities in a balanced chemical equation.

Key Concept Video
Reaction Stoichiometry

For the sake of simplicity, this recipe omits liquid ingredients.

The concepts of stoichiometry are similar to the concepts we use in following a cooking recipe. Calculating the amount of carbon dioxide produced by the combustion of a given amount of a fossil fuel is similar to calculating the number of pancakes that we can make from a given number of eggs. For example, suppose we use the following pancake recipe:

1 cup flour + 2 eggs + $\frac{1}{2}$ tsp baking powder ⟶ 5 pancakes

| 1 cup flour | 2 eggs | $\frac{1}{2}$ tsp baking powder | 5 pancakes |

▲ A recipe gives numerical relationships between the ingredients and the number of pancakes.

The recipe shows the numerical relationships between the pancake ingredients. It says that if we have 2 eggs—and enough of everything else—we can make 5 pancakes. We can write this relationship as a ratio:

| 2 eggs | 5 pancakes |

2 eggs : 5 pancakes

What if we have 8 eggs? Assuming that we have enough of everything else, how many pancakes can we make? Using the preceding ratio as a conversion factor, we can determine that 8 eggs are sufficient to make 20 pancakes.

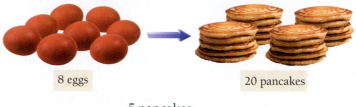

| 8 eggs | 20 pancakes |

$$8 \text{ eggs} \times \frac{5 \text{ pancakes}}{2 \text{ eggs}} = 20 \text{ pancakes}$$

The pancake recipe contains numerical conversion factors between the pancake ingredients and the number of pancakes. Other conversion factors from this recipe include:

1 cup flour : 5 pancakes

$\frac{1}{2}$ tsp baking powder : 5 pancakes

The recipe also gives us relationships among the ingredients themselves. For example, how much baking powder is required to go with 3 cups of flour? From the recipe:

1 cup flour : $\frac{1}{2}$ tsp baking powder

With this ratio, we can form the conversion factor to calculate the appropriate amount of baking powder.

$$3 \text{ cups flour} \times \frac{\frac{1}{2} \text{ tsp baking powder}}{1 \text{ cup flour}} = \tfrac{3}{2} \text{ tsp baking powder}$$

8.3 Making Molecules: Mole-to-Mole Conversions

▶ Carry out mole-to-mole conversions between reactants and products in a balanced chemical equation.

A balanced chemical equation is like a "recipe" for how reactants combine to form products. For example, the following equation shows how hydrogen and nitrogen combine to form ammonia (NH_3):

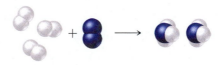

$$3\,H_2(g) + N_2(g) \longrightarrow 2\,NH_3(g)$$

The balanced equation shows that 3 H_2 molecules react with 1 N_2 molecule to form 2 NH_3 molecules. We can express these relationships as the following ratios:

3 H_2 molecules : 1 N_2 molecule : 2 NH_3 molecules

Since we do not ordinarily deal with individual molecules, we can express the same ratios in moles.

3 mol H_2 : 1 mol N_2 : 2 mol NH_3

If we have 3 mol of N_2, and more than enough H_2, how much NH_3 can we make? We first sort the information in the problem.

GIVEN: 3 mol N_2

FIND: mol NH_3

SOLUTION MAP

We then strategize by drawing a solution map that begins with mol N_2 and ends with mol NH_3. The conversion factor comes from the balanced chemical equation.

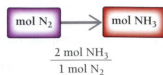

$$\frac{2 \text{ mol } NH_3}{1 \text{ mol } N_2}$$

RELATIONSHIPS USED

1 mol N_2 : 2 mol NH_3 (from balanced equation)

SOLUTION

We can then do the conversion.

$$3 \text{ mol } N_2 \times \frac{2 \text{ mol } NH_3}{1 \text{ mol } N_2} = 6 \text{ mol } NH_3$$

We have enough N_2 to make 6 mol of NH_3.

EXAMPLE **8.1**   **Mole-to-Mole Conversions**

Sodium chloride, NaCl, forms in this reaction between sodium and chlorine.

$$2 \, Na(s) + Cl_2(g) \longrightarrow 2 \, NaCl(s)$$

How many moles of NaCl result from the complete reaction of 3.4 mol of Cl_2? Assume that there is more than enough Na.

SORT You are given the number of moles of a reactant (Cl_2) and asked to find the number of moles of product (NaCl) that will form if the reactant completely reacts.	**GIVEN:** 3.4 mol Cl_2 **FIND:** mol NaCl
STRATEGIZE Draw the solution map beginning with moles of chlorine and using the stoichiometric conversion factor to calculate moles of sodium chloride. The conversion factor comes from the balanced chemical equation.	**SOLUTION MAP** 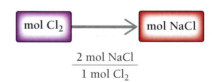 $$\frac{2 \text{ mol NaCl}}{1 \text{ mol } Cl_2}$$ **RELATIONSHIPS USED** 1 mol Cl_2 : 2 mol NaCl (from balanced chemical equation)
SOLVE Follow the solution map to solve the problem.	**SOLUTION** $$3.4 \text{ mol } Cl_2 \frac{2 \text{ mol NaCl}}{1 \text{ mol } Cl_2} = 6.8 \text{ mol NaCl}$$ There is enough Cl_2 to produce 6.8 mol of NaCl.
CHECK Check your answer. Are the units correct? Does the answer make physical sense?	The answer has the correct units, moles. The answer is reasonable because each mole of Cl_2 makes two moles of NaCl.

▶ SKILLBUILDER 8.1 | **Mole-to-Mole Conversions**

Water forms when hydrogen gas reacts explosively with oxygen gas according to the balanced equation:

$$O_2(g) + 2 \, H_2(g) \longrightarrow 2 \, H_2O(g)$$

How many mol of H_2O result from the complete reaction of 24.6 mol of O_2? Assume that there is more than enough H_2.

▶ **FOR MORE PRACTICE**   Example 8.8; Problems 15, 16, 17, 18.

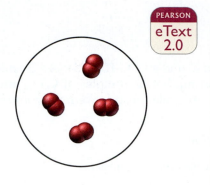

CONCEPTUAL ✔ CHECKPOINT 8.1

Methane (CH_4) undergoes combustion according to this reaction.

$$CH_4(g) + 2\,O_2(g) \longrightarrow CO_2(g) + 2\,H_2O(g)$$

If the figure shown in the left margin represents the amount of oxygen available to react, which of the following figures best represents the amount of CH_4 required to completely react with all of the oxygen?

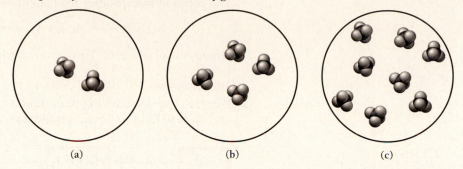

(a)	(b)	(c)

8.4 Making Molecules: Mass-to-Mass Conversions

▶ Carry out mass-to-mass conversions between reactants and products in a balanced chemical equation and molar masses.

In Chapter 6, we learned how a chemical *formula* contains conversion factors for converting between moles of a compound and moles of its constituent elements. In this chapter, we have seen how a chemical *equation* contains conversion factors between moles of reactants and moles of products. However, we are often interested in relationships between *mass* of reactants and *mass* of products. For example, we might want to know the mass of carbon dioxide emitted by an automobile per kilogram of gasoline used. Or we might want to know the mass of each reactant required to obtain a certain mass of a product in a synthesis reaction.

These calculations are similar to calculations covered in Section 6.5, where we converted between mass of a compound and mass of a constituent element. The general outline for these types of calculations is:

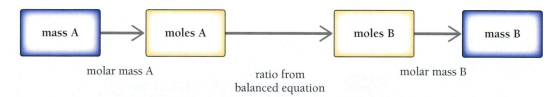

$$\boxed{\text{mass A}} \longrightarrow \boxed{\text{moles A}} \longrightarrow \boxed{\text{moles B}} \longrightarrow \boxed{\text{mass B}}$$

molar mass A   ratio from balanced equation   molar mass B

where A and B are two different substances involved in the reaction. We use the molar mass of A to convert from mass of A to moles of A. We use the ratio from the balanced equation to convert from moles of A to moles of B, and we use the molar mass of B to convert moles of B to mass of B.

For example, suppose we want to calculate the mass of CO_2 emitted upon the combustion of 5.0×10^2 g of pure octane. The balanced chemical equation for octane combustion is:

$$2\,C_8H_{18}(l) + 25\,O_2(g) \longrightarrow 16\,CO_2(g) + 18\,H_2O(g)$$

We begin by sorting the information in the problem.

GIVEN: 5.0×10^2 g C_8H_{18}

FIND: g CO_2

Notice that we are given g C_8H_{18} and asked to find g CO_2. The balanced chemical equation, however, gives us a relationship between moles of C_8H_{18} and moles of CO_2. Consequently, before using that relationship, we must convert from grams to moles.

The solution map follows the general outline:

$$\text{mass A} \longrightarrow \text{moles A} \longrightarrow \text{moles B} \longrightarrow \text{mass B}$$

where A is octane and B is carbon dioxide.

SOLUTION MAP

We strategize by drawing the solution map, which begins with mass of octane and ends with mass of carbon dioxide.

$$\text{g } C_8H_{18} \longrightarrow \text{mol } C_8H_{18} \longrightarrow \text{mol } CO_2 \longrightarrow \text{g } CO_2$$

$$\frac{1 \text{ mol } C_8H_{18}}{114.3 \text{ g } C_8H_{18}} \qquad \frac{16 \text{ mol } CO_2}{2 \text{ mol } C_8H_{18}} \qquad \frac{44.01 \text{ g } CO_2}{1 \text{ mol } CO_2}$$

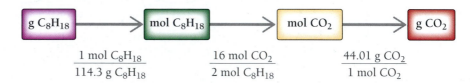

RELATIONSHIPS USED

2 mol C_8H_{18} : 16 mol CO_2 (from chemical equation)

molar mass C_8H_{18} = 114.3 g/mol

molar mass CO_2 = 44.01 g/mol

SOLUTION

We then follow the solution map to solve the problem, beginning with g C_8H_{18} and canceling units to arrive at g CO_2.

$$5.0 \times 10^2 \text{ g } C_8H_{18} \times \frac{1 \text{ mol } C_8H_{18}}{114.3 \text{ g } C_8H_{18}} \times \frac{16 \text{ mol } CO_2}{2 \text{ mol } C_8H_{18}} \times \frac{44.01 \text{ g } CO_2}{1 \text{ mol } CO_2} = 1.5 \times 10^3 \text{ g } CO_2$$

Upon combustion, 5.0×10^2 g of octane produces 1.5×10^3 g of carbon dioxide.

PEARSON eText 2.0

CONCEPTUAL CHECKPOINT 8.2

Consider the reaction $A + 2\,B \rightarrow 3\,C$. If the molar mass of C is twice the molar mass of A, what mass of C is produced by the complete reaction of 10.0 g A?

(a) 10.0 g

(b) 30.0 g

(c) 60.0 g

Interactive Worked Example Video 8.2

EXAMPLE **8.2**   **Mass-to-Mass Conversions**

In photosynthesis, plants convert carbon dioxide and water into glucose ($C_6H_{12}O_6$) according to the reaction:

$$6\ CO_2(g) + 6\ H_2O(l) \xrightarrow{\text{sunlight}} 6\ O_2(g) + C_6H_{12}O_6(aq)$$

How many grams of glucose can be synthesized from 58.5 g of CO_2? Assume that there is more than enough water present to react with all of the CO_2.

SORT	
You are given the mass of carbon dioxide and asked to find the mass of glucose that can form if the carbon dioxide completely reacts.	**GIVEN:** 58.5 g CO_2 **FIND:** g $C_6H_{12}O_6$

STRATEGIZE	
The solution map uses the general outline: $\quad$ mass A $\longrightarrow$ moles A $\longrightarrow$ $\qquad$ moles B $\longrightarrow$ mass B where A is carbon dioxide and B is glucose. The main conversion factor is the stoichiometric relationship between moles of carbon dioxide and moles of glucose. This conversion factor comes from the balanced equation. The other conversion factors are the molar masses of carbon dioxide and glucose.	**SOLUTION MAP** 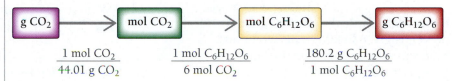 **RELATIONSHIPS USED** $\quad$ 6 mol CO_2 : 1 mol $C_6H_{12}O_6$ (from balanced chemical equation) $\quad$ molar mass CO_2 = 44.01 g/mol $\quad$ molar mass $C_6H_{12}O_6$ = 180.2 g/mol

SOLVE	
Follow the solution map to solve the problem. Begin with grams of carbon dioxide and multiply by the appropriate factors to arrive at grams of glucose.	**SOLUTION** $58.5\ \cancel{\text{g } CO_2} \times \dfrac{1\ \cancel{\text{mol } CO_2}}{44.01\ \cancel{\text{g } CO_2}} \times \dfrac{1\ \cancel{\text{mol } C_6H_{12}O_6}}{6\ \cancel{\text{mol } CO_2}} \times \dfrac{180.2\ \text{g } C_6H_{12}O_6}{1\ \cancel{\text{mol } C_6H_{12}O_6}}$ $= 39.9\ \text{g } C_6H_{12}O_6$

CHECK	
Are the units correct? Does the answer make physical sense?	The units, g $C_6H_{12}O_6$, are correct. The magnitude of the answer seems reasonable because it is of the same order of magnitude as the given mass of carbon dioxide. An answer that is orders of magnitude different would immediately be suspect.

▶ **SKILLBUILDER 8.2** | **Mass-to-Mass Conversions**

Magnesium hydroxide, the active ingredient in milk of magnesia, neutralizes stomach acid, primarily HCl, according to the reaction:

$$Mg(OH)_2(aq) + 2\ HCl(aq) \longrightarrow 2\ H_2O(l) + MgCl_2(aq)$$

How much HCl in grams can be neutralized by 5.50 g of $Mg(OH)_2$?

▶ **FOR MORE PRACTICE**   Example 8.9; Problems 31, 32, 33, 34.

EXAMPLE **8.3** **Mass-to-Mass Conversions**

One of the components of acid rain (rain that becomes acidified due to air pollution) is nitric acid, which forms when NO_2, a pollutant, reacts with oxygen and rainwater according to the following simplified reaction:

$$4\,NO_2(g) + O_2(g) + 2\,H_2O(l) \longrightarrow 4\,HNO_3(aq)$$

Assuming that there is more than enough O_2 and H_2O, how much HNO_3 in kilograms forms from 1.5×10^3 kg of NO_2 pollutant?

SORT	
You are given the mass of nitrogen dioxide (a reactant) and asked to find the mass of nitric acid that can form if the nitrogen dioxide completely reacts.	**GIVEN:** 1.5×10^3 kg NO_2 **FIND:** kg HNO_3

STRATEGIZE

The solution map follows the general format of:

mass $\longrightarrow$ moles $\longrightarrow$

           moles $\longrightarrow$ mass

However, because the original quantity of NO_2 is given in kilograms, you must first convert to grams. The final quantity is requested in kilograms, so you must convert back to kilograms at the end. The main conversion factor is the stoichiometric relationship between moles of nitrogen dioxide and moles of nitric acid. This conversion factor comes from the balanced equation. The other conversion factors are the molar masses of nitrogen dioxide and nitric acid and the relationship between kilograms and grams.

SOLUTION MAP

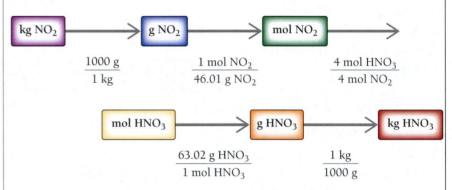

RELATIONSHIPS USED

4 mol NO_2 : 4 mol HNO_3 (from balanced chemical equation)

molar mass NO_2 = 46.01 g/mol

molar mass HNO_3 = 63.02 g/mol

1 kg = 1000 g

SOLVE

Follow the solution map to solve the problem. Begin with kilograms of nitrogen dioxide and multiply by the appropriate conversion factors to arrive at kilograms of nitric acid.

SOLUTION

$$1.5 \times 10^3 \text{ kg } NO_2 \times \frac{1000 \text{ g}}{1 \text{ kg}} \times \frac{1 \text{ mol } NO_2}{46.01 \text{ g } NO_2} \times \frac{4 \text{ mol } HNO_3}{4 \text{ mol } NO_2} \times$$

$$\frac{63.02 \text{ g } HNO_3}{1 \text{ mol } HNO_3} \times \frac{1 \text{ kg}}{1000 \text{ g}} = 2.1 \times 10^3 \text{ kg } HNO_3$$

CHECK

Are the units correct? Does the answer make physical sense?

The units, kg HNO_3, are correct. The magnitude of the answer seems reasonable because it is of the same order of magnitude as the given mass of nitrogen dioxide. An answer that is orders of magnitude different would immediately be suspect.

▶ **SKILLBUILDER 8.3** | **Mass-to-Mass Conversions**

Another component of acid rain is sulfuric acid, which forms when SO_2, also a pollutant, reacts with oxygen and rainwater according to the following reaction:

$$2\,SO_2(g) + O_2(g) + 2\,H_2O(l) \longrightarrow 2\,H_2SO_4(aq)$$

Assuming that there is more than enough O_2 and H_2O, how much H_2SO_4 in kilograms forms from 2.6×10^3 kg of SO_2?

▶ **FOR MORE PRACTICE** Problems 35, 36, 37, 38.

8.5 More Pancakes: Limiting Reactant, Theoretical Yield, and Percent Yield

▶ Calculate limiting reactant, theoretical yield, and percent yield in a balanced chemical equation.

PEARSON eText 2.0

Key Concept Video
Limiting Reactant, Theoretical Yield, and Percent Yield

The term *limiting reagent* is sometimes used in place of limiting reactant.

Let's return to our pancake analogy to understand two more concepts important in reaction stoichiometry: limiting reactant and percent yield. Recall our pancake recipe:

$$1 \text{ cup flour} + 2 \text{ eggs} + \tfrac{1}{2} \text{ tsp baking powder} \longrightarrow 5 \text{ pancakes}$$

Suppose we have 3 cups flour, 10 eggs, and 4 tsp baking powder. How many pancakes can we make? We have enough flour to make:

$$3 \text{ cups flour} \times \frac{5 \text{ pancakes}}{1 \text{ cup flour}} = 15 \text{ pancakes}$$

We have enough eggs to make:

$$10 \text{ eggs} \times \frac{5 \text{ pancakes}}{2 \text{ eggs}} = 25 \text{ pancakes}$$

We have enough baking powder to make:

$$4 \text{ tsp baking powder} \times \frac{5 \text{ pancakes}}{\tfrac{1}{2} \text{ tsp baking powder}} = 40 \text{ pancakes}$$

We have enough flour for 15 pancakes, enough eggs for 25 pancakes, and enough baking powder for 40 pancakes. Consequently, unless we get more ingredients, *we can make only 15 pancakes*. The amount of flour we have *limits* the number of pancakes we can make. If this were a chemical reaction, the flour would be the *limiting reactant*, the reactant that limits the amount of product in a chemical reaction. Notice that the **limiting reactant** is simply the reactant that makes *the least amount of product*. If this were a chemical reaction, 15 pancakes would be the **theoretical yield**, the amount of product that can be made in a chemical reaction based on the amount of limiting reactant.

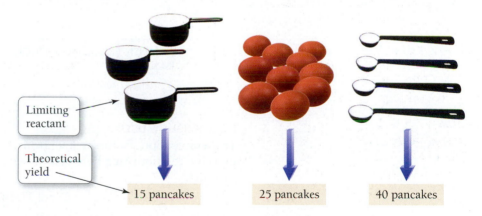

Limiting reactant

Theoretical yield

15 pancakes · 25 pancakes · 40 pancakes

▶ If this were a chemical reaction, the flour would be the limiting reactant and 15 pancakes would be the theoretical yield.

Let us carry this analogy one step further. Suppose we go on to make our pancakes. We accidentally burn three of them and one falls on the floor. So even though we had enough flour for 15 pancakes, we finished with only 11 pancakes. If this were a chemical reaction, the 11 pancakes would be our **actual yield**, the amount of product actually produced by a chemical reaction. Finally, our **percent yield**, the percentage of the theoretical yield that was actually attained, would be:

$$\text{percent yield} = \frac{11 \text{ pancakes}}{15 \text{ pancakes}} \times 100\% = 73\%$$

The actual yield of a chemical reaction, which must be determined experimentally, often depends in various ways on the reaction conditions. We will explore some of the factors involved in Chapter 15.

Since four of the pancakes were ruined, we ended up with only 73% of our theoretical yield. In a chemical reaction, the actual yield is almost always less than 100% because at least some of the product does not form or is lost in the process of recovering it (in analogy to some of the pancakes being burned).

To summarize:

- **Limiting reactant (or limiting reagent)**—the reactant that is completely consumed in a chemical reaction.
- **Theoretical yield**—the amount of product that can be made in a chemical reaction based on the amount of limiting reactant.
- **Actual yield**—the amount of product actually produced by a chemical reaction.
- **Percent yield** $= \dfrac{\textbf{Actual yield}}{\textbf{Theoretical yield}} \times \textbf{100\%}$

Consider this reaction:

$$Ti(s) + 2\,Cl_2(g) \longrightarrow TiCl_4(s)$$

If we begin with 1.8 mol of titanium and 3.2 mol of chlorine, what is the limiting reactant and theoretical yield of $TiCl_4$ in mol? We begin by sorting the information in the problem according to our standard problem-solving procedure.

GIVEN: 1.8 mol Ti
3.2 mol Cl_2

FIND: limiting reactant
theoretical yield

SOLUTION MAP

As in our pancake analogy, we determine the limiting reactant by calculating how much product can be made from each reactant. The reactant that makes the *least amount of product* is the limiting reactant.

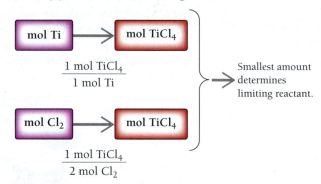

RELATIONSHIPS USED

The conversion factors come from the balanced chemical equation and give the relationships between moles of each of the reactants and moles of product.

$$1 \text{ mol Ti} : 1 \text{ mol TiCl}_4$$

$$2 \text{ mol Cl}_2 : 1 \text{ mol TiCl}_4$$

SOLUTION

$$1.8 \text{ mol Ti} \times \frac{1 \text{ mol TiCl}_4}{1 \text{ mol Ti}} = 1.8 \text{ mol TiCl}_4$$

$$3.2 \text{ mol Cl}_2 \times \frac{1 \text{ mol TiCl}_4}{2 \text{ mol Cl}_2} = 1.6 \text{ mol TiCl}_4$$

Limiting reactant

Least amount of product

In many industrial applications, the more costly reactant or the reactant that is most difficult to remove from the product mixture is chosen to be the limiting reactant.

Since the 3.2 mol of Cl_2 make the least amount of $TiCl_4$, Cl_2 is the limiting reactant. Notice that we began with more moles of Cl_2 than Ti, but because the reaction requires 2 Cl_2 for each Ti, Cl_2 is still the limiting reactant. The theoretical yield is 1.6 mol of $TiCl_4$.

EXAMPLE **8.4** | **Limiting Reactant and Theoretical Yield from Initial Moles of Reactants**

Consider this reaction:

$$2\,Al(s) + 3\,Cl_2(g) \longrightarrow 2\,AlCl_3(s)$$

If you begin with 0.552 mol of aluminum and 0.887 mol of chlorine, what is the limiting reactant and theoretical yield of $AlCl_3$ in moles?

SORT You are given the number of moles of aluminum and chlorine and asked to find the limiting reactant and theoretical yield of aluminum chloride.	**GIVEN:** 0.552 mol Al 0.887 mol Cl_2 **FIND:** limiting reactant theoretical yield of $AlCl_3$
STRATEGIZE Draw a solution map that shows how to get from moles of each reactant to moles of $AlCl_3$. The reactant that makes the *least amount of AlCl$_3$* is the limiting reactant. The conversion factors are the stoichiometric relationships (from the balanced equation).	**SOLUTION MAP** 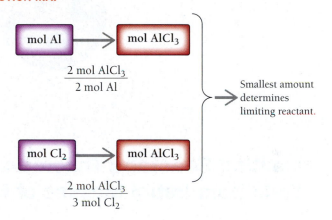 **RELATIONSHIPS USED** 2 mol Al : 2 mol $AlCl_3$ (from balanced equation) 3 mol Cl_2 : 2 mol $AlCl_3$ (from balanced equation)
SOLVE Follow the solution map to solve the problem.	**SOLUTION** $$0.552\ \text{mol Al} \times \dfrac{2\ \text{mol AlCl}_3}{2\ \text{mol Al}} = 0.552\ \text{mol AlCl}_3$$ Limiting reactant Least amount of product $$0.887\ \text{mol Cl}_2 \times \dfrac{2\ \text{mol AlCl}_3}{3\ \text{mol Cl}_2} = 0.591\ \text{mol AlCl}_3$$ Because the 0.552 mol of Al makes the least amount of $AlCl_3$, Al is the limiting reactant. The theoretical yield is 0.552 mol of $AlCl_3$.
CHECK Are the units correct? Does the answer make physical sense?	The units, mol $AlCl_3$, are correct. The magnitude of the answer seems reasonable because it is of the same order of magnitude as the given number of moles of Al and Cl_2. An answer that is orders of magnitude different would immediately be suspect.

▶ **SKILLBUILDER 8.4** | **Limiting Reactant and Theoretical Yield from Initial Moles of Reactants**

Consider the reaction:

$$2\,Na(s) + F_2(g) \longrightarrow 2\,NaF(s)$$

If you begin with 4.8 mol of sodium and 2.6 mol of fluorine, what is the limiting reactant and theoretical yield of NaF in mol?

▶ **FOR MORE PRACTICE**   Problems 43, 44, 45, 46, 47, 48, 49, 50.

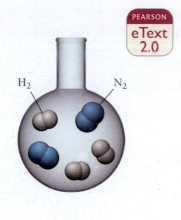

H₂   N₂

CONCEPTUAL ✓ CHECKPOINT 8.3

Consider the reaction:

$$N_2(g) + 3\,H_2(g) \longrightarrow 2\,NH_3(g)$$

If the flask in the left margin represents the mixture before the reaction, which flask represents the products after the limiting reactant has completely reacted?

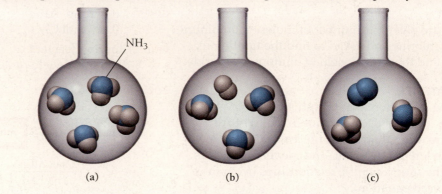

NH₃

(a)   (b)   (c)

8.6 Limiting Reactant, Theoretical Yield, and Percent Yield from Initial Masses of Reactants

▶ Calculate limiting reactant, theoretical yield, and percent yield in a balanced chemical equation.

When working in the laboratory, we normally measure the initial amounts of reactants in grams. To find limiting reactants and theoretical yields from initial masses, we must add two steps to our calculations. Consider, for example, the synthesis reaction:

$$2\,Na(s) + Cl_2(g) \longrightarrow 2\,NaCl(s)$$

If we have 53.2 g of Na and 65.8 g of Cl₂, what is the limiting reactant and theoretical yield? We begin by sorting the information in the problem.

GIVEN: 53.2 g Na
65.8 g Cl₂

FIND: limiting reactant
theoretical yield

SOLUTION MAP

Again, we find the limiting reactant by calculating how much product can be made from each reactant. Since we are given the initial amounts in grams, we must first convert to moles. After we convert to moles of product, we convert back to grams of product. The reactant that makes the *least amount of product* is the limiting reactant.

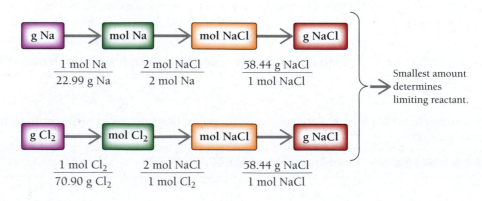

g Na → mol Na → mol NaCl → g NaCl
$\dfrac{1\ \text{mol Na}}{22.99\ \text{g Na}}$ $\dfrac{2\ \text{mol NaCl}}{2\ \text{mol Na}}$ $\dfrac{58.44\ \text{g NaCl}}{1\ \text{mol NaCl}}$

Smallest amount → determines limiting reactant.

g Cl₂ → mol Cl₂ → mol NaCl → g NaCl
$\dfrac{1\ \text{mol Cl}_2}{70.90\ \text{g Cl}_2}$ $\dfrac{2\ \text{mol NaCl}}{1\ \text{mol Cl}_2}$ $\dfrac{58.44\ \text{g NaCl}}{1\ \text{mol NaCl}}$

RELATIONSHIPS USED

From the balanced chemical equation, we know:

$$2\,mol\,Na : 2\,mol\,NaCl$$

$$1\,mol\,Cl_2 : 2\,mol\,NaCl$$

We also use these molar masses:

$$molar\ mass\ Na = \frac{22.99\ g\ Na}{1\ mol\ Na}$$

$$molar\ mass\ Cl_2 = \frac{70.90\ g\ Cl_2}{1\ mol\ Cl_2}$$

$$molar\ mass\ NaCl = \frac{58.44\ g\ NaCl}{1\ mol\ NaCl}$$

SOLUTION

Beginning with the actual amounts of each reactant, we follow the solution map to calculate how much product can be made from each.

> We can also find the limiting reactant by calculating the number of moles of NaCl (rather than grams) that can be made from each reactant. However, since theoretical yields are normally calculated in grams, we take the calculation all the way to grams to determine limiting reactant.

$$53.2\ g\ Na \times \frac{1\ mol\ Na}{22.99\ g\ Na} \times \frac{2\ mol\ NaCl}{2\ mol\ Na} \times \frac{58.44\ g\ NaCl}{1\ mol\ NaCl} = 135\ g\ NaCl$$

$$65.8\ g\ Cl_2 \times \frac{1\ mol\ Cl_2}{70.90\ g\ Cl_2} \times \frac{2\ mol\ NaCl}{1\ mol\ Cl_2} \times \frac{58.44\ g\ NaCl}{1\ mol\ NaCl} = 108\ g\ NaCl$$

Limiting reactant

Least amount of product

Since Cl_2 makes the least amount of product, it is the limiting reactant. Notice that the limiting reactant is not necessarily the reactant with the least mass. In this case, we had fewer grams of Na than Cl_2, yet Cl_2 was the limiting reactant because it made less NaCl. The theoretical yield is therefore 108 g of NaCl, the amount of product possible based on the limiting reactant.

> The actual yield is always less than the theoretical yield because at least a small amount of product is usually lost or does not form during a reaction.

Now suppose that when the synthesis is carried out, the actual yield of NaCl is 86.4 g. What is the percent yield? The percent yield is:

$$percent\ yield = \frac{actual\ yield}{theoretical\ yield} \times 100\% = \frac{86.4\ g}{108\ g} \times 100\% = 80.0\%$$

CONCEPTUAL ✅ **CHECKPOINT 8.4**

Consider the reaction A + 2 B ⟶ 3 C. The molar mass of B is twice the molar mass of A. Equal masses of A and B are in a reaction vessel. Which reactant, A or B, is the limiting reactant?

EXAMPLE **8.5** **Finding Limiting Reactant and Theoretical Yield**

Ammonia, NH_3, can be synthesized by this reaction:

$$2\ NO(g) + 5\ H_2(g) \longrightarrow 2\ NH_3(g) + 2\ H_2O(g)$$

What maximum amount of ammonia in grams can be synthesized from 45.8 g of NO and 12.4 g of H_2?

SORT	
You are given the masses of two reactants and asked to find the maximum mass of ammonia that forms. Although this problem does not specifically ask for the limiting reactant, you must know it to determine the theoretical yield, which is the maximum amount of ammonia that can be synthesized.	**GIVEN:** 45.8 g NO, 12.4 g H_2 **FIND:** maximum amount of NH_3 in g (this is the theoretical yield)

STRATEGIZE

Identify the limiting reactant by calculating how much product can be made from each reactant. The reactant that makes the *least amount of product* is the limiting reactant. The mass of ammonia formed by the limiting reactant is the maximum amount of ammonia that can be synthesized.

SOLUTION MAP

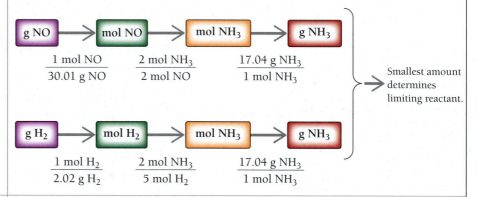

$$\boxed{\text{g NO}} \rightarrow \boxed{\text{mol NO}} \rightarrow \boxed{\text{mol NH}_3} \rightarrow \boxed{\text{g NH}_3}$$

$$\frac{1\ \text{mol NO}}{30.01\ \text{g NO}} \qquad \frac{2\ \text{mol NH}_3}{2\ \text{mol NO}} \qquad \frac{17.04\ \text{g NH}_3}{1\ \text{mol NH}_3}$$

$$\boxed{\text{g H}_2} \rightarrow \boxed{\text{mol H}_2} \rightarrow \boxed{\text{mol NH}_3} \rightarrow \boxed{\text{g NH}_3}$$

$$\frac{1\ \text{mol H}_2}{2.02\ \text{g H}_2} \qquad \frac{2\ \text{mol NH}_3}{5\ \text{mol H}_2} \qquad \frac{17.04\ \text{g NH}_3}{1\ \text{mol NH}_3}$$

Smallest amount determines limiting reactant.

The main conversion factors come from the stoichiometric relationship between moles of each reactant and moles of ammonia. The other conversion factors are the molar masses of nitrogen monoxide, hydrogen gas, and ammonia.

RELATIONSHIPS USED

2 mol NO : 2 mol NH_3 5 mol H_2 : 2 mol NH_3

$$\text{molar mass NO} = \frac{30.01\ \text{g NO}}{1\ \text{mol NO}} \qquad \text{molar mass H}_2 = \frac{2.02\ \text{g H}_2}{1\ \text{mol H}_2}$$

$$\text{molar mass NH}_3 = \frac{17.04\ \text{g NH}_3}{1\ \text{mol NH}_3}$$

SOLVE

Follow the solution map, beginning with the actual amount of each reactant given, to calculate the amount of product that can be made from each reactant.

SOLUTION

$$45.8\ \text{g NO} \times \frac{1\ \text{mol NO}}{30.01\ \text{g NO}} \times \frac{2\ \text{mol NH}_3}{2\ \text{mol NO}} \times \frac{17.04\ \text{g NH}_3}{1\ \text{mol NH}_3} = 26.0\ \text{g NH}_3$$

Limiting reactant   Least amount of product

$$12.4\ \text{g H}_2 \times \frac{1\ \text{mol H}_2}{2.02\ \text{g H}_2} \times \frac{2\ \text{mol NH}_3}{5\ \text{mol H}_2} \times \frac{17.04\ \text{g NH}_3}{1\ \text{mol NH}_3} = 41.8\ \text{g NH}_3$$

There is enough NO to make 26.0 g of NH_3 and enough H_2 to make 41.8 g of NH_3. Therefore, NO is the limiting reactant, and the maximum amount of ammonia that can possibly be made is 26.0 g, which is the theoretical yield.

CHECK Are the units correct? Does the answer make physical sense?	The units of the answer, g NH_3, are correct. The magnitude of the answer seems reasonable because it is of the same order of magnitude as the given masses of NO and H_2. An answer that is orders of magnitude different would immediately be suspect.

▶ **SKILLBUILDER 8.5** | **Finding Limiting Reactant and Theoretical Yield**

Ammonia can also be synthesized by this reaction:

$$3 H_2(g) + N_2(g) \longrightarrow 2 NH_3(g)$$

What maximum amount of ammonia in grams can be synthesized from 25.2 g of N_2 and 8.42 g of H_2?

▶ **SKILLBUILDER PLUS**   What maximum amount of ammonia in kilograms can be synthesized from 5.22 kg of H_2 and 31.5 kg of N_2?

▶ **FOR MORE PRACTICE**   Problems 55, 56, 57, 58.

EXAMPLE **8.6**   **Finding Limiting Reactant, Theoretical Yield, and Percent Yield**

Consider this reaction:

$$Cu_2O(s) + C(s) \longrightarrow 2 Cu(s) + CO(g)$$

When 11.5 g of C reacts with 114.5 g of Cu_2O, 87.4 g of Cu are obtained. Determine the limiting reactant, theoretical yield, and percent yield.

SORT You are given the mass of the reactants, carbon and copper(I) oxide, as well as the mass of copper formed by the reaction. You are asked to find the limiting reactant, theoretical yield, and percent yield.	**GIVEN:** 11.5 g C 114.5 g Cu_2O 87.4 g Cu produced **FIND:** limiting reactant theoretical yield percent yield
STRATEGIZE The solution map shows how to find the mass of Cu formed by the initial masses of Cu_2O and C. The reactant that makes the *least amount of product* is the limiting reactant and determines the theoretical yield.	**SOLUTION MAP** 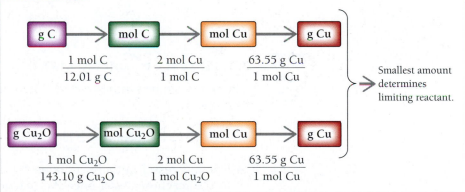
The main conversion factors are the stoichiometric relationships between moles of each reactant and moles of copper. The other conversion factors are the molar masses of copper(I) oxide, carbon, and copper.	**RELATIONSHIPS USED** 1 mol Cu_2O : 2 mol Cu 1 mol C : 2 mol Cu molar mass Cu_2O = 143.10 g/mol molar mass C = 12.01 g/mol molar mass Cu = 63.55 g/mol

continued on page 264 ▶

continued from page 263

SOLVE	SOLUTION
Follow the solution map, beginning with the actual given amount of each reactant, to calculate the amount of product that can be made from each reactant. Since Cu_2O makes the least amount of product, Cu_2O is the limiting reactant. The theoretical yield is then the amount of product made by the limiting reactant. The percent yield is the actual yield (87.4 g Cu) divided by the theoretical yield (101.7 g Cu) multiplied by 100%.	$$11.5 \text{ g C} \times \frac{1 \text{ mol C}}{12.01 \text{ g C}} \times \frac{2 \text{ mol Cu}}{1 \text{ mol C}} \times \frac{63.55 \text{ g Cu}}{1 \text{ mol Cu}} = 122 \text{ g Cu}$$ $$114.5 \text{ g Cu}_2\text{O} \times \frac{1 \text{ mol Cu}_2\text{O}}{143.10 \text{ g Cu}_2\text{O}} \times \frac{2 \text{ mol Cu}}{1 \text{ mol Cu}_2\text{O}} \times \frac{63.55 \text{ g Cu}}{1 \text{ mol Cu}} = 101.7 \text{ g Cu}$$ Limiting reactant Least amount of product theoretical yield = 101.7 g Cu $$\text{percent yield} = \frac{\text{actual yield}}{\text{theoretical yield}} \times 100\%$$ $$= \frac{87.4 \text{ g}}{101.7 \text{ g}} \times 100\% = 85.9\%$$
CHECK Are the units correct? Does the answer make physical sense?	The theoretical yield has the right units (g Cu). The magnitude of the theoretical yield seems reasonable because it is of the same order of magnitude as the given masses of C and Cu_2O. The theoretical yield is reasonable because it is less than 100%. Any calculated theoretical yield above 100% is incorrect.

▶ **SKILLBUILDER 8.6** | **Finding Limiting Reactant, Theoretical Yield, and Percent Yield**

This reaction is used to obtain iron from iron ore:

$$Fe_2O_3(s) + 3\,CO(g) \longrightarrow 2\,Fe(s) + 3\,CO_2(g)$$

The reaction of 185 g of Fe_2O_3 with 95.3 g of CO produces 87.4 g of Fe. Determine the limiting reactant, theoretical yield, and percent yield.

▶ **FOR MORE PRACTICE** Example 8.10; Problems 61, 62, 63, 64, 65, 66.

CONCEPTUAL ✔ **CHECKPOINT 8.5**

Ammonia can be synthesized by the reaction of nitrogen monoxide and hydrogen gas.

$$2\,NO(g) + 5\,H_2(g) \longrightarrow 2\,NH_3(g) + 2\,H_2O(g)$$

A reaction vessel initially contains 4.0 mol of NO and 15.0 mol of H_2. What is in the reaction vessel once the reaction has occurred to the fullest extent possible?

(a) 2 mol NO; 5 mol H_2; 2 mol NH_3; and 2 mol H_2O

(b) 0 mol NO; 0 mol H_2; 6 mol NH_3; and 6 mol H_2O

(c) 2 mol NO; 0 mol H_2; 4 mol NH_3; and 2 mol H_2O

(d) 0 mol NO; 5 mol H_2; 4 mol NH_3; and 4 mol H_2O

8.7 Enthalpy: A Measure of the Heat Evolved or Absorbed in a Reaction

▶ Calculate the amount of thermal energy emitted or absorbed by a chemical reaction.

Chapter 3 (see Section 3.9) describes how chemical reactions can be *exothermic* (in which case they *emit* thermal energy when they occur) or *endothermic* (in which case they *absorb* thermal energy when they occur). The *amount* of thermal energy emitted or absorbed by a chemical reaction, under conditions of constant pressure (which are common for most everyday reactions), can be quantified with a function called **enthalpy**. Specifically, we define a quantity called the **enthalpy of reaction** (ΔH_{rxn}) as the amount of thermal energy (or heat) that is emitted or absorbed when a reaction occurs at constant pressure.

Sign of ΔH_{rxn}

The *sign* of ΔH_{rxn} (positive or negative) depends on the *direction* in which thermal energy flows when the reaction occurs. If thermal energy flows out of the reaction and into the surroundings (as in an exothermic reaction), then ΔH_{rxn} is negative.

For example, we can specify the enthalpy of reaction for the combustion of CH_4, the main component in natural gas, as:

$$CH_4(g) + 2\,O_2(g) \longrightarrow CO_2(g) + 2\,H_2O(g) \qquad \Delta H_{rxn} = -802.3\ kJ$$

EVERYDAY CHEMISTRY

Bunsen Burners

In the laboratory, we often use Bunsen burners as heat sources. These burners are normally fueled by methane. The balanced equation for methane (CH_4) combustion is:

$$CH_4(g) + 2\,O_2(g) \longrightarrow CO_2(g) + 2\,H_2O(g)$$

Most Bunsen burners have a mechanism to adjust the amount of air (and therefore of oxygen) that is mixed with the methane. If you light the burner with the air completely closed off, you get a yellow, smoky flame that is not very hot. As you increase the amount of air going into the burner,

the flame becomes bluer, less smoky, and hotter. When you reach the optimum adjustment, the flame has a sharp, inner blue triangle, no smoke, and is hot enough to melt glass easily. Continuing to increase the air beyond this point causes the flame to become cooler again and may actually extinguish it.

B8.1 CAN YOU ANSWER THIS? *Can you use the concepts from this chapter to explain the changes in the Bunsen burner flame as the air intake is adjusted?*

| (a) No air | (b) Small amount of air | (c) Optimum | (d) Too much air |

▲ Bunsen burner flame at various stages of air-intake adjustment.

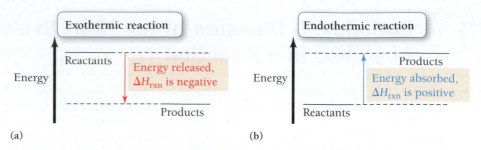

▲ FIGURE 8.3 **Exothermic and endothermic reactions** **(a)** In an exothermic reaction, energy is released into the surroundings. **(b)** In an endothermic reaction, energy is absorbed from the surroundings.

This reaction is exothermic and therefore has a negative enthalpy of reaction. The magnitude of ΔH_{rxn} tells us that 802.3 kJ of heat are emitted when 1 mol of CH_4 reacts with 2 mol of O_2.

If, by contrast, thermal energy flows into the reaction and out of the surroundings (as in an endothermic reaction), then ΔH_{rxn} is positive. For example, we specify the enthalpy of reaction for the reaction between nitrogen and oxygen gas to form nitrogen monoxide as:

$$N_2(g) + O_2(g) \longrightarrow 2\,NO(g) \qquad \Delta H_{rxn} = +182.6\,kJ$$

This reaction is endothermic and therefore has a positive enthalpy of reaction. When 1 mol of N_2 reacts with 1 mol of O_2, 182.6 kJ of heat are absorbed from the surroundings.

We can think of the energy of a chemical system in the same way that we think about the balance in a checking account. Energy flowing *out* of the chemical system is like a withdrawal and carries a negative sign as shown in ▲ FIGURE 8.3a. Energy flowing *into* the system is like a deposit and carries a positive sign as shown in ▲ FIGURE 8.3b.

Stoichiometry of ΔH_{rxn}

The amount of heat emitted or absorbed when a chemical reaction occurs depends on the *amounts* of reactants that actually react. As we have just seen, we usually specify ΔH_{rxn} in combination with the balanced chemical equation for the reaction. The magnitude of ΔH_{rxn} is for the stoichiometric amounts of reactants and products for the reaction *as written*.

For example, the balanced equation and ΔH_{rxn} for the combustion of propane (the fuel used in LP gas) is:

$$C_3H_8(g) + 5\,O_2(g) \longrightarrow 3\,CO_2(g) + 4\,H_2O(g) \qquad \Delta H_{rxn} = -2044\,kJ$$

This means that when 1 mol of C_3H_8 reacts with 5 mol of O_2 to form 3 mol of CO_2 and 4 mol of H_2O, 2044 kJ of heat are emitted. We can write these relationships in the same way that we express stoichiometric relationships: as ratios between two quantities. For the reactants in this reaction, we write:

$$1 \text{ mol } C_3H_8 : -2044 \text{ kJ} \quad \text{or} \quad 5 \text{ mol } O_2 : -2044 \text{ kJ}$$

The ratios mean that 2044 kJ of thermal energy are evolved when 1 mol of C_3H_8 and 5 mol of O_2 completely react. We can use these ratios to construct conversion factors between amounts of reactants or products and the quantity of heat emitted (for exothermic reactions) or absorbed (for endothermic reactions). To find out how much heat is emitted upon the combustion of a certain mass in grams of C_3H_8, we use the following solution map:

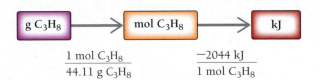

We use the molar mass to convert between grams and moles, and the stoichiometric relationship between moles of C_3H_8 and kilojoules to convert between moles and kilojoules, as shown in Example 8.7.

PEARSON
eText
2.0
**Interactive
Worked Example
Video 8.7**

EXAMPLE **8.7** | **Stoichiometry Involving ΔH_{rxn}**

An LP gas tank in a home barbecue contains 1.18×10^4 g of propane (C_3H_8). Calculate the heat (in kJ) associated with the complete combustion of all of the propane in the tank.

$$C_3H_8(g) + 5\,O_2(g) \longrightarrow 3\,CO_2(g) + 4\,H_2O(g) \qquad \Delta H_{rxn} = -2044 \text{ kJ}$$

SORT You are given the mass of propane and asked to find the heat evolved (in kJ) in its combustion.	**GIVEN:** 1.18×10^4 g C_3H_8 **FIND:** kJ
STRATEGIZE Start with the given mass of propane and use its molar mass to find the number of moles. Use the stoichiometric relationship between moles of propane and kilojoules of heat to find the heat evolved.	**SOLUTION MAP** 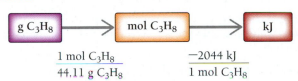 **RELATIONSHIPS USED** $\quad$ 1 mol C_3H_8 : -2044 kJ (from balanced equation) $\quad$ molar mass $C_3H_8 = 44.11$ g/mol
SOLVE Follow the solution map to solve the problem. Begin with 11.8×10^4 g C_3H_8 and multiply by the appropriate conversion factors to arrive at kJ.	**SOLUTION** $$1.18 \times 10^4 \text{ g } C_3H_8 \times \frac{1 \text{ mol } C_3H_8}{44.11 \text{ g } C_3H_8} \times \frac{-2044 \text{ kJ}}{1 \text{ mol } C_3H_8} = -5.47 \times 10^5 \text{ kJ}$$
CHECK Check your answer. Are the units correct? Does the answer make physical sense?	The units, kJ, are correct. The answer is negative, as it should be when heat is evolved by a reaction.

▶ **SKILLBUILDER 8.7** | **Stoichiometry Involving ΔH**

Ammonia reacts with oxygen according to the equation:

$$4\,NH_3(g) + 5\,O_2(g) \longrightarrow 4\,NO(g) + 6\,H_2O \qquad \Delta H_{rxn} = -906 \text{ kJ}$$

Calculate the heat (in kJ) associated with the complete reaction of 155 g of NH_3.

▶ **SKILLBUILDER PLUS** What mass of butane in grams is necessary to produce 1.5×10^3 kJ of heat? What mass of CO_2 is produced?

$$C_4H_{10}(g) + \tfrac{13}{2}\,O_2(g) \longrightarrow 4\,CO_2(g) + 5\,H_2O(g) \qquad \Delta H_{rxn} = -2658 \text{ kJ}$$

▶ **FOR MORE PRACTICE** Example 8.11; Problems 71, 72, 73, 74, 75, 76.

PEARSON eText 2.0

CONCEPTUAL ✔ **CHECKPOINT 8.6**

Consider the generic reaction:

$$2\,A + 3\,B \longrightarrow 2\,C \qquad \Delta H_{rxn} = -100 \text{ kJ}$$

If a reaction mixture initially contains 5 mol of A and 6 mol of B, how much heat (in kJ) will have evolved once the reaction has occurred to the greatest extent possible?

(a) 100 kJ　　　　**(b)** 150 kJ　　　　**(c)** 200 kJ　　　　**(d)** 300 kJ

Chapter 8 in Review

MasteringChemistry™ provides end-of-chapter exercises, feedback-enriched tutorial problems, animations, and interactive activities to encourage problem solving practice and deeper understanding of key concepts and topics.

Self-Assessment Quiz

PEARSON eText 2.0

Q1. Sulfur and fluorine react to form sulfur hexafluoride according to the reaction shown here. How many mol of F_2 are required to react completely with 2.55 mol of S?

$$S(s) + 3\,F_2(g) \longrightarrow SF_6(g)$$

(a) 0.85 mol F_2 　　　**(b)** 2.55 mol F_2
(c) 7.65 mol F_2 　　　**(d)** 15.3 mol F_2

Q2. Hydrogen chloride gas and oxygen gas react to form gaseous water and chlorine gas according to the reaction shown here.

$$4\,HCl(g) + O_2(g) \longrightarrow 2\,H_2O(g) + 2\,Cl_2(g)$$

If the first image below represents the amount of HCl available for the reaction, which image represents the amount of oxygen required to react completely with the amount of available HCl?

$= O_2$
$= HCl$

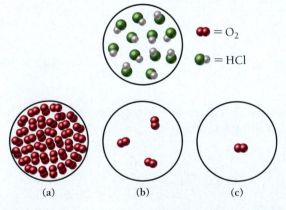

　(a)　　　　　(b)　　　　　(c)

Q3. Sodium reacts with fluorine to form sodium fluoride. What mass of sodium fluoride forms from the complete reaction of 12.5 g of fluorine with enough sodium to completely react with it?

$$2\,Na(s) + F_2(g) \longrightarrow 2\,NaF(s)$$

(a) 0.658 g NaF 　　　**(b)** 13.8 g NaF
(c) 6.91 g NaF 　　　**(d)** 27.6 g NaF

Q4. Consider the hypothetical reaction shown here. If 11 mol of A combine with 16 mol of B and the reaction occurs to the greatest extent possible, how many mol of C form?

$$3\,A + 4\,B \longrightarrow 2\,C$$

(a) 7.3 mol C 　　　　**(b)** 8.0 mol C
(c) 17 mol C 　　　　**(d)** 32 mol C

Q5. Consider the generic reaction:

$$2\,A + 3\,B + C \longrightarrow 2\,D$$

A reaction mixture contains 6 mol A, 8 mol B, and 10 mol C. What is the limiting reactant?
(a) A 　　**(b)** B 　　**(c)** C 　　**(d)** D

Q6. Methanol (CH_3OH) reacts with oxygen to form carbon dioxide and water according to the reaction shown here.

$$2\,CH_3OH(g) + 3\,O_2(g) \longrightarrow 2\,CO_2(g) + 4\,H_2O(g)$$

If the first image below represents a reaction mixture of methanol and oxygen, which image represents the reaction mixture after the reaction has occurred to the maximum extent possible?

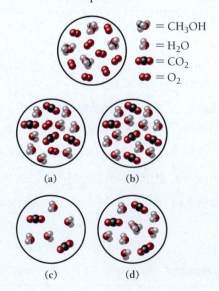

$= CH_3OH$
$= H_2O$
$= CO_2$
$= O_2$

　(a)　　　　　(b)

　(c)　　　　　(d)

Q7. Sodium and chlorine react to form sodium chloride.

$$2\,Na(s) + Cl_2(g) \longrightarrow 2\,NaCl(s)$$

What is the theoretical yield of sodium chloride for the reaction of 55.0 g Na with 67.2 g Cl_2?
(a) 1.40×10^2 g NaCl   **(b)** 111 g NaCl
(c) 55.4 g NaCl   **(d)** 222 g NaCl

Q8. A reaction has a theoretical yield of 22.8 g. When the reaction is carried out, 15.1 g of the product is obtained. What is the percent yield?
(a) 151%
(b) 66.2%
(c) 344%
(d) 88.2%

Q9. Titanium can be obtained from its oxide by the reaction shown here. When 42.0 g of TiO_2 react with 11.5 g C, 18.7 g Ti are obtained. What is the percent yield for the reaction?

$$TiO_2(s) + 2\,C(s) \longrightarrow Ti(s) + CO(g)$$

(a) 74.2%
(b) 81.6%
(c) 122%
(d) 41%

Q10. Which statement best describes an exothermic reaction?
(a) An exothermic reaction gives off heat.
(b) An exothermic reaction absorbs heat.
(c) An exothermic reaction produces only small amounts of products.
(d) none of the above

Q11. Consider the generic reaction:

$$A + 2\,B \longrightarrow AB_2 \quad \Delta H_{rxn} = -155\,kJ$$

If a reaction mixture contains 5 mol A and 8 mol B, how much heat is emitted or absorbed once the reaction has occurred to the greatest extent possible?
(a) 775 kJ emitted
(b) 775 kJ absorbed
(c) 620 kJ emitted
(d) 620 kJ absorbed

Q12. Hydrogen gas reacts with oxygen to form water.

$$2\,H_2(g) + O_2(g) \longrightarrow 2\,H_2O(g) \quad \Delta H = -483.5\,kJ$$

Determine the minimum mass of hydrogen gas required to produce 226 kJ of heat.
(a) 8.63 g
(b) 1.88 g
(c) 0.942 g
(d) 0.935 g

Answers: 1:c, 2:b, 3:d, 4:a, 5:b, 6:a, 7:b, 8:b, 9:b, 10:a, 11:c, 12:b

Chemical Principles

Relevance

Stoichiometry

A balanced chemical equation indicates quantitative relationships between the amounts of reactants and products. For example, the reaction $2\,H_2 + O_2 \longrightarrow 2\,H_2O$ tells us that 2 mol of H_2 reacts with 1 mol of O_2 to form 2 mol of H_2O. We can use these relationships to calculate quantities such as the amount of product possible with a certain amount of reactant, or the amount of one reactant required to completely react with a certain amount of another reactant. The quantitative relationship between reactants and products in a chemical reaction is reaction stoichiometry.

Reaction stoichiometry is important because we often want to know the numerical relationship between the reactants and products in a chemical reaction. For example, we might want to know how much carbon dioxide, a greenhouse gas, is formed when a certain amount of a particular fossil fuel burns.

Limiting Reactant, Theoretical Yield, and Percent Yield

The limiting reactant in a chemical reaction is the reactant that limits the amount of product that can be made. The theoretical yield in a chemical reaction is the amount of product that can be made based on the amount of the limiting reactant. The actual yield in a chemical reaction is the amount of product actually produced. The percent yield in a chemical reaction is the actual yield divided by theoretical yield times 100%.

Calculations of limiting reactant, theoretical yield, and percent yield are central to chemistry because they allow for quantitative understanding of chemical reactions. Just as we need to know relationships between ingredients to follow a recipe, so we must know relationships between reactants and products to carry out a chemical reaction. The percent yield in a chemical reaction is often used as a measure of the success of the reaction. Imagine following a recipe and making only 1% of the final product—your cooking would be a failure. Similarly, low percent yields in chemical reactions are usually considered poor, and high percent yields are considered good.

Chemical Principles

Relevance

Enthalpy of Reaction

The amount of heat released or absorbed by a chemical reaction under conditions of constant pressure is the enthalpy of reaction (ΔH_{rxn}).

The enthalpy of reaction describes the relationship between the amount of reactant that undergoes reaction and the amount of thermal energy produced. This is important, for example, in determining quantities such as the amount of fuel needed to produce a given amount of energy.

Chemical Skills

Examples

LO: Carry out mole-to-mole conversions between reactants and products in a balanced chemical equation (Section 8.3).

EXAMPLE **8.8**   **Mole-to-Mole Conversions**

How many mol of sodium oxide can be synthesized from 4.8 mol of sodium? Assume that more than enough oxygen is present. The balanced equation is:

$$4\,Na(s) + O_2(g) \longrightarrow 2\,Na_2O(s)$$

SORT

You are given the number of moles of sodium and asked to find the number of moles of sodium oxide formed by the reaction.

GIVEN: 4.8 mol Na

FIND: mol Na_2O

STRATEGIZE

Draw a solution map beginning with the number of moles of the given substance and then use the conversion factor from the balanced chemical equation to determine the number of moles of the substance you are trying to find.

SOLUTION MAP

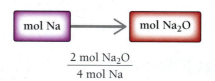

$$\frac{2\;mol\;Na_2O}{4\;mol\;Na}$$

RELATIONSHIPS USED

$4\,mol\,Na : 2\,mol\,Na_2O$

SOLVE

Follow the solution map to get to the number of moles of the substance you are trying to find.

CHECK

Are the units correct? Does the answer make physical sense?

SOLUTION

$$4.8\;\cancel{mol\;Na} \times \frac{2\;mol\;Na_2O}{4\;\cancel{mol\;Na}} = 2.4\;mol\;Na_2O$$

The units of the answer, mol Na_2O, are correct. The magnitude of the answer seems reasonable because it is of the same order of magnitude as the given number of mol of Na.

LO: Carry out mass-to-mass conversions between reactants and products in a balanced chemical equation and molar masses (Section 8.4).

EXAMPLE **8.9** | **Mass-to-Mass Conversions**

How many grams of sodium oxide can be synthesized from 17.4 g of sodium? Assume that more than enough oxygen is present. The balanced equation is:

$$4\,Na(s) + O_2(g) \longrightarrow 2\,Na_2O(s)$$

GIVEN: 17.4 g Na

FIND: g Na_2O

SOLUTION MAP

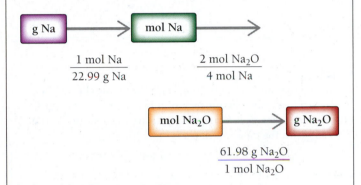

RELATIONSHIPS USED

4 mol Na : 2 mol Na_2O (from balanced equation)

molar mass Na = 22.99 g/mol

molar mass Na_2O = 61.98 g/mol

SORT
You are given the mass of sodium and asked to find the mass of sodium oxide that forms upon reaction.

STRATEGIZE
Draw the solution map by beginning with the mass of the given substance. Convert to moles using the molar mass and then convert to moles of the substance you are trying to find, using the conversion factor obtained from the balanced chemical equation.

Finally, convert to the mass of the substance you are trying to find, using its molar mass.

SOLVE
Follow the solution map and calculate the answer by beginning with the mass of the given substance and multiplying by the appropriate conversion factors to determine the mass of the substance you are trying to find.

SOLUTION

$$17.4\ g\,\cancel{Na} \times \frac{1\ \cancel{mol\,Na}}{22.99\ g\,\cancel{Na}} \times \frac{2\ \cancel{mol\,Na_2O}}{4\ \cancel{mol\,Na}} \times$$

$$\frac{61.98\ g\,Na_2O}{1\ \cancel{mol\,Na_2O}} = 23.5\ g\,Na_2O$$

CHECK
Are the units correct? Does the answer make physical sense?

The units of the answer, g Na_2O, are correct. The magnitude of the answer seems reasonable because it is of the same order of magnitude as the given number of mol of Na.

LO: Calculate limiting reactant, theoretical yield, and percent yield in a balanced chemical equation (Sections 8.5, 8.6).

EXAMPLE **8.10** **Limiting Reactant, Theoretical Yield, and Percent Yield**

10.4 g of As reacts with 11.8 g of S to produce 14.2 g of As_2S_3. Find the limiting reactant, theoretical yield, and percent yield for this reaction. The balanced chemical equation is:

$$2\,As(s) + 3S(l) \longrightarrow As_2S_3(s)$$

SORT

You are given the masses of arsenic and sulfur as well as the mass of arsenic sulfide formed by the reaction. You are asked to find the limiting reactant, theoretical yield, and percent yield.

GIVEN: 10.4 g As

11.8 g S

14.2 g As_2S_3

FIND: limiting reactant

theoretical yield

percent yield

STRATEGIZE

The solution map for limiting-reactant problems shows how to convert from the mass of each of the reactants to the mass of the product for each reactant. These are mass-to-mass conversions with the basic outline of

$$mass \longrightarrow moles \longrightarrow moles \longrightarrow mass$$

The reactant that forms the least amount of product is the limiting reactant.

The conversion factors you need are the stoichiometric relationships between each of the reactants and the product. You also need the molar masses of each reactant and product.

SOLUTION MAP

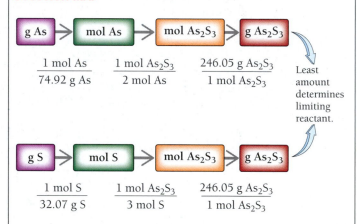

RELATIONSHIPS USED

2 mol As : 1 mol As_2S_3

3 mol S : 1 mol As_2S_3

molar mass As = 74.92 g/mol

molar mass S = 32.07 g/mol

molar mass As_2S_3 = 246.05 g/mol

SOLVE

To calculate the amount of product formed by each reactant, begin with the given amount of each reactant and multiply by the appropriate conversion factors, as shown in the solution map, to arrive at the mass of product for each reactant. The reactant that forms the least amount of product is the limiting reactant.

SOLUTION

$$10.4 \text{ g As} \times \frac{1 \text{ mol As}}{74.92 \text{ g As}} \times \frac{1 \text{ mol As}_2\text{S}_3}{2 \text{ mol As}} \times \frac{246.05 \text{ g As}_2\text{S}_3}{1 \text{ mol As}_2\text{S}_3}$$

Limiting reactant

$$= \boxed{17.1 \text{ g As}_2\text{S}_3}$$

Least amount of product

$$11.8 \text{ g S} \times \frac{1 \text{ mol S}}{32.07 \text{ g S}} \times \frac{1 \text{ mol As}_2\text{S}_3}{3 \text{ mol S}} \times \frac{246.05 \text{ g As}_2\text{S}_3}{1 \text{ mol As}_2\text{S}_3}$$

$$= 30.2 \text{ g As}_2\text{S}_3$$

The limiting reactant is As.

The theoretical yield is the amount of product formed by the limiting reactant.

The percent yield is the actual yield divided by the theoretical yield times 100%.

CHECK

Check your answer. Are the units correct? Does the answer make physical sense?

The theoretical yield is 17.1 g of As_2S_3.

$$\text{percent yield} = \frac{\text{actual yield}}{\text{theoretical yield}} \times 100\%$$

$$= \frac{14.2\,\text{g}}{17.1\,\text{g}} \times 100\% = 83.0\%$$

The percent yield is 83.0%.

The theoretical yield has the right units (g As_2S_3). The magnitude of the theoretical yield seems reasonable because it is of the same order of magnitude as the given masses of As and S. The theoretical yield is reasonable because it is less than 100%. Any calculated theoretical yield above 100% would be suspect.

LO: Calculate the amount of thermal energy emitted or absorbed by a chemical reaction (Section 8.7).

SORT

You are given the mass of methane and asked to find the quantity of heat in kJ emitted upon combustion.

Draw the solution map by beginning with the mass of the given substance. Convert to moles using molar mass and to kJ using ΔH_{rxn}.

SOLVE

Follow the solution map to solve the problem. Begin with the mass of the given substance and multiply by the appropriate conversion factors to arrive at kJ. A negative answer indicates that heat is evolved into the surroundings. A positive answer indicates that heat is absorbed from the surroundings.

CHECK

Are the units correct? Does the answer make physical sense?

EXAMPLE **8.11** **Stoichiometry Involving ΔH_{rxn}**

Calculate the heat evolved (in kJ) upon complete combustion of 25.0 g of methane (CH_4).

$$CH_4(g) + 2O_2(g) \longrightarrow CO_2(g) + 2H_2O(g)$$
$$\Delta H_{rxn} = -802\,\text{kJ}$$

GIVEN: 25 g CH_4

FIND: kJ

SOLUTION MAP

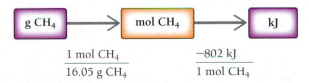

$$\frac{1\text{ mol }CH_4}{16.05\text{ g }CH_4} \qquad \frac{-802\text{ kJ}}{1\text{ mol }CH_4}$$

RELATIONSHIPS USED

1 mol CH_4 : -802 kJ (from balanced equation)

molar mass CH_4 = 16.05 g/mol

SOLUTION

$$25.0\text{ g }CH_4 \times \frac{1\text{ mol }CH_4}{16.05\text{ g }CH_4} \times \frac{-802\text{ kJ}}{1\text{ mol }CH_4} = -1.25 \times 10^3\text{ kJ}$$

The units, kJ, are correct. The answer is negative, as it should be since heat is evolved by the reaction.

Key Terms

actual yield **[8.5]**

climate change **[8.1]**

enthalpy **[8.7]**

enthalpy of reaction (ΔH_{rxn}) **[8.7]**

greenhouse gases **[8.1]**

limiting reactant **[8.5]**

percent yield **[8.5]**

stoichiometry **[8.1]**

theoretical yield **[8.5]**

Exercises

Questions

1. Why is reaction stoichiometry important? Cite some examples in your answer.
2. Nitrogen and hydrogen can react to form ammonia:

$$N_2(g) + 3H_2(g) \longrightarrow 2NH_3(g)$$

   (a) Write ratios showing the relationships between moles of each of the reactants and products in the reaction.
   (b) How many molecules of H_2 are required to completely react with two molecules of N_2?
   (c) How many moles of H_2 are required to completely react with 2 mol of N_2?

3. Write the conversion factor that you would use to convert from moles of Cl_2 to moles of NaCl in the reaction:

$$2Na(s) + Cl_2(g) \longrightarrow 2NaCl(s)$$

4. What is wrong with this statement in reference to the reaction in the previous problem? "2 g of Na react with 1 g of Cl_2 to form 2 g of NaCl." Correct the statement to make it true.
5. What is the general form of the solution map for problems in which you are given the mass of a reactant in a chemical reaction and asked to find the mass of the product that can be made from the given amount of reactant?
6. Consider the recipe for making tomato and garlic pasta.

   2 cups noodles + 12 tomatoes + 3 cloves garlic $\longrightarrow$
   4 servings pasta

   If you have 7 cups of noodles, 27 tomatoes, and 9 cloves of garlic, how many servings of pasta can you make? Which ingredient limits the amount of pasta that it is possible to make?

7. In a chemical reaction, what is the limiting reactant?
8. In a chemical reaction, what is the theoretical yield?
9. In a chemical reaction, what are the actual yield and the percent yield?
10. If you are given a chemical equation and specific amounts for each reactant in grams, how do you determine the maximum amount of product that can be made?
11. Consider the generic chemical reaction:

$$A + 2B \longrightarrow C + D$$

   Suppose you have 12 g of A and 24 g of B. Which statement is true?
   (a) A will definitely be the limiting reactant.
   (b) B will definitely be the limiting reactant.
   (c) A will be the limiting reactant if its molar mass is less than B.
   (d) A will be the limiting reactant if its molar mass is greater than B.

12. Consider the generic chemical equation:

$$A + B \longrightarrow C$$

   Suppose 25 g of A were allowed to react with 8 g of B. Analysis of the final mixture showed that A was completely used up and 4 g of B remained. What was the limiting reactant?
13. What is the enthalpy of reaction (ΔH_{rxn})? Why is this quantity important?
14. Explain the relationship between the sign of ΔH_{rxn} and whether a reaction is exothermic or endothermic.

Problems

MOLE-TO-MOLE CONVERSIONS

15. Consider the generic chemical reaction:

$$A + 2B \longrightarrow C$$

   How many moles of C are formed upon complete reaction of:
   (a) 2 mol of A
   (b) 2 mol of B
   (c) 3 mol of A
   (d) 3 mol of B

16. Consider the generic chemical reaction:

$$2A + 3B \longrightarrow 3C$$

   How many moles of B are required to completely react with:
   (a) 6 mol of A
   (b) 2 mol of A
   (c) 7 mol of A
   (d) 11 mol of A

17. For the reaction shown, calculate how many moles of NO_2 form when each amount of reactant completely reacts.

$$2N_2O_5(g) \longrightarrow 4NO_2(g) + O_2(g)$$

   (a) 1.3 mol N_2O_5
   (b) 5.8 mol N_2O_5
   (c) 4.45×10^3 mol N_2O_5
   (d) 1.006×10^{-3} mol N_2O_5

18. For the reaction shown, calculate how many moles of NH_3 form when each amount of reactant completely reacts.

$$3N_2H_4(l) \longrightarrow 4NH_3(g) + N_2(g)$$

   (a) 5.3 mol N_2H_4
   (b) 2.28 mol N_2H_4
   (c) 5.8×10^{-2} mol N_2H_4
   (d) 9.76×10^7 mol N_2H_4

19. Dihydrogen monosulfide reacts with sulfur dioxide according to the balanced equation:

$$2\,H_2S(g) + SO_2(g) \longrightarrow 3\,S(s) + 2\,H_2O(g)$$

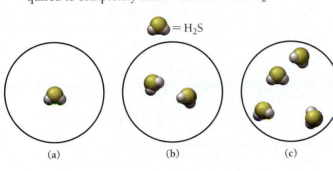

= SO_2

If the first figure represents the amount of SO_2 available to react, which figure best represents the amount of H_2S required to completely react with all of the SO_2?

= H_2S

(a)   (b)   (c)

20. Chlorine gas reacts with fluorine gas according to the balanced equation:

$$Cl_2(g) + 3\,F_2(g) \longrightarrow 2\,ClF_3(g)$$

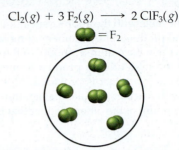

= F_2

If the first figure represents the amount of fluorine available to react, and assuming that there is more than enough chlorine, which figure best represents the amount of chlorine trifluoride that would form upon complete reaction of all of the fluorine?

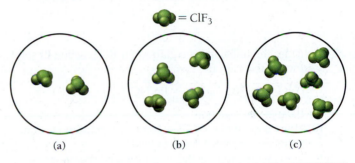

= ClF_3

(a)   (b)   (c)

21. For each reaction, calculate how many moles of product form when 1.75 mol of the reactant in color completely reacts. Assume there is more than enough of the other reactant.
(a) $H_2(g) + Cl_2(g) \longrightarrow 2\,HCl(g)$
(b) $2\,H_2(g) + O_2(g) \longrightarrow 2\,H_2O(l)$
(c) $2\,Na(s) + O_2(g) \longrightarrow Na_2O_2(s)$
(d) $2\,S(s) + 3\,O_2(g) \longrightarrow 2\,SO_3(g)$

22. For each reaction, calculate how many moles of the product form when 0.112 mol of the reactant in color completely reacts. Assume there is more than enough of the other reactant.
(a) $2\,Ca(s) + O_2(g) \longrightarrow 2\,CaO(s)$
(b) $4\,Fe(s) + 3\,O_2(g) \longrightarrow 2\,Fe_2O_3(s)$
(c) $4\,K(s) + O_2(g) \longrightarrow 2\,K_2O(s)$
(d) $4\,Al(s) + 3\,O_2(g) \longrightarrow 2\,Al_2O_3(s)$

23. For the reaction shown, calculate how many moles of each product form when the given amount of each reactant completely reacts. Assume there is more than enough of the other reactant.

$$2\,PbS(s) + 3\,O_2(g) \longrightarrow 2\,PbO(s) + 2\,SO_2(g)$$

(a) 2.4 mol PbS
(b) 2.4 mol O_2
(c) 5.3 mol PbS
(d) 5.3 mol O_2

24. For the reaction shown, calculate how many moles of each product form when the given amount of each reactant completely reacts. Assume there is more than enough of the other reactant.

$$C_3H_8(g) + 5\,O_2(g) \longrightarrow 3\,CO_2(g) + 4\,H_2O(g)$$

(a) 4.6 mol C_3H_8
(b) 4.6 mol O_2
(c) 0.0558 mol C_3H_8
(d) 0.0558 mol O_2

25. Consider the balanced equation:

$$2\,N_2H_4(g) + N_2O_4(g) \longrightarrow 3\,N_2(g) + 4\,H_2O(g)$$

Complete the table with the appropriate number of moles of reactants and products. If the number of moles of a reactant is provided, fill in the required amount of the other reactant, as well as the moles of each product formed. If the number of moles of a product is provided, fill in the required amount of each reactant to make that amount of product, as well as the amount of the other product that is made.

mol N_2H_4	mol N_2O_4	mol N_2	mol H_2O
	2		
6			
			8
	5.5		
3			
		12.4	

26. Consider the balanced equation:

$$SiO_2(s) + 3C(s) \longrightarrow SiC(s) + 2CO(g)$$

Complete the table with the appropriate number of moles of reactants and products. If the number of moles of a reactant is provided, fill in the required amount of the other reactant, as well as the moles of each product formed. If the number of moles of a product is provided, fill in the required amount of each reactant to make that amount of product, as well as the amount of the other product that is made.

mol SiO₂	mol C	mol SiC	mol CO
	6		
3	___	___	
___			10
___	9.5	___	___
3.2	___	___	___

27. Consider the unbalanced equation for the combustion of butane:

$$C_4H_{10}(g) + O_2(g) \longrightarrow CO_2(g) + H_2O(g)$$

Balance the equation and determine how many moles of O_2 are required to react completely with 4.9 mol of C_4H_{10}.

28. Consider the unbalanced equation for the neutralization of acetic acid:

$$HC_2H_3O_2(aq) + Ca(OH)_2(aq) \longrightarrow H_2O(l) + Ca(C_2H_3O_2)_2(aq)$$

Balance the equation and determine how many moles of $Ca(OH)_2$ are required to completely neutralize 1.07 mol of $HC_2H_3O_2$.

29. Consider the unbalanced equation for the reaction of solid lead with silver nitrate:

$$Pb(s) + AgNO_3(aq) \longrightarrow Pb(NO_3)_2(aq) + Ag(s)$$

(a) Balance the equation.
(b) How many moles of silver nitrate are required to completely react with 9.3 mol of lead?
(c) How many moles of Ag are formed by the complete reaction of 28.4 mol of Pb?

30. Consider the unbalanced equation for the reaction of aluminum with sulfuric acid:

$$Al(s) + H_2SO_4(aq) \longrightarrow Al_2(SO_4)_3(aq) + H_2(g)$$

(a) Balance the equation.
(b) How many moles of H_2SO_4 are required to completely react with 8.3 mol of Al?
(c) How many moles of H_2 are formed by the complete reaction of 0.341 mol of Al?

MASS-TO-MASS CONVERSIONS

31. For the reaction shown, calculate how many grams of oxygen form when each quantity of reactant completely reacts.

$$2HgO(s) \longrightarrow 2Hg(l) + O_2(g)$$

(a) 2.13 g HgO
(b) 6.77 g HgO
(c) 1.55 kg HgO
(d) 3.87 mg HgO

32. For the reaction shown, calculate how many grams of oxygen form when each quantity of reactant completely reacts.

$$2KClO_3(s) \longrightarrow 2KCl(s) + 3O_2(g)$$

(a) 2.72 g KClO₃
(b) 0.361 g KClO₃
(c) 83.6 kg KClO₃
(d) 22.4 mg KClO₃

33. For each of the reactions, calculate how many grams of the product form when 2.4 g of the reactant in color completely reacts. Assume there is more than enough of the other reactant.
(a) $2Na(s) + Cl_2(g) \longrightarrow 2NaCl(s)$
(b) $CaO(s) + CO_2(g) \longrightarrow CaCO_3(s)$
(c) $2Mg(s) + O_2(g) \longrightarrow 2MgO(s)$
(d) $Na_2O(s) + H_2O(l) \longrightarrow 2NaOH(aq)$

34. For each of the reactions, calculate how many grams of the product form when 17.8 g of the reactant in color completely reacts. Assume there is more than enough of the other reactant.
(a) $Ca(s) + Cl_2(g) \longrightarrow CaCl_2(s)$
(b) $2K(s) + Br_2(l) \longrightarrow 2KBr(s)$
(c) $4Cr(s) + 3O_2(g) \longrightarrow 2Cr_2O_3(s)$
(d) $2Sr(s) + O_2(g) \longrightarrow 2SrO(s)$

35. For the reaction shown, calculate how many grams of each product form when the given amount of each reactant completely reacts to form products. Assume there is more than enough of the other reactant.

$$2Al(s) + Fe_2O_3(s) \longrightarrow Al_2O_3(s) + 2Fe(l)$$

(a) 4.7 g Al
(b) 4.7 g Fe₂O₃

36. For the reaction shown, calculate how many grams of each product form when the given amount of each reactant completely reacts to form products. Assume there is more than enough of the other reactant.

$$2HCl(aq) + Na_2CO_3(aq) \longrightarrow 2NaCl(aq) + H_2O(l) + CO_2(g)$$

(a) 10.8 g HCl
(b) 10.8 g Na₂CO₃

37. Consider the balanced equation for the combustion of methane, a component of natural gas:

$$CH_4(g) + 2\,O_2(g) \longrightarrow CO_2(g) + 2\,H_2O(g)$$

Complete the table with the appropriate masses of reactants and products. If the mass of a reactant is provided, fill in the mass of other reactants required to completely react with the given mass, as well as the mass of each product formed. If the mass of a product is provided, fill in the required masses of each reactant to make that amount of product, as well as the mass of the other product that forms.

Mass CH_4	Mass O_2	Mass CO_4	Mass H_2O
___	2.57 g	___	___
22.32 g	___	___	___
___	___	___	11.32 g
___	___	2.94 g	___
3.18 kg	___	___	___
___	___	2.35×10^3 kg	___

38. Consider the balanced equation for the combustion of butane, a fuel often used in lighters:

$$2\,C_4H_{10}(g) + 13\,O_2(g) \longrightarrow 8\,CO_2(g) + 10\,H_2O(g)$$

Complete the table showing the appropriate masses of reactants and products. If the mass of a reactant is provided, fill in the mass of other reactants required to completely react with the given mass, as well as the mass of each product formed. If the mass of a product is provided, fill in the required masses of each reactant to make that amount of product, as well as the mass of the other product that forms.

Mass C_4H_{10}	Mass O_2	Mass CO_2	Mass H_2O
___	1.11 g	___	___
5.22 g	___	___	___
___	___	10.12 g	___
___	___	___	9.04 g
232 mg	___	___	___
___	___	118 mg	___

39. For each acid–base reaction, calculate how many grams of acid are necessary to completely react with and neutralize 2.5 g of the base.
(a) $HCl(aq) + NaOH(aq) \longrightarrow H_2O(l) + NaCl(aq)$
(b) $2\,HNO_3(aq) + Ca(OH)_2(aq) \longrightarrow$
$\qquad\qquad 2\,H_2O(l) + Ca(NO_3)_2(aq)$
(c) $H_2SO_4(aq) + 2\,KOH(aq) \longrightarrow 2\,H_2O(l) + K_2SO_4(aq)$

40. For each precipitation reaction, calculate how many grams of the first reactant are necessary to completely react with 17.3 g of the second reactant.
(a) $2\,KI(aq) + Pb(NO_3)_2(aq) \longrightarrow PbI_2(s) + 2\,KNO_3(aq)$
(b) $Na_2CO_3(aq) + CuCl_2(aq) \longrightarrow CuCO_3(s) + 2\,NaCl(aq)$
(c) $K_2SO_4(aq) + Sr(NO_3)_2(aq) \longrightarrow SrSO_4(s) + 2\,KNO_3(aq)$

41. Sulfuric acid can dissolve aluminum metal according to the reaction:

$$2\,Al(s) + 3\,H_2SO_4(aq) \longrightarrow Al_2(SO_4)_3(aq) + 3\,H_2(g)$$

Suppose you wanted to dissolve an aluminum block with a mass of 22.5 g. What minimum amount of H_2SO_4 in grams do you need? How many grams of H_2 gas will be produced by the complete reaction of the aluminum block?

42. Hydrochloric acid can dissolve solid iron according to the reaction:

$$Fe(s) + 2\,HCl(aq) \longrightarrow FeCl_2(aq) + H_2(g)$$

What minimum mass of HCl in grams dissolves a 2.8-g iron bar on a padlock? How much H_2 is produced by the complete reaction of the iron bar?

LIMITING REACTANT, THEORETICAL YIELD, AND PERCENT YIELD

43. Consider the generic chemical equation:

$$2\,A + 4\,B \longrightarrow 3\,C$$

What is the limiting reactant when each of the initial quantities of A and B is allowed to react?
(a) 2 mol A; 5 mol B
(b) 1.8 mol A; 4 mol B
(c) 3 mol A; 4 mol B
(d) 22 mol A; 40 mol B

44. Consider the generic chemical equation:

$$A + 3\,B \longrightarrow C$$

What is the limiting reactant when each of the initial quantities of A and B is allowed to react?
(a) 1 mol A; 4 mol B
(b) 2 mol A; 3 mol B
(c) 0.5 mol A; 1.6 mol B
(d) 24 mol A; 75 mol B

45. Determine the theoretical yield of C when each of the initial quantities of A and B is allowed to react in the generic reaction:

$$A + 2\,B \longrightarrow 3\,C$$

(a) 1 mol A; 1 mol B
(b) 2 mol A; 2 mol B
(c) 1 mol A; 3 mol B
(d) 32 mol A; 68 mol B

46. Determine the theoretical yield of C when each of the initial quantities of A and B is allowed to react in the generic reaction:

$$2\,A + 3\,B \longrightarrow 2\,C$$

(a) 2 mol A; 4 mol B
(b) 3 mol A; 3 mol B
(c) 5 mol A; 6 mol B
(d) 4 mol A; 5 mol B

47. For the reaction shown, find the limiting reactant for each of the initial quantities of reactants.

$$2 K(s) + Cl_2(g) \longrightarrow 2 KCl(s)$$

(a) 1 mol K; 1 mol Cl_2
(b) 1.8 mol K; 1 mol Cl_2
(c) 2.2 mol K; 1 mol Cl_2
(d) 14.6 mol K; 7.8 mol Cl_2

48. For the reaction shown, find the limiting reactant for each of the initial quantities of reactants.

$$4 Cr(s) + 3 O_2(g) \longrightarrow 2 Cr_2O_3(s)$$

(a) 1 mol Cr; 1 mol O_2
(b) 4 mol Cr; 2.5 mol O_2
(c) 12 mol Cr; 10 mol O_2
(d) 14.8 mol Cr; 10.3 mol O_2

49. For the reaction shown, calculate the theoretical yield of product in moles for each of the initial quantities of reactants.

$$2 Mn(s) + 3 O_2(g) \longrightarrow 2 MnO_3(s)$$

(a) 2 mol Mn; 2 mol O_2
(b) 4.8 mol Mn; 8.5 mol O_2
(c) 0.114 mol Mn; 0.161 mol O_2
(d) 27.5 mol Mn; 43.8 mol O_2

50. For the reaction shown, calculate the theoretical yield of the product in moles for each of the initial quantities of reactants.

$$Ti(s) + 2 Cl_2(g) \longrightarrow TiCl_4(s)$$

(a) 2 mol Ti; 2 mol Cl_2
(b) 5 mol Ti; 9 mol Cl_2
(c) 0.483 mol Ti; 0.911 mol Cl_2
(d) 12.4 mol Ti; 15.8 mol Cl_2

51. Consider the generic reaction between reactants A and B:

$$3 A + 4 B \longrightarrow 2 C$$

If a reaction vessel initially contains 9 mol A and 8 mol B, how many moles of A, B, and C will be in the reaction vessel after the reactants have reacted as much as possible? (Assume 100% actual yield.)

52. Consider the reaction between reactants S and O_2:

$$2 S(s) + 3 O_2(g) \longrightarrow 2 SO_3(g)$$

If a reaction vessel initially contains 5 mol S and 9 mol O_2, how many moles of S, O_2, and SO_3 will be in the reaction vessel after the reactants have reacted as much as possible? (Assume 100% actual yield.)

53. Consider the reaction:

$$4 HCl(g) + O_2(g) \longrightarrow 2 H_2O(g) + 2 Cl_2(g)$$

Each molecular diagram represents an initial mixture of the reactants. How many molecules of Cl_2 are formed by complete reaction in each case? (Assume 100% actual yield.)

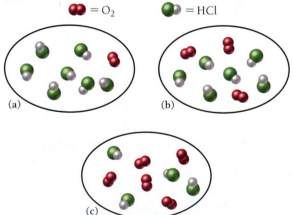

54. Consider the reaction:

$$2 CH_3OH(g) + 3 O_2(g) \longrightarrow 2 CO_2(g) + 4 H_2O(g)$$

Each molecular diagram represents an initial mixture of the reactants. How many CO_2 molecules are formed by complete reaction in each case? (Assume 100% actual yield.)

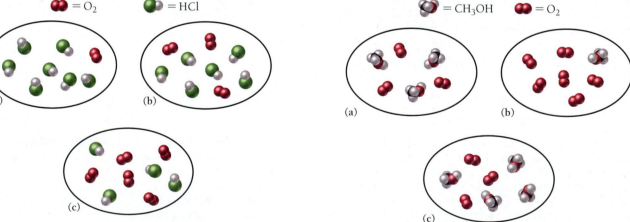

55. For the reaction shown, find the limiting reactant for each of the initial quantities of reactants.

$$2 Li(s) + F_2(g) \longrightarrow 2 LiF(s)$$

(a) 1.0 g Li; 1.0 g F_2
(b) 10.5 g Li; 37.2 g F_2
(c) 2.85×10^3 g Li; 6.79×10^3 g F_2

56. For the reaction shown, find the limiting reactant for each of the initial quantities of reactants.

$$4 Al(s) + 3 O_2(g) \longrightarrow 2 Al_2O_3(s)$$

(a) 1.0 g Al; 1.0 g O_2
(b) 2.2 g Al; 1.8 g O_2
(c) 0.353 g Al; 0.482 g O_2

57. For the reaction shown, calculate the theoretical yield of the product in grams for each of the initial quantities of reactants.

$$2\,Al(s) + 3\,Cl_2(g) \longrightarrow 2\,AlCl_3(s)$$

(a) 1.0 g Al; 1.0 g Cl_2
(b) 5.5 g Al; 19.8 g Cl_2
(c) 0.439 g Al; 2.29 g Cl_2

58. For the reaction shown, calculate the theoretical yield of the product in grams for each of the initial quantities of reactants.

$$Ti(s) + 2\,F_2(g) \longrightarrow TiF_4(s)$$

(a) 1.0 g Ti; 1.0 g F_2
(b) 4.8 g Ti; 3.2 g F_2
(c) 0.388 g Ti; 0.341 g F_2

59. If the theoretical yield of a reaction is 24.8 g and the actual yield is 18.5 g, what is the percent yield?

60. If the theoretical yield of a reaction is 0.118 g and the actual yield is 0.104 g, what is the percent yield?

61. Consider the reaction between calcium oxide and carbon dioxide:

$$CaO(s) + CO_2(g) \longrightarrow CaCO_3(s)$$

A chemist allows 14.4 g of CaO and 13.8 g of CO_2 to react. When the reaction is finished, the chemist collects 19.4 g of $CaCO_3$. Determine the limiting reactant, theoretical yield, and percent yield for the reaction.

62. Consider the reaction between sulfur trioxide and water:

$$SO_3(g) + H_2O(l) \longrightarrow H_2SO_4(aq)$$

A chemist allows 61.5 g of SO_3 and 11.2 g of H_2O to react. When the reaction is finished, the chemist collects 54.9 g of H_2SO_4. Determine the limiting reactant, theoretical yield, and percent yield for the reaction.

63. Consider the reaction between NiS_2 and O_2:

$$2\,NiS_2(s) + 5\,O_2(g) \longrightarrow 2\,NiO(s) + 4\,SO_2(g)$$

When 11.2 g of NiS_2 react with 5.43 g of O_2, 4.86 g of NiO are obtained. Determine the limiting reactant, theoretical yield of NiO, and percent yield for the reaction.

64. Consider the reaction between HCl and O_2:

$$4\,HCl(g) + O_2(g) \longrightarrow 2\,H_2O(l) + 2\,Cl_2(g)$$

When 63.1 g of HCl react with 17.2 g of O_2, 49.3 g of Cl_2 are collected. Determine the limiting reactant, theoretical yield of Cl_2, and percent yield for the reaction.

65. Lead ions can be precipitated from solution with NaCl according to the reaction:

$$Pb^{2+}(aq) + 2\,NaCl(aq) \longrightarrow PbCl_2(s) + 2\,Na^+(aq)$$

When 135.8 g of NaCl are added to a solution containing 195.7 g of Pb^{2+}, a $PbCl_2$ precipitate forms. The precipitate is filtered and dried and found to have a mass of 252.4 g. Determine the limiting reactant, theoretical yield of $PbCl_2$, and percent yield for the reaction.

66. Magnesium oxide can be produced by heating magnesium metal in the presence of oxygen. The balanced equation for the reaction is:

$$2\,Mg(s) + O_2(g) \longrightarrow 2\,MgO(s)$$

When 10.1 g of Mg react with 10.5 g of O_2, 11.9 g of MgO are collected. Determine the limiting reactant, theoretical yield, and percent yield for the reaction.

67. Consider the reaction between TiO_2 and C:

$$TiO_2(s) + 2\,C(s) \longrightarrow Ti(s) + 2\,CO(g)$$

A reaction vessel initially contains 10.0 g of each of the reactants. Calculate the masses of TiO_2, C, Ti, and CO that will be in the reaction vessel after the reactants have reacted as much as possible. Assume 100% yield. *Hint:* The limiting reactant is completely consumed, but the reactant in excess is not. Use the amount of limiting reactant to determine the amount of products that form and the amount of the reactant in excess that remains after complete reaction.

68. Consider the reaction between N_2H_4 and N_2O_4:

$$2\,N_2H_4(g) + N_2O_4(g) \longrightarrow 3\,N_2(g) + 4\,H_2O(g)$$

A reaction vessel initially contains 27.5 g N_2H_4 and 74.9 g of N_2O_4. Calculate the masses of N_2H_4, N_2O_4, N_2, and H_2O that will be in the reaction vessel after the reactants have reacted as much as possible. Assume 100% yield. *Hint:* The limiting reactant is completely consumed, but the reactant in excess is not. Use the amount of limiting reactant to determine the amount of products that form and the amount of the reactant in excess that remains after complete reaction.

ENTHALPY AND STOICHIOMETRY OF ΔH_{rxn}

69. Classify each process as exothermic or endothermic and indicate the sign of ΔH_{rxn}.
(a) butane gas burning in a lighter
(b) the reaction that occurs in the chemical cold packs used to ice athletic injuries
(c) the burning of wax in a candle

70. Classify each process as exothermic or endothermic and indicate the sign of ΔH_{rxn}.
(a) ice melting
(b) a sparkler burning
(c) acetone evaporating from skin

71. Consider the generic reaction:

$$A + 2B \longrightarrow C \quad \Delta H_{rxn} = -55 \text{ kJ}$$

Determine the amount of heat emitted when each amount of reactant completely reacts (assume that there is more than enough of the other reactant).
(a) 1 mol A
(b) 2 mol A
(c) 1 mol B
(d) 2 mol B

72. Consider the generic reaction:

$$2A + 3B \longrightarrow C \quad \Delta H_{rxn} = -125 \text{ kJ}$$

Determine the amount of heat emitted when each amount of reactant completely reacts (assume that there is more than enough of the other reactant).
(a) 2 mol A
(b) 3 mol A
(c) 3 mol B
(d) 5 mol B

73. Consider the equation for the combustion of acetone (C_3H_6O), the main ingredient in nail polish remover:

$$C_3H_6O(l) + 4O_2(g) \longrightarrow 3CO_2(g) + 3H_2O(g)$$
$$\Delta H_{rxn} = -1790 \text{ kJ}$$

If a bottle of nail polish remover contains 155 g of acetone, how much heat is released by its complete combustion?

74. The equation for the combustion of CH_4 (the main component of natural gas) is shown below. How much heat is produced by the complete combustion of 237 g of CH_4?

$$CH_4(g) + 2O_2(g) \longrightarrow CO_2(g) + 2H_2O(g)$$
$$\Delta H_{rxn} = -802.3 \text{ kJ}$$

75. Octane (C_8H_{18}) is a component of gasoline that burns according to the equation:

$$C_8H_{18}(l) + \tfrac{25}{2}O_2(g) \longrightarrow 8CO_2(g) + 9H_2O(g)$$
$$\Delta H_{rxn} = -5074.1 \text{ kJ}$$

What mass of octane (in g) is required to produce 1.55×10^3 kJ of heat?

76. The evaporation of water is endothermic:

$$H_2O(l) \longrightarrow H_2O(g) \quad \Delta H_{rxn} = +44.01 \text{ kJ}$$

What minimum mass of water (in g) has to evaporate to absorb 175 kJ of heat?

Cumulative Problems

77. Consider the reaction:

$$2N_2(g) + 5O_2(g) + 2H_2O(g) \longrightarrow 4HNO_3(g)$$

If a reaction mixture contains 28 g of N_2, 150 g of O_2, and 36 g of H_2O, what is the limiting reactant? (Try to do this problem in your head without any written calculations.)

78. Consider the reaction:

$$2CO(g) + O_2(g) \longrightarrow 2CO_2(g)$$

If a reaction mixture contains 28 g of CO and 32 g of O_2, what is the limiting reactant? (Try to do this problem in your head without any written calculations.)

79. A solution contains an unknown mass of dissolved barium ions. When sodium sulfate is added to the solution, a white precipitate forms. The precipitate is filtered and dried and found to have a mass of 258 mg. What mass of barium was in the original solution? (Assume that all of the barium was precipitated out of solution by the reaction.)

80. A solution contains an unknown mass of dissolved silver ions. When potassium chloride is added to the solution, a white precipitate forms. The precipitate is filtered and dried and found to have a mass of 212 mg. What mass of silver was in the original solution? (Assume that all of the silver was precipitated out of solution by the reaction.)

81. Sodium bicarbonate is often used as an antacid to neutralize excess hydrochloric acid in an upset stomach. How much hydrochloric acid in grams can be neutralized by 3.5 g of sodium bicarbonate? (*Hint:* Begin by writing a balanced equation for the reaction between aqueous sodium bicarbonate and aqueous hydrochloric acid.)

82. Toilet bowl cleaners often contain hydrochloric acid to dissolve the calcium carbonate deposits that accumulate within a toilet bowl. How much calcium carbonate in grams can be dissolved by 5.8 g of HCl? (*Hint:* Begin by writing a balanced equation for the reaction between hydrochloric acid and calcium carbonate.)

83. The combustion of gasoline produces carbon dioxide and water. Assume gasoline to be pure octane (C_8H_{18}) and calculate how many kilograms of carbon dioxide are added to the atmosphere per 1.0 kg of octane burned. (*Hint:* Begin by writing a balanced equation for the combustion reaction.)

84. Many home barbecues are fueled with propane gas (C_3H_8). How much carbon dioxide in kilograms is produced upon the complete combustion of 18.9 L of propane (approximate contents of one 5-gal tank)? Assume that the density of the liquid propane in the tank is 0.621 g/mL. (*Hint:* Begin by writing a balanced equation for the combustion reaction.)

85. A hard-water solution contains 4.8 g of calcium chloride. How much sodium phosphate in grams should be added to the solution to completely precipitate all of the calcium?

86. Magnesium ions can be precipitated from seawater by the addition of sodium hydroxide. How much sodium hydroxide in grams must be added to a sample of seawater to completely precipitate the 88.4 mg of magnesium present?

87. Hydrogen gas can be prepared in the laboratory by a single-displacement reaction in which solid zinc reacts with hydrochloric acid. How much zinc in grams is required to make 14.5 g of hydrogen gas through this reaction?

88. Sodium peroxide (Na_2O_2) reacts with water to form sodium hydroxide and oxygen gas. Write a balanced equation for the reaction and determine how much oxygen in grams is formed by the complete reaction of 35.23 g of Na_2O_2.

89. Ammonium nitrate reacts explosively upon heating to form nitrogen gas, oxygen gas, and gaseous water. Write a balanced equation for this reaction and determine how much oxygen in grams is produced by the complete reaction of 1.00 kg of ammonium nitrate.

90. Pure oxygen gas can be prepared in the laboratory by the decomposition of solid potassium chlorate to form solid potassium chloride and oxygen gas. How much oxygen gas in grams can be prepared from 45.8 g of potassium chlorate?

91. Aspirin can be made in the laboratory by reacting acetic anhydride ($C_4H_6O_3$) with salicylic acid ($C_7H_6O_3$) to form aspirin ($C_9H_8O_4$) and acetic acid ($C_2H_4O_2$). The balanced equation is:

$$C_4H_6O_3 + C_7H_6O_3 \longrightarrow C_9H_8O_4 + C_2H_4O_2$$

In a laboratory synthesis, a student begins with 5.00 mL of acetic anhydride (density = 1.08 g/mL) and 2.08 g of salicylic acid. Once the reaction is complete, the student collects 2.01 g of aspirin. Determine the limiting reactant, theoretical yield of aspirin, and percent yield for the reaction.

92. The combustion of liquid ethanol (C_2H_5OH) produces carbon dioxide and water. After 3.8 mL of ethanol (density = 0.789 g/mL) is allowed to burn in the presence of 12.5 g of oxygen gas, 3.10 mL of water (density = 1.00 g/mL) is collected. Determine the limiting reactant, theoretical yield of H_2O, and percent yield for the reaction. (*Hint*: Write a balanced equation for the combustion of ethanol.)

93. Urea (CH_4N_2O), a common fertilizer, can be synthesized by the reaction of ammonia (NH_3) with carbon dioxide:

$$2\,NH_3(aq) + CO_2(aq) \longrightarrow CH_4N_2O(aq) + H_2O(l)$$

An industrial synthesis of urea produces 87.5 kg of urea upon reaction of 68.2 kg of ammonia with 105 kg of carbon dioxide. Determine the limiting reactant, theoretical yield of urea, and percent yield for the reaction.

94. Silicon, which occurs in nature as SiO_2, is the material from which most computer chips are made. If SiO_2 is heated until it melts into a liquid, it reacts with solid carbon to form liquid silicon and carbon monoxide gas. In an industrial preparation of silicon, 52.8 kg of SiO_2 reacts with 25.8 kg of carbon to produce 22.4 kg of silicon. Determine the limiting reactant, theoretical yield, and percent yield for the reaction.

95. The ingestion of lead from food, water, or other environmental sources can cause lead poisoning, a serious condition that affects the central nervous system, causing symptoms such as distractibility, lethargy, and loss of motor function. Lead poisoning is treated with chelating agents, substances that bind to lead and allow it to be eliminated in the urine. A modern chelating agent used for this purpose is succimer ($C_4H_6O_4S_2$). Suppose you are trying to determine the appropriate dose for succimer treatment of lead poisoning. Assume that a patient's blood lead levels are 0.550 mg/L, that total blood volume is 5.0 L, and that 1 mol of succimer binds 1 mol of lead. What minimum mass of succimer in milligrams is needed to bind all of the lead in this patient's bloodstream?

96. An emergency breathing apparatus placed in mines or caves works via the chemical reaction:

$$4\,KO_2(s) + 2\,CO_2(g) \longrightarrow 2\,K_2CO_3(s) + 3\,O_2(g)$$

If the oxygen supply becomes limited or if the air becomes poisoned, a worker can use the apparatus to breathe while exiting the mine. Notice that the reaction produces O_2, which can be breathed, and absorbs CO_2, a product of respiration. What minimum amount of KO_2 is required for the apparatus to produce enough oxygen to allow the user 15 minutes to exit the mine in an emergency? Assume that an adult consumes approximately 4.4 g of oxygen in 15 minutes of normal breathing.

97. The propane fuel (C_3H_8) used in gas barbecues burns according to the equation:

$$C_3H_8(g) + 5\,O_2(g) \longrightarrow 3\,CO_2(g) + 4\,H_2O(g)$$
$$\Delta H_{rxn} = -2044\ \text{kJ}$$

If a pork roast must absorb 1.6×10^3 kJ to fully cook, and if only 10% of the heat produced by the barbecue is actually absorbed by the roast, what mass of CO_2 is emitted into the atmosphere during the grilling of the pork roast?

98. Charcoal is primarily carbon. Determine the mass of CO_2 produced by burning enough carbon to produce 5.00×10^2 kJ of heat.

$$C(s) + O_2(g) \longrightarrow CO_2(g) \qquad \Delta H_{rxn} = -393.5\ \text{kJ}$$

Highlight Problems

99. A loud classroom demonstration involves igniting a hydrogen-filled balloon. The hydrogen within the balloon reacts explosively with oxygen in the air to form water according to this reaction:

$$2 H_2(g) + O_2(g) \longrightarrow 2 H_2O(g)$$

If the balloon is filled with a mixture of hydrogen and oxygen, the explosion is even louder than if the balloon is filled with only hydrogen; the intensity of the explosion depends on the relative amounts of oxygen and hydrogen within the balloon. Consider the molecular views representing different amounts of hydrogen and oxygen in four different balloons. Based on the balanced chemical equation, which balloon will make the loudest explosion?

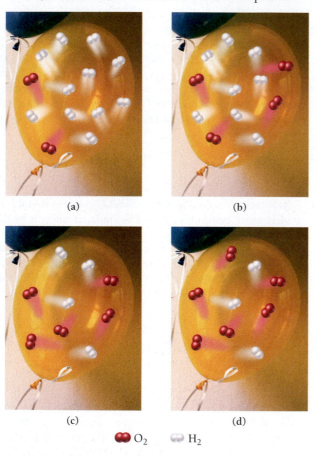

O_2 H_2

100. A hydrochloric acid solution will neutralize a sodium hydroxide solution. Consider the molecular views showing one beaker of HCl and four beakers of NaOH. Which NaOH beaker will just neutralize the HCl beaker? Begin by writing a balanced chemical equation for the neutralization reaction.

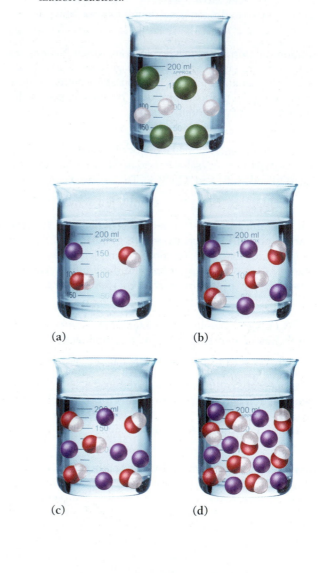

101. Scientists have grown progressively more worried about the potential for climate change caused by increasing atmospheric carbon dioxide levels. The world burns the fossil fuel equivalent of approximately 9.0×10^{12} kg of petroleum per year. Assume that all of this fossil fuel is in the form of octane (C_8H_{18}) and calculate how much CO_2 in kilograms is produced by world fossil fuel combustion per year. (*Hint*: Begin by writing a balanced equation for the combustion of octane.) If the atmosphere currently contains approximately 3.0×10^{15} kg of CO_2, how long will it take for the world's fossil fuel combustion to double the amount of atmospheric carbon dioxide?

102. Lakes that have been acidified by acid rain can be neutralized by the addition of limestone ($CaCO_3$). How much limestone in kilograms would be required to completely neutralize a 5.2×10^9-L lake containing 5.0×10^{-3} g of H_2SO_4 per liter?

Questions for Group Work

Discuss these questions with the group and record your consensus answer.

103. What volume of air is needed to burn an entire 55-L (approximately 15-gal) tank of gasoline? Assume that the gasoline is pure octane, C_8H_{18}. *Hint*: Air is 20% oxygen, 1 mol of a gas occupies about 25 L at room temperature, and the density of octane is 0.70 g/cm^3.

104. Have each member of your group choose a precipitation reaction from Chapter 7 and write a limiting reagent problem based on it. Provide the masses of reactants and the product. Trade problems within your group and solve them by determining the limiting reagent, the theoretical yield, and the percent yield.

105. Consider the combustion of propane:

$$C_3H_8(g) + O_2(g) \longrightarrow CO_2(g) + H_2O(g)$$

(a) Balance the reaction.
(b) Divide all coefficients by the coefficient on propane, so that you have the reaction for the combustion of 1 mol of propane.
(c) ΔH_{rxn} for the combustion of one mole of propane is -2219 kJ. What mass of propane would you need to burn to generate 5.0 MJ of heat?
(d) If propane costs about \$0.67/L and has a density of 2.01 g/cm^3, how much would it cost to generate 5.0 MJ of heat by burning propane?

Data Interpretation and Analysis

106. A chemical reaction in which reactants A and B form the product C is studied in the laboratory. The researcher carries out the reaction with differing relative amounts of reactants and measures the amount of product produced. Examine the given tabulated data from the experiment and answer the questions.

(a) For which experiments is A the limiting reactant?
(b) For which experiments is B the limiting reactant?
(c) The molar mass of A is 50.0 g/mol, and the molar mass of B is 75.0 g/mol. What are the coefficients of A and B in the balanced chemical equation?
(d) The molar mass of C is 88.0 g/mol. What is the coefficient of C in the balanced chemical equation?
(e) Calculate an average percent yield for the reaction.

Experiment #	Mass A (g)	Mass B (g)	Mass C Obtalned (g)
1	2.51	7.54	3.76
2	5.03	7.51	7.43
3	7.55	7.52	11.13
4	12.53	7.49	14.84
5	15.04	7.47	14.94
6	19.98	7.51	15.17
7	20.04	9.95	19.31
8	20.02	12.55	24.69

Answers to Skillbuilder Exercises

Skillbuilder 8.1................. 49.2 mol H_2O
Skillbuilder 8.2................. 6.88 g HCl
Skillbuilder 8.3................. 4.0×10^3 kg H_2SO_4
Skillbuilder 8.4................. Limiting reactant is Na; theoretical yield is 4.8 mol of NaF
Skillbuilder 8.5................. 30.7 g NH_3
Skillbuilder Plus, p. 263... 29.4 kg NH_3

Skillbuilder 8.6................. Limiting reactant is CO; theoretical yield = 127 g Fe; percent yield = 68.8%
Skillbuilder 8.7................. -2.06×10^3 kJ
Skillbuilder Plus, p. 267... 33 g C_4H_{10} necessary; 99 g CO_2 produced

Answers to Conceptual Checkpoints

8.1 (a) Because the reaction requires two O_2 molecules to react with 1 CH_4 molecule, and there are four O_2 molecules available to react, two CH_4 molecules are required for complete reaction.

8.2 (c) One mole of A produces three moles of C. If the molar mass of C is three times the molar mass of A, then the mass of C produced is six times the mass of A that reacts.

8.3 (c) Hydrogen is the limiting reactant. The reaction mixture contains three H_2 molecules; therefore, two NH_3 molecules will form when the reactants have reacted as completely as possible. Nitrogen is in excess, and there is one leftover nitrogen molecule.

8.4 B is the limiting reactant because since B has a higher molar mass, a reaction mixture containing equal masses of both A and B has fewer moles of B. Since two moles of B are required to react with one mole of A and since fewer moles of B are present, B is the limiting reactant.

8.5 (d) NO is the limiting reactant. The reaction mixture initially contains 4 mol NO; therefore, 10 mol of H_2O will be consumed, leaving 5 mol H_2 unreacted. The products will be 4 mol NH_3 and 4 mol H_2O.

8.6 (c) B is the limiting reactant. If 4 mol B react, then 200 kJ of heat is produced.

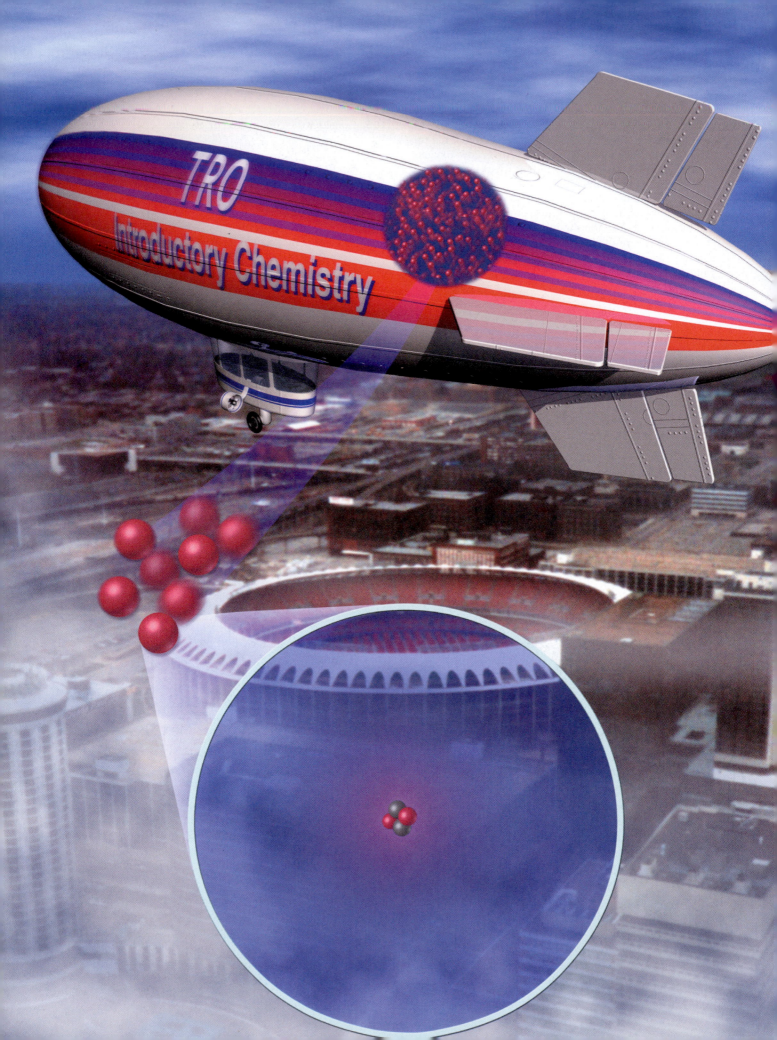

9 Electrons in Atoms and the Periodic Table

Anyone who is not shocked by quantum mechanics has not understood it.
—Niels Bohr (1885–1962)

9.1 Blimps, Balloons, and Models of the Atom

▲ The *Hindenburg* was filled with hydrogen, a reactive and flammable gas.
Question: What makes hydrogen reactive?

You have probably seen one of the Goodyear blimps floating in the sky. A Goodyear blimp is often present at championship sporting events such as the Rose Bowl, the Indy 500, or the U.S. Open golf tournament. The blimp's inherent stability allows cameras to provide spectacular views of the world below for television and film.

The Goodyear blimp is similar to a large balloon. Unlike airplanes, which must be moving fast to stay in flight, a blimp or *airship* floats in air because it is filled with a gas that is less dense than air. The Goodyear blimp is filled with helium. Other airships in history, however, have used hydrogen for buoyancy. For example, the *Hindenburg*—the largest airship ever constructed—was filled with hydrogen. Hydrogen, a reactive and flammable gas, turned out to be a poor choice. On May 6, 1937, while landing in New Jersey after its first transatlantic crossing, the *Hindenburg* burst into flames. The fire destroyed the airship, killing 36 of its 97 passengers. Apparently, as the *Hindenburg* was landing, leaking hydrogen gas ignited, resulting in the explosion that destroyed the ship. A similar accident cannot happen to the Goodyear blimp because it uses helium—an inert gas—for buoyancy. A spark or even a flame would actually be *extinguished* by helium.

Why is helium inert? What is it about helium *atoms* that makes helium *gas* inert? By contrast, why is hydrogen so reactive? Recall from Chapter 5 that elemental hydrogen exists as a diatomic element. Hydrogen atoms are so reactive that they react with each other to form hydrogen molecules. What is it about hydrogen atoms that makes them so reactive? What is the difference between hydrogen and helium that accounts for their different reactivities?

◀ Modern blimps are filled with helium, an inert gas. The nucleus of the helium atom (inset) has two protons, so the neutral helium atom has two electrons—a highly stable configuration. In this chapter, we learn about models that explain the inertness of helium and the reactivity of other elements.

When we examine the properties of hydrogen and helium, we make observations about nature. Mendeleev's periodic law, first discussed in Chapter 4, summarizes the results of many similar observations on the properties of elements:

When the elements are arranged in order of increasing atomic number, certain sets of properties recur periodically.

The reactivity exhibited by hydrogen (the first element in Group 1A) is also seen in other Group 1A elements, such as lithium and sodium. Likewise, the inertness of helium (a Group 8A element) is seen in other Group 8A elements such as neon and argon and the other noble gases. In this chapter, we consider models that help explain the observed behaviors of groups of elements such as the Group 1A metals and the noble gases. We examine two important models in particular—the **Bohr model** and the **quantum-mechanical model**—that propose explanations for the periodic law. These models explain how electrons exist in atoms and how those electrons affect the chemical and physical properties of elements.

We have already learned much about the behavior of elements in this book. We know, for example, that sodium tends to form 1+ ions and that fluorine tends to form 1− ions. We know that some elements are metals and that others are nonmetals. And we know that the noble gases are, in general, chemically inert and that the alkali metals are chemically reactive. But we have not yet explored *why*. The models in this chapter explain why.

When the Bohr model and the quantum-mechanical model were developed in the early 1900s, they caused a revolution in the physical sciences, changing our fundamental view of matter at its most basic level. The scientists who devised these models—including Niels Bohr, Erwin Schrödinger, and Albert Einstein—were bewildered by their own discoveries. Bohr claimed, "Anyone who is not shocked by quantum mechanics has not understood it." Schrödinger lamented, "I don't like it, and I am sorry I ever had anything to do with it." Einstein disbelieved it, insisting that "God does not play dice with the universe." However, the quantum-mechanical model has such explanatory power that it is rarely questioned today. It forms the basis of the modern periodic table and our understanding of chemical bonding. Its applications include lasers, computers, and semiconductor devices, and it has led us to discover new ways to design drugs that cure disease. The quantum-mechanical model for the atom is, in many ways, the foundation of modern chemistry.

> The periodic law stated here is a modification of Mendeleev's original formulation. Mendeleev listed elements in order of increasing *mass*; today we list them in order of increasing *atomic number*.

Alkali metals

Noble gases

1 1A	18 8A
1 H	
3 Li	2 He
11 Na	10 Ne
19 K	18 Ar
37 Rb	36 Kr
55 Cs	54 Xe
87 Fr	86 Rn

▲ The noble gases are chemically inert, and the alkali metals are chemically reactive. Why? (Hydrogen is a Group 1 element, but it is not considered an alkali metal.)

▲ Niels Bohr (left) and Erwin Schrödinger (right), along with Albert Einstein, played a role in the development of quantum mechanics, yet they were bewildered by their own findings.

9.2 Light: Electromagnetic Radiation

▶ Understand and explain the nature of electromagnetic radiation.

Before we explore models of the atom, we must understand a few things about light, because observations of the interaction of light with atoms helped to shape these models. Light is familiar to all of us—we see the world by it—but what is light? Unlike most of what we have encountered so far in this book, light is not matter—it has no mass. Light is a form of **electromagnetic radiation**, a

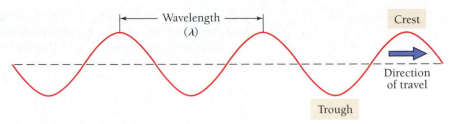

▲ When a water surface is disturbed, waves are created that radiate outward from the site.

▲ FIGURE 9.1 **Wavelength**   The wavelength of light (λ) is the distance between adjacent wave crests.

type of energy that travels through space at a constant speed of 3.0×10^8 m/s (186,000 mi/s). At this speed, a flash of light generated at the equator can travel around the world in one-seventh of a second. This extremely fast speed is part of the reason that we *see* a firework in the sky before we *hear* the sound of its explosion. The light from the exploding firework reaches our eyes almost instantaneously. The sound, traveling much more slowly, takes longer.

Before the advent of quantum mechanics, light was described exclusively as a wave of electromagnetic energy traveling through space. You are probably familiar with water waves (think of the waves created by a rock dropped into a still pond), or you may have created a wave on a rope by moving the end of the rope up and down in a quick motion. In either case, the wave carries energy as it moves through the water or along the rope.

> The Greek letter *lambda* (λ) is pronounced "lam-duh."

Waves are generally characterized by **wavelength** (λ), the distance between adjacent wave crests (▲ FIGURE 9.1). For visible light, wavelength determines color. For example, orange light has a longer wavelength than blue light. White light, as produced by the sun or by a light bulb, contains a spectrum of wavelengths and therefore a spectrum of color. We see these colors—red, orange, yellow, green, blue, indigo, and violet—in a rainbow or when white light is passed through a prism (▼ FIGURE 9.2). Red light, with a wavelength of 750 nm (nanometers), has the longest wavelength of visible light. Violet light, with a wavelength of 400 nm, has the shortest ($1\,\text{nm} = 10^{-9}$ m). The presence of color in white light is responsible for the colors we see in our everyday vision. For example, a red shirt is red because it reflects red light (▼ FIGURE 9.3). Our eyes see only the reflected light, making the shirt appear red.

> Helpful mnemonic: ROY G BIV—Red, Orange, Yellow, Green, Blue, Indigo, Violet

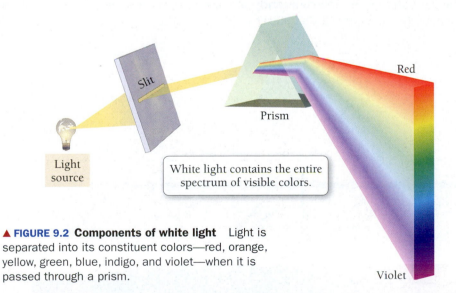

White light contains the entire spectrum of visible colors.

▲ FIGURE 9.2 **Components of white light**   Light is separated into its constituent colors—red, orange, yellow, green, blue, indigo, and violet—when it is passed through a prism.

▲ FIGURE 9.3 **Color in objects**   A red shirt appears red because it absorbs all colors except red, which it reflects.

> The Greek letter *nu* (ν) is pronounced "noo."

Light waves are also often characterized by **frequency** (ν), the number of cycles or crests that pass through a stationary point in one second. Wavelength and frequency are inversely related—the shorter the wavelength, the higher the frequency. Blue light, for example, has a higher frequency than red light.

In the early twentieth century, scientists such as Albert Einstein discovered that the results of certain experiments could be explained only by describing light, not as waves, but as particles. In this description, the light leaving a flashlight, for example, is viewed as a stream of particles. A particle of light is called a **photon**, and we can think of a photon as a single packet of light energy. The amount of energy carried in the packet depends on the wavelength of the light—the shorter the wavelength, the greater the energy. Therefore, violet light (shorter wavelength) carries more energy per photon than red light (longer wavelength).

To summarize:

- Electromagnetic radiation is a form of energy that travels through space at a constant speed of 3.0×10^8 m/s and can exhibit wavelike or particle-like properties.
- The wavelength of electromagnetic radiation determines the amount of energy carried by one of its photons. The shorter the wavelength, the greater the energy of each photon.
- The frequency and energy of electromagnetic radiation are inversely related to its wavelength.

PEARSON
eText
2.0

CONCEPTUAL ✔ **CHECKPOINT 9.1**

Which wavelength of light has the highest frequency?

(a) 350 nm   (b) 500 nm   (c) 750 nm

9.3 The Electromagnetic Spectrum

▶ Predict the relative wavelength, energy, and frequency of different types of light.

Electromagnetic radiation ranges in wavelength from 10^{-16} m (gamma rays) to 10^6 m (radio waves). Visible light composes only a tiny portion of that range. The entire range of electromagnetic radiation is called the **electromagnetic spectrum**. ▼ FIGURE 9.4 shows the electromagnetic spectrum, with short-wavelength, high-frequency radiation on the right and long-wavelength, low-frequency radiation on the left. Visible light is the small sliver in the middle.

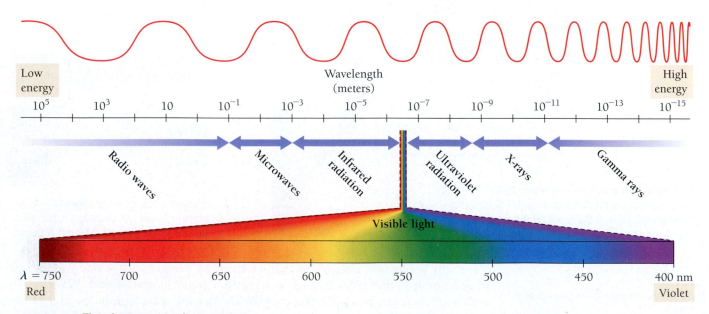

▲ **FIGURE 9.4 The electromagnetic spectrum**

Remember that the energy carried per photon is greater for short-wavelength light than for long-wavelength light. The shortest wavelength (and therefore most energetic) photons are those of **gamma rays**, shown on the far right of Figure 9.4. Gamma rays are produced by the sun, by stars, and by certain unstable atomic nuclei on Earth. Excessive human exposure to gamma rays is dangerous because the high energy of gamma-ray photons can damage biological molecules.

Next on the electromagnetic spectrum (to the left in Figure 9.4), with longer wavelengths (and lower energy) than gamma rays, are **X-rays**, familiar to us from their medical use. X-rays pass through many substances that block visible light and are therefore used to image internal bones and organs. Like gamma-ray photons, X-ray photons carry enough energy to damage biological molecules. While several yearly exposures to X-rays are relatively harmless, excessive exposure to X-rays increases cancer risk.

Sandwiched between X-rays and visible light in the electromagnetic spectrum is **ultraviolet** or **UV light**, most familiar to us as the component of sunlight that produces a sunburn or suntan. Though not as energetic as gamma-ray or X-ray photons, ultraviolet photons still carry enough energy to damage biological molecules. Excessive exposure to ultraviolet light increases the risk of skin cancer and cataracts and causes premature wrinkling of the skin. Next on the spectrum is **visible light**, ranging from violet (shorter wavelength, higher energy) to red (longer wavelength, lower energy). Photons of visible light do not damage biological molecules. They do, however, cause molecules in our eyes to rearrange, which sends a signal to our brains that results in vision.

Infrared light is next, with even longer wavelengths than visible light. The heat we feel when we place a hand near a hot object is infrared light. All warm objects, including human bodies, emit infrared light. While infrared light is invisible to our eyes, infrared sensors can detect it and are often used in night-vision technology to "see" in the dark. In the infrared region of the spectrum, warm objects—such as human bodies—glow, much as a light bulb glows in the visible region of the spectrum.

Beyond infrared light, at longer wavelengths still, are **microwaves**, used for radar and in microwave ovens. Although microwave light has longer wavelengths—and therefore lower energy per photon—than visible or infrared light, it is efficiently absorbed by water and therefore heats substances that contain water. For this reason substances that contain water, such as food, are warmed when placed in a microwave oven, but substances that do not contain water, such as plate and cups, are not.

The longest wavelengths of light are **radio waves**, which are used to transmit the signals used by AM and FM radio, cellular telephones, television, and other forms of communication.

Some dishes contain substances that absorb microwave radiation, but most do not.

Normal photograph

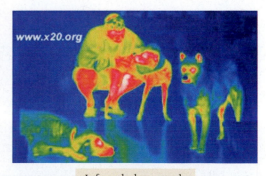

Infrared photograph

▲ Warm objects, such as human or animal bodies, give off infrared light that is easily detected with an infrared camera. In the infrared photograph, the warmest areas appear as red and the coolest as dark blue.

CHEMISTRY AND HEALTH

Radiation Treatment for Cancer

X-rays and gamma rays are sometimes called ionizing radiation because the high energy in their photons can ionize atoms and molecules. When ionizing radiation interacts with biological molecules, it can permanently change or even destroy them. Consequently we normally try to limit our exposure to ionizing radiation. However, doctors can use ionizing radiation to destroy molecules within unwanted cells such as cancer cells.

In radiation therapy (or radiotherapy), medical professionals aim X-ray or gamma-ray beams at cancerous tumors. The ionizing radiation damages the molecules within the tumor's cells that carry genetic information—information necessary for the cell to grow and divide—and the cells die or stop dividing. Ionizing radiation also damages molecules within healthy cells; however, cancerous cells divide more quickly than healthy cells, making cancerous cells more susceptible to genetic damage. Nonetheless, healthy cells often inadvertently sustain damage during treatments, resulting in side effects for patients such as fatigue, skin lesions, and hair loss. Doctors try to minimize the exposure of healthy cells by appropriate shielding and by targeting the tumor from multiple directions, minimizing the exposure of healthy cells while maximizing the exposure of cancerous cells (▶ FIGURE 9.5).

Another side effect of exposing healthy cells to radiation is that they too may become cancerous. In this way, a treatment for cancer may cause cancer. So why do we continue to use it? Radiation therapy, as most other disease therapies, has associated risks. However, we take risks all the time, many for lesser reasons. For example, every time we drive a car, we risk injury or even death. Why? Because we perceive the benefit—such as getting to the grocery store to buy food—to be worth the risk. The situation is similar in cancer therapy or any other therapy for that matter. The benefit of cancer therapy (possibly curing a cancer that will certainly kill) is worth the risk (a slight increase in the chance of developing a future cancer).

B9.1 CAN YOU ANSWER THIS? *Why is visible light not used to destroy cancerous tumors?*

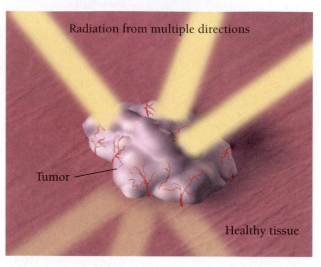

▲ **FIGURE 9.5 Radiation therapy** By targeting the tumor from various different directions, radiologists attempt to limit damage to healthy tissue.

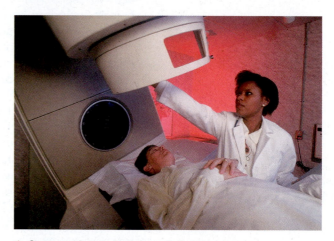

▲ Cancer patient undergoing radiation therapy.

EXAMPLE **9.1**   **Wavelength, Energy, and Frequency**

Arrange these three types of electromagnetic radiation—visible light, X-rays, and microwaves—in order of increasing:

(a) wavelength   **(b)** frequency   **(c)** energy per photon

	SOLUTION
(a) wavelength Figure 9.4 indicates that X-rays have the shortest wavelength, followed by visible light and then microwaves.	X-rays, visible light, microwaves
(b) frequency Since frequency and wavelength are inversely proportional—the longer the wavelength, the shorter the frequency—the ordering with respect to frequency is exactly the reverse of the ordering with respect to wavelength.	microwaves, visible light, X-rays

(c) energy per photon Energy per photon decreases with increasing wavelength but increases with increasing frequency; therefore, the ordering with respect to energy per photon is the same as frequency.	microwaves, visible light, X-rays

▶ **SKILLBUILDER 9.1 | Wavelength, Energy, and Frequency**

Arrange these colors of visible light—green, red, and blue—in order of increasing:

(a) wavelength   **(b)** frequency   **(c)** energy per photon

▶ **FOR MORE PRACTICE**   Example 9.9; Problems 31, 32, 33, 34, 35, 36, 37, 38.

CONCEPTUAL ✔ CHECKPOINT **9.2**

Yellow light has a longer wavelength than violet light. Therefore:

(a) Yellow light has more energy per photon than violet light.

(b) Yellow light has less energy per photon than violet light.

(c) Both yellow light and violet light have the same energy per photon.

9.4   The Bohr Model: Atoms with Orbits

▶ Understand and explain the key characteristics of the Bohr model of the atom.

When an atom absorbs energy—in the form of heat, light, or electricity—it often reemits that energy as light. For example, a neon sign is composed of one or more glass tubes filled with gaseous neon atoms. When an electrical current passes through the tube, the neon atoms absorb some of the electrical energy and reemit it as the familiar red light of a neon sign (◀ **FIGURE 9.6**).

The absorption and emission of light by atoms is due to the interaction of the light with the electrons in the atom. If the atoms in the tube are different, the electrons have different energies, and the emitted light is a different color. In other words, atoms of each unique element emit light of a unique color (or unique wavelength). Mercury atoms, for example, emit light that appears blue, hydrogen atoms emit light that appears pink (▼ **FIGURE 9.7**), and helium atoms emit light that appears yellow-orange.

▲ **FIGURE 9.6 A neon sign**   Neon atoms inside a glass tube absorb electrical energy and reemit the energy as light.

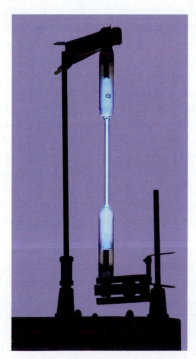
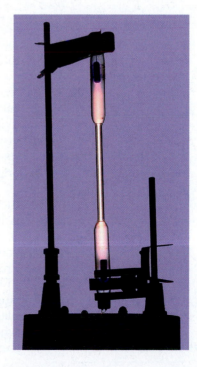

▶ **FIGURE 9.7 Light emission by different elements**   Light emitted from a mercury lamp (left) appears blue, and light emitted from a hydrogen lamp (right) appears pink.

▶ **FIGURE 9.8 Emission spectra**
A white-light spectrum is continuous, with some radiation emitted at every wavelength. The emission spectrum of an individual element, however, includes only certain specific wavelengths. (The different wavelengths appear as *lines* because the light from the source passes through a slit before entering the prism.) Each element produces its own unique and distinctive emission spectrum.

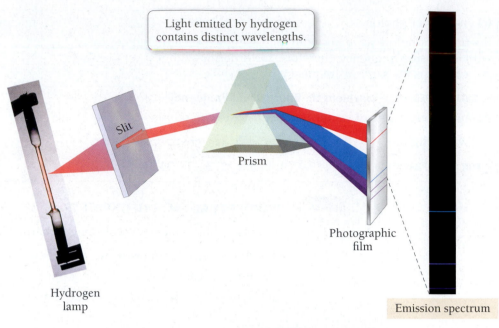

Light emitted by hydrogen contains distinct wavelengths.

Slit

Prism

Photographic film

Hydrogen lamp

Emission spectrum

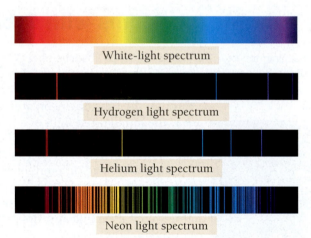

White-light spectrum

Hydrogen light spectrum

Helium light spectrum

Neon light spectrum

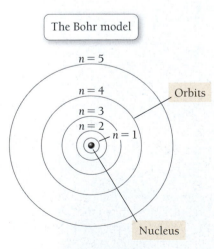

The Bohr model

$n = 5$

$n = 4$

$n = 3$

$n = 2$

$n = 1$

Orbits

Nucleus

▲ **FIGURE 9.9 Bohr orbits**

Closer inspection of the light emitted by hydrogen, helium, and neon atoms reveals that the light contains several distinct colors or wavelengths. Just as we can separate the white light from a light bulb into its constituent wavelengths by passing it through a prism, so we can separate the light emitted by glowing hydrogen, helium, or neon into its constituent wavelengths (▲ **FIGURE 9.8**) by passing it through a prism. The result is an **emission spectrum**. Notice the differences between a white-light spectrum and the emission spectra of hydrogen, helium, and neon. The white-light spectrum is *continuous*, meaning that the light intensity is uninterrupted or smooth across the entire visible range—there is some radiation at all wavelengths, with no gaps. The emission spectra of hydrogen, helium, and neon, however, are not continuous. They consist of bright spots or lines at specific wavelengths with complete darkness in between.

Since the emission of light in atoms is related to the motion of electrons within the atoms, a model for how electrons exist in atoms must account for these spectra. A major challenge in developing a model for electrons in atoms is the discrete or bright-line nature of the emission spectra. Why do atoms, when excited with energy, emit light only at particular wavelengths? Why do they *not* emit a continuous spectrum? Niels Bohr developed a simple model to explain these results. In his model, now called the Bohr model, electrons travel around the nucleus in circular orbits that are similar to planetary orbits around the sun. However, unlike planets revolving around the sun—which can theoretically orbit at any distance whatsoever from the sun—electrons in the Bohr model can orbit only at *specific, fixed* distances from the nucleus (◀ **FIGURE 9.9**).

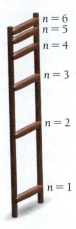

▲ **FIGURE 9.10   The Bohr energy ladder**   Bohr orbits are like steps on a ladder. It is possible to stand on one step or another, but impossible to stand between steps.

The radii of Bohr orbits become increasingly further apart as *n* increases, but the energy levels get increasingly closer together.

The Bohr model is still important because it provides a logical foundation to the quantum-mechanical model and reveals the historical development of scientific understanding.

The *energy* of each Bohr orbit, specified by a **quantum number** $n = 1, 2, 3...$, is also fixed, or **quantized**. The energy of each orbit increases with increasing value of *n*, but the energy levels become more closely spaced as *n* increases. Bohr orbits are like steps of a ladder (◀ **FIGURE 9.10**), each at a specific distance from the nucleus and each at a specific energy. Just as it is impossible to stand *between steps* on a ladder, so it is impossible for an electron to exist *between orbits* in the Bohr model. An electron in an $n = 3$ orbit, for example, is farther from the nucleus and has more energy than an electron in an $n = 2$ orbit. And an electron cannot exist at an intermediate distance or energy between the two orbits—the orbits are *quantized*. As long as an electron remains in a given orbit, it does not absorb or emit light, and its energy remains fixed and constant.

When an atom absorbs energy, an electron in one of these fixed orbits is *excited* or promoted to an orbit that is farther away from the nucleus (▼ **FIGURE 9.11**) and therefore higher in energy (this is analogous to moving up a step on the ladder). However, in this new configuration, the atom is less stable, and the electron quickly falls back or *relaxes* to a lower-energy orbit (this is analogous to moving down a step on the ladder). As it does so, it releases a photon of light containing the precise amount of energy—called a **quantum** of energy—that corresponds to the energy difference between the two orbits.

Since the amount of energy in a photon is directly related to its wavelength, the photon has a specific wavelength. *Consequently, the light emitted by excited atoms consists of specific lines at specific wavelengths, each corresponding to a specific transition between two orbits.* For example, the line at 486 nm in the hydrogen emission spectrum corresponds to an electron relaxing from the $n = 4$ orbit to the $n = 2$ orbit (▼ **FIGURE 9.12**). In the same way, the line at 657 nm (longer wavelength and therefore lower energy) corresponds to an electron relaxing from the $n = 3$ orbit to the $n = 2$ orbit. Notice that transitions between orbits that are closer together produce lower-energy (and therefore longer-wavelength) light than transitions between orbits that are farther apart.

The great success of the Bohr model of the atom was that it predicted the lines of the hydrogen emission spectrum. However, it failed to predict the emission spectra of other elements that contained more than one electron. For this, and other reasons, the Bohr model was replaced with a more sophisticated model called the quantum-mechanical or wave-mechanical model.

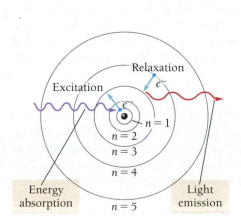

▲ **FIGURE 9.11   Excitation and emission**   When a hydrogen atom absorbs energy, an electron is excited to a higher-energy orbit. The electron then relaxes back to a lower-energy orbit, emitting a photon of light.

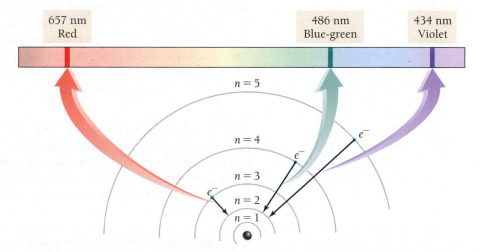

▲ **FIGURE 9.12   Hydrogen emission lines**   The 657-nm line of the hydrogen emission spectrum corresponds to an electron relaxing from the $n = 3$ orbit to the $n = 2$ orbit. The 486-nm line corresponds to an electron relaxing from the $n = 4$ orbit to the $n = 2$ orbit, and the 434-nm line corresponds to an electron relaxing from $n = 5$ to $n = 2$.

To summarize:

- Electrons exist in quantized orbits at specific, fixed energies and specific, fixed distances from the nucleus.
- When energy is put into an atom, electrons are excited to higher-energy orbits.
- When an electron in an atom relaxes (or falls) from a higher-energy orbit to a lower-energy orbit, the atom emits light.
- The energy (and therefore the wavelength) of the emitted light corresponds to the energy difference between the two orbits in the transition. Since these energies are fixed and discrete, the energy (and therefore the wavelength) of the emitted light is fixed and discrete.

CONCEPTUAL ✔ **CHECKPOINT 9.3**

In one transition, an electron in a hydrogen atom falls from the $n = 3$ level to the $n = 2$ level. In a second transition, an electron in a hydrogen atom falls from the $n = 2$ level to the $n = 1$ level. Compared to the radiation emitted by the first of these transitions, the radiation emitted by the second has:

(a) a lower frequency

(b) a smaller energy per photon

(c) a shorter wavelength

(d) a longer wavelength

9.5   The Quantum-Mechanical Model: Atoms with Orbitals

▶ Understand and explain the key characteristics of the quantum-mechanical model of the atom.

The quantum-mechanical model of the atom replaced the Bohr model in the early twentieth century. In the quantum-mechanical model, *Bohr orbits* are replaced with *quantum-mechanical* **orbitals**. Orbitals are different from orbits in that they represent, not specific paths that electrons follow, but probability maps that show a statistical distribution of where the electron is likely to be found. The idea of an orbital is not easy to visualize. Quantum mechanics revolutionized physics and chemistry because in the quantum-mechanical model, electrons *do not* behave like particles flying through space. We cannot, in general, describe their exact paths. An orbital is a probability map that shows where the electron is *likely* to be found when the atom is probed; it does not represent the exact path that an electron takes as it travels through space.

Baseball Paths and Electron Probability Maps

To understand orbitals, let's contrast the behavior of a baseball with that of an electron. Imagine a baseball thrown from the pitcher's mound to a catcher at home plate (▶ FIGURE 9.13). The baseball's path can easily be traced as it travels from the pitcher to the catcher. The catcher can watch the baseball as it travels through the air, and he can predict exactly where the baseball will cross over home plate. He can even place his mitt in the correct place to catch it. *This sequence of events would be impossible for an electron.* Like photons, electrons exhibit a *wave–particle duality*; sometimes they act as particles, and other times as waves. This duality leads to behavior that makes it impossible to trace an electron's path. If an electron were "thrown" from the pitcher's mound to home plate, it would land in a different place every time, *even if it were thrown in exactly the same way.* Baseballs have predictable paths—electrons do not.

▶ **FIGURE 9.13 Baseballs follow predictable paths** A baseball follows a well-defined path as it travels from the pitcher to the catcher.

▲ **FIGURE 9.14 Electrons are unpredictable** To describe the behavior of a "pitched" electron, we would have to construct a probability map of where it would cross home plate.

In the quantum-mechanical world of the electron, the catcher cannot know exactly where the electron would cross the plate for any given throw. He has no way of putting his mitt in the right place to catch it. However, if the catcher were able to keep track of hundreds of electron throws, he could observe a reproducible, statistical pattern of where the electron crosses the plate. He could even draw maps in the strike zone showing the probability of an electron crossing a certain area (◀ **FIGURE 9.14**). These maps are called *probability maps*.

From Orbits to Orbitals

In the Bohr model, an *orbit* is a circular path—analogous to a baseball's path—that shows the electron's motion around an atomic nucleus. In the quantum-mechanical model, an *orbital* is a probability map, analogous to the probability map drawn by our catcher. It shows the relative likelihood of the electron being found at various locations when the atom is probed. Just as the Bohr model has different orbits with different radii, the quantum-mechanical model has different orbitals with different shapes.

9.6 Quantum-Mechanical Orbitals and Electron Configurations

▶ Write electron configurations and orbital diagrams for atoms.

In the Bohr model of the atom, a single quantum number (n) specifies each orbit. In the quantum-mechanical model, a number and a letter specify an orbital (or orbitals). In this section, we examine quantum-mechanical orbitals and electron configurations. An electron configuration is a compact way to specify the occupation of quantum-mechanical orbitals by electrons.

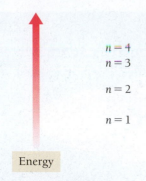

$n = 4$
$n = 3$

$n = 2$

$n = 1$

Energy

▲ **FIGURE 9.15 Principal quantum numbers** The principal quantum numbers ($n = 1, 2, 3 \ldots$) determine the energy of the hydrogen quantum-mechanical orbitals.

This analogy is purely hypothetical. It is impossible to photograph electrons in this way.

Quantum-Mechanical Orbitals

The lowest-energy orbital in the quantum-mechanical model—analogous to the $n = 1$ orbit in the Bohr model—is the *1s orbital*. We specify it by the number 1 and the letter s. The number is the **principal quantum number** (*n*) and specifies the **principal shell** of the orbital. The higher the principal quantum number, the higher the energy of the orbital. The possible principal quantum numbers are $n = 1, 2, 3 \ldots$, with energy increasing as *n* increases (◄ **FIGURE 9.15**). Because the 1s orbital has the lowest possible principal quantum number, it is in the lowest-energy shell and has the lowest possible energy.

The letter indicates the **subshell** of the orbital and specifies its shape. The possible letters are *s*, *p*, *d*, and *f*, and each letter corresponds to a different shape. For example, orbitals within the *s* subshell have a spherical shape. Unlike the $n = 1$ Bohr orbit, which shows the electron's circular path, the 1s quantum-mechanical orbital is a three-dimensional probability map. We sometimes represent orbitals with dots (▼ **FIGURE 9.16**), where the dot density is proportional to the probability of finding the electron.

We can understand the dot representation of an orbital better with another analogy. Imagine you could take a photograph of an electron in an atom every second for 10 or 15 minutes. One second the electron is very close to the nucleus; the next second it is farther away and so on. Each photo shows a dot representing the electron's position relative to the nucleus at that time. If we took hundreds of photos and superimposed all of them, we would have an image like Figure 9.16—a statistical representation of where the electron is found. Notice that the dot density for the 1s orbital is greatest near the nucleus and decreases farther away from the nucleus. This means that the electron is more likely to be found close to the nucleus than far away from it.

Orbitals can also be represented as geometric shapes that encompass most of the volume where the electron is likely to be found. For example, we represent the 1s orbital as a sphere (▼ **FIGURE 9.17**) that encompasses the volume within which the electron is found 90% of the time. If we superimpose the dot representation of the 1s orbital on the shape representation (▼ **FIGURE 9.18**), we can see that most of the dots are within the sphere, indicating that the electron is most likely to be found within the sphere when it is in the 1s orbital.

The single electron of an undisturbed hydrogen atom at room temperature is in the 1s orbital. This is the **ground state**, or lowest energy state, of the hydrogen atom. However, like the Bohr model, the quantum-mechanical model allows electrons to transition to higher-energy orbitals upon the absorption of energy. What are these higher-energy orbitals? What do they look like?

Dot representation
of 1s orbital

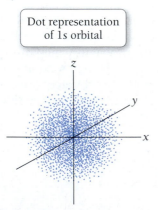

Shape representation
of 1s orbital

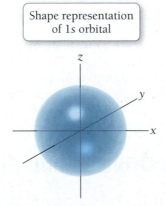

Both representations
of 1s superimposed

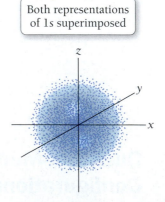

▲ **FIGURE 9.16 1s orbital** The dot density in this plot is proportional to the probability of finding the electron. The greater dot density near the middle indicates a higher probability of finding the electron near the nucleus.

▲ **FIGURE 9.17 Shape representation of the 1s orbital** Because the distribution of electron density around the nucleus in Figure 9.16 is symmetrical—the same in all directions—we can represent the 1s orbital as a sphere.

▲ **FIGURE 9.18 Orbital shape and dot representation for the 1s orbital** The shape representation of the 1s orbital superimposed on the dot density representation. We can see that when the electron is in the 1s orbital, it is most likely to be found within the sphere.

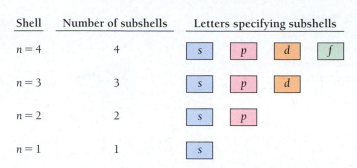

Shell	Number of subshells	Letters specifying subshells			
$n = 4$	4	s	p	d	f
$n = 3$	3	s	p	d	
$n = 2$	2	s	p		
$n = 1$	1	s			

▲ **FIGURE 9.19 Subshells**   The number of subshells in a given principal shell is equal to the value of n.

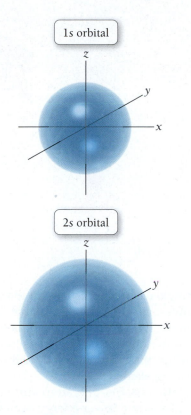

1s orbital

2s orbital

▲ **FIGURE 9.20 The 2s orbital**   The 2s orbital is similar to the 1s orbital, but larger in size.

The next orbitals in the quantum-mechanical model are those with principal quantum number $n = 2$. Unlike the $n = 1$ principal shell, which contains only one subshell (specified by s), the $n = 2$ principal shell contains two subshells, specified by s and p.

The number of subshells in a given principal shell is equal to the value of n. Therefore, the $n = 1$ principal shell has one subshell, the $n = 2$ principal shell has two subshells, and so on (▲ **FIGURE 9.19**). The s subshell contains the 2s orbital, which is higher in energy than the 1s orbital and slightly larger (◄ **FIGURE 9.20**), but otherwise similar in shape. The p subshell contains three 2p orbitals. The three 2p orbitals all have the same dumbbell-like shape but each has a different orientation (▼ **FIGURE 9.21**).

The next principal shell, $n = 3$, contains three subshells specified by s, p, and d. The s and p subshells contain the 3s and 3p orbitals, similar in shape to the 2s and 2p orbitals, but slightly larger and higher in energy. The 3d subshell contains the five d orbitals shown in ▶ **FIGURE 9.22**, on the next page. The next principal shell, $n = 4$, contains four subshells specified by s, p, d, and f. The s, p, and d subshells for the $n = 4$ principal shell are similar to those in $n = 3$. The 4f subshell contains seven orbitals (called the 4f orbitals), whose shape we do not consider in this book.

The 2p orbitals

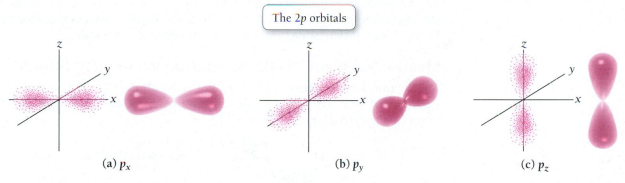

(a) p_x (b) p_y (c) p_z

▲ **FIGURE 9.21 The 2p orbitals**   This figure shows both the dot representation (left) and shape representation (right) for each p orbital.

PEARSON
eText
2.0

CONCEPTUAL ✓ CHECKPOINT 9.4

Which subshells are in the $n = 3$ principal shell?

(a) s subshell (only)

(b) s and p subshells (only)

(c) s, p, and d subshells (only)

PEARSON
eText
2.0

CONCEPTUAL ✓ CHECKPOINT 9.5

How many orbitals are in the p subshell?

(a) 1   **(b)** 3   **(c)** 5

▶ **FIGURE 9.22 The 3d orbitals** This figure shows both the dot representation (left) and shape representation (right) for each *d* orbital.

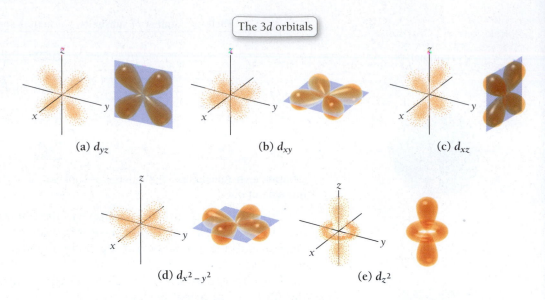

The 3d orbitals

(a) d_{yz} (b) d_{xy} (c) d_{xz}

(d) $d_{x^2-y^2}$ (e) d_{z^2}

As we have already discussed, hydrogen's single electron is usually in the 1s orbital because electrons generally occupy the lowest-energy orbital available. In hydrogen, the rest of the orbitals are normally empty. However, the absorption of energy by a hydrogen atom can cause the electron to jump (or make a transition) from the 1s orbital to a higher-energy orbital. When the electron is in a higher-energy orbital, we say that the hydrogen atom is in an **excited state**.

Because of their higher energy, excited states are unstable, and an electron in a higher-energy orbital will usually fall (or relax) back to a lower-energy orbital. In the process the electron emits energy, often in the form of light. As in the Bohr model, the energy difference between the two orbitals involved in the transition determines the wavelength of the emitted light (the greater the energy difference, the shorter the wavelength). The quantum-mechanical model predicts the bright-line spectrum of hydrogen as well as the Bohr model. However, unlike the Bohr model, it also predicts the bright-line spectra of other elements.

Electron Configurations: How Electrons Occupy Orbitals

An **electron configuration** illustrates the occupation of orbitals by electrons for a particular atom. For example, the electron configuration for a ground-state (or lowest energy) hydrogen atom is:

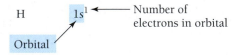

The electron configuration tells us that hydrogen's single electron is in the 1s orbital.

Another way to represent this information is with an **orbital diagram**, which gives similar information but shows the electrons as arrows in a box representing the orbital. The orbital diagram for a ground-state hydrogen atom is:

The box represents the 1s orbital, and the arrow within the box represents the electron in the 1s orbital. In orbital diagrams, the direction of the arrow (pointing up or pointing down) represents **electron spin**, a fundamental property of electrons. All electrons have spin. The **Pauli exclusion principle** states that *orbitals may hold no more than two electrons with opposing spins*. We symbolize this as two arrows pointing in opposite directions:

⇅

► **FIGURE 9.23 Energy ordering of orbitals for multi-electron atoms**
Different subshells within the same principal shell have different energies.

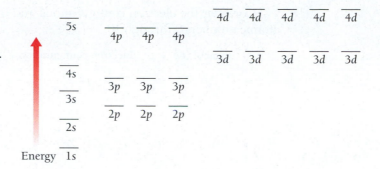

A helium atom, for example, has two electrons. The electron configuration and orbital diagram for helium are:

Electron configuration	Orbital diagram
He $\quad 1s^2$	⇅
	1s

Since we know that electrons occupy the lowest-energy orbitals available, and since we know that only two electrons (with opposing spins) are allowed in each orbital, we can continue to build ground-state electron configurations for the rest of the elements as long as we know the energy ordering of the orbitals. ▲ **FIGURE 9.23** shows the energy ordering of a number of orbitals for multi-electron atoms.

Notice that, for multi-electron atoms (in contrast to hydrogen which has only one electron), the subshells within a principal shell *do not* have the same energy. In elements other than hydrogen, the energy ordering is not determined by the principal quantum number alone. For example, in multi-electron atoms, the $4s$ subshell is lower in energy than the $3d$ subshell, even though its principal quantum number is higher. Using this relative energy ordering, we can write ground-state electron configurations and orbital diagrams for other elements.

In multi-electron atoms, the subshells within a principal shell do not have the same energy because of electron–electron interactions.

For lithium, which has three electrons, the electron configuration and orbital diagram are:

Electron configuration	Orbital diagram
Li $\quad 1s^2 2s^1$	⇅ ↑
	1s 2s

Remember that the number of electrons in an atom is equal to its atomic number.

For carbon, which has six electrons, the electron configuration and orbital diagram are:

Electron configuration	Orbital diagram
C $\quad 1s^2 2s^2 2p^2$	⇅ ⇅ ↑ ↑ ☐
	1s 2s 2p

Notice that the $2p$ electrons occupy the p orbitals (of equal energy) singly rather than pairing in one orbital. This is the result of **Hund's rule**, which states that *when filling orbitals of equal energy, electrons fill them singly first, with parallel spins*.

Before we write electron configurations for other elements, let's summarize what we have learned so far:

- Electrons occupy orbitals so as to minimize the energy of the atom; therefore, lower-energy orbitals fill before higher-energy orbitals. Orbitals fill in the following order: $1s\ 2s\ 2p\ 3s\ 3p\ 4s\ 3d\ 4p\ 5s\ 4d\ 5p\ 6s$ (◄ **FIGURE 9.24**).

- Orbitals can hold no more than two electrons each. When two electrons occupy the same orbital, they must have opposing spins. This is known as the Pauli exclusion principle.

- When orbitals of identical energy are available, these are first occupied singly with parallel spins rather than in pairs. This is known as Hund's rule.

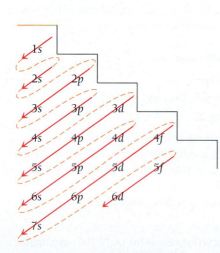

▲ **FIGURE 9.24 Orbital filling order** The arrows indicate the order in which orbitals fill.

Consider the electron configurations and orbital diagrams for elements with atomic numbers 3 through 10:

Symbol (#e^-)	Electron configuration	Orbital diagram
Li (3)	$1s^2 2s^1$	↑↓ ↑ 1s 2s
Be (4)	$1s^2 2s^2$	↑↓ ↑↓ 1s 2s
B (5)	$1s^2 2s^2 2p^1$	↑↓ ↑↓ ↑ 1s 2s 2p
C (6)	$1s^2 2s^2 2p^2$	↑↓ ↑↓ ↑ ↑ 1s 2s 2p
N (7)	$1s^2 2s^2 2p^3$	↑↓ ↑↓ ↑ ↑ ↑ 1s 2s 2p
O (8)	$1s^2 2s^2 2p^4$	↑↓ ↑↓ ↑↓ ↑ ↑ 1s 2s 2p
F (9)	$1s^2 2s^2 2p^5$	↑↓ ↑↓ ↑↓ ↑↓ ↑ 1s 2s 2p
Ne (10)	$1s^2 2s^2 2p^6$	↑↓ ↑↓ ↑↓ ↑↓ ↑↓ 1s 2s 2p

Notice how the p orbitals fill. As a result of Hund's rule, the p orbitals fill with single electrons before they fill with paired electrons. The electron configuration of neon represents the complete filling of the $n = 2$ principal shell. When writing electron configurations for elements beyond neon—or beyond any other noble gas—we often abbreviate the electron configuration of the previous noble gas by the symbol for the noble gas in brackets.

For example, the electron configuration of sodium is:

$$\text{Na} \qquad 1s^2 2s^2 2p^6 3s^1$$

We can write this using the noble gas core notation as:

$$\text{Na} \qquad [\text{Ne}]3s^1$$

where [Ne] represents $1s^2 2s^2 2p^6$, the electron configuration for neon.

To write an electron configuration for an element, we first find its atomic number from the periodic table—this number equals the number of electrons in the neutral atom. Then we use the order of filling from Figure 9.23 or 9.24 to distribute the electrons in the appropriate orbitals. Remember that each orbital can hold a maximum of two electrons. Consequently:

- the s subshell has only one orbital and therefore can hold only two electrons.
- the p subshell has three orbitals and therefore can hold six electrons.
- the d subshell has five orbitals and therefore can hold ten electrons.
- the f subshell has seven orbitals and therefore can hold 14 electrons.

EXAMPLE **9.2** | **Electron Configurations**

Write electron configurations for each element.

(a) Mg   (b) S   (c) Ga

	SOLUTION
(a) Magnesium has 12 electrons. Distribute two of these into the 1s orbital, two into the 2s orbital, six into the 2p orbitals, and two into the 3s orbital. You can also write the electron configuration more compactly using the noble gas core notation. For magnesium, use [Ne] to represent $1s^2 2s^2 2p^6$.	Mg $1s^2 2s^2 2p^6 3s^2$ *or* Mg $[Ne]3s^2$
(b) Sulfur has 16 electrons. Distribute two of these into the 1s orbital, two into the 2s orbital, six into the 2p orbitals, two into the 3s orbital, and four into the 3p orbitals. You can write the electron configuration more compactly by using [Ne] to represent $1s^2 2s^2 2p^6$.	S $1s^2 2s^2 2p^6 3s^2 3p^4$ *or* S $[Ne]3s^2 3p^4$
(c) Gallium has 31 electrons. Distribute two of these into the 1s orbital, two into the 2s orbital, six into the 2p orbitals, two into the 3s orbital, six into the 3p orbitals, two into the 4s orbital, ten into the 3d orbitals, and one into the 4p orbitals. Notice that the d subshell has five orbitals and can therefore hold 10 electrons. You can write the electron configuration more compactly by using [Ar] to represent $1s^2 2s^2 2p^6 3s^2 3p^6$.	Ga $1s^2 2s^2 2p^6 3s^2 3p^6 4s^2 3d^{10} 4p^1$ *or* Ga $[Ar]4s^2 3d^{10} 4p^1$

▶ **SKILLBUILDER 9.2** | **Electron Configurations**

Write electron configurations for each element.

(a) Al   (b) Br   (c) Sr

▶ **SKILLBUILDER PLUS**

Write electron configurations for each ion. (*Hint:* To determine the number of electrons to include in the electron configuration of an ion, add or subtract electrons as needed to account for the charge of the ion.)

(a) Al^{3+} (b) Cl^- (c) O^{2-}

▶ **FOR MORE PRACTICE**   Problems 49, 50, 53, 54, 55, 56.

EXAMPLE **9.3**   **Writing Orbital Diagrams**

Write an orbital diagram for silicon.

SOLUTION

Since silicon is atomic number 14, it has 14 electrons. Draw a box for each orbital, putting the lowest-energy orbital (1s) on the far left and proceeding to orbitals of higher energy to the right.

    1s   2s   2p   3s   3p

Distribute the 14 electrons into the orbitals, allowing a maximum of two electrons per orbital and remembering Hund's rule. The complete orbital diagram is:

Si   [↑↓] [↑↓] [↑↓][↑↓][↑↓] [↑↓] [↑][↑][ ]

    1s   2s   2p   3s   3p

▶ **SKILLBUILDER 9.3** | **Writing Orbital Diagrams**

Write an orbital diagram for argon.

▶ **FOR MORE PRACTICE**   Example 9.10; Problems 51, 52.

CONCEPTUAL ✔ **CHECKPOINT 9.6**

Which pair of elements has the same *total* number of electrons in its p orbitals?

(a) Na and K

(b) K and Kr

(c) P and N

(d) Ar and Ca

9.7 Electron Configurations and the Periodic Table

▶ Identify valence electrons and core electrons.

▶ Write electron configurations for elements based on their positions in the periodic table.

Valence electrons are the electrons in the outermost principal shell (the principal shell with the highest principal quantum number, n). These electrons are important because, as we will see in Chapter 10, they are held most loosely and are most easily lost or shared; therefore they are involved in chemical bonding. Electrons that are *not* in the outermost principal shell are **core electrons**. For example, silicon, with the electron configuration of $1s^2 2s^2 2p^6 3s^2 3p^2$, has four valence electrons (those in the $n = 3$ principal shell) and ten core electrons.

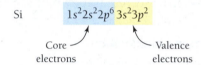

Si $\qquad 1s^2 2s^2 2p^6\ 3s^2 3p^2$

Core electrons $\qquad$ Valence electrons

EXAMPLE **9.4** | **Valence Electrons and Core Electrons**

Write an electron configuration for selenium and identify the valence electrons and the core electrons.

SOLUTION

Write the electron configuration for selenium by determining the total number of electrons from selenium's atomic number (34) and distributing them into the appropriate orbitals.

$$\text{Se} \quad 1s^2 2s^2 2p^6 3s^2 3p^6 4s^2 3d^{10} 4p^4$$

The valence electrons are those in the outermost principal shell. For selenium, the outermost principal shell is the $n = 4$ shell, which contains six electrons (two in the $4s$ orbital and four in the three $4p$ orbitals). All other electrons, including those in the $3d$ orbitals, are core electrons.

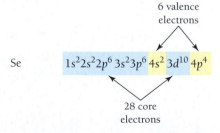

6 valence electrons

Se $\qquad 1s^2 2s^2 2p^6\ 3s^2 3p^6\ 4s^2\ 3d^{10}\ 4p^4$

28 core electrons

▶ **SKILLBUILDER 9.4** | **Valence Electrons and Core Electrons**

Write an electron configuration for chlorine and identify the valence electrons and core electrons.

▶ **FOR MORE PRACTICE** Example 9.11; Problems 57, 58, 61, 62.

▶ FIGURE 9.25 Outer electron configurations of the first 18 elements

1A	2A	3A	4A	5A	6A	7A	8A
1 H $1s^1$							2 He $1s^2$
3 Li $2s^1$	4 Be $2s^2$	5 B $2s^22p^1$	6 C $2s^22p^2$	7 N $2s^22p^3$	8 O $2s^22p^4$	9 F $2s^22p^5$	10 Ne $2s^22p^6$
11 Na $3s^1$	12 Mg $3s^2$	13 Al $3s^23p^1$	14 Si $3s^23p^2$	15 P $3s^23p^3$	16 S $3s^23p^4$	17 Cl $3s^23p^5$	18 Ar $3s^23p^6$

▲ FIGURE 9.25 shows the first 18 elements in the periodic table with an outer electron configuration listed below each one. As we move across a row, the orbitals are simply filling in the correct order. As we move down a column, the highest principal quantum number increases, but the number of electrons in each subshell remains the same. Consequently, the elements within a column (or family) all have the same number of valence electrons and similar outer electron configurations.

A similar pattern exists for the entire periodic table (▼ FIGURE 9.26). Notice that, because of the filling order of orbitals, we can divide the periodic table into blocks representing the filling of particular subshells.

- The first two columns on the left side of the periodic table are the *s* block with outer electron configurations of ns^1 (first column) and ns^2 (second column).

- The six columns on the right side of the periodic table are the *p* block with outer electron configurations of: ns^2np^1, ns^2np^2, ns^2np^3, ns^2np^4, ns^2np^5 (halogens), and ns^2np^6 (noble gases).

- The transition metals are the *d* block.

- The lanthanides and actinides (also called the inner transition metals) are the *f* block.

Notice that, except for helium, the number of valence electrons for any main-group element is equal to the group number of its column. For example, we can tell that chlorine has seven valence electrons because it is in the column with group

Groups

s-block elements   *p*-block elements

d-block elements   *f*-block elements

Period	1A	2A	3B	4B	5B	6B	7B	8B	8B	8B	1B	2B	3A	4A	5A	6A	7A	8A
	1	2	3	4	5	6	7	8	9	10	11	12	13	14	15	16	17	18
1	1 H $1s^1$																	2 He $1s^2$
2	3 Li $2s^1$	4 Be $2s^2$											5 B $2s^22p^1$	6 C $2s^22p^2$	7 N $2s^22p^3$	8 O $2s^22p^4$	9 F $2s^22p^5$	10 Ne $2s^22p^6$
3	11 Na $3s^1$	12 Mg $3s^2$											13 Al $3s^23p^1$	14 Si $3s^23p^2$	15 P $3s^23p^3$	16 S $3s^23p^4$	17 Cl $3s^23p^5$	18 Ar $3s^23p^6$
4	19 K $4s^1$	20 Ca $4s^2$	21 Sc $4s^23d^1$	22 Ti $4s^23d^2$	23 V $4s^23d^3$	24 Cr $4s^13d^5$	25 Mn $4s^23d^5$	26 Fe $4s^23d^6$	27 Co $4s^23d^7$	28 Ni $4s^23d^8$	29 Cu $4s^13d^{10}$	30 Zn $4s^23d^{10}$	31 Ga $4s^24p^1$	32 Ge $4s^24p^2$	33 As $4s^24p^3$	34 Se $4s^24p^4$	35 Br $4s^24p^5$	36 Kr $4s^24p^6$
5	37 Rb $5s^1$	38 Sr $5s^2$	39 Y $5s^24d^1$	40 Zr $5s^24d^2$	41 Nb $5s^14d^4$	42 Mo $5s^14d^5$	43 Tc $5s^24d^5$	44 Ru $5s^14d^7$	45 Rh $5s^14d^8$	46 Pd $4d^{10}$	47 Ag $5s^14d^{10}$	48 Cd $5s^24d^{10}$	49 In $5s^25p^1$	50 Sn $5s^25p^2$	51 Sb $5s^25p^3$	52 Te $5s^25p^4$	53 I $5s^25p^5$	54 Xe $5s^25p^6$
6	55 Cs $6s^1$	56 Ba $6s^2$	57 La $6s^25d^1$	72 Hf $6s^25d^2$	73 Ta $6s^25d^3$	74 W $6s^25d^4$	75 Re $6s^25d^5$	76 Os $6s^25d^6$	77 Ir $6s^25d^7$	78 Pt $6s^15d^9$	79 Au $6s^15d^{10}$	80 Hg $6s^25d^{10}$	81 Tl $6s^26p^1$	82 Pb $6s^26p^2$	83 Bi $6s^26p^3$	84 Po $6s^26p^4$	85 At $6s^26p^5$	86 Rn $6s^26p^6$
7	87 Fr $7s^1$	88 Ra $7s^2$	89 Ac $7s^26d^1$	104 Rf $7s^26d^2$	105 Db $7s^26d^3$	106 Sg $7s^26d^4$	107 Bh	108 Hs	109 Mt	110 Ds	111 Rg	112 Cn	113 Nh	114 Fl	115 Mc	116 Lv	117 Ts	118 Og

Lanthanides

58 Ce $6s^25d^14f^1$	59 Pr $6s^24f^3$	60 Nd $6s^24f^4$	61 Pm $6s^24f^5$	62 Sm $6s^24f^6$	63 Eu $6s^24f^7$	64 Gd $6s^24f^75d$	65 Tb $6s^24f^9$	66 Dy $6s^24f^{10}$	67 Ho $6s^24f^{11}$	68 Er $6s^24f^{12}$	69 Tm $6s^24f^{13}$	70 Yb $6s^24f^{14}$	71 Lu $6s^24f^{14}5d^1$

Actinides

90 Th $7s^26d^2$	91 Pa $7s^25f^26d^1$	92 U $7s^25f^36d^1$	93 Np $7s^25f^46d^1$	94 Pu $7s^25f^6$	95 Am $7s^25f^7$	96 Cm $7s^25f^76d^1$	97 Bk $7s^25f^9$	98 Cf $7s^25f^{10}$	99 Es $7s^25f^{11}$	100 Fm $7s^25f^{12}$	101 Md $7s^25f^{13}$	102 No $7s^25f^{14}$	103 Lr $7s^25f^{14}6d^1$

▲ FIGURE 9.26 Outer electron configurations of the elements

Remember that main-group elements are those in the two far-left columns (1A, 2A) and the six far-right columns (3A–8A) of the periodic table (see Section 4.6).

number 7A. The row number in the periodic table is equal to the number of the highest principal shell (n value). For example, since chlorine is in row 3, its highest principal shell is the $n = 3$ shell.

The transition metals have electron configurations with trends that differ somewhat from main-group elements. As we move across a row in the d block, the d orbitals are filling (see Figure 9.26). However, the principal quantum number of the d orbital being filled across each row in the transition series is equal to the row number minus one (in the fourth row, the $3d$ orbitals fill; in the fifth row, the $4d$ orbitals fill; and so on). For the first transition series, the outer configuration is $4s^2 3d^x$ (x = number of d electrons) with two exceptions: Cr is $4s^1 3d^5$ and Cu is $4s^1 3d^{10}$. These exceptions occur because a half-filled d subshell and a completely filled d subshell are particularly stable. Otherwise, the number of outershell electrons in a transition series does not change as we move across a period. In other words, *the transition series represents the filling of core orbitals, and the number of outershell electrons is mostly constant.*

We can now see that the organization of the periodic table allows us to write the electron configuration for any element based simply on its position in the periodic table. For example, suppose we want to write an electron configuration for P. The inner electrons of P are those of the noble gas that precedes P in the periodic table, Ne. So we can represent the inner electrons with [Ne]. We obtain the outer electron configuration by tracing the elements between Ne and P and assigning electrons to the appropriate orbitals (▼ **FIGURE 9.27**). Remember that the highest n value is given by the row number (3 for phosphorus). So we begin with [Ne], then add in the two $3s$ electrons as we trace across the s block, followed by three $3p$ electrons as we trace across the p block to P, which is in the third column of the p block. The electron configuration is:

$$P \qquad [Ne]3s^2 3p^3$$

Notice that P is in column 5A and therefore has five valence electrons and an outer electron configuration of $ns^2 np^3$.

To summarize writing an electron configuration for an element based on its position in the periodic table:

- The inner electron configuration for any element is the electron configuration of the noble gas that immediately precedes that element in the periodic table. We represent the inner configuration with the symbol for the noble gas in brackets.
- We can determine the outer electrons from the element's position within a particular block (s, p, d, or f) in the periodic table. We trace the elements between the preceding noble gas and the element of interest, and assign electrons to the appropriate orbitals.
- The highest principal quantum number (highest n value) is equal to the row number of the element in the periodic table.
- For any element containing d electrons, the principal quantum number (n value) of the outermost d electrons is equal to the row number of the element minus 1.

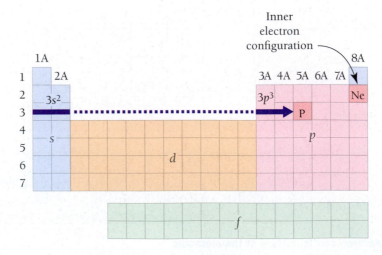

▶ **FIGURE 9.27 Electron configuration of phosphorus** Determining the electron configuration for P from its position in the periodic table.

Interactive
Worked Example
Video 9.5

EXAMPLE **9.5**   **Writing Electron Configurations from the Periodic Table**

Write an electron configuration for arsenic based on its position in the periodic table.

SOLUTION

The noble gas that precedes arsenic in the periodic table is argon, so the inner electron configuration is [Ar]. Obtain the outer electron configuration by tracing the elements between Ar and As and assigning electrons to the appropriate orbitals.

Remember that the highest n value is given by the row number (4 for arsenic). So, begin with [Ar], then add in the two $4s$ electrons as you trace across the s block, followed by ten $3d$ electrons as you trace across the d block (the n value for d subshells is equal to the row number minus one), and finally the three $4p$ electrons as you trace across the p block to As, which is in the third column of the p block.

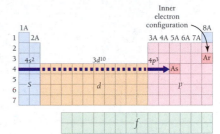

The electron configuration is:

$$\text{As} \qquad [\text{Ar}]4s^2 3d^{10} 4p^3$$

▶ **SKILLBUILDER 9.5 | Writing Electron Configurations from the Periodic Table**

Use the periodic table to determine the electron configuration for tin.

▶ **FOR MORE PRACTICE**   Example 9.12; Problems 65, 66, 67, 68.

CONCEPTUAL ✔ CHECKPOINT 9.7

Which element has the *fewest* valence electrons?

(a) B   **(b)** Ca   **(c)** O   **(d)** K   **(e)** Ga

9.8 The Explanatory Power of the Quantum-Mechanical Model

▶ Explain why the chemical properties of elements are largely determined by the number of valence electrons they contain.

Noble gases
18 8A
2 He $1s^2$
10 Ne $2s^2 2p^6$
18 Ar $3s^2 3p^6$
36 Kr $4s^2 4p^6$
54 Xe $5s^2 5p^6$
86 Rn $6s^2 6p^6$

▶ **FIGURE 9.28 Electron configurations of the noble gases**   The noble gases (except for helium) all have eight valence electrons and completely full outer principal shells.

At the beginning of this chapter, we asserted that the quantum-mechanical model explains the chemical properties of the elements such as the inertness of helium, the reactivity of hydrogen, and the periodic law. We can now see why: *The chemical properties of elements are largely determined by the number of valence electrons they contain.* The properties of elements vary in a periodic fashion because the number of valence electrons is periodic.

Because elements within a column in the periodic table have the same number of valence electrons, they also have similar chemical properties. The noble gases, for example, all have eight valence electrons, except for helium, which has two (◀ **FIGURE 9.28**). Although we don't get into the quantitative (or numerical) aspects of the quantum-mechanical model in this book, calculations show that atoms with eight valence electrons (or two for helium) are particularly low in energy, and therefore stable. The noble gases are indeed chemically stable and thus relatively inert or nonreactive as accounted for by the quantum model.

Elements with electron configurations close to the noble gases are the most reactive because they can attain noble gas electron configurations by losing or gaining a small number of electrons. Alkali metals (Group 1) are among the most reactive metals since their outer electron

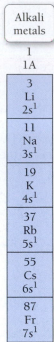

▲ **FIGURE 9.29 Electron configurations of the alkali metals** The alkali metals all have ns^1 electron configurations and are therefore one electron beyond a noble gas configuration. In their reactions, they tend to lose that electron, forming 1+ ions and attaining a noble gas configuration.

> Atoms and/or ions that share the same electron configuration are termed **isoelectronic**.

configuration (ns^1) is one electron beyond a noble gas configuration (◄ **FIGURE 9.29**). If an alkali metal can react to lose its ns^1 electron, it attains a noble gas configuration. This explains why—as we learned in Chapter 4—the Group 1A metals tend to form 1+ cations. As an example, consider the electron configuration of sodium:

$$\text{Na} \qquad 1s^2 2s^2 2p^6 3s^1$$

In reactions, sodium loses its 3s electron, forming a 1+ ion with the electron configuration of neon:

$$\text{Na}^+ \qquad 1s^2 2s^2 2p^6$$

$$\text{Ne} \qquad 1s^2 2s^2 2p^6$$

Similarly, alkaline earth metals, with an outer electron configuration of ns^2, also tend to be reactive metals. Each alkaline earth metal loses its two ns^2 electrons to form a 2+ cation (▼ **FIGURE 9.30**). For example, consider magnesium:

$$\text{Mg} \qquad 1s^2 2s^2 2p^6 3s^2$$

In reactions, magnesium loses its two 3s electrons, forming a 2+ ion with the electron configuration of neon:

$$\text{Mg}^{2+} \qquad 1s^2 2s^2 2p^6$$

On the other side of the periodic table, halogens are among the most reactive nonmetals because of their $ns^2 np^5$ electron configurations (▼ **FIGURE 9.31**). Each halogen is only one electron away from a noble gas configuration and tends to react to gain that one electron, forming a 1− ion. For example, consider fluorine:

$$\text{F} \qquad 1s^2 2s^2 2p^5$$

In reactions, fluorine gains one additional 2p electron, forming a 1− ion with the electron configuration of neon:

$$\text{F}^- \qquad 1s^2 2s^2 2p^6$$

The elements that form predictable ions are shown in ▶ **FIGURE 9.32** (first introduced in Chapter 4). Notice how the charge of these ions reflects their electron configurations—these elements form ions with noble gas electron configurations.

◄ **FIGURE 9.30 Electron configurations of the alkaline earth metals** The alkaline earth metals all have ns^2 electron configurations and are therefore two electrons beyond a noble gas configuration. In their reactions, they tend to lose two electrons, forming 2+ ions and attaining a noble gas configuration.

▶ **FIGURE 9.31 Electron configurations of the halogens** The halogens all have $ns^2 np^5$ electron configurations and are therefore one electron short of a noble gas configuration. In their reactions, they tend to gain one electron, forming 1− ions and attaining a noble gas configuration.

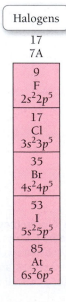

1																	8
	2									3	4	5	6	7			
Li^+													N^{3-}	O^{2-}	F^-		
Na^+	Mg^{2+}									Al^{3+}				S^{2-}	Cl^-		
K^+	Ca^{2+}									Ga^{3+}				Se^{2-}	Br^-		
Rb^+	Sr^{2+}				Most transition metals form					In^{3+}				Te^{2-}	I^-		
Cs^+	Ba^{2+}				cations with various charges												

▲ **FIGURE 9.32 Elements that form predictable ions**

CONCEPTUAL ✓ CHECKPOINT 9.8

Shown here is the electron configuration of calcium:

$$Ca \qquad 1s^2 2s^2 2p^6 3s^2 3p^6 4s^2$$

In its reactions, calcium tends to form the Ca^{2+} ion. Which electrons are lost upon ionization?

(a) all of the $4s$ electrons
(b) two of the $3p$ electrons
(c) all of the $3s$ electrons
(d) the $1s$ electrons

9.9   Periodic Trends: Atomic Size, Ionization Energy, and Metallic Character

▶ Identify and understand periodic trends in atomic size, ionization energy, and metallic character.

The quantum-mechanical model also explains other periodic trends such as atomic size, ionization energy, and metallic character. We examine these trends individually in this section of the chapter.

Atomic Size

The **atomic size** of an atom is determined by the distance between its outermost electrons and its nucleus. As we move across a period in the periodic table, we know that electrons occupy orbitals with the same principal quantum number, n. Since the principal quantum number largely determines the size of an orbital, electrons are therefore filling orbitals of approximately the same size, and we might expect atomic size to remain constant across a period. However, with each step across a period, the number of protons in the nucleus also increases. This increase in the number of protons results in a greater pull on the electrons from the nucleus, causing atomic size to actually decrease. Therefore:

As we move to the right across a period, or row, in the periodic table, atomic size decreases, as shown in ▶ **FIGURE 9.33**, on the next page.

As we move down a column in the periodic table, the highest principal quantum number, n, increases. Because the size of an orbital increases with increasing principal quantum number, the electrons that occupy the outermost orbitals are farther from the nucleus as we move down a column. Therefore:

As we move down a column, or family, in the periodic table, atomic size increases, as shown in Figure 9.33.

▶ **FIGURE 9.33 Periodic properties: atomic size** Atomic size decreases as we move to the right across a period and increases as we move down a column in the periodic table.

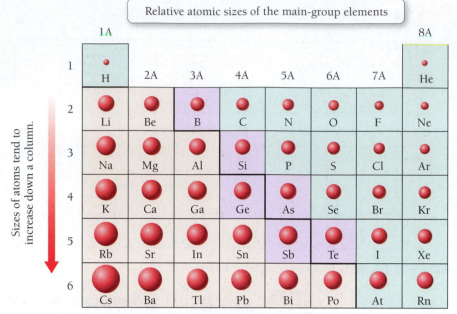

Relative atomic sizes of the main-group elements

Sizes of atoms tend to increase down a column.

Sizes of atoms tend to decrease across a period.

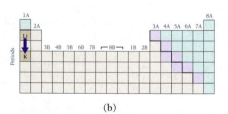

(a)

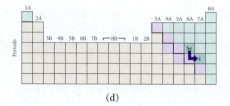

(b)

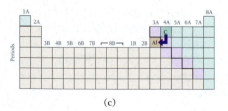

(c)

(d)

PEARSON eText 2.0

Interactive Worked Example Video 9.6

EXAMPLE **9.6** **Atomic Size**

Choose the larger atom in each pair.

(a) C or O   **(b)** Li or K   **(c)** C or Al   **(d)** Se or I

SOLUTION

(a) C or O
Carbon atoms are larger than O atoms because, as you trace the path between C and O on the periodic table, you move to the right within the same period. Atomic size decreases as you go to the right.

(b) Li or K
Potassium atoms are larger than Li atoms because, as you trace the path between Li and K on the periodic table, you move down a column. Atomic size increases as you go down a column.

(c) C or Al
Aluminum atoms are larger than C atoms because, as you trace the path between C and Al on the periodic table, you move down a column (atomic size increases) and then to the left across a period (atomic size increases). These effects add together for an overall increase.

(d) Se or I
Based on periodic properties alone, you cannot tell which atom is larger because as you trace the path between Se and I, you go down a column (atomic size increases) and then to the right across a period (atomic size decreases). These effects tend to cancel one another.

▶ **SKILLBUILDER 9.6 | Atomic Size**

Choose the larger atom in each pair.
(a) Pb or Po   **(b)** Rb or Na   **(c)** Sn or Bi   **(d)** F or Se

▶ **FOR MORE PRACTICE** Example 9.13a; Problems 81, 82, 83, 84.

CHEMISTRY AND HEALTH

Pumping Ions: Atomic Size and Nerve Impulses

No matter what you are doing at this moment, tiny pumps in each of the trillions of cells that make up your body are hard at work. These pumps, located in the cell membrane, move a number of different ions into and out of the cell. The most important of these ions are sodium (Na^+) and potassium (K^+), which happen to be pumped in opposite directions. Sodium ions are pumped *out of cells*, while potassium ions are pumped *into cells*. The result is a *chemical gradient* for each ion: The concentration of sodium is higher outside the cell than within, while exactly the opposite is true for potassium.

The ion pumps within the cell membrane are analogous to water pumps in a high-rise building that pump water against the force of gravity to a tank on the roof. Other structures within the membrane, called ion channels, are like the building's faucets. When they open momentarily, bursts of sodium and potassium ions, driven by their concentration gradients, flow back across the membrane—sodium flowing in and potassium flowing out. These ion pulses are the basis for the transmission of nerve signals in the brain, heart, and throughout the body. Consequently, every move you make or every thought you have is mediated by the flow of these ions.

How do the pumps and channels differentiate between sodium and potassium ions? How do the ion pumps selectively move sodium out of the cell and potassium into the cell? To answer this question, we must examine the sodium and potassium ions more closely. In what ways do they differ? Both are cations of Group I metals. All Group I metals tend to lose one electron to form cations with 1+ charge, so the magnitude of the charge cannot be the decisive factor. But potassium (atomic number 19) lies directly below sodium in the periodic table (atomic number 11) and based on periodic properties potassium is therefore larger than sodium. The potassium ion has a radius of 133 pm, while the sodium ion has a radius of 95 pm. (Recall from Chapter 2 that $1\,pm = 10^{-12}\,m$.) The pumps and channels within cell membranes are so sensitive that they distinguish between the sizes of these two ions and selectively allow only one or the other to pass. The result is the transmission of nerve signals that allows you to read this page.

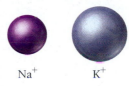

Na$^+$ K$^+$

B9.2 CAN YOU ANSWER THIS? *Other ions, including calcium and magnesium, are also important to nerve signal transmission. Arrange these four ions in order of increasing size: K^+, Na^+, Mg^{2+}, and Ca^{2+}.*

Ionization Energy

The **ionization energy** of an atom is the energy required to remove an electron from the atom in the gaseous state. The ionization of sodium, for example, is represented with the equation:

$$Na \ + \ \text{ionization energy} \ \longrightarrow \ Na^+ \ + \ 1e^-$$

Based on what we know about electron configurations, what can we predict about ionization energy trends? Would it take more or less energy to remove an electron from Na than from Cl? We know that Na has an outer electron configuration of $3s^1$ and Cl has an outer electron configuration of $3s^2 3p^5$. Since removing an electron from Na gives it a noble gas configuration—and removing an electron from Cl does not—we would expect sodium to have a lower ionization energy, and that is the case. It is easier to remove an electron from sodium than it is from chlorine. We can generalize this idea in this statement:

As we move across a period, or row, to the right in the periodic table, ionization energy increases (▶ **FIGURE 9.34**, on the next page).

What happens to ionization energy as we move down a column? As we have learned, the principal quantum number, n, increases as we move down a column. Within a given subshell, orbitals with higher principal quantum numbers are larger than orbitals with smaller principal quantum numbers. Consequently, electrons in the outermost principal shell are farther away from the positively charged nucleus—and therefore are held less tightly—as we move down a column. This

▶ **FIGURE 9.34 Periodic properties: ionization energy**
Ionization energy increases as we move to the right across a period and decreases as we move down a column in the periodic table.

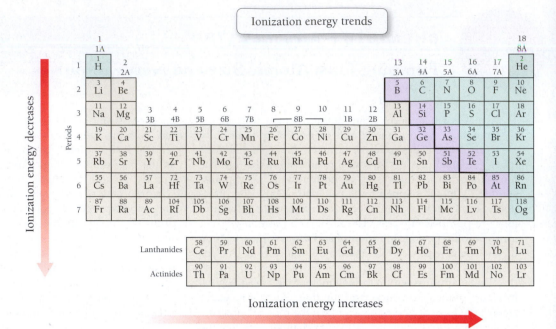

Ionization energy trends

Ionization energy decreases

Ionization energy increases

results in a lower ionization energy (if the electron is held less tightly, it is easier to pull away) as we move down a column. Therefore:

As we move down a column (or family) in the periodic table, ionization energy decreases (see Figure 9.34).

Notice that the trends in ionization energy are consistent with the trends in atomic size. Smaller atoms are more difficult to ionize because their electrons are held more tightly. Therefore, as we move across a period, atomic size decreases and ionization energy increases. Similarly, as we move down a column, atomic size increases and ionization energy decreases since electrons are farther from the nucleus and are therefore less tightly held.

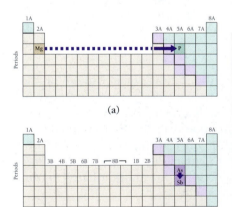

(a)

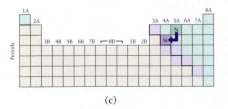

(b)

(c)

EXAMPLE **9.7** | **Ionization Energy**

Choose the element with the higher ionization energy from each pair.

(a) Mg or P   **(b)** As or Sb   **(c)** N or Si   **(d)** O or Cl

SOLUTION

(a) Mg or P
P has a higher ionization than Mg because, as you trace the path between Mg and P on the periodic table, you move to the right within the same period. Ionization energy increases as you go to the right.

(b) As or Sb
As has a higher ionization energy than Sb because, as you trace the path between As and Sb on the periodic table, you move down a column. Ionization energy decreases as you go down a column.

(c) N or Si
N has a higher ionization energy than Si because, as you trace the path between N and Si on the periodic table, you move down a column (ionization energy decreases) and then to the left across a period (ionization energy decreases). These effects sum together for an overall decrease.

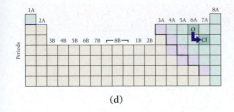

(d)

(d) O or Cl

Based on periodic properties alone, you cannot tell which has a higher ionization energy because, as you trace the path between O and Cl you move down a column (ionization energy decreases) and then to the right across a period (ionization energy increases). These effects tend to cancel.

▶ **SKILLBUILDER 9.7 | Ionization Energy**

Choose the element with the higher ionization energy from each pair.
(a) Mg or Sr   **(b)** In or Te   **(c)** C or P   **(d)** F or S

▶ **FOR MORE PRACTICE**   Example 9.13b; Problems 77, 78, 79, 80.

Metallic Character

As we learned in Chapter 4, metals tend to lose electrons in their chemical reactions, while nonmetals tend to gain electrons. As we move across a period in the periodic table, ionization energy increases, which means that electrons are less likely to be lost in chemical reactions. Consequently:

> As we move across a period, or row, to the right in the periodic table, **metallic character** decreases (▼ **FIGURE 9.35**).

As we move down a column in the periodic table, ionization energy decreases, making electrons more likely to be lost in chemical reactions. Consequently:

> As we move down a column, or family, in the periodic table, metallic character increases (see Figure 9.35).

These trends, based on the quantum-mechanical model, explain the distribution of metals and nonmetals that we were introduced to in Chapter 4. Metals are found toward the left side of the periodic table and nonmetals (with the exception of hydrogen) toward the upper right.

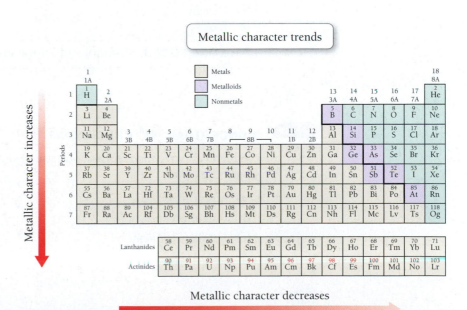

▲ **FIGURE 9.35 Periodic properties: metallic character**   Metallic character decreases as we move to the right across a period and increases as we move down a column in the periodic table.

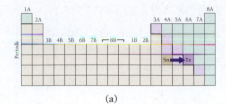

(a)

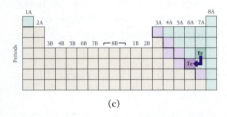

(b)

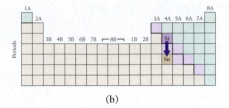

(c)

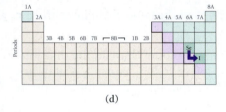

(d)

EXAMPLE **9.8** **Metallic Character**

Choose the more metallic element from each pair.

(a) Sn or Te
(b) Si or Sn
(c) Br or Te
(d) Se or I

SOLUTION

(a) Sn or Te
Sn is more metallic than Te because, as you trace the path between Sn and Te on the periodic table, you move to the right within the same period. Metallic character decreases as you go to the right.

(b) Si or Sn
Sn is more metallic than Si because, as you trace the path between Si and Sn on the periodic table, you move down a column. Metallic character increases as you go down a column.

(c) Br or Te
Te is more metallic than Br because, as you trace the path between Br and Te on the periodic table, you move down a column (metallic character increases) and then to the left across a period (metallic character increases). These effects add together for an overall increase.

(d) Se or I
Based on periodic properties alone, you cannot tell which is more metallic because, as you trace the path between Se and I, you go down a column (metallic character increases) and then to the right across a period (metallic character decreases). These effects tend to cancel.

▶ SKILLBUILDER 9.8 | **Metallic Character**

Choose the more metallic element from each pair.

(a) Ge or In
(b) Ga or Sn
(c) P or Bi
(d) B or N

▶ FOR MORE PRACTICE   Example 9.13; Problems 85, 86, 87, 88.

CONCEPTUAL ☑ CHECKPOINT 9.9

Which property *increases* as you move from left to right across a row in the periodic table?

(a) atomic size

(b) ionization energy

(c) metallic character

Chapter 9 in Review

MasteringChemistry™ provides end-of-chapter exercises, feedback-enriched tutorial problems, animations, and interactive activities to encourage problem solving practice and deeper understanding of key concepts and topics.

Self-Assessment Quiz

PEARSON
eText
2.0

Q1. Which set of wavelengths for light are arranged in order of increasing frequency?
(a) 250 nm; 300 nm; 350 nm
(b) 350 nm; 300 nm; 250 nm
(c) 300 nm; 350 nm; 250 nm
(d) 300 nm; 250 nm; 350 nm

Q2. Which of the listed types of electromagnetic radiation has the longest wavelength?
(a) ultraviolet   **(b)** X-ray
(c) infrared   **(d)** microwaves

Q3. Which electron transition in the Bohr model would produce light with the longest wavelength?
(a) $n = 2 \longrightarrow n = 1$ **(b)** $n = 3 \longrightarrow n = 1$
(c) $n = 4 \longrightarrow n = 1$ **(d)** $n = 5 \longrightarrow n = 1$

Q4. What is the electron configuration of arsenic (As)?
(a) $[Ar]4s^2 4p^3$ **(b)** $[Ar]4s^2 4d^{10} 4p^3$
(c) $[Ar]4s^2 3d^6 4p^3$ **(d)** $[Ar]4s^2 3d^{10} 4p^3$

Q5. Which orbital diagram corresponds to phosphorus (P)?

(a)   1s   2s   2p   3s   3p

(b)   1s   2s   2p

(c)   1s   2s   2p   3s   3p

(d)   1s   2s   2p   3s   3p

Q6. How many valence electrons does tellurium (Te) have?
(a) 5   **(b)** 6   **(c)** 16   **(d)** 52

Q7. The element sulfur forms an ion with what charge?
(a) 2−   **(b)** 1−   **(c)** 1+   **(d)** 2+

Q8. Order the elements Sr, Ca, and Se in order of decreasing atomic size.
(a) Se > Sr > Ca   **(b)** Ca > Se > Sr
(c) Sr > Ca > Se   **(d)** Se > Ca > Sr

Q9. Which of the listed elements has the highest ionization energy?
(a) Sn   **(b)** S   **(c)** Si   **(d)** F

Q10. Which of the listed elements is most metallic?
(a) Al   **(b)** N   **(c)** P   **(d)** O

Q11. Which property decreases as you move down a column in the periodic table?
(a) atomic size
(b) ionization energy
(c) metallic character
(d) none of the above (all increase as you move down a column).

Q12. When aluminum forms an ion, it loses electrons. How many electrons does it lose, and which orbitals do the electrons come from?
(a) one electron from the 3s orbital
(b) two electrons: one from the 3s orbital and one from the 2s orbital
(c) three electrons: two from the 3s orbital and one from the 3p orbital
(d) five electrons from the 3p orbital

Answers: 1:b, 2:d, 3:a, 4:d, 5:c, 6:b, 7:a, 8:c, 9:d, 10:a, 11:b, 12:c

Chemical Principles

Light

Light is electromagnetic radiation, energy that travels through space at a constant speed of 3.0×10^8 m/s (186,000 mi/s) and exhibits both wavelike and particle-like behavior. Particles of light are called photons. The wave nature of light is characterized by its wavelength, the distance between adjacent crests in the wave. The wavelength of light is inversely proportional to both the frequency—the number of cycles that pass a stationary point in one second—and the energy of a photon. Electromagnetic radiation ranges in wavelength from 10^{-16} m (gamma rays) to 10^6 m (radio waves). In between these lie X-rays, ultraviolet light, visible light, infrared light, and microwaves.

Relevance

Light enables us to see the world. However, we see only visible light, a small sliver in the center of the electromagnetic spectrum. We use other forms of electromagnetic radiation for cancer therapy, X-ray imaging, night vision, microwave cooking, and communications. Light is also important to many chemical processes. We can learn about the electronic structure of atoms, for example, by examining their interaction with light.

The Bohr Model

The emission spectrum of hydrogen, consisting of bright lines at specific wavelengths, is explained by the Bohr model for the hydrogen atom. In this model, electrons occupy circular orbits at specific fixed distances from the nucleus. Each orbit is specified by a quantum number (n), which also specifies the orbit's energy. While an electron is in a given orbit, its energy remains constant. When an electron jumps between orbits, a quantum of energy is absorbed or emitted. Since the difference in energy between orbits is fixed, the energy emitted or absorbed is also fixed. Emitted energy is carried away in the form of a photon of specific wavelength.

The Bohr model was a first attempt to explain the bright-line spectra of atoms. While it does predict the spectrum of the hydrogen atom, it fails to predict the spectra of other atoms and was consequently replaced by the quantum-mechanical model.

The Quantum-Mechanical Model

The quantum-mechanical model for the atom describes electron orbitals, which are electron probability maps that show the relative probability of finding an electron in various places surrounding the atomic nucleus. Orbitals are specified with a number (n), called the principal quantum number, and a letter. The principal quantum number ($n = 1, 2, 3 \ldots$) specifies the principal shell, and the letter (s, p, d, or f) specifies the subshell of the orbital. In the hydrogen atom, the energy of orbitals depends only on n. In multi-electron atoms, the energy ordering is $1s\ 2s\ 2p\ 3s\ 3p\ 4s\ 3d\ 4p\ 5s\ 4d\ 5p\ 6s$.

An electron configuration indicates which orbitals are occupied for a particular atom. Orbitals are filled in order of increasing energy and obey the Pauli exclusion principle (each orbital can hold a maximum of two electrons with opposing spins) and Hund's rule (electrons occupy orbitals of identical energy singly before pairing).

The quantum-mechanical model changed the way we view nature. Before the quantum-mechanical model, electrons were viewed as small particles, much like any other particle. Electrons were expected to follow the normal laws of motion, just as a baseball does. However, the electron, with its wavelike properties, does not follow these laws. Instead, electron motion is describable only through probabilistic predictions. Quantum theory single-handedly changed the predictability of nature at its most fundamental level.

The quantum-mechanical model of the atom predicts and explains many of the chemical properties we learned about in earlier chapters.

The Periodic Table

Elements in the same column of the periodic table have similar outer electron configurations and the same number of valence electrons (electrons in the outermost principal shell), and therefore similar chemical properties. We divide the periodic table into blocks (s block, p block, d block, and f block) in which particular sublevels are filled. As we move across a period to the right in the periodic table, atomic size decreases, ionization energy increases, and metallic character decreases. As we move down a column in the periodic table, atomic size increases, ionization energy decreases, and metallic character increases.

The periodic law exists because the number of valence electrons is periodic, and valence electrons determine chemical properties. Quantum theory also predicts that atoms with eight outershell electrons (or two for helium) are particularly stable, thus explaining the inertness of the noble gases. Atoms without noble gas configurations undergo chemical reactions to attain them, explaining the reactivity of the alkali metals and the halogens as well as the tendency of several families to form ions with certain charges.

Chemical Skills

Examples

LO: Predict the relative wavelength, energy, and frequency of different types of light (Section 9.3).

- Figure 9.4 includes relative wavelengths.
- Energy per photon increases with decreasing (shorter) wavelength.
- Frequency increases with decreasing (shorter) wavelength.

EXAMPLE **9.9** **Predicting Relative Wavelength, Energy, and Frequency of Light**

Which type of light—infrared or ultraviolet—has the longer wavelength? Higher frequency? Higher energy per photon?

SOLUTION

Infrared light has the longer wavelength (see Figure 9.4). Ultraviolet light has the higher frequency and the higher energy per photon.

LO: Write electron configurations and orbital diagrams for atoms (Section 9.6).

To write electron configurations, determine the number of electrons in the atom from the element's atomic number and then follow these rules:

- Electrons occupy orbitals so as to minimize the energy of the atom; therefore, lower-energy orbitals fill before higher-energy orbitals. Orbitals fill in the order: $1s$ $2s$ $2p$ $3s$ $3p$ $4s$ $3d$ $4p$ $5s$ $4d$ $5p$ $6s$ (Figure 9.24). The s subshells hold up to two electrons, p subshells hold up to six, d subshells hold up to ten, and f subshells hold up to 14.

- Orbitals can hold no more than two electrons each. When two electrons occupy the same orbital, they must have opposing spins.

- When orbitals of identical energy are available, these are first occupied singly with parallel spins rather than in pairs.

EXAMPLE **9.10** **Writing Electron Configurations and Orbital Diagrams**

Write an electron configuration and orbital diagram (outer electrons only) for germanium.

SOLUTION

Germanium is atomic number 32; therefore, it has 32 electrons.

Electron Configuration

$$\text{Ge} \qquad 1s^2 2s^2 2p^6 3s^2 3p^6 4s^2 3d^{10} 4p^2$$

or

$$\text{Ge} \qquad [\text{Ar}]4s^2 3d^{10} 4p^2$$

Orbital Diagram (Outer Electrons)

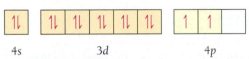

4s 3d 4p

LO: Identify valence electrons and core electrons (Section 9.7).

- Valence electrons are the electrons in the outermost principal energy shell (the principal shell with the highest principal quantum number).

- Core electrons are electrons that are not in the outermost principal shell.

EXAMPLE **9.11** **Identifying Valence Electrons and Core Electrons**

Identify the valence electrons and core electrons in the electron configuration of germanium (given in Example 9.10).

SOLUTION

$$\text{Ge} \qquad 1s^2 2s^2 2p^6 3s^2 3p^6 \; 4s^2 \; 3d^{10} \; 4p^2$$

28 core electrons   4 valence electrons

LO: Write electron configurations for elements based on their positions in the periodic table (Section 9.7).

- The inner electron configuration for any element is the electron configuration of the noble gas that immediately precedes that element in the periodic table. Represent the inner configuration with the symbol for the noble gas in brackets.

- The outer electrons can be determined from the element's position within a particular block (s, p, d, or f) in the periodic table. Trace the elements between the preceding noble gas and the element of interest and assign electrons to the appropriate orbitals. Figure 9.26 shows the outer electron configuration based on the position of an element in the periodic table.

- The highest principal quantum number (highest n value) is equal to the row number of the element in the periodic table.

- The principal quantum number (n value) of the outermost d electrons for any element containing d electrons is equal to the row number of the element minus 1.

EXAMPLE **9.12** **Writing an Electron Configuration for an Element Based on Its Position in the Periodic Table**

Write an electron configuration for iodine based on its position in the periodic table.

SOLUTION

The inner configuration for I is [Kr].

Begin with the [Kr] inner electron configuration. As you trace from Kr to I, add two $5s$ electrons, ten $4d$ electrons, and five $5p$ electrons. The overall configuration is:

$$\text{I} \qquad [\text{Kr}]5s^2 4d^{10} 5p^5$$

LO: Identify and understand periodic trends in atomic size, ionization energy, and metallic character (Section 9.9).

On the periodic table:

- Atomic size decreases as you move to the right and increases as you move down.

- Ionization energy increases as you move to the right and decreases as you move down.

- Metallic character decreases as you move to the right and increases as you move down.

EXAMPLE **9.13** **Periodic Trends: Atomic Size, Ionization Energy, and Metallic Character**

Arrange Si, In, and S in order of **(a)** increasing atomic size, **(b)** increasing ionization energy, and **(c)** increasing metallic character.

SOLUTION

(a) S, Si, In
(b) In, Si, S
(c) S, Si, In

Key Terms

atomic size [9.9]
Bohr model [9.1]
core electrons [9.7]
electromagnetic
   radiation [9.2]
electromagnetic
   spectrum [9.3]
electron configuration [9.6]
electron spin [9.6]
emission spectrum (plural,
   emission *spectra*) [9.4]

excited state [9.6]
frequency (ν) [9.2]
gamma ray [9.3]
ground state [9.6]
Hund's rule [9.6]
infrared light [9.3]
ionization energy [9.9]
metallic character [9.9]
microwaves [9.3]
orbital [9.5]
orbital diagram [9.6]

Pauli exclusion principle [9.6]
photon [9.2]
principal quantum
   number [9.6]
principal shell [9.6]
quantized [9.4]
quantum (plural,
   quanta) [9.4]
quantum-mechanical
   model [9.1]

quantum number [9.4]
radio waves [9.3]
subshell [9.6]
ultraviolet (UV)
   light [9.3]
valence electron [9.7]
visible light [9.3]
wavelength (λ) [9.2]
X-rays [9.3]

Exercises

Questions

1. When were the Bohr model and the quantum-mechanical model for the atom developed? What purpose do these models serve?
2. What is light? How fast does light travel?
3. What is white light? Colored light?
4. Explain, in terms of absorbed and reflected light, why a blue object appears blue.
5. What is the relationship between the wavelength of light and the amount of energy carried by its photons? How are wavelength and frequency of light related?
6. List some sources of gamma rays.
7. How are X-rays used?
8. Why should excess exposure to gamma rays and X-rays be avoided?
9. Why should excess exposure to ultraviolet light be avoided?
10. What objects emit infrared light? What technology exploits this?
11. Why do microwave ovens heat food but tend not to heat the dish the food is on?
12. What type of electromagnetic radiation is used in communications devices such as cellular telephones?
13. Describe the Bohr model for the hydrogen atom.

14. What is an emission spectrum? Use the Bohr model to explain why the emission spectrum of the hydrogen atom consists of distinct lines at specific wavelengths.
15. Explain the difference between a Bohr orbit and a quantum-mechanical orbital.
16. What is the difference between the ground state of an atom and an excited state of an atom?
17. Explain how the motion of an electron is different from the motion of a baseball. What is a probability map?
18. Why do quantum-mechanical orbitals have "fuzzy" boundaries?
19. List the four possible subshells in the quantum-mechanical model, the number of orbitals in each subshell, and the maximum number of electrons that can be contained in each subshell.
20. List the quantum-mechanical orbitals through 5*s*, in the correct energy order for multi-electron atoms.
21. What is the Pauli exclusion principle? Why is it important when writing electron configurations?
22. What is Hund's rule? Why is it important when writing orbital diagrams?
23. Within an electron configuration, what do symbols such as [Ne] and [Kr] represent?

24. Explain the difference between valence electrons and core electrons.

25. Identify each block in the blank periodic table.
(a) *s* block   (b) *p* block
(c) *d* block   (d) *f* block

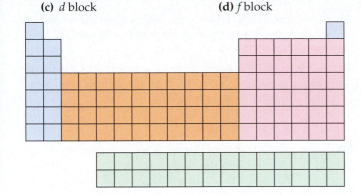

26. List some examples of the explanatory power of the quantum-mechanical model.

27. Explain why Group 1 elements tend to form 1+ ions and Group 7 elements tend to form 1− ions.

28. Explain the periodic trends in each chemical property.
(a) ionization energy
(b) atomic size
(c) metallic character

Problems

WAVELENGTH, ENERGY, AND FREQUENCY OF ELECTROMAGNETIC RADIATION

29. How long does it take light to travel:
(a) 1.0 ft (report answer in nanoseconds)
(b) 2462 mi, the distance between Los Angeles and New York (report answer in milliseconds)
(c) 4.5 billion km, the average separation between the sun and Neptune (report answer in hours and minutes)

30. How far does light travel in each time period?
(a) 1.0 s
(b) 1.0 day
(c) 1.0 yr

31. Which type of electromagnetic radiation has the longest wavelength?
(a) visible   (b) ultraviolet
(c) infrared   (d) X-ray

32. Which type of electromagnetic radiation has the shortest wavelength?
(a) radio waves   (b) microwaves
(c) infrared   (d) ultraviolet

33. List the types of electromagnetic radiation in order of increasing energy per photon.
(a) radio waves   (b) microwaves
(c) infrared   (d) ultraviolet

34. List the types of electromagnetic radiation in order of decreasing energy per photon.
(a) gamma rays   (b) radio waves
(c) microwaves   (d) visible light

35. List two types of electromagnetic radiation with frequencies higher than visible light.

36. List two types of electromagnetic radiation with frequencies lower than infrared light.

37. List these three types of radiation—infrared, X-ray, and radio waves—in order of:
(a) increasing energy per photon
(b) increasing frequency
(c) increasing wavelength

38. List these three types of electromagnetic radiation—visible, gamma rays, and microwaves—in order of:
(a) decreasing energy per photon
(b) decreasing frequency
(c) decreasing wavelength

THE BOHR MODEL

39. Bohr orbits have fixed _____ and fixed _____ .

40. In the Bohr model, what happens when an electron makes a transition between orbits?

41. Two of the emission wavelengths in the hydrogen emission spectrum are 410 nm and 434 nm. One of these is due to the $n = 6$ to $n = 2$ transition, and the other is due to the $n = 5$ to $n = 2$ transition. Which wavelength corresponds to which transition?

42. Two of the emission wavelengths in the hydrogen emission spectrum are 656 nm and 486 nm. One of these is due to the $n = 4$ to $n = 2$ transition, and the other is due to the $n = 3$ to $n = 2$ transition. Which wavelength corresponds to which transition?

THE QUANTUM-MECHANICAL MODEL

43. Sketch the $1s$ and $2p$ orbitals. How do the $2s$ and $3p$ orbitals differ from the $1s$ and $2p$ orbitals?

44. Sketch the $3d$ orbitals. How do the $4d$ orbitals differ from the $3d$ orbitals?

45. Which electron is, on average, closer to the nucleus: an electron in a $2s$ orbital or an electron in a $3s$ orbital?

46. Which electron is, on average, farther from the nucleus: an electron in a $3p$ orbital or an electron in a $4p$ orbital?

47. According to the quantum-mechanical model for the hydrogen atom, which electron transition produces light with longer wavelength: $2p$ to $1s$ or $3p$ to $1s$?

48. According to the quantum-mechanical model for the hydrogen atom, which transition produces light with longer wavelength: $3p$ to $2s$ or $4p$ to $2s$?

ELECTRON CONFIGURATIONS

49. Write full electron configurations for each element.
  (a) Sr (b) Ge (c) Li (d) Kr

50. Write full electron configurations for each element.
  (a) N (b) Mg (c) Ar (d) Se

51. Write full orbital diagrams and indicate the number of unpaired electrons for each element.
  (a) He (b) B (c) Li (d) N

52. Write full orbital diagrams and indicate the number of unpaired electrons for each element.
  (a) F (b) C (c) Ne (d) Be

53. Write electron configurations for each element. Use the symbol of the previous noble gas in brackets to represent the core electrons.
  (a) Ga (b) As (c) Rb (d) Sn

54. Write electron configurations for each element. Use the symbol of the previous noble gas in brackets to represent the core electrons.
  (a) Te (b) Br (c) I (d) Cs

55. Write electron configurations for each transition metal.
  (a) Zn (b) Cu (c) Zr (d) Fe

56. Write electron configurations for each transition metal.
  (a) Mn (b) Ti (c) Cd (d) V

VALENCE ELECTRONS AND CORE ELECTRONS

57. Write full electron configurations and indicate the valence electrons and the core electrons for each element.
  (a) Kr (b) Ge (c) Cl (d) Sr

58. Write full electron configurations and indicate the valence electrons and the core electrons for each element.
  (a) Sb (b) N (c) B (d) K

59. Write orbital diagrams for the valence electrons and indicate the number of unpaired electrons for each element.
  (a) Br (b) Kr (c) Na (d) In

60. Write orbital diagrams for the valence electrons and indicate the number of unpaired electrons for each element.
  (a) Ne (b) I (c) Sr (d) Ge

61. How many valence electrons are in each element?
  (a) O (b) S (c) Br (d) Rb

62. How many valence electrons are in each element?
  (a) Ba (b) Al (c) Be (d) Se

ELECTRON CONFIGURATIONS AND THE PERIODIC TABLE

63. List the outer electron configuration for each column in the periodic table.
  (a) 1A (b) 2A (c) 5A (d) 7A

64. List the outer electron configuration for each column in the periodic table.
  (a) 3A (b) 4A (c) 6A (d) 8A

65. Use the periodic table to write electron configurations for each element.
  (a) Al (b) Be (c) In (d) Zr

66. Use the periodic table to write electron configurations for each element.
  (a) Tl (b) Co (c) Ba (d) Sb

67. Use the periodic table to write electron configurations for each element.
  (a) Sr (b) Y (c) Ti (d) Te

68. Use the periodic table to write electron configurations for each element.
  (a) Se (b) Sn (c) Pb (d) Cd

69. How many $2p$ electrons are in an atom of each element?
  (a) C (b) N (c) F (d) P

70. How many $3d$ electrons are in an atom of each element?
  (a) Fe (b) Zn (c) K (d) As

71. List the number of elements in periods 1 and 2 of the periodic table. Why does each period have a different number of elements?

72. List the number of elements in periods 3 and 4 of the periodic table. Why does each period have a different number of elements?

73. Name the element in the third period (row) of the periodic table with:
(a) three valence electrons
(b) a total of four $3p$ electrons
(c) six $3p$ electrons
(d) two $3s$ electrons and no $3p$ electrons

74. Name the element in the fourth period of the periodic table with:
(a) five valence electrons
(b) a total of four $4p$ electrons
(c) a total of three $3d$ electrons
(d) a complete outer shell

75. Use the periodic table to identify the element with each electron configuration.
(a) $[Ne]3s^23p^5$
(b) $[Ar]4s^23d^{10}4p^1$
(c) $[Ar]4s^23d^6$
(d) $[Kr]5s^1$

76. Use the periodic table to identify the element with each electron configuration.
(a) $[Ne]3s^1$
(b) $[Kr]5s^24d^{10}$
(c) $[Xe]6s^2$
(d) $[Kr]5s^24d^{10}5p^3$

PERIODIC TRENDS

77. Choose the element with the higher ionization energy from each pair.
(a) As or Bi   (b) As or Br   (c) S or I   (d) S or Sb

78. Choose the element with the higher ionization energy from each pair.
(a) Al or In   (b) Cl or Sb   (c) K or Ge   (d) S or Se

79. Arrange the elements in order of increasing ionization energy: Te, Pb, Cl, S, Sn.

80. Arrange the elements in order of increasing ionization energy: Ga, In, F, Si, N.

81. Choose the element with the larger atoms from each pair.
(a) Al or In
(b) Si or N
(c) P or Pb
(d) C or F

82. Choose the element with the larger atoms from each pair.
(a) Sn or Si
(b) Br or Ga
(c) Sn or Bi
(d) Se or Sn

83. Arrange these elements in order of increasing atomic size: Ca, Rb, S, Si, Ge, F.

84. Arrange these elements in order of increasing atomic size: Cs, Sb, S, Pb, Se.

85. Choose the more metallic element from each pair.
(a) Sr or Sb
(b) As or Bi
(c) Cl or O
(d) S or As

86. Choose the more metallic element from each pair.
(a) Sb or Pb
(b) K or Ge
(c) Ge or Sb
(d) As or Sn

87. Arrange these elements in order of increasing metallic character: Fr, Sb, In, S, Ba, Se.

88. Arrange these elements in order of increasing metallic character: Sr, N, Si, P, Ga, Al.

Cumulative Problems

89. What is the maximum number of electrons that can occupy the $n = 3$ quantum shell?

90. What is the maximum number of electrons that can occupy the $n = 4$ quantum shell?

91. Use the electron configurations of the alkaline earth metals to explain why they tend to form 2+ ions.

92. Use the electron configuration of oxygen to explain why it tends to form a 2− ion.

93. Write the electron configuration for each ion. What do all of the electron configurations have in common?
(a) Ca^{2+} (b) K^+ (c) S^{2-} (d) Br^-

94. Write the electron configuration for each ion. What do all of the electron configurations have in common?
(a) F^- (b) P^{3-} (c) Li^+ (d) Al^{3+}

95. Examine Figure 4.12, which shows the division of the periodic table into metals, nonmetals, and metalloids. Use what you know about electron configurations to explain these divisions.

96. Examine Figure 4.14, which shows the elements that form predictable ions. Use what you know about electron configurations to explain these trends.

97. Identify what is wrong with each electron configuration and write the correct ground-state (or lowest energy) configuration based on the number of electrons.
(a) $1s^32s^32p^9$
(b) $1s^22s^22p^62d^4$
(c) $1s^21p^5$
(d) $1s^22s^22p^83s^23p^1$

98. Identify what is wrong with each electron configuration and write the correct ground-state (or lowest energy) configuration based on the number of electrons.
(a) $1s^42s^42p^{12}$
(b) $1s^22s^22p^63s^23p^63d^{10}$
(c) $1s^22p^63s^2$
(d) $1s^22s^22p^63s^23p^64s^24d^{10}4p^3$

99. Bromine is a highly reactive liquid, while krypton is an inert gas. Explain this difference based on their electron configurations.

100. Potassium is a highly reactive metal, while argon is an inert gas. Explain this difference based on their electron configurations.

101. Based on periodic trends, which one of these elements would you expect to be most easily oxidized: Ge, K, S, or N?

102. Based on periodic trends, which one of these elements would you expect to be most easily reduced: Ca, Sr, P, or Cl?

103. When an electron makes a transition from the $n = 3$ to the $n = 2$ hydrogen atom Bohr orbit, the energy difference between these two orbits (3.0×10^{-19} J) is emitted as a photon of light. The relationship between the energy of a photon and its wavelength is given by $E = hc/\lambda$, where E is the energy of the photon in J, h is Planck's constant (6.626×10^{-34} J·s), and c is the speed of light (3.00×10^8 m/s). Find the wavelength of light emitted by hydrogen atoms when an electron makes this transition.

104. When an electron makes a transition from the $n = 4$ to the $n = 2$ hydrogen atom Bohr orbit, the energy difference between these two orbits (4.1×10^{-19} J) is emitted as a photon of light. The relationship between the energy of a photon and its wavelength is given by $E = hc/\lambda$, where E is the energy of the photon in J, h is Planck's constant (6.626×10^{-34} J·s), and c is the speed of light (3.00×10^8 m/s). Find the wavelength of light emitted by hydrogen atoms when an electron makes this transition.

105. The distance from the sun to Earth is 1.496×10^8 km. How long does it take light to travel from the sun to Earth?

106. The nearest star is Alpha Centauri, at a distance of 4.3 light-years from Earth. A light-year is the distance that light travels in one year (365 days). How far away, in kilometers, is Alpha Centauri from Earth?

107. The wave nature of matter was first proposed by Louis de Broglie, who suggested that the wavelength (λ) of a particle was related to its mass (m) and its velocity (v) by the equation: $\lambda = h/mv$, where h is Planck's constant (6.626×10^{-34} J·s). Calculate the de Broglie wavelength of: (a) a 0.0459 kg golf ball traveling at 95 m/s; (b) an electron traveling at 3.88×10^6 m/s. Can you explain why the wave nature of matter is significant for the electron but not for the golf ball? (*Hint:* Express mass in kilograms.)

108. The particle nature of light was first proposed by Albert Einstein, who suggested that light could be described as a stream of particles called photons. A photon of wavelength λ has an energy (E) given by the equation: $E = hc/\lambda$, where E is the energy of the photon in J, h is Planck's constant (6.626×10^{-34} J·s), and c is the speed of light (3.00×10^8 m/s). Calculate the energy of 1 mol of photons with a wavelength of 632 nm.

109. You learned in this chapter that ionization generally increases as you move from left to right across the periodic table. However, consider the following data, which shows the ionization energies of the period 2 and 3 elements:

Group	Period 2 Elements	Ionization Energy (kJ/mol)	Period 3 Elements	Ionization Energy (kJ/mol)
1A	Li	520	Na	496
2A	Be	899	Mg	738
3A	B	801	Al	578
4A	C	1086	Si	786
5A	N	1402	P	1012
6A	O	1314	S	1000
7A	F	1681	Cl	1251
8A	Ne	2081	Ar	1521

Notice that the increase is not uniform. In fact, ionization energy actually decreases a bit in going from elements in group 2A to 3A and then again from 5A to 6A. Use what you know about electron configurations to explain why these dips in ionization energy exist.

110. When atoms lose more than one electron, the ionization energy to remove the second electron is always more than the ionization energy to remove the first. Similarly, the ionization energy to remove the third electron is more than the second and so on. However, the increase in ionization energy upon the removal of subsequent electrons is not necessarily uniform. For example, consider the first three ionization energies of magnesium:

First ionization energy	738 kJ/mol
Second ionization energy	1450 kJ/mol
Third ionization energy	7730 kJ/mol

The second ionization energy is roughly twice the first ionization energy, but then the third ionization energy is over five times the second. Use the electron configuration of magnesium to explain why this is so. Would you expect the same behavior in sodium? Why or why not?

Highlight Problems

111. Excessive exposure to sunlight increases the risk of skin cancer because some of the photons have enough energy to break chemical bonds in biological molecules. These bonds require approximately 250−800 kJ/mol of energy to break. The energy of a single photon is given by $E = hc/\lambda$, where E is the energy of the photon in J, h is Planck's constant (6.626×10^{-34} J·s), and c is the speed of light (3.00×10^8 m/s). Determine which kinds of light contain enough energy to break chemical bonds in biological molecules by calculating the total energy in 1 mol of photons for light of each wavelength.

(a) infrared light (1500 nm)
(b) visible light (500 nm)
(c) ultraviolet light (150 nm)

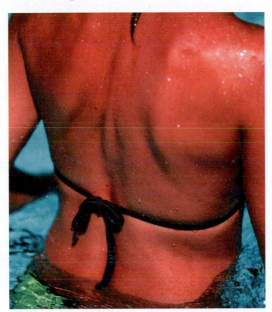

112. The quantum-mechanical model, besides revolutionizing chemistry, shook the philosophical world because of its implications regarding determinism. Determinism is the idea that the outcomes of future events are determined by preceding events. The trajectory of a baseball, for example, is deterministic; that is, its trajectory—and therefore its landing place—is determined by its position, speed, and direction of travel. Before quantum mechanics, most scientists thought that fundamental particles—such as electrons and protons—also behaved deterministically. The implication of this belief was that the entire universe must behave deterministically—the future must be determined by preceding events. Quantum mechanics challenged this reasoning because fundamental particles do not behave deterministically—their future paths are not determined by preceding events. Some scientists struggled with this idea. Einstein himself refused to believe it, stating, "God does not play dice with the universe." Explain what Einstein meant by this statement.

▲ "God does not play dice with the universe."

Questions for Group Work

Discuss these questions with the group and record your consensus answer.

113. Sketch the following orbitals (including the x, y, and z axes): $1s$, $2p_x$, $3d_{xy}$, $3d_{z^2}$.

114. Draw the best periodic table you can from memory (do not look at a table to do this). You do not need to label the elements, but you should put the correct number of elements in each block. After your group agrees that the group has done its best, spend exactly three minutes comparing your table with Figure 9.26. Make a second periodic table from memory. Is it better than your first one?

115. Play the following game to memorize the order in which orbitals fill. Go around your group and have each group member say the name of the next orbital to fill and the maximum number of electrons it can hold ("$1s$ two," "$2s$ two," "$2p$ six,". . .). If a group member gets stuck, other members can help, referring to Figure 9.24 or 9.26 if necessary. However, when anyone gets stuck, the next player starts back at "$1s$ two." Keep going until each group member can list the correct sequence up to "$6s$ two."

116. Using grammatically correct sentences, describe the periodic trends for atomic size, ionization energy, and metallic character.

Data Interpretation and Analysis

117. The first graph shown here is of the first ionization energies (the energy associated with removing an electron) of the period 3 elements. The second graph shows the electron affinities (the energy associated with gaining an electron) of the period 3 elements. Refer to the graphs to answer the questions.

(a) Notice that the ionization energies are positive and the electron affinities are negative. Explain the significance of this difference.

(b) Describe the general trend in period 3 first ionization energies as you move from left to right across the periodic table. Explain why this trend occurs. (*Hint:* Consider the trend in atomic size as you move from left to right across the periodic table.)

(c) The trend in first ionization energy has two exceptions: one at Al and another at S. Write the electron configurations of Mg, Al, P, and S and refer to them to explain the exceptions.

(d) Describe the general trend in period 3 electron affinities as you move from left to right across the periodic table. Explain why this trend occurs. (*Hint:* Consider the trend in atomic size as you move from left to right across the periodic table.)

(e) The trend in electron affinities has exceptions. Write the electron configurations of Si and P and explain why the electron affinity for Si is more exothermic than that of P.

(f) Determine the overall energy change for removing one electron from Na and adding that electron to Cl. Is the exchange of the electron exothermic or endothermic?

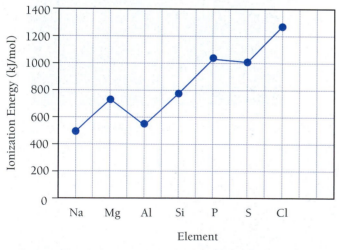

▲ **First Ionization Energies of Period 3 Elements**

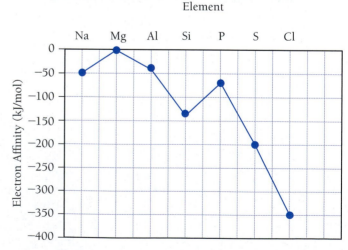

▲ **Electron Affinities of Period 3 Elements**

Answers to Skillbuilder Exercises

Skillbuilder 9.1................. (a) blue, green, red
(b) red, green, blue
(c) red, green, blue

Skillbuilder 9.2
(a) Al $1s^22s^22p^63s^23p^1$ or [Ne]$3s^23p^1$
(b) Br $1s^22s^22p^63s^23p^64s^23d^{10}4p^5$ or [Ar]$4s^23d^{10}4p^5$
(c) Sr $1s^22s^22p^63s^23p^64s^23d^{10}4p^65s^2$ or [Kr]$5s^2$

Skillbuilder Plus, p. 301
Subtract one electron for each unit of positive charge. Add one electron for each unit of negative charge.
(a) Al^{3+} $1s^22s^22p^6$
(b) Cl$^-$ $1s^22s^22p^63s^23p^6$
(c) O^{2-} $1s^22s^22p^6$

Skillbuilder 9.3

Ar

       1s   2s   2p   3s   3p

Skillbuilder 9.4

Cl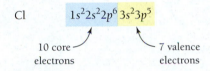

     10 core   7 valence
     electrons   electrons

Skillbuilder 9.5................. [Kr]$5s^24d^{10}5p^2$
Skillbuilder 9.6................. (a) Pb
(b) Rb
(c) cannot determine based on periodic properties
(d) Se

Skillbuilder 9.7................. (a) Mg
(b) Te
(c) cannot determine based on periodic properties
(d) F

Skillbuilder 9.8................. (a) In
(b) cannot determine based on periodic properties
(c) Bi
(d) B

Answers to Conceptual Checkpoints

9.1 (a) Wavelength and frequency are inversely related. Therefore, the shortest wavelength has the highest frequency.

9.2 (b) Wavelength and energy per photon are inversely related. Since yellow light has a longer wavelength, it has less energy per photon than violet light.

9.3 (c) The higher-energy levels are more closely spaced than the lower ones, so the difference in energy between $n = 2$ and $n = 1$ is greater than the difference in energy between $n = 3$ and $n = 2$. The photon emitted when an electron falls from $n = 2$ to $n = 1$ therefore carries more energy, corresponding to radiation with a shorter wavelength and higher frequency.

9.4 (c) The $n = 3$ principal shell contains three subshells: s, p, and d.

9.5 (b) The p subshell contains three orbitals.

9.6 (d) Both have six electrons in $2p$ orbitals and six electrons in $3p$ orbitals.

9.7 (d) The outermost principal shell for K is $n = 4$, which contains only a single valence electron, $4s^1$.

9.8 (a) Calcium loses its $4s$ electron and attains a noble gas configuration (that of Ar).

9.9 (b) Ionization energy increases as you move from left to right across a row in the periodic table. Both atomic size and metallic character decrease as you move from left to right across a row.

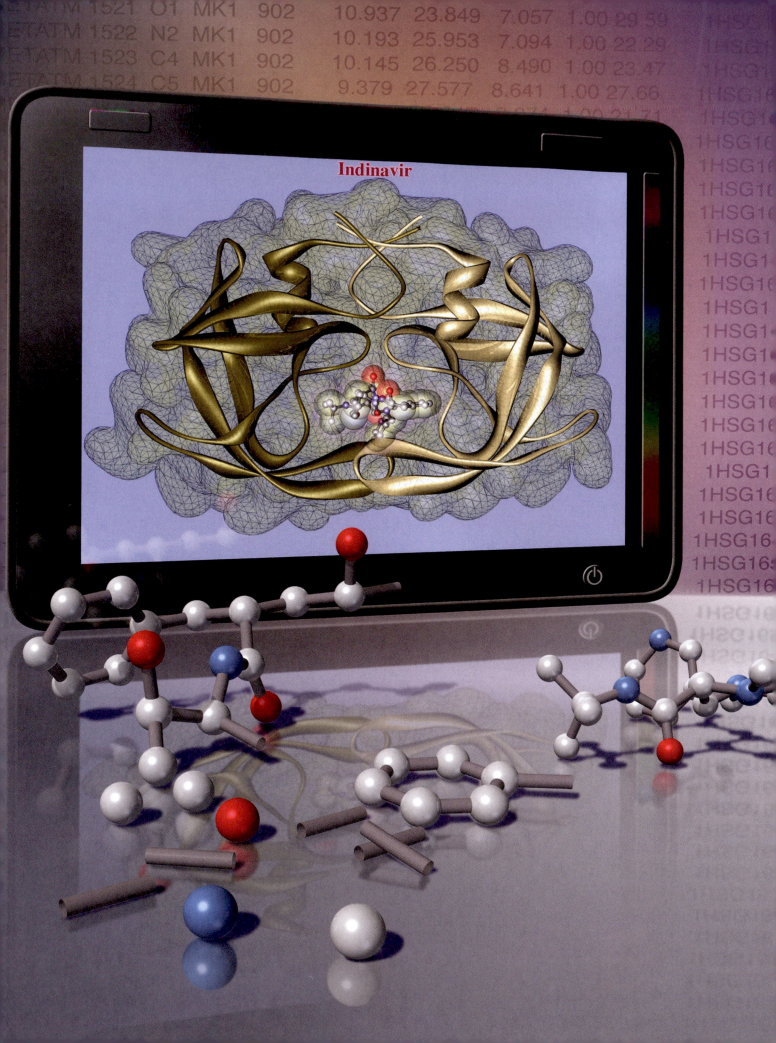

10 Chemical Bonding

The fascination of a growing science lies in the work of the pioneers at the very borderland of the unknown, but to reach this frontier one must pass over well traveled roads.

—Gilbert N. Lewis (1875–1946)

10.1 Bonding Models and AIDS Drugs

We will discuss proteins in more detail in Chapter 19.

In 1989, researchers discovered the structure of a molecule called HIV-protease. HIV-protease is a protein (a class of biological molecules) synthesized by the human immunodeficiency virus (HIV), which causes AIDS. HIV-protease is crucial to the virus's ability to replicate itself. Without HIV-protease, HIV could not spread in the human body because the virus could not copy itself, and AIDS would not develop.

With knowledge of the HIV-protease structure, drug companies set out to design a molecule that would disable the protease by attaching to the working part of the molecule (called the *active site*). To design such a molecule, researchers used **bonding theories**—models that predict how atoms bond together to form molecules—to simulate how potential drug molecules would interact with the protease molecule. By the early 1990s, these companies had developed several effective drug molecules. Since these molecules inhibit the action of HIV-protease, they are called *protease inhibitors*. In human trials, protease inhibitors in combination with other drugs decrease the viral count in HIV-infected individuals to undetectable levels. Although these drugs do not cure AIDS, HIV-infected individuals who regularly take their medication can now expect nearly normal life spans. The use of protease inhibitors has since been expanded to include treatment for Hepatitis C with similarly excellent results.

Bonding theories are central to chemistry because they predict how atoms bond together to form compounds. They predict which combinations of atoms form compounds and which combinations do not. Bonding theories predict why salt is $NaCl$ and not $NaCl_2$ and why water is H_2O and not H_3O. Bonding theories also explain the shapes of molecules, which in turn determine many of their physical and chemical properties.

The bonding theory you will learn in this chapter is the **Lewis model**, named after G. N. Lewis (1875–1946), the American chemist who developed it. In this

◀ The gold-colored structure on the tablet screen is a representation of HIV-protease. The molecule shown in the center is Indinavir, a protease inhibitor.

model, we represent electrons as dots and draw *dot structures* or *Lewis structures* to represent molecules. These structures, which are fairly simple to draw, have tremendous predictive power. It takes just a few minutes to apply the Lewis model to determine whether a particular set of atoms will form a stable molecule and what that molecule might look like. Although modern chemists also use more advanced bonding theories to better predict molecular properties, the Lewis model remains the simplest method for making quick, everyday predictions about molecules.

10.2 Representing Valence Electrons with Dots

▶ Write Lewis structures for elements.

As we discussed in Chapter 9, valence electrons are the electrons in the outermost principal shell. Since valence electrons are most important in bonding, the Lewis model focuses on these. In the Lewis model, the valence electrons of main-group elements are represented as dots surrounding the symbol of the element. The result is a **Lewis structure**, or **dot structure**. For example, the electron configuration of O is:

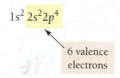

and its Lewis structure is:

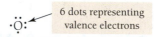

Remember, the number of valence electrons for any main-group element is equal to the group number of the element (except helium, which has two valence electrons but is in Group 8A).

Each dot represents a valence electron. We place the dots around the element's symbol with a maximum of two dots per side. Although the exact location of dots is not critical, in this book we fill in the dots singly first and then pair them (with the exception of helium, described shortly).

The Lewis structures for the period 2 elements are:

$$ \text{Li·} \quad \text{·Be·} \quad \text{·}\overset{\cdot}{\text{B}}\text{·} \quad \text{·}\overset{\cdot}{\underset{\cdot}{\text{C}}}\text{·} \quad \text{·}\overset{\cdot\cdot}{\underset{\cdot}{\text{N}}}\text{:} \quad \text{·}\overset{\cdot\cdot}{\underset{\cdot\cdot}{\text{O}}}\text{:} \quad \text{:}\overset{\cdot\cdot}{\underset{\cdot\cdot}{\text{F}}}\text{:} \quad \text{:}\overset{\cdot\cdot}{\underset{\cdot\cdot}{\text{Ne}}}\text{:} $$

Lewis structures allow us to easily see the number of valence electrons in an atom. Atoms with eight valence electrons—which are particularly stable—are easily identified because they have eight dots, an **octet**.

Helium is somewhat of an exception. Its electron configuration and Lewis structure are:

$$ 1s^2 \quad \text{He:} $$

The Lewis structure of helium contains two paired dots (a **duet**). For helium, a duet represents a stable electron configuration.

In the Lewis model, a **chemical bond** involves the sharing or transfer of electrons to attain stable electron configurations for the bonding atoms. If the electrons are transferred, the bond is an **ionic bond**. If the electrons are shared, the bond is a **covalent bond**. In either case, the bonding atoms attain stable electron configurations. As we have seen, a stable configuration usually consists of eight electrons in the outermost or valence shell. This observation leads to the **octet rule**:

In chemical bonding, atoms transfer or share electrons to obtain outer shells with eight electrons.

The octet rule generally applies to all main-group elements except hydrogen and helium. Each of these elements achieves stability when it has two electrons (a duet) in its outermost shell.

EXAMPLE **10.1** | **Writing Lewis Structures for Elements**

Write the Lewis structure of phosphorus.

Since phosphorus is in Group 5A in the periodic table, it has five valence electrons. Represent these as five dots surrounding the symbol for phosphorus.	**SOLUTION** $\cdot \overset{\displaystyle \cdot}{\underset{\displaystyle \cdot}{P}} :$

▶ **SKILLBUILDER 10.1** | **Writing Lewis Structures for Elements**

Write the Lewis structure of Mg.

▶ **FOR MORE PRACTICE** Example 10.12; Problems 25, 26.

CONCEPTUAL ✓ CHECKPOINT 10.1

Which two elements have the most similar Lewis structures?

(**a**) C and Si   (**b**) O and P   (**c**) Li and F   (**d**) S and Br

10.3 Lewis Structures of Ionic Compounds: Electrons Transferred

▶ Write Lewis structures for ionic compounds.

▶ Use the Lewis model to predict the chemical formula of an ionic compound.

Recall from Chapter 5 that when metals bond with nonmetals, electrons are transferred from the metal to the nonmetal. The metal becomes a cation and the nonmetal becomes an anion. The attraction between the cation and the anion results in an ionic compound. In the Lewis model, we represent this by moving electron dots from the metal to the nonmetal. For example, the Lewis structures for potassium and chlorine are:

$$K \cdot \quad : \overset{\displaystyle \cdot \cdot}{\underset{\displaystyle \cdot \cdot}{Cl}} :$$

When potassium and chlorine bond, potassium transfers its valence electron to chlorine.

$$K \cdot \quad : \overset{\displaystyle \cdot \cdot}{\underset{\displaystyle \cdot \cdot}{Cl}} : \quad \longrightarrow \quad K^+ \; [: \overset{\displaystyle \cdot \cdot}{\underset{\displaystyle \cdot \cdot}{Cl}} :]^-$$

Recall from Section 4.7 that atoms that lose electrons become positively charged and atoms that gain electrons become negatively charged.

The transfer of the electron gives chlorine an octet (shown as eight dots around chlorine) and leaves potassium with an octet in the previous principal shell, which is now the valence shell. Because the potassium lost an electron, it becomes positively charged, while the chlorine, which gained an electron, becomes negatively charged. We usually write the Lewis structure of an anion in brackets with the charge in the upper right corner (outside the brackets). The positive and negative charges attract one another, forming the compound KCl.

EXAMPLE **10.2** | **Writing Ionic Lewis Structures**

Write the Lewis structure of the compound MgO.

Draw the Lewis structures of magnesium and oxygen by drawing two dots around the symbol for magnesium and six dots around the symbol for oxygen.	**SOLUTION** $\cdot Mg \cdot \quad \cdot \overset{\displaystyle \cdot \cdot}{\underset{\displaystyle \cdot}{O}} :$
In MgO, magnesium loses its two valence electrons, resulting in a 2+ charge, and oxygen gains two electrons, attaining a 2− charge and an octet.	$Mg^{2+} \; [: \overset{\displaystyle \cdot \cdot}{\underset{\displaystyle \cdot \cdot}{O}} :]^{2-}$

▶ **SKILLBUILDER 10.2** | **Writing Ionic Lewis Structures**

Write the Lewis structure of the compound NaBr.

▶ **FOR MORE PRACTICE** Example 10.13; Problems 37, 38.

> Recall from Section 5.4 that ionic compounds do not exist as distinct molecules, but rather as part of a large three-dimensional array (or lattice) of alternating cations and anions.

The Lewis model predicts the correct chemical formulas for ionic compounds. For the compound that forms between K and Cl, for example, the Lewis model predicts one potassium cation to every chlorine anion, KCl. As another example, consider the ionic compound formed between sodium and sulfur. The Lewis structures for sodium and sulfur are:

$$\text{Na} \cdot \quad \cdot \ddot{\text{S}} :$$

Notice that sodium must lose its one valence electron to obtain an octet (in the previous principal shell), while sulfur must gain two electrons to obtain an octet. Consequently, the compound that forms between sodium and sulfur requires two sodium atoms to every one sulfur atom. The Lewis structure is:

$$\text{Na}^+ \quad [: \ddot{\text{S}} :]^{2-} \quad \text{Na}^+$$

The two sodium atoms each lose their single valence electron, while the sulfur atom gains two electrons and obtains an octet. The correct chemical formula is Na_2S.

EXAMPLE 10.3

Using the Lewis Model to Predict the Chemical Formula of an Ionic Compound

Use the Lewis model to predict the formula of the compound that forms between calcium and chlorine.

	SOLUTION
Draw the Lewis structures of calcium and chlorine by drawing two dots around the symbol for calcium and seven dots around the symbol for chlorine.	$\cdot \text{Ca} \cdot \quad : \ddot{\text{Cl}} :$
Calcium must lose its two valence electrons (to effectively attain an octet in its previous principal shell), while chlorine needs to gain only one electron to obtain an octet. Consequently, the compound that forms between Ca and Cl has two chlorine atoms to every one calcium atom.	$[: \ddot{\text{Cl}} :]^- \quad \text{Ca}^{2+} \quad [: \ddot{\text{Cl}} :]^-$ The formula is therefore $CaCl_2$.

▶ **SKILLBUILDER 10.3** | **Using the Lewis Model to Predict the Chemical Formula of an Ionic Compound**

Use the Lewis model to predict the formula of the compound that forms between magnesium and nitrogen.

▶ **FOR MORE PRACTICE** Example 10.14; Problems 39, 40, 41, 42.

CONCEPTUAL ✔ CHECKPOINT 10.2

Which nonmetal forms an ionic compound with aluminum that has the formula Al_2X_3 (where X represents the nonmetal)?

(a) Cl   (b) S   (c) N   (d) C

10.4 Covalent Lewis Structures: Electrons Shared

▶ Write Lewis structures for covalent compounds.

Recall from Chapter 5 that when nonmetals bond with other nonmetals, a molecular compound results. Molecular compounds contain covalent bonds in which electrons are shared between atoms rather than transferred. Electrons are normally shared in pairs to form single, double, or triple bonds.

Single Bonds

In the Lewis model, we represent a single covalent bond by allowing neighboring atoms to share a pair of valence electrons to attain an octet (or duet for hydrogen). For example, hydrogen and oxygen have the Lewis structures:

$$\text{H} \cdot \quad \cdot \ddot{\text{O}} :$$

In water, hydrogen and oxygen share their electrons so that each hydrogen atom has a duet and the oxygen atom has an octet.

$$\text{H} \!:\! \overset{..}{\text{O}} \!:\! \text{H}$$

The shared electrons—those that appear in the space between the two atoms—count toward the octets (or duets) of *both of the atoms*.

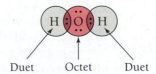

Duet   Octet   Duet

Electrons that are shared between two atoms are **bonding pair** electrons, while those that are only on one atom are **lone pair** (or nonbonding) electrons.

Bonding
pair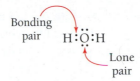
Lone
pair

A bonding pair of electrons is often represented by a dash to emphasize that it is a chemical bond.

$$\text{H} \!-\! \overset{..}{\text{O}} \!-\! \text{H}$$

Remember that each dash represents a *pair* of shared electrons.

The Lewis model also explains why the halogens form diatomic molecules. Consider the Lewis structure of chlorine:

$$: \overset{..}{\underset{..}{\text{Cl}}} :$$

If two Cl atoms pair, they can each attain an octet:

$$: \overset{..}{\underset{..}{\text{Cl}}} \!:\! \overset{..}{\underset{..}{\text{Cl}}} : \quad or \quad : \overset{..}{\underset{..}{\text{Cl}}} \!-\! \overset{..}{\underset{..}{\text{Cl}}} :$$

When we examine elemental chlorine, we find that it indeed exists as a diatomic molecule, just as the Lewis model predicts. The same is true for the other halogens.

Similarly, the Lewis model predicts that hydrogen, which has this Lewis structure:

$$\text{H} \cdot$$

should exist as H_2. When two hydrogen atoms share their valence electrons, they each have a duet, a stable configuration for hydrogen.

$$\text{H} \!:\! \text{H} \quad or \quad \text{H} \!-\! \text{H}$$

Again, the Lewis model prediction is correct. In nature, elemental hydrogen exists as H_2 molecules.

Double and Triple Bonds

In the Lewis model, atoms can share more than one electron pair to attain an octet. For example, we know from Chapter 5 that oxygen exists as the diatomic molecule, O_2. The Lewis structure of an oxygen atom is:

$$\cdot \overset{..}{\text{O}} :$$

If we pair two oxygen atoms and then try to write the Lewis structure, we do not have enough electrons to give each O atom an octet.

$$: \overset{..}{\text{O}} \!:\! \overset{..}{\underset{..}{\text{O}}} :$$

However, we can convert a lone pair into an additional bonding pair by moving it into the bonding region.

$$: \overset{..}{\text{O}} \!:\! \overset{..}{\underset{..}{\text{O}}} : \longrightarrow : \overset{..}{\text{O}} \!::\! \overset{..}{\text{O}} : \quad or \quad : \overset{..}{\text{O}} \!=\! \overset{..}{\text{O}} :$$

Each oxygen atom now has an octet because the additional bonding pair counts toward the octet of both oxygen atoms.

Octet ———→  ←——— Octet

When two atoms share two electron pairs, the resulting bond is a **double bond**. In general, double bonds are shorter and stronger than single bonds. For example, the distance between oxygen nuclei in an oxygen–oxygen double bond is 121 pm. In a single bond, it is 148 pm.

Two atoms can also share three electron pairs. Consider the Lewis structure of N_2. Since each N atom has five valence electrons, the Lewis structure for N_2 has ten electrons. A first attempt at writing the Lewis structure looks like this:

$$:\ddot{N}:\ddot{N}:$$

As with O_2, we do not have enough electrons to satisfy the octet rule for both N atoms. However, if we convert two additional lone pairs into bonding pairs, each nitrogen atom has an octet.

$$:\ddot{N}:\ddot{N}: \longrightarrow :N:::N: \quad or \quad :N \equiv N:$$

The resulting bond is a **triple bond**. Triple bonds are even shorter and stronger than double bonds. The distance between nitrogen nuclei in a nitrogen–nitrogen triple bond is 110 pm. In a double bond, the distance is 124 pm. When we examine nitrogen in nature, we find that it exists as a diatomic molecule with a very strong short bond between the two nitrogen atoms. The bond is so strong that it is difficult to break, making N_2 a relatively unreactive molecule.

PEARSON eText 2.0

CONCEPTUAL ✔ CHECKPOINT 10.3

How many bonding electrons are in the Lewis structure of O_2?

(a) 2   (b) 4   (c) 6

10.5 Writing Lewis Structures for Covalent Compounds

▶ Write Lewis structures for covalent compounds.

When guessing at skeletal structures, we put the less metallic elements in terminal positions and the more metallic elements in central positions. Halogens, which are among the least metallic elements, are almost always terminal.

Nonterminal hydrogen atoms exist in some compounds. However, they are rare and beyond the scope of this text.

To write the Lewis structure for a covalent compound, follow these steps:

1. **Write the correct skeletal structure for the molecule.** The skeletal structure shows the relative positions of the atoms and does not include electrons, but it must have the atoms in the correct positions. For example, you *cannot* write the Lewis structure for water if you start with the hydrogen atoms next to each other and the oxygen atom at the end (H H O). In nature, oxygen is the central atom, and the hydrogen atoms are **terminal atoms** (at the ends). The correct skeletal structure is H O H.

   The only way to absolutely know the correct skeletal structure for any molecule is to examine its structure in nature. However, you can write likely skeletal structures by remembering two guidelines. First, *hydrogen atoms are always terminal*. Since hydrogen requires only a duet, it is never a central atom because central atoms must be able to form at least two bonds and hydrogen can form only one. Second, *many molecules tend to be symmetrical*, so when a molecule contains several atoms of the same type, these tend to be in terminal positions. *This symmetry guideline, however, has many exceptions.* In cases where the skeletal structure is unclear, this text provides the correct skeletal structure.

2. **Calculate the total number of electrons for the Lewis structure by summing the valence electrons of each atom in the molecule.** Remember that the number of valence electrons for any main-group element is equal to its group number in the periodic table. **If you are writing a Lewis structure for a**

1 pm = 10^{-12} m

polyatomic ion, you must consider the charge of the ion when calculating the total number of electrons. Add one electron for each negative charge and subtract one electron for each positive charge.

3. **Distribute the electrons among the atoms, giving octets (or duets for hydrogen) to as many atoms as possible.** Begin by placing two electrons between each pair of atoms. These are the minimal number of bonding electrons. Then distribute the remaining electrons, first to terminal atoms and then to the central atom, giving octets to as many atoms as possible.

4. **If any atoms lack an octet, form double or triple bonds as necessary to give them octets.** Do this by moving lone electron pairs from terminal atoms into the bonding region with the central atom.

A brief version of this procedure is presented in the left column. In the center and right columns, Examples 10.4 and 10.5 illustrate the procedure.

PEARSON eText 2.0 | Interactive Worked Example Video 10.4

Writing Lewis Structures for Covalent Compounds	EXAMPLE **10.4** Write the Lewis structure for CO_2.	EXAMPLE **10.5** Write the Lewis structure for CCl_4.
1. Write the correct skeletal structure for the molecule.	**SOLUTION** Following the symmetry guideline, write: O C O	**SOLUTION** Following the symmetry guideline, write: Cl Cl C Cl Cl
2. Calculate the total number of electrons for the Lewis structure by summing the valence electrons of each atom in the molecule.	Total number of electrons for Lewis structure = $\left(\begin{matrix}\text{\# valence}\\\text{e}^- \text{ for C}\end{matrix}\right) + 2\left(\begin{matrix}\text{\# valence}\\\text{e}^- \text{ for O}\end{matrix}\right)$ $= 4 + 2(6)$ $= 16$	Total number of electrons for Lewis structure = $\left(\begin{matrix}\text{\# valence}\\\text{e}^- \text{ for C}\end{matrix}\right) + 4\left(\begin{matrix}\text{\# valence}\\\text{e}^- \text{ for Cl}\end{matrix}\right)$ $= 4 + 4(7)$ $= 32$
3. Distribute the electrons among the atoms, giving octets (or duets for hydrogen) to as many atoms as possible. Begin with the bonding electrons, proceed to lone pairs on terminal atoms, and finally go to lone pairs on the central atom.	Work with bonding electrons first. O:C:O (4 of 16 electrons used) Proceed to lone pairs on terminal atoms next. :Ö:C:Ö: (16 of 16 electrons used)	Work with bonding electrons first. Cl Cl:C:Cl Cl (8 of 32 electrons used) Proceed to lone pairs on terminal atoms next. :Cl: :Cl:C:Cl: :Cl: (32 of 32 electrons used)
4. If any atoms lack octets, form double or triple bonds as necessary to give them octets.	Move lone pairs from the oxygen atoms to bonding regions to form double bonds. :Ö:C:Ö: ⟶ :O::C::O: ▶ **SKILLBUILDER 10.4** | Write the Lewis structure for CO.	Since all of the atoms have octets, the Lewis structure is complete. ▶ **SKILLBUILDER 10.5** | Write the Lewis structure for H_2CO. ▶ **FOR MORE PRACTICE** Example 10.15; Problems 47, 48, 49, 50, 51, 52.

Writing Lewis Structures for Polyatomic Ions

We write Lewis structures for polyatomic ions by following the same procedure, but we pay special attention to the charge of the ion when calculating the number of electrons for the Lewis structure. We add one electron for each negative charge and subtract one electron for each positive charge. We normally show the Lewis structure for a polyatomic ion within brackets and write the charge of the ion in the upper right corner. For example, suppose we want to write the Lewis structure for the CN^- ion. We begin by writing the skeletal structure:

CN

Next we calculate the total number of electrons for the Lewis structure by summing the number of valence electrons for each atom and adding one for the negative charge.

Total number of electrons
for Lewis structure $= (\text{\# valence } e^- \text{ in C}) + (\text{\# valence } e^- \text{ in N}) + 1$

$$= 4 + 5 + 1$$

Add one e^- to account
for 1− charge of ion.

$$= 10$$

We then place two electrons between each pair of atoms

C:N   (2 of 10 electrons used)

and distribute the remaining electrons.

:C̈:N̈:   (10 of 10 electrons used)

Since neither of the atoms has octets, we move two lone pairs into the bonding region to form a triple bond, giving both atoms octets. We also enclose the Lewis structure in brackets and write the charge of the ion in the upper right corner.

[:C:::N:][−]   or   [:C≡N:][−]

PEARSON eText 2.0

CONCEPTUAL ✔ CHECKPOINT 10.4

How many electrons are there in the Lewis structure of OH^-?

(a) 6   (b) 7   (c) 8   (d) 9

EXAMPLE 10.6   Writing Lewis Structures for Polyatomic Ions

Write the Lewis structure for the NH_4^+ ion.

Begin by writing the skeletal structure. Hydrogen atoms must be terminal, and following the guideline of symmetry, the nitrogen atom should be in the middle surrounded by four hydrogen atoms.	**SOLUTION** H H N H H
Calculate the total number of electrons for the Lewis structure by summing the number of valence electrons for each atom and subtracting one for the positive charge.	Total number of electrons for Lewis structure $= 5 + 4 - 1 = 8$ $4 \times (\text{\# valence } e^- \text{ in H})$ # valence e^- in N Subtract 1 e^- to account for 1+ charge of ion.
Next, place two electrons between each pair of atoms.	H H:N̈:H (8 of 8 electrons used) H

Since the nitrogen atom has an octet and all of the hydrogen atoms have duets, the placement of electrons is complete. Write the entire Lewis structure in brackets indicating the charge of the ion in the upper right corner.

$$\left[\begin{array}{c} H \\ H\!:\!N\!:\!H \\ H \end{array}\right]^{+} \quad or \quad \left[\begin{array}{c} H \\ | \\ H\!-\!N\!-\!H \\ | \\ H \end{array}\right]^{+}$$

▶ **SKILLBUILDER 10.6** | **Writing Lewis Structures for Polyatomic Ions**

Write the Lewis structure for the ClO^{-} ion.

▶ **FOR MORE PRACTICE** Problems 55bcd, 56abc, 57, 58.

CONCEPTUAL ✔ CHECKPOINT 10.5

Which two species have the same number of lone electron pairs in their Lewis structures?

(a) H_2O and H_3O^+

(b) NH_3 and H_3O^+

(c) NH_3 and CH_4

(d) NH_3 and NH_4^+

Exceptions to the Octet Rule

Lewis model predictions are often correct, but exceptions exist. For example, if we try to write the Lewis structure for NO, which has 11 electrons, the best we can do is:

$$:\!\overset{.}{N}\!:\!:\!\overset{..}{O}\!: \quad or \quad :\!\overset{.}{N}\!=\!\overset{..}{O}\!:$$

The nitrogen atom does not have an octet, so this is not a great Lewis structure. However, NO exists in nature. Why does the Lewis model not account for the existence of NO? As with any simple theory, the Lewis model is not sophisticated enough to be correct every time. It is impossible to write good Lewis structures for molecules with odd numbers of electrons, yet some of these molecules exist in nature. In such cases, we write the best Lewis structure that we can. Another significant exception to the octet rule is boron, which tends to form compounds with only six electrons around B, rather than eight. For example, BF_3 and BH_3—which both exist in nature—each lack an octet for B.

$$\begin{array}{c} :\!\overset{..}{F}\!: \\ :\!\overset{..}{F}\!:\!B\!:\!\overset{..}{F}\!: \end{array} \qquad \begin{array}{c} H \\ H\!:\!\overset{..}{B}\!:\!H \end{array}$$

A third type of exception to the octet rule is also common. A number of molecules, such as SF_6 and PCl_5, have more than eight electrons around a central atom in their Lewis structures.

$$\begin{array}{c} :\!\overset{..}{F}\!: \\ :\!\overset{..}{F}\!\diagdown\!|\!\diagup\!\overset{..}{F}\!: \\ S \\ :\!\overset{..}{F}\!\diagup\!|\!\diagdown\!\overset{..}{F}\!: \\ :\!\overset{..}{F}\!: \end{array} \qquad \begin{array}{c} :\!\overset{..}{Cl}\!: \\ :\!\overset{..}{Cl}\!\diagdown\!|\!\diagup\!\overset{..}{Cl}\!: \\ P \\ :\!\overset{..}{Cl}\!. \quad .\!\overset{..}{Cl}\!: \end{array}$$

We often refer to these as *expanded octets*. Expanded octets can form for period 3 elements and beyond. Beyond mentioning them, we do not cover expanded octets in this book. In spite of these exceptions, the Lewis model remains a powerful and simple way to understand chemical bonding.

10.6 Resonance: Equivalent Lewis Structures for the Same Molecule

▶ Write resonance structures.

Key Concept Video
Resonance and Formal Charge

When writing Lewis structures, we may find that, for some molecules, we can write more than one good Lewis structure. For example, consider writing a Lewis structure for SO_2. We begin with the skeletal structure:

$$O\ S\ O$$

We then sum the valence electrons.

Total number of electrons for Lewis structure

$= (\#\text{ valence e}^-\text{ in S}) + 2(\#\text{ valence e}^-\text{ in O})$

$= 6 + 2(6)$

$= 18$

We next place two electrons between each pair of atoms

$$O:S:O \quad \text{(4 of 18 electrons used)}$$

and distribute the remaining electrons, first to terminal atoms

$$:\ddot{O}:S:\ddot{O}: \quad \text{(16 of 18 electrons used)}$$

and finally to the central atom.

$$:\ddot{O}:\ddot{S}:\ddot{O}: \quad \text{(18 of 18 electrons used)}$$

Since the central atom lacks an octet, we move one lone pair from an oxygen atom into the bonding region to form a double bond, giving all of the atoms octets.

$$:\ddot{O}::\ddot{S}:\ddot{O}: \quad or \quad :\ddot{O}\!=\!\ddot{S}\!-\!\ddot{O}:$$

However, we could have formed the double bond with the other oxygen atom.

$$:\ddot{O}\!-\!\ddot{S}\!=\!\ddot{O}:$$

These two Lewis structures are equally correct. In cases such as this—where we can write two or more equivalent (or nearly equivalent) Lewis structures for the same molecule—we find that the molecule exists in nature as an average or intermediate between the two Lewis structures. Both of the two Lewis structures for SO_2 predict that SO_2 would contain two different kinds of bonds (one double bond and one single bond). However, when we examine SO_2 in nature, we find that both of the bonds are equivalent and intermediate in strength and length between a double bond and single bond.

We address this in the Lewis model by representing the molecule with both structures, called **resonance structures**, with a double-headed arrow between them.

$$:\ddot{O}\!=\!\ddot{S}\!-\!\ddot{O}: \longleftrightarrow :\ddot{O}\!-\!\ddot{S}\!=\!\ddot{O}:$$

The true structure of SO_2 is intermediate between these two resonance structures and is called a *resonance hybrid*. Resonance structures always have the same skeletal structure (the atoms are in the same relative positions); only the distribution of electron dots differs between them.

EXAMPLE **10.7** | **Writing Resonance Structures**

Write the Lewis structure for the NO_3^- ion. Include resonance structures.

Begin by writing the skeletal structure. Applying the guideline of symmetry, make the three oxygen atoms terminal.	**SOLUTION** O O N O
Sum the valence electrons (adding one electron to account for the 1− charge) to determine the total number of electrons in the Lewis structure.	Total number of electrons for Lewis structure $= 5 + 3(6) + 1 = 24$ 3 × (# valence e⁻ in O) # valence e⁻ in N Add one e⁻ to account for negative charge of ion.
Place two electrons between each pair of atoms.	 O O:N:O (6 of 24 electrons used)
Distribute the remaining electrons, first to the terminal atoms.	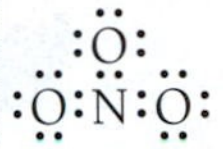 :Ö: :Ö:N:Ö: (24 of 24 electrons used)
Since there are no electrons remaining to complete the octet of the central atom, form a double bond by moving a lone pair from one of the oxygen atoms into the bonding region with nitrogen. Enclose the structure in brackets and write the charge at the upper right.	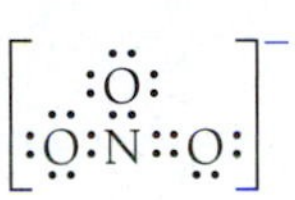 $\left[\begin{array}{c}:\ddot{O}:\\:\ddot{O}:N::\ddot{O}:\end{array}\right]^-$ or $\left[\begin{array}{c}:\ddot{O}:\\:\ddot{O}-N=\ddot{O}:\end{array}\right]^-$ 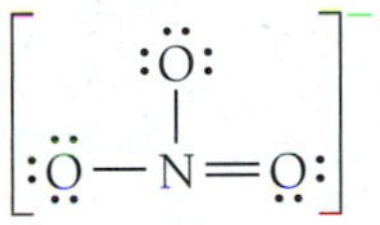
Notice that you can form the double bond with either of the other two oxygen atoms as well.	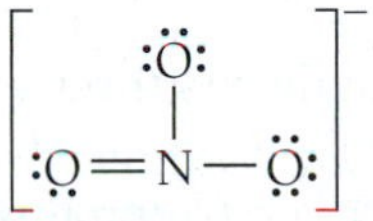 $\left[\begin{array}{c}:\ddot{O}:\\:\ddot{O}=N-\ddot{O}:\end{array}\right]^-$ or $\left[\begin{array}{c}:\ddot{O}\\:\ddot{O}-N-\ddot{O}:\end{array}\right]^-$ 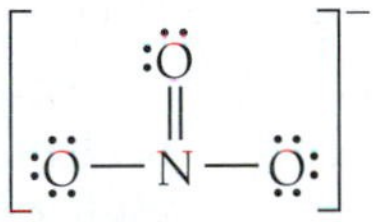
Since the three Lewis structures are equally correct, write the three structures as resonance structures.	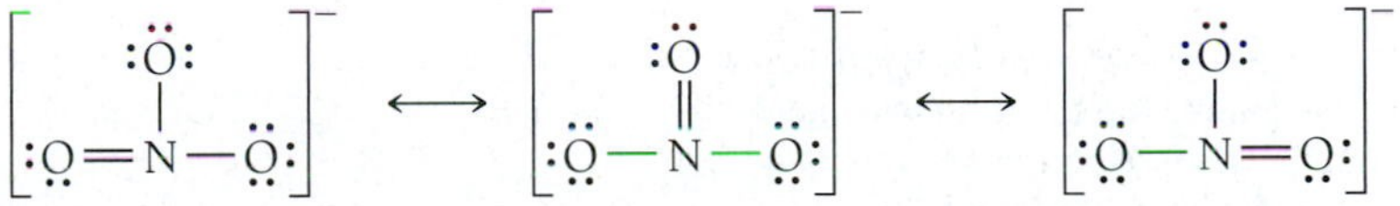 $\left[\begin{array}{c}:\ddot{O}:\\:\ddot{O}=N-\ddot{O}:\end{array}\right]^- \longleftrightarrow \left[\begin{array}{c}:\ddot{O}\\:\ddot{O}-N-\ddot{O}:\end{array}\right]^- \longleftrightarrow \left[\begin{array}{c}:\ddot{O}:\\:\ddot{O}-N=\ddot{O}:\end{array}\right]^-$

▶ **SKILLBUILDER 10.7** | **Writing Resonance Structures**

Write the Lewis structure for the NO_2^- ion. Include resonance structures.

▶ **FOR MORE PRACTICE**   Example 10.16; Problems 55, 56, 57, 58.

PEARSON eText 2.0

CONCEPTUAL ✔ CHECKPOINT 10.6

Which one of the structures that follow is NOT a resonance structure of the Lewis structure for N_2O shown here?

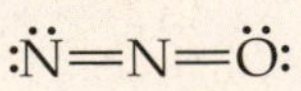

$:\ddot{N}=N=\ddot{O}:$

(a) 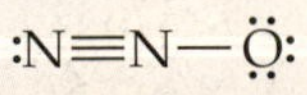 $:N\equiv N-\ddot{O}:$

(b) 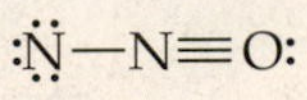 $:\ddot{N}-N\equiv O:$

(c) 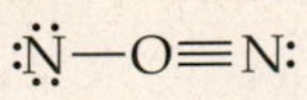 $:\ddot{N}-O\equiv N:$

10.7 Predicting the Shapes of Molecules

▶ Predict the shapes of molecules.

We can use the Lewis model, in combination with **valence shell electron pair repulsion (VSEPR) theory**, to predict the shapes of molecules. VSEPR theory is based on the idea that **electron groups**—lone pairs, single bonds, or multiple bonds—repel each other. This repulsion between the negative charges of electron groups on the central atom determines the geometry of the molecule. For example, consider CO_2, which has the Lewis structure:

$$\ddot{\text{O}}=\text{C}=\ddot{\text{O}}:$$

The geometry of CO_2 is determined by the repulsion between the two electron groups (the two double bonds) on the central carbon atom. These two electron groups get as far away from each other as possible, resulting in a bond angle of 180° and a **linear** geometry for CO_2.

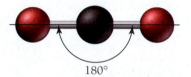

180°

As another example, consider the molecule H_2CO. Its Lewis structure is:

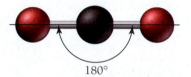

$$\overset{\displaystyle :\text{O}:}{\underset{\displaystyle |}{\overset{\displaystyle \|}{\text{H}-\text{C}-\text{H}}}}$$

This molecule has three electron groups around the central atom. These three electron groups get as far away from each other as possible, resulting in a bond angle of 120° and a **trigonal planar** geometry.

> The angles shown here for **H₂CO** are approximate. The **C=O** double bond contains more electron density than do **C—H** single bonds, resulting in a slightly greater repulsion; thus the HCH bond angle is actually 116°, and the HCO bond angles are actually 122°.

~120°   ~120°

~120°

> A tetrahedron is a geometric shape with four triangular faces.

If a molecule has four electron groups around the central atom, as CH_4 does, it has a **tetrahedral** geometry with bond angles of 109.5°.

> **CH₄** is shown here with both a ball-and-stick model (left) and a space-filling model (right). Although space-filling models more closely portray molecules, ball-and-stick models are often used to clearly illustrate molecular geometries.

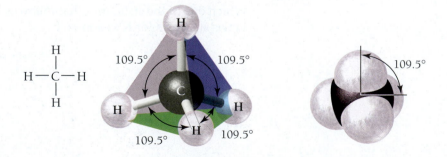

$$\overset{\displaystyle \text{H}}{\underset{\displaystyle \text{H}}{\overset{\displaystyle |}{\underset{\displaystyle |}{\text{H}-\text{C}-\text{H}}}}}$$

109.5°   109.5°

109.5°   109.5°

109.5°

The mutual repulsion of the four electron groups causes the tetrahedral shape; the tetrahedron allows the maximum separation among the four groups. When we write the structure of CH_4 on paper, it may seem that the molecule should be square planar, with bond angles of 90°. However, in three dimensions the electron groups can get farther away from each other by forming the tetrahedral geometry.

Each of the molecules in the preceding examples has only bonding groups of electrons around the central atom. What happens in molecules with lone pairs around the central atom? These lone pairs also repel other electron groups. For example, consider the NH_3 molecule:

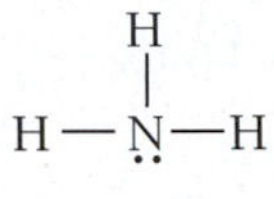

$$H - \underset{..}{N} - H$$

(with an H above N)

The four electron groups (one lone pair and three bonding pairs) get as far away from each other as possible. If we look only at the electrons, we find that the **electron geometry**—the geometrical arrangement of the electron groups—is tetrahedral.

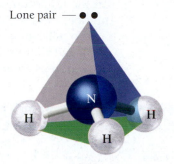

Lone pair

However, the **molecular geometry**—the geometrical arrangement of the atoms— is **trigonal pyramidal**.

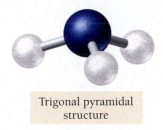

Trigonal pyramidal
structure

Notice that, although the electron geometry and the molecular geometry are different, the electron geometry is relevant to the molecular geometry. In other words, the lone pair exerts its influence on the bonding pairs.

Consider one last example, H_2O. Its Lewis structure is:

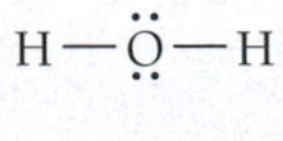

$$H - \underset{..}{\overset{..}{O}} - H$$

The bond angles in NH_3 and H_2O are actually a few degrees smaller than the ideal tetrahedral angles because lone pairs exert a slightly greater repulsion than bonding pairs.

Since it has four electron groups, its electron geometry is also tetrahedral.

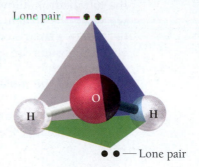

However, its molecular geometry is **bent**.

Bent structure

Table 10.1 summarizes the electron and molecular geometry of a molecule based on the total number of electron groups, the number of bonding groups, and the number of lone pairs.

To determine the geometry of any molecule, we use the procedure presented in the left column of Examples 10.8 and 10.9. As usual, the two examples of applying the steps are in the center and right columns.

TABLE 10.1 Electron and Molecular Geometries

Electron Groups*	Bonding Groups	Lone Pairs	Electron Geometry	Angle between Electron Groups**	Molecular Geometry	Example
2	2	0	linear	180°	linear	$:\ddot{O}=C=\ddot{O}:$
3	3	0	trigonal planar	120°	trigonal planar	$\ddot{O}:$ ‖ $H-C-H$
3	2	1	trigonal planar	120°	bent	$:\ddot{O}=\ddot{S}-\ddot{O}:$
4	4	0	tetrahedral	109.5°	tetrahedral	H ｜ $H-C-H$ ｜ H
4	3	1	tetrahedral	109.5°	trigonal pyramidal	$H-\ddot{N}-H$ ｜ H
4	2	2	tetrahedral	109.5°	bent	$H-\ddot{\underset{..}{O}}-H$

* Count only electron groups around the *central* atom. Each of the following is considered one electron group: a lone pair, a single bond, a double bond, and a triple bond.
** Angles listed here are idealized. Actual angles in specific molecules may vary by several degrees. For example, the bond angles in ammonia are 107° and the bond angle in water is 104.5°.

Predicting Geometry Using VSEPR Theory

	EXAMPLE **10.8**	EXAMPLE **10.9**
	Predict the electron and molecular geometry of PCl_3.	Predict the electron and molecular geometry of the $[NO_3]^-$ ion.
1. Draw a Lewis structure for the molecule.	**SOLUTION** PCl_3 has 26 electrons. $$:\ddot{C}l:$$ $$:\ddot{C}l:\ddot{P}:\ddot{C}l:$$	**SOLUTION** $[NO_3]^-$ has 24 electrons. $$\left[\begin{array}{c} :\ddot{O}: \\ :\ddot{O}:N::\ddot{O}: \end{array} \right]^-$$
2. Determine the total number of electron groups around the central atom. Lone pairs, single bonds, double bonds, and triple bonds each count as one group.	The central atom (P) has four electron groups.	The central atom (N) has three electron groups (the double bond counts as one group).
3. Determine the number of bonding groups and the number of lone pairs around the central atom. These should sum to the result from Step 2. Bonding groups include single bonds, double bonds, and triple bonds.	 Lone pair Three of the four electron groups around P are bonding groups, and one is a lone pair.	$$\left[\begin{array}{c} :\ddot{O}: \\ :\ddot{O}:N::\ddot{O}: \end{array} \right]^-$$ No lone pairs All three of the electron groups around N are bonding groups.
4. Refer to Table 10.1 to determine the electron geometry and molecular geometry.	The electron geometry is tetrahedral (four electron groups), and the molecular geometry — the shape of the molecule — is trigonal pyramidal (four electron groups, three bonding groups, and one lone pair).	The electron geometry is trigonal planar (three electron groups), and the molecular geometry — the shape of the molecule — is trigonal planar (three electron groups, three bonding groups, and no lone pairs).
	▶ **SKILLBUILDER 10.8**	Predict the molecular geometry of ClNO (N is the central atom).	▶ **SKILLBUILDER 10.9**	Predict the molecular geometry of the $SO_3{}^{2-}$ ion. ▶ **FOR MORE PRACTICE** Example 10.17; Problems 65, 66, 69, 70, 73, 74.

CONCEPTUAL ✔ CHECKPOINT **10.7**

Which condition necessarily leads to a molecular geometry that is identical to the electron geometry?

(a) The presence of a double bond between the central atom and a terminal atom.

(b) The presence of two or more identical terminal atoms bonded to the central atom.

(c) The presence of one or more lone pairs on the central atom.

(d) The absence of any lone pairs on the central atom.

Representing Molecular Geometries on Paper

Because molecular geometries are three-dimensional, they are often difficult to represent on two-dimensional paper. Many chemists use this notation for bonds to indicate three-dimensional structures on two-dimensional paper:

Straight line	*Hashed lines*	*Wedge*
Bond in plane of paper	Bond projecting into the paper	Bond projecting out of the paper

The major molecular geometries used in this book are shown here using this notation:

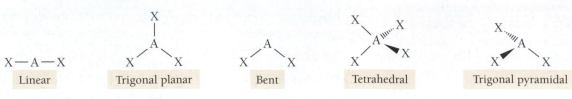

| Linear | Trigonal planar | Bent | Tetrahedral | Trigonal pyramidal |

CHEMISTRY AND HEALTH

Fooled by Molecular Shape

Artificial sweeteners, such as aspartame (Nutrasweet™), taste sweet but have few or no calories. Why? Because taste and caloric value are entirely separate properties of foods.

The caloric value of a food depends on the amount of energy released when the food is metabolized. Sucrose (table sugar) is metabolized by oxidation to carbon dioxide and water:

$$C_{12}H_{22}O_{11} + 6\,O_2 \longrightarrow 12\,CO_2 + 11\,H_2O$$
$$\Delta H = -5644 \text{ kJ}$$

When your body metabolizes one mole of sucrose, it obtains 5644 kJ of energy. Some artificial sweeteners, such as saccharin, are not metabolized at all—they just pass through the body unchanged—and therefore have no caloric value. Other artificial sweeteners, such as aspartame, are metabolized but have a much lower caloric content (for a given amount of sweetness) than sucrose.

The *taste* of a food is independent of its metabolism. The sensation of taste originates in the tongue, where specialized cells called taste cells act as highly sensitive and specific molecular detectors. These cells can distinguish the sugar molecules from the thousands of different types of molecules present in a mouthful of food. The main basis for this discrimination is the molecule's *shape*.

The surface of a taste cell contains specialized protein molecules called taste receptors. Each particular *tastant*—a molecule that you can taste—fits snugly into a special pocket on the taste receptor protein called the *active site*, just as a key fits into a lock (see Section 15.12). For example, a sugar molecule fits only into the active site of the sugar receptor protein called Tlr3. When the sugar molecule (the key) enters the active site (the lock), the different subunits of the Tlr3 protein split apart. This split causes a series of events that results in transmission of a nerve signal, which reaches the brain and registers a sweet taste.

Artificial sweeteners taste sweet because they fit into the receptor pocket that normally binds sucrose. In fact, both aspartame and saccharin bind to the active site in the Tlr3 protein more strongly than sugar does! For this reason, artificial sweeteners are "sweeter than sugar." It takes 200 times as much sucrose as aspartame to trigger the same amount of nerve signal transmission from taste cells.

This type of lock-and-key fit between the active site of a protein and a particular molecule is important not only to taste but to many other biological functions as well. For example, immune response, the sense of smell, and many types of drug action all depend on shape-specific interactions between molecules and proteins. The ability of scientists to determine the shapes of key biological molecules is largely responsible for the revolution in biology that has occurred over the last 50 years.

B10.1 CAN YOU ANSWER THIS? *Proteins are long-chain molecules in which each link is an amino acid. The simplest amino acid is glycine, which has this structure:*

Determine the geometry around each interior atom in the glycine structure and make a three-dimensional sketch of the molecule.

10.8 Electronegativity and Polarity: Why Oil and Water Don't Mix

▶ Determine whether a molecule is polar.

▲ **FIGURE 10.1 Oil and water don't mix**   Question: Why not?

The representation for depicting electron density in this figure is introduced in Section 9.6.

The value of electronegativity is assigned using a relative scale on which fluorine, the most electronegative element, has an electronegativity of 4.0. All other electronegativities are defined relative to fluorine.

If we combine oil and water in a container, they separate into distinct regions (◀ **FIGURE 10.1**). Why? Something about water molecules causes them to bunch together into one region, expelling the oil molecules into a separate region. What is that something? We can begin to understand the answer by examining the Lewis structure of water.

$$H—\overset{..}{\underset{..}{O}}—H$$

The two bonds between O and H each consist of an electron pair—two electrons shared between the oxygen atom and the hydrogen atom. The oxygen and hydrogen atoms each donate one electron to this electron pair; however, like most children, they don't share them equally. The oxygen atom takes more than its fair share of the electron pair.

Electronegativity

The ability of an element to attract electrons within a covalent bond is **electronegativity.** Oxygen is more electronegative than hydrogen, which means that, on average, shared electrons are more likely to be found near the oxygen atom than near the hydrogen atom. Consider one of the two OH bonds:

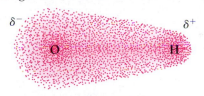

Dipole moment

Since the electron pair is unequally shared (with oxygen getting the larger share), the oxygen atom has a partial negative charge, symbolized by $\delta-$ (delta minus). The hydrogen atom (which gets the smaller share) has a partial positive charge, symbolized by $\delta+$ (delta plus). The result of this uneven electron sharing is a **dipole moment,** a separation of charge within the bond. We call covalent bonds that have a dipole moment **polar covalent bonds**. The magnitude of the dipole moment, and therefore the degree of polarity of the bond, depend on the electronegativity difference between the two elements in the bond and the length of the bond. For a fixed bond length, the greater the electronegativity difference, the greater the dipole moment and the more polar the bond.

▼ **FIGURE 10.2** shows the relative electronegativities of the elements. Notice that electronegativity increases as we move toward the right across a period in the

▶ **FIGURE 10.2 Electronegativity of the elements**   Linus Pauling introduced the scale shown in this figure. He arbitrarily set the electronegativity of fluorine at 4.0 and calculated all other values relative to fluorine.

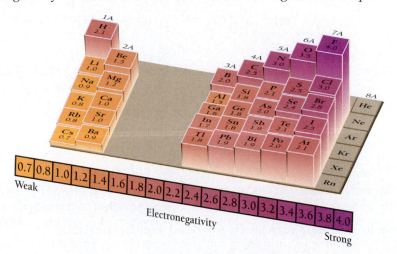

▲ **FIGURE 10.3 Pure covalent bonding** In Cl_2, the two Cl atoms share the electrons evenly. This is a pure covalent bond.

▲ **FIGURE 10.4 Ionic bonding** In NaCl, Na completely transfers an electron to Cl. This is an ionic bond.

> The degree of bond polarity is a continuous function. The guidelines given here are approximate.

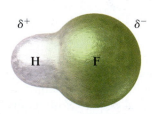

▲ **FIGURE 10.5 Polar covalent bonding** In HF, the electrons are shared, but the shared electrons are more likely to be found on F than on H. The bond is polar covalent.

periodic table and decreases as we move down a column in the periodic table. If two elements with identical electronegativities form a covalent bond, they share the electrons equally, and there is no dipole moment. For example, the chlorine molecule, composed of two chlorine atoms (which of course have identical electronegativities), has a pure covalent bond in which electrons are evenly shared (◄ **FIGURE 10.3**). The bond has no dipole moment, and the molecule is **nonpolar**.

If there is a large electronegativity difference between the two elements in a bond, such as normally occurs between a metal and a nonmetal, the electron is completely transferred and the bond is ionic. For example, sodium and chlorine form an ionic bond (◄ **FIGURE 10.4**).

If there is an intermediate electronegativity difference between the two elements, such as between two different nonmetals, the bond is polar covalent. For example, HF forms a polar covalent bond (▼ **FIGURE 10.5**).

Table 10.2 and ▼ **FIGURE 10.6** summarize these concepts.

TABLE 10.2 The Effect of Electronegativity Difference on Bond Type

Electronegativity Difference (ΔEN)	Bond Type	Example
zero (0–0.4)	pure covalent	Cl_2
intermediate (0.4–2.0)	polar covalent	HF
large (2.0+)	ionic	NaCl

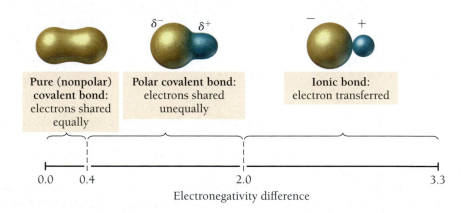

▲ **FIGURE 10.6 The continuum of bond types** The electronegativity difference between two bonded atoms determines the type of bond (pure covalent, polar covalent, or ionic).

EXAMPLE 10.10   Classifying Bonds as Pure Covalent, Polar Covalent, or Ionic

Is the bond formed between each pair of atoms pure covalent, polar covalent, or ionic?

(a) Sr and F   **(b)** N and Cl   **(c)** N and O

SOLUTION

(a) In Figure 10.2, find the electronegativity of Sr (1.0) and of F (4.0). The electronegativity difference (ΔEN) is:

$$\Delta EN = 4.0 - 1.0 = 3.0$$

Refer to Table 10.2 and classify this bond as ionic.

(b) In Figure 10.2, find the electronegativity of N (3.0) and of Cl (3.0). The electronegativity difference (ΔEN) is:

$$\Delta EN = 3.0 - 3.0 = 0$$

Refer to Table 10.2 and classify this bond as pure covalent.

(c) In Figure 10.2, find the electronegativity of N (3.0) and of O (3.5). The electronegativity difference (ΔEN) is:

$$\Delta EN = 3.5 - 3.0 = 0.5$$

Refer to Table 10.2 and classify this bond as polar covalent.

▶ **SKILLBUILDER 10.10** | **Classifying Bonds as Pure Covalent, Polar Covalent, or Ionic**

Is the bond formed between each pair of atoms pure covalent, polar covalent, or ionic?

(a) I and I   (b) Cs and Br   (c) P and O

▶ **FOR MORE PRACTICE** Problems 81, 82.

CONCEPTUAL ✔ CHECKPOINT 10.8

Which bond would you expect to be more polar: the bond in HCl or the bond in HBr?

Polar Bonds and Polar Molecules

Does the presence of one or more polar bonds in a molecule always result in a polar molecule? The answer is no. A **polar molecule** is one with polar bonds that add together—they do not cancel each other—to form a net dipole moment. For diatomic molecules, we can readily tell polar molecules from nonpolar ones. If a diatomic molecule contains a polar bond, then the molecule is polar. However, for molecules with more than two atoms, it is more difficult to tell polar molecules from nonpolar ones because two or more polar bonds may cancel one another.

For example, consider carbon dioxide:

$$:\ddot{O}=C=\ddot{O}:$$

Each C=O *bond* is polar because the difference in electronegativity between oxygen and carbon is 1.0. However, since CO_2 has a linear geometry, the dipole moment of one bond completely cancels the dipole moment of the other and the *molecule* is nonpolar. We can understand this with an analogy. Imagine each polar bond to be a rope pulling on the central atom. In CO_2 we can see how the two ropes pulling in opposing directions cancel each other's effect:

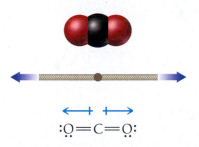

$$:\underset{..}{O}=C=\underset{..}{O}:$$

We can also represent polar bonds with arrows (or *vectors*) that point in the direction of the negative pole and have a plus sign at the positive pole (as we just saw for carbon dioxide). If the arrows (or vectors) point in exactly opposing directions as in carbon dioxide, the dipole moments cancel.

Water, on the other hand, has two dipole moments that do not cancel. If we imagine each bond as a rope pulling on oxygen, we see that, because of the angle between the bonds, the pulls of the two ropes do not cancel as shown at right.

In the vector representation of a dipole moment, the vector points in the direction of the atom with the partial negative charge.

$$\delta^+ \qquad \delta^-$$

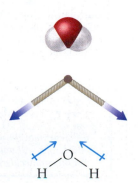

TABLE 10.3 Common Cases of Adding Dipole Moments to Determine Whether a Molecule Is Polar

Nonpolar

Two identical polar bonds pointing in opposite directions will cancel. The molecule is nonpolar.

Nonpolar

Three identical polar bonds at 120° from each other will cancel. The molecule is nonpolar.

Polar

Three polar bonds in a trigonal pyramidal arrangement (109.5°) will not cancel. The molecule is polar.

Polar

Two polar bonds with an angle of less than 180° between them will not cancel. The molecule is polar.

Nonpolar

Four identical polar bonds in a tetrahedral arrangement (109.5° from each other) will cancel. The molecule is nonpolar.

Note: In all cases where the polar bonds cancel, the bonds are assumed to be identical. If one or more of the bonds are different than the other(s), the bonds will not cancel and the molecule is polar.

Consequently, water is a polar molecule. We can use symmetry as a guide to determine whether a molecule containing polar bonds is indeed polar. Highly symmetric molecules tend to be nonpolar even if they have polar bonds because the bond dipole moments (or the pulls of the ropes) tend to cancel. Asymmetric molecules that contain polar bonds tend to be polar because the bond dipole moments (or the pulls of the ropes) tend not to cancel. Table 10.3 summarizes some common cases.

In summary, to determine whether a molecule is polar:

- **Determine whether the molecule contains polar bonds.** A bond is polar if the two bonding atoms have different electronegativities. If there are no polar bonds, the molecule is nonpolar.

- **Determine whether the polar bonds add together to form a net dipole moment.** We must first use VSEPR theory to determine the geometry of the molecule. Then we visualize each bond as a rope pulling on the central atom. Is the molecule highly symmetrical? Do the pulls of the ropes cancel? If so, there is no net dipole moment and the molecule is nonpolar. If the molecule is asymmetrical and the pulls of the rope do not cancel, the molecule is polar.

PEARSON eText 2.0 | Interactive Worked Example Video 10.11

EXAMPLE **10.11** | **Determining Whether a Molecule Is Polar**

Is NH_3 polar?

	SOLUTION
Begin by drawing the Lewis structure of NH_3. Since N and H have different electronegativities, the bonds are polar.	
The geometry of NH_3 is trigonal pyramidal (four electron groups, three bonding groups, one lone pair). Draw a three-dimensional picture of NH_3 and imagine each bond as a rope that is being pulled. The pulls of the ropes do not cancel and the molecule is polar.	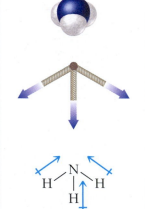 NH_3 is polar

▶ **SKILLBUILDER 10.11 | Determining Whether a Molecule Is Polar**

Determine whether CH_4 is polar.

▶ **FOR MORE PRACTICE** Example 10.18; Problems 89, 90, 91, 92.

Polarity is important because polar molecules tend to behave differently than nonpolar molecules. Water and oil do not mix, for example, because water molecules are polar and the molecules that compose oil are generally nonpolar. Polar molecules interact strongly with other polar molecules because the positive end of one molecule is attracted to the negative end of another, just as the south pole of a

EVERYDAY CHEMISTRY

How Soap Works

Imagine eating a greasy cheeseburger without flatware or napkins. By the end of the meal, your hands are coated with grease and oil. If you try to wash them with only water, they remain greasy. However, if you add a little soap, the grease washes away. Why? As we learned previously, water molecules are polar and the molecules that compose grease and oil are nonpolar. As a result, water and grease repel each other.

The molecules that compose soap, however, have a special structure that allows them to interact strongly with both water and grease. One end of a soap molecule is polar, while the other end is nonpolar.

The polar head of a soap molecule strongly attracts water molecules, while the nonpolar tail strongly attracts grease and oil molecules. Soap is a sort of molecular liaison, one end interacting with water and the other end interacting with grease. Soap therefore allows water and grease to mix, removing the grease from your hands and washing it down the drain.

B10.2 CAN YOU ANSWER THIS? *Consider this detergent molecule. Which end do you think is polar? Which end is nonpolar?*

Soap molecule

Polar head
attracts water

Nonpolar tail
attracts grease

$$CH_3(CH_2)_{11}OCH_2CH_2OH$$

▶ **FIGURE 10.7 Dipole–dipole attraction** Just as the north pole of one magnet is attracted to the south pole of another, so the positive end of one molecule with a dipole is attracted to the negative end of another molecule with a dipole.

The positive end of one molecule is attracted to the negative end of another molecule.

magnet is attracted to the north pole of another magnet (▲ **FIGURE 10.7**). A mixture of polar and nonpolar molecules is similar to a mixture of small magnetic and nonmagnetic particles. The magnetic particles clump together, excluding the nonmagnetic ones and separating into distinct regions (◀ **FIGURE 10.8**). Similarly, the polar water molecules attract one another, forming regions from which the nonpolar oil molecules are excluded (▼ **FIGURE 10.9**).

▲ **FIGURE 10.8 Magnetic and nonmagnetic particles** Magnetic particles (the colored marbles) attract one another, excluding nonmagnetic particles (the clear marbles). This behavior is analogous to that of polar and nonpolar molecules.

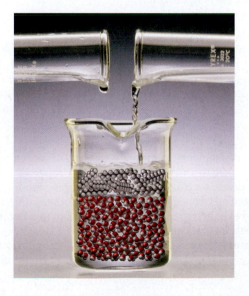

◀ **FIGURE 10.9 Polar and nonpolar molecules** A mixture of polar and nonpolar molecules, like a mixture of magnetic and nonmagnetic particles, separates into distinct regions because the polar molecules attract one another, excluding the nonpolar ones.
Question: Can you think of some examples of this behavior?

Chapter 10 in Review

MasteringChemistry™ provides end-of-chapter exercises, feedback-enriched tutorial problems, animations, and interactive activities to encourage problem solving practice and deeper understanding of key concepts and topics.

Self-Assessment Quiz

Q1. Which pair of elements has the most similar Lewis structures?
(a) N and S
(b) F and Ar
(c) Cl and Ar
(d) O and S

Q2. What is the Lewis structure for the compound that forms between K and S?

(a) $K—\ddot{\underset{..}{S}}—K$
(b) $K^+ [:\ddot{\underset{..}{S}}:]^-$

(c) $K^+ [:\ddot{\underset{..}{S}}:]^{2-} K^+$
(d) $[:\ddot{\underset{..}{S}}:]^{2-} K^+ [:\ddot{\underset{..}{S}}:]^{2-}$

Q3. Use the Lewis model to predict the correct formula for the compound that forms between K and S.
(a) KS
(b) K_2S
(c) KS_2
(d) K_2S_2

Q4. What is the correct Lewis structure for H_2CS?

(a) $H—H—\overset{\displaystyle :\ddot{S}:}{\underset{\displaystyle }{\overset{|}{C}}}—\ddot{\underset{..}{S}}:$
(b) $H—\overset{\displaystyle :\ddot{S}:}{\underset{\displaystyle }{\overset{|}{C}}}—H$

(c) $H=\overset{\displaystyle :\ddot{S}:}{\underset{\displaystyle }{\overset{|}{C}}}—H$
(d) $H—\overset{\displaystyle :\ddot{S}}{\underset{\displaystyle }{\overset{\|}{C}}}—H$

Q5. How many electron dots are in the Lewis structure of NO_2^-?
(a) 17
(b) 18
(c) 19
(d) 20

Q6. Which compound has two or more resonance structures?
(a) NO_2^-
(b) CO_2
(c) NH_4^+
(d) CCl_4

Q7. What is the molecular geometry of PBr_3?
(a) Bent
(b) Tetrahedral
(c) Trigonal pyramidal
(d) Linear

Q8. What is the molecular geometry of N_2O? (Nitrogen is the central atom.)
(a) Bent
(b) Tetrahedral
(c) Trigonal pyramidal
(d) Linear

Q9. Which bond is polar?
(a) A bond between C and S
(b) A bond between Br and Br
(c) A bond between C and O
(d) A bond between B and H

Q10. Which molecule is polar?
(a) SCl_2
(b) CS_2
(c) CF_4
(d) $SiCl_4$

Answers: 1:d, 2:c, 3:b, 4:d, 5:b, 6:a, 7:c, 8:d, 9:c, 10:a

Chemical Principles

Relevance

The Lewis Model

The Lewis model is a model for chemical bonding. According to the Lewis model, chemical bonds form when atoms transfer valence electrons (ionic bonding) or share valence electrons (covalent bonding) to attain noble gas electron configurations. In the Lewis model, we represent valence electrons as dots surrounding the symbol for an element. When two or more elements bond together, the dots are transferred or shared so that every atom attains eight dots (an octet), or two dots (a duet) in the case of hydrogen.

Bonding theories predict what combinations of elements will form stable compounds, and we can use them to predict the properties of those compounds. For example, pharmaceutical companies use bonding theories when they design drug molecules to interact with a specific part of a protein molecule.

Molecular Shapes

We can predict the shapes of molecules by combining the Lewis model with valence shell electron pair repulsion (VSEPR) theory. In this model, electron groups—lone pairs, single bonds, double bonds, and triple bonds—around the central atom repel one another and determine the geometry of the molecule.

Molecular shapes determine many of the properties of compounds. Water's bent geometry, for example, is the reason it is a liquid at room temperature instead of a gas. Its geometry is also the reason ice floats and snowflakes have hexagonal patterns.

Electronegativity and Polarity

Electronegativity refers to the relative ability of elements to attract electrons within a chemical bond. Electronegativity increases as we move to the right across a period in the periodic table and decreases as we move down a column. When two nonmetal atoms of different electronegativities form a covalent bond, the electrons in the bond are not evenly shared and the bond is polar. In diatomic molecules, a polar bond results in a polar molecule. In molecules with more than two atoms, polar bonds may cancel, forming a nonpolar molecule, or they may sum, forming a polar molecule.

The polarity of a molecule influences many of its properties such as whether it is a solid, liquid, or gas at room temperature and whether it mixes with other compounds. Oil and water, for example, do not mix because water is polar while oil is nonpolar.

Chemical Skills

Examples

LO: Write Lewis structures for elements (Section 10.2).

The Lewis structure of any element is the symbol for the element with the valence electrons represented as dots drawn around the element. The number of valence electrons is equal to the group number of the element (for main-group elements).

EXAMPLE 10.12 | **Lewis Structures for Elements**

Draw the Lewis structure of sulfur.

SOLUTION

Since S is in Group 6A, it has six valence electrons. Draw these as dots surrounding its symbol, S.

$$\cdot \ddot{\text{S}} \colon$$

LO: Write Lewis structures for ionic compounds (Section 10.3).

In an ionic Lewis structure, the metal loses all of its valence electrons to the nonmetal, which attains an octet. We place the nonmetal in brackets with the charge in the upper right corner.

EXAMPLE 10.13 | **Writing Lewis Structures of Ionic Compounds**

Write the Lewis structure for lithium bromide.

SOLUTION

$$\text{Li}^+ \; [\,\colon\!\ddot{\text{Br}}\colon\,]^-$$

LO: Use the Lewis model to predict the chemical formula of an ionic compound (Section 10.3).

To determine the chemical formula of an ionic compound, write the Lewis structures of each of the elements. Then choose the correct number of each type of atom so that the metal atom(s) lose all of their valence electrons and the nonmetal atom(s) attain an octet.

EXAMPLE 10.14 | **Using the Lewis Model to Predict the Chemical Formula of an Ionic Compound**

Use the Lewis model to predict the formula for the compound that forms between potassium and sulfur.

SOLUTION

The Lewis structures of K and S are:

$$\text{K}\cdot \qquad \cdot \ddot{\text{S}}\colon$$

Potassium must lose one electron, and sulfur must gain two. Consequently, there are two potassium atoms for every sulfur atom. The Lewis structure is:

$$\text{K}^+ \, [\,\colon\!\ddot{\ddot{\text{S}}}\colon\,]^{2-} \text{K}^+$$

The correct formula is K_2S.

LO: Write Lewis structures for covalent compounds (Sections 10.4, 10.5).

To write covalent Lewis structures, follow these steps:

1. **Write the correct skeletal structure for the molecule.** Hydrogen atoms are always terminal, halogens are usually terminal, and many molecules tend to be symmetrical.

2. **Calculate the total number of electrons for the Lewis structure by summing the valence electrons of each atom in the molecule.** Remember that the number of valence electrons for any main-group element is equal to its group number in the periodic table. For polyatomic ions, add one electron for each negative charge and subtract one electron for each positive charge.

EXAMPLE 10.15 | **Writing Lewis Structures for Covalent Compounds**

Write the Lewis structure for CS_2.

SOLUTION

S C S

$$
\begin{aligned}
\text{Total e}^- &= 1 \times (\text{\# valence e}^- \text{ in C}) \\
&\quad + 2 \times (\text{\# valence e}^- \text{ in S}) \\
&= 4 + 2(6) \\
&= 16
\end{aligned}
$$

3. **Distribute the electrons among the atoms, giving octets (or duets for hydrogen) to as many atoms as possible.** Begin by placing two electrons between each pair of atoms. These are the bonding electrons. Then distribute the remaining electrons, first to terminal atoms and then to the central atom.

4. **If any atoms lack an octet, form double or triple bonds as necessary to give them octets.** Do this by moving lone electron pairs from terminal atoms into the bonding region with the central atom.

$$\ddot{S}{:}C{:}\ddot{S} \quad (\text{4 of 16 } e^- \text{ used})$$

$$:\!\ddot{S}\!:\!C\!:\!\ddot{S}\!: \quad (\text{16 of 16 } e^- \text{ used})$$

$$:\!\ddot{S}\!::\!C\!::\!\ddot{S}\!: \quad \text{or} \quad :\!\ddot{S}\!=\!C\!=\!\ddot{S}\!:$$

LO: Write resonance structures (Section 10.6).

When we can write two or more equivalent (or nearly equivalent) Lewis structures for a molecule, the true structure is an average between these. Represent this by writing all of the correct structures (called resonance structures) with double-headed arrows between them.

EXAMPLE 10.16 Writing Resonance Structures

Write resonance structures for SeO_2.

SOLUTION

You can write the Lewis structure for SeO_2 by following the steps for writing covalent Lewis structures. You can write two equally correct structures, so draw them both as resonance structures.

$$:\!\ddot{O}\!-\!\ddot{Se}\!=\!\ddot{O}\!: \quad \longleftrightarrow \quad :\!\ddot{O}\!=\!\ddot{Se}\!-\!\ddot{O}\!:$$

LO: Predict the shapes of molecules (Section 10.7).

To determine the shape of a molecule, follow these steps:

1. **Draw the Lewis structure for the molecule.**

2. **Determine the total number of electron groups around the central atom.** Lone pairs, single bonds, double bonds, and triple bonds each count as one group.

3. **Determine the number of bonding groups and the number of lone pairs around the central atom.** These should sum to the result from Step 2. Bonding groups include single bonds, double bonds, and triple bonds.

4. **Refer to Table 10.1 to determine the electron geometry and molecular geometry.**

EXAMPLE 10.17 Predicting the Shapes of Molecules

Predict the geometry of SeO_2.

SOLUTION

The Lewis structure for SeO_2 (as you determined in Example 10.16) is composed of the following two resonance structures.

$$:\!\ddot{O}\!-\!\ddot{Se}\!=\!\ddot{O}\!: \quad \longleftrightarrow \quad :\!\ddot{O}\!=\!\ddot{Se}\!-\!\ddot{O}\!:$$

Either of the resonance structures will give the same geometry.
  Total number of electron groups = 3
  Number of bonding groups = 2
  Number of lone pairs = 1

  Electron geometry = Trigonal planar
  Molecular geometry = Bent

LO: Determine whether a molecule is polar (Section 10.8).

- **Determine whether the molecule contains polar bonds.** A bond is polar if the two bonding atoms have different electronegativities. If there are no polar bonds, the molecule is nonpolar.

- **Determine whether the polar bonds add together to form a net dipole moment.** Use VSEPR theory to determine the geometry of the molecule. Then visualize each bond as a rope pulling on the central atom. Is the molecule highly symmetrical? Do the pulls of the ropes cancel? If so, there is no net dipole moment and the molecule is nonpolar. If the molecule is asymmetrical and the pulls of the rope do not cancel, the molecule is polar.

EXAMPLE 10.18 Determining Whether a Molecule Is Polar

Determine whether SeO_2 is polar.

SOLUTION

Se and O are nonmetals with different electronegativities (2.4 for Se and 3.5 for O). Therefore, the Se—O bonds are polar.
  As you determined in Example 10.17, the geometry of SeO_2 is bent.

The polar bonds do not cancel but rather sum to give a net dipole moment. Therefore the molecule is polar.

Key Terms

bent [10.7]
bonding pair [10.4]
bonding theory [10.1]
chemical bond [10.2]
covalent bond [10.2]
dipole moment [10.8]
dot structure [10.2]
double bond [10.4]

duet [10.2]
electron geometry [10.7]
electron group [10.7]
electronegativity [10.8]
ionic bond [10.2]
Lewis model [10.1]
Lewis structure [10.2]
linear [10.7]

lone pair [10.4]
molecular geometry [10.7]
nonpolar [10.8]
octet [10.2]
octet rule [10.2]
polar covalent bond [10.8]
polar molecule [10.8]
resonance structures [10.6]

terminal atom [10.5]
tetrahedral [10.7]
trigonal planar [10.7]
trigonal pyramidal [10.7]
triple bond [10.4]
valence shell electron pair
repulsion (VSEPR) theory
[10.7]

Exercises

Questions

1. Why are bonding theories important? Cite some examples of what bonding theories can predict.
2. Write the electron configurations for Ne and Ar. How many valence electrons does each element have?
3. In the Lewis model, what is an octet? What is a duet? What is a chemical bond?
4. What is the difference between ionic bonding and covalent bonding?
5. How can the Lewis model be used to determine the formula of ionic compounds? You may explain this with an example.
6. What is the difference between lone pair and bonding pair electrons?
7. How are double and triple bonds physically different from single bonds?
8. What is the procedure for writing a covalent Lewis structure?
9. How do you determine the number of electrons that go into the Lewis structure of a molecule?
10. How do you determine the number of electrons that go into the Lewis structure of a polyatomic ion?
11. Why does the octet rule have exceptions? List some examples.

12. What are resonance structures? Why are they necessary?
13. Explain how VSEPR theory predicts the shapes of molecules.
14. If all of the electron groups around a central atom are bonding groups (that is, there are no lone pairs), what is the molecular geometry for:
    (a) two electron groups
    (b) three electron groups
    (c) four electron groups
15. Give the bond angles for each of the geometries in the preceding question.
16. What is the difference between electron geometry and molecular geometry in VSEPR theory?
17. What is electronegativity?
18. What is the most electronegative element on the periodic table?
19. What is a polar covalent bond?
20. What is a dipole moment?
21. What happens if you try to mix a polar liquid with a nonpolar one?
22. If a molecule has polar bonds, is the molecule itself polar? Why or why not?

Problems

WRITING LEWIS STRUCTURES FOR ELEMENTS

23. Write an electron configuration for each element and the corresponding Lewis structure. Indicate which electrons in the electron configuration are included in the Lewis structure.
    (a) N
    (b) C
    (c) Cl
    (d) Ar

24. Write an electron configuration for each element and the corresponding Lewis structure. Indicate which electrons in the electron configuration are included in the Lewis structure.
    (a) Li
    (b) P
    (c) F
    (d) Ne

25. Write the Lewis structure for each element.
    (a) I
    (b) S
    (c) Ge
    (d) Ca

26. Write the Lewis structure for each element.
    (a) Kr
    (b) P
    (c) B
    (d) Na

27. Write a generic Lewis structure for the halogens. Do the halogens tend to gain or lose electrons in chemical reactions? How many?

28. Write a generic Lewis structure for the alkali metals. Do the alkali metals tend to gain or lose electrons in chemical reactions? How many?

29. Write a generic Lewis structure for the alkaline earth metals. Do the alkaline earth metals tend to gain or lose electrons in chemical reactions? How many?

30. Write a generic Lewis structure for the elements in the oxygen family (Group 6A). Do the elements in the oxygen family tend to gain or lose electrons in chemical reactions? How many?

31. Write the Lewis structure for each ion.
  (a) Al^{3+}
  (b) Mg^{2+}
  (c) Se^{2-}
  (d) N^{3-}

32. Write the Lewis structure for each ion.
  (a) Sr^{2+}
  (b) S^{2-}
  (c) Li^+
  (d) Cl^-

33. Indicate the noble gas that has the same Lewis structure as each ion.
  (a) Br^-
  (b) O^{2-}
  (c) Rb^+
  (d) Ba^{2+}

34. Indicate the noble gas that has the same Lewis structure as each ion.
  (a) Se^{2-}
  (b) I^-
  (c) Sr^{2+}
  (d) F^-

LEWIS STRUCTURES FOR IONIC COMPOUNDS

35. Is each compound best represented by an ionic or a covalent Lewis structure?
  (a) SF_6
  (b) $MgCl_2$
  (c) $BrCl$
  (d) K_2S

36. Is each compound best represented by an ionic or a covalent Lewis structure?
  (a) NO
  (b) CO_2
  (c) Rb_2O
  (d) Al_2S_3

37. Write the Lewis structure for each ionic compound.
  (a) NaF
  (b) CaO
  (c) $SrBr_2$
  (d) K_2O

38. Write the Lewis structure for each ionic compound.
  (a) SrO
  (b) Li_2S
  (c) CaI_2
  (d) RbF

39. Use the Lewis model to determine the formula for the compound that forms from each pair of atoms.
  (a) Ca and S
  (b) Mg and Br
  (c) Cs and I
  (d) Ca and N

40. Use the Lewis model to determine the formula for the compound that forms from each pair of atoms.
  (a) Al and S
  (b) Na and S
  (c) Sr and Se
  (d) Ba and F

41. Draw the Lewis structure for the ionic compound that forms from Mg and each atom.
  (a) F
  (b) O
  (c) N

42. Draw the Lewis structure for the ionic compound that forms from Al and each atom.
  (a) F
  (b) O
  (c) N

43. Determine what is wrong with each ionic Lewis structure and write the correct structure.

  (a) $[Cs:]^+ [:\ddot{C}l:]^-$

  (b) $Ba^+ [:\ddot{O}:]^-$

  (c) $Ca^{2+} [:\ddot{I}:]^-$

44. Determine what is wrong with each ionic Lewis structure and write the correct structure.

  (a) $[:\ddot{O}:]^{2-} Na^+ [:\ddot{O}:]^{2-}$

  (b) $Mg:\ddot{O}:$

  (c) $[Li:]^+ [:\ddot{S}:]^-$

LEWIS STRUCTURES FOR COVALENT COMPOUNDS

45. Use the Lewis model to explain why each element exists as a diatomic molecule.
(a) hydrogen
(b) iodine
(c) nitrogen
(d) oxygen

46. Use the Lewis model to explain why the compound that forms between hydrogen and sulfur has the formula H_2S. Would you expect HS to be stable? H_3S?

47. Write the Lewis structure for each molecule.
(a) PH_3
(b) SCl_2
(c) F_2
(d) HI

48. Write the Lewis structure for each molecule.
(a) CH_4
(b) NF_3
(c) OF_2
(d) H_2O

49. Write the Lewis structure for each molecule.
(a) O_2
(b) CO
(c) HONO (N is central; H bonded to one of the O atoms)
(d) SO_2

50. Write the Lewis structure for each molecule.
(a) N_2O (oxygen is terminal)
(b) SiH_4
(c) CI_4
(d) Cl_2CO (carbon is central)

51. Write the Lewis structure for each molecule.
(a) C_2H_2
(b) C_2H_4
(c) N_2H_2
(d) N_2H_4

52. Write the Lewis structure for each molecule.
(a) H_2CO (carbon is central)
(b) H_3COH (carbon and oxygen are both central)
(c) H_3COCH_3 (oxygen is between the two carbon atoms)
(d) H_2O_2

53. Determine what is wrong with each Lewis structure and write the correct structure.

(a) 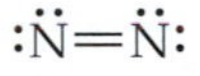 $:\ddot{N}=\ddot{N}:$

(b) 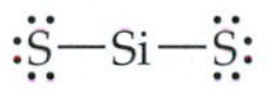 $:\ddot{S}-Si-\ddot{S}:$

(c) 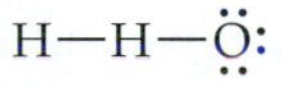 $H-H-\ddot{O}:$

(d) 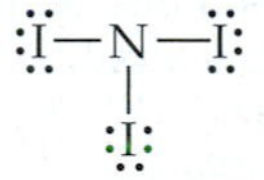 $:\ddot{I}-N-\ddot{I}:$
     $\qquad\quad |$
     $\qquad\;\;:\ddot{I}:$

54. Determine what is wrong with each Lewis structure and write the correct structure.

(a) 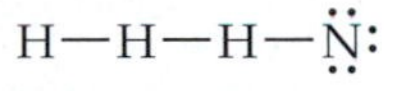 $H-H-H-\ddot{N}:$

(b) 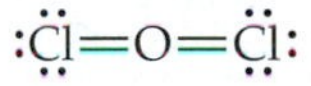 $:\ddot{Cl}=O=\ddot{Cl}:$

(c) 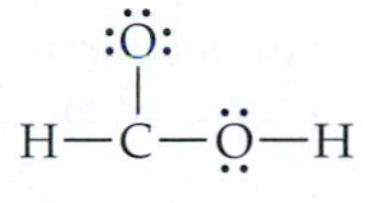 $\qquad\quad :\ddot{O}:$
    $\qquad\quad\;\;|$
    $H-C-\ddot{O}-H$

(d) 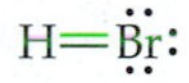 $H=\ddot{Br}:$

55. Write the Lewis structure for each molecule or ion. Include resonance structures if necessary.
(a) SeO_2
(b) CO_3^{2-}
(c) ClO^-
(d) ClO_2^-

56. Write the Lewis structure for each molecule or ion. Include resonance structures if necessary.
(a) ClO_3^-
(b) ClO_4^-
(c) NO_3^-
(d) SO_3

57. Write the Lewis structure for each ion. Include resonance structures if necessary.
(a) PO_4^{3-}
(b) CN^-
(c) NO_2^-
(d) SO_3^{2-}

58. Write the Lewis structure for each ion. Include resonance structures if necessary.
(a) SO_4^{2-}
(b) HSO_4^- (S is central; H is attached to one of the O atoms)
(c) NH_4^+
(d) BrO_2^- (Br is central)

59. Write the Lewis structure for each molecule. These molecules do not follow the octet rule.
(a) BCl_3
(b) NO_2
(c) BH_3

60. Write the Lewis structure for each molecule. These molecules do not follow the octet rule.
(a) BBr_3
(b) NO

PREDICTING THE SHAPES OF MOLECULES

61. Determine the number of electron groups around the central atom for each molecule.
(a) OF_2
(b) NF_3
(c) CS_2
(d) CH_4

62. Determine the number of electron groups around the central atom for each molecule.
(a) CH_2Cl_2
(b) SBr_2
(c) H_2S
(d) PCl_3

63. Determine the number of bonding groups and the number of lone pairs for each of the molecules in Problem 61. The sum of these should equal your answers to Problem 61.

64. Determine the number of bonding groups and the number of lone pairs for each of the molecules in Problem 62. The sum of these should equal your answers to Problem 62.

65. Determine the molecular geometry of each molecule.
(a) CBr_4
(b) H_2CO
(c) CS_2
(d) BH_3

66. Determine the molecular geometry of each molecule.
(a) SiO_2
(b) BF_3
(c) $CFCl_3$ (carbon is central)
(d) H_2CS (carbon is central)

67. Determine the bond angles for each molecule in Problem 65.

68. Determine the bond angles for each molecule in Problem 66.

69. Determine the electron and molecular geometries of each molecule.
(a) N_2O (oxygen is terminal)
(b) SO_2
(c) H_2S
(d) PF_3

70. Determine the electron and molecular geometries of each molecule. (*Hint:* Determine the geometry around each of the two central atoms.)
(a) C_2H_2 (skeletal structure HCCH)
(b) C_2H_4 (skeletal structure H_2CCH_2)
(c) C_2H_6 (skeletal structure H_3CCH_3)

71. Determine the bond angles for each molecule in Problem 69.

72. Determine the bond angles for each molecule in Problem 70.

73. Determine the electron and molecular geometries of each molecule. For molecules with two central atoms, indicate the geometry about each central atom.
(a) N_2
(b) N_2H_2 (skeletal structure HNNH)
(c) N_2H_4 (skeletal structure H_2NNH_2)

74. Determine the electron and molecular geometries of each molecule. For molecules with more than one central atom, indicate the geometry about each central atom.
(a) CH_3OH (skeletal structure H_3COH)
(b) H_3COCH_3 (skeletal structure H_3COCH_3)
(c) H_2O_2 (skeletal structure HOOH)

75. Determine the molecular geometry of each polyatomic ion.
(a) CO_3^{2-}
(b) ClO_2^-
(c) NO_3^-
(d) NH_4^+

76. Determine the molecular geometry of each polyatomic ion.
(a) ClO_4^-
(b) BrO_2^-
(c) NO_2^-
(d) SO_4^{2-}

ELECTRONEGATIVITY AND POLARITY

77. Refer to Figure 10.2 to determine the electronegativity of each element.
(a) Mg
(b) Si
(c) Br

78. Refer to Figure 10.2 to determine the electronegativity of each element.
(a) F
(b) C
(c) S

79. List these elements in order of decreasing electronegativity: Rb, Si, Cl, Ca, Ga.

80. List these elements in order of increasing electronegativity: Ba, N, F, Si, Cs.

81. Refer to Figure 10.2 to find the electronegativity difference between each pair of elements; then refer to Table 10.2 to classify the bonds that occur between them as pure covalent, polar covalent, or ionic.
  (a) Mg and Br
  (b) Cr and F
  (c) Br and Br
  (d) Si and O

82. Refer to Figure 10.2 to find the electronegativity difference between each pair of elements; then refer to Table 10.2 to classify the bonds that occur between them as pure covalent, polar covalent, or ionic.
  (a) K and Cl
  (b) N and N
  (c) C and S
  (d) C and Cl

83. Arrange these diatomic molecules in order of increasing bond polarity: ICl, HBr, H_2, CO.

84. Arrange these diatomic molecules in order of decreasing bond polarity: HCl, NO, F_2, HI.

85. Classify each diatomic molecule as polar or nonpolar.
  (a) CO
  (b) O_2
  (c) F_2
  (d) HBr

86. Classify each diatomic molecule as polar or nonpolar.
  (a) I_2
  (b) NO
  (c) HCl
  (d) N_2

87. For each polar molecule in Problem 85 draw the molecule and indicate the positive and negative ends of the dipole moment.

88. For each polar molecule in Problem 86 draw the molecule and indicate the positive and negative ends of the dipole moment.

89. Classify each molecule as polar or nonpolar.
  (a) CS_2
  (b) SO_2
  (c) CH_4
  (d) CH_3Cl

90. Classify each molecule as polar or nonpolar.
  (a) H_2CO
  (b) CH_3OH
  (c) CH_2Cl_2
  (d) CO_2

91. Classify each molecule as polar or nonpolar.
  (a) BH_3
  (b) $CHCl_3$
  (c) C_2H_2
  (d) NH_3

92. Classify each molecule as polar or nonpolar.
  (a) N_2H_2
  (b) H_2O_2
  (c) CF_4
  (d) NO_2

Cumulative Problems

93. Write electron configurations and Lewis structures for each element. Indicate which of the electrons in the electron configuration are shown in the Lewis structure.
  (a) Ca
  (b) Ga
  (c) As
  (d) I

94. Write electron configurations and Lewis structures for each element. Indicate which of the electrons in the electron configuration are shown in the Lewis structure.
  (a) Rb
  (b) Ge
  (c) Kr
  (d) Se

95. Determine whether each compound is ionic or covalent and write the appropriate Lewis structure.
  (a) K_2S
  (b) CHFO (carbon is central)
  (c) MgSe
  (d) PBr_3

96. Determine whether each compound is ionic or covalent and write the appropriate Lewis structure.
  (a) HCN
  (b) ClF
  (c) MgI_2
  (d) CaS

97. Write the Lewis structure for $OCCl_2$ (carbon is central) and determine whether the molecule is polar. Draw the three-dimensional structure of the molecule.

98. Write the Lewis structure for CH_3COH and determine whether the molecule is polar. Draw the three-dimensional structure of the molecule. The skeletal structure is:

$$
\begin{array}{ccc}
\text{H} & \text{O} & \\
\text{H} \quad \text{C} & \text{C} & \text{H} \\
\text{H} & &
\end{array}
$$

99. Write the Lewis structure for acetic acid (a component of vinegar) CH_3COOH, and draw the three-dimensional sketch of the molecule. Its skeletal structure is:

$$
\begin{array}{cccc}
\text{H} & \text{O} & & \\
\text{H} \quad \text{C} & \text{C} & \text{O} & \text{H} \\
\text{H} & & &
\end{array}
$$

100. Write the Lewis structure for benzene, C_6H_6, and draw a three-dimensional sketch of the molecule. The skeletal structure is the ring shown here. (*Hint:* The Lewis structure consists of two resonance structures.)

$$
\begin{array}{ccc}
& \text{H} & \\
& \text{C} & \\
\text{HC} & & \text{CH} \\
\text{HC} & & \text{CH} \\
& \text{C} & \\
& \text{H} &
\end{array}
$$

101. Consider the neutralization reaction.

$$HCl(aq) + NaOH(aq) \longrightarrow H_2O(l) + NaCl(aq)$$

Write the reaction showing the Lewis structures of each of the reactants and products.

102. Consider the precipitation reaction.

$$Pb(NO_3)_2(aq) + 2\ LiCl(aq) \longrightarrow PbCl_2(s) + 2\ LiNO_3(aq)$$

Write the reaction showing the Lewis structures of each of the reactants and products.

103. Consider the redox reaction.

$$2\ K(s) + Cl_2(g) \longrightarrow 2\ KCl(s)$$

Draw the Lewis structure for each reactant and product and determine which reactant was oxidized and which one was reduced.

104. Consider the redox reaction.

$$Ca(s) + Br_2(g) \longrightarrow CaBr_2(s)$$

Draw the Lewis structure for each reactant and product and determine which reactant was oxidized and which one was reduced.

105. Each compound listed contains both ionic and covalent bonds. Write the ionic Lewis structure for each one including the covalent structure for the polyatomic ion. Write resonance structures if necessary.
(a) KOH
(b) KNO_3
(c) LiIO
(d) $BaCO_3$

106. Each of the compounds listed contains both ionic and covalent bonds. Write an ionic Lewis structure for each one, including the covalent structure for the polyatomic ion. Write resonance structures if necessary.
(a) $RbIO_2$
(b) $Ca(OH)_2$
(c) NH_4Cl
(d) $Sr(CN)_2$

107. Each molecule listed contains an expanded octet (10 or 12 electrons) around the central atom. Write the Lewis structure for each molecule.
(a) PF_5
(b) SF_4
(c) SeF_4

108. Each molecule listed contains an expanded octet (10 or 12 electrons) around the central atom. Write the Lewis structure for each molecule.
(a) ClF_5
(b) SF_6
(c) IF_5

109. Formic acid is responsible for the sting you feel when stung by fire ants. By mass, formic acid is 26.10% C, 4.38% H, and 69.52% O. The molar mass of formic acid is 46.02 g/mol. Find the molecular formula of formic acid and draw its Lewis structure.

110. Diazomethane has the following composition by mass: 28.57% C, 4.80% H, and 66.64% N. The molar mass of diazomethane is 42.04 g/mol. Find the molecular formula of diazomethane and draw its Lewis structure.

111. Free radicals are molecules that contain an odd number of valence electrons and therefore contain an unpaired electron in their Lewis structure. Write the best possible Lewis structure for the free radical HOO. Does the Lewis model predict that HOO is stable? Predict its geometry.

112. Free radicals (as explained in the previous problem) are molecules that contain an odd number of valence electrons. Write the best possible Lewis structure for the free radical CH_3. Predict its geometry.

Highlight Problems

113. Some theories on aging suggest that free radicals cause a variety of diseases and aging. Free radicals (as explained in Problems 111 and 112) are molecules or ions containing an unpaired electron. As you know from the Lewis model, such molecules are not chemically stable and quickly react with other molecules. Free radicals may attack molecules within the cell, such as DNA, changing them and causing cancer or other diseases. Free radicals may also attack molecules on the surfaces of cells, making them appear foreign to the body's immune system. The immune system then attacks the cell and destroys it, weakening the body. Draw the Lewis structure for each of these free radicals, which have been implicated in theories of aging.

(a) O_2^-

(b) O^-

(c) OH

(d) CH_3OO (unpaired electron on terminal oxygen)

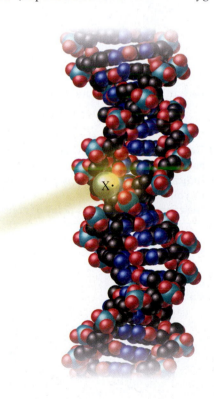

▲ Free radicals, molecules containing unpaired electrons (represented here as X·), may attack biological molecules such as the DNA molecule depicted here.

114. Free radicals (see Problem 113) are important in many environmentally significant reactions. For example, photochemical smog, which forms as a result of the action of sunlight on air pollutants, is formed in part by these two steps:

$$NO_2 \xrightarrow{\text{UV light}} NO + O$$

$$O + O_2 \longrightarrow O_3$$

The product of this reaction, ozone, is a pollutant in the lower atmosphere. Ozone is an eye and lung irritant and also accelerates the weathering of rubber products. Write Lewis structures for each of the reactants and products in the preceding reactions.

▲ Ozone damages rubber products.

115. Examine the formulas and space-filling models of the molecules shown here. Determine whether the structure is correct. If the structure is incorrect, sketch the correct structure.

(a) H_2Se

(b) CSe_2

(c) PCl_3

(d) CF_2Cl_2

Questions for Group Work

Discuss these questions with the group and record your consensus answer.

116. Draw the Lewis dot structure for the atoms Al and O. Use the Lewis model to determine the formula for the compound these atoms form.

117. Draft a list of step-by-step instructions for writing a correct Lewis dot structure for any molecule or polyatomic ion.

118. For each of the following molecules:

$$CS_2 \quad NCl_3 \quad CF_2 \quad CH_2F_2$$

(a) Draw the Lewis dot structure.
(b) Determine the molecular geometry and draw it as accurately as you can.
(c) Indicate the polarity of any polar bonds with the δ symbol.
(d) Classify the molecule as polar or nonpolar.

Data Interpretation and Analysis

119. The VSEPR model is useful in predicting bond angles for many compounds. Consider the tabulated data for bond angles in related species and answer the questions.

Bond Angles in NO_2 and Associated Ions	
Species	**Bond Angle**
NO_2	134°
NO_2^+	180°
NO_2^-	115°

(a) Draw Lewis structures for all of the species in the table.
(b) Use the Lewis structures from part a) to explain the observed bond angles in NO_2 and its associated ions.

Answers to Skillbuilder Exercises

Skillbuilder 10.1 .... $\cdot \text{Mg} \cdot$

Skillbuilder 10.2 .... Na^+ $[:\ddot{\text{Br}}:]^-$

Skillbuilder 10.3 .... Mg_3N_2

Skillbuilder 10.4 .... $:\text{C}\equiv\text{O}:$

Skillbuilder 10.5 ....
$$\begin{array}{c} \ddot{\text{O}}: \\ \parallel \\ \text{H}-\text{C}-\text{H} \end{array}$$

Skillbuilder 10.6 .... $[:\ddot{\text{Cl}}-\ddot{\text{O}}:]^-$

Skillbuilder 10.7

$$[:\ddot{\text{O}}=\ddot{\text{N}}-\ddot{\text{O}}:]^- \longleftrightarrow [:\ddot{\text{O}}-\ddot{\text{N}}=\ddot{\text{O}}:]^-$$

Skillbuilder 10.8 .... bent

Skillbuilder 10.9 .... trigonal pyramidal

Skillbuilder 10.10 .... (a) pure covalent
(b) ionic
(c) polar covalent

Skillbuilder 10.11 .... CH_4 is nonpolar

Answers to Conceptual Checkpoints

10.1 (a) C and Si both have four dots in their Lewis structure because they are both in the same column in the periodic table.

10.2 (b) Aluminum must lose its three valence electrons to obtain an octet. Sulfur must gain two electrons to obtain an octet. Therefore, two Al atoms are required for every three S atoms.

10.3 (b) The Lewis structure of O_2 has one double bond that contains four electrons (all of them bonding electrons); therefore, the number of bonding electrons is four.

10.4 (c) The Lewis structure of OH^- has eight electrons: six from oxygen, one from hydrogen, and one from the negative charge.

10.5 (b) Both NH_3 and H_3O^+ have one lone electron pair.

10.6 (c) Resonance structures must have the same skeletal structure. Structure c differs from the other structures and is therefore not a resonance structure.

10.7 (d) If there are no lone pairs on the central atom, all of its valence electrons are involved in bonds, so the molecular geometry must be the same as the electron geometry.

10.8 The bond in H—Cl is more polar than the bond in H—Br because Cl is more electronegative than Br, so the electronegativity difference between H and Cl is greater than the difference between H and Br.

11 Gases

The generality of men are so accustomed to judge of things by their senses that, because the air is invisible, they ascribe but little to it, and think it but one removed from nothing.

—Robert Boyle (1627–1691)

11.1 Extra-Long Straws

Like too many kids, I grew up preferring fast food to home cooking. My favorite stunt at the burger restaurant was drinking my orange soda from an extra-long straw that I pieced together from several smaller straws. I would pinch the end of one straw and squeeze it into the end of another. By attaching several straws together in this way, I could put my orange soda on the floor and drink it while standing on my chair (for some reason, my parents did not appreciate my scientific curiosity). I sometimes planned ahead and brought duct tape to the restaurant to form extra-tight seals between adjacent straws. My brother and I would compete to see who could make the longest working straw. Since I was older, I usually won.

I often wondered how long the straw could be if I made perfect seals between the straws. Could I drink my orange soda from a cup on the ground while I sat in my tree house? Could I drink it from the top of a ten-story building? It seemed to me that I could, but I was wrong. Even if the extended straw had perfect seals and rigid walls, and even if I could create a perfect vacuum (the absence of all air), I could never suck my orange soda from a straw longer than about 10.3 m (34 ft). Why?

Straws work because sucking creates a *pressure* difference between the inside of the straw and the outside. We define pressure and its units more thoroughly later in this chapter; for now, think of pressure as the force exerted per unit area by gas molecules as they collide with the surfaces around them (◀ **FIGURE 11.1**). Just as a ball exerts a force when it bounces against a wall, so a molecule exerts a force when it collides with

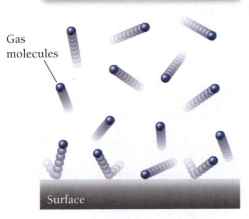

Pressure results from the collisions between gas molecules and the surrounding surfaces.

Gas molecules

Surface

▲ **FIGURE 11.1 Gas pressure** Pressure is the force exerted by gas molecules as they collide with the surfaces around them.

◀ When we drink from a straw, we remove some of the molecules from inside the straw. This creates a pressure difference between the inside of the straw and the outside of the straw that results in the liquid being pushed up the straw. The pushing is done by molecules in the atmosphere—primarily nitrogen and oxygen—as shown in this illustration.

A Newton (N) is a unit of force equal to 1 kg m s^{-2}.

a surface. The result of many of these collisions is pressure. The total amount of pressure exerted by a gas sample depends on several factors, including the concentration of gas molecules in the sample. On Earth at sea level, the gas molecules in our atmosphere exert an average pressure of 101,325 N/m^2 or, in English units, 14.7 lb/in^2.

When we put a straw in a glass of soda, the pressure inside and outside the straw are the same, so the soda does not rise within the straw (▼ FIGURE 11.2a). When we suck on the straw, we remove some of the air molecules from inside the straw, reducing the number of collisions that occur and therefore lowering the pressure inside the straw (▼ FIGURE 11.2b). However, the pressure outside the straw remains the same. This causes a pressure differential—the pressure outside the straw becomes greater than the pressure inside the straw. The greater external pressure pushes the liquid up the straw.

How high can this greater external pressure push the liquid up the straw? If we formed a perfect vacuum within the straw, the pressure outside of the straw at sea level would be enough to push the orange soda (which is mostly water) to a total height of about 10.3 m (▼ FIGURE 11.3). This is because a 10.3-m column of water exerts the same pressure—101,325 N/m^2 or 14.7 lb/in.^{2}—as the gas molecules in our atmosphere. In other words, the orange soda can rise up the straw until the pressure exerted by its weight equals the pressure exerted by the molecules in our atmosphere. When the two pressures are equal, the liquid stops rising (just like a two-arm balance stops moving when the masses on the two arms are equal).

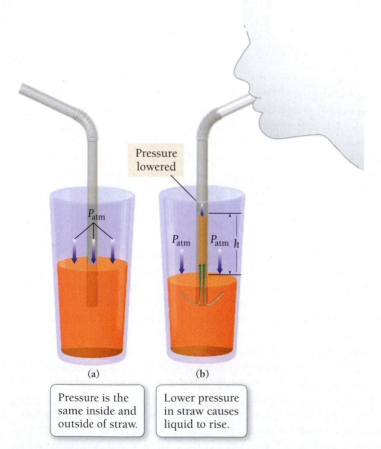

(a)

(b)

Pressure is the same inside and outside of straw.

Lower pressure in straw causes liquid to rise.

▲ FIGURE 11.2 Sipping soda (a) When a straw is put into a glass of orange soda, the pressure inside and outside the straw is the same, so the liquid levels inside and outside the straw are the same. (b) When a person sucks on the straw, the pressure inside the straw decreases. The greater pressure on the surface of the liquid outside the straw pushes the liquid up the straw.

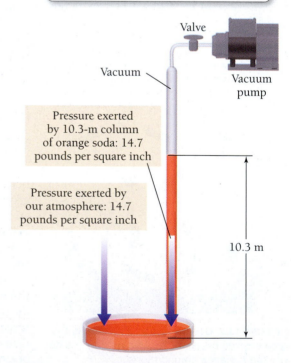

The pressure exerted by the atmosphere is equal to the pressure exerted by a 10.3-m column of water.

Valve

Vacuum

Vacuum pump

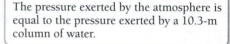

Pressure exerted by 10.3-m column of orange soda: 14.7 pounds per square inch

Pressure exerted by our atmosphere: 14.7 pounds per square inch

10.3 m

▲ FIGURE 11.3 Atmospheric pressure Even if we form a perfect vacuum, atmospheric pressure can only push orange soda to a total height of about 10 m. This is because a column of water (or soda) 10.3 m high exerts the same pressure (14.7 lb/in.^{2}) as the gas molecules in our atmosphere.

11.2 Kinetic Molecular Theory: A Model for Gases

▶ Describe how kinetic molecular theory predicts the main properties of a gas.

In prior chapters we have seen the importance of models or theories in understanding nature. A simple model for understanding the behavior of gases is the **kinetic molecular theory**. This model predicts the correct behavior for most gases under many conditions. Like other models, the kinetic molecular theory is not perfect and breaks down under certain conditions. In this book, however, we focus on conditions where it works well.

▶ **FIGURE 11.4 Simplified representation of an ideal gas** In reality, the spaces between the gas molecules would be larger in relation to the size of the molecules than shown here.

Kinetic molecular theory

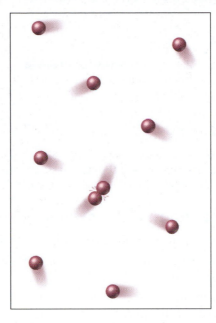

1. Collection of particles in constant motion

2. No attractions or repulsions between particles; collisions like billiard ball collisions

3. A lot of space between the particles compared to the size of the particles themselves

4. The speed of the particles increases with increasing temperature

Gases are compressible.

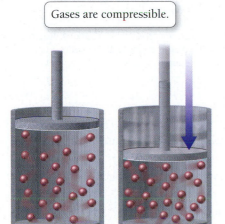

Gas

▲ **FIGURE 11.5 Compressibility of gases** Gases are compressible because there is so much empty space between gas particles.

Kinetic molecular theory makes the following assumptions (▲ **FIGURE 11.4**):

1. A gas is a collection of particles (molecules or atoms) in constant, straight-line motion.
2. Gas particles do not attract or repel each other—they do not interact. The particles collide with each other and with the surfaces around them, but they bounce back from these collisions like idealized billiard balls.
3. There is a lot of space between gas particles compared with the size of the particles themselves.
4. The average kinetic energy—energy due to motion (see Chapter 3)—of gas particles is proportional to the temperature of the gas in kelvins. This means that as the temperature increases, the particles move faster and therefore have more energy.

Kinetic molecular theory is consistent with, and indeed predicts, the properties of gases. Recall from Section 3.3 that gases:

- are compressible
- assume the shape and volume of their container
- have low densities in comparison with liquids and solids

Liquids are not compressible.

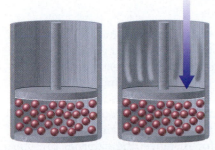

Liquid

▲ **FIGURE 11.6 Incompressibility of liquids** Liquids are not compressible because there is so little space between the liquid particles.

Gases are compressible because the atoms or molecules that compose them have a lot of space between them. When we apply external pressure to a gas sample, we force the atoms or molecules closer together, compressing the gas. Liquids and solids, in contrast, are not compressible because the atoms or molecules composing them are already in close contact—they cannot be forced any closer together. We can witness the compressibility of a gas, for example, by pushing a piston into a cylinder containing a gas. The piston goes down (◀ **FIGURE 11.5**) in response to the external pressure. If the cylinder is filled with a liquid or a solid, the piston does not move when pushed (◀ **FIGURE 11.6**).

▲ FIGURE 11.7 **A gas assumes the shape of its container** Since the attractions between molecules in a gas are negligible, and since the particles are in constant motion, a gas expands to fill the volume of its container.

Gases assume the shape and volume of their container because gaseous atoms or molecules are in constant, straight-line motion. In contrast to a solid or liquid, whose atoms or molecules interact with one another, the atoms or molecules in a gas do not interact with one another (or more precisely, their interactions are negligible). They simply move in straight lines, colliding with each other and with the walls of their container. As a result, they fill the entire container, collectively assuming its shape (◄ FIGURE 11.7).

Gases have a low density in comparison with solids and liquids because there is so much empty space between the atoms or molecules in a gas. For example, if the water in a 350-mL (12-oz) soda can were converted to steam (gaseous water), the steam would occupy a volume of 595 L (the equivalent of 1700 soda cans).

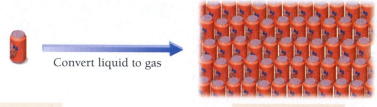

Convert liquid to gas

(1 can of soda)

(1700 cans of soda)

▲ If all of the water in a 12-oz (350-mL) can of orange soda were converted to gaseous steam (at 1 atm pressure and 100 °C), the steam would occupy a volume equal to 1700 soda cans. (Only a few of the 1700 cans are shown in the figure due to space constraints.)

PEARSON eText 2.0

CONCEPTUAL ✓ CHECKPOINT 11.1

Which statement is not consistent with the assumptions of kinetic molecular theory?

(a) As temperature increases, the particles that compose a gas move faster (on average).

(b) The size of the particles in a gas should have a dramatic effect on the properties of the gas.

(c) Gases have lower densities than solids and liquids.

11.3 Pressure: The Result of Constant Molecular Collisions

▶ Identify and explain the relationship between pressure, force, and area.

▶ Convert among pressure units.

Pressure is the result of the constant collisions between the atoms or molecules in a gas and the surfaces around them. Because of pressure, we can drink from straws, inflate basketballs, and move air into and out of our lungs. Variation in pressure in Earth's atmosphere creates wind, and changes in pressure help predict weather. Pressure is all around us and even inside us. The pressure exerted by a gas sample is defined as the force per unit area that results from the collisions of gas particles with surrounding surfaces:

$$\text{pressure} = \frac{\text{force}}{\text{area}}$$

The pressure exerted by a gas depends on several factors, including the number of gas particles in a given volume (▶ FIGURE 11.8). The fewer the gas particles, the lower the pressure. Pressure decreases, for example, with increasing altitude. As we climb a mountain or ascend in an airplane, there are fewer molecules per unit volume in air and the pressure consequently drops. For this reason, most airplane cabins are artificially pressurized (see the *Everyday Chemistry* box on page 366).

You may feel the effect of a drop in pressure as a pain in your ears. (▶ FIGURE 11.9). When you climb a mountain, for instance, the external pressure (that pressure that surrounds you) drops while the pressure within your ear cavities (the internal pressure) remains the same. This creates an imbalance—the lower external pressure causes your eardrum to bulge outward, causing pain. With time and a yawn or two, the excess air within your ears' cavities escapes, equalizing the internal and external pressure and relieving the pain.

▲ **FIGURE 11.8 Pressure** Since pressure is a result of collisions between gas particles and the surfaces around them, the amount of pressure increases when the number of particles in a given volume increases (assuming constant temperature).

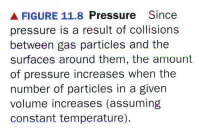

▲ **FIGURE 11.10 The mercury barometer** Average atmospheric pressure at sea level pushes a column of mercury to a height of 760 mm (29.92 in.). Question: What happens to the height of the mercury column if the external pressure decreases? increases?

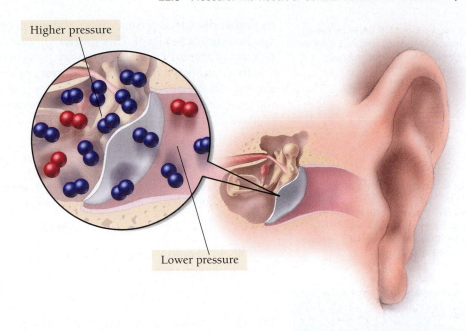

▲ **FIGURE 11.9 Pressure imbalance** The pain you feel in your ears upon climbing a mountain or ascending in an airplane is caused by an imbalance of pressure between the cavities inside your ear and the outside air.

CONCEPTUAL ✓ CHECKPOINT 11.2

Which sample of gas has the *lowest* pressure? Assume that all of the particles are identical and that the three samples are at the same temperature.

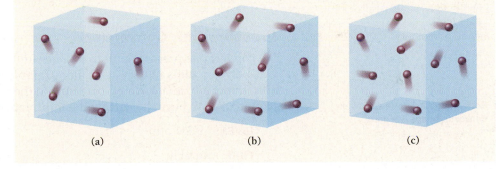

(a)   (b)   (c)

Pressure Units

The most common unit of pressure is the **atmosphere (atm)**, the average pressure at sea level. The SI unit of pressure is the **pascal (Pa)**, defined as 1 newton (N) per square meter.

$$1 \, Pa = 1 \, N/m^2$$

The pascal is a much smaller unit of pressure; 1 atm is equal to 101,325 Pa.

$$1 \, atm = 101,325 \, Pa$$

The pressure in a fully inflated mountain bike tire is about 6 atm, and the pressure on top of Mount Everest is about 0.31 atm.

A third unit of pressure, the **millimeter of mercury (mm Hg)**, originates from how pressure is measured with a barometer (◄ FIGURE 11.10). A barometer is an evacuated glass tube whose tip is submerged in a pool of mercury. Recall from our discussion in Section 11.1 that liquid is pushed up an evacuated tube by the atmospheric pressure on the liquid's surface. We learned that water is pushed up to a height of 10.3 m by the average pressure at sea level. Mercury, however, with

> Since mercury is 13.5 times as dense as water, it is pushed up 1/13.5 times as high as water by atmospheric pressure.

its higher density, is pushed up to a height of only 0.760 m, or 760 mm, by the average pressure at sea level. This shorter length—0.760 m instead of 10.3 m—makes a column of mercury a convenient way to measure pressure.

In a barometer, the mercury column rises or falls with changes in atmospheric pressure. If the pressure increases, the level of mercury within the column rises. If the pressure decreases, the level of mercury within the column falls. Since 1 atm of pressure pushes a column of mercury to a height of 760 mm, 1 atm and 760 mm Hg are equal.

$$1 \text{ atm} = 760 \text{ mm Hg}$$

A millimeter of mercury is also called a **torr**, after Italian physicist Evangelista Torricelli (1608–1647), who invented the barometer.

$$1 \text{ mm Hg} = 1 \text{ torr}$$

> Inches of mercury is still a widely used unit in weather reports. You may have heard a weather forecaster say, "The barometer is 30.07 and rising," meaning that the atmospheric pressure is currently 30.07 in. Hg.

Other common units of pressure include inches of mercury (in. Hg) and **pounds per square inch (psi)**.

$$1 \text{ atm} = 29.92 \text{ in. Hg} \qquad 1 \text{ atm} = 14.7 \text{ psi}$$

Table 11.1 lists the common units of pressure.

TABLE 11.1 Common Units of Pressure

Unit	Average Air Pressure at Sea Level
pascal (Pa)	101,325 Pa
atmosphere (atm)	1 atm
millimeter of mercury (mm Hg)	760 mm Hg (exact)
torr (torr)	760 torr (exact)
pounds per square inch (psi)	14.7 psi
inches of mercury (in. Hg)	29.92 in. Hg

PEARSON
eText
2.0

CONCEPTUAL ✔ CHECKPOINT 11.3

In place of mercury, a liquid that is about twice as dense as water is used in a barometer. With this barometer, normal atmospheric pressure would be about:

(a) 0.38 m   **(b)** 1.52 m   **(c)** 5.15 m   **(d)** 20.6 m

> See Section 2.6 for a review on converting between units.

Pressure Unit Conversion

We convert one pressure unit to another in the same way that we learned to convert between units in Chapter 2. For example, suppose we want to convert 0.311 atm (the approximate average pressure at the top of Mount Everest) to millimeters of mercury. We begin by sorting the information in the problem statement.

GIVEN: 0.311 atm

FIND: mm Hg

SOLUTION MAP
We then strategize by building a solution map that shows how to convert from atm to mm Hg.

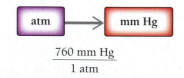

$$\frac{760 \text{ mm Hg}}{1 \text{ atm}}$$

RELATIONSHIPS USED

$$1 \text{ atm} = 760 \text{ mm Hg (from Table 11.1)}$$

SOLUTION
To solve, we begin with the given value (0.311 atm) and convert it to mm Hg.

$$0.311 \text{ atm} \times \frac{760 \text{ mm Hg}}{1 \text{ atm}} = 236 \text{ mm Hg}$$

EXAMPLE **11.1**   **Converting between Pressure Units**

A high-performance road bicycle tire is inflated to a total pressure of 125 psi. What is this pressure in millimeters of mercury?

SORT	GIVEN: 125 psi
You are given a pressure in psi and asked to convert it to mm Hg.	FIND: mm Hg

STRATEGIZE	SOLUTION MAP
Begin the solution map with the given units of psi. Use conversion factors to convert first to atm and then to mm Hg.	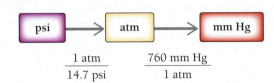

RELATIONSHIPS USED

   1 atm = 14.7 psi (Table 11.1)

   760 mm Hg = 1 atm (Table 11.1)

SOLVE	SOLUTION
Follow the solution map to solve the problem.	$125 \text{ psi} \times \dfrac{1 \text{ atm}}{14.7 \text{ psi}} \times \dfrac{760 \text{ mm Hg}}{1 \text{ atm}} = 6.46 \times 10^3 \text{ mm Hg}$

CHECK	
Check your answer. Are the units correct? Does the answer make physical sense?	The answer has the correct units, mm Hg. The answer is reasonable because the mm Hg is a smaller unit than psi; therefore, the value of the pressure in mm Hg should be greater than the value of the same pressure in psi.

▶ **SKILLBUILDER 11.1** | **Converting between Pressure Units**

Convert a pressure of 173 in. Hg into pounds per square inch.

▶ **SKILLBUILDER PLUS**   Convert a pressure of 23.8 in. Hg into kilopascals.

▶ **FOR MORE PRACTICE**   Example 11.13; Problems 23, 24, 25, 26, 29, 30, 31, 32.

11.4 Boyle's Law: Pressure and Volume

▶ Restate and apply Boyle's law.

The pressure of a gas sample depends, in part, on its volume. If the temperature and the amount of gas are constant, the pressure of a gas sample *increases* when the volume *decreases* and *decreases* when the volume *increases*. A simple hand pump (such as the kind you use to pump up a basketball or a bicycle tire) works on this principle. A hand pump is basically a cylinder equipped with a moveable piston (▶ FIGURE 11.11, on the next page). The volume in the cylinder increases when you pull the handle up (the upstroke) and decreases when you push the handle down (the downstroke). On the upstroke, the *increasing* volume causes a *decrease* in the internal pressure (the pressure within the pump's cylinder). This, in turn, draws air into the pump's cylinder through a one-way valve. On the downstroke, the *decreasing* volume causes an *increase* in the internal pressure. This increase forces the air out of the pump, through a different one-way valve, and into the tire or whatever else you are inflating.

The relationships between gas properties—such as the relationship between pressure and volume—are described by gas laws. These laws show how a change in one of these properties affects one or more of the others. The relationship between volume and pressure was discovered by Robert Boyle (1627–1691) and is called **Boyle's law**.

Boyle's law: The volume of a gas and its pressure are inversely proportional.

Boyle's law assumes constant temperature and a constant number of gas particles.

$$V \propto \frac{1}{P} \qquad \propto \text{ means "proportional to"}$$

EVERYDAY CHEMISTRY

Airplane Cabin Pressurization

Most commercial airplanes fly at elevations between 25,000 and 40,000 ft. At these elevations, atmospheric pressure is below 0.50 atm, much less than the normal atmospheric pressure to which our bodies are accustomed. The physiological effects of these lowered pressures—and the correspondingly lowered oxygen levels (see Section 11.9)—include dizziness, headache, shortness of breath, and even unconsciousness. Consequently, commercial airplanes pressurize the air in their cabins. If, for some reason, an airplane cabin should lose its pressurization, passengers are directed to breathe oxygen through an oxygen mask.

Cabin air pressurization is accomplished as part of the cabin's overall air circulation system. As air flows into the plane's jet engines, the large turbines at the front of the engines compress it. Most of this compressed (or pressurized) air exits out the back of the engines, creating the thrust that drives the plane forward. However, some of the pressurized air is directed into the cabin, where it is cooled and mixed with existing cabin air. This air is then circulated through the cabin through the overhead vents. The air leaves the cabin through ducts that direct it into the lower portion of the airplane. About half of this exiting air is mixed with incoming, pressurized air to circulate again. The other half is vented out of the plane through an outflow valve. The outflow valve is adjusted to maintain the desired cabin pressure. Federal regulations require that cabin pressure in commercial airliners be greater than the equivalent of outside air pressure at 8000 ft. Some newer jets, such as the Boeing 787 Dreamliner, pressurize their cabins at an equivalent pressure of 6000 ft for greater passenger comfort.

B11.1 CAN YOU ANSWER THIS? *Atmospheric pressure at elevations of 8000 ft averages about 0.72 atm. Convert this pressure to millimeters of mercury, inches of mercury, and pounds per square inch. Would a cabin pressurized at 500 mm Hg meet federal standards?*

◄ Commercial airplane cabins must be pressurized to a pressure greater than the equivalent atmospheric pressure at an elevation of 8000 ft.

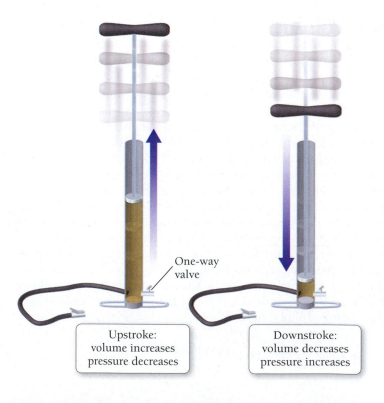

One-way valve

Upstroke: volume increases pressure decreases

Downstroke: volume decreases pressure increases

◄ **FIGURE 11.11 Operation of a hand pump**

If two quantities are inversely proportional, increasing one decreases the other (► **FIGURE 11.12**). As we saw for the hand pump, when the volume of a gas sample is decreased, its pressure increases and vice versa. Kinetic molecular theory explains the observed change in pressure. If the volume of a gas sample is decreased, the same number of gas particles is crowded into a smaller volume, resulting in more collisions with the walls of the container and therefore increasing the pressure (► **FIGURE 11.13**).

Scuba divers learn about Boyle's law during certification courses because it explains why ascending too quickly toward the surface is dangerous. For every 10 m of depth that a diver descends in water, he experiences an additional 1 atm of pressure due to the weight of the water above him (► **FIGURE 11.14**). The pressure regulator used in scuba diving delivers air at a pressure that matches the external pressure; otherwise the diver could not inhale the air (see the *Everyday Chemistry* box on page 370). For example, when a diver is at 20 m of depth, the

▶ **FIGURE 11.12 Volume versus pressure** **(a)** We can use a J-tube to measure the volume of a gas at different pressures. Adding mercury to the J-tube increases the height (*h*) of the mercury in the column, which causes the pressure on the gas sample to increase and its volume to decrease. **(b)** A plot of the volume of a gas as a function of pressure.

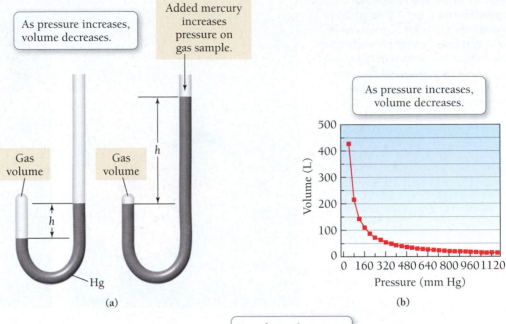

As pressure increases, volume decreases.

Added mercury increases pressure on gas sample.

Gas volume

Gas volume

h

h

Hg

(a)

As pressure increases, volume decreases.

500
400
300
200
100
0

Volume (L)

0 160 320 480 640 800 960 1120

Pressure (mm Hg)

(b)

▶ **FIGURE 11.13 Volume versus pressure: a molecular view** As the volume of a sample of gas is decreased, the number of collisions between the gas molecules and each square meter of the container increases. This raises the pressure exerted by the gas.

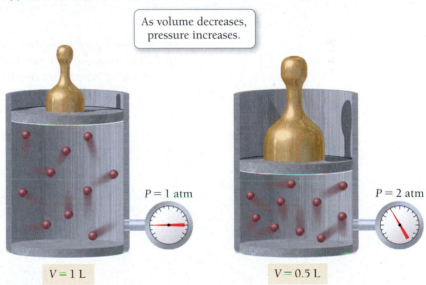

As volume decreases, pressure increases.

$P = 1$ atm

$V = 1$ L

$P = 2$ atm

$V = 0.5$ L

Depth = 0 m
$P = 1$ atm

Depth = 20 m
$P = 3$ atm

regulator delivers air at a pressure of 3 atm to match the 3 atm of pressure around the diver—1 atm due to normal atmospheric pressure and 2 additional atmospheres due to the weight of the water at 20 m (▶ **FIGURE 11.15**, on the next page).

Suppose that a diver inhales a lungful of 3-atm air and swims quickly to the surface (where the pressure drops to 1 atm) while holding his breath. What happens to the volume of air in his lungs? Since the pressure decreases by a factor of 3, the volume of the air in his lungs increases by a factor of 3, severely damaging his lungs and possibly killing him. The volume increase in the diver's lungs would be so great that the diver would not be able to hold his breath all the way to the surface—the air would force itself out of his mouth. So the most important rule in diving is *never hold your breath*. Divers must ascend slowly and breathe continuously, allowing the regulator to bring the air pressure in their lungs back to 1 atm by the time they reach the surface.

◀ **FIGURE 11.14 Pressure at depth** For every 10 m of depth, a diver experiences an additional 1 atm of pressure due to the weight of the water surrounding him. At 20 m, the diver experiences a total pressure of 3 atm (1 atm from atmospheric pressure plus an additional 2 atm from the weight of the water).

▶ **FIGURE 11.15 The dangers of decompression** **(a)** A diver at 20 m experiences an external pressure of 3 atm and breathes air pressurized at 3 atm. **(b)** If the diver shoots toward the surface with lungs full of 3 atm air, his lungs will expand as the external pressure drops to 1 atm.

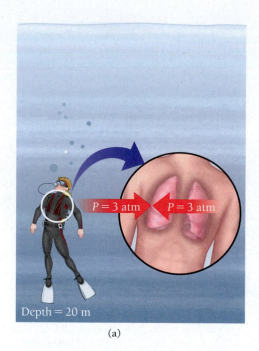
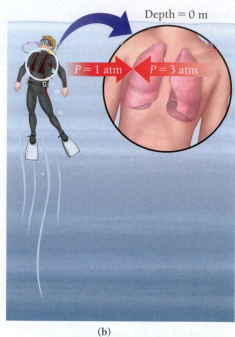

(a)

(b)

We can use Boyle's law to calculate the volume of a gas following a pressure change or the pressure of a gas following a volume change *as long as the temperature and the amount of gas remain constant*. For these calculations, we write Boyle's law in a slightly different way:

$$\text{Since } V \propto \frac{1}{P}\text{, then } V = \frac{\text{constant}}{P}$$

If we multiply both sides by P, we get:

$$PV = \text{constant}$$

This relationship is true because if the pressure increases, the volume decreases, but the product $P \times V$ is always equal to the same constant. For two different sets of conditions, we can say that

$$P_1V_1 = \text{constant} = P_2V_2\text{, or}$$

$$P_1V_1 = P_2V_2$$

where P_1 and V_1 are the initial pressure and volume of the gas, and P_2 and V_2 are the final volume and pressure. For example, suppose we want to calculate the pressure of a gas that was initially at 765 mm Hg and 1.78 L and later compressed to 1.25 L. We first sort the information in the problem.

GIVEN: $P_1 = 765$ mm Hg
$V_1 = 1.78$ L
$V_2 = 1.25$ L

FIND: P_2

Based on Boyle's law, and before doing any calculations, do you expect P_2 to be greater than or less than P_1?

SOLUTION MAP
We then draw a solution map showing how the equation takes us from the given quantities (what we have) to the find quantity (what we want to find).

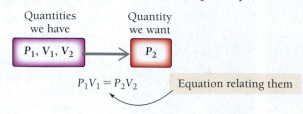

RELATIONSHIPS USED

$$P_1V_1 = P_2V_2 \quad \text{(Boyle's law, presented in this section)}$$

SOLUTION

We then solve the equation for the quantity we are trying to find (P_2).

$$P_1V_1 = P_2V_2$$

$$P_2 = \frac{P_1V_1}{V_2}$$

Lastly, we substitute the numerical values into the equation and calculate the answer.

$$P_2 = \frac{P_1V_1}{V_2} = \frac{(765 \text{ mm Hg})(1.78 \text{ L})}{1.25 \text{ L}}$$

$$= 1.09 \times 10^3 \text{ mm Hg}$$

EXAMPLE **11.2**   **Boyle's Law**

A cylinder equipped with a moveable piston has an applied pressure of 4.0 atm and a volume of 6.0 L. What is the volume of the cylinder if the applied pressure is decreased to 1.0 atm?

SORT	GIVEN:
You are given an initial pressure, an initial volume, and a final pressure. You are asked to find the final volume.	$P_1 = 4.0$ atm $V_1 = 6.0$ L $P_2 = 1.0$ atm FIND: V_2

STRATEGIZE

Draw a solution map beginning with the given quantities. Boyle's law shows the relationship necessary to determine the quantity you need to find.

SOLUTION MAP

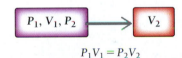

$$P_1V_1 = P_2V_2$$

RELATIONSHIPS USED

$$P_1V_1 = P_2V_2 \qquad \text{(Boyle's law, presented in this section)}$$

SOLVE

Solve the equation for the quantity you are trying to find (V_2), and substitute the numerical quantities into the equation to calculate the answer.

SOLUTION

$$P_1V_1 = P_2V_2$$

$$V_2 = \frac{V_1P_1}{P_2}$$

$$= \frac{(6.0 \text{ L})(4.0 \text{ atm})}{1.0 \text{ atm}}$$

$$= 24 \text{ L}$$

CHECK

Check your answer. Are the units correct? Does the answer make physical sense?

The answer has units of volume (L) as expected. The answer is reasonable because you expect the volume to increase as the pressure decreases.

▶ **SKILLBUILDER 11.2 | Boyle's Law**

A snorkeler takes a syringe filled with 16 mL of air from the surface, where the pressure is 1.0 atm, to an unknown depth. The volume of the air in the syringe at this depth is 7.5 mL. What is the pressure at this depth? If pressure increases by an additional 1 atm for every 10 m of depth, how deep is the snorkeler?

▶ **FOR MORE PRACTICE** Example 11.14; Problems 33, 34, 35, 36.

CONCEPTUAL ✓ CHECKPOINT 11.4

A flask contains a gas sample at pressure x. If the volume of the container triples at constant temperature and constant amount of gas, what is the pressure after the volume change?

(a) $3x$ **(b)** $\frac{1}{3}x$ **(c)** $9x$

EVERYDAY CHEMISTRY

Extra-long Snorkels

Several episodes of *The Flintstones* featured Fred Flintstone and Barney Rubble snorkeling. Their snorkels, however, were not the modern kind, but long reeds that stretched from the surface of the water down many meters. Fred and Barney swam around in the deep water while breathing air provided to them by these extra-long snorkels. Would this work? Why do people bother with scuba diving equipment if they could simply use 10-m snorkels like Fred and Barney did?

When we breathe, we expand the volume of our lungs, lowering the pressure within them (Boyle's law). Air from outside our lungs then flows into them. Extra-long snorkels, such as those used by Fred and Barney, do not work because of the pressure caused by water at depth. A diver at 10 m experiences a pressure of 2 atm that compresses the air in his lungs to a pressure of 2 atm. If the diver had a snorkel that went to the surface—where the air pressure is 1 atm—air would flow out of his lungs, not into them, making it impossible to breathe.

B11.2 CAN YOU ANSWER THIS? *Suppose a diver takes a balloon with a volume of 2.5 L from the surface, where the pressure is 1.0 atm, to a depth of 20 m, where the pressure is 3.0 atm. What would happen to the volume of the balloon? What if the end of the balloon was on a long tube that went to the surface and was attached to another balloon, as shown in the drawing? Which way would air flow as the diver descends?*

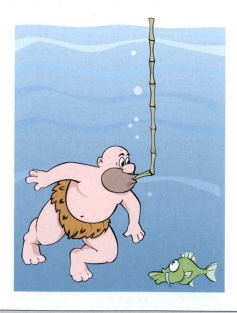

◀ This cartoon caveman would be unable to use a reed to breathe air from the surface because the pressure at depth would push air out of his lungs.

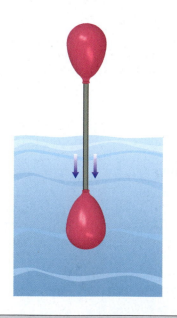

▶ If one end of a long tube with balloons tied on both ends is submerged in water, in which direction does air flow?

11.5 Charles's Law: Volume and Temperature

▶ Restate and apply Charles's law.

Recall from Section 2.10 that density = mass/volume. If the volume increases and the mass remains constant, the density must decrease.

Have you ever noticed that hot air rises? You may have walked upstairs in your house and noticed it feeling warmer than the ground floor. Or you may have witnessed a hot-air balloon take flight. The air that fills a hot-air balloon is warmed with a burner, which causes the balloon to rise in the cooler air around it. Why does hot air rise? Hot air rises because the volume of a gas sample at constant pressure increases with increasing temperature. As long as the amount of gas (and

▲ Heating the air in a balloon makes it expand (Charles's law). As the volume occupied by the hot air increases, its density decreases, allowing the balloon to float in the cooler, denser air that surrounds it.

⎡ The extrapolated line in Figure 11.16 cannot be measured experimentally because all gases condense into liquids before −273 °C is reached.

⎡ Section 3.10 summarizes the three different temperature scales.

⎡ Charles's law assumes constant pressure and a constant amount of gas.

therefore its mass) remains constant, warming decreases its density because density is mass divided by volume. A lower-density gas floats in a higher-density gas just as wood floats in water.

Suppose we keep the pressure of a gas sample constant and measure its volume at a number of different temperatures. The results of a number of such measurements are shown in ▼ **FIGURE 11.16**. The plot reveals the relationship between volume and temperature: The volume of a gas increases with increasing temperature. Note also that temperature and volume are *linearly related*. If two variables are linearly related, plotting one against the other produces a straight line.

We can predict an important property of matter by extending the line on our plot in Figure 11.16 backward from the lowest measured point—a process called *extrapolation*. Our extrapolated line shows that the gas should have a zero volume at −273 °C. Recall from Chapter 3 that −273 °C corresponds to 0 K, the coldest possible temperature. Our extrapolated line shows that below −273 °C, our gas would have a negative volume, which is physically impossible. For this reason, we refer to 0 K as **absolute zero**—colder temperatures do not exist.

The first person to carefully quantify the relationship between the volume of a gas and its temperature was J. A. C. Charles (1746–1823), a French mathematician and physicist. Charles was among the first people to ascend in a hydrogen-filled balloon. The law he formulated is called **Charles's law**.

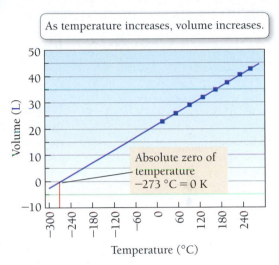

As temperature increases, volume increases.

▲ **FIGURE 11.16 Volume versus temperature** The volume of a gas increases linearly with increasing temperature. Question: How does this graph demonstrate that −273 °C is the coldest possible temperature?

Charles's law: The volume (V) of a gas and its Kelvin temperature (T) are directly proportional.

$$V \propto T$$

If two variables are directly proportional, increasing one by some factor increases the other by the same factor. For example, when the temperature of a gas sample (in kelvins) is doubled, its volume doubles; when the temperature is tripled, its volume triples; and so on. The observed relationship between the temperature and volume of a gas follows from kinetic molecular theory. If the temperature of a gas sample increases, the gas particles move faster (they have more kinetic energy), and if the pressure is to remain constant, the volume must increase (▶ **FIGURE 11.17**, on the next page).

You can experience Charles's law directly by holding a partially inflated balloon over a warm toaster. As the air in the balloon warms, you can feel the balloon expanding. Alternatively, you can put an inflated balloon in the freezer or take it outside on a very cold day (below freezing) and see that it becomes smaller as it cools.

We can use Charles's law to calculate the volume of a gas following a temperature change or the temperature of a gas following a volume change as *long as the pressure and the amount of gas are constant*. For these calculations, we express Charles's law in a different way.

▲ A partially inflated balloon held over a warm toaster expand as the air within the balloon warms.

Since $V \propto T$, then $V = \text{constant} \times T$

▶ **FIGURE 11.17 Volume versus temperature: a molecular view** If a balloon is moved from an ice-water bath into a boiling-water bath, the gas molecules inside it move faster (they have more kinetic energy) due to the increased temperature. If the external pressure remains constant, the molecules will expand the balloon and collectively occupy a larger volume.

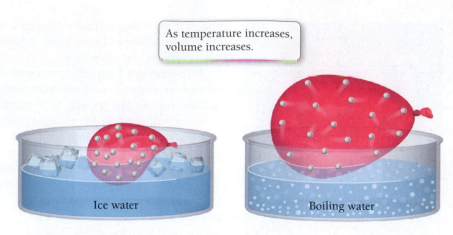

As temperature increases, volume increases.

Ice water

Boiling water

If we divide both sides by T, we get:

$$\frac{V}{T} = \text{constant}$$

If the temperature increases, the volume increases in direct proportion so that the quotient, V/T, is always equal to the same constant. So, for two different measurements, we can say that:

$$\frac{V_1}{T_1} = \text{constant} = \frac{V_2}{T_2}, \text{ or}$$

$$\frac{V_1}{T_1} = \frac{V_2}{T_2}$$

where V_1 and T_1 are the initial volume and temperature of the gas and V_2 and T_2 are the final volume and temperature. *All temperatures must be expressed in kelvins.*

For example, suppose we have a 2.37-L sample of a gas at 298 K that is then heated to 354 K with no change in pressure. To determine the final volume of the gas, we begin by sorting the information in the problem statement.

Based on Charles's law, and before doing any calculations, do you expect V_2 to be greater than or less than V_1?

GIVEN: $T_1 = 298$ K

$V_1 = 2.37$ L

$T_2 = 354$ K

FIND: V_2

SOLUTION MAP

We then strategize by building a solution map that shows how the equation takes us from the given quantities to the unknown quantity.

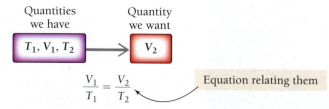

Quantities we have

Quantity we want

T_1, V_1, T_2 → V_2

$\frac{V_1}{T_1} = \frac{V_2}{T_2}$

Equation relating them

RELATIONSHIPS USED

$$\frac{V_1}{T_1} = \frac{V_2}{T_2} \quad \text{(Charles's law, presented in this section)}$$

SOLUTION

We then solve the equation for the quantity we are trying to find (V_2).

$$\frac{V_1}{T_1} = \frac{V_2}{T_2}$$

$$V_2 = \frac{V_1}{T_1} T_2$$

Lastly, we substitute the numerical values into the equation and calculate the answer.

$$V_2 = \frac{V_1}{T_1} T_2$$

$$= \frac{2.37 \text{ L}}{298 \text{ K}} 354 \text{ K}$$

$$= 2.82 \text{ L}$$

EXAMPLE **11.3** | **Charles's Law**

A sample of gas has a volume of 2.80 L at an unknown temperature. When the sample is submerged in ice water at $t = 0\,°C$, its volume decreases to 2.57 L. What was its initial temperature (in kelvins and in Celsius)? Assume a constant pressure. (To distinguish between the two temperature scales, use t for temperature in °C and T for temperature in K.)

SORT You are given an initial volume, a final volume, and a final temperature. You are asked to find the initial temperature in both kelvins (T_1) and degrees Celsius (t_1).	**GIVEN:** $V_1 = 2.80$ L $V_2 = 2.57$ L $t_2 = 0\,°C$ **FIND:** T_1 and t_1
STRATEGIZE Draw a solution map beginning with the given quantities. Charles's law shows the relationship necessary to get to the find quantity.	**SOLUTION MAP** 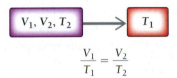 $$\frac{V_1}{T_1} = \frac{V_2}{T_2}$$ **RELATIONSHIPS USED** $\dfrac{V_1}{T_1} = \dfrac{V_2}{T_2}$ (Charles's law, presented in this section)
SOLVE Solve the equation for the quantity you are trying to find (T_1). Before you substitute in the numerical values, convert the temperature to kelvins. *Remember, you must always work gas law problems using Kelvin temperatures.* Once you have converted the temperature to kelvins, substitute into the equation to find T_1. Convert the temperature to degrees Celsius to find t_1.	**SOLUTION** $\dfrac{V_1}{T_1} = \dfrac{V_2}{T_2}$ $T_1 = \dfrac{V_1}{V_2} T_2$ $T_2 = 0 + 273 = 273$ K $T_1 = \dfrac{V_1}{V_2} T_2$ $ = \dfrac{2.80 \text{ L}}{2.57 \text{ L}} 273$ K $ = 297$ K $t_1 = 297 - 273 = 24\,°C$
CHECK Check your answer. Are the units correct? Does the answer make physical sense?	The answers have the correct units, K and °C. The answer is reasonable because the initial volume was larger than the final volume; therefore, the initial temperature must be higher than the final temperature.

▶ **SKILLBUILDER 11.3** | **Charles's Law**

A gas in a cylinder with a moveable piston has an initial volume of 88.2 mL and is heated from 35 °C to 155 °C. What is the final volume of the gas in milliliters?

▶ **FOR MORE PRACTICE**   Problems 39, 40, 41, 42.

CONCEPTUAL ✓ CHECKPOINT 11.5

A volume of gas is confined to a cylinder with a freely moveable piston at one end. If you apply enough heat to double the Kelvin temperature of the gas, what happens? (Assume constant pressure.)

(a) The volume doubles.

(b) The volume remains the same.

(c) The volume decreases to half the initial volume.

11.6 The Combined Gas Law: Pressure, Volume, and Temperature

▶ Restate and apply the combined gas law.

Boyle's law shows how P and V are related at constant temperature, and Charles's law shows how V and T are related at constant pressure. But what if two of these variables change at once? For example, what happens to the volume of a gas if both its pressure and its temperature change?

Since volume is inversely proportional to pressure ($V \propto 1/P$) and directly proportional to temperature ($V \propto T$), we can write:

$$V \propto \frac{T}{P} \quad \text{or} \quad \frac{PV}{T} = \text{constant}$$

For a sample of gas under two different sets of conditions we use the **combined gas law**.

> The combined gas law encompasses both Boyle's law and Charles's law, and we can use it in place of them. If one physical property (P, V, or T) is constant, it cancels out of our calculations when we use the combined gas law.

The combined gas law: $\dfrac{P_1 V_1}{T_1} = \dfrac{P_2 V_2}{T_2}$

The combined gas law applies only when the *amount* of gas is constant. We must express the temperature (as with Charles's law) in kelvins.

Suppose a person carries a cylinder with a moveable piston that has an initial volume of 3.65 L up a mountain. The pressure at the bottom of the mountain is 755 mm Hg, and the temperature is 302 K. The pressure at the top of the mountain is 687 mm Hg, and the temperature is 291 K. What is the volume of the cylinder at the top of the mountain? We begin by sorting the information in the problem statement.

GIVEN: $P_1 = 755$ mm Hg $\quad T_2 = 291$ K
$V_1 = 3.65$ L $\quad P_2 = 687$ mm Hg
$T_1 = 302$ K

FIND: V_2

SOLUTION MAP

We strategize by building a solution map that shows how the combined gas law equation allows us to determine the find quantity from the given quantities.

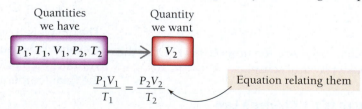

RELATIONSHIPS USED

$$\frac{P_1 V_1}{T_1} = \frac{P_2 V_2}{T_2} \quad \text{(combined gas law, presented in this section)}$$

SOLUTION

We then solve the equation for the quantity we are trying to find (V_2).

$$\frac{P_1 V_1}{T_1} = \frac{P_2 V_2}{T_2}$$

$$V_2 = \frac{P_1 V_1 T_2}{T_1 P_2}$$

Lastly, we substitute in the appropriate values and calculate the answer.

$$V_2 = \frac{P_1 V_1 T_2}{T_1 P_2}$$

$$= \frac{755 \text{ mm Hg} \times 3.65 \text{ L} \times 291 \text{ K}}{302 \text{ K} \times 687 \text{ mm Hg}}$$

$$= 3.87 \text{ L}$$

EXAMPLE **11.4** | **The Combined Gas Law**

A sample of gas has an initial volume of 158 mL at a pressure of 735 mm Hg and a temperature of 34 °C. If the gas is compressed to a volume of 108 mL and heated to a temperature of 85 °C, what is its final pressure in millimeters of mercury?

SORT You are given the initial pressure, temperature, and volume as well as the final temperature and volume. You are asked to find the final pressure.	**GIVEN:** $P_1 = 735$ mm Hg $\quad t_1 = 34$ °C $\qquad t_2 = 85$ °C $\quad V_1 = 158$ mL $\qquad V_2 = 108$ mL **FIND:** P_2
STRATEGIZE Draw a solution map beginning with the given quantities. The combined gas law shows the relationship necessary to get to the find quantity.	**SOLUTION MAP** $\boxed{P_1, T_1, V_1, T_2, V_2} \longrightarrow \boxed{P_2}$ $$\frac{P_1 V_1}{T_1} = \frac{P_2 V_2}{T_2}$$ **RELATIONSHIPS USED** $\dfrac{P_1 V_1}{T_1} = \dfrac{P_2 V_2}{T_2}$ (combined gas law, presented in this section)
SOLVE Solve the equation for the quantity you are trying to find (P_2). Before you substitute in the numerical values, you must convert the temperatures to kelvins. Once you have converted the temperature to kelvins, substitute into the equation to find P_2.	**SOLUTION** $$\frac{P_1 V_1}{T_1} = \frac{P_2 V_2}{T_2}$$ $$P_2 = \frac{P_1 V_1 T_2}{T_1 V_2}$$ $T_1 = 34 + 273 = 307$ K $T_2 = 85 + 273 = 358$ K $$P_2 = \frac{735 \text{ mm Hg} \times 158 \text{ mL} \times 358 \text{ K}}{307 \text{ K} \times 108 \text{ mL}}$$ $$= 1.25 \times 10^3 \text{ mm Hg}$$
CHECK Check your answer. Are the units correct? Does the answer make physical sense?	The answer has the correct units, mm Hg. The answer is reasonable because the decrease in volume and the increase in temperature should result in a pressure that is higher than the initial pressure.

continued on page 376 ▶

continued from page 375

> ▶ **SKILLBUILDER 11.4** | **The Combined Gas Law**
>
> A balloon has a volume of 3.7 L at a pressure of 1.1 atm and a temperature of 30°C. If the balloon is submerged in water to a depth where the pressure is 4.7 atm and the temperature is 15°C, what is its volume (assume that any changes in pressure caused by the skin of the balloon are negligible)?
>
> ▶ **FOR MORE PRACTICE** Example 11.15; Problems 51, 52, 53, 54, 55, 56.

PEARSON
eText
2.0

CONCEPTUAL ✔ CHECKPOINT 11.6

A volume of gas is confined to a container. If you apply enough heat to double the Kelvin temperature of the gas and expand the size of the container to twice its initial volume, what happens to the pressure?

(a) The pressure doubles.

(b) The pressure decreases to half of its initial value.

(c) The pressure is the same as its initial value.

11.7 Avogadro's Law: Volume and Moles

▶ Restate and apply Avogadro's law.

So far, we have learned how V, P, and T are interrelated, but we have considered only a constant amount of a gas. What happens when the amount of gas changes? If we make several measurements of the volume of a gas sample (at constant temperature and pressure) while varying the number of moles in the sample, our results are similar to those shown in ▼ **FIGURE 11.18**. We can see that the relationship between volume and number of moles is linear. An extrapolation to zero moles shows a zero volume, as we might expect. This relationship was first stated formally by Amedeo Avogadro (1776–1856) and is called **Avogadro's law**.

Avogadro's law assumes constant temperature and pressure.

> **Avogadro's law:** The volume of a gas and the amount of the gas in moles (n) are directly proportional.

$$V \propto n$$

Since $V \propto n$, then $V/n =$ constant. If the number of moles increases, then the volume increases in direct proportion so that the quotient, V/n, is always equal to the same constant. Thus, for two different measurements, we can say that
$$\frac{V_1}{n_1} = \text{constant} = \frac{V_2}{n_2} \text{ or } \frac{V_1}{n_1} = \frac{V_2}{n_2}.$$

When the amount of gas in a sample increases, its volume increases in direct proportion, which is yet another prediction of kinetic molecular theory. If the number of gas particles increases at constant pressure and temperature, the particles must occupy more volume.

We experience Avogadro's law when we inflate a balloon, for example. With each exhaled breath, we add more gas particles to the inside of the balloon, increasing its volume (▼ **FIGURE 11.19**). We can use Avogadro's law to calculate the volume of a gas following a change in the amount of the gas *as long as the pressure and temperature of the gas are constant.* For these calculations, Avogadro's law is expressed as:

$$\frac{V_1}{n_1} = \frac{V_2}{n_2}$$

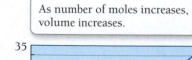
As number of moles increases, volume increases.

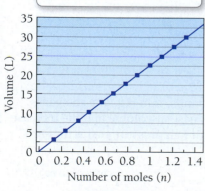
Number of moles (*n*)

◀ **FIGURE 11.18 Volume versus number of moles** The volume of a gas sample increases linearly with the number of moles in the sample.

▶ **FIGURE 11.19 Blow-up** As you exhale into a balloon, you add gas molecules to the inside of the balloon, increasing its volume.

where V_1 and n_1 are the initial volume and number of moles of the gas and V_2 and n_2 are the final volume and number of moles. In calculations, we use Avogadro's law in a manner similar to the other gas laws, as shown in Example 11.5.

EXAMPLE **11.5** | **Avogadro's Law**

A 4.8-L sample of helium gas contains 0.22 mol of helium. How many additional moles of helium gas should you add to the sample to obtain a volume of 6.4 L? Assume constant temperature and pressure.

SORT You are given an initial volume, an initial number of moles, and a final volume. You are (essentially) asked to find the final number of moles.	**GIVEN:** $V_1 = 4.8$ L $\quad\quad\quad n_1 = 0.22$ mol $\quad\quad\quad V_2 = 6.4$ L **FIND:** n_2
STRATEGIZE Draw a solution map beginning with the given quantities. Avogadro's law addresses the relationship necessary to determine the find quantity.	**SOLUTION MAP** 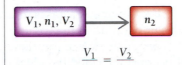 $$\dfrac{V_1}{n_1} = \dfrac{V_2}{n_2}$$ **RELATIONSHIPS USED** $\dfrac{V_1}{n_1} = \dfrac{V_2}{n_2}$ (Avogadro's law, presented in this section)
SOLVE Solve the equation for the quantity you are trying to find (n_2), and substitute the appropriate quantities to calculate n_2. Since the balloon already contains 0.22 mol, subtract this quantity from the final number of moles to determine how much you must add.	**SOLUTION** $$\dfrac{V_1}{n_1} = \dfrac{V_2}{n_2}$$ $$n_2 = \dfrac{V_2}{V_1}\, n_1$$ $$= \dfrac{6.4 \; \cancel{L}}{4.8 \; \cancel{L}} \, 0.22 \text{ mol}$$ $$= 0.29 \text{ mol}$$ mol to add $= 0.29$ mol $- 0.22$ mol $= 0.07$ mol
CHECK Check your answer. Are the units correct? Does the answer make physical sense?	The answer has the correct units, moles. The answer is reasonable because the increase in the number of moles is proportional to the given increase in the volume.

▶ **SKILLBUILDER 11.5 | Avogadro's Law**

A chemical reaction occurring in a cylinder equipped with a moveable piston produces 0.58 mol of a gaseous product. If the cylinder contained 0.11 mol of gas before the reaction and had an initial volume of 2.1 L, what was its volume after the reaction?

▶ **FOR MORE PRACTICE** Problems 45, 46, 47, 48.

PEARSON
eText
2.0

CONCEPTUAL ✔ **CHECKPOINT 11.7**

If each gas sample has the same temperature and pressure, which has the greatest volume?

(a) 1 g O_2

(b) 1 g Ar

(c) 1 g H_2

11.8 The Ideal Gas Law: Pressure, Volume, Temperature, and Moles

▶ Restate and apply the ideal gas law.

The relationships covered so far can be combined into a single law that encompasses all of them. So far, we know that:

$$V \propto \frac{1}{P} \quad \text{(Boyle's law)}$$

$$V \propto T \quad \text{(Charles's law)}$$

$$V \propto n \quad \text{(Avogadro's law)}$$

Combining these three expressions, we arrive at:

$$V \propto \frac{nT}{P}$$

The volume of a gas is directly proportional to the number of moles of gas and the temperature of the gas and is inversely proportional to the pressure of the gas. We can replace the proportional sign with an equal sign by adding R, a proportionality constant called the **ideal gas constant**.

$$V = \frac{RnT}{P}$$

Rearranging, we get the **ideal gas law**:

The ideal gas law: $\quad PV = nRT$

The value of R, the ideal gas constant, is:

R can be expressed in other units, but its numerical value will then be different.

$$R = 0.0821 \frac{\text{L} \cdot \text{atm}}{\text{mol} \cdot \text{K}}$$

The ideal gas law contains within it the simple gas laws. For example, recall that Boyle's law states that $V \propto 1/P$ when the amount of gas (n) and the temperature of the gas (T) are kept constant. To derive Boyle's law, we can rearrange the ideal gas law as follows:

$$PV = nRT$$

First, divide both sides by P.

$$V = \frac{nRT}{P}$$

Then put the variables that are constant in parentheses.

$$V = (nRT)\frac{1}{P}$$

Since n and T are constant in this case and since R is always a constant:

$$V = (\text{Constant}) \times \frac{1}{P}$$

which gives us Boyle's law $\left(V \propto \frac{1}{P}\right)$.

The ideal gas law also shows how other pairs of variables are related. For example, from Charles's law we know that volume is proportional to temperature at constant pressure and a constant number of moles. But what if we heat a sample of gas at constant *volume* and a constant number of moles? This question applies to the warning labels on aerosol cans such as hair spray or deodorants. These labels warn the user against excessive heating or incineration of the can, even after the contents are used up. Why? An aerosol can that appears empty actually contains a fixed amount of gas trapped in a fixed volume. What would happen if we heated the can? Let's rearrange the ideal gas law to clearly see the relationship between pressure and temperature at constant volume and a constant number of moles:

$$PV = nRT$$

If we divide both sides by V, we get:

$$P = \frac{nRT}{V}$$

$$P = \left(\frac{nR}{V}\right)T$$

Since n and V are constant and since R is always a constant:

$$P = \text{Constant} \times T$$

> The relationship between pressure and temperature is also known as Gay-Lussac's law.

As the temperature of a fixed amount of gas in a fixed volume increases, the pressure increases. In an aerosol can, this pressure increase can cause the can to explode, which is why aerosol cans should not be heated or incinerated. Table 11.2 summarizes the relationships between all of the simple gas laws and the ideal gas law.

TABLE 11.2   Relationships between Simple Gas Laws and Ideal Gas Law

Variable Quantities	Constant Quantities	Ideal Gas Law in Form of Variables-Constant	Simple Gas Law	Name of Simple Law
V and P	n and T	$PV = nRT$	$P_1V_1 = P_2V_2$	Boyle's law
V and T	n and P	$\frac{V}{T} = \frac{nR}{P}$	$\frac{V_1}{T_1} = \frac{V_2}{T_2}$	Charles's law
P and T	n and V	$\frac{P}{T} = \frac{nR}{V}$	$\frac{P_1}{T_1} = \frac{P_2}{T_2}$	Gay-Lussac's law
P and n	V and T	$\frac{P}{n} = \frac{RT}{V}$	$\frac{P_1}{n_1} = \frac{P_2}{n_2}$	
V and n	T and P	$\frac{V}{n} = \frac{RT}{P}$	$\frac{V_1}{n_1} = \frac{V_2}{n_2}$	Avogadro's law

We can use the ideal gas law to determine the value of any one of the four variables (P, V, n, or T) given the other three. However, each of the quantities in the ideal gas law *must be expressed* in the units within R.

- Pressure (**P**) must be expressed in atmospheres.
- Volume (**V**) must be expressed in liters.
- Amount of gas (**n**) must be expressed in moles.
- Temperature (**T**) must be expressed in kelvins.

For example, suppose we want to know the pressure of 0.18 mol of a gas in a 1.2-L flask at 298 K. We begin by sorting the information in the problem statement.

GIVEN: $n = 0.18$ mol
$V = 1.2$ L
$T = 298$ K

FIND: P

SOLUTION MAP

We strategize by drawing a solution map that shows how the ideal gas law takes us from the given quantities to the find quantity.

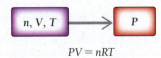

$$PV = nRT$$

RELATIONSHIPS USED

$PV = nRT$ (ideal gas law, presented in this section)

SOLUTION

We solve the equation for the quantity we are trying to find (in this case, P).

$$PV = nRT$$

$$P = \frac{nRT}{V}$$

Next we substitute in the numerical values and calculate the answer.

$$P = \frac{0.18 \cancel{\text{mol}} \times 0.0821\dfrac{\cancel{L} \cdot \text{atm}}{\cancel{\text{mol}} \cdot \cancel{K}} \times 298 \cancel{K}}{1.2 \cancel{L}}$$

$$= 3.7 \text{ atm}$$

Notice that all units cancel except the units of the quantity we need (atm).

PEARSON
eText 2.0
Interactive
Worked Example
Video 11.6

EXAMPLE **11.6**   **The Ideal Gas Law**

Calculate the volume occupied by 0.845 mol of nitrogen gas at a pressure of 1.37 atm and a temperature of 315 K.

SORT	
You are given the number of moles, the pressure, and the temperature of a gas sample. You are asked to find the volume.	**GIVEN:** $n = 0.845$ mol $P = 1.37$ atm $T = 315$ K **FIND:** V

STRATEGIZE	
Draw a solution map beginning with the given quantities. The ideal gas law shows the relationship necessary to determine the find quantity.	**SOLUTION MAP** 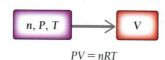 $PV = nRT$ **RELATIONSHIPS USED** $PV = nRT$ (ideal gas law, presented in this section)

SOLVE	
Solve the equation for the quantity you are trying to find (V) and substitute the appropriate quantities to calculate V.	**SOLUTION** $PV = nRT$ $V = \dfrac{nRT}{P}$ $V = \dfrac{0.845 \cancel{\text{mol}} \times 0.0821\frac{L \cdot \text{atm}}{\cancel{\text{mol}} \cdot \cancel{K}} \times 315 \cancel{K}}{1.37 \cancel{\text{atm}}} = 16.0 \text{ L}$

CHECK	
Check your answer. Are the units correct? Does the answer make physical sense?	The answer has the correct units for volume, liters. The *value* of the answer is a bit more difficult to judge. However, at standard temperature and pressure ($T = 0\,°C$ or 273.15 K and $P = 1$ atm), 1 mol gas occupies 22.4 L (see Section 11.10). Therefore, an answer of 16.0 L seems reasonable for the volume of 0.85 mol of gas under conditions that are not too far from standard temperature and pressure.

▶ **SKILLBUILDER 11.6 | The Ideal Gas Law**

An 8.5-L tire is filled with 0.55 mol of gas at a temperature of 305 K. What is the pressure of the gas in the tire?

▶ **FOR MORE PRACTICE**   Example 11.16; Problems 59, 60, 61, 62.

If the units given in an ideal gas law problem are different from those of the ideal gas constant (atm, L, mol, and K), we must convert to the correct units before we substitute into the ideal gas equation, as demonstrated in Example 11.7.

EXAMPLE **11.7**   **The Ideal Gas Law Requiring Unit Conversion**

Calculate the number of moles of gas in a basketball inflated to a total pressure of 24.2 psi with a volume of 3.2 L at 25 °C.

SORT	**GIVEN:** $P = 24.2$ psi
You are given the pressure, the volume, and the temperature of a gas sample. You are asked to find the number of moles.	$V = 3.2$ L
	$t = 25$ °C
	FIND: n

STRATEGIZE	**SOLUTION MAP**
Draw a solution map beginning with the given quantities. The ideal gas law shows the relationship necessary to determine the find quantity.	
	$PV = nRT$
	RELATIONSHIPS USED
	$PV = nRT$ (ideal gas law, presented in this section)

SOLVE	**SOLUTION**
Solve the equation for the quantity you are trying to find (n).	$PV = nRT$
	$n = \dfrac{PV}{RT}$
Before substituting into the equation, convert P and t into the correct units. (Since 1.6462 atm is an intermediate answer, mark the least significant digit, but don't round until the end.)	$P = 24.2 \text{ psi} \times \dfrac{1 \text{ atm}}{14.7 \text{ psi}} = 1.6\underline{4}62 \text{ atm}$
	$T = t + 273$
	$= 25 + 273 = 298 \text{ K}$
Finally, substitute into the equation to calculate n.	$n = \dfrac{1.6\underline{4}62 \text{ atm} \times 3.2 \text{ L}}{0.0821\dfrac{\text{L} \cdot \text{atm}}{\text{mol} \cdot \text{K}} \times 298 \text{ K}}$
	$= 0.22 \text{ mol}$

CHECK	
Check your answer. Are the units correct? Does the answer make physical sense?	The answer has the correct units, moles. The *value* of the answer is a bit more difficult to judge. Again, it is helpful to know that at standard temperature and pressure ($T = 0$ °C or 273.15 K and $P = 1$ atm), 1 mol of gas occupies 22.4 L (see Check step in Example 11.6). A 3.2-L sample of gas at standard temperature and pressure (STP) would contain about 0.15 mol; therefore, at a greater pressure, the sample should contain a bit more than 0.15 mol, which is consistent with the answer.

▶ **SKILLBUILDER 11.7 | The Ideal Gas Law Requiring Unit Conversion**

How much volume does 0.556 mol of gas occupy when its pressure is 715 mm Hg and its temperature is 58 °C?

▶ **SKILLBUILDER PLUS**

Find the pressure in millimeters of mercury of a 0.133-g sample of helium gas at 32 °C contained in a 648-mL container.

▶ **FOR MORE PRACTICE**   Problems 63, 64, 67, 68.

Determining Molar Mass of a Gas from the Ideal Gas Law

We can use the ideal gas law in combination with mass measurements to calculate the molar mass of a gas. For example, a sample of gas has a mass of 0.136 g. Its volume is 0.112 L at a temperature of 298 K and a pressure of 1.06 atm. With this information, we can find its molar mass.

We begin by sorting the information given in the problem.

GIVEN: $m = 0.136$ g $V = 0.112$ L

$T = 298$ K $P = 1.06$ atm

FIND: molar mass (g/mol)

SOLUTION MAP

We strategize by drawing a solution map, which in this case has two parts. In the first part, we use P, V, and T to find the number of moles of gas. In the second part, we use the number of moles of gas and the given mass to find the molar mass.

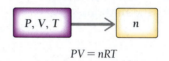

$$PV = nRT$$

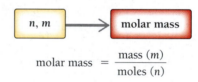

$$\text{molar mass} = \frac{\text{mass } (m)}{\text{moles } (n)}$$

RELATIONSHIPS USED

$$PV = nRT \quad \text{(ideal gas law, presented in this section)}$$

$$\text{molar mass} = \frac{\text{mass}}{\text{moles}} \quad \text{(definition of molar mass from Section 6.3)}$$

SOLUTION

$$PV = nRT$$

$$n = \frac{PV}{RT}$$

$$= \frac{1.06 \ \cancel{atm} \times 0.112 \ \cancel{L}}{0.0821 \dfrac{\cancel{L} \cdot \cancel{atm}}{mol \cdot \cancel{K}} \times 298 \ \cancel{K}}$$

$$= 4.8\underline{5}25 \times 10^{-3} \ \text{mol}$$

$$\text{molar mass} = \frac{\text{mass } (m)}{\text{moles } (n)}$$

$$= \frac{0.136 \ \text{g}}{4.8\underline{5}25 \times 10^{-3} \ \text{mol}}$$

$$= 28.0 \ \text{g/mol}$$

EXAMPLE **11.8**   **Molar Mass, the Ideal Gas Law, and Mass Measurement**

A sample of gas has a mass of 0.311 g. Its volume is 0.225 L at a temperature of 55 °C and a pressure of 886 mm Hg. Determine its molar mass.

SORT	**GIVEN:**
You are given the mass, the volume, the temperature, and the pressure of a gas sample. You are asked to find the molar mass of the gas.	$m = 0.311$ g $V = 0.225$ L $t = 55$ °C $P = 886$ mm Hg **FIND:** molar mass (g/mol)

STRATEGIZE

In the first part of the solution map, use the ideal gas law to find the number of moles of gas from the other given quantities.

In the second part, use the number of moles from the first part, as well as the given mass, to find the molar mass.

SOLUTION MAP

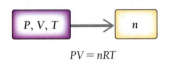

$PV = nRT$

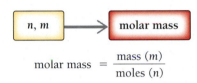

$$\text{molar mass} = \frac{\text{mass } (m)}{\text{moles } (n)}$$

RELATIONSHIPS USED

$PV = nRT$ (ideal gas law, presented in this section)

$$\text{molar mass} = \frac{\text{mass } (m)}{\text{moles } (n)} \quad \text{(definition of molar mass from Section 6.3)}$$

SOLVE

First, solve the ideal gas law for n.

Before substituting into the equation, convert the pressure to atm and temperature to K.

Substitute into the equation to calculate n, the number of moles.

Finally, use the number of moles just found and the given mass (m) to find the molar mass.

SOLUTION

$$PV = nRT$$
$$n = \frac{PV}{RT}$$
$$P = 886 \ \cancel{\text{mm Hg}} \times \frac{1 \ \text{atm}}{760 \ \cancel{\text{mm Hg}}} = 1.1658 \ \text{atm}$$
$$T = 55\,°C + 273 = 328 \ K$$
$$n = \frac{1.1658 \ \cancel{\text{atm}} \times 0.225 \ \cancel{L}}{0.0821 \ \frac{\cancel{L} \cdot \cancel{\text{atm}}}{\text{mol} \cdot \cancel{K}} \times 328 \ \cancel{K}}$$
$$= 9.7406 \times 10^{-3} \ \text{mol}$$
$$\text{molar mass} = \frac{\text{mass } (m)}{\text{moles } (n)}$$
$$= \frac{0.311 \ \text{g}}{9.7406 \times 10^{-3} \ \text{mol}}$$
$$= 31.9 \ \text{g/mol}$$

CHECK

Check your answer. Are the units correct? Does the answer make physical sense?

The answer has the correct units, g/mol. The answer is reasonable because its value is within the range of molar masses for common compounds.

▶ **SKILLBUILDER 11.8 | Molar Mass Using the Ideal Gas Law and Mass Measurement**

A sample of gas has a mass of 827 mg. Its volume is 0.270 L at a temperature of 88 °C and a pressure of 975 mm Hg. Find its molar mass.

▶ **FOR MORE PRACTICE** Problems 69, 70, 71, 72.

Ideal and Nonideal Gas Behavior

Although a complete derivation is beyond the scope of this book, the ideal gas law follows directly from the kinetic molecular theory of gases. Consequently, the ideal gas law holds only under conditions where the kinetic molecular theory holds. The ideal gas law works exactly only for gases that are acting ideally (▼ FIGURE 11.20), which means that (a) the volume of the gas particles is small compared to the space between them, and (b) the forces between the gas particles are not significant. These assumptions break down (▼ FIGURE 11.21) under conditions of high pressure (because the space between particles is no longer much larger than the size of the particles themselves) or low temperatures (because the gas particles move so slowly that their interactions become significant). For all of the problems encountered in this book, you may assume ideal gas behavior.

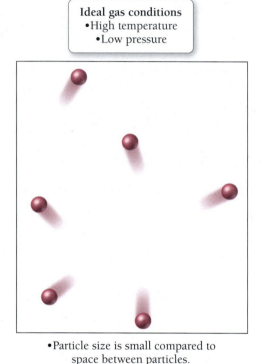

Ideal gas conditions
- High temperature
- Low pressure

- Particle size is small compared to space between particles.
- Interactions between particles are insignificant.

▲ **FIGURE 11.20 Conditions for ideal gas behavior** At high temperatures and low pressures, the assumptions of the kinetic molecular theory apply.

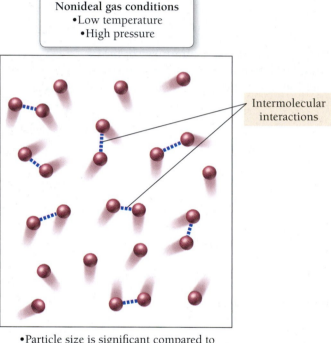

Nonideal gas conditions
- Low temperature
- High pressure

Intermolecular interactions

- Particle size is significant compared to space between particles.
- Interactions between particles are significant.

▲ **FIGURE 11.21 Conditions for nonideal gas behavior** At low temperatures and high pressures, the assumptions of kinetic molecular theory do not apply.

11.9 Mixtures of Gases

▶ Restate and apply Dalton's law of partial pressures.

Many gas samples are not pure but consist of mixtures of gases. The air in our atmosphere, for example, is a mixture containing 78% nitrogen, 21% oxygen, 0.9% argon, 0.04% carbon dioxide (Table 11.3), and a few other gases in smaller amounts.

According to the kinetic molecular theory, each of the components in a gas mixture acts independently of the others. For example, the nitrogen molecules in air exert a certain pressure—78% of the total pressure—that is independent of the presence of the other gases in the mixture. Likewise, the oxygen molecules in air exert a certain pressure—21% of the total pressure—that is also independent of the presence of the other gases in the mixture. The pressure due to any individual component in a gas mixture is the **partial pressure** of that component. The partial pressure of any component is that component's fractional composition times the total pressure of the mixture (▼ FIGURE 11.22).

partial pressure of component:

= fractional composition of component × total pressure

For example, the partial pressure of nitrogen (P_{N_2}) in air at 1.0 atm is:

$$P_{N_2} = 0.78 \times 1.0 \text{ atm}$$
$$= 0.78 \text{ atm}$$

Similarly, the partial pressure of oxygen in air at 1.0 atm is:

$$P_{O_2} = 0.21 \times 1.0 \text{ atm}$$
$$= 0.21 \text{ atm}$$

The sum of the partial pressures of each component in a gas mixture must equal the total pressure, as expressed by **Dalton's law of partial pressures**:

Dalton's law of partial pressures:

$$P_{tot} = P_a + P_b + P_c + \ldots$$

where P_{tot} is the total pressure and $P_a, P_b, P_c, \ldots$ are the partial pressures of the components.

For 1 atm air:

$$P_{tot} = P_{N_2} + P_{O_2} + P_{Ar}$$
$$P_{tot} = 0.78 \text{ atm} + 0.21 \text{ atm} + 0.01 \text{ atm}$$
$$= 1.00 \text{ atm}$$

TABLE 11.3 Composition of Dry Air	
Gas	**Percent by Volume (%)**
nitrogen (N_2)	78
oxygen (O_2)	21
argon (Ar)	0.9
carbon dioxide (CO_2)	0.04

The fractional composition is the percent composition divided by 100.

Gas mixture (80% He ●, 20% Ne ●)
$P_{tot} = 1.0 \text{ atm}$
$P_{He} = 0.80 \text{ atm}$
$P_{Ne} = 0.20 \text{ atm}$

◀ **FIGURE 11.22 Partial pressures** A gas mixture at a total pressure of 1.0 atm consisting of 80% helium and 20% neon has a helium partial pressure of 0.80 atm and a neon partial pressure of 0.20 atm.

EXAMPLE **11.9** | **Total Pressure and Partial Pressure**

A mixture of helium, neon, and argon has a total pressure of 558 mm Hg. If the partial pressure of helium is 341 mm Hg and the partial pressure of neon is 112 mm Hg, what is the partial pressure of argon?

You are given the total pressure of a gas mixture and the partial pressures of two (of its three) components. You are asked to find the partial pressure of the third component.	**GIVEN:** $P_{tot} = 558$ mm Hg $P_{He} = 341$ mm Hg $P_{Ne} = 112$ mm Hg **FIND:** P_{Ar}
To solve this problem, solve Dalton's law for the partial pressure of argon, and substitute the correct values to calculate it.	**SOLUTION** $P_{tot} = P_{He} + P_{Ne} + P_{Ar}$ $P_{Ar} = P_{tot} - P_{He} - P_{Ne}$ $\quad = 558 \text{ mm Hg} - 341 \text{ mm Hg} - 112 \text{ mm Hg}$ $\quad = 105 \text{ mm Hg}$

▶ **SKILLBUILDER 11.9 | Total Pressure and Partial Pressure**

A sample of hydrogen gas is mixed with water vapor. The mixture has a total pressure of 745 torr, and the water vapor has a partial pressure of 24 torr. What is the partial pressure of the hydrogen gas?

▶ **FOR MORE PRACTICE** Example 11.17; Problems 73, 74, 75, 76.

Partial Pressure and Physiology

Human lungs have evolved to breathe oxygen at a partial pressure of $P_{O_2} = 0.21$ atm. If the total pressure decreases—as happens when we climb a mountain, for example—the partial pressure of oxygen also decreases. For example, on top of Mount Everest, where the total pressure is 0.311 atm, the partial pressure of oxygen is only 0.065 atm. As we learned earlier, low oxygen levels can have negative physiological effects, a condition called **hypoxia**, or oxygen starvation. Mild hypoxia causes dizziness, headache, and shortness of breath. Severe hypoxia, which occurs when P_{O_2} drops below 0.1 atm, may cause unconsciousness or even death. For this reason, climbers hoping to summit Mount Everest carry oxygen to breathe.

▲ Mountain climbers on Mount Everest require oxygen because the pressure is so low that the lack of oxygen causes hypoxia, a condition that in severe cases can be fatal.

High oxygen levels can also have negative physiological effects. Scuba divers, as we discussed in Section 11.4, breathe pressurized air. At 30 m, a scuba diver breathes air at a total pressure of 4.0 atm, making P_{O_2} about 0.84 atm. This increased partial pressure of oxygen causes a higher density of oxygen molecules in the lungs (▼ FIGURE 11.23), which results in a higher concentration of oxygen in body tissues. When P_{O_2} increases beyond 1.4 atm, the increased oxygen concentration in body tissues causes a condition called **oxygen toxicity**, which is characterized by muscle twitching, tunnel vision, and convulsions (▶ FIGURE 11.24). Divers who venture too deep without proper precautions have drowned because of oxygen toxicity.

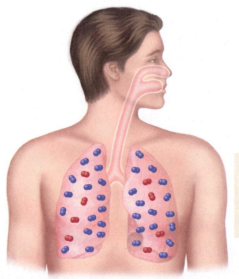

Surface

$P_{tot} = 1$ atm

$P_{N_2} = 0.78$ atm

$P_{O_2} = 0.21$ atm

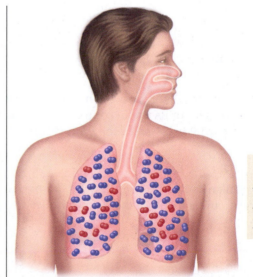

30 m

$P_{tot} = 4$ atm

$P_{N_2} = 3.12$ atm

$P_{O_2} = 0.84$ atm

▲ **FIGURE 11.23 Too much of a good thing**   When a person is breathing compressed air, there is a larger partial pressure of oxygen in the lungs. A large oxygen partial pressure in the lungs results in a larger amount of oxygen in bodily tissues. When the oxygen partial pressure increases beyond 1.4 atm, oxygen toxicity results. (In this figure, the red molecules are oxygen and the blue ones are nitrogen.)

▶ **FIGURE 11.24 Oxygen partial pressure limits** The partial pressure of oxygen in air at sea level is 0.21 atm. If this pressure drops by 50%, fatal hypoxia can result. High oxygen levels can also be harmful, but only if the partial pressure of oxygen increases by a factor of 7 or more.

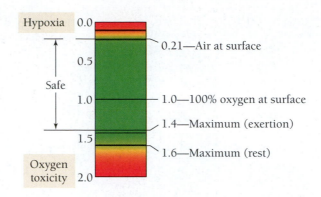

A second problem associated with breathing pressurized air is the increase in nitrogen in the lungs. At 30 m a scuba diver breathes nitrogen at $P_{N_2} = 3.1$ atm, which causes an increase in nitrogen concentration in bodily tissues and fluids. When P_{N_2} increases beyond about 4 atm, a condition called **nitrogen narcosis**, which is referred to as *rapture of the deep*, results. Divers describe this condition as feeling intoxicated. A diver breathing compressed air at 60 m feels as if he has had too much wine.

To avoid oxygen toxicity and nitrogen narcosis, deep-sea divers—those venturing beyond 50 m—breathe specialized mixtures of gases. One common mixture is called heliox, a mixture of helium and oxygen. These mixtures usually contain a smaller percentage of oxygen than would be found in air, thereby lowering the risk of oxygen toxicity. Heliox also contains helium instead of nitrogen, eliminating the risk of nitrogen narcosis.

EXAMPLE **11.10** | **Partial Pressure, Total Pressure, and Percent Composition**

Calculate the partial pressure of oxygen that a diver breathes with a heliox mixture containing 2.0% oxygen at a depth of 90 m where the total pressure is 10.0 atm.

You are given the percent oxygen in the mixture and the total pressure. You are asked to find the partial pressure of oxygen.	**GIVEN:** O_2 percent = 2.0% $\qquad\quad$ $P_{tot} = 10.0$ atm **FIND:** P_{O_2}
The partial pressure of a component in a gas mixture is equal to the fractional composition of the component multiplied by the total pressure. Calculate the fractional composition of O_2 by dividing the percent composition by 100. Calculate the partial pressure of O_2 by multiplying the fractional composition by the total pressure.	**SOLUTION** partial pressure of component $\quad$ = fractional composition of component $\times$ total pressure fractional composition of $O_2 = \dfrac{2.0}{100} = 0.020$ $P_{O_2} = 0.020 \times 10.0$ atm $= 0.20$ atm

▶ **SKILLBUILDER 11.10 | Partial Pressure, Total Pressure, and Percent Composition**

A diver breathing heliox with an oxygen composition of 5.0% wants to adjust the total pressure so that $P_{O_2} = 0.21$ atm. What must the total pressure be?

▶ **FOR MORE PRACTICE** Problems 79, 80, 81, 82.

Collecting Gases over Water

When the product of a chemical reaction is gaseous, it is often collected by the displacement of water. For example, suppose the following reaction is used as a source of hydrogen gas:

$$Zn(s) + 2\,HCl(aq) \longrightarrow ZnCl_2(aq) + H_2(g)$$

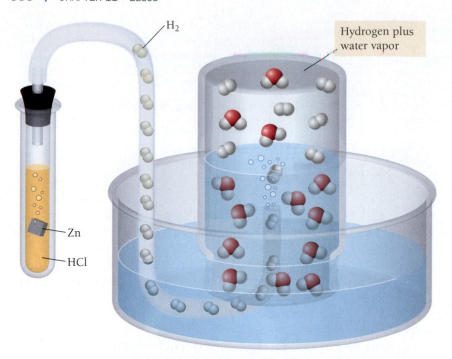

▲ **FIGURE 11.25 Vapor pressure** When a gas from a chemical reaction is collected through water, water molecules mix with the gas molecules. The pressure of water vapor in the final mixture is the vapor pressure of water at the temperature at which the gas is collected.

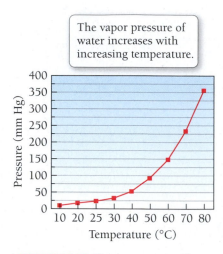

The vapor pressure of water increases with increasing temperature.

▲ **FIGURE 11.26 Vapor pressure of water as a function of temperature**

> We will cover vapor pressure in detail in Chapter 12.

TABLE 11.4 Vapor Pressure of Water versus Temperature

Temperature (°C)	Pressure (mm Hg)
10 °C	9.2
20 °C	17.5
25 °C	23.8
30 °C	31.8
40 °C	55.3
50 °C	92.5
60 °C	149.4
70 °C	233.7
80 °C	355.1

As the hydrogen gas forms, it bubbles through the water and gathers in the collection flask (▲ **FIGURE 11.25**). However, the hydrogen gas collected in this way is not pure but is mixed with water vapor because some water molecules evaporate and become mixed with the hydrogen molecules. The partial pressure of water in the mixture is called its **vapor pressure** (Table 11.4 and ▲ **FIGURE 11.26**). Vapor pressure depends on temperature. It increases with increasing temperature because higher temperatures cause more water molecules to evaporate.

Suppose we collect the hydrogen gas over water at a total pressure of 758 mm Hg and a temperature of 25 °C. What is the partial pressure of the hydrogen gas? We know that the total pressure is 758 mm Hg and that the partial pressure of water is 23.8 mm Hg (its vapor pressure at 25 °C).

$$P_{tot} = P_{H_2} + P_{H_2O}$$
$$758 \text{ mm Hg} = P_{H_2} + 23.8 \text{ mm Hg}$$

Therefore,

$$P_{H_2} = 758 \text{ mm Hg} - 23.8 \text{ mm Hg}$$
$$= 734 \text{ mm Hg}$$

The partial pressure of the hydrogen in the mixture is 734 mm Hg.

11.10 Gases in Chemical Reactions

▶ Apply the principles of stoichiometry to chemical reactions involving gases.

Chapter 8 describes how we can use the coefficients in chemical equations as conversion factors between moles of reactants and moles of products in a chemical reaction. We can use these conversion factors to determine, for example, the amount of product obtained in a chemical reaction based on a given amount of reactant or the amount of one reactant we need to completely react with a given amount of another reactant. The general solution map for these kinds of calculations is:

$$\text{moles A} \longrightarrow \text{moles B}$$

where A and B are two different substances involved in the reaction and the conversion factor between them comes from the stoichiometric coefficients in the balanced chemical equation.

In reactions involving gaseous reactants or products, the amount of gas is often specified in terms of its volume at a given temperature and pressure. In cases like this, we can use the ideal gas law to convert pressure, volume, and temperature to moles.

$$n = \frac{PV}{RT}$$

We can then use the stoichiometric coefficients to convert to other quantities in the reaction. For example, consider the reaction for the synthesis of ammonia.

$$3\,H_2(g) + N_2(g) \longrightarrow 2\,NH_3(g)$$

How many moles of NH_3 are formed by the complete reaction of 2.5 L of hydrogen at 381 K and 1.32 atm? Assume that there is more than enough N_2.

We begin by sorting the information in the problem statement.

GIVEN: $V = 2.5\,L$
$T = 381\,K$
$P = 1.32$ atm (of H_2)

FIND: mol NH_3

SOLUTION MAP

We strategize by drawing a solution map. The solution map for this problem is similar to the solution maps for other stoichiometric problems (see Sections 8.3 and 8.4). However, we first use the ideal gas law to find mol H_2 from P, V, and T. Then we use the stoichiometric coefficients from the equation to convert mol H_2 to mol NH_3.

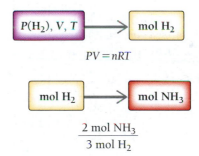

RELATIONSHIPS USED

$PV = nRT$ (ideal gas law, Section 11.8)

3 mol H_2:2 mol NH_3 (from balanced equation given in problem)

SOLUTION

We first solve the ideal gas equation for n.

$$PV = nRT$$

$$n = \frac{PV}{RT}$$

Then we substitute in the appropriate values.

$$n = \frac{1.32\ \cancel{atm} \times 2.5\ \cancel{L}}{0.0821\ \dfrac{\cancel{L} \cdot \cancel{atm}}{mol \cdot \cancel{K}} \times 381\ \cancel{K}}$$

$$= 0.1055\ mol\ H_2$$

Next, we convert mol H_2 to mol NH_3.

$$0.1055 \ \cancel{\text{mol } H_2} \times \frac{2 \ \text{mol } NH_3}{3 \ \cancel{\text{mol } H_2}} = 0.070 \ \text{mol } NH_3$$

There is enough H_2 to form 0.070 mol NH_3.

Interactive
Worked Example
Video 11.11

EXAMPLE **11.11**   **Gases in Chemical Reactions**

How many liters of oxygen gas form when 294 g of $KClO_3$ completely react in this reaction (which is used in the ignition of fireworks)?

$$2 \ KClO_3(s) \longrightarrow 2 \ KCl(s) + 3 \ O_2(g)$$

Assume that the oxygen gas is collected at $P = 755$ mm Hg and $T = 305$ K.

SORT	GIVEN:
You are given the mass of a reactant in a chemical reaction. You are asked to find the volume of a gaseous product at a given pressure and temperature.	294 g $KClO_3$ $P = 755$ mm Hg (of oxygen gas) $T = 305$ K FIND: volume of O_2 in liters

STRATEGIZE

The solution map has two parts. In the first part, convert from g $KClO_3$ to mol $KClO_3$ and then to mol O_2.

In the second part, use mol O_2 as n in the ideal gas law to find the volume of O_2.

You need the molar mass of $KClO_3$ and the stoichiometric relationship between $KClO_3$ and O_2 (from the balanced chemical equation). You also use the ideal gas law.

SOLUTION MAP

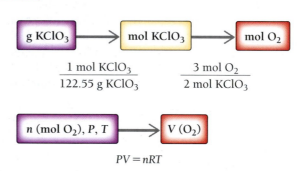

$PV = nRT$

RELATIONSHIPS USED

1 mol $KClO_3$ = 122.55 g (molar mass of $KClO_3$)

2 mol $KClO_3$: 3 mol O_2 (from balanced equation given in problem)

$PV = nRT$ (ideal gas law, Section 11.8)

SOLVE

Begin by converting mass $KClO_3$ to mol $KClO_3$ and then to mol O_2.

Then solve the ideal gas equation for V.

Before substituting the values into this equation, convert the pressure to atm.

Finally, substitute the given quantities along with the number of moles just calculated to calculate the volume.

SOLUTION

$$294 \ \cancel{\text{g } KClO_3} \times \frac{1 \ \cancel{\text{mol } KClO_3}}{122.55 \ \cancel{\text{g } KClO_3}} \times \frac{3 \ \text{mol } O_2}{2 \ \cancel{\text{mol } KClO_3}} = 3.60 \ \text{mol } O_2$$

$$PV = nRT$$

$$V = \frac{nRT}{P}$$

$$P = 755 \ \cancel{\text{mm Hg}} \times \frac{1 \ \text{atm}}{760 \ \cancel{\text{mm Hg}}} = 0.99342 \ \text{atm}$$

$$V = \frac{3.60 \ \cancel{\text{mol}} \times 0.0821 \ \dfrac{L \cdot \cancel{\text{atm}}}{\cancel{\text{mol}} \cdot \cancel{K}} \times 305 \ \cancel{K}}{0.99342 \ \cancel{\text{atm}}}$$

$$= 90.7 \ L$$

CHECK	The answer has the correct units, liters. The *value* of the answer is a bit more difficult to judge. Again, it is helpful to know that at standard temperature and pressure ($T = 0\,°C$ or 273.15 K and $P = 1$ atm), 1 mol of gas occupies 22.4 L (see Check step in Example 11.6). A 90.7-L sample of gas at STP would contain about 4 mol; since we started with a little more than 2 mol $KClO_3$, and since 2 mol $KClO_3$ forms 3 mol O_2, an answer that corresponds to about 4 mol O_2 is reasonable.
Check your answer. Are the units correct? Does the answer make physical sense?	

▶ **SKILLBUILDER 11.11** | **Gases in Chemical Reactions**

In this reaction, 4.58 L of O_2 is formed at 745 mm Hg and 308 K. How many grams of Ag_2O decomposed?

$$2\,Ag_2O(s) \longrightarrow 4\,Ag(s) + O_2(g)$$

▶ **FOR MORE PRACTICE**   Problems 89, 90, 91, 92, 93, 94.

Molar Volume at Standard Temperature and Pressure

Recall from the Check step of Example 11.6 that the volume occupied by 1 mol of gas at $0\,°C$ (273.15 K) and 1 atm is 22.4 L. These conditions are called **standard temperature and pressure (STP)**, and the volume occupied by 1 mol of gas under these conditions is called the **molar volume** of an ideal gas at STP. Using the ideal gas law, we can confirm that the molar volume at STP is 22.4 L:

$$V = \frac{nRT}{P}$$

$$= \frac{1.00\ \text{mol} \times 0.0821\ \dfrac{\text{L}\cdot\text{atm}}{\text{mol}\cdot\text{K}} \times 273\ \text{K}}{1.00\ \text{atm}}$$

$$= 22.4\ \text{L}$$

Under standard conditions, therefore, we can use this ratio as a conversion factor.

$$1\ \text{mol} : 22.4\ \text{L}$$

> The molar volume of 22.4 L applies only at STP.

One mole of any gas at standard temperature and pressure (STP) occupies 22.4 L.

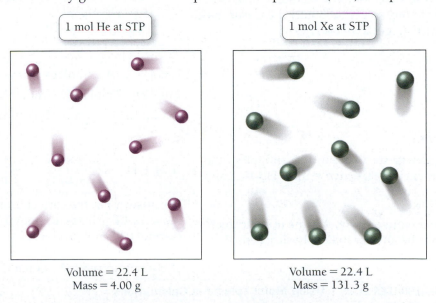

1 mol He at STP	1 mol Xe at STP
Volume = 22.4 L Mass = 4.00 g	Volume = 22.4 L Mass = 131.3 g

For example, suppose we wanted to calculate the number of liters of CO_2 gas that forms at STP when 0.879 mole of $CaCO_3$ undergoes this reaction:

$$CaCO_3(s) \longrightarrow CaO(s) + CO_2(g)$$

We begin by sorting the information in the problem statement.

GIVEN: 0.879 mol $CaCO_3$

FIND: $CO_2(g)$ in liters

SOLUTION MAP

We strategize by drawing a solution map that shows how to convert from mol $CaCO_3$ to mol CO_2 to L CO_2 using the molar volume at STP.

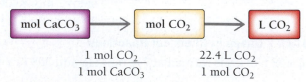

$$\dfrac{1 \text{ mol } CO_2}{1 \text{ mol } CaCO_3} \qquad \dfrac{22.4 \text{ L } CO_2}{1 \text{ mol } CO_2}$$

RELATIONSHIPS USED

1 mol $CaCO_3$: 1 mol CO_2 (from balanced equation given in problem)

1 mol = 22.4 L (at STP) (molar volume at STP, presented in this section)

SOLUTION

$$0.879 \text{ mol } \cancel{CaCO_3} \times \frac{1 \text{ mol } \cancel{CO_2}}{1 \text{ mol } \cancel{CaCO_3}} \times \frac{22.4 \text{ L } CO_2}{1 \text{ mol } \cancel{CO_2}} = 19.7 \text{ L } CO_2$$

EXAMPLE **11.12** | **Using Molar Volume in Calculations**

How many grams of water form when 1.24 L of H_2 gas at STP completely reacts with O_2?

$$2 H_2(g) + O_2(g) \longrightarrow 2 H_2O(g)$$

SORT You are given the volume of a reactant at STP and asked to find the mass of the product formed.	**GIVEN:** 1.24 L H_2 **FIND:** g H_2O
STRATEGIZE In the solution map, use the molar volume to convert from volume H_2 to mol H_2. Then use the stoichiometric relationship to convert to mol H_2O and finally the molar mass of H_2O to get to mass H_2O.	**SOLUTION MAP** 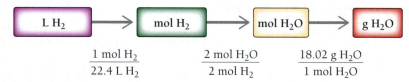 $\dfrac{1 \text{ mol } H_2}{22.4 \text{ L } H_2} \qquad \dfrac{2 \text{ mol } H_2O}{2 \text{ mol } H_2} \qquad \dfrac{18.02 \text{ g } H_2O}{1 \text{ mol } H_2O}$ **RELATIONSHIPS USED** 1 mol = 22.4 L (molar volume at STP, presented in this section) 2 mol H_2 : 2 mol H_2O (from balanced equation given in problem) 18.02 g H_2O = 1 mol H_2O (molar mass of H_2O)
SOLVE Begin with the volume of H_2 and follow the solution map to arrive at mass H_2O in grams.	**SOLUTION** $1.24 \text{ L } \cancel{H_2} \times \dfrac{1 \text{ mol } \cancel{H_2}}{22.4 \text{ L } \cancel{H_2}} \times \dfrac{2 \text{ mol } \cancel{H_2O}}{2 \text{ mol } \cancel{H_2}} \times \dfrac{18.02 \text{ g } H_2O}{1 \text{ mol } \cancel{H_2O}} = 0.998 \text{ g } H_2O$
CHECK Check your answer. Are the units correct? Does the answer make physical sense?	The answer has the correct units, g H_2O. The answer is reasonable because 1.24 L of a gas is about 0.05 mol of reactant (at STP) and 1 g H_2O is about 0.05 mol product (1 g/18 g/mol ≈ 0.05 mol). Since the reaction produces 1 mol H_2O to 1 mol H_2, you expect the number of moles of H_2O produced to be equal to the number of moles of H_2 that react.

▶ **SKILLBUILDER 11.12 | Using Molar Volume in Calculations**

How many liters of oxygen (at STP) are required to form 10.5 g of H_2O?

$$2 H_2(g) + O_2(g) \longrightarrow 2 H_2O(g)$$

▶ **FOR MORE PRACTICE** Problems 95, 96, 97, 98.

CHEMISTRY IN THE ENVIRONMENT

Air Pollution

All major cities in the world have polluted air. This pollution comes from a number of sources, including electricity generation, motor vehicles, and industrial waste. While there are many different kinds of air pollutants, the major gaseous air pollutants include the following:

Sulfur dioxide (SO_2)—Sulfur dioxide is primarily a by-product of electricity generation and industrial metal refining. SO_2 is a lung and eye irritant that affects the respiratory system. SO_2 is also one of the main precursors of acid rain.

Carbon monoxide (CO)—Carbon monoxide is formed by the incomplete combustion of fossil fuels (petroleum, natural gas, and coal). It is emitted mainly by motor vehicles. CO displaces oxygen in the blood and causes the heart and lungs to work harder. At high levels, CO can cause sensory impairment, decreased thinking ability, unconsciousness, and even death.

Ozone (O_3)—Ozone in the upper atmosphere is a normal part of our environment. Upper atmospheric ozone filters out part of the harmful UV light contained in sunlight. Lower-atmospheric or *ground-level ozone*, on the other hand, is a pollutant that results from the action of sunlight on motor vehicle emissions. Ground-level ozone is an eye and lung irritant. Prolonged exposure to ozone has been shown to permanently damage the lungs.

Nitrogen dioxide (NO_2)—Nitrogen dioxide is emitted by motor vehicles and by electricity generation plants. It is an orange-brown gas that causes the dark haze often seen over polluted cities. NO_2 is an eye and lung irritant and a precursor of acid rain.

In the United States, the U.S. Environmental Protection Agency (EPA) has set standards for these pollutants. Beginning in the 1970s, the U.S. Congress passed the Clean Air Act and its amendments, requiring U.S. cities to reduce their pollution and maintain levels below the limits set by the EPA. As a result of this legislation, pollutant levels in U.S. cities have decreased significantly over the last 40 years, even as the number of vehicles has increased. For example, according to the EPA, the levels of all four of the previously mentioned pollutants in major U.S. cities decreased during 1980–2010. Table 11.5 lists the amounts of these decreases.

TABLE 11.5 Changes in Pollutant Levels for Major U.S. Cities, 1980–2014	
Pollutant	**Change, 1980–2014**
SO_2	−80%
CO	−85%
O_3	−33%
NO_2	−60%

(*Source*: U.S. EPA Air Quality Trends)

As you can see from Table 11.5, much progress has been made. The pollutant levels in most cities are now at or below what the EPA considers safe. These trends demonstrate that good legislation can clean up our environment.

B11.3 CAN YOU ANSWER THIS?   *Calculate the amount (in grams) of SO_2 emitted when 1.0 kg of coal containing 4.0% S by mass is completely burned. At STP, what volume in liters does this SO_2 occupy?*

▲ Air pollution plagues most large cities.

Chapter 11 in Review

MasteringChemistry™ provides end-of-chapter exercises, feedback-enriched tutorial problems, animations, and interactive activities to encourage problem solving practice and deeper understanding of key concepts and topics.

Self-Assessment Quiz
PEARSON eText 2.0

Q1. According to kinetic molecular theory, the particles that compose a gas:
(a) attract one another strongly
(b) are packed closely together—nearly touching one another
(c) have an average kinetic energy proportional to the temperature in kelvins
(d) all of the above

Q2. Convert 558 mm Hg to atm.
(a) 0.734 atm
(b) 4.24×10^5 atm
(c) 38.0 atm
(d) 18.6 atm

Q3. If you double the volume of a constant amount of gas at a constant temperature, what happens to the pressure?
(a) The pressure doubles.
(b) The pressure quadruples.
(c) The pressure does not change.
(d) The pressure is one-half the initial pressure.

Q4. A 2.55-L gas sample in a cylinder with a freely moveable piston is initially at 25.0 °C. What is the volume of the gas when it is heated to 145 °C (at constant pressure)?
(a) 1.82 L
(b) 3.58 L
(c) 0.440 L
(d) 14.8 L

Q5. A gas sample has an initial volume of 55.2 mL, an initial temperature of 35.0 °C, and an initial pressure of 735 mm Hg. The volume is decreased to 48.8 mL and the temperature is increased to 72.5 °C. What is the final pressure?
(a) 933 mm Hg
(b) 186 mm Hg
(c) 1.72×10^3 mm Hg
(d) 401 mm Hg

Q6. Each gas sample has the same temperature and pressure. Which sample occupies the greatest volume?
(a) 4.0 g He
(b) 40.0 g Ne
(c) 40.0 g Ar
(d) 84.0 g Kr

Q7. Find the volume occupied by 22.0 g of helium gas at 25.0 °C and 1.18 atm of pressure.
(a) 123 L
(b) 9.57 L
(c) 114 L
(d) 159 L

Q8. A 2.82-g sample of an unknown gas occupies 525 mL at 25.0 °C and a pressure of 1.00 atm. Find the molar mass of the gas.
(a) 11.0 g/mol
(b) 0.011 g/mol
(c) 90.9 g/mol
(d) 131 g/mol

Q9. A mixture of three gases containing an equal number of moles of each gas has a total pressure of 9.0 atm. What is the partial pressure of each gas?
(a) 1.0 atm
(b) 3.0 atm
(c) 9.0 atm
(d) 27.0 atm

Q10. A gas mixture containing only helium and neon is 22.1% neon (by volume) and has a total pressure of 748 mm Hg. What is the partial pressure of neon?
(a) 165 mm Hg
(b) 1.69×10^4 mm Hg
(c) 583 mm Hg
(d) 22.1 mm Hg

Q11. Aluminum reacts with chlorine gas to form aluminum chloride.

$$2\,Al(s) + 3\,Cl_2(g) \longrightarrow 2\,AlCl_3(s)$$

What minimum volume of chlorine gas (at 298 K and 225 mm Hg) is required to completely react with 7.85 g of aluminum?
(a) 36.0 L
(b) 24.0 L
(c) 0.0474 L
(d) 16.0 L

Q12. A gas sample at STP contains 1.15 g oxygen and 1.55 g nitrogen. What is the volume of the gas sample?
(a) 1.26 L
(b) 2.04 L
(c) 4.08 L
(d) 61.0 L

Answers: 1:c, 2:a, 3:d, 4:b, 5:a, 6:b, 7:c, 8:d, 9:b, 10:a, 11:a, 12:b

Chemical Principles

Kinetic Molecular Theory
The kinetic molecular theory is a model for gases. In this model, gases are composed of widely spaced, noninteracting particles whose average kinetic energy depends on temperature.

Pressure
Pressure is the force per unit area that results from the collision of gas particles with surfaces. The SI unit of pressure is the pascal, but we often express pressure in other units such as atmospheres, millimeters of mercury, torr, pounds per square inch, and inches of mercury.

Relevance

The kinetic molecular theory predicts many of the properties of gases, including their low density in comparison to solids, their compressibility, and their tendency to assume the shape and volume of their container. The kinetic molecular theory also predicts the ideal gas law.

Pressure is a fundamental property of a gas. It allows tires to be inflated and makes it possible to drink from straws.

Simple Gas Laws

The simple gas laws show how one property of a gas varies with another property of the same gas.

volume (*V*) and pressure (*P*)

$$V \propto \frac{1}{P} \quad \text{(Boyle's law)}$$

volume (*V*) and temperature (*T*)

$$V \propto T \quad \text{(Charles's law)}$$

volume (*V*) and moles (*n*)

$$V \propto n \quad \text{(Avogadro's law)}$$

Each of the simple gas laws allows us to see how two properties of a gas are interrelated. They are also useful in calculating how one property of a gas changes when another does. We can use Boyle's law, for example, to calculate how the volume of a gas will change in response to a change in pressure, or vice versa.

The Combined Gas Law

The combined gas law joins Boyle's law and Charles's law.

$$\frac{P_1 V_1}{T_1} = \frac{P_2 V_2}{T_2}$$

We use the combined gas law to calculate how one property of a gas (pressure, volume, or temperature) changes when two other properties change at the same time.

The Ideal Gas Law

The ideal gas law combines the four properties of a gas—pressure, volume, temperature, and number of moles—in a single equation showing their interrelatedness.

$$PV = nRT$$

We can use the ideal gas law to find any one of the four properties of a gas if we know the other three.

Mixtures of Gases

The pressure of an individual component in a mixture of gases is its partial pressure, and we define it as the fractional composition of the component multiplied by the total pressure:

partial pressure of component = fractional
composition of component × total pressure

Dalton's law states that the total pressure of a mixture of gases is equal to the sum of the partial pressures of its components.

$$P_{tot} = P_a + P_b + P_c + \ldots$$

Since many gases are not pure but mixtures of several components, it is useful to know how each component contributes to the properties of the entire mixture. The concepts of partial pressure are relevant to deep-sea diving, for example, and to collecting gases over water, where water vapor mixes with the gas being collected.

Gases in Chemical Reactions

Coefficients in a balanced chemical equation provide conversion factors among moles of reactants and products in the reaction. For gases, the amount of a reactant or product is often specified by the volume of reactant or product at a given temperature and pressure. The ideal gas law is then used to convert from these quantities to moles of reactant or product. Alternatively, at standard temperature and pressure, volume can be converted directly to moles with the equality:

$$1 \text{ mol} = 22.4 \text{ L (at STP)}$$

Reactions involving gases are common in chemistry. For example, many atmospheric reactions—some of which are important to the environment—occur as gaseous reactions.

Chemical Skills

Examples

SORT
You are given a pressure in inches of mercury and asked to convert the units to torr.

STRATEGIZE
Begin with the quantity you are given and multiply by the appropriate conversion factor(s) to determine the quantity you are trying to find.

SOLVE
Follow the solution map to solve the problem.

CHECK
Are the units correct? Does the answer make physical sense?

EXAMPLE **11.13** **Pressure Unit Conversion**

Convert 18.4 in. Hg to torr.

GIVEN: 18.4 in. Hg

FIND: torr

SOLUTION MAP

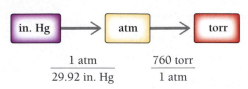

$$\frac{1 \text{ atm}}{29.92 \text{ in. Hg}} \qquad \frac{760 \text{ torr}}{1 \text{ atm}}$$

RELATIONSHIPS USED

$$1 \text{ atm} = 29.92 \text{ in. Hg (Table 11.1)}$$
$$760 \text{ torr} = 1 \text{ atm} \qquad \text{(Table 11.1)}$$

SOLUTION

$$18.4 \text{ in. Hg} \times \frac{1 \text{ atm}}{29.92 \text{ in. Hg}} \times \frac{760 \text{ torr}}{1 \text{ atm}} = 467 \text{ torr}$$

The units (torr) are correct. The value of the answer is reasonable because the torr is a smaller unit than in. Hg; therefore, the pressure in torr should be larger than in in. Hg.

LO: Restate and apply Boyle's law (Section 11.4).

SORT
You are given the initial and final pressures and the initial volume. You are asked to find the final volume.

STRATEGIZE
Calculations involving the simple gas laws usually consist of finding one of the initial or final conditions given the other initial and final conditions. In this case, use Boyle's law to find V_2 given P_1, V_1, and P_2.

SOLVE
Solve Boyle's law for V_2 and substitute the correct variables to calculate its value.

CHECK
Are the units correct? Does the answer make physical sense?

EXAMPLE **11.14** **Simple Gas Laws**

A gas has a volume of 5.7 L at a pressure of 3.2 atm. What is its volume at 4.7 atm? (Assume constant temperature.)

GIVEN: $P_1 = 3.2 \text{ atm}$
$\qquad\quad V_1 = 5.7 \text{ L}$
$\qquad\quad P_2 = 4.7 \text{ atm}$

FIND: V_2

SOLUTION MAP

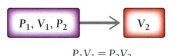

$$P_1 V_1 = P_2 V_2$$

RELATIONSHIPS USED

$$P_1 V_1 = P_2 V_2 \quad \text{(Boyle's law, Section 11.4)}$$

SOLUTION

$$P_1 V_1 = P_2 V_2$$
$$V_2 = \frac{P_1}{P_2} V_1$$
$$= \frac{3.2 \text{ atm}}{4.7 \text{ atm}} 5.7 \text{ L}$$
$$= 3.9 \text{ L}$$

The units (L) are correct. The value is reasonable because as pressure increases volume should decrease.

LO: Restate and apply the combined gas law (Section 11.6).

| EXAMPLE **11.15** | **The Combined Gas Law** |

A sample of gas has an initial volume of 2.4 L at a pressure of 855 mm Hg and a temperature of 298 K. If the gas is heated to a temperature of 387 K and expanded to a volume of 4.1 L, what is its final pressure in millimeters of mercury?

SORT
You are given the initial and final volume and temperature, and you are also given the initial pressure. You are asked to find the final pressure.

GIVEN: $P_1 = 855$ mm Hg

$V_1 = 2.4$ L

$T_1 = 298$ K

$V_2 = 4.1$ L

$T_2 = 387$ K

FIND: P_2

STRATEGIZE
Problems involving the combined gas law usually consist of finding one of the initial or final conditions given the other initial and final conditions. In this case, you can use the combined gas law to find the unknown quantity, P_2.

SOLUTION MAP

$$\boxed{P_1, V_1, T_1, V_2, T_2} \longrightarrow \boxed{P_2}$$

$$\frac{P_1 V_1}{T_1} = \frac{P_2 V_2}{T_2}$$

RELATIONSHIPS USED

$$\frac{P_1 V_1}{T_1} = \frac{P_2 V_2}{T_2} \quad \text{(combined gas law, Section 11.6)}$$

SOLVE
Solve the combined gas law for the quantity you are trying to find, in this case P_2, and substitute the known quantities to calculate the value of P_2.

SOLUTION

$$\frac{P_1 V_1}{T_1} = \frac{P_2 V_2}{T_2}$$

$$P_2 = \frac{P_1 V_1 T_2}{T_1 V_2}$$

$$= \frac{855 \text{ mm Hg} \times 2.4 \text{ L} \times 387 \text{ K}}{298 \text{ K} \times 4.1 \text{ L}}$$

$$= 6.5 \times 10^2 \text{ mm Hg}$$

CHECK
Are the units correct? Does the answer make physical sense?

The units, mm Hg, are correct. The value of the answer makes sense because the volume increase is proportionally more than the temperature decrease; therefore, you would expect the pressure to decrease.

LO: Restate and apply the ideal gas law (Section 11.8).

| EXAMPLE **11.16** | **The Ideal Gas Law** |

Calculate the pressure exerted by 1.2 mol of gas in a volume of 28.2 L and at a temperature of 334 K.

SORT
You are given the number of moles of a gas, its volume, and its temperature. You are asked to find its pressure.

GIVEN: $n = 1.2$ mol

$V = 28.2$ L

$T = 334$ K

FIND: P

STRATEGIZE

Calculations involving the ideal gas law often involve finding one of the four variables (P, V, n, or T) given the other three. In this case, you are asked to find P. Use the given variables and the ideal gas law to arrive at P.

SOLVE

Solve the ideal gas law equation for P and substitute the given variables to calculate P.

CHECK

Are the units correct? Does the answer make physical sense?

SOLUTION MAP

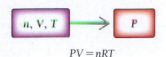

$$PV = nRT$$

RELATIONSHIPS USED

$$PV = nRT \text{ (ideal gas law, Section 11.8)}$$

SOLUTION

$$PV = nRT$$

$$P = \frac{nRT}{V}$$

$$= \frac{1.2 \text{ mol} \times 0.0821 \frac{\text{L} \cdot \text{atm}}{\text{mol} \cdot \text{K}} \times 334 \text{ K}}{28.2 \text{ L}}$$

$$= 1.2 \text{ atm}$$

The units (atm) are correct units for pressure. The *value* of the answer is a bit more difficult to judge. However, knowing that at standard temperature and pressure ($T = 0$ °C or 273.15 K and $P = 1$ atm) 1 mol of gas occupies 22.4 L, you can see that the answer is reasonable because you have a bit more than one mole of a gas at a temperature not too far from standard temperature. The volume of the gas is a bit higher than 22.4 L; therefore you might expect the pressure to be a bit higher than 1 atm.

LO: Restate and apply Dalton's law of partial pressures (Section 11.9).

You are given the partial pressures of three gases in a mixture and asked to find the total pressure.

Use Dalton's law of partial pressures ($P_{tot} = P_a + P_b + P_c + \ldots$) to solve the problem. Sum the partial pressures to obtain the total pressure.

EXAMPLE 11.17 | Total Pressure and Partial Pressure

A mixture of three gases has the partial pressures:

$$P_{CO_2} = 289 \text{ mm Hg}$$

$$P_{O_2} = 342 \text{ mm Hg}$$

$$P_{N_2} = 122 \text{ mm Hg}$$

What is the total pressure of the mixture?

GIVEN: $P_{CO_2} = 289 \text{ mm Hg}$
$P_{O_2} = 342 \text{ mm Hg}$
$P_{N_2} = 122 \text{ mm Hg}$

FIND: P_{tot}

SOLUTION

$$P_{tot} = P_{CO_2} + P_{O_2} + P_{N_2}$$

$$= 289 \text{ mm Hg} + 342 \text{ mm Hg} + 122 \text{ mm Hg}$$

$$= 753 \text{ mm Hg}$$

Key Terms

absolute zero [11.5]
atmosphere (atm) [11.3]
Avogadro's law [11.7]
Boyle's law [11.4]
Charles's law [11.5]
combined gas
    law [11.6]

Dalton's law of partial
    pressures [11.9]
hypoxia [11.9]
ideal gas constant (R) [11.8]
ideal gas law [11.8]
kinetic molecular theory
    [11.2]

millimeter of mercury
    (mm Hg) [11.3]
molar volume [11.10]
nitrogen narcosis [11.9]
oxygen toxicity [11.9]
partial pressure [11.9]
pascal (Pa) [11.3]

pounds per square inch
    (psi) [11.3]
pressure [11.3]
standard temperature and
    pressure (STP) [11.10]
torr [11.3]
vapor pressure [11.9]

Exercises

Questions

1. What is pressure?
2. Explain how drinking from a straw works. What causes the beverage to go up the straw? Is there an upper limit to how long a straw can theoretically be and still work as a drinking straw?
3. What are the main assumptions of kinetic molecular theory?
4. Describe the main properties of a gas. How are these predicted by kinetic molecular theory?
5. Why do we experience pain in our ears during changes in altitude?
6. What units are used to measure pressure?
7. What is Boyle's law? Explain Boyle's law from the perspective of kinetic molecular theory.
8. Explain why scuba divers should not hold their breath as they ascend to the surface.
9. Why would it be impossible to breathe air through an extra-long snorkel (greater than a couple of meters) while swimming underwater?
10. What is Charles's law? Explain Charles's law from the perspective of kinetic molecular theory.
11. Explain why hot-air balloons float above the ground.
12. What is the combined gas law? When is it useful?
13. What is Avogadro's law? Explain Avogadro's law from the perspective of kinetic molecular theory.
14. What is the ideal gas law? When is it useful?
15. Under what conditions is the ideal gas law most accurate? Under what conditions does the ideal gas law break down? Why?
16. What is partial pressure?
17. What is Dalton's law?
18. Describe hypoxia and oxygen toxicity.
19. Why do deep-sea divers breathe a mixture of helium and oxygen?
20. When a gas is collected over water, is the gas pure? Why or why not?
21. What is vapor pressure?
22. What is standard temperature and pressure (STP)? What is the molar volume of a gas at STP?

Problems

CONVERTING BETWEEN PRESSURE UNITS

23. Convert each measurement to atm.
    (a) 1277 mm Hg
    (b) 2.38×10^5 Pa
    (c) 127 psi
    (d) 455 torr

24. Convert each measurement to atm.
    (a) 921 torr
    (b) 4.8×10^4 Pa
    (c) 87.5 psi
    (d) 34.22 in. Hg

25. Perform each conversion.
    (a) 2.3 atm to torr
    (b) 4.7×10^{-2} atm to millimeters of mercury
    (c) 24.8 psi to millimeters of mercury
    (d) 32.84 in. Hg to torr

26. Perform each conversion.
    (a) 1.06 atm to millimeters of mercury
    (b) 95,422 Pa to millimeters of mercury
    (c) 22.3 psi to torr
    (d) 35.78 in. Hg to millimeters of mercury

27. Complete the table.

Pascals	Atmospheres	Millimeters of Mercury	Torr	Pounds per Square Inch
882 Pa	_____	6.62 mm Hg	_____	_____
_____	0.558 atm	_____	_____	_____
_____	_____	_____	_____	24.8 psi
_____	_____	_____	764 torr	_____
_____	_____	249 mm Hg	_____	_____

28. Complete the table.

Pascals	Atmospheres	Millimeters of Mercury	Torr	Pounds per Square Inch
_____	1.91 atm	_____	1.45×10^3 torr	_____
1.15×10^4 Pa	_____	_____	_____	_____
_____	_____	_____	721 torr	_____
_____	_____	109 mm Hg	_____	_____
_____	_____	_____	_____	38.9 psi

29. The pressure in Denver, Colorado (5280-ft elevation), averages about 24.9 in. Hg. Convert this pressure to:
    (a) atmospheres
    (b) millimeters of mercury
    (c) pounds per square inch
    (d) pascals

30. The pressure on top of Mount Everest averages about 235 mm Hg. Convert this pressure to:
    (a) torr
    (b) pounds per square inch
    (c) inches of mercury
    (d) atmospheres

| CHAPTER 11 Gases

31. The U.S. record for highest recorded barometric pressure is 31.85 in. Hg, set in 1989 in Northway, Alaska. Convert this pressure to:
(a) millimeters of mercury
(b) atmospheres
(c) torr
(d) kilopascals

32. The world record for lowest pressure (at sea level) was 652 mm Hg, inside of Typhoon Tip, 300 miles west of Guam on October 12, 1979. Convert this pressure to:
(a) torr
(b) atmospheres
(c) inches of mercury
(d) pounds per square inch

SIMPLE GAS LAWS

33. A sample of gas has an initial volume of 3.95 L at a pressure of 705 mm Hg. If the volume of the gas is increased to 5.38 L, what is the pressure? (Assume constant temperature.)

34. A sample of gas has an initial volume of 22.8 L at a pressure of 1.65 atm. If the sample is compressed to a volume of 10.7 L, what is its pressure? (Assume constant temperature.)

35. A snorkeler with a lung capacity of 6.3 L inhales a lungful of air at the surface, where the pressure is 1.0 atm. The snorkeler then descends to a depth of 25 m, where the pressure increases to 3.5 atm. What is the capacity of the snorkeler's lungs at this depth? (Assume constant temperature.)

36. A scuba diver with a lung capacity of 5.2 L inhales at a depth of 45 m and a pressure of 5.5 atm. If the diver were to ascend to the surface (where the pressure is 1.0 atm) while holding her breath, to what volume would the air in her lungs expand? (Assume constant temperature.)

37. Use Boyle's law to complete the table (assume temperature and number of moles of gas to be constant).

P_1	V_1	P_2	V_2
755 mm Hg	2.85 L	885 mm Hg	_____
_____	1.33 L	4.32 atm	2.88 L
192 mm Hg	382 mL	_____	482 mL
2.11 atm	_____	3.82 atm	125 mL

38. Use Boyle's law to complete the table (assume temperature and number of moles of gas to be constant).

P_1	V_1	P_2	V_2
_____	1.90 L	4.19 atm	1.09 L
755 mm Hg	118 mL	709 mm Hg	_____
2.75 atm	6.75 mL	_____	49.8 mL
343 torr	_____	683 torr	8.79 L

39. A balloon with an initial volume of 3.2 L at a temperature of 299 K is warmed to 376 K. What is its volume at 376 K?

40. A dramatic classroom demonstration involves cooling a balloon from room temperature (298 K) to liquid nitrogen temperature (77 K). If the initial volume of the balloon is 2.7 L, what is its volume after it cools? (Ignore the effect of any gases that might liquefy upon cooling.)

41. A 48.3-mL sample of gas in a cylinder equipped with a piston is warmed from 22 °C to 87 °C. What is its volume at the final temperature? (*Hint:* Use the combined gas law and make sure all temperatures are in kelvins.)

42. A syringe containing 1.55 mL of oxygen gas is cooled from 95.3 °C to 0.0 °C. What is the final volume of oxygen gas?

43. Use Charles's law to complete the table (assume pressure and number of moles of gas to be constant). (*Hint:* For calculations using Charles's law, temperatures must be in kelvins.)

V_1	T_1	V_2	T_2
1.08 L	25.4 °C	1.33 L	_____
_____	77 K	228 mL	298 K
115 cm³	_____	119 cm³	22.4 °C
232 L	18.5 °C	_____	96.2 °C

44. Use Charles's law to complete the table (assume pressure and number of moles of gas to be constant).

V_1	T_1	V_2	T_2
119 L	10.5 °C	_____	112.3 °C
_____	135 K	176 mL	315 K
2.11 L	15.4 °C	2.33 L	_____
15.4 cm³	_____	19.2 cm³	10.4 °C

45. A 0.12-mol sample of nitrogen gas occupies a volume of 2.55 L. What is the volume of 0.32 mol of nitrogen gas under the same conditions?

46. A 0.48-mol sample of helium gas occupies a volume of 11.7 L. What is the volume of 0.72 mol of helium gas under the same conditions?

47. A balloon contains 0.128 mol of gas and has a volume of 2.76 L. If an additional 0.073 mol of gas is added to the balloon, what is its final volume? (*Hint:* The final number of moles is the sum of the initial number and the amount added.)

48. A cylinder with a moveable piston contains 0.87 mol of gas and has a volume of 334 mL. What will its volume be if an additional 0.22 mol of gas is added to the cylinder? (*Hint:* The final number of moles is the sum of the initial number and the amount added.)

49. Use Avogadro's law to complete the table (assume pressure and temperature to be constant).

V_1	n_1	V_2	n_2
38.5 mL	1.55×10^{-3} mol	49.4 mL	_____
_____	1.37 mol	26.8 L	4.57 mol
11.2 L	0.628 mol	_____	0.881 mol
422 mL	_____	671 mL	0.0174 mol

50. Use Avogadro's law to complete the table (assume pressure and temperature to be constant).

V_1	n_1	V_2	n_2
25.2 L	5.05 mol	_____	3.03 mol
_____	1.10 mol	414 mL	0.913 mol
8.63 L	0.0018 mol	10.9 L	_____
53 mL	_____	13 mL	2.61×10^{-4} mol

THE COMBINED GAS LAW

51. A sample of gas with an initial volume of 28.4 L at a pressure of 725 mm Hg and a temperature of 305 K is compressed to a volume of 14.8 L and warmed to a temperature of 375 K. What is the final pressure of the gas?

52. A cylinder with a moveable piston contains 218 mL of nitrogen gas at a pressure of 1.32 atm and a temperature of 298 K. What must the final volume be for the pressure of the gas to be 1.55 atm at a temperature of 335 K?

53. A scuba diver takes a 2.8-L balloon from the surface, where the pressure is 1.0 atm and the temperature is 34 °C, to a depth of 25 m, where the pressure is 3.5 atm and the temperature is 18 °C. What is the volume of the balloon at this depth?

54. A bag of potato chips contains 585 mL of air at 25 °C and a pressure of 765 mm Hg. Assuming the bag does not break, what will be its volume at the top of a mountain where the pressure is 442 mm Hg and the temperature is 5.0 °C?

55. A gas sample with a volume of 5.3 L has a pressure of 735 mm Hg at 28 °C. What is the pressure of the sample if the volume remains at 5.3 L but the temperature rises to 86 °C?

56. The total pressure in a 11.7-L automobile tire is 44 psi at 11°C. How much does the pressure in the tire rise if its temperature increases to 37 °C and the volume remains at 11.7 L?

57. Use the combined gas law to complete the table (assume the number of moles of gas to be constant).

P_1	V_1	T_1	P_2	V_2	T_2
1.21 atm	1.58 L	12.2 °C	1.54 atm	_____	32.3 °C
721 torr	141 mL	135 K	801 torr	152 mL	_____
5.51 atm	0.879 L	22.1 °C	_____	1.05 L	38.3 °C

58. Use the combined gas law to complete the table (assume the number of moles of gas to be constant).

P_1	V_1	T_1	P_2	V_2	T_2
1.01 atm	_____	2.7 °C	0.54 atm	0.58 L	42.3 °C
123 torr	41.5 mL	_____	626 torr	36.5 mL	205 K
_____	1.879 L	20.8 °C	0.412 atm	2.05 L	48.1 °C

THE IDEAL GAS LAW

59. What is the volume occupied by 0.255 mol of helium gas at 1.25 atm and 305 K?

60. What is the pressure in a 20.0-L cylinder filled with 0.683 mol of nitrogen gas at 325 K?

61. A cylinder contains 28.5 L of oxygen gas at a pressure of 1.8 atm and a temperature of 298 K. How many moles of gas are in the cylinder?

62. What is the temperature of 0.52 mol of gas at a pressure of 1.3 atm and a volume of 11.8 L?

63. A cylinder contains 11.8 L of air at a total pressure of 43.2 psi and a temperature of 25 °C. How many moles of gas does the cylinder contain? (*Hint:* You must convert each quantity into the correct units (L, atm, mol, and K) before substituting into the ideal gas law.)

64. What is the pressure in millimeters of mercury of 0.0115 mol of helium gas with a volume of 214 mL at 45 °C? (*Hint:* You must convert each quantity into the correct units (L, atm, mol, and K) before substituting into the ideal gas law.)

65. Use the ideal gas law to complete the table.

P	V	n	T
1.05 atm	1.19 L	0.112 mol	_____
112 torr	_____	0.241 mol	304 K
_____	28.5 mL	1.74×10^{-3} mol	25.4 °C
0.559 atm	0.439 L	_____	255 K

66. Use the ideal gas law to complete the table.

P	V	n	T
2.39 atm	1.21 L	_____	205 K
512 torr	_____	0.741 mol	298 K
0.433 atm	0.192 L	0.0131 mol	_____
_____	20.2 mL	5.71×10^{-3} mol	20.4 °C

67. How many moles of gas must be forced into a 3.5-L ball to give it a gauge pressure of 9.4 psi at 25 °C? The gauge pressure is relative to atmospheric pressure. Assume that atmospheric pressure is 14.7 psi so that the total pressure in the ball is 24.1 psi.

68. How many moles of gas must be forced into a 4.8-L tire to give it a gauge pressure of 32.4 psi at 25 °C? The gauge pressure is relative to atmospheric pressure. Assume that atmospheric pressure is 14.7 psi so that the total pressure in the tire is 47.1 psi.

69. An experiment shows that a 248-mL gas sample has a mass of 0.433 g at a pressure of 745 mm Hg and a temperature of 28 °C. What is the molar mass of the gas?

70. An experiment shows that a 113-mL gas sample has a mass of 0.171 g at a pressure of 721 mm Hg and a temperature of 32 °C. What is the molar mass of the gas?

71. A sample of gas has a mass of 38.8 mg. Its volume is 224 mL at a temperature of 55 °C and a pressure of 886 torr. Find the molar mass of the gas.

72. A sample of gas has a mass of 0.555 g. Its volume is 117 mL at a temperature of 85 °C and a pressure of 753 mm Hg. Find the molar mass of the gas.

PARTIAL PRESSURE

73. A gas mixture contains each gas at the indicated partial pressure.

N_2	217 torr
O_2	106 torr
He	248 torr

What is the total pressure of the mixture?

74. A gas mixture contains each gas at the indicated partial pressure.

CO_2	422 mm Hg
Ar	102 mm Hg
O_2	165 mm Hg
H_2	52 mm Hg

What is the total pressure of the mixture?

75. A heliox deep-sea diving mixture delivers an oxygen partial pressure of 0.30 atm when the total pressure is 11.0 atm. What is the partial pressure of helium in this mixture?

76. A mixture of helium, nitrogen, and oxygen has a total pressure of 752 mm Hg. The partial pressures of helium and nitrogen are 234 mm Hg and 197 mm Hg, respectively. What is the partial pressure of oxygen in the mixture?

77. The hydrogen gas formed in a chemical reaction is collected over water at 30 °C at a total pressure of 732 mm Hg. What is the partial pressure of the hydrogen gas collected in this way?

78. The oxygen gas emitted from an aquatic plant during photosynthesis is collected over water at 25 °C and a total pressure of 753 torr. What is the partial pressure of the oxygen gas?

79. A gas mixture contains 78% nitrogen and 22% oxygen. If the total pressure is 1.12 atm, what are the partial pressures of each component?

80. An air sample contains 0.038% CO_2. If the total pressure is 758 mm Hg, what is the partial pressure of CO_2?

81. A heliox deep-sea diving mixture contains 4.0% oxygen and 96.0% helium. What is the partial pressure of oxygen when this mixture is delivered at a total pressure of 8.5 atm?

82. A scuba diver breathing normal air descends to 100 m of depth, where the total pressure is 11 atm. What is the partial pressure of oxygen that the diver experiences at this depth? Is the diver in danger of experiencing oxygen toxicity?

MOLAR VOLUME

83. Calculate the volume of each gas sample at STP.
   (a) 22.5 mol Cl_2
   (b) 3.6 mol nitrogen
   (c) 2.2 mol helium
   (d) 27 mol CH_4

84. Calculate the volume of each gas sample at STP.
   (a) 21.2 mol N_2O
   (b) 0.215 mol CO
   (c) 0.364 mol CO_2
   (d) 8.6 mol C_2H_6

85. Calculate the volume of each gas sample at STP.
 (a) 73.9 g N_2
 (b) 42.9 g O_2
 (c) 148 g NO_2
 (d) 245 mg CO_2

86. Calculate the volume of each gas sample at STP.
 (a) 48.9 g He
 (b) 45.2 g Xe
 (c) 48.2 mg Cl_2
 (d) 3.83 kg SO_2

87. Calculate the mass of each gas sample at STP.
 (a) 178 mL CO_2
 (b) 155 mL O_2
 (c) 1.25 L SF_6

88. Calculate the mass of each gas sample at STP.
 (a) 5.82 L NO
 (b) 0.324 L N_2
 (c) 139 cm^3 Ar

GASES IN CHEMICAL REACTIONS

89. Consider the chemical reaction:

$$C(s) + H_2O(g) \longrightarrow CO(g) + H_2(g)$$

How many liters of hydrogen gas are formed from the complete reaction of 1.07 mol of C? Assume that the hydrogen gas is collected at 1.0 atm and 315 K.

90. Consider the chemical reaction:

$$2\,H_2O(l) \longrightarrow 2\,H_2(g) + O_2(g)$$

How many moles of H_2O are required to form 1.3 L of O_2 at 325 K and 0.988 atm?

91. CH_3OH can be synthesized by the reaction:

$$CO(g) + 2\,H_2(g) \longrightarrow CH_3OH(g)$$

How many liters of H_2 gas, measured at 748 mm Hg and 86 °C, are required to synthesize 0.55 mol of CH_3OH? How many liters of CO gas, measured under the same conditions, are required?

92. Oxygen gas reacts with powdered aluminum according to the reaction:

$$4\,Al(s) + 3\,O_2(g) \longrightarrow 2\,Al_2O_3(s)$$

How many liters of O_2 gas, measured at 782 mm Hg and 25 °C, are required to completely react with 2.4 mol of Al?

93. Nitrogen reacts with powdered aluminum according to the reaction:

$$2\,Al(s) + N_2(g) \longrightarrow 2\,AlN(s)$$

How many liters of N_2 gas, measured at 892 torr and 95 °C, are required to completely react with 18.5 g of Al?

94. Sodium reacts with chlorine gas according to the reaction:

$$2\,Na(s) + Cl_2(g) \longrightarrow 2\,NaCl(s)$$

What volume of Cl_2 gas, measured at 687 torr and 35 °C, is required to form 28 g of NaCl?

95. How many grams of NH_3 form when 24.8 L of $H_2(g)$ (measured at STP) reacts with N_2 to form NH_3 according to this reaction?

$$N_2(g) + 3\,H_2(g) \longrightarrow 2\,NH_3(g)$$

96. Lithium reacts with nitrogen gas according to the reaction:

$$6\,Li(s) + N_2(g) \longrightarrow 2\,Li_3N(s)$$

How many grams of lithium are required to completely react with 58.5 mL of N_2 gas measured at STP?

97. How many grams of calcium are consumed when 156.8 mL of oxygen gas, measured at STP, reacts with calcium according to this reaction?

$$2\,Ca(s) + O_2(g) \longrightarrow 2\,CaO(s)$$

98. How many grams of magnesium oxide form when 14.8 L of oxygen gas, measured at STP, completely reacts with magnesium metal according to this reaction?

$$2\,Mg(s) + O_2(g) \longrightarrow 2\,MgO(s)$$

Cumulative Problems

99. Use the ideal gas law to show that the molar volume of a gas at STP is 22.4 L.

100. Use the ideal gas law to show that 28.0 g of nitrogen gas and 4.00 g of helium gas occupy the same volume at any temperature and pressure.

101. The mass of an evacuated 255-mL flask is 143.187 g. The mass of the flask filled with 267 torr of an unknown gas at 25 °C is 143.289 g. Calculate the molar mass of the unknown gas.

102. A 118-mL flask is evacuated, and its mass is measured as 97.129 g. When the flask is filled with 768 torr of helium gas at 35 °C, it is found to have a mass of 97.171 g. Is the gas pure helium?

103. A gaseous compound containing hydrogen and carbon is decomposed and found to contain 82.66% carbon and 17.34% hydrogen by mass. The mass of 158 mL of the gas, measured at 556 mm Hg and 25 °C, is 0.275 g. What is the molecular formula of the compound?

104. A gaseous compound containing hydrogen and carbon is decomposed and found to contain 85.63% C and 14.37% H by mass. The mass of 258 mL of the gas, measured at STP, is 0.646 g. What is the molecular formula of the compound?

105. The reaction between zinc and hydrochloric acid is carried out as a source of hydrogen gas in the laboratory:

$$Zn(s) + 2 HCl(aq) \longrightarrow ZnCl_2(aq) + H_2(g)$$

If 325 mL of hydrogen gas is collected over water at 25 °C at a total pressure of 748 mm Hg, how many grams of Zn reacted?

106. Consider the reaction:

$$2 NiO(s) \longrightarrow 2 Ni(s) + O_2(g)$$

If O_2 is collected over water at 40 °C and a total pressure of 745 mm Hg, what volume of gas will be collected for the complete reaction of 24.78 g of NiO?

107. How many grams of hydrogen are collected in a reaction where 1.78 L of hydrogen gas is collected over water at a temperature of 40 °C and a total pressure of 748 torr?

108. How many grams of oxygen are collected in a reaction where 235 mL of oxygen gas is collected over water at a temperature of 25 °C and a total pressure of 697 torr?

109. The decomposition of a silver oxide sample forms 15.8 g of $Ag(s)$:

$$2 Ag_2O(s) \longrightarrow 4 Ag(s) + O_2(g)$$

What total volume of O_2 gas forms if it is collected over water at a temperature of 25 °C and a total pressure of 752 mm Hg?

110. The following reaction consumes 2.45 kg of $CO(g)$:

$$CO(g) + H_2O(g) \longrightarrow CO_2(g) + H_2(g)$$

How many total liters of gas are formed if the products are collected at STP?

111. When hydrochloric acid is poured over a sample of sodium bicarbonate, 28.2 mL of carbon dioxide gas is produced at a pressure of 0.954 atm and a temperature of 22.7 °C. Write an equation for the gas-evolution reaction and determine how much sodium bicarbonate reacted.

112. When hydrochloric acid is poured over potassium sulfide, 42.9 mL of hydrogen sulfide gas is produced at a pressure of 752 torr and a temperature of 25.8 °C. Write an equation for the gas-evolution reaction and determine how much potassium sulfide (in grams) reacted.

113. Consider the reaction:

$$2 SO_2(g) + O_2(g) \longrightarrow 2 SO_3(s)$$

(a) If 285.5 mL of SO_2 is allowed to react with 158.9 mL of O_2 (both measured at STP), what are the limiting reactant and the theoretical yield of SO_3?
(b) If 2.805 g of SO_3 is collected (measured at STP), what is the percent yield for the reaction?

114. Consider the reaction:

$$P_4(s) + 6 H_2(g) \longrightarrow 4 PH_3(g)$$

(a) If 88.6 L of $H_2(g)$, measured at STP, is allowed to react with 158.3 g of P_4, what is the limiting reactant?
(b) If 48.3 L of PH_3, measured at STP, forms, what is the percent yield?

115. Consider the reaction for the synthesis of nitric acid:

$$3 NO_2(g) + H_2O(l) \longrightarrow 2 HNO_3(aq) + NO(g)$$

(a) If 12.8 L of $NO_2(g)$, measured at STP, is allowed to react with 14.9 g of water, find the limiting reagent and the theoretical yield of HNO_3 in grams.
(b) If 14.8 g of HNO_3 forms, what is the percent yield?

116. Consider the reaction for the production of NO_2 from NO:

$$2 NO(g) + O_2(g) \longrightarrow 2 NO_2(g)$$

(a) If 84.8 L of $O_2(g)$, measured at 35 °C and 632 mm Hg, is allowed to react with 158.2 g of NO, find the limiting reagent.
(b) If 97.3 L of NO_2 forms, measured at 35 °C and 632 mm Hg, what is the percent yield?

117. Ammonium carbonate decomposes upon heating according to the balanced equation:

$$(NH_4)_2CO_3(s) \longrightarrow 2 NH_3(g) + CO_2(g) + H_2O(g)$$

Calculate the total volume of gas produced at 22 °C and 1.02 atm by the complete decomposition of 11.83 g of ammonium carbonate.

118. Ammonium nitrate decomposes explosively upon heating according to the balanced equation:

$$2 NH_4NO_3(s) \longrightarrow 2 N_2(g) + O_2(g) + 4 H_2O(g)$$

Calculate the total volume of gas (at 25 °C and 748 mm Hg) produced by the complete decomposition of 1.55 kg of ammonium nitrate.

119. A mixture containing 235 mg of helium and 325 mg of neon has a total pressure of 453 torr. What is the partial pressure of helium in the mixture?

120. A mixture containing 4.33 g of CO_2 and 3.11 g of CH_4 has a total pressure of 1.09 atm. What is the partial pressure of CO_2 in the mixture?

121. Consider the reaction:

$$2\,SO_2(g) + O_2(g) \longrightarrow 2\,SO_3(g)$$

A reaction flask initially contains 0.10 atm of SO_2 and 0.10 atm of O_2. What is the total pressure in the flask once the limiting reactant is completely consumed? Assume a constant temperature and volume and a 100% reaction yield.

122. Consider the reaction:

$$CO(g) + 2\,H_2(g) \longrightarrow CH_3OH(g)$$

A reaction flask initially contains 112 torr of CO and 282 torr of H_2. The reaction is allowed to proceed until the pressure stops changing, at which point the total pressure is 196 torr. Determine the percent yield for the reaction. Assume that temperature is constant and that no other reactions occur other than the one indicated.

Highlight Problems

123. Which gas sample has the greatest pressure? Assume they are all at the same temperature. Explain.

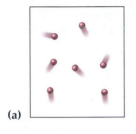

(a)

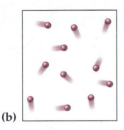

(b)

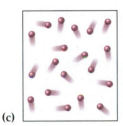

(c)

124. This image represents a sample of gas at a pressure of 1 atm, a volume of 1 L, and a temperature of 25 °C. Draw a similar picture showing what happens if the volume is reduced to 0.5 L and the temperature is increased to 250 °C. What happens to the pressure?

$V = 1.0$ L
$T = 25\ °C$
$P = 1.0$ atm

125. Automobile airbags inflate following a serious impact. The impact triggers the chemical reaction:

$$2\,NaN_3(s) \longrightarrow 2\,Na(s) + 3\,N_2(g)$$

If an automobile airbag has a volume of 11.8 L, how much NaN_3 in grams is required to fully inflate the airbag upon impact? Assume STP conditions.

FO4305OZ02

126. Olympic cyclists fill their tires with helium to make them lighter. Calculate the mass of air in an air-filled tire and the mass of helium in a helium-filled tire. What is the mass difference between the two? Assume that the volume of the tire is 855 mL, that it is filled with a total pressure of 125 psi, and that the temperature is 25 °C. Also, assume an average molar mass for air of 28.8 g/mol.

127. In a common classroom demonstration, a balloon is filled with air and submerged into liquid nitrogen. The balloon contracts as the gases within the balloon cool. Suppose the balloon initially contains 2.95 L of air at 25.0 °C and a pressure of 0.998 atm. Calculate the expected volume of the balloon upon cooling to −196 °C (the boiling point of liquid nitrogen). When the demonstration is carried out, the actual volume of the balloon decreases to 0.61 L. How well does the observed volume of the balloon compare to your calculated value? Can you explain the difference?

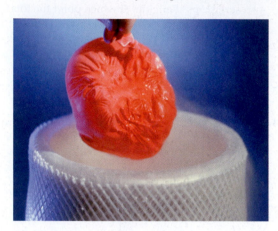

128. Aerosol cans carry clear warnings against incineration because of the high pressures that can develop upon heating. Suppose a can contains a residual amount of gas at a pressure of 755 mm Hg and 25 °C. What would the pressure be if the can were heated to 1155 °C?

Questions for Group Work

Discuss these questions with the group and record your consensus answer.

129. Complete the table.

Variables Related	Name of Law	Proportionality Expression	Equality Expression	Held Constant
V, P	Boyle's law	$V \propto 1/P$	$P_1V_1 = P_2V_2$	n, T
_____	_____	$V \propto T$	$V_1/T_1 = V_2/T_2$	n, P
V, n	Avogadro's law	_____	_____	P, T
_____	Gay-Lussac's law	_____	$P_1/T_1 = P_2/T_2$	_____
P, V, T	_____	_____	_____	_____

130. A chemical reaction produces 10.4 g of CO_2. What volume does this gas occupy at 1.2 atm and 29 °C? Assign one member of your group to check all units and another to check all significant figures.

131. A 14.22-g aluminum soda can reacts with hydrochloric acid according to the reaction:

$$2\,Al(s) \;+\; 6\,HCl(aq) \longrightarrow 3\,H_2(g) \;+\; 2\,AlCl_3(aq)$$

The hydrogen gas is collected by the displacement of water. What volume of hydrogen is collected if the external pressure is 749 mm Hg and the temperature is 31 °C, assuming that there is no water vapor present? What is the vapor pressure of water at this temperature? What volume do the hydrogen and water vapor occupy under these conditions?

Data Interpretation and Analysis

132. When fuels are burned in the presence of air, such as in an automobile engine, some of the nitrogen in the air oxidizes to form nitrogen oxide gases such as NO and NO_2 (known collectively as NO_x). The U.S. Environmental Protection Agency (EPA) sets standards for air quality of several pollutants including NO_2. According to the EPA, NO_2 levels in U.S. cities are not to exceed a yearly average of 53 parts per billion by volume (ppbv) nor a 1-hour average of 100 ppbv.

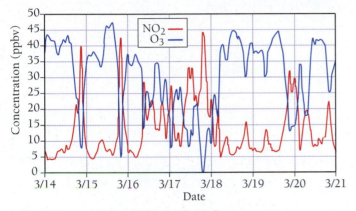

▲ Concentration of NO_2 and O_3 over Seven Days

Source: http://www.cas.manchester.ac.uk/resprojects/holmemoss/results/fig2/

Another pollutant associated with automobile exhaust is ozone (O_3). The EPA standard for ozone is an 8-hour average of 70 ppbv. Breathing air with elevated levels of NO_2 or O_3 can cause asthma and other respiratory problems. This graph shows the average concentration of nitrogen dioxide (NO_2) and ozone (O_3) gases in units of parts per billion by volume (ppbv) of four sampling sites over a period of one week near Manchester, England (March 1999).

Study the graph and answer the following questions:

(a) Do the concentrations of NO_2 or O_3 exceed the U.S. standards set by the EPA?
(b) What type of relationship exists between nitrogen dioxide and ozone between March 14 and March 16?
(c) Since the volume of a gas is directly proportional to the number of moles of the gas at constant temperature and volume, what is the mole-to-mole ratio of O_3 consumed to NO_2 produced?
(d) The following chemical equations model the interactions of nitrogen dioxide gas and ozone gas. Can this set of equations account for the trends observed in the graph? Explain your answer.

$$\text{i. } N_2 + O_2 \longrightarrow 2\,NO$$
$$NO + O_3 \longrightarrow NO_2 + O_2 + \text{light}$$

Answers to Skillbuilder Exercises

Skillbuilder 11.1	85.0 psi
Skillbuilder Plus, p. 365	80.6 kPa
Skillbuilder 11.2	$P_2 = 2.1$ atm; depth is approximately 11 m
Skillbuilder 11.3	123 mL
Skillbuilder 11.4	0.82 L
Skillbuilder 11.5	13 L
Skillbuilder 11.6	1.6 atm

Skillbuilder 11.7	16.1 L
Skillbuilder Plus, p. 381	977 mm Hg
Skillbuilder 11.8	70.8 g/mol
Skillbuilder 11.9	721 torr
Skillbuilder 11.10	$P_{tot} = 4.2$ atm
Skillbuilder 11.11	82.3 g
Skillbuilder 11.12	6.53 L O_2

Answers to Conceptual Checkpoints

11.1 (b) Since the size of the particles is small compared to the space between them, a change in the size of the particle should not dramatically change the properties of the gas.

11.2 (a) Since all the particles are identical and since (a) has the smallest number of particles per unit volume, it will have the lowest pressure.

11.3 (c) Atmospheric pressure will support a column of water 10.3 m in height. If the liquid in a barometer were twice as dense as water, a column of the liquid would be twice as heavy and the pressure it exerted at its base would be twice as great. Therefore, atmospheric pressure would be able to support a column only half as high.

11.4 (b) Since the volume triples, and since according to Boyle's law the volume and pressure are inversely proportional, the pressure will decrease by a factor of three.

11.5 (a) At constant pressure, the volume of the gas will be proportional to the temperature—if the Kelvin temperature doubles, the volume will double.

11.6 (c) Doubling the temperature in kelvins doubles the pressure, but doubling the volume halves the pressure. The net result is that the pressure is the same as its initial value.

11.7 (c) Since hydrogen gas has the lowest molar mass of the set, 1 g of hydrogen gas will have the greatest number of moles and therefore the greatest volume.

12 Liquids, Solids, and Intermolecular Forces

It will be found that everything depends on the composition of the forces with which the particles of matter act upon one another; and from these forces . . . all phenomena of nature take their origin.

—Roger Joseph Boscovich (1711–1787)

12.1 Spherical Water

When a solid object is dropped into still water, a water jet—a small column of rising water—often forms immediately following impact. You can see this yourself by dropping a coin (flat side down) into a glass of water. Watch carefully at the instant the coin falls beneath the water surface and you will see the characteristic water jet splashing upward. The jet forms because the solid object forms a cylindrical crater in the water. As the cylinder fills back in, the collapsing walls of the crater collide at its center. The water has nowhere to go but up. These water jets have been the subject of photographers and videographers striving to capture water's stunning performance. Many of these images also show a small droplet of water that forms at the top of the water jet as the column begins to fall back down. The small droplet is nearly perfectly spherical, as you can see in the chapter-opening image. Why does the water form this perfect sphere at that instant? The reason is the main topic of this chapter: intermolecular forces, the attractive forces that exist among the particles that compose matter.

After the jet rises and begins to fall, the column of water is in free fall, and the distorting effects of gravity are absent. This allows the intermolecular forces between the water molecules to determine the shape of the falling water. These forces cause attractions between water molecules, much like the attractions in a collection of small magnets. These attractions hold water together as a liquid (instead of a gas) at room temperature; they also cause a sample of water in free fall to clump together into a sphere, which you can clearly see at the top of the water jet. The sphere is the geometrical shape with the lowest surface area to volume ratio. By forming a sphere, the water molecules maximize their interaction with one another because a sphere minimizes the number of molecules at the surface of the liquid, where fewer interactions occur (compared to the interior of the liquid).

◀ The spherical shape of the water drop in this image is caused by intermolecular forces, attractive forces that exist among the water molecules.

Intermolecular forces exist, not only among water molecules, but among all particles that compose matter. These forces are responsible for the very existence of liquids and solids. The state of a sample of matter—solid, liquid, or gas—depends on the magnitude of intermolecular forces relative to the amount of thermal energy in the sample. Recall from Section 3.10 that the molecules and atoms that compose matter are in constant random motion that increases with increasing temperature. The energy associated with this motion is **thermal energy**. The weaker the intermolecular forces relative to thermal energy, the more likely it is that the sample will be gaseous. The stronger the intermolecular forces relative to thermal energy, the more likely it is that the sample will be liquid or solid.

12.2 Properties of Liquids and Solids

▶ Describe the properties of solids and liquids and relate them to their constituent atoms and molecules.

We are all familiar with solids and liquids. Water, gasoline, rubbing alcohol, and fingernail-polish remover are all common liquids. Ice, dry ice, and diamond are familiar solids. In contrast to gases—in which molecules or atoms are separated by large distances—the molecules or atoms that compose liquids and solids are in close contact with one another (▼ FIGURE 12.1).

The difference between solids and liquids is in the freedom of movement of the constituent molecules or atoms. In liquids, even though the atoms or molecules are in close contact, they are still free to move around each other. In solids, the atoms or molecules are fixed in their positions, although thermal energy causes them to vibrate about a fixed point. These molecular properties of solids and liquids result in characteristic macroscopic properties.

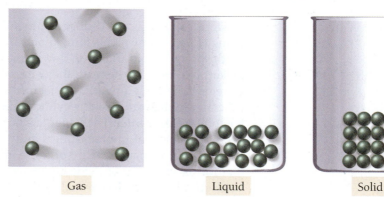

▶ **FIGURE 12.1 Gas, liquid, and solid states**

Gas   Liquid   Solid

Properties of Liquids

- High densities in comparison to gases.
- Indefinite shape; liquids assume the shape of their container.
- Definite volume; liquids are not easily compressed.

Properties of Solids

- High densities in comparison to gases.
- Definite shape; solids do not assume the shape of their container.
- Definite volume; solids are not easily compressed.
- May be crystalline (ordered) or amorphous (disordered).

Table 12.1 summarizes these properties, as well as the properties of gases for comparison.

TABLE 12.1	**Properties of the States of Matter**				
Phase	Density	Shape	Volume	Strength of Intermolecular Forces[a]	Example
gas	low	indefinite	indefinite	weak	carbon dioxide gas (CO_2)
liquid	high	indefinite	definite	moderate	liquid water (H_2O)
solid	high	definite	definite	strong	sugar ($C_{12}H_{22}O_{11}$)

[a]Relative to thermal energy.

Liquids have high densities in comparison to gases because the atoms or molecules that compose liquids are much closer together. The density of liquid water, for example, is 1.0 g/cm^3 (at 25 °C), while the density of gaseous water at 100 °C and 1 atm is 0.59 g/L, or 5.9×10^{-4} g/cm^3. Liquids assume the shape of their containers because the atoms or molecules that compose them are free to flow. When we pour water into a flask, the water flows and assumes the shape of the flask (▼ FIGURE 12.2). Liquids are not easily compressed because the molecules or atoms that compose them are in close contact—they cannot be pushed closer together.

Like liquids, solids have high densities in comparison to gases because the atoms or molecules that compose solids are also close together. The densities of solids are usually just slightly greater than those of the corresponding liquids. A major exception is water, whose solid (ice) is slightly less dense than liquid water. Solids have a definite shape because, in contrast to liquids or gases, the molecules or atoms that compose solids are fixed in place (▼ FIGURE 12.3). Each molecule or atom in a solid only vibrates about a fixed point. Like liquids, solids have a definite volume and cannot be compressed because the molecules or atoms composing them are in close contact. As described in Section 3.3, solids may be *crystalline*, in which case the atoms or molecules that compose them arrange themselves in a well-ordered, three-dimensional array, or they may be *amorphous*, in which case the atoms or molecules that compose them have no long-range order.

As we will see in Section 12.8, ice is less dense than liquid water because water expands when it freezes due to its unique crystalline structure.

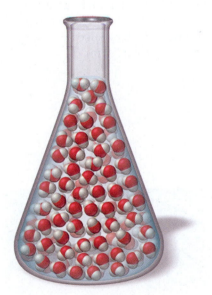

▲ FIGURE 12.2 **A liquid assumes the shape of its container** Because the molecules in liquid water are free to move around each other, they flow and assume the shape of their container.

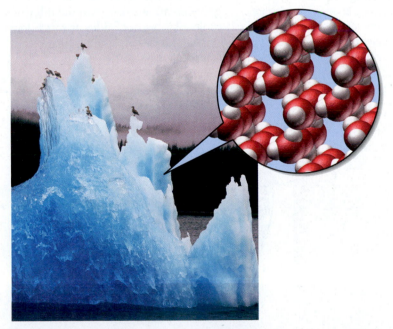

▲ FIGURE 12.3 **Solids have a definite shape** In a solid such as ice, the molecules are fixed in place. However, they vibrate about fixed points.

PEARSON eText 2.0

CONCEPTUAL ✔ CHECKPOINT 12.1

A substance has a definite shape and definite volume. What is the state of the substance?

(a) solid   **(b)** liquid   **(c)** gas

12.3 Intermolecular Forces in Action: Surface Tension and Viscosity

▶ Describe how surface tension and viscosity are manifestations of the intermolecular forces in liquids.

The most important manifestation of intermolecular forces is the very existence of molecular liquids and solids. Without intermolecular forces, molecular solids and liquids would not exist (they would be gaseous). In liquids, we can observe several other manifestations of intermolecular forces, including surface tension and viscosity.

Surface Tension

An angler delicately casts a small metal hook (with a few feathers and strings attached to make it look like an insect) onto the surface of a moving stream. The hook floats on the surface of the water and attracts trout (◀ FIGURE 12.4). The hook floats because of **surface tension**, the tendency of liquids to minimize their surface area. This tendency causes liquids to have a sort of "skin" that resists penetration. For the hook to sink into the water, the water's surface area would have to increase slightly. The increase is resisted because molecules at the surface interact with fewer neighbors than those in the interior of the liquid (▼ FIGURE 12.5). Because molecules at the surface have fewer interactions with other molecules, they are inherently less stable than those in the interior. Consequently, liquids tend to minimize the number of molecules at the surface, which results in surface tension. You can observe surface tension by carefully placing a paper clip on the surface of water (▼ FIGURE 12.6). The paper clip, even though it is denser than water, floats on the surface of the water. A slight tap on the clip overcomes surface tension and causes the clip to sink. Surface tension increases with increasing intermolecular forces. You can't float a paper clip on gasoline, for example, because the intermolecular forces among the molecules composing gasoline are weaker than the intermolecular forces among water molecules. The gas molecules are not under as much tension, so they do not form a "skin."

▲ **FIGURE 12.4 Floating flies** Even though fly-fishing lures are denser than water, they float on the surface of a stream or lake because of surface tension.

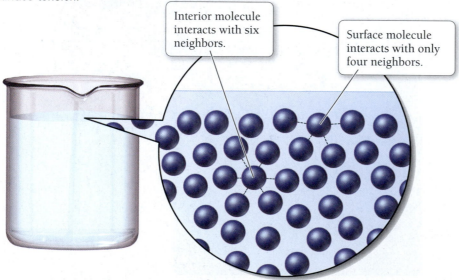

Interior molecule interacts with six neighbors.

Surface molecule interacts with only four neighbors.

▲ **FIGURE 12.5 Origin of surface tension** Molecules at the surface of a liquid interact with fewer molecules than those in the interior; the lower number of interactions makes the surface molecules less stable.

▲ **FIGURE 12.6 Surface tension at work** A paper clip will float on water if it is carefully placed on the surface of the water. It is held up by surface tension.

Viscosity

Another manifestation of intermolecular forces is **viscosity**, the resistance of a liquid to flow. Liquids that are viscous flow more slowly than liquids that are not viscous. For example, motor oil is more viscous than gasoline, and maple syrup is more viscous than water (◄ FIGURE 12.7). Viscosity is greater in substances with stronger intermolecular forces because molecules cannot move around each other as freely, hindering flow. Long molecules, such as the hydrocarbons in motor oil, also tend to form viscous liquids because of molecular entanglement (the long chainlike molecules get tangled together).

◄ **FIGURE 12.7 Viscosity** Maple syrup is more viscous than water because the syrup molecules interact strongly and so cannot flow past one another easily.

12.4 Evaporation and Condensation

▶ Describe and explain the processes of evaporation and condensation.

▶ Use the heat of vaporization in calculations.

Leave a glass of water in the open for several days and the water level within the glass slowly drops. Why? The first reason is that water molecules at the surface of the liquid—which experience fewer attractions to neighboring molecules and are therefore held less tightly—can break away from the rest of the liquid. The second reason is that all of the molecules in the liquid have a *distribution of kinetic energy* at any given temperature (▼ FIGURE 12.8). At any given moment, some molecules in the liquid are moving faster than the average (higher energy), and others are moving more slowly (lower energy). Some of the molecules that are moving faster have enough energy to break free from the surface, resulting in **evaporation** or **vaporization**, a physical change in which a substance converts from its liquid state to its gaseous state (▼ FIGURE 12.9).

If we spill the same amount of water (as was in the glass) on a table, it evaporates more quickly, probably within a few hours. Why? The surface area of the spilled water is greater, leaving more molecules susceptible to evaporation. If we warm the water in the glass, it also evaporates more quickly because the greater thermal energy causes a greater fraction of molecules to have enough energy to break away from the surface (see Figure 12.8). If we fill the glass with rubbing

> At a higher temperature, the fraction of molecules with enough energy to escape increases.

> Molecules on the surface are held less tightly than those in the interior so the most energetic can break away into the gas state.

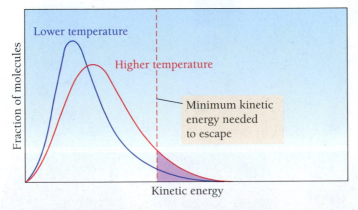

▲ **FIGURE 12.8 Energy distribution** At a given temperature, a sample of molecules or atoms will have a distribution of kinetic energies. The fraction of molecule having enough energy to escape is shown in purple.

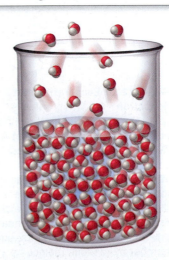

▲ **FIGURE 12.9 Evaporation**

In evaporation or vaporization, a substance is converted from its liquid state into its gaseous state.

alcohol instead of water, the liquid again evaporates more quickly because the intermolecular forces between the alcohol molecules are weaker than the intermolecular forces between water molecules. In general, the rate of vaporization increases with:

- Increasing surface area
- Increasing temperature
- Decreasing strength of intermolecular forces

Liquids that evaporate easily are **volatile**, while those that do not vaporize easily are **nonvolatile**. Rubbing alcohol, for example, is more volatile than water. Motor oil at room temperature is virtually nonvolatile.

If we leave water in a *closed* container, its level remains constant because the molecules that leave the liquid are trapped in the air space above the water. These gaseous molecules bounce off the walls of the container and eventually hit the surface of the water again and recondense. **Condensation** is a physical change in which a substance converts from its gaseous state to its liquid state.

Evaporation and condensation are opposites: Evaporation is a liquid turning into a gas, and condensation is a gas turning into a liquid. When we initially put liquid water into a closed container, more evaporation happens than condensation because there are so few gaseous water molecules in the space above the water (▼ FIGURE 12.10a). However, as the number of gaseous water molecules increases, the rate of condensation also increases (▼ FIGURE 12.10b). At the point where the rates of condensation and evaporation become equal (▼ FIGURE 12.10c), **dynamic equilibrium** is reached and the number of gaseous water molecules above the liquid remains constant.

Dynamic equilibrium is so named because both condensation and evaporation of individual molecules continue, but at the same rate.

The **vapor pressure** of a liquid is the partial pressure of its vapor in dynamic equilibrium with its liquid. At 25 °C, water's vapor pressure is 23.8 mm Hg. Vapor pressure increases with:

Table 11.4 lists the vapor pressure of water at different temperatures.

- Increasing temperature
- Decreasing strength of intermolecular forces

Vapor pressure is independent of surface area because an increase in surface area at equilibrium equally affects the rate of evaporation and the rate of condensation.

▶ FIGURE 12.10 Evaporation and condensation (a) When water is first put into a closed container, water molecules begin to evaporate. (b) As the number of gaseous molecules increases, some of the molecules begin to collide with the liquid and are recaptured—that is, they recondense into liquid. (c) When the rate of evaporation equals the rate of condensation, dynamic equilibrium occurs, and the number of gaseous molecules remains constant.

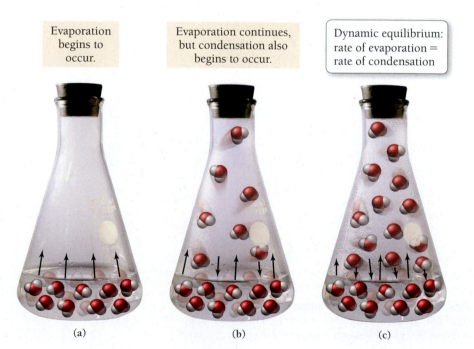

Evaporation begins to occur.

Evaporation continues, but condensation also begins to occur.

Dynamic equilibrium: rate of evaporation = rate of condensation

(a)   (b)   (c)

Boiling

As you increase the temperature of water in an open container, the increasing thermal energy causes molecules to leave the surface and vaporize at a faster and faster rate. At the **boiling point**—the temperature at which the vapor pressure of a liquid is equal to the pressure above it—the thermal energy is enough for molecules within the *interior* of the liquid (not just those at the surface) to break free into the gas phase (▼ FIGURE 12.11). Water's **normal boiling point**—its boiling point at a pressure of 1 atmosphere—is 100 °C. When a sample of water reaches 100 °C, you can see bubbles form within the liquid. These bubbles are pockets of gaseous water. The bubbles quickly rise to the surface of the liquid, and the water molecules that were in the bubble leave as gaseous water, or steam.

Once the boiling point of a liquid is reached, additional heating only causes more rapid boiling; it does not raise the temperature of the liquid above its boiling point (▼ FIGURE 12.12). A mixture of boiling water *and* steam always has a temperature of 100 °C (at 1 atm pressure). Only after all the water has been converted to steam can the temperature of the steam rise beyond 100 °C.

> Sometimes you see bubbles begin to form in hot water below 100 °C. These bubbles are dissolved air—not gaseous water—leaving the liquid. Dissolved air comes out of water as you heat it because the solubility of a gas in a liquid decreases with increasing temperature (Section 13.4).

▶ **FIGURE 12.11 Boiling** During boiling, thermal energy is enough to cause water molecules in the interior of the liquid to become gaseous, forming bubbles containing gaseous water molecules.

▶ **FIGURE 12.12 Heating curve during boiling** The temperature of water as it is heated from room temperature to its boiling point. During boiling, the temperature remains at 100 °C until all the liquid is evaporated.

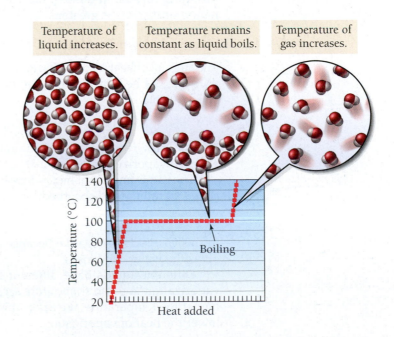

Temperature of liquid increases.   Temperature remains constant as liquid boils.   Temperature of gas increases.

CONCEPTUAL ✔ **CHECKPOINT 12.2**

The gas over a rapidly boiling pot of water is sampled and analyzed. Which substance composes a large fraction of the gas sample?

(a) $H_2(g)$ **(b)** $H_2O(g)$ **(c)** $O_2(g)$ **(d)** $H_2O_2(g)$

Energetics of Evaporation and Condensation

> In an endothermic process, heat is absorbed; in an exothermic process, heat is released.

Evaporation is *endothermic*—when a liquid is converted into a gas, it absorbs heat because energy is required to break molecules away from the rest of the liquid. Imagine a collection of water molecules in the liquid state. As the water evaporates, it cools—typical of endothermic processes—because only the fastest-moving molecules break away, which leaves the slower-moving cooler molecules behind. Under ordinary conditions, the slight decrease in the temperature of water as it evaporates is counteracted by thermal energy transfer from the surroundings, which warms the water back up. However, if the evaporating water were thermally isolated from the surroundings, it would continue to cool down as it evaporated.

You can observe the endothermic nature of evaporation by turning off the heat beneath a boiling pot of water; it quickly stops boiling as the heat lost due to vaporization causes the water to cool below its boiling point. Our bodies use the endothermic nature of evaporation for cooling. When we overheat, we sweat, causing our skin to be covered with liquid water. As this water evaporates it absorbs heat from our bodies, cooling us down. A fan intensifies the cooling effect because it blows newly vaporized water away from the skin, allowing more sweat to vaporize and cause even more cooling. High humidity, however, slows down evaporation, preventing cooling. When the air already contains high amounts of water vapor, sweat does not evaporate as easily, making our cooling system less efficient.

Condensation, the opposite of evaporation, is *exothermic*—heat is released when a gas condenses to a liquid. If you have ever accidentally put your hand above a steaming kettle, you may have experienced a *steam burn*. As the steam condenses to a liquid on your skin, it releases heat, causing a severe burn. The exothermic nature of condensation is also the reason that winter overnight temperatures in coastal cities, which tend to have water vapor in the air, do not get as low as in deserts, which tend to have dry air. As the air temperature in a coastal city drops, water condenses out of the air, releasing heat and preventing the temperature from dropping further. In deserts, there is little moisture in the air to condense, so the temperature drop is greater.

Heat of Vaporization

The amount of heat required to vaporize one mole of liquid is the **heat of vaporization** (ΔH_{vap}). The heat of vaporization of water at its normal boiling point (100 °C) is 40.7 kJ/mole.

$$H_2O(l) \longrightarrow H_2O(g) \quad \Delta H = +40.7 \text{ kJ (at 100 °C)}$$

ΔH is positive because vaporization is endothermic; energy must be added to the water to vaporize it.

The same amount of heat is involved when 1 mol of gas condenses, but *the heat is emitted* rather than absorbed.

$$H_2O(g) \longrightarrow H_2O(l) \quad \Delta H = -40.7 \text{ kJ (at 100 °C)}$$

In this case, ΔH is negative because condensation is exothermic; energy is given off as the water condenses.

Different liquids have different heats of vaporization (Table 12.2). Heats of vaporization are also *temperature dependent* (they change with temperature). The higher the temperature, the easier it is to vaporize a given liquid and therefore the lower the heat of vaporization.

TABLE 12.2 Heats of Vaporization of Several Liquids at Their Boiling Points and at 25 °C

Liquid	Chemical Formula	Normal Boiling Point (°C)	Heat of Vaporization (kJ/mol) at Boiling Point	Heat of Vaporization (kJ/mol) at 25 °C
water	H_2O	100.0	40.7	44.0
isopropyl alcohol (rubbing alcohol)	C_3H_8O	82.3	39.9	45.4
acetone	C_3H_6O	56.1	29.1	31.0
diethyl ether	$C_4H_{10}O$	34.5	26.5	27.1

We can use the heat of vaporization of a liquid to calculate the amount of heat energy required to vaporize a given amount of that liquid. To do so, we use the heat of vaporization as a conversion factor between moles of the liquid and the amount of heat required to vaporize it. For example, suppose we want to calculate the amount of heat required to vaporize 25.0 g of water at its boiling point. We begin by sorting the information in the problem statement.

GIVEN: 25.0 g H_2O

FIND: heat (kJ)

SOLUTION MAP
We then strategize by building a solution map that begins with the mass of water and ends with the energy required to vaporize it.

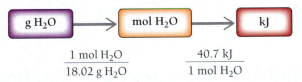

$$\dfrac{1 \text{ mol } H_2O}{18.02 \text{ g } H_2O} \qquad \dfrac{40.7 \text{ kJ}}{1 \text{ mol } H_2O}$$

RELATIONSHIPS USED

$$\Delta H_{vap} = 40.7 \text{ kJ/mol at } 100 \text{ °C (Table 12.2)}$$
$$1 \text{ mol } H_2O = 18.02 \text{ g } H_2O \text{ (molar mass of water)}$$

SOLUTION

PEARSON
eText 2.0
Interactive
Worked Example
Video 12.1

$$25.0 \text{ g } H_2O \times \dfrac{1 \text{ mol } H_2O}{18.02 \text{ g } H_2O} \times \dfrac{40.7 \text{ kJ}}{1 \text{ mol } H_2O} = 56.5 \text{ kJ}$$

EXAMPLE **12.1** | **Using the Heat of Vaporization in Calculations**

Calculate the amount of water in grams that can be vaporized at its boiling point with 155 kJ of heat.

SORT
You are given the number of kilojoules of heat energy and asked to find the mass of water that can be vaporized with the given amount of energy.

GIVEN: 155 kJ

FIND: g H_2O

STRATEGIZE
Draw the solution map beginning with the energy in kilojoules and converting to moles of water and then to grams of water.

SOLUTION MAP

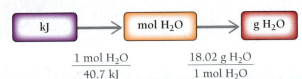

$$\dfrac{1 \text{ mol } H_2O}{40.7 \text{ kJ}} \qquad \dfrac{18.02 \text{ g } H_2O}{1 \text{ mol } H_2O}$$

RELATIONSHIPS USED

$$\Delta H_{vap} = 40.7 \text{ kJ/mol at } 100 \text{ °C} \quad \text{(Table 12.2)}$$

$$18.02 \text{ g } H_2O = 1 \text{ mol } H_2O \text{ (molar mass of water)}$$

continued on page 418 ▶

continued from page 417

SOLVE Follow the solution map to solve the problem.	SOLUTION $$155 \text{ kJ} \times \frac{1 \text{ mol } H_2O}{40.7 \text{ kJ}} \times \frac{18.02 \text{ g}}{1 \text{ mol } H_2O} = 68.6 \text{ g}$$
CHECK Check your answer. Are the units correct? Does the answer make physical sense?	The units (g) are correct. The magnitude of the answer makes sense because each mole of water absorbs about 40 kJ of energy upon vaporization. Therefore 155 kJ should vaporize close to 4 mol of water, which is consistent with the answer (4 mol of water has a mass of about 72 g).

▶ **SKILLBUILDER 12.1** | **Using the Heat of Vaporization in Calculations**

Calculate the amount of heat in kilojoules required to vaporize 2.58 kg of water at its boiling point.

▶ **SKILLBUILDER PLUS**

A drop of water weighing 0.48 g condenses on the surface of a 55-g block of aluminum that is initially at 25 °C. If the heat released during condensation goes only toward heating the metal, what is the final temperature in Celsius of the metal block? (The specific heat capacity of aluminum is listed in Table 3.4 and is 0.903 J/g °C. Use 40.7 kJ/mol as the heat of vaporization of water.)

▶ **FOR MORE PRACTICE**   Example 12.7; Problems 49, 50, 51, 52, 53, 54.

PEARSON eText 2.0

CONCEPTUAL ✔ **CHECKPOINT 12.3**

When water condenses on a small metal block, what would you expect to happen to the temperature of the metal block?

12.5 Melting, Freezing, and Sublimation

▶ Describe the processes of melting, freezing, and sublimation.

▶ Use the heat of fusion in calculations.

As the temperature of a solid increases, thermal energy causes the molecules and atoms composing the solid to vibrate faster. At the **melting point**, atoms and molecules have enough thermal energy to overcome the intermolecular forces that hold them at their stationary points, and the solid turns into a liquid. The melting point of ice, for example, is 0 °C. Once the melting point of a solid is reached, additional heating only causes more rapid melting; it does not raise the temperature of the solid above its melting point (▼ FIGURE 12.13). Only after all of the ice has melted does additional heating raise the temperature of the liquid water past 0 °C. A mixture of water *and* ice always has a temperature of 0 °C (at 1 atm pressure).

▶ **FIGURE 12.13 Heating curve during melting** A graph of the temperature of ice as it is heated from 220 °C to 35 °C. During melting, the temperature of the solid and the liquid remains at 0 °C until the entire solid is melted.

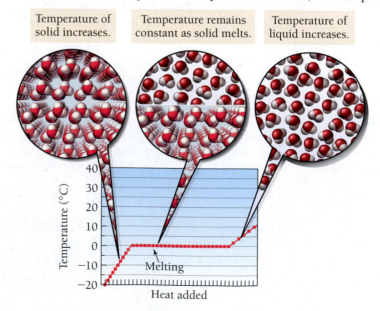

Temperature of solid increases.

Temperature remains constant as solid melts.

Temperature of liquid increases.

Melting

Heat added

Energetics of Melting and Freezing

The most common way to cool down a drink is to drop several ice cubes into it. As the ice melts, the drink cools because melting is endothermic—heat is absorbed when a solid is converted into a liquid. The melting ice absorbs heat from the liquid in the drink and cools the liquid. Melting is endothermic because energy is required to partially overcome the attractions between molecules in the solid and free them into the liquid state.

Freezing, the opposite of melting, is exothermic—heat is released when a liquid freezes into a solid. For example, as water in a freezer turns into ice, it releases heat, which must be removed by the refrigeration system of the freezer. If the refrigeration system did not remove the heat, the water would not completely freeze into ice. The heat released as it began to freeze would warm the freezer, preventing further freezing.

Heat of Fusion

The amount of heat required to melt 1 mol of a solid is the **heat of fusion** ($\Delta H^{\circ}_{\text{fus}}$). The heat of fusion for water is 6.02 kJ/mol.

$$H_2O(s) \longrightarrow H_2O(l) \quad \Delta H = +6.02 \text{ kJ}$$

ΔH is positive because melting is endothermic; energy must be added to the ice to melt it.

The same amount of heat is involved when 1 mol of liquid water freezes, but the heat is emitted rather than absorbed.

$$H_2O(l) \longrightarrow H_2O(s) \quad \Delta H = -6.02 \text{ kJ}$$

In this case, ΔH is negative because freezing is exothermic; energy is given off as the water freezes.

Different substances have different heats of fusion (Table 12.3, on the next page). Notice that, in general, the heat of fusion is significantly less than the heat of vaporization. It takes less energy to melt 1 mol of ice than it does to vaporize 1 mol of liquid water. Why? Vaporization requires complete separation of one molecule from another, so the intermolecular forces must be completely overcome. Melting, on the other hand, requires that intermolecular forces be only partially overcome, allowing molecules to move around one another while still remaining in contact.

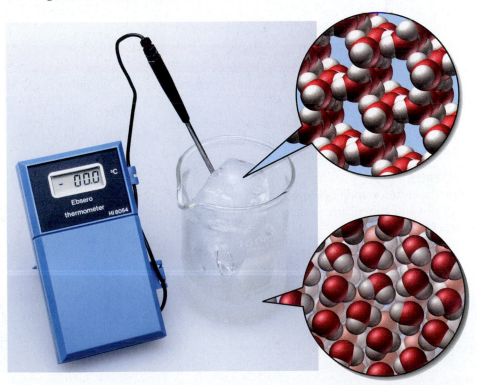

▶ When ice melts, water molecules break free from the solid structure and become liquid. As long as ice and water are both present, the temperature is 0.0 °C.

TABLE 12.3 Heats of Fusion of Several Substances

Liquid	Chemical Formula	Melting Point (°C)	Heat of Fusion (kJ/mol)
water	H_2O	0.00	6.02
isopropyl alcohol (rubbing alcohol)	C_3H_8O	−89.5	5.37
acetone	C_3H_6O	−94.8	5.69
diethyl ether	$C_4H_{10}O$	−116.3	7.27

We can use the heat of fusion to calculate the amount of heat energy required to melt a given amount of a solid. The heat of fusion is a conversion factor between moles of a solid and the amount of heat required to melt it. For example, suppose we want to calculate the amount of heat required to melt 25.0 g of ice (at 0 °C). We first sort the information in the problem statement.

GIVEN: 25.0 g H_2O

FIND: heat (kJ)

SOLUTION MAP

We then draw the solution map, beginning with the mass of water and ending with the energy required to melt it.

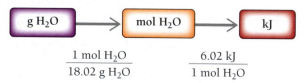

$$\frac{1 \text{ mol } H_2O}{18.02 \text{ g } H_2O} \qquad \frac{6.02 \text{ kJ}}{1 \text{ mol } H_2O}$$

RELATIONSHIPS USED

$$\Delta H_{fus} = 6.02 \text{ kJ/mol (Table 12.3)}$$
$$1 \text{ mol } H_2O = 18.02 \text{ g } H_2O \text{ (molar mass of water)}$$

SOLUTION

$$25.0 \text{ g } H_2O \times \frac{1 \text{ mol } H_2O}{18.02 \text{ g } H_2O} \times \frac{6.02 \text{ kJ}}{1 \text{ mol } H_2O} = 8.35 \text{ kJ}$$

EXAMPLE **12.2** **Using the Heat of Fusion in Calculations**

Calculate the amount of ice in grams that, upon melting (at 0 °C), absorbs 237 kJ of heat.

SORT
You are given the number of kilojoules of heat energy and asked to find the mass of ice that absorbs the given amount of energy upon melting.

GIVEN: 237 kJ

FIND: g H_2O (ice)

STRATEGIZE
Draw the solution map beginning with the energy in kilojoules and converting to moles of water and then to grams of water.

SOLUTION MAP

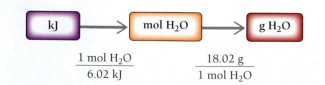

$$\frac{1 \text{ mol } H_2O}{6.02 \text{ kJ}} \qquad \frac{18.02 \text{ g}}{1 \text{ mol } H_2O}$$

RELATIONSHIPS USED

$$\Delta H_{fus} = 6.02 \text{ kJ/mol (Table 12.3)}$$
$$1 \text{ mol } H_2O = 18.02 \text{ g } H_2O \text{ (molar mass of water)}$$

SOLVE	SOLUTION
Follow the solution map to solve the problem.	$237 \text{ kJ} \times \dfrac{1 \text{ mol H}_2\text{O}}{6.02 \text{ kJ}} \times \dfrac{18.02 \text{ g}}{1 \text{ mol H}_2\text{O}} = 709 \text{ g}$
CHECK Check your answer. Are the units correct? Does the answer make physical sense?	The units (g) are correct. The magnitude of the answer makes sense because each mole of water absorbs about 6 kJ of energy upon melting. Therefore, 237 kJ should melt close to 40 mol of water, which is consistent with the answer (40 mol of water has a mass of about 720 g).

▶ **SKILLBUILDER 12.2** | **Using the Heat of Fusion in Calculations**

Calculate the amount of heat absorbed when a 15.5-g ice cube melts (at 0 °C).

▶ **SKILLBUILDER PLUS**

A 5.6-g ice cube (at 0 °C) is placed into 195 g of water initially at 25 °C. If the heat absorbed for melting the ice comes only from the 195 g of water, what is the temperature change of the 195 g of water?

▶ **FOR MORE PRACTICE** Example 12.8; Problems 57, 58, 59, 60.

CONCEPTUAL ✔ **CHECKPOINT 12.4**

This diagram shows a heating curve for ice beginning at −25 °C and ending at 125 °C. Correlate sections i, ii, and iii with the correct states of water.

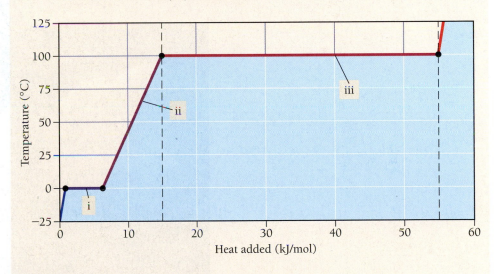

(a) i—solid, ii—liquid, iii—gas

(b) i—solid and liquid, ii—liquid, iii—liquid and gas

(c) i—liquid, ii—liquid and gas, iii—gas

(d) i—solid and liquid, ii—liquid and gas , iii—gas

Sublimation

Sublimation is a physical change in which a substance changes from its solid state directly to its gaseous state. When a substance sublimes, molecules leave the surface of the solid, where they are held less tightly than in the interior, and become gaseous. For example, dry ice, which is solid carbon dioxide, does not melt under atmospheric pressure (at any temperature). At −78 °C, the CO_2 molecules have enough energy to leave the surface of the dry ice and become gaseous. Regular ice

▲ Dry ice is solid carbon dioxide. The solid does not melt but rather sublimes. It transforms directly from solid carbon dioxide to gaseous carbon dioxide.

slowly sublimes at temperatures below 0 °C. You can observe the sublimation of ice in cold climates; ice or snow laying on the ground gradually disappears, even if the temperature remains below 0 °C. Similarly, ice cubes left in the freezer for a long time slowly become smaller, even though the freezer is always below 0 °C. In both cases, the ice is subliming, turning directly into water vapor.

Ice also sublimes out of frozen foods. You can see this in food that is frozen in an airtight plastic bag for a long time. The ice crystals that form in the bag are water that has sublimed out of the food and redeposited on the surface of the bag. For this reason, food that remains frozen for too long becomes dried out. This can be avoided to some degree by freezing foods to colder temperatures (below 0 °C), a process called deep-freezing. The colder temperature lowers the rate of sublimation and preserves the food longer.

CONCEPTUAL ✔ CHECKPOINT 12.5

Solid carbon dioxide (dry ice) can be depicted as follows:

Which image best represents the dry ice after it has sublimed?

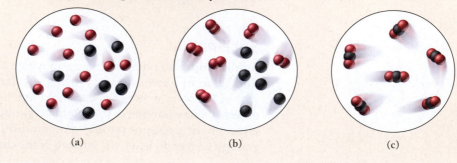

(a)   (b)   (c)

12.6 Types of Intermolecular Forces: Dispersion, Dipole–Dipole, Hydrogen Bonding, and Ion–Dipole

► Compare and contrast four types of intermolecular forces: dispersion, dipole–dipole, hydrogen bonds, and ion–dipole.

► Determine the types of intermolecular forces in compounds.

► Use intermolecular forces to determine relative melting and/or boiling points.

The strength of the intermolecular forces between the molecules or atoms that compose a substance determines the state—solid, liquid, or gas—of the substance at room temperature. Strong intermolecular forces tend to result in liquids and solids (with high melting and boiling points). Weak intermolecular forces tend to result in gases (with low melting and boiling points). In this book, we focus on four fundamental types of intermolecular forces. In order of increasing strength, they are the dispersion force, the dipole–dipole force, the hydrogen bond, and the ion–dipole force.

Dispersion Force

The default intermolecular force, present in all molecules and atoms, is the **dispersion force** (also called the *London force*). Dispersion forces are caused by fluctuations in the electron distribution within molecules or atoms. Since all atoms and molecules have electrons, they all have dispersion forces. The electrons in an atom or a molecule may, at any one instant, be unevenly distributed. For example, imagine a frame-by-frame movie of a helium atom in which each "frame" captures the position of the helium atom's two electrons (▼ **FIGURE 12.14**). In any one frame, the electrons are not symmetrically arranged around the nucleus. In Frame 3, for example, helium's two electrons are on the left side of the helium atom. The left side then acquires a slightly negative charge (δ^-). The right side of the atom, which is void of electrons, acquires a slightly positive charge (δ^+).

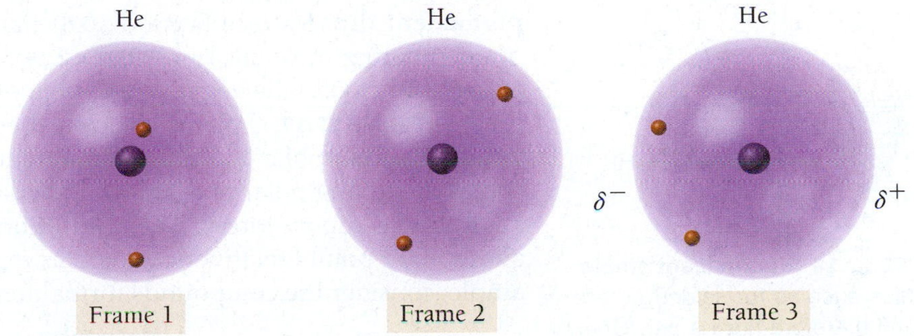

The nature of dispersion forces was first recognized by Fritz W. London (1900–1954), a German American physicist.

► **FIGURE 12.14 Instantaneous dipoles** Random fluctuations in the electron distribution of a helium atom cause instantaneous dipoles to form.

He   He   He

Frame 1   Frame 2   Frame 3

This fleeting charge separation is called an **instantaneous dipole** (or *temporary dipole*). An instantaneous dipole on one helium atom induces an instantaneous dipole on its neighboring atoms because the positive end of the instantaneous dipole attracts electrons in the neighboring atoms (▼ **FIGURE 12.15**). The dispersion force occurs as neighboring atoms attract one another—the positive end of one instantaneous dipole attracts the negative end of another. The dipoles responsible for the dispersion force are transient, constantly appearing and disappearing in response to fluctuations in electron clouds.

► **FIGURE 12.15 Dispersion force** An instantaneous dipole on any one helium atom induces instantaneous dipoles on neighboring atoms. The neighboring atoms then attract one another. This attraction is called the dispersion force.

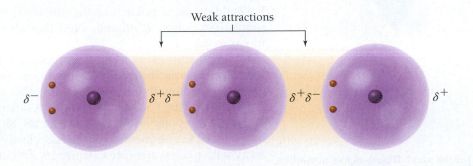

Weak attractions

To *polarize* means to form a dipole moment.

TABLE 12.4 Noble Gas Boiling Points

Noble Gas	Molar Mass (g/mol)	Boiling Point (K)
He	4.00	4.2 K
Ne	20.18	27 K
Ar	39.95	87 K
Kr	83.80	120 K
Xe	131.29	165 K

The magnitude of the dispersion force depends on how easily the electrons in the atom or molecule move or *polarize* in response to an instantaneous dipole, which in turn depends on the size of the electron cloud. A larger electron cloud results in a greater dispersion force because the electrons are held less tightly by the nucleus and therefore can polarize more easily. If all other variables are constant, the dispersion force increases with increasing molar mass. For example, consider the boiling points of the noble gases listed in Table 12.4. As the molar mass of the noble gases increases, their boiling points increase. While molar mass alone does not determine the magnitude of the dispersion force, it can be useful as a guide when comparing dispersion forces within a family of similar elements or compounds.

EXAMPLE 12.3 | Dispersion Forces

Which halogen, Cl_2 or I_2, has the higher boiling point?

SOLUTION

The molar mass of Cl_2 is 70.90 g/mol, and the molar mass of I_2 is 253.80 g/mol. Since I_2 has the higher molar mass, it has stronger dispersion forces and therefore the higher boiling point.

▶ **SKILLBUILDER 12.3 | Dispersion Forces**

Which hydrocarbon, CH_4 or C_2H_6, has the higher boiling point?

▶ **FOR MORE PRACTICE** Problems 69, 70.

See Section 10.8 to review how to determine whether a molecule is polar.

Dipole–Dipole Force

The **dipole–dipole force** exists in all polar molecules. Polar molecules have **permanent dipoles** (see Section 10.8) that interact with the permanent dipoles of neighboring molecules (◀ FIGURE 12.16). The positive end of one permanent dipole is attracted to the negative end of another; this attraction is the dipole–dipole force (◀ FIGURE 12.17). Polar molecules, therefore, have higher melting and boiling points than nonpolar molecules of similar molar mass. Remember that all molecules (including polar ones) have dispersion forces. In addition, polar molecules have dipole–dipole forces. These additional attractive forces raise their melting and boiling points relative to nonpolar molecules of similar molar mass. For example, consider the compounds formaldehyde and ethane:

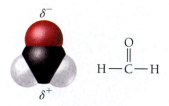

▲ **FIGURE 12.16 A permanent dipole** Molecules such as formaldehyde are polar and therefore have a permanent dipole.

The positive end of a polar molecule is attracted to the negative end of neighboring molecule.

▲ **FIGURE 12.17 Dipole–dipole attraction**

Name	Formula	Molar mass (g/mol)	Structure	Boiling point (°C)	Melting point (°C)
formaldehyde	CH_2O	30.0	H—C(=O)—H	−19.5	−92
ethane	C_2H_6	30.1	H—C—C—H	−89	−183

Formaldehyde is polar and therefore has a higher melting point and boiling point than nonpolar ethane, even though the two compounds have the same molar mass.

The polarity of molecules composing liquids is also important in determining a liquid's **miscibility**—its ability to mix without separating into two phases. In general, polar liquids are miscible with other polar liquids but are not miscible with nonpolar liquids. For example, water, a polar liquid, is not miscible with pentane (C_5H_{12}), a nonpolar liquid (▶ FIGURE 12.18). Similarly, water and oil (also nonpolar) do not mix. Consequently, oily hands or oily stains on clothes cannot be washed away with plain water (see Chapter 10, *Everyday Chemistry: How Soap Works*).

► **FIGURE 12.18 Polar and nonpolar compounds** **(a)** Pentane, a nonpolar compound, does not mix with water, a polar compound. **(b)** For the same reason, the oil and vinegar (vinegar is largely a water solution of acetic acid) in salad dressing tend to separate into distinct layers. **(c)** An oil spill from a tanker demonstrates dramatically that petroleum and seawater are not miscible.

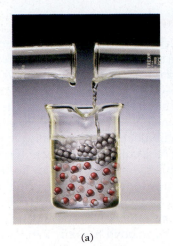  

(a) (b) (c)

 Interactive Worked Example Video 12.4

EXAMPLE **12.4**   **Dipole–Dipole Forces**

Determine whether each molecule has dipole–dipole forces.

(a) CO_2 **(b)** CH_2Cl_2 **(c)** CH_4

SOLUTION

A molecule has dipole–dipole forces if it is polar. To find out whether a molecule is polar, you must:

1. determine whether the molecule contains polar bonds, and
2. determine whether the polar bonds add together to form a net dipole moment (Section 10.8).

(a) The electronegativities of carbon and oxygen are 2.5 and 3.5, respectively (see Figure 10.2), so CO_2 has polar bonds. The geometry of CO_2 is linear. Consequently, the polar bonds cancel; the molecule is not polar and does not have dipole–dipole forces.	 O=C=O Nonpolar; no dipole–dipole forces
(b) The electronegativities of C, H, and Cl are 2.5, 2.1, and 3.5, respectively. Consequently, CH_2Cl_2 has two polar bonds (C—Cl) and two bonds that are nearly nonpolar (C—H). The geometry of CH_2Cl_2 is tetrahedral. Since the C—Cl bonds and the C—H bonds are different, they do not cancel but sum to a net dipole moment. Therefore, the molecule is polar and has dipole–dipole forces.	 Polar; dipole–dipole forces
(c) Since the electronegativities of C and H are 2.5 and 2.1, respectively, the C—H bonds are nearly nonpolar. In addition, because the geometry of the molecule is tetrahedral, any slight polarities that the bonds might have will cancel. CH_4 is therefore nonpolar and does not have dipole–dipole forces.	 Nonpolar; no dipole–dipole forces

► **SKILLBUILDER 12.4 | Dipole–Dipole Forces**

Determine whether each molecule has dipole–dipole forces.

(a) CI_4 **(b)** CH_3Cl **(c)** HCl

► **FOR MORE PRACTICE**   Problems 61, 62, 65, 66.

Hydrogen Bonding

Polar molecules containing hydrogen atoms bonded directly to fluorine, oxygen, or nitrogen exhibit an additional intermolecular force called a **hydrogen bond**. HF, NH_3, and H_2O, for example, all undergo hydrogen bonding. A hydrogen bond is a sort of *super* dipole–dipole force. The large electronegativity difference between hydrogen and these electronegative elements, as well as the small size of these atoms (which allows neighboring molecules to get very close to each other), gives rise to a strong attraction between the H in each of these molecules and the F, O, or N on neighboring molecules. This attraction between a hydrogen atom and an electronegative atom is the hydrogen bond. For example, in HF the hydrogen is strongly attracted to the fluorine on neighboring molecules (◀ FIGURE 12.19).

Do not confuse hydrogen bonds with chemical bonds. Chemical bonds occur between *individual atoms within a molecule* and are generally much stronger than hydrogen bonds. A hydrogen bond has only 2 to 5% the strength of a typical covalent chemical bond. Hydrogen bonds—like dispersion forces and dipole–dipole forces—are intermolecular forces that occur *between molecules*. In liquid water, for example, the hydrogen bonds are transient, constantly forming, breaking, and re-forming as water molecules move within the liquid. Hydrogen bonds are, however, strong intermolecular forces. Substances composed of molecules that form hydrogen bonds have much higher melting and boiling points than you would predict based on molar mass. For example, consider the two compounds, methanol and ethane:

Hydrogen on each molecule is strongly attracted to fluorine on its neighbor.

δ^+ δ^- δ^+ δ^- δ^+ δ^-

H—F ····· H—F ····· H—F

▲ FIGURE 12.19 **The hydrogen bond** The intermolecular attraction of a hydrogen atom to an electronegative atom is a *hydrogen bond*.

Name	Formula	Molar mass (g/mol)	Structure	Boiling Point (°C)	Melting Point (°C)
methanol	CH_3OH	32.0	H—C—O—H (with H above and H below C)	64.7	−97.6
ethane	C_2H_6	30.1	H—C—C—H (with H above and below each C)	−89	−183

Since methanol contains hydrogen directly bonded to oxygen, its molecules have hydrogen bonding as an intermolecular force. The hydrogen that is directly bonded to oxygen is strongly attracted to the oxygen on neighboring molecules (◀ FIGURE 12.20). This strong attraction makes the boiling point of methanol 64.7 °C. Consequently, methanol is a liquid at room temperature. Water is another good example of a molecule with hydrogen bonding as an intermolecular force (▼ FIGURE 12.21). The boiling point of water (100 °C) is remarkably high for a molecule with such a low molar mass (18.02 g/mol). Hydrogen bonding is important in biological molecules. The shapes of proteins and nucleic acids are largely

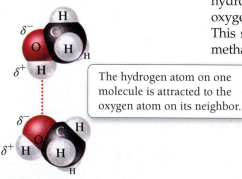

The hydrogen atom on one molecule is attracted to the oxygen atom on its neighbor.

▲ FIGURE 12.20 **Hydrogen bonding in methanol** Since methanol contains hydrogen atoms directly bonded to oxygen, methanol molecules form hydrogen bonds to one another.

The hydrogen atoms on each water molecule are attracted to the oxygen atoms on its neighbors.

◀ FIGURE 12.21 **Hydrogen bonding in water** Water molecules form strong hydrogen bonds with one another.

influenced by hydrogen bonding; for example, the two halves of DNA are held together by hydrogen bonds (see the *Chemistry and Health* box on p. 428).

EXAMPLE **12.5**   **Hydrogen Bonding**

One of these compounds is a liquid at room temperature. Which one and why?

$$
\begin{array}{ccc}
\text{O} & \text{H} & \\
\parallel & | & \\
\text{H—C—H} & \text{H—C—F} & \text{H—O—O—H} \\
& | & \\
& \text{H} & \\
\text{Formaldehyde} & \text{Fluoromethane} & \text{Hydrogen peroxide}
\end{array}
$$

SOLUTION

The three compounds have similar molar masses.

formaldehyde	30.03 g/mol
fluoromethane	34.04 g/mol
hydrogen peroxide	34.02 g/mol

Therefore, the strengths of their dispersion forces are similar. All three compounds are also polar, so they have dipole–dipole forces. Hydrogen peroxide, however, is the only compound that also contains H bonded directly to F, O, or N. Therefore, it also has hydrogen bonding and is most likely to have the highest boiling point of the three. Since the problem stated that only one of the compounds was a liquid, we can safely assume that hydrogen peroxide is the liquid. Note that although fluoromethane *contains* both H and F, H is not *directly bonded* to F, so fluoromethane does not have hydrogen bonding as an intermolecular force. Similarly, although formaldehyde *contains* both H and O, H is not *directly bonded* to O, so formaldehyde does not have hydrogen bonding either.

▶ **SKILLBUILDER 12.5 | Hydrogen Bonding**
Which has the higher boiling point, HF or HCl? Why?

▶ **FOR MORE PRACTICE**   Examples 12.9, 12.10; Problems 71, 72, 73, 74, 75, 76.

> In some cases, hydrogen bonding can occur between one molecule in which H is directly bonded to F, O, or N and another molecule containing an electronegative atom. (For an example, see the box, *Chemistry and Health*, in this section.)

CONCEPTUAL ✔ CHECKPOINT 12.6

Three molecular compounds A, B, and C have nearly identical molar masses. Substance A is nonpolar, substance B is polar, and substance C undergoes hydrogen bonding. What is most likely to be the relative order of their boiling points?

(a) A < B < C   **(b)** C < B < A   **(c)** B < C < A

Ion–Dipole Force

> The positive sodium ions interact with the negative ends of water molecules, while the negative chloride ions interact with the positive ends of water molecules.

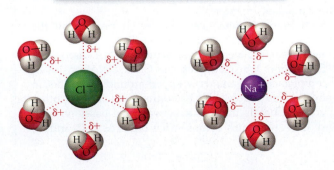

◀ **FIGURE 12.22 Ion–dipole forces**

The **ion–dipole force** occurs in mixtures of ionic compounds and polar compounds; it is especially important in aqueous solutions of ionic compounds. For example, when sodium chloride is mixed with water, the sodium and chloride ions interact with water molecules via ion–dipole forces, as shown in ◀ **FIGURE 12.22**. The positive sodium ions interact with the negative poles of water molecules, while the negative chloride ions interact with the positive poles. Ion–dipole forces are the strongest of the four types of intermolecular forces discussed and are responsible for the ability of ionic substances to form solutions with water. We will discuss aqueous solutions more thoroughly in Chapter 13.

Table 12.5 summarizes the different types of intermolecular forces. Remember that dispersion forces, the weakest kind of intermolecular force, are present in all molecules and atoms and increase with increasing molar mass. These forces are weak in small molecules, but they become substantial in molecules with high molar masses. Dipole–dipole forces are present only in polar molecules. Hydrogen bonds are present in molecules containing hydrogen bonded directly to fluorine, oxygen, or nitrogen, and ion–dipole forces occur in mixtures of ionic compounds and polar compounds.

TABLE 12.5 Types of Intermolecular Forces			
Type of Force	**Relative Strength**	**Present in**	**Example**
dispersion force (or London force)	weak, but increases with increasing molar mass	all atoms and molecules	H_2 H_2
dipole–dipole force	moderate	only polar molecules	δ^+ δ^- δ^+ δ^- HCl HCl
hydrogen bond	strong	molecules containing H bonded directly to F, O, or N	δ^+ δ^- δ^+ δ^- HF HF
ion–dipole force	very strong	mixtures of ionic compounds and polar compounds	

CHEMISTRY AND HEALTH

Hydrogen Bonding in DNA

DNA is a long chainlike molecule that acts as a blueprint for living organisms. Copies of DNA are passed from parent to offspring, which is why we inherit traits from our parents. A DNA molecule is composed of thousands of repeating units called *nucleotides* (▶ FIGURE 12.23). Each nucleotide contains one of four different bases: adenine, thymine, cytosine, and guanine (abbreviated A, T, C, and G, respectively). The order of these bases in DNA encodes the instructions that specify how proteins—the workhorse molecules in living organisms—are made in each cell of the body. Proteins determine virtually all human characteristics,

including how we look, how we fight infections, and even how we behave. Consequently, human DNA is a blueprint for how humans are made.

Each time a human cell divides, it must copy the blueprint—which means replicating its DNA. The replicating mechanism is related to the structure of DNA, discovered in 1953 by James Watson and Francis Crick. DNA consists of two complementary strands wrapped around each other in the now famous double helix. Each strand is held to the other by hydrogen bonds that occur between the bases on each strand. DNA replicates because each base (A, T, C, and G)

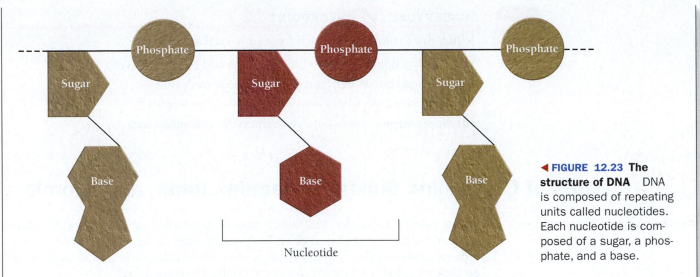

◀ **FIGURE 12.23 The structure of DNA**   DNA is composed of repeating units called nucleotides. Each nucleotide is composed of a sugar, a phosphate, and a base.

Nucleotide

has a complementary partner with which it hydrogen-bonds (▼ **FIGURE 12.24**). Adenine **(a)** hydrogen-bonds with thymine (T), and cytosine (C) hydrogen-bonds with guanine (G). The hydrogen bonds are so specific that each base will pair only with its complementary partner. When a cell is going to divide, the DNA unzips across the hydrogen bonds that run along its length. Then new nucleotides, containing bases complementary to the bases in each half, add along each of

the halves, forming hydrogen bonds with their complement. The result is two identical copies of the original DNA (see Chapter 19).

B12.1 CAN YOU ANSWER THIS?   *Why would dispersion forces not work as a way to hold the two halves of DNA together? Why would covalent bonds not work?*

▶ **FIGURE 12.24 Hydrogen bonding in DNA**   The two halves of the DNA double helix are held together by hydrogen bonds.

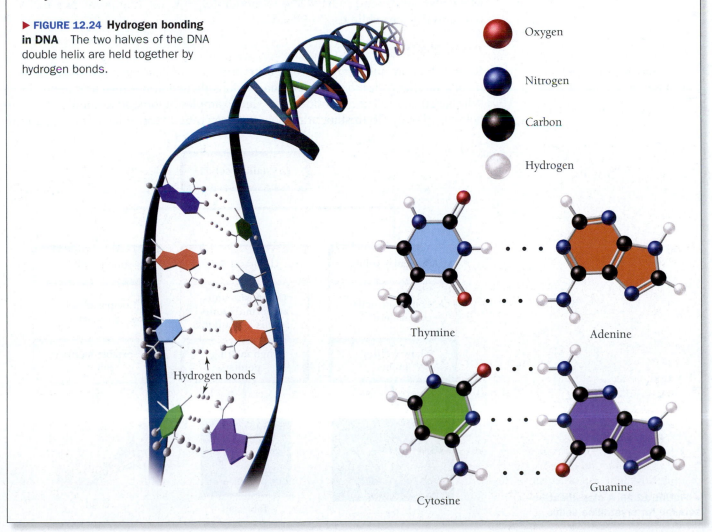

CONCEPTUAL ✓ **CHECKPOINT 12.7**

When dry ice sublimes, which forces are overcome?

(a) chemical bonds between carbon atoms and oxygen atoms

(b) hydrogen bonds between carbon dioxide molecules

(c) dispersion forces between carbon dioxide molecules

(d) dipole–dipole forces between carbon dioxide molecules

12.7 Types of Crystalline Solids: Molecular, Ionic, and Atomic

▶ Identify types of crystalline solids.

As we learned in Section 12.2, solids may be crystalline (showing a well-ordered array of atoms or molecules) or amorphous (having no long-range order). We divide crystalline solids into three categories—molecular, ionic, and atomic—based on the individual units that compose the solid (▼ **FIGURE 12.25**).

Molecular Solids

Molecular solids are solids whose composite units are *molecules*. Ice (solid H_2O) and dry ice (solid CO_2) are examples of molecular solids. Molecular solids are held together by the kinds of intermolecular forces—dispersion forces, dipole–dipole forces, and hydrogen bonding—discussed in Section 12.6. For example, ice is held together by hydrogen bonds, and dry ice is held together by dispersion forces. Molecular solids tend to have low to moderately low melting points; ice melts at 0 °C and dry ice sublimes at −78.5 °C.

Ionic Solids

See Section 5.4 for a complete description of the formula unit.

Ionic solids are solids whose composite units are *formula units*, the smallest electrically neutral collection of cations and anions that compose the compound. Table salt (NaCl) and calcium fluoride (CaF_2) are examples of ionic solids. Ionic solids are held together by electrostatic attractions between cations and anions. For example, in

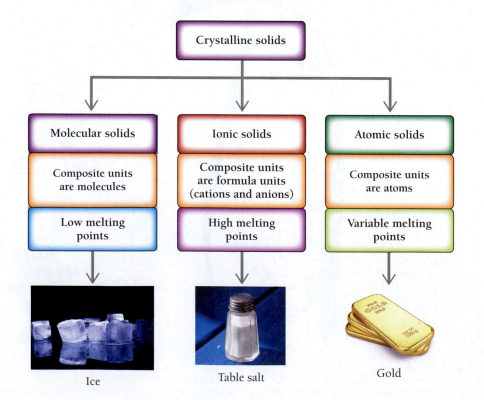

▶ **FIGURE 12.25 A classification scheme for crystalline solids**

NaCl, the attraction between the Na$^+$ cation and the Cl$^-$ anion holds the solid lattice together because the lattice is composed of alternating Na$^+$ cations and Cl$^-$ anions in a three-dimensional array. In other words, the forces that hold ionic solids together are actual ionic bonds. Because ionic bonds are much stronger than any of the intermolecular forces discussed previously, ionic solids tend to have much higher melting points than molecular solids. For example, sodium chloride melts at 801 °C, while carbon disulfide CS$_2$—a molecular solid with a higher molar mass—melts at −110 °C.

Atomic Solids

Atomic solids are solids whose composite units are *individual atoms*. Diamond (C), iron (Fe), and solid xenon (Xe) are examples of atomic solids. We divide atomic solids into three categories—**covalent atomic solids**, **nonbonding atomic solids**, and **metallic atomic solids**—each held together by a different kind of force (▼ FIGURE 12.26).

▶ **FIGURE 12.26 A classification scheme for atomic solids**

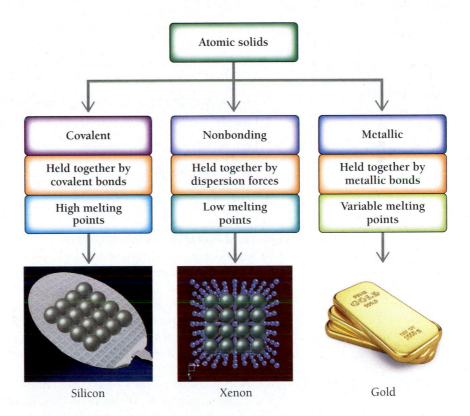

Silicon   Xenon   Gold

Covalent atomic solids, such as diamond, are held together by covalent bonds. In diamond (▼ FIGURE 12.27), each carbon atom forms four covalent bonds to four other carbon atoms in a tetrahedral geometry. This structure extends throughout the entire crystal, so that a diamond crystal can be thought of as a giant molecule held together by these covalent bonds. Since covalent bonds are very strong, covalent atomic solids have high melting points. Diamond is estimated to melt at about 3800 °C.

Nonbonding atomic solids, such as solid xenon, are held together by relatively weak dispersion forces. Xenon atoms have stable electron configurations and therefore do not form covalent bonds with each other. Consequently, solid xenon, like other nonbonding atomic solids, has a very low melting point (about −112 °C).

◀ **FIGURE 12.27 Diamond: a covalent atomic solid**
In diamond, carbon atoms form covalent bonds in a three-dimensional hexagonal pattern.

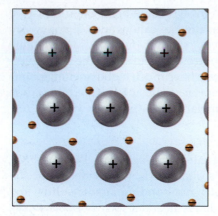

▲ FIGURE 12.28 **Structure of a metallic atomic solid** In the simplest model of a metal, each atom donates one or more electrons to an "electron sea." The metal consists of the metal cations in a negatively charged electron sea.

Metallic atomic solids, such as iron, have variable melting points. Metals are held together by metallic bonds that, in the simplest model, consist of positively charged ions in a sea of electrons (◄ FIGURE 12.29). Metallic bonds are of varying strengths. Some metals, such as mercury, have melting points below room temperature, and other metals, such as iron, have relatively high melting points (iron melts at 1809 °C).

EXAMPLE **12.6**	**Identifying Types of Crystalline Solids**

Identify each solid as molecular, ionic, or atomic.

(a) $CaCl_2(s)$　　**(b)** $Co(s)$　　**(c)** $CS_2(s)$

SOLUTION

(a) $CaCl_2$ is an ionic compound (metal and nonmetal) and therefore forms an ionic solid ($CaCl_2$ melts at 772 °C).

(b) Co is a metal and therefore forms a metallic atomic solid (Co melts at 1768 °C).

(c) CS_2 is a molecular compound (nonmetal bonded to a nonmetal) and therefore forms a molecular solid (CS_2 melts at −110 °C).

▶ **SKILLBUILDER 12.6 | Identifying Types of Crystalline Solids**

Identify each solid as molecular, ionic, or atomic.

(a) $NH_3(s)$　　　**(b)** $CaO(s)$　　　**(c)** $Kr(s)$

▶ **FOR MORE PRACTICE** Problems 79, 80, 81, 82.

12.8 Water: A Remarkable Molecule

▶ Describe the properties that make water unique among molecules.

▲ FIGURE 12.29 **The water molecule**

Water reaches its maximum density at 4.0 °C.

Water is the most common and important liquid on Earth. It fills our oceans, lakes, and streams. In its solid form, it covers nearly an entire continent (Antarctica), as well as large regions around the North Pole, and caps our tallest mountains. In its gaseous form, it humidifies our air. We drink water, we sweat water, and we excrete bodily wastes dissolved in water. Indeed, the majority of our body mass *is* water. Life is impossible without water, and in most places on Earth where liquid water exists, life exists. Evidence of water on Mars—that existed either in the past or exists in the present—has fueled hopes of finding life or evidence of life there.

Among liquids, water is unique. It has a low molar mass (18.02 g/mol), yet is a liquid at room temperature. No other compound of similar molar mass even comes close to being a liquid at room temperature. For example, nitrogen (28.02 g/mol) and carbon dioxide (44.01 g/mol) are both gases at room temperature. Water's relatively high boiling point (for its low molar mass) can be understood by examining the structure of the water molecule (◄ FIGURE 12.29). The bent geometry of the water molecule and the highly polar nature of the O—H bonds result in a molecule with a significant dipole moment. Water's two O—H bonds (hydrogen directly bonded to oxygen) allow water molecules to form strong hydrogen bonds with other water molecules, resulting in a relatively high boiling point. Water's high polarity also allows it to dissolve many other polar and ionic compounds. Consequently, water is the main solvent of living organisms, transporting nutrients and other important compounds throughout the body.

The way water freezes is also unique. Unlike other substances, which contract upon freezing, water expands upon freezing. This seemingly trivial property has significant consequences. Because liquid water expands when it freezes, ice is less dense than liquid water. Consequently, ice cubes and icebergs both float. The frozen layer of ice at the surface of a winter lake insulates the water in the lake from further freezing. If this ice layer were to sink, it would kill bottom-dwelling aquatic life and possibly allow the lake to freeze solid, eliminating virtually all aquatic life in the lake.

The expansion of water upon freezing, however, is one reason that most organisms do not survive freezing. When the water within a cell freezes, it expands and often ruptures the cell, just as water freezing within a pipe bursts the pipe. Many foods, especially those with high water content, do not survive freezing very well either. Have you ever tried, for example, to freeze a vegetable? Try putting lettuce or spinach in the freezer. When you defrost it, it will be limp and damaged. The frozen food industry gets around this problem by *flash-freezing* vegetables and other foods. In this process, foods are frozen instantaneously, preventing water molecules from settling into their preferred crystalline structure. Consequently, the water does not expand very much, and the food remains largely undamaged.

▲ Lettuce does not survive freezing because the expansion of water upon freezing ruptures the cells within the lettuce leaf.

CHEMISTRY IN THE ENVIRONMENT

Water Pollution and the Flint River Water Crisis

Water quality is critical to human health. Many human diseases—especially in developing nations—are caused by poor water quality. Several kinds of pollutants, including biological and chemical contaminants, can enter water supplies. Biological contaminants are microorganisms that cause diseases such as hepatitis, cholera, dysentery, and typhoid. Water containing biological contaminants poses an immediate danger to human health and should not be consumed. Drinking water in developed nations is usually treated to kill microorganisms. Most biological contaminants can be eliminated from untreated water by boiling.

Chemical contaminants get into drinking water from sources such as industrial dumping, pesticide and fertilizer use, and household dumping. These contaminants include organic compounds, such as carbon tetrachloride and dioxin, and inorganic elements and compounds such as mercury, lead, and nitrates. Since many chemical contaminants are neither volatile nor alive like biological contaminants, they are *not* eliminated through boiling.

The Environmental Protection Agency (EPA), under the Safe Drinking Water Act of 1974 and its amendments, sets standards that specify the maximum contamination level (MCL) for nearly 100 biological and chemical contaminants in water. Water providers that serve more than 25 people must periodically test the water they deliver to their consumers for these contaminants. If levels exceed the standards set by the EPA, the water provider must notify the consumer and take appropriate measures to remove the contaminant from the water. According to the EPA, if water comes from a provider that serves more than 25 people, it should be safe to consume over a lifetime. If it is not safe to drink for a short period of time, consumers are notified. However, a failure of this system in 2014 in Flint, Michigan, has raised questions about water regulation.

In April of 2014, in an effort to save money, officials in Flint, Michigan, changed their water source from Lake Huron to the Flint River. In subsequent months, residents began complaining about the quality of the water. Routine monitoring of the tap water in select homes, however, did not reveal any problem because samples were collected only after preflushing taps (allowing the water to run for a time) before collecting samples. A Virginia Tech professor and his students independently tested the water coming from the city's taps and gathered much different data by analyzing the water that initially came from taps (so-called first draw samples). They discovered lead levels that exceeded the EPA maximum contaminant level standard.

The lead in the city's tap water was a direct result of the switch from Lake Huron water to the Flint River water, which was highly corrosive. As the Flint River water flowed through service lines and home pipes, many of which contained lead, the water became contaminated with lead. The contaminated water was ultimately consumed by many residents. The results of the independent study caused a national scandal and resulted in felony charges against several officials involved in the crisis. The municipality has since switched back to the Lake Huron water source, which is much less corrosive. Continued reliable monitoring of the lead levels in the water has shown lead levels that are back below the FDA maximum contaminant level.

B12.2 CAN YOU ANSWER THIS? *Suppose a sample of water is contaminated by a nonvolatile contaminant such as lead. Why doesn't boiling eliminate the contaminant?*

▲ Safe drinking water has a major effect on public health and the spread of disease. In many parts of the world, the water supply is unsafe to drink. In the United States the Environmental Protection Agency (EPA) is charged with maintaining water safety.

Chapter **12** in Review

MasteringChemistry™ provides end-of-chapter exercises, feedback-enriched tutorial problems, animations, and interactive activities to encourage problem solving practice and deeper understanding of key concepts and topics.

Self-Assessment Quiz

Q1. The first diagram shown here represents liquid water. Which of the diagrams that follow best represents the water after it has boiled?

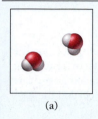  

   (a)   (b)   (c)

Q2. Which change affects the vapor pressure of a liquid?
(a) pouring the liquid into a different container
(b) increasing the surface area of the liquid
(c) increasing the temperature of the liquid
(d) all of the above

Q3. How much heat is required to vaporize 38.5 g of acetone (C_3H_6O) at 25 °C? The heat of vaporization for acetone at this temperature is 31.0 kJ/mol.
(a) 31.0 kJ   (b) 1194 kJ
(c) 0.0214 kJ   (d) 20.5 kJ

Q4. How many 20.0-g ice cubes are required to absorb 47.0 kJ from a glass of water upon melting?
(a) 6   (b) 7   (c) 127   (d) 140

Q5. A sample of ice is heated past its melting point, and its temperature is monitored. The graph shows the results. What is the first point on the graph where the sample contains no solid ice?

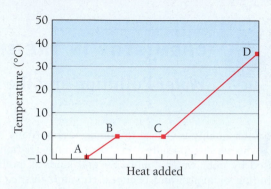

(a) A   (b) B   (c) C   (d) D

Q6. Which halogen has the highest boiling point?
(a) F_2 (b) Cl_2 (c) Br_2 (d) I_2

Q7. Which substance has dipole–dipole forces?
(a) OF_2 (b) CBr_4 (c) CS_2 (d) Br_2

Q8. Which substance exhibits hydrogen bonding?
(a) CF_4 (b) CH_3OH (c) H_2 (d) PH_3

Q9. Which substance is an ionic solid?
(a) Ti(s)   (b) CaO(s)   (c) $CO_2(s)$ (d) $H_2O(s)$

Q10. Which substance would you expect to have the highest melting point?
(a) CH_3CH_3 (b) CH_3CH_2Cl
(c) CH_4 (d) CH_3CH_2OH

Answers: 1:a, 2:c, 3:d, 4:b, 5:c, 6:d, 7:a, 8:b, 9:b, 10:d

Chemical Principles

Properties of Liquids
- High densities in comparison to gases.
- Indefinite shape; they assume the shape of their container.
- Definite volume; they are not easily compressed.

Properties of Solids
- High densities in comparison to gases.
- Definite shape; they do not assume the shape of their container.
- Definite volume; they are not easily compressed.
- May be crystalline (ordered) or amorphous (disordered).

Relevance

Common liquids include water, acetone (fingernail-polish remover), and rubbing alcohol. Water is the most common and most important liquid on Earth. It is difficult to imagine life without water.

Much of the matter we encounter is solid. Common solids include ice, dry ice, and diamond. Understanding the properties of solids involves understanding the particles that compose them and how those particles interact.

Manifestations of Intermolecular Forces

Surface tension—the tendency for liquids to minimize their surface area—is a direct result of intermolecular forces. Viscosity—the resistance of liquids to flow—is another result of intermolecular forces. Both surface tension and viscosity increase with greater intermolecular forces.

Many insects can walk on water due to surface tension. The viscosity of a liquid is one of its defining properties and is important in applications such as automobile lubrication; the viscosity of a motor oil must be high enough to coat an engine's surfaces, but not so high that it can't flow to remote parts of the engine.

Evaporation and Condensation

Evaporation or vaporization—an endothermic physical change—is the conversion of a liquid to a gas. Condensation—an exothermic physical change—is the conversion of a gas to a liquid. When the rates of evaporation and condensation in a liquid/gas sample are equal, dynamic equilibrium is reached and the partial pressure of the gas at that point is its vapor pressure. When the vapor pressure equals the external pressure, the boiling point is reached. At the boiling point, thermal energy causes molecules in the interior of the liquid, as well as those at the surface, to convert to gas, resulting in the bubbling. We can calculate the heat absorbed or emitted during evaporation and condensation (respectively) using the heat of vaporization.

Evaporation is the body's natural cooling system. When we get overheated, we sweat; the sweat then evaporates and cools us. Evaporation and condensation both play roles in moderating climate. Humid areas, for example, cool less at night because as the temperature drops, water condenses out of the air, releasing heat and preventing a further temperature drop.

Melting and Freezing

Melting—an endothermic physical change—is the conversion of a solid to a liquid, and freezing—an exothermic physical change—is the conversion of a liquid to a solid. We can calculate the heat absorbed or emitted during melting and freezing (respectively) using the heat of fusion.

We use the melting of solid ice, for example, to cool drinks when we place ice cubes in them. Since melting is endothermic, it absorbs heat from the liquid and cools it.

Types of Intermolecular Forces

The four main types of intermolecular forces are:

Dispersion forces—Dispersion forces occur between all molecules and atoms due to instantaneous fluctuations in electron charge distribution. The strength of the dispersion force increases with increasing molar mass.

Dipole–dipole forces—Dipole–dipole forces exist between molecules that are polar. Consequently, polar molecules have higher melting and boiling points than nonpolar molecules of similar molar mass.

Hydrogen bonding—Hydrogen bonding exists between molecules that have H bonded directly to F, O, or N. Hydrogen bonds are stronger than dispersion forces or dipole–dipole forces.

Ion–dipole forces—The ion–dipole force occurs in mixtures of ionic compound and polar compounds.

The type of intermolecular force present in a substance determines many of the properties of the substance. The stronger the intermolecular force, for example, the greater the melting and boiling points of the substance. In addition, the miscibility of liquids—their ability to mix without separating—depends on the relative kinds of intermolecular forces present within them. In general, polar liquids are miscible with other polar liquids, but not with nonpolar liquids. Hydrogen bonding is important in many biological molecules such as proteins and DNA. Ion–dipole forces are important in mixtures of ionic compounds and water.

Types of Crystalline Solids

We divide crystalline solids into three categories based on the individual units composing the solid:

Molecular solids—Molecules are the composite units of molecular solids, which are held together by dispersion forces, dipole–dipole forces, or hydrogen bonding.

Ionic solids—Formula units (the smallest electrically neutral collection of cations and anions) are the composite units of ionic solids. They are held together by the electrostatic attractions that occur between cations and anions.

Atomic solids—Atoms are the composite units of atomic solids, which are held together by different forces depending on the particular solid.

Solids have different properties depending on their individual units and the forces that hold those units together. Molecular solids tend to have low melting points. Ionic solids tend to have intermediate to high melting points. Atomic solids have varied melting points, depending on the particular solid.

Water

Water is a unique molecule. Because of its strong hydrogen bonding, water is a liquid at room temperature. Unlike most liquids, water expands when it freezes. In addition, water is highly polar, making it a good solvent for many polar substances.

Water is critical to life. On Earth, wherever there is water, there is life. Water acts as a solvent and transport medium, and virtually all the chemical reactions on which life depends take place in aqueous solution. The expansion of water upon freezing allows life within frozen lakes to survive the winter. The ice on top of the lake acts as insulation, protecting the rest of the lake (and the life within it) from freezing.

Chemical Skills

LO: Use the heat of vaporization in calculations (Section 12.4).

SORT
You are given the mass of water and asked to find the amount of heat required to vaporize it.

STRATEGIZE
To calculate the amount of heat required to vaporize a given amount of a substance, first convert the given amount of the substance to moles and then use the heat of vaporization as a conversion factor to get to kilojoules. For vaporization, the heat is always absorbed. For condensation, follow the same procedure, but the heat is always emitted.

SOLVE
Follow the solution map to solve the problem.

CHECK
Check your answer. Are the units correct? Does the answer make physical sense?

Examples

EXAMPLE **12.7**   **Heat of Vaporization in Calculations**

Calculate the amount of heat required to vaporize 84.8 g of water at its boiling point.

GIVEN: 84.8 g H_2O

FIND: heat (kJ)

SOLUTION MAP

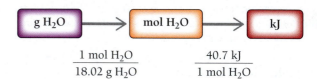

$$\frac{1\ mol\ H_2O}{18.02\ g\ H_2O}\qquad\qquad \frac{40.7\ kJ}{1\ mol\ H_2O}$$

RELATIONSHIPS USED

$\Delta H_{vap} = 40.7$ kJ/mol   at 100 °C   (Table 12.2)

1 mol H_2O = 18.02 g H_2O (molar mass of water)

SOLUTION

$$84.8\ \cancel{g\ H_2O} \times \frac{1\ \cancel{mol\ H_2O}}{18.02\ \cancel{g\ H_2O}} \times \frac{40.7\ kJ}{1\ \cancel{mol\ H_2O}} = 192\ kJ$$

The units (kJ) are correct. The magnitude of the answer makes sense because each mole of water absorbs about 40 kJ of energy upon vaporization. Therefore, 84.8 g (between 4 and 5 mol) should absorb between 160 and 200 kJ upon vaporization.

LO: Use the heat of fusion in calculations (Section 12.5).

EXAMPLE **12.8**   **Using Heat of Fusion in Calculations**

Calculate the amount of heat emitted when 12.4 g of water freezes to solid ice.

GIVEN: 12.4 g H_2O

FIND: heat (kJ)

SORT
You are given the mass of water and asked to find the amount of heat emitted when it freezes.

STRATEGIZE

Use the heat of fusion as a conversion factor between moles of a substance and the amount of heat required to melt it. To calculate the amount of heat required to melt a given amount of a substance, first convert the given amount of the substance to moles and then use the heat of fusion as a conversion factor to get to kilojoules. For melting, the heat is always absorbed. For freezing, follow the same procedure, but the heat is always emitted.

SOLVE

Follow the solution map to solve the problem.

CHECK

Check your answer. Are the units correct? Does the answer make physical sense?

SOLUTION MAP

g H₂O $\longrightarrow$ mol H₂O $\longrightarrow$ kJ

$$\frac{1 \text{ mol H}_2\text{O}}{18.02 \text{ g H}_2\text{O}} \qquad \frac{6.02 \text{ kJ}}{1 \text{ mol H}_2\text{O}}$$

RELATIONSHIPS USED

ΔH_{fus} = 6.02 kJ/mol   (Table 12.3)

1 mol H₂O = 18.02 g H₂O (molar mass of water)

SOLUTION

$$12.4 \text{ g H}_2\text{O} \times \frac{1 \text{ mol H}_2\text{O}}{18.02 \text{ g H}_2\text{O}} \times \frac{6.02 \text{ kJ}}{1 \text{ mol H}_2\text{O}} = 4.14 \text{ kJ}$$

The heat emitted is 4.14 kJ.

The units (kJ) are correct. The magnitude of the answer makes sense because each mole of water emits about 6 kJ of energy upon freezing. Therefore, 12.2 g (less than 1 mol) should emit less than 6 kJ upon freezing.

LO: Determine the types of intermolecular forces in compounds (Section 12.6).

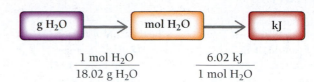

EXAMPLE **12.9** **Determining the Type of Intermolecular Forces in a Compound**

Determine the types of intermolecular forces present in each substance.

(a) N₂
(b) CO
(c) NH₃

All substances exhibit dispersion forces. Polar substances—those whose molecules have polar bonds that add to net dipole moment—also exhibit dipole–dipole forces. Substances whose molecules contain H bonded directly to F, O, or N exhibit hydrogen bonding as well.

SOLUTION

(a) N₂ is nonpolar and therefore has only dispersion forces.
(b) CO is polar and therefore has dipole–dipole forces (in addition to dispersion forces).
(c) NH₃ has hydrogen bonding (in addition to dispersion forces and dipole–dipole forces).

LO: Use intermolecular forces to determine melting and/or boiling points (Section 12.6).

EXAMPLE **12.10** **Using Intermolecular Forces to Determine Melting and/or Boiling Points**

Arrange each group of compounds in order of increasing boiling point.

(a) F₂, Cl₂, Br₂
(b) HF, HCl, HBr

SOLUTION

To determine relative boiling points and melting points among compounds, you must evaluate the types of intermolecular forces that each compound exhibits. Dispersion forces are the weakest kind of intermolecular force, but they increase with increasing molar mass. Dipole–dipole forces are stronger than dispersion forces. If two compounds have similar molar mass, but one is polar, it will have higher melting and boiling points. Hydrogen bonds are a stronger type of intermolecular force. Substances that exhibit hydrogen bonding will have much higher boiling and melting points than substances without hydrogen bonding, even if the substance without hydrogen bonding is of higher molar mass.

(a) Since these all have only dispersion forces, and since they are similar substances (all halogens), the strength of the dispersion force will increase with increasing molar mass. Therefore, the correct order is F₂ < Cl₂ < Br₂.

(b) Since HF has hydrogen bonding, it has the highest boiling point. Between HCl and HBr, HBr (because of its higher molar mass) has a higher boiling point. Therefore, the correct order is HCl < HBr < HF.

Key Terms

atomic solid [12.7]
boiling point [12.4]
condensation [12.4]
covalent atomic solid [12.7]
dipole–dipole force [12.6]
dispersion force [12.6]
dynamic equilibrium [12.4]
evaporation [12.4]
heat of fusion (ΔH_{fus}) [12.5]

heat of vaporization (ΔH_{vap})
    [12.4]
hydrogen bond [12.6]
instantaneous (temporary)
    dipole [12.6]
intermolecular forces [12.1]
ion–dipole force [12.6]
ionic solid [12.7]
melting point [12.5]

metallic atomic solid [12.7]
miscibility [12.6]
molecular solid [12.7]
nonbonding atomic solid
    [12.7]
nonvolatile [12.4]
normal boiling point [12.4]
permanent dipole [12.6]
sublimation [12.5]

surface tension [12.3]
thermal energy [12.1]
vaporization [12.4]
vapor pressure [12.4]
viscosity [12.3]
volatile [12.4]

Exercises

Questions

1. What are intermolecular forces? Why are intermolecular forces important?
2. Why are water droplets spherical?
3. What determines whether a substance is a solid, liquid, or gas?
4. What are the properties of liquids? Explain the properties of liquids in terms of the molecules or atoms that compose them.
5. What are the properties of solids? Explain the properties of solids in terms of the molecules or atoms that compose them.
6. What is the difference between a crystalline solid and an amorphous solid?
7. What is surface tension? How does it depend on intermolecular forces?
8. What is viscosity? How does it depend on intermolecular forces?
9. What is evaporation? Condensation?
10. Why does a glass of water evaporate more slowly in the glass than if you spilled the same amount of water on a table?
11. Explain the difference between evaporation below the boiling point of a liquid and evaporation at the boiling point of a liquid.
12. What is the boiling point of a liquid? What is the normal boiling point?
13. Acetone evaporates more quickly than water at room temperature. What can you say about the relative strength of the intermolecular forces in the two compounds? Which substance is more volatile?
14. Explain condensation and dynamic equilibrium.
15. What is the vapor pressure of a substance? How does it depend on the temperature and strength of intermolecular forces?
16. Explain how sweat cools the body.
17. Explain why a steam burn from gaseous water at 100 °C is worse than a water burn involving the same amount of liquid water at 100 °C.

18. Explain what happens when a liquid boils.
19. Explain why the water in a cup placed in a small ice chest (without a refrigeration mechanism) initially at −5 °C does *not* freeze.
20. Explain how ice cubes cool down beverages.
21. Is the melting of ice endothermic or exothermic? What is the sign of ΔH for the melting of ice? For the freezing of water?
22. Is the boiling of water endothermic or exothermic? What is the sign of ΔH for the boiling of water? For the condensation of steam?
23. What are dispersion forces? How does the strength of dispersion forces relate to molar mass?
24. What are dipole–dipole forces? How can you tell whether a compound has dipole–dipole forces?
25. What is hydrogen bonding? How can you tell whether a compound has hydrogen bonding?
26. What are ion–dipole forces? What kinds of substances contain ion–dipole forces?
27. List the four types of intermolecular forces discussed in this chapter in order of increasing relative strength.
28. What is a molecular solid? What kinds of forces hold molecular solids together?
29. How do the melting points of molecular solids relate to those of other types of solids?
30. What is an ionic solid? What kinds of forces hold ionic solids together?
31. How do the melting points of ionic solids relate to those of other types of solids?
32. What is an atomic solid? What are the properties of atomic solids?
33. In what ways is water unique?
34. How would ice be different if it were denser than water? How would that affect aquatic life in cold-climate lakes?

Problems

EVAPORATION, CONDENSATION, MELTING, AND FREEZING

35. Which evaporates more quickly: 55 mL of water in a beaker with a diameter of 4.5 cm or 55 mL of water in a dish with a diameter of 12 cm? Why?

36. Two samples of pure water of equal volume are put into separate dishes and kept at room temperature for several days. The water in the first dish is completely vaporized after 2.8 days, while the water in the second dish takes 8.3 days to completely evaporate. What can you conclude about the two dishes?

37. One milliliter of water is poured onto one hand, and one milliliter of acetone (fingernail-polish remover) is poured onto the other. As they evaporate, they both feel cool. Which one feels cooler and why? (*Hint:* Which substance is more volatile?)

38. Spilling water over your skin on a hot day will cool you down. Spilling vegetable oil over your skin on a hot day will not. Explain the difference.

39. Several ice cubes are placed in a beaker on a lab bench, and their temperature, initially at −5.0 °C, is monitored. Explain what happens to the temperature as a function of time. Make a sketch of how the temperature might change with time. (Assume that the lab is at 25 °C.)

40. Water is put into a beaker and heated with a Bunsen burner. The temperature of the water, initially at 25 °C, is monitored. Explain what happens to the temperature as a function of time. Make a sketch of how the temperature might change with time. (Assume that the Bunsen burner is hot enough to heat the water to its boiling point.)

41. Which causes a more severe burn: spilling 0.50 g of 100 °C water on your hand or allowing 0.50 g of 100 °C steam to condense on your hand? Why?

42. The nightly winter temperature drop in a seaside town is usually less than that in nearby towns that are farther inland. Explain.

43. When a plastic bag containing a water and ice mixture is placed in an ice chest initially at −8 °C, the temperature of the ice chest goes up. Why?

44. The refrigeration mechanism in a freezer with an automatic ice maker runs extensively each time ice forms from liquid water in the freezer. Why?

45. An ice chest is filled with 3.5 kg of ice at 0 °C. A second ice chest is filled with 3.5 kg of water at 0 °C. After several hours, which ice chest is colder? Why?

46. Why does 50 g of water initially at 0 °C warm more quickly than 50 g of an ice/water mixture initially at 0 °C?

47. In Denver, Colorado, water boils at 95 °C. Explain.

48. At the top of Mount Everest, water boils at 70 °C. Explain.

HEAT OF VAPORIZATION AND HEAT OF FUSION

49. How much heat is required to vaporize 33.8 g of water at 100 °C?

50. How much heat is required to vaporize 43.9 g of acetone at its boiling point?

51. How much heat does your body lose when 2.8 g of sweat evaporates from your skin at 25 °C? (Assume that the sweat is only water.)

52. How much heat does your body lose when 4.86 g of sweat evaporates from your skin at 25 °C? (Assume that the sweat is only water.)

53. How much heat is emitted when 4.25 g of water condenses at 25 °C?

54. How much heat is emitted when 65.6 g of isopropyl alcohol condenses at 25 °C?

55. The human body obtains 835 kJ of energy from a chocolate chip cookie. If this energy were used to vaporize water at 100 °C, how many grams of water could be vaporized? Assume that the density of water is 1.0 g/mL. (*Hint:* Begin by using the enthalpy of vaporization of water to convert between the given number of kilojoules and moles of water.)

56. The human body obtains 1078 kJ from a candy bar. If this energy were used to vaporize water at 100 °C, how much water in liters could be vaporized? Assume that the density of water is 1.0 g/mL. (*Hint:* Begin by using the enthalpy of vaporization of water to convert between the given number of kilojoules and moles of water.)

57. How much heat is required to melt 37.4 g of ice at 0 °C?

58. How much heat is required to melt 23.9 g of solid diethyl ether (at its melting point)?

59. How much energy is released when 34.2 g of water freezes?

60. How much energy is released when 2.55 kg of diethyl ether freezes?

61. How much heat is required to convert 2.55 g of water at 28.0 °C to steam at 100.0 °C?

62. How much heat is required to convert 5.88 g of ice at −12.0 °C to water at 25.0 °C? (The heat capacity of ice is 2.09 J/g °C.)

INTERMOLECULAR FORCES

63. What kinds of intermolecular forces are present in each substance?
  (a) Kr (b) N_2 (c) CO (d) HF

64. What kinds of intermolecular forces are present in each substance?
  (a) HCl (b) H_2O (c) Br_2 (d) He

65. What kinds of intermolecular forces are present in each substance?
  (a) NCl_3 (trigonal pyramidal)
  (b) NH_3 (trigonal pyramidal)
  (c) SiH_4 (tetrahedral)
  (d) CCl_4 (tetrahedral)

66. What kinds of intermolecular forces are present in each substance?
  (a) O_3
  (b) HBr
  (c) CH_3OH
  (d) I_2

67. What kinds of intermolecular forces are present in a mixture of potassium chloride and water?

68. What kinds of intermolecular forces are present in a mixture of calcium bromide and water?

69. Which substance has the highest boiling point? Why? *Hint:* They are all nonpolar.
  (a) CH_4 (b) CH_3CH_3
  (c) $CH_3CH_2CH_3$ (d) $CH_3CH_2CH_2CH_3$

70. Which noble gas has the highest boiling point? Why?
  (a) Kr (b) Xe (c) Rn

71. One of these two substances is a liquid at room temperature and the other one is a gas. Which one is the liquid and why?
  CH_3OH CH_3SH

72. One of these two substances is a liquid at room temperature and the other one is a gas. Which one is the liquid and why?
  CH_3OCH_3 CH_3CH_2OH

73. A flask containing a mixture of $NH_3(g)$ and $CH_4(g)$ is cooled. At $-33.3\,°C$ a liquid begins to form in the flask. What is the liquid?

74. Explain why CS_2 is a liquid at room temperature while CO_2 is a gas.

75. Are $CH_3CH_2CH_2CH_2CH_3$ and H_2O miscible?

76. Are CH_3OH and H_2O miscible?

77. Determine whether a homogeneous solution forms when each pair of substances is mixed.
  (a) CCl_4 and H_2O
  (b) Br_2 and CCl_4
  (c) CH_3CH_2OH and H_2O

78. Determine whether a homogeneous solution forms when each pair of substances is mixed.
  (a) $CH_3CH_2CH_2CH_2CH_3$ and $CH_3CH_2CH_2CH_2CH_2CH_3$
  (b) CBr_4 and H_2O
  (c) Cl_2 and H_2O

TYPES OF SOLIDS

79. Identify each solid as molecular, ionic, or atomic.
  (a) Ar(s) (b) $H_2O(s)$ (c) $K_2O(s)$ (d) Fe(s)

80. Identify each solid as molecular, ionic, or atomic.
  (a) $CaCl_2(s)$ (b) $CO_2(s)$ (c) Ni(s) (d) $I_2(s)$

81. Identify each solid as molecular, ionic, or atomic.
  (a) $H_2S(s)$ (b) KCl(s) (c) $N_2(s)$ (d) $NI_3(s)$

82. Identify each solid as molecular, ionic, or atomic.
  (a) $SF_6(s)$ (b) C(s) (c) $MgCl_2(s)$ (d) Ti(s)

83. Which solid has the highest melting point? Why?
  (a) Ar(s) (b) $CCl_4(s)$
  (c) LiCl(s) (d) $CH_3OH(s)$

84. Which solid has the highest melting point? Why?
  (a) C (s, diamond) (b) Kr(s)
  (c) NaCl(s) (d) $H_2O(s)$

85. For each pair of solids, determine which solid has the higher melting point and explain why.
  (a) Ti(s) and Ne(s) (b) $H_2O(s)$ and $H_2S(s)$
  (c) Kr(s) and Xe(s) (d) NaCl(s) and $CH_4(s)$

86. For each pair of solids, determine which solid has the higher melting point and explain why.
  (a) Fe(s) and $CCl_4(s)$ (b) KCl(s) or HCl(s)
  (c) $TiO_2(s)$ or HOOH(s)

87. List these substances in order of increasing boiling point:
  H_2O, Ne, NH_3, NaF, SO_2

88. List these substances in order of decreasing boiling point:
  CO_2, Ne, CH_3OH, KF

Cumulative Problems

89. Ice actually has negative caloric content. How much energy, in each of the following units, does your body lose from eating (and therefore melting) 78 g of ice?
(a) joules
(b) kilojoules
(c) calories (1 cal = 4.18 J)
(d) nutritional Calories or capital "C" Calories (1000 cal = 1 Cal)

90. Ice has negative caloric content. How much energy, in each of the following units, does your body lose from eating (and therefore melting) 145 g of ice?
(a) joules
(b) kilojoules
(c) calories (1 cal = 4.18 J)
(d) nutritional Calories or capital "C" calories (1000 cal = 1 Cal)

91. An 8.5-g ice cube is placed into 255 g of water. Calculate the temperature change in the water upon the complete melting of the ice. *Hint:* Determine how much heat is absorbed by the melting ice and then use $q = mC\Delta T$ to calculate the temperature change of the 255 g of water.

92. A 14.7-g ice cube is placed into 324 g of water. Calculate the temperature change in the water upon complete melting of the ice. *Hint:* Determine how much heat is absorbed by the melting ice and then use $q = mC\Delta T$ to calculate the temperature change of the 324 g of water.

93. How much ice in grams would have to melt to lower the temperature of 352 mL of water from 25 °C to 0 °C? (Assume that the density of water is 1.0 g/mL.)

94. How much ice in grams would have to melt to lower the temperature of 55.8 g of water from 55.0 °C to 0 °C? (Assume that the density of water is 1.0 g/mL.)

95. How much heat in kilojoules is evolved in converting 1.00 mol of steam at 145 °C to ice at −50.0 °C? The specific heat capacity of steam is 1.84 J/g °C and that of ice is 2.09 J/g °C.

96. How much heat in kilojoules is required to warm 10.0 g of ice, initially at −10.0 °C, to steam at 110.0 °C? The specific heat capacity of ice is 2.09 J/g °C and that of steam is 1.84 J/g °C.

97. Draw a Lewis structure for each molecule and determine its molecular geometry. What kind of intermolecular forces are present in each substance?
(a) H_2Se
(b) SO_2
(c) $CHCl_3$
(d) CO_2

98. Draw a Lewis structure for each molecule and determine its molecular geometry. What kind of intermolecular forces are present in each substance?
(a) BCl_3 (remember that B is a frequent exception to the octet rule)
(b) HCOH (carbon is central; each H and O bonded directly to C)
(c) CS_2
(d) NCl_3

99. The melting point of ionic solids depends on the magnitude of the electrostatic attractions that hold the solid together. Draw ionic Lewis structures for NaF and MgO. Which do you think has the higher melting point?

100. Draw ionic Lewis structures for KF and CaO. Use the information and the method in the previous problem to predict which of these two ionic solids has the higher melting point.

101. Explain the observed trend in the melting points of the alkyl halides. Why is HF atypical?

Compound	Melting Point
HI	−50.8 °C
HBr	−88.5 °C
HCl	−114.8 °C
HF	−83.1 °C

102. Explain the observed trend in the boiling points of the compounds listed. Why is H_2O atypical?

Compound	Boiling Point
H_2Te	−2 °C
H_2Se	−41.5 °C
H_2S	−60.7 °C
H_2O	+100 °C

103. An ice cube at 0.00 °C with a mass of 23.5 g is placed into 550.0 g of water, initially at 28.0 °C, in an insulated container. Assuming that no heat is lost to the surroundings, what is the temperature of the entire water sample after all of the ice has melted?

104. If 1.10 g of steam at 100.0 °C condenses into 38.5 g of water, initially at 27.0 °C, in an insulated container, what is the final temperature of the entire water sample? Assume no loss of heat into the surroundings.

Highlight Problems

105. Consider the molecular view of water shown here. Pick a molecule in the interior and draw a line to each of its direct neighbors. Pick a molecule near the edge (analogous to a molecule on the surface in three dimensions) and do the same. Which molecule has the most neighbors? Which molecule is more likely to evaporate?

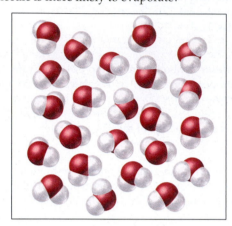

106. Water does not easily remove grease from dirty hands because grease is nonpolar and water is polar; therefore they are immiscible. The addition of soap, however, results in the removal of the grease. Examine the structure of soap shown here and explain how soap works.

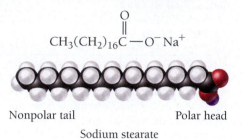

Nonpolar tail   Polar head

Sodium stearate
a soap

107. One prediction of global warming is the melting of global ice, which may result in coastal flooding. A criticism of this prediction is that the melting of icebergs does not increase ocean levels any more than the melting of ice in a glass of water increases the level of liquid in the glass.
  (a) Is this a valid criticism? Does the melting of an ice cube in a cup of water raise the level of the liquid in the cup? Why or why not?
  A response to this criticism is that scientists are not worried about rising ocean levels due to melting icebergs; rather, scientists are worried about rising ocean levels due to melting ice sheets that sit on the continent of Antarctica.
  (b) Would the melting of the ice sheets increase ocean levels? Why or why not?

108. Explain why rubbing alcohol feels cold when applied to the skin.

Questions for Group Work

Discuss these questions with the group and record your consensus answer.

109. Consider the following compounds.

Acetone   Butane   1–Propanol

  (a) Calculate the molar mass of each compound.
  (b) Redraw each Lewis structure and indicate polar bonds with $\delta+$ and $\delta-$.
  (c) Indicate which of these molecules would be considered polar, and why.
  (d) Indicate which of these molecules can hydrogen-bond, and why.
  (e) Identify the predominant intermolecular force in each molecule.
  (f) Rank the molecules in order of increasing boiling point.
  (g) Why is your answer to part a important in answering part f?

110. Look up the boiling points of carbon monoxide and carbon dioxide. Which intermolecular force would you cite to account for the difference? Explain.

111. The existence of "triads" of elements was a big clue that eventually led to the discovery of the periodic table. A triad is a group of three elements for which the middle element has properties that are the average of the properties of the first and third elements. Some triads include: (O, S, Se); (Cl, Br, I); and (Li, Na, K). Verify that the atomic weight of the middle element in each triad is approximately the average of the other two. Explain why the melting points of the compounds formed by combining each element with hydrogen (e.g., HCl, HBr, and HI) might be expected to show the same trend.

112. How much sweat (in mL) would you have to evaporate per hour to remove the same amount of heat a 100 W light bulb produces? (1 W = 1 J/s.)

Data Interpretation and Analysis

113. The boiling points and melting points of several hydrocarbons (compounds composed of carbon and hydrogen) are tabulated here. Examine the data and answer the questions that follow.

Name	Formula	Boiling point	Melting point
methane	CH_4	$-161.5\ °C$	$-182\ °C$
ethane	C_2H_6	$-88.6\ °C$	$-183\ °C$
propane	C_3H_8	$-42.1\ °C$	$-188\ °C$
butane	C_4H_{10}	$-0.5\ °C$	$-140\ °C$
pentane	C_5H_{12}	$36.0\ °C$	$-130\ °C$
hexane	C_6H_{14}	$68.7\ °C$	$-95\ °C$

(a) Which of the hydrocarbons in the table are liquids at room temperature?

(b) Describe the trends in the boiling points and melting points of these hydrocarbons. Which of the two trends (the trend in boiling points or the trend in melting points) is more regular?

(c) Which types of intermolecular forces are present in these hydrocarbons? Explain the trend in boiling points based on their intermolecular forces.

(d) Prepare a graph of boiling point versus molar mass for these hydrocarbons.

(e) Use the graph from part d to predict the boiling point of heptane (C_7H_{16}).

Answers to Skillbuilder Exercises

Skillbuilder 12.1 .... $5.83 \times 10^3\ kJ$
Skillbuilder Plus, p. 418... $47\ °C$
Skillbuilder 12.2 .... $5.18\ kJ$
Skillbuilder Plus, p. 421... $-2.3\ °C$
Skillbuilder 12.3 .... C_2H_6
Skillbuilder 12.4 .... **(a)** no dipole–dipole forces
(b) yes, it has dipole–dipole forces
(c) yes, it has dipole–dipole forces

Skillbuilder 12.5 .... HF, because it has hydrogen bonding as an intermolecular force
Skillbuilder 12.6 .... **(a)** molecular
(b) ionic
(c) atomic

Answers to Conceptual Checkpoints

12.1 (a) The substance has definite volume *and* a definite shape, so it must be a solid.

12.2 (b) Because boiling is a physical rather than a chemical change, the water molecules (H_2O) undergo no chemical alteration—they merely change from the liquid to the gaseous state.

12.3 The temperature of the metal block increases. Condensation is exothermic, so it gives off heat to the metal block and raises its temperature.

12.4 (b) The temperature remains flat during a state transition; therefore, since the water starts as ice, the first flat section (i) is the melting of the ice to liquid water. The increasing temperature section (ii) is the water warming to its boiling point. The subsequent flat section (iii) is the boiling of the liquid water to gaseous steam.

12.5 (c) Since sublimation is a physical change, the carbon dioxide molecules do not decompose into other molecules or atoms; they simply change state from the solid to the gaseous state.

12.6 (a) All three compounds have nearly identical molar masses, so the strength of the dispersion forces is similar in all three. A is nonpolar; it has only dispersion forces and has the lowest boiling point. B is polar; it has dipole–dipole forces in addition and therefore has the next highest boiling point. C has hydrogen bonding in addition; it therefore has the highest boiling point.

12.7 (c) The chemical bonds between carbon and oxygen atoms are not broken by changes of state such as sublimation. Because carbon dioxide contains no hydrogen atoms, it cannot undergo hydrogen bonding, and because the molecule is nonpolar, it does not experience dipole–dipole interactions.

13 Solutions

The goal of science is to make sense of the diversity of nature.

—John Barrow (b. 1952)

13.1 Tragedy in Cameroon

▲ Cameroon is in West Africa.

On August 22, 1986, most people living near Lake Nyos in Cameroon, West Africa, began their day in an ordinary way. Unfortunately, the day ended in tragedy. On that evening, a large cloud of carbon dioxide gas, burped up from the depths of Lake Nyos, killed more than 1700 people and about 3000 head of cattle. Survivors tell of smelling rotten eggs, feeling a warm sensation, and then losing consciousness. Two years before that, a similar tragedy occurred in Lake Monoun, just 60 miles away, killing 37 people. In the wake of these events, scientists have taken steps to prevent these lakes from burping again.

Lake Nyos is a water-filled volcanic crater. Some 50 miles beneath the surface of the lake, molten volcanic rock (magma) produces carbon dioxide gas that seeps into the lake through the volcano's plumbing system. The carbon dioxide then mixes with the lake water. However, as we will see later in this chapter, the concentration of a gas (such as carbon dioxide) that can build up in water increases with increasing pressure. The great pressure at the bottom of the deep lake allows the concentration of carbon dioxide to become very high (just as the pressure in a soda can allows the concentration of carbon dioxide in an unopened soda to be very high). Over time, the carbon dioxide and water mixture at the bottom of the lake became so concentrated that—either because of the high concentration itself or because of some other natural trigger, such as a landslide—some gaseous carbon dioxide escaped. The rising bubbles disrupted the lake water, causing the highly concentrated carbon dioxide and water mixture at the bottom of the lake to rise, which lowered the pressure on the mixture. The drop in pressure on the mixture released more carbon dioxide bubbles just as the

◀ Late in the summer of 1986, carbon dioxide bubbled out of Lake Nyos and flowed into the adjacent valley. The carbon dioxide came from the bottom of the lake where it was held in a mixture with water by the pressure above it. When the layers in the lake were disturbed, the carbon dioxide came out of the water due to the decrease in pressure—with lethal consequences.

Carbon dioxide, a colorless and odorless gas, displaced the air in low-lying regions surrounding Lake Nyos, leaving no oxygen for the inhabitants to breathe. The rotten-egg smell was an indication of the presence of additional sulfur-containing gases.

drop in pressure upon opening a soda can releases carbon dioxide bubbles. This in turn caused further churning and more carbon dioxide release. Since carbon dioxide is more dense than air, once freed from the lake, it traveled down the sides of the volcano and into the nearby valley, displacing air and asphyxiating many of the local residents.

In efforts to prevent these events from occurring again—by 2001, carbon dioxide concentrations had already returned to dangerously high levels—scientists built a piping system to slowly vent carbon dioxide from the lake bottom. Since 2001, this system has gradually been releasing the carbon dioxide into the atmosphere, preventing a repeat of the tragedy.

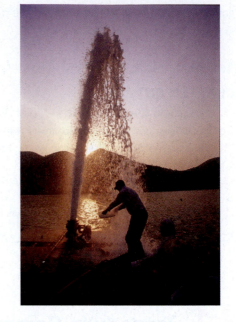

▶ Engineers watch as the carbon dioxide vented from the bottom of Lake Nyos creates a geyser. The controlled release of carbon dioxide from the lake bed is designed to prevent future catastrophes like the one that killed more than 1700 people in 1986.

13.2 Solutions: Homogeneous Mixtures

▶ Define solution, solute, and solvent.

The carbon dioxide and water mixture at the bottom of Lake Nyos is a **solution**, a homogeneous mixture of two or more substances. Solutions are common—most of the liquids and gases that we encounter every day are actually solutions. When most people think of a solution, they think of a solid dissolved in water. The ocean, for example, is a solution of salt and other solids dissolved in water. Blood plasma (blood that has had blood cells removed from it) is a solution of several solids (as well as some gases) dissolved in water. In addition to these solutions, many other kinds exist. A solution may be composed of a gas and a liquid (like the carbon dioxide and water of Lake Nyos), a liquid and another liquid, a solid and a gas, or other combinations (see Table 13.1).

The most common solutions are those containing a solid, a liquid, or a gas and water. These are *aqueous solutions*—they are critical to life and are the main focus of this chapter. Common examples of aqueous solutions include sugar water and salt-water, both solutions of solids and water. Similarly, ethyl alcohol—the alcohol in alcoholic beverages—readily mixes with water to form a solution of a liquid with water, and we have already discussed an example of a gas-and-water solution in Lake Nyos.

Aqueous comes from the Latin word *aqua*, which means "water."

TABLE 13.1 Common Types of Solutions			
Solution Phase	**Solute Phase**	**Solvent Phase**	**Example**
gaseous solutions	gas	gas	air (mainly oxygen and nitrogen)
liquid solutions	gas	liquid	soda water (CO_2 and water)
	liquid	liquid	vodka (ethanol and water)
	solid	liquid	seawater (salt and water)
solid solutions	solid	solid	brass (copper and zinc) and other alloys

In a solid/liquid solution, the liquid is usually considered the solvent, regardless of the relative proportions of the components.

See Sections 10.8 and 12.6 to review the concept of polarity.

A solution has at least two components. The majority component is usually called the **solvent**, and the minority component is called the **solute**. In a carbon-dioxide-and-water solution, carbon dioxide is the solute and water is the solvent. In a salt-and-water solution, salt is the solute and water is the solvent. Because water is so abundant on Earth, it is a common solvent. However, other solvents are often used in the laboratory, in industry, and even in the home, especially to form solutions with nonpolar solutes. For example, you may use paint thinner, a nonpolar solvent, to remove grease from a dirty bicycle chain or from ball bearings. The paint thinner dissolves (that is, it forms a solution with) the grease, removing it from the metal.

In general, polar solvents dissolve polar or ionic solutes, and nonpolar solvents dissolve nonpolar solutes. We describe this tendency with the rule *like dissolves like*. This statement means that similar kinds of solvents dissolve similar kinds of solutes. Table 13.2 lists some common polar and nonpolar laboratory solvents.

TABLE 13.2 Common Laboratory Solvents	
Common Polar Solvents	**Common Nonpolar Solvents**
water (H_2O)	hexane (C_6H_{14})
acetone (CH_3COCH_3)	diethyl ether ($CH_3CH_2OCH_2CH_3$)
methyl alcohol (CH_3OH)	toluene (C_7H_8)

PEARSON
eText
2.0

CONCEPTUAL ✔ **CHECKPOINT 13.1**

Which compound would you expect to be *least* soluble in water?
(a) CCl_4 (b) CH_3Cl (c) NH_3 (d) KF

13.3 Solutions of Solids Dissolved in Water: How to Make Rock Candy

▶ Relate the solubility of solids in water to temperature.

We have already discussed several examples of solutions of a solid dissolved in water. The ocean, for example, is a solution of salt and other solids dissolved in water. A sweetened cup of coffee is a solution of sugar and other solids dissolved in water. Blood plasma is a solution of several solids (and some gases) dissolved in water. Not all solids, however, dissolve in water. We already know that nonpolar solids—such as lard and shortening—do not dissolve in water. Solids such as calcium carbonate and sand do not dissolve either.

When a solid is put into water, there is competition between the attractive forces that hold the solid together (the *solute–solute* interactions) and the attractive forces occurring between the water molecules and the particles that compose the solid (the *solvent–solute* interactions). The solvent–solute interactions are usually intermolecular forces of the type discussed in Chapter 12. For example, when sodium chloride is put into water, there is competition between the mutual attraction of Na^+ cations and Cl^- anions and the ion–dipole forces between Na^+ or Cl^- and water molecules as shown in the margin. For sodium ions, the attraction is between the positive charge of the sodium ion and the negative side of water's dipole moment as shown in ▶ **FIGURE 13.1**, on the next page (see Section 10.8 to review dipole moment). For chloride ions, the attraction is between the negative charge of the chloride ion and the positive side of water's dipole moment. In the

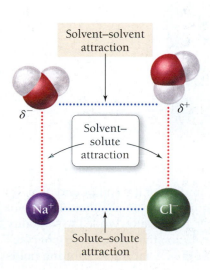

◀ When NaCl is put into water, the attraction between water molecules and Na^+ and Cl^- ions (solvent–solute attraction) overcomes the attraction between Na^+ and Cl^- ions (solute–solute attraction).

▶ **FIGURE 13.1 How an ionic solid dissolves in water**

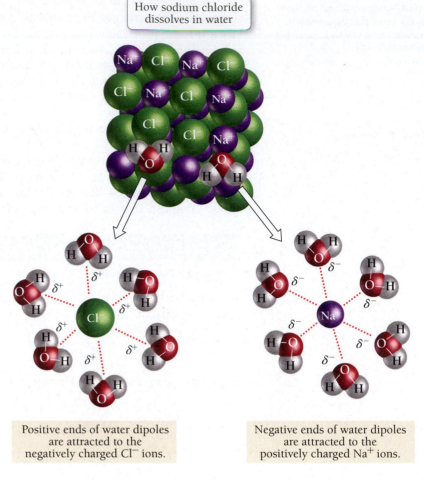

How sodium chloride dissolves in water

Positive ends of water dipoles are attracted to the negatively charged Cl⁻ ions.

Negative ends of water dipoles are attracted to the positively charged Na⁺ ions.

case of NaCl, the attraction to water wins, and sodium chloride dissolves (▼ **FIGURE 13.2**). In contrast, in the case of calcium carbonate ($CaCO_3$), the attraction between Ca^{2+} ions and CO_3^{2-} ions wins and calcium carbonate does not dissolve in water.

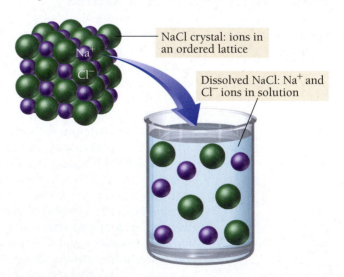

NaCl crystal: ions in an ordered lattice

Dissolved NaCl: Na⁺ and Cl⁻ ions in solution

▶ **FIGURE 13.2 A sodium chloride solution** In a solution of NaCl, the Na⁺ and Cl⁻ ions are dispersed in the water.

Solubility and Saturation

The **solubility** of a compound is defined as the amount of the compound, usually in grams, that dissolves in a certain amount of liquid. For example, the solubility of sodium chloride in water at 25 °C is 36 g NaCl per 100 g water, while the solubility of calcium carbonate in water is close to zero. A solution that contains 36 g of NaCl per 100 g water is a *saturated* sodium chloride solution. A **saturated solution** holds the maximum amount of solute under the solution conditions. If additional solute is added to a saturated solution, it will not dissolve. An **unsaturated solution** is holding less than the maximum amount of solute. If additional solute is added to an unsaturated solution, it will dissolve. A **supersaturated solution** holds more than the normal maximum amount of solute. The solute will normally *precipitate* from (come out of) a supersaturated solution. As the carbon dioxide and water solution rose from the bottom of Lake Nyos, for example, the solution became supersaturated because of the drop in pressure. The excess gas came out of the solution and rose to the surface of the lake, where the gas was emitted into the surrounding air.

> Supersaturated solutions can form under special circumstances, such as the sudden release in pressure that occurs when a soda is opened.

(a)   (b)   (c)

▶ A supersaturated solution holds more than the normal maximum amount of solute. In some cases, such as the sodium acetate solution pictured here, a supersaturated solution may be temporarily stable. Any disturbance, however, such as dropping in a small piece of solid sodium acetate (a), causes the solid to come out of solution (b, c).

Recall the solubility rules from Chapter 7 (Section 7.5, Table 7.2), which provide a qualitative description of the solubility of ionic solids. Molecular solids may also be soluble in water depending on whether the solid is polar. Table sugar ($C_{12}H_{22}O_{11}$), for example, is polar and soluble in water. Nonpolar solids, such as lard and vegetable shortening, are usually insoluble in water.

Electrolyte Solutions: Dissolved Ionic Solids

A sugar solution (containing a molecular solid) and a salt solution (containing an ionic solid) are very different, as shown in ▼ FIGURE 13.3. In a salt solution the dissolved particles are ions, while in a sugar solution the dissolved particles are molecules. The ions in the salt solution are mobile charged particles and can therefore conduct electricity. As described in Section 7.5, a solution containing a solute that dissociates into ions is an **electrolyte solution**. The sugar solution contains dissolved sugar molecules and cannot conduct electricity; it is a **nonelectrolyte solution**. In general, soluble ionic solids form electrolyte solutions, while soluble molecular solids form nonelectrolyte solutions.

> NaCl forms a strong electrolyte solution (Section 7.5). Both strong and weak electrolyte solutions are covered in Chapter 14.

Dissolved ions (NaCl)   Dissolved molecules (sugar)

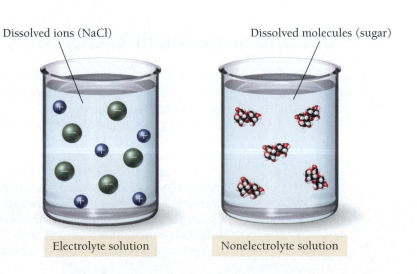

Electrolyte solution   Nonelectrolyte solution

▶ **FIGURE 13.3 Electrolyte and nonelectrolyte solutions**
Electrolyte solutions contain dissolved ions (charged particles) and therefore conduct electricity. Nonelectrolyte solutions contain dissolved molecules (neutral particles) and therefore do not conduct electricity.

▶ **FIGURE 13.4 Solubility of some ionic solids as a function of temperature**

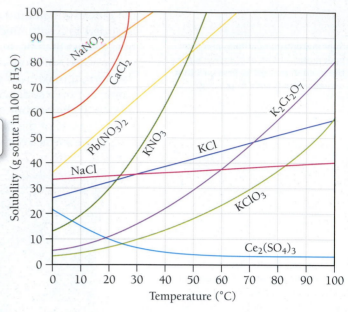

The solubility of *solids* in water generally increases with increasing temperature.

▲ Rock candy is composed of sugar crystals that form through recrystallization.

How Solubility Varies with Temperature

Have you ever noticed how much easier it is to dissolve sugar in hot tea than in cold tea? The solubility of solids in water can be highly dependent on temperature. In general, the solubility of *solids* in water increases with increasing temperature (▲ **FIGURE 13.4**). For example, the solubility of potassium nitrate (KNO_3) at 20 °C is about 30 g KNO_3 per 100 g of water. However, at 50 °C, its solubility rises to 88 g KNO_3 per 100 g of water.

A common way to purify a solid is **recrystallization**. Recrystallization involves putting the solid into heated water or some other solvent with an elevated temperature. Enough solid is added to the solvent to create a saturated solution at the elevated temperature. As the solution cools, the solubility decreases, causing some of the solid to precipitate from solution. If the solution cools slowly, the solid will form crystals as it comes out. The crystalline structure tends to reject impurities, resulting in a purer solid.

You can use recrystallization to make rock candy. To make rock candy, prepare a saturated sucrose (table sugar) solution at an elevated temperature. Dangle a string in the solution, and let it to cool and stand for several days. As the solution cools, it becomes supersaturated and sugar crystals grow on the string. After several days, beautiful and sweet crystals, or "rocks," of sugar cover the string, ready to be admired and eaten.

13.4 Solutions of Gases in Water: How Soda Pop Gets Its Fizz

▶ Relate the solubility of gases in liquids to temperature and pressure.

The water at the bottom of Lake Nyos and a can of soda pop are both examples of solutions in which a gas (carbon dioxide) is dissolved in a liquid (water). Most liquids exposed to air contain some dissolved gases. Lake water and seawater, for example, contain dissolved oxygen necessary for the survival of fish. Our blood contains dissolved nitrogen, oxygen, and carbon dioxide. Even tap water contains dissolved atmospheric gases.

We can see the dissolved gases in ordinary tap water when we heat it on a stove. Before the water reaches its boiling point, small bubbles develop in the water. These bubbles are dissolved air (mostly nitrogen and oxygen) coming out of solution. Once the water boils, the bubbling becomes more vigorous—these larger bubbles are composed of water vapor. The dissolved air comes out of solution upon heating because—unlike solids, whose solubility *increases* with increasing

Cold soda pop: carbon dioxide more likely to stay in solution

Warm soda pop: carbon dioxide more likely to bubble out of solution

▲ Warm soda pop fizzes more than cold soda pop because the solubility of the dissolved carbon dioxide decreases with increasing temperature.

temperature—the solubility of gases in water *decreases* with increasing temperature. As the temperature of the water rises, the solubility of the dissolved nitrogen and oxygen decreases and these gases come out of solution, forming small bubbles around the bottom of the pot.

The decrease in the solubility of gases with increasing temperature is the reason that warm soda pop bubbles more than cold soda pop and also the reason that warm soda goes flat faster than cold soda. The carbon dioxide comes out of solution faster (bubbles more) at room temperature than at cooler temperature because the gas is less soluble at room temperature.

The solubility of gases also depends on pressure. The higher the pressure above a liquid, the more soluble the gas is in the liquid (▼ **FIGURE 13.5**), a relationship known as **Henry's law**.

The solubility of a gas in a liquid increases with increasing pressure.

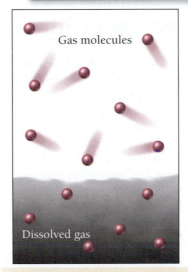
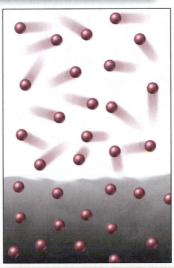

Gas molecules

Dissolved gas

Gas at low pressure over a liquid

Gas at high pressure over a liquid

▶ **FIGURE 13.5 Pressure and solubility**   The higher the pressure above a liquid, the more soluble the gas is in the liquid.

In a can of soda pop and at the bottom of Lake Nyos, high pressure maintains the carbon dioxide in solution. In soda pop, the pressure is provided by a large amount of carbon dioxide gas that is pumped into the can before sealing it. When we open the can, we release the pressure and the solubility of carbon dioxide decreases, resulting in bubbling (▼ **FIGURE 13.6**). The bubbles are formed by the carbon

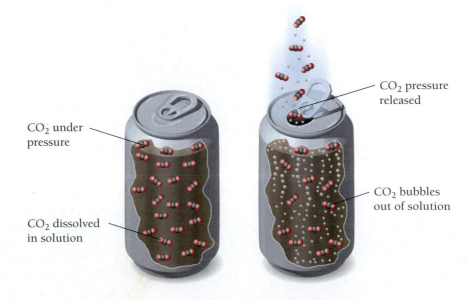

CO_2 pressure released

CO_2 under pressure

CO_2 dissolved in solution

CO_2 bubbles out of solution

▶ **FIGURE 13.6 Pop! Fizz!**   A can of soda pop is pressurized with carbon dioxide. When the can is opened, the pressure is released, lowering the solubility of carbon dioxide in the solution and causing it to come out of solution as bubbles.

dioxide gas as it escapes. In Lake Nyos, the mass of the lake water provides the pressure by pushing down on the carbon-dioxide–rich water at the bottom of the lake. When the stratification (or layering) of the lake is disturbed, the pressure on the carbon dioxide solution decreases and the solubility of carbon dioxide decreases, resulting in the release of excess carbon dioxide gas.

CONCEPTUAL ✔ CHECKPOINT 13.2

A solution is saturated in both nitrogen gas (N_2) and potassium chloride (KCl) at 75 °C. What happens when the solution is cooled to room temperature?

(a) Some nitrogen gas bubbles out of solution.

(b) Some potassium chloride precipitates out of solution.

(c) Both (a) and (b) happen.

(d) Nothing happens.

13.5 Specifying Solution Concentration: Mass Percent

▶ Calculate mass percent.
▶ Use mass percent in calculations.

As we have seen, the amount of solute in a solution is an important property of the solution. For example, the amount of carbon dioxide in the water at the bottom of Lake Nyos is an important predictor of when the deadly event may repeat itself. A **dilute solution** is one containing small amounts of solute relative to solvent. If the water at the bottom of Lake Nyos were a dilute carbon dioxide solution, it would pose little threat. A **concentrated solution** is one containing large amounts of solute relative to solvent. If the carbon dioxide in the water at the bottom of Lake Nyos becomes concentrated (through the continual feeding of carbon dioxide from magma into the lake), it becomes a threat. A common method of reporting solution concentration is *mass percent*.

Mass Percent

> Also in common use are *parts per million* (ppm), the number of grams of solute per 1 million g of solution, and *parts per billion* (ppb), the number of grams of solute per 1 billion g of solution.

Mass percent is the number of grams of solute per 100 g of solution. A solution with a concentration of 14% by mass, for example, contains 14 g of solute per 100 g of solution. To calculate mass percent, we divide the mass of the solute by the mass of the solution (solute *and* solvent) and multiply by 100%.

$$\text{mass percent} = \frac{\text{mass solute}}{\text{mass solute} + \text{mass solvent}} \times 100\%$$

> Note that the denominator is the mass of *solution*, not the mass of solvent.

Suppose we want to calculate the mass percent of NaCl in a solution containing 15.3 g of NaCl and 155.0 g of water. We begin by sorting the information in the problem statement.

GIVEN: 15.3 g NaCl
155.0 g H_2O

FIND: mass percent

SOLUTION

To solve this problem, we substitute the correct values into the mass percent equation just presented.

$$\text{mass percent} = \frac{\text{mass solute}}{\text{mass solute} + \text{mass solvent}} \times 100\%$$

$$= \frac{15.3 \text{ g}}{15.3 \text{ g} + 155.0 \text{ g}} \times 100\%$$

$$= \frac{15.3 \text{ g}}{170.3 \text{ g}} \times 100\%$$

$$= 8.98\%$$

The solution is 8.98% NaCl by mass.

EXAMPLE **13.1** | **Calculating Mass Percent**

Calculate the mass percent of a solution containing 27.5 g of ethanol (C_2H_6O) and 175 mL of H_2O. (Assume that the density of water is 1.00 g/mL.)

Begin by setting up the problem. You are given the mass of ethanol and the volume of water and asked to find the mass percent of the solution.	**GIVEN:** 27.5 g C_2H_6O 175 mL H_2O $d_{H_2O} = \dfrac{1.00 \text{ g}}{mL}$ **FIND:** mass percent
To find the mass percent, substitute into the equation for mass percent. You need the mass of ethanol and the mass of water. Obtain the mass of water from the volume of water by using the density as a conversion factor.	**SOLUTION** $\text{mass percent} = \dfrac{\text{mass solute}}{\text{mass solute } + \text{ mass solvent}} \times 100\%$ $\text{mass } H_2O = 175 \text{ mL } H_2O \times \dfrac{1.00 \text{ g}}{mL} = 175 \text{ g}$
Finally, substitute the correct quantities into the equation and calculate the mass percent.	$\text{mass percent} = \dfrac{\text{mass solute}}{\text{mass solute } + \text{ mass solvent}} \times 100\%$ $= \dfrac{27.5 \text{ g}}{27.5 \text{ g } + 175 \text{ g}}$ $= \dfrac{27.5 \text{ g}}{202.5 \text{ g}} \times 100\%$ $= 13.6\%$

▶ **SKILLBUILDER 13.1** | **Calculating Mass Percent**

Calculate the mass percent of a sucrose solution containing 11.3 g of sucrose and 412.1 mL of water. (Assume that the density of water is 1.00 g/mL.)

▶ **FOR MORE PRACTICE** Example 13.11; Problems 41, 42, 43, 44, 45, 46.

Using Mass Percent in Calculations

We can use the mass percent of a solution as a conversion factor between mass of the solute and mass of the solution. The key to using mass percent in this way is writing it as a fraction.

$$\text{mass percent} = \frac{\text{g solute}}{100 \text{ g solution}}$$

A solution containing 3.5% sodium chloride, for example, has the following conversion factor:

$$\frac{3.5 \text{ g NaCl}}{100 \text{ g solution}} \quad \text{converts g solution} \longrightarrow \text{g NaCl}$$

This conversion factor converts from grams of solution to grams of NaCl. If we want to go the other way, we invert the conversion factor:

$$\frac{100 \text{ g solution}}{3.5 \text{ g NaCl}} \quad \text{converts g NaCl} \longrightarrow \text{g solution}$$

To use mass percent as a conversion factor, let's consider a water sample from the bottom of Lake Nyos containing 8.5% carbon dioxide by mass. We can determine how much carbon dioxide in grams is contained in 28.6 L of the water solution. (Assume that the density of the solution is 1.03 g/mL.) We begin by sorting the information in the problem statement.

GIVEN: 8.5% CO_2 by mass

28.6 L solution

$$d = \frac{1.03 \text{ g}}{\text{mL}}$$

FIND: g CO_2

SOLUTION MAP

We strategize by drawing a solution map that begins with L solution and shows the conversion to mL solution and then to g solution using density. Then we proceed from g solution to g CO_2, using the mass percent (expressed as a fraction) as a conversion factor.

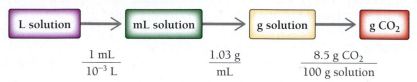

$$\frac{1 \text{ mL}}{10^{-3} \text{ L}} \qquad \frac{1.03 \text{ g}}{\text{mL}} \qquad \frac{8.5 \text{ g } CO_2}{100 \text{ g solution}}$$

RELATIONSHIPS USED

$$\frac{8.5 \text{ g } CO_2}{100 \text{ g solution}} \text{ (given mass percent, written as a fraction)}$$

$$\frac{1.03 \text{ g}}{\text{mL}} \text{ (given density of the solution)}$$

$$1 \text{ mL} = 10^{-3} \text{ L} \quad \text{(from Table 2.2)}$$

SOLUTION

We follow the solution map to calculate the answer.

$$28.6 \text{ L solution} \times \frac{1 \text{ mL}}{10^{-3} \text{ L}} \times \frac{1.03 \text{ g}}{\text{mL}} \times \frac{8.5 \text{ g } CO_2}{100 \text{ g solution}} = 2.5 \times 10^3 \text{ g } CO_2$$

In this example, we used mass percent to convert from a given amount of *solution* to the amount of *solute* present in the solution. In Example 13.2, we use mass percent to convert from a given amount of *solute* to the amount of *solution* containing that solute.

PEARSON
eText
2.0

Interactive
Worked Example
Video 13.2

EXAMPLE **13.2** | **Using Mass Percent in Calculations**

A soft drink contains 11.5% sucrose ($C_{12}H_{22}O_{11}$) by mass. What volume of the soft drink solution in milliliters contains 85.2 g of sucrose? (Assume a density of 1.04 g/mL.)

SORT	
You are given the concentration of sucrose in a soft drink and a mass of sucrose. You are asked to find the volume of the soft drink that contains the given mass of sucrose.	GIVEN: 11.5% $C_{12}H_{22}O_{11}$ by mass 85.2 g $C_{12}H_{22}O_{11}$ $d = \dfrac{1.04 \text{ g}}{\text{mL}}$ FIND: mL solution (soft drink)

STRATEGIZE	
Draw a solution map to convert from g solute ($C_{12}H_{22}O_{11}$) to g solution using the mass percent in fractional form as the conversion factor. Convert to mL using the density.	SOLUTION MAP 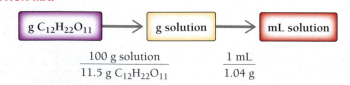 $\dfrac{100 \text{ g solution}}{11.5 \text{ g } C_{12}H_{22}O_{11}} \qquad \dfrac{1 \text{ mL}}{1.04 \text{ g}}$ **RELATIONSHIPS USED** $\dfrac{11.5 \text{ g } C_{12}H_{22}O_{11}}{100 \text{ g solution}}$ (given mass percent, written as a fraction) $\dfrac{1.04 \text{ g}}{\text{mL}}$ (given density of solution)

SOLVE	SOLUTION
Follow the solution map to solve the problem.	$85.2 \text{ g } C_{12}H_{22}O_{11} \times \dfrac{100 \text{ g solution}}{11.5 \text{ g } C_{12}H_{22}O_{11}} \times \dfrac{1 \text{ mL}}{1.04 \text{ g}} = 712 \text{ mL solution}$
CHECK Check your answer. Are the units correct? Does the answer make physical sense?	The units (mL solution) are correct. The magnitude of the answer makes sense because each 100 mL of solution contains 11.5 g sucrose; therefore, 712 mL should contain a bit more than 77 g, which is close to the given amount of 85.2 g.

▶ **SKILLBUILDER 13.2 | Using Mass Percent in Calculations**

How much sucrose ($C_{12}H_{22}O_{11}$) in grams is contained in 355 mL (12 oz) of the soft drink in Example 13.2?

▶ **FOR MORE PRACTICE** Example 13.12; Problems 47, 48, 49, 50, 51, 52.

13.6 Specifying Solution Concentration: Molarity

▶ Calculate molarity.

▶ Use molarity in calculations.

▶ Calculate ion concentration.

> Note that molarity is abbreviated with a capital M.

A second way to express solution concentration is **molarity** (M), defined as the number of moles of solute per liter of solution. We calculate the molarity of a solution as follows:

$$\text{molarity (M)} = \frac{\text{moles solute}}{\text{liters solution}}$$

Note that molarity is moles of solute per liter of *solution*, not per liter of solvent. To make a solution of a specified molarity, we put the solute into a flask and then add water to the desired volume of solution. For example, to make 1.00 L of a 1.00 M NaCl solution, we add 1.00 mol of NaCl to a flask and then add water to make 1.00 L of solution (▼ FIGURE 13.7). We *do not* combine 1.00 mol of NaCl with 1.00 L of water because that would result in a total volume exceeding 1.00 L and therefore a molarity of less than 1.00 M.

To calculate molarity, we divide the number of moles of the solute by the volume of the solution (solute *and* solvent) in liters. For example, to calculate the molarity of a sucrose ($C_{12}H_{22}O_{11}$) solution made with 1.58 mol of sucrose diluted to a total volume of 5.0 L of solution, we begin by sorting the information in the problem statement.

How to prepare a 1.00 molar NaCl solution

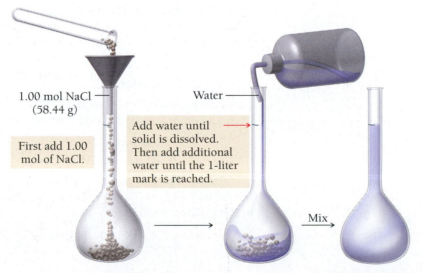

1.00 mol NaCl (58.44 g)

First add 1.00 mol of NaCl.

Water

Add water until solid is dissolved. Then add additional water until the 1-liter mark is reached.

Mix

A 1.00 molar NaCl solution

▶ **FIGURE 13.7 Making a solution of specific molarity** To make 1.00 L of a 1.00 M NaCl solution, we add 1.00 mol (58.44 g) of sodium chloride to a flask and then dilute to 1.00 L of total volume. **Question: What would happen if we added 1 L of water to 1 mol of sodium chloride? Would the resulting solution be 1 M?**

GIVEN: 1.58 mol $C_{12}H_{22}O_{11}$
5.0 L solution

FIND: molarity (M)

SOLUTION

We substitute the correct values into the equation for molarity and calculate the answer.

$$\text{molarity (M)} = \frac{\text{moles solute}}{\text{liters solution}}$$

$$= \frac{1.58 \text{ mol } C_{12}H_{22}O_{11}}{5.0 \text{ L solution}}$$

$$= 0.32 \text{ M}$$

Interactive
Worked Example
Video 13.3

EXAMPLE **13.3** **Calculating Molarity**

Calculate the molarity of a solution made by putting 15.5 g NaCl into a beaker and adding water to make 1.50 L of NaCl solution.

You are given the mass of sodium chloride (the solute) and the volume of solution. You are asked to find the molarity of the solution.	**GIVEN:** 15.5 g NaCl 1.50 L solution **FIND:** molarity (M)
To calculate molarity, substitute the correct values into the equation and calculate the answer. You must first convert the amount of NaCl from grams to moles using the molar mass of NaCl.	**SOLUTION** $\text{mol NaCl} = 15.5 \text{ g NaCl} \times \dfrac{1 \text{ mol NaCl}}{58.44 \text{ g NaCl}} = 0.2652 \text{ mol NaCl}$ $\text{molarity (M)} = \dfrac{\text{moles solute}}{\text{liters solution}}$ $= \dfrac{0.2652 \text{ mol NaCl}}{1.50 \text{ L solution}}$ $= 0.177 \text{ M}$

▶ **SKILLBUILDER 13.3 | Calculating Molarity**

Calculate the molarity of a solution made by putting 55.8 g of $NaNO_3$ into a beaker and diluting to 2.50 L.

▶ **FOR MORE PRACTICE**   Example 13.13; Problems 59, 60, 61, 62, 63, 64.

Using Molarity in Calculations

We can use the molarity of a solution as a conversion factor between moles of the solute and liters of the solution. For example, a 0.500 M NaCl solution contains 0.500 mol NaCl for every liter of solution.

$$\frac{0.500 \text{ mol NaCl}}{\text{L solution}} \quad \text{converts L solution} \longrightarrow \text{mol NaCl}$$

This conversion factor converts from liters of solution to moles of NaCl. If we want to go the other way, we invert the conversion factor.

$$\frac{\text{L solution}}{0.500 \text{ mol NaCl}} \quad \text{converts mol NaCl} \longrightarrow \text{L solution}$$

For example, to determine how many grams of sucrose ($C_{12}H_{22}O_{11}$) are contained in 1.72 L of 0.758 M sucrose solution, we begin by sorting the information in the problem statement.

GIVEN: 0.758 M $C_{12}H_{22}O_{11}$
1.72 L solution

FIND: g $C_{12}H_{22}O_{11}$

SOLUTION MAP

We strategize by drawing a solution map that begins with L solution and shows the conversion to moles of sucrose using the molarity, and then the conversion to mass of sucrose using the molar mass.

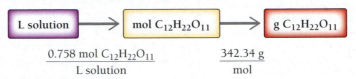

$$\underbrace{\frac{0.758 \text{ mol } C_{12}H_{22}O_{11}}{\text{L solution}}}_{} \qquad \underbrace{\frac{342.34 \text{ g}}{\text{mol}}}_{}$$

RELATIONSHIPS USED

$$\frac{0.758 \text{ mol } C_{12}H_{22}O_{11}}{\text{L solution}} \text{ (given molarity of solution, written out as a fraction)}$$

$$1 \text{ mol } C_{12}H_{22}O_{11} = 342.34 \text{ g (molar mass of sucrose)}$$

SOLUTION

We then follow the solution map to calculate the answer.

$$1.72 \text{ L solution} \times \frac{0.758 \text{ mol } C_{12}H_{22}O_{11}}{\text{L solution}}$$

$$\times \frac{342.34 \text{ g } C_{12}H_{22}O_{11}}{\text{mol } C_{12}H_{22}O_{11}} = 446 \text{ g } C_{12}H_{22}O_{11}$$

In this example, we used molarity to convert from a given amount of *solution* to the amount of *solute* in that solution. In Example 13.4, we use molarity to convert from a given amount of *solute* to the amount of *solution* containing that solute.

PEARSON
eText 2.0 Interactive Worked Example Video 13.4

EXAMPLE **13.4** | **Using Molarity in Calculations**

How many liters of a 0.114 M NaOH solution contains 1.24 mol of NaOH?

SORT	**GIVEN:** 0.114 M NaOH
You are given the molarity of an NaOH solution and the number of moles of NaOH. You are asked to find the volume of solution that contains the given number of moles.	1.24 mol NaOH
	FIND: L solution

STRATEGIZE

The solution map begins with mol NaOH and shows the conversion to liters of solution using the molarity as a conversion factor.

SOLUTION MAP

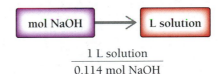

$$\frac{1 \text{ L solution}}{0.114 \text{ mol NaOH}}$$

RELATIONSHIP USED

$$\frac{0.114 \text{ mol NaOH}}{\text{L solution}} \text{ (given molarity of solution, written out as a fraction)}$$

SOLVE

Solve the problem by following the solution map.

SOLUTION

$$1.24 \text{ mol NaOH} \times \frac{1 \text{ L solution}}{0.114 \text{ mol NaOH}} = 10.9 \text{ L solution}$$

CHECK

Check your answer. Are the units correct? Does the answer make physical sense?

The units (L solution) are correct. The magnitude of the answer makes sense because each L of solution contains a little more than 0.10 mol; therefore, about 10 L contains a little more than 1 mol.

▶ **SKILLBUILDER 13.4** | **Using Molarity in Calculations**

How much of a 0.225 M KCl solution contains 55.8 g of KCl?

▶ **FOR MORE PRACTICE** Example 13.14; Problems 65, 66, 67, 68, 69, 70.

Ion Concentrations

> When an ionic compound dissolves in solution, some of the cations and anions may pair up, so that the actual concentrations of the ions are lower than what we would expect if we assume complete dissociation occurred.

The reported concentration of a solution containing a *molecular* compound usually reflects the concentration of the solute as it actually exists in solution. For example, a 1.0 M glucose ($C_6H_{12}O_6$) solution indicates that the solution contains 1.0 mol of $C_6H_{12}O_6$ per liter of solution. However, the reported concentration of solution containing an *ionic* compound reflects the concentration of the solute *before it is dissolved in solution*. For example, a 1.0 M $CaCl_2$ solution contains 1.0 mol of Ca^{2+} per liter and 2.0 mol of Cl^- per liter. The concentration of the individual ions present in a solution containing an ionic compound can usually be approximated from the overall concentration as we demonstrate in Example 13.5.

EXAMPLE **13.5** | **Calculating Ion Concentration**

Determine the molar concentrations of Na^+ and PO_4^{3-} in a 1.50 M Na_3PO_4 solution.

You are given the concentration of an ionic solution and asked to find the concentrations of the component ions.	**GIVEN:** 1.50 M Na_3PO_4 **FIND:** molarity (M) of Na^+ and PO_4^{3-}
A formula unit of Na_3PO_4 contains 3 Na^+ ions (as indicated by the subscript), so the concentration of Na^+ is three times the concentration of Na_3PO_4. Since the same formula unit contains one PO_4^{3-} ion, the concentration of PO_4^{3-} is equal to the concentration of Na_3PO_4.	**SOLUTION** molarity of Na^+ = 3(1.50 M) = 4.50 M molarity of PO_4^{3-} = 1.50 M

▶ **SKILLBUILDER 13.5** | **Ion Concentration**

Determine the molar concentrations of Ca^{2+} and Cl^- in a 0.75 M $CaCl_2$ solution.

▶ **FOR MORE PRACTICE** Problems 77, 78, 79, 80.

CONCEPTUAL ✓ **CHECKPOINT 13.3**

A solution is 0.15 M in K_2SO_4. What is the concentration of K^+ in solution?

(a) 0.075 M

(b) 0.15 M

(c) 0.30 M

(d) 0.45 M

13.7 Solution Dilution

▶ Use the dilution equation in calculations.

> When diluting acids, always add the concentrated acid to the water. *Never add water to concentrated acid solutions.*

To save space in laboratory storerooms, solutions are often stored in concentrated forms called **stock solutions.** For example, hydrochloric acid is typically stored as a 12 M stock solution. However, many lab procedures call for much less concentrated hydrochloric acid solutions, so chemists must dilute the stock solution to the required concentration. We normally do this by diluting a certain amount of the stock solution with water. How do we determine how much of the stock solution to use? The easiest way to solve this problem is to use the dilution equation:

$$M_1V_1 = M_2V_2$$

where M_1 and V_1 are the molarity and volume of the initial concentrated solution and M_2 and V_2 are the molarity and volume of the final diluted solution.

This equation works because the molarity multiplied by the volume gives the number of moles of solute ($M \times V = $ mol), which is the same in both solutions.

For example, suppose a laboratory procedure calls for 5.00 L of a 1.50 M KCl solution. How should we prepare this solution from a 12.0 M stock solution? We begin by sorting the information in the problem statement.

> The equation $M_1V_1 = M_2V_2$ applies only to solution dilution, NOT to stoichiometry.

GIVEN: $M_1 = 12.0$ M

$M_2 = 1.50$ M

$V_2 = 5.00$ L

FIND: V_1

SOLUTION

We solve the solution dilution equation for V_1 (the volume of the stock solution required for the dilution) and substitute in the correct values to calculate V_1.

$$M_1V_1 = M_2V_2$$

$$V_1 = \frac{M_2V_2}{M_1}$$

$$= \frac{1.50 \, \frac{\text{mol}}{\text{L}} \times 5.00 \, \text{L}}{12.0 \, \frac{\text{mol}}{\text{L}}}$$

$$= 0.625 \, \text{L}$$

We can therefore make the solution by diluting 0.625 L of the stock solution to a total volume of 5.00 L (V_2). The resulting solution is 1.50 M in KCl (▶ **FIGURE 13.8**, on the next page).

EXAMPLE **13.6** | **Solution Dilution**

To what volume should you dilute 0.100 L of a 15 M NaOH solution to obtain a 1.0 M NaOH solution?	
You are given the initial volume and concentration of an NaOH solution and a final concentration. You are asked to find the volume required to dilute the initial solution to the given final concentration.	**GIVEN:** $V_1 = 0.100$ L $M_1 = 15$ M $M_2 = 1.0$ M **FIND:** V_2
Solve the solution dilution equation for V_2 (the volume of the final solution) and substitute the required quantities to calculate V_2. You can make the solution by diluting 0.100 L of the stock solution to a total volume of 1.5 L (V_2). The resulting solution has a concentration of 1.0 M.	**SOLUTION** $M_1V_1 = M_2V_2$ $V_2 = \dfrac{M_1V_1}{M_2}$ $= \dfrac{15 \, \frac{\text{mol}}{\text{L}} \times 0.100 \, \text{L}}{1.0 \, \frac{\text{mol}}{\text{L}}}$ $= 1.5$ L

▶ **SKILLBUILDER 13.6** | **Solution Dilution**

How much 6.0 M NaNO$_3$ solution should you use to make 0.585 L of a 1.2 M NaNO$_3$ solution?

▶ **FOR MORE PRACTICE** Example 13.15; Problems 81, 82, 83, 84, 85, 86, 87, 88.

▶ **FIGURE 13.8 Making a solution by dilution of a more concentrated solution**

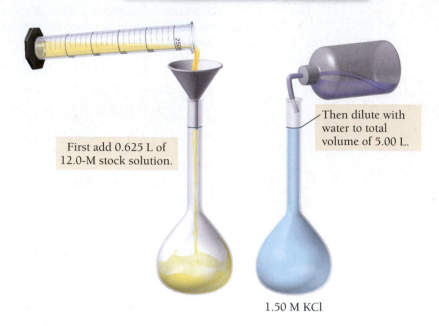

How to make 5.00 L of a 1.50-M KCl solution from a 12.0-M stock solution

First add 0.625 L of 12.0-M stock solution.

Then dilute with water to total volume of 5.00 L.

1.50 M KCl

$$M_1V_1 = M_2V_2$$

$$\frac{12.0 \text{ mol}}{\cancel{L}} \times 0.625\cancel{L} = \frac{1.50 \text{ mol}}{\cancel{L}} \times 5.00\cancel{L}$$

$$7.50 \text{ mol} = 7.50 \text{ mol}$$

PEARSON eText 2.0

CONCEPTUAL ✔ CHECKPOINT 13.4

What is the molarity of a solution in which 100.0 mL of 1.0 M KCl is diluted to 1.0 L?

(a) 0.10 M

(b) 1.0 M

(c) 10.0 M

13.8 Solution Stoichiometry

▶ Use volume and concentration to calculate the number of moles of reactants or products and then use stoichiometric coefficients to convert to other quantities in a reaction.

See Sections 8.2 through 8.4 for a review of reaction stoichiometry.

As we discussed in Chapter 7, many chemical reactions take place in aqueous solutions. Precipitation reactions, neutralization reactions, and gas-evolution reactions, for example, all occur in aqueous solutions. Chapter 8 described how we use the coefficients in chemical equations as conversion factors between moles of reactants and moles of products in stoichiometric calculations. We often use these conversion factors to determine, for example, the amount of product obtained in a chemical reaction based on a given amount of reactant or the amount of one reactant needed to completely react with a given amount of another reactant. The general solution map for these kinds of calculations is:

moles A → moles B

where A and B are two different substances involved in the reaction and the conversion factor between them comes from the stoichiometric coefficients in the balanced chemical equation.

In reactions involving aqueous reactant and products, it is often convenient to specify the amount of reactants or products in terms of their volume and concentration. We can use the volume and concentration to calculate the number of moles of reactants or products, and then use the stoichiometric coefficients to convert to other quantities in the reaction. The general solution map for these kinds of calculations is:

where the conversions between volume and moles are achieved using the molarities of the solutions.

For example, consider the reaction for the neutralization of sulfuric acid.

$$H_2SO_4(aq) + 2\,NaOH(aq) \longrightarrow Na_2SO_4(aq) + 2\,H_2O(l)$$

How much 0.125 M NaOH solution do we need to completely neutralize 0.225 L of 0.175 M H_2SO_4 solution? We begin by sorting the information in the problem statement.

GIVEN: 0.225 L H_2SO_4 solution
0.175 M H_2SO_4
0.125 M NaOH

FIND: L NaOH solution

SOLUTION MAP

We strategize by drawing a solution map similar to those for other stoichiometric problems. We first use the volume and molarity of H_2SO_4 solution to get mol H_2SO_4. Then we use the stoichiometric coefficients from the equation to convert mol H_2SO_4 to mol NaOH. Finally, we use the molarity of NaOH to calculate L NaOH solution.

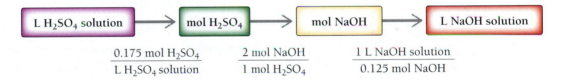

RELATIONSHIPS USED

$$M(H_2SO_4) = \frac{0.175 \text{ mol } H_2SO_4}{\text{L } H_2SO_4 \text{ solution}} \quad \text{(given molarity of } H_2SO_4 \text{ solution, written out as a fraction)}$$

$$M(NaOH) = \frac{0.125 \text{ mol NaOH}}{\text{L NaOH solution}} \quad \text{(given molarity of NaOH solution, written out as a fraction)}$$

1 mol H_2SO_4:2 mol NaOH (stoichiometric relationship between H_2SO_4 and NaOH, from balanced chemical equation)

SOLUTION

To solve the problem, we follow the solution map and calculate the answer.

$$0.225 \text{ L } H_2SO_4 \text{ solution} \times \frac{0.175 \text{ mol } H_2SO_4}{\text{L } H_2SO_4 \text{ solution}} \times \frac{2 \text{ mol NaOH}}{1 \text{ mol } H_2SO_4}$$

$$\times \frac{1 \text{ L NaOH solution}}{0.125 \text{ mol NaOH}} = 0.630 \text{ L NaOH solution}$$

It will take 0.630 L of the NaOH solution to completely neutralize the H_2SO_4.

Interactive
Worked Example
Video 13.7

EXAMPLE **13.7** | **Solution Stoichiometry**

Consider the precipitation reaction.

$$2\ KI(aq) + Pb(NO_3)_2(aq) \longrightarrow PbI_2(s) + 2\ KNO_3(aq)$$

How much 0.115 M KI solution in liters will completely precipitate the Pb^{2+} in 0.104 L of 0.225 M $Pb(NO_3)_2$ solution?

SORT	
You are given the concentration of a reactant, KI, in a chemical reaction. You are also given the volume and concentration of a second reactant, $Pb(NO_3)_2$. You are asked to find the volume of the first reactant that completely reacts with the given amount of the second.	**GIVEN:** 0.115 M KI 0.104 L $Pb(NO_3)_2$ solution 0.225 M $Pb(NO_3)_2$ **FIND:** L KI solution

STRATEGIZE

The solution map for this problem is similar to the solution maps for other stoichiometric problems. First use the volume and molarity of $Pb(NO_3)_2$ solution to determine mol $Pb(NO_3)_2$. Then use the stoichiometric coefficients from the equation to convert mol $Pb(NO_3)_2$ to mol KI. Finally, use mol KI to find L KI solution.

SOLUTION MAP

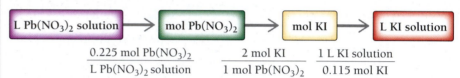

RELATIONSHIPS USED

$M\ KI = \dfrac{0.115\ mol\ KI}{L\ KI\ solution}$ (given molarity of KI solution, written out as a fraction)

$M\ Pb(NO_3)_2 = \dfrac{0.225\ mol\ Pb(NO_3)_2}{L\ Pb(NO_3)_2\ solution}$ (given molarity of $Pb(NO_3)_2$ solution, written out as a fraction)

2 mol KI:1 mol $Pb(NO_3)_2$ (stoichiometric relationship between KI and $Pb(NO_3)_2$, from balanced chemical equation)

SOLVE

Follow the solution map to solve the problem. Begin with volume of $Pb(NO_3)_2$ solution and cancel units to arrive at volume of KI solution.

SOLUTION

$0.104\ \cancel{L\ Pb(NO_3)_2\ solution} \times \dfrac{0.225\ \cancel{mol\ Pb(NO_3)_2}}{\cancel{L\ Pb(NO_3)_2\ solution}} \times \dfrac{2\ \cancel{mol\ KI}}{\cancel{mol\ Pb(NO_3)}}$

$\times \dfrac{L\ KI\ solution}{0.115\ \cancel{mol\ KI}} = 0.407\ L\ KI\ solution$

CHECK

Check your answer. Are the units correct? Does the answer make physical sense?

The units (L KI solution) are correct. The magnitude of the answer makes sense because the lead nitrate solution is about twice as concentrated as the potassium iodide solution and 2 mol of potassium iodide are required to react with 1 mol of lead(II) nitrate. Therefore, you would expect the volume of the potassium solution required to completely react with a given volume of the $Pb(NO_3)_2$ solution to be about four times as much.

▶ **SKILLBUILDER 13.7 | Solution Stoichiometry**

How many milliliters of 0.112 M Na_2CO_3 will completely react with 27.2 mL of 0.135 M HNO_3 according to the reaction?

$$2\ HNO_3(aq) + Na_2CO_3(aq) \longrightarrow H_2O(l) + CO_2(g) + 2\ NaNO_3(aq)$$

▶ **SKILLBUILDER PLUS** A 25.0-mL sample of HNO_3 solution requires 35.7 mL of 0.108 M Na_2CO_3 to completely react with all of the HNO_3 in the solution. What is the concentration of the HNO_3 solution?

▶ **FOR MORE PRACTICE** Example 13.16; Problems 89, 90, 91, 92.

CONCEPTUAL ✔ **CHECKPOINT 13.5**

Consider the following reaction occurring in aqueous solution.

$$A(aq) + 2\,B(aq) \longrightarrow \text{Products}$$

What volume of a 0.100 M solution of B is required to completely react with 50.0 mL of a 0.200-M solution of A?

(a) 25.0 mL

(b) 50.0 mL

(c) 100.0 mL

(d) 200.0 mL

13.9 Freezing Point Depression and Boiling Point Elevation: Making Water Freeze Colder and Boil Hotter

▶ Calculate molality.

▶ Calculate freezing points and boiling points for solutions.

▲ Sprinkling salt on icy roads lowers the freezing point of water, so the ice melts even if the temperature is below 0 °C.

Have you ever wondered why salt is added to ice in an ice-cream maker? Or why salt is scattered on icy roads? Salt actually lowers the melting point of ice. A salt-and-water solution remains a liquid even below 0 °C. By adding salt to ice in the ice-cream maker, you form a mixture of ice, salt, and water that can reach a temperature of about −10 °C, which is cold enough to freeze the cream. On the road, salt allows the ice to melt, even if the ambient temperature is below freezing.

Adding a nonvolatile solute—a solute that does not readily evaporate—to a liquid extends the temperature range over which the liquid remains a liquid. The solution has a lower melting point and a higher boiling point than the pure liquid; these effects are called **freezing point depression** and **boiling point elevation**. Freezing point depression and boiling point elevation depend only on the number of solute particles *in solution*, not on the type of solute particles. Properties such as these—which depend on the number of dissolved solute particles and not on the type of solute particles—are **colligative properties**.

Freezing Point Depression

The freezing point of a solution containing a nonvolatile solute is lower than the freezing point of the pure solvent. For example, antifreeze, which is added to engine coolant to prevent it from freezing in cold climates, is an aqueous solution of ethylene glycol ($C_2H_6O_2$). The ethylene glycol lowers the freezing point of the aqueous solution. The more concentrated the solution is, the lower the freezing point becomes. For freezing point depression and boiling point elevation, the concentration of the solution is usually expressed in **molality (*m*)**, the number of moles of solute per kilogram of solvent.

Molality is abbreviated with a lowercase italic *m*, while molarity is abbreviated with a capital M.

$$\text{molality } (m) = \frac{\text{moles solute}}{\text{kilograms solvent}}$$

Notice that molality is defined with respect to kilograms of *solvent*, not kilograms of *solution*.

EXAMPLE **13.8** | **Calculating Molality**

Calculate the molality of a solution containing 17.2 g of ethylene glycol ($C_2H_6O_2$) dissolved in 0.500 kg of water.

You are given the mass of ethylene glycol in grams and the mass of the solvent in kilograms. You are asked to find the molality of the resulting solution.	**GIVEN:** 17.2 g $C_2H_6O_2$ 0.500 kg H_2O **FIND:** molality (m)
To calculate molality, substitute the correct values into the equation and calculate the answer. First convert the amount of $C_2H_6O_2$ from grams to moles using the molar mass of $C_2H_6O_2$.	**SOLUTION** $$\text{mol } C_2H_6O_2 = 17.2 \text{ g } C_2H_6O_2 \times \frac{1 \text{ mol } C_2H_6O_2}{62.08 \text{ g } C_2H_6O_2}$$ $$= 0.2771 \text{ mol } C_2H_6O_2$$ $$\text{molality } (m) = \frac{\text{moles solute}}{\text{kilograms solvent}}$$ $$= \frac{0.2771 \text{ mol } C_2H_6O_2}{0.500 \text{ kg } H_2O}$$ $$= 0.554 \text{ } m$$

▶ **SKILLBUILDER 13.8** | **Calculating Molality**

Calculate the molality (m) of a sucrose ($C_{12}H_{22}O_{11}$) solution containing 50.4 g sucrose and 0.332 kg of water.

▶ **FOR MORE PRACTICE** Example 13.17; Problems 97, 98, 99, 100.

CONCEPTUAL ✔ CHECKPOINT 13.6

A laboratory procedure calls for a 2.0 molal aqueous solution. A student accidentally makes a 2.0 molar solution. The solution made by the student is:

(a) too concentrated

(b) too dilute

(c) just right

(d) It depends on the molar mass of the solute.

With an understanding of molality, we can now quantify freezing point depression. The amount that the freezing point of a solution is lowered by a particular amount of solute is given by the following equation:

Freezing Point Depression of a Solution

$$\Delta T_f = m \times K_f$$

where

- ΔT_f is the change in temperature of the freezing point in °C (from the freezing point of the pure solvent).

- m is the molality of the solution in $\dfrac{\text{mol solute}}{\text{kg solvent}}$.

- K_f is the freezing point depression constant for the solvent.

For water:

$$K_f = 1.86 \frac{\text{°C kg solvent}}{\text{mol solute}}$$

Calculating the freezing point of a solution involves substituting into the given equation, as Example 13.9 demonstrates.

▲ Ethylene glycol is the chief component of antifreeze, which keeps engine coolant from freezing in winter or boiling over in summer.

The equations for freezing point depression and boiling point elevation given in this section apply only to nonelectrolyte solutions.

Different solvents have different values of K_f.

EVERYDAY CHEMISTRY

Antifreeze in Frogs

Wood frogs *(Rana sylvatica)* look like most other frogs. They are a few inches long and have characteristic greenish-brown skin. However, wood frogs survive cold winters in a remarkable way—they partially freeze. In its frozen state, the frog has no heartbeat, no blood circulation, no breathing, and no brain activity. Within one to two hours of thawing, however, these vital functions return, and the frog hops off to find food. How is this possible?

Most cold-blooded animals cannot survive freezing temperatures because the water within their cells freezes. As described in Section 12.8, when water freezes, it expands, irreversibly damaging cells. When the wood frog hibernates for the winter, however, it secretes large amounts of glucose into its blood and into the interior of its cells. When the temperature drops below freezing, extracellular bodily fluids, such as those in the frog's abdominal cavity, freeze solid. Fluids within the frogs' cells, however, remain liquid because the high glucose concentration lowers their freezing point. In other words, the concentrated glucose solution within the cells acts as antifreeze, preventing the water within from freezing and allowing the frog to survive.

▲ The wood frog survives cold winters by partially freezing. The fluids in frog cells are protected by a high concentration of glucose that acts as antifreeze, lowering their freezing point so that the intracellular fluids remain liquid to temperatures as low as −8 °C.

B13.1 CAN YOU ANSWER THIS?   *The wood frog can survive at body temperatures as low as −8.0 °C. Calculate the molality of a glucose solution ($C_6H_{12}O_6$) required to lower the freezing point of water to −8.0 °C.*

EXAMPLE **13.9** | **Freezing Point Depression**

Calculate the freezing point of a 1.7 *m* ethylene glycol solution.

You are given the molality of an aqueous solution and asked to find the freezing point depression. You will need the freezing point depression equation provided in this section.	**GIVEN:** 1.7 *m* solution **FIND:** freezing point
To solve this problem, substitute the values into the equation for freezing point depression and calculate ΔT_f.	**SOLUTION** $\Delta T_f = m \times K_f$ $= 1.7 \dfrac{\text{mol solute}}{\text{kg solvent}} \times 1.86 \dfrac{°C \text{ kg solvent}}{\text{mol solute}}$ $= 3.2 \, °C$
The actual freezing point is the freezing point of pure water (0.00 °C) − ΔT_f.	freezing point = 0.00 °C − 3.2 °C $= -3.2 \, °C$

▶ **SKILLBUILDER 13.9** | **Freezing Point Depression**

Calculate the freezing point of a 2.6 *m* sucrose solution.

▶ **FOR MORE PRACTICE**   Example 13.18; Problems 101, 102.

Boiling Point Elevation

The boiling point of a solution containing a nonvolatile solute is higher than the boiling point of the pure solvent. In automobiles, antifreeze not only prevents the freezing of coolant within engine blocks in cold climates, but it also prevents the boiling of engine coolant in hot climates. The amount that the boiling point is raised for solutions is given by the following equation:

Boiling Point Elevation of a Solution

$$\Delta T_b = m \times K_b$$

where

- ΔT_b is change in temperature of the boiling point in °C (from the boiling point of the pure solvent).

- m is the molality of the solution in $\dfrac{\text{mol solute}}{\text{kg solvent}}$.

- K_b is the boiling point elevation constant for the solvent.

For water:

$$K_b = 0.512 \, \frac{\text{°C kg solvent}}{\text{mol solute}}$$

Different solvents have different values of K_b.

We calculate the boiling point of solutions by substituting into the preceding equation, as Example 13.10 demonstrates.

EXAMPLE **13.10** | **Boiling Point Elevation**

Calculate the boiling point of a 1.7 m ethylene glycol solution.

You are given the molality of an aqueous solution and asked to find the boiling point.	**GIVEN:** 1.7 m solution **FIND:** boiling point
To solve this problem, substitute the values into the equation for boiling point elevation and calculate ΔT_b.	**SOLUTION** $\Delta T_b = m \times K_b$ $= 1.7 \, \dfrac{\text{mol solute}}{\text{kg solvent}} \times 0.512 \, \dfrac{\text{°C kg solvent}}{\text{mol solute}}$ $= 0.87 \, \text{°C}$
The actual boiling point of the solution is the boiling point of pure water (100.00 °C) plus ΔT_b.	boiling point $= 100.00 \, \text{°C} + 0.87 \, \text{°C}$ $= 100.87 \, \text{°C}$

▶ **SKILLBUILDER 13.10** | **Boiling Point Elevation**

Calculate the boiling point of a 3.5 m glucose solution.

▶ **FOR MORE PRACTICE** Problems 103, 104, 105, 106.

PEARSON eText 2.0

CONCEPTUAL ✓ **CHECKPOINT 13.7**

Which solution has the highest boiling point?

(a) 0.50 m $C_{12}H_{22}O_{11}$
(b) 0.50 m $C_6H_{12}O_6$
(c) 0.50 m $C_2H_6O_2$
(d) All of these solutions will have the same boiling point.

13.10 Osmosis: Why Drinking Saltwater Causes Dehydration

▶ Summarize and explain the process of osmosis.

Humans adrift at sea are surrounded by water, yet drinking that water only accelerates their dehydration. Why? Saltwater causes dehydration because of **osmosis**, the flow of solvent from a less concentrated solution to a more concentrated solution. Solutions containing a high concentration of solute draw solvent from solutions containing a lower concentration of solute. In other words, aqueous solutions with high concentrations of solute, such as seawater, are actually *thirsty solutions*—they draw water away from other, less concentrated solutions, including those in the human body (▼ **FIGURE 13.9**).

▶ **FIGURE 13.9 Seawater is a *thirsty* solution**

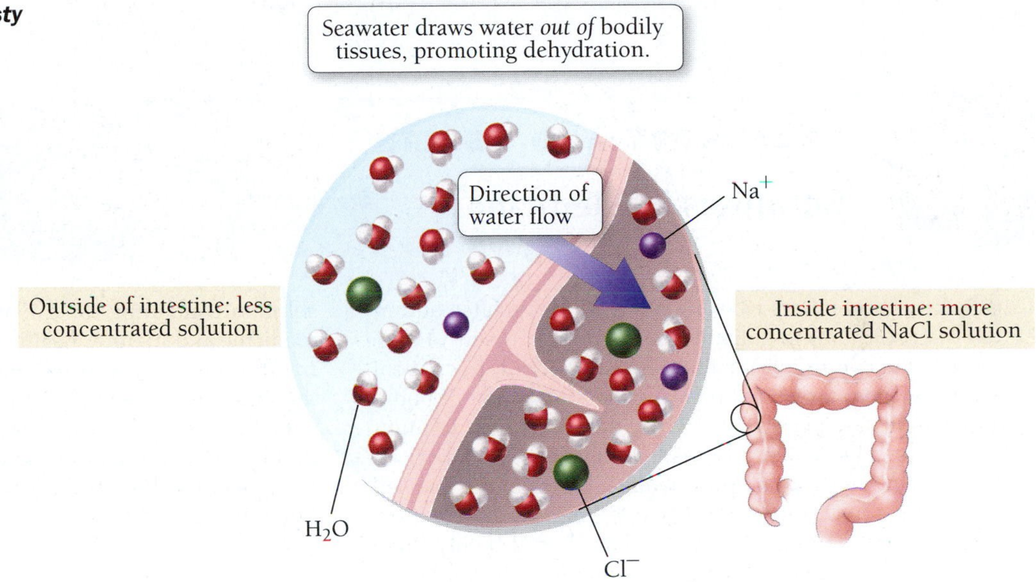

Seawater draws water *out of* bodily tissues, promoting dehydration.

Direction of water flow

Na⁺

Outside of intestine: less concentrated solution

Inside intestine: more concentrated NaCl solution

H_2O

Cl⁻

▼ **FIGURE 13.10** shows an osmosis cell. The left side of the cell contains a concentrated saltwater solution, and the right side of the cell contains pure water. A **semipermeable membrane**—a membrane that allows some substances to pass

Osmosis Cell: Water flows toward the more concentrated solution.

Initial

At equilibrium

Osmotic pressure

Semipermeable membrane

Water molecules

Solute particles

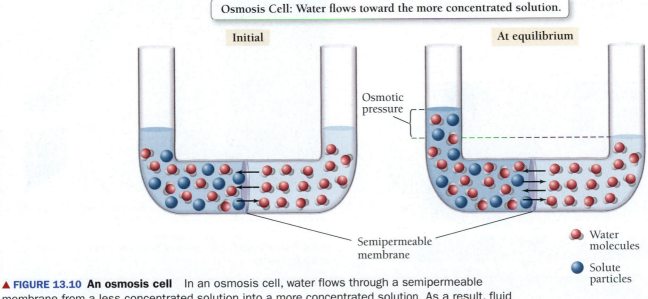

▲ **FIGURE 13.10 An osmosis cell**   In an osmosis cell, water flows through a semipermeable membrane from a less concentrated solution into a more concentrated solution. As a result, fluid rises in one side of the tube until the weight of the excess fluid creates enough pressure to stop the flow. This pressure is the osmotic pressure of the solution.

through but not others—separates the two halves of the cell. Through osmosis, water flows from the pure-water side of the cell through the semipermeable membrane into the saltwater side. Over time, the water level on the left side of the cell rises while the water level on the right side of the cell falls. This continues until the pressure created by the weight of the water on the left side is enough to stop the osmotic flow. The pressure required to stop the osmotic flow is the **osmotic pressure** of the solution. Osmotic pressure—like freezing point depression and boiling point elevation—is a colligative property; it depends only on the concentration of the solute particles, not on the type of solute. The more concentrated the solution, the greater its osmotic pressure.

The **membranes** of living cells act as semipermeable membranes. Consequently, if you put a living cell into seawater, it loses water through osmosis and

CHEMISTRY AND HEALTH

Solutions in Medicine

Doctors and others working in health fields often administer solutions to patients. The osmotic pressure of these solutions is controlled for the desired effect on the patient. Solutions having osmotic pressures less than that of bodily fluids are called *hypoosmotic*. These solutions tend to pump water into cells. When a human cell is placed in a hypoosmotic solution—such as pure water—water enters the cell, sometimes causing it to burst (see Figure 13.11b). Solutions having osmotic pressures greater than that of bodily fluids are called *hyperosmotic*. These solutions tend to take water out of cells and tissues. When a human cell is placed in a hyperosmotic solution, it typically shrivels as it loses water to the surrounding solution (see Figure 13.11c).

Intravenous solutions—those that are administered directly into a patient's veins—must have osmotic pressure equal to that of bodily fluids. These solutions are called *isoosmotic*. When a patient is given an IV in a hospital, the majority of the fluid is usually an isoosmotic saline solution—a

solution containing 0.9 g NaCl per 100 mL of solution. In medicine and in other health-related fields, solution concentrations are often reported in units that indicate the mass of the solute in a given volume of solution. Also common is *percent mass to volume*—which is the mass of the solute in grams divided by volume of the solution in milliliters times 100%. In these units, the concentration of an isoosmotic saline solution is 0.9% mass/volume.

B13.2 CAN YOU ANSWER THIS? *An isoosmotic sucrose ($C_{12}H_{22}O_{11}$) solution has a concentration of 0.30 M. Calculate its concentration in percent mass to volume.*

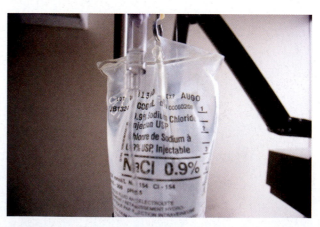

▲ Intravenous fluids consist mostly of isoosmotic saline solutions with an osmotic pressure equal to that of bodily fluids.
Question: Why would it be dangerous to administer intravenous fluids that do not have an osmotic pressure comparable to that of bodily fluids?

▶ **FIGURE 13.11 Red blood cells in solutions of different concentration** (a) When the solute concentration of the surrounding fluid is equal to that within the cell, there is no net osmotic flow, and the red blood cell exhibits its typical shape. (b) When a cell is placed in pure water, osmotic flow of water into the cell causes it to swell up. Eventually it may burst. (c) When a cell is placed in a concentrated solution, osmosis draws water out of the cell, distorting its normal shape.

Normal red blood cell

(a)

Red blood cell in pure water: water flows into cell

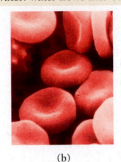

(b)

Red blood cell in concentrated solution: water flows out of cell

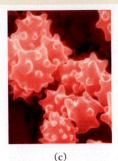

(c)

becomes dehydrated. ▲ **FIGURE 13.11** shows red blood cells in solutions of various concentrations. The cells in Figure 13.11a, immersed in a solution with the same solute concentration as the cell interior, have the normal red blood cell shape. The cells in Figure 13.11b, in pure water, are swollen. Because the solute concentration within the cells is higher than that of the surrounding fluid, osmosis has pulled water across the membrane *into* the cells. The cells in Figure 13.11c, in a solution more concentrated than the cell interior, are starting to shrivel as osmosis draws water *out of* the cells. Similarly, if you drink seawater, the seawater actually draws water out of your body as it passes through your stomach and intestines. All of that extra water in your intestine promotes dehydration of bodily tissues and diarrhea. Consequently, seawater should never be consumed.

Chapter **13** in Review

MasteringChemistry™ provides end-of-chapter exercises, feedback-enriched tutorial problems, animations, and interactive activities to encourage problem solving practice and deeper understanding of key concepts and topics.

Self-Assessment Quiz

PEARSON
eText
2.0

Q1. Which compound forms an electrolyte solution when dissolved in water?
(a) KBr
(b) Br_2
(c) CH_3OH
(d) $C_6H_{12}O_6$ (glucose)

Q2. A solution is saturated in O_2 gas and KNO_3 at room temperature. What happens if the solution is warmed to 75 °C?
(a) Solid KNO_3 precipitates out of solution.
(b) Gaseous O_2 bubbles out of solution.
(c) Solid KNO_3 precipitates out of solution *and* gaseous O_2 bubbles out of solution.
(d) Nothing happens (both O_2 and KNO_3 remain in solution).

Q3. What is the mass percent concentration of a solution containing 25.0 g NaCl and 155.0 mL of water? Assume a density of 1.00 g/mL for water.
(a) 2.76%
(b) 6.20%
(c) 13.9%
(d) 16.1%

Q4. A glucose solution is 3.25% glucose by mass and has a density of 1.03 g/mL. What mass of glucose is contained in 58.2 mL of this solution?
(a) 18.4 g
(b) 184 g
(c) 195 g
(d) 1.95 g

Q5. What is the molarity of a solution containing 11.2 g $Ca(NO_3)_2$ in 175 mL of solution?
(a) 64.0 M
(b) 0.390 M
(c) 0.064 M
(d) 1.96×10^3 M

Q6. What mass of glucose ($C_6H_{12}O_6$) is contained in 75.0 mL of a 1.75 M glucose solution?
(a) 23.6 g
(b) 7.72 g
(c) 0.131 g
(d) 75.0 g

Q7. What is the molar concentration of potassium ions in a 0.250 M K_2SO_4 solution?
(a) 0.125 M
(b) 0.250 M
(c) 0.500 M
(d) 0.750 M

Q8. What volume of a 3.50 M Na_3PO_4 solution should you use to make 1.50 L of a 2.55 M Na_3PO_4 solution?
(a) 0.917 L
(b) 13.4 L
(c) 2.06 L
(d) 1.09 L

Q9. Potassium iodide reacts with lead(II) nitrate in the following precipitation reaction:

$$2 \text{ KI}(aq) + \text{Pb(NO}_3)_2(aq) \longrightarrow 2 \text{ KNO}_3(aq) + \text{PbI}_2(s)$$

What minimum volume of 0.200 M potassium iodide solution is required to completely precipitate all of the lead in 155.0 mL of a 0.112 M lead(II) nitrate solution?

(a) 348 mL   (b) 86.8 mL
(c) 43.4 mL   (d) 174 mL

Q10. A solution contains 22.4 g glucose ($C_6H_{12}O_6$) dissolved in 500.0 g of water. What is the molality of the solution?

(a) 0.238 *m*   (b) 0.249 *m*
(c) 44.8 *m*   (d) 4.03 *m*

Q11. Calculate the freezing point of a 1.30 *m* sucrose ($C_{12}H_{22}O_{11}$) solution.

(a) 2.42 °C
(b) −2.42 °C
(c) 0.472 °C
(d) −0.472 °C

Q12. What mass of ethylene glycol ($C_2H_6O_2$) must be added to 250.0 g of water to obtain a solution with a boiling point of 102.5 °C?

(a) 75.7 g
(b) 3.11×10^3 g
(c) 1.22 g
(d) 0.0197 g

Answers: 1:a, 2:b, 3:c, 4:d, 5:b, 6:a, 7:c, 8:d, 9:d, 10:b, 11:b, 12:a

Chemical Principles

Relevance

Solutions

A solution is a homogeneous mixture with two or more components. The solvent is the majority component, and the solute is the minority component. Water is the solvent in aqueous solutions.

Solutions are all around us—most of the fluids that we encounter every day are solutions. Common solutions include seawater (solid and liquid), soda pop (gas and liquid), alcoholic beverages such as vodka (liquid and liquid), air (gas and gas), and blood (solid, gas, and liquid).

Solid-and-Liquid Solutions

The solubility—the amount of solute that dissolves in a certain amount of solvent—of solids in liquids increases with increasing temperature. Recrystallization involves dissolving a solid into hot solvent to saturation and then allowing it to cool. As the solution cools, it becomes supersaturated and the solid crystallizes.

Solutions of solids dissolved in liquids, such as seawater, coffee, and sugar water, are important both in chemistry and in everyday life. Recrystallization is used extensively in the laboratory to purify solids.

Gas-and-Liquid Solutions

The solubility of gases in liquids decreases with increasing temperature but increases with increasing pressure.

The temperature and pressure dependence of gas solubility is the reason that soda pop fizzes when you open the can and the reason that warm soda goes flat.

Solution Concentration

We use solution concentration to specify how much of the solute is present in a given amount of solution. Three common ways to express solution concentration are mass percent, molarity, and molality.

$$\text{mass percent} = \frac{\text{mass solute}}{\text{mass solute} + \text{mass solvent}} \times 100\%$$

$$\text{molarity (M)} = \frac{\text{moles solute}}{\text{liters solution}}$$

$$\text{molality (}m\text{)} = \frac{\text{moles solute}}{\text{kilograms solvent}}$$

Solution concentration is useful in converting between amounts of solute and solution. Mass percent and molarity are the most common concentration units. We use molality to quantify colligative properties such as freezing point depression and boiling point elevation.

Solution Dilution

We can solve solution dilution problems using the following equation:

$$M_1 V_1 = M_2 V_2$$

As many solutions are stored in concentrated form, it is often necessary to dilute them to a desired concentration.

Freezing Point Depression and Boiling Point Elevation

A nonvolatile solute will extend the liquid temperature range of a solution relative to the pure solvent. The freezing point of a solution is lower than the freezing point of the pure solvent, and the boiling point of a solution is higher than the boiling point of the pure solvent. These relationships are quantified by the following equations:

freezing point depression:

$$\Delta T_f = m \times K_f$$

boiling point elevation:

$$\Delta T_b = m \times K_b$$

We add salt to ice in ice-cream makers and use it to melt ice on roads in frigid weather. The salt lowers the freezing point of water, allowing the cream within the ice-cream maker to freeze and the ice on icy roads to melt. We use antifreeze in the cooling systems of cars both to lower the freezing point of the coolant in winter and to raise its boiling point in summer.

Osmosis

Osmosis is the flow of water from a low-concentration solution to a high-concentration solution through a semipermeable membrane.

Osmosis is the reason drinking seawater causes dehydration. As seawater goes through the stomach and intestines, it draws water away from the body through osmosis, resulting in diarrhea and dehydration. To avoid damage to body tissues, transfused fluids must always be isoosmotic with body fluids. Most transfused fluids consist in whole or part of 0.9% mass/volume saline solution.

Chemical Skills

Examples

LO: Calculate mass percent (Section 13.5).

You are given the mass of the solute and the solvent and asked to find the concentration of the solution in mass percent.

To calculate mass percent concentration, divide the mass of the solute by the mass of the solution (solute and solvent) and multiply by 100%.

EXAMPLE **13.11** | **Calculating Mass Percent**

Find the mass percent concentration of a solution containing 19 g of solute and 158 g of solvent.

GIVEN: 19 g solute
158 g solvent

FIND: mass percent

SOLUTION

$$\text{mass percent} = \frac{\text{mass solute}}{\text{mass solute} + \text{mass solvent}} \times 100\%$$

$$\text{mass percent} = \frac{19\ g}{19\ g + 158\ g} \times 100\%$$

$$\text{mass percent} = \frac{19\ g}{177\ g} \times 100\%$$

$$= 11\%$$

LO: Use mass percent in calculations (Section 13.5).

EXAMPLE **13.12** | **Using Mass Percent in Calculations**

How much KCl in grams is in 0.337 L of a 5.80% mass percent KCl solution? (Assume that the density of the solution is 1.05 g/mL.)

SORT

You are given the volume of a potassium chloride solution and its mass percent concentration. You are asked to find the mass of potassium chloride.

GIVEN: 5.80% KCl by mass
0.337 L solution

$$d = \frac{1.05\ g}{mL}$$

FIND: g KCl

STRATEGIZE

Draw a solution map. Begin with the given volume of the solution in L and convert to mL. Use the density to find the mass of the solution. Finally, use the mass percent to get to the mass of potassium chloride.

SOLUTION MAP

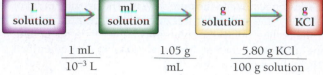

$$\frac{1 \text{ mL}}{10^{-3} \text{ L}} \qquad \frac{1.05 \text{ g}}{\text{mL}} \qquad \frac{5.80 \text{ g KCl}}{100 \text{ g solution}}$$

RELATIONSHIPS USED

$\dfrac{5.80 \text{ g KCl}}{100 \text{ g solution}}$ (given mass percent, written as a fraction)

$\dfrac{1.05 \text{ g}}{\text{mL}}$ (given density)

$1 \text{ mL} = 10^{-3} \text{ L}$ (Table 2.2)

SOLVE

Follow your solution map to calculate the answer.

SOLUTION

$$0.337 \text{ L solution} \times \frac{1 \text{ mL}}{10^{-3} \text{ L}} \times \frac{1.05 \text{ g}}{\text{mL}} \times \frac{5.80 \text{ g KCl}}{100 \text{ g solution}}$$

$$= 20.5 \text{ g KCl}$$

CHECK

Check your answer. Are the units correct? Does the answer make physical sense?

The units (g KCl) are correct. The magnitude of the answer makes sense because 0.337 L is a bit more than 300 g of solution. Each 100 g of solution contains about 6 g KCl. Therefore, the answer should be a bit more than 18 g.

LO: Calculate molarity (Section 13.6).

You are given the number of moles of potassium chloride and the volume of solution. You are asked to find the molarity.

To calculate molarity, divide the number of moles of solute by the volume of the solution in liters.

EXAMPLE 13.13 | **Calculating Molarity**

Calculate the molarity of a KCl solution containing 0.22 mol of KCl in 0.455 L of solution.

GIVEN: 0.22 mol KCl

0.455 L solution

FIND: molarity (M)

SOLUTION

$$\text{molarity (M)} = \frac{0.22 \text{ mol KCl}}{0.455 \text{ L solution}}$$

$$= 0.48 \text{ M}$$

LO: Use molarity in calculations (Section 13.6).

SORT

You are given the volume and molarity of a potassium chloride solution and asked to find the mass of potassium chloride contained in the solution.

STRATEGIZE

Draw a solution map beginning with liters of solution and converting to moles of solute using the molarity as a conversion factor. Then convert to grams using the molar mass.

EXAMPLE 13.14 | **Using Molarity in Calculations**

How much KCl in grams is contained in 0.488 L of 1.25 M KCl solution?

GIVEN: 1.25 M KCl

0.488 L solution

FIND: g KCl

SOLUTION MAP

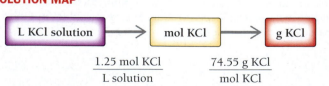

$$\frac{1.25 \text{ mol KCl}}{\text{L solution}} \qquad \frac{74.55 \text{ g KCl}}{\text{mol KCl}}$$

RELATIONSHIPS USED

$$\frac{1.25 \text{ mol KCl}}{\text{L solution}} \text{ (given molarity, written as a fraction)}$$

$$1 \text{ mol} = 74.55 \text{ g (molar mass KCl)}$$

SOLVE
Follow your solution map to calculate the answer.

SOLUTION

$$0.488 \text{ L solution} \times \frac{1.25 \text{ mol KCl}}{\text{L solution}} \times \frac{74.55 \text{ g KCl}}{\text{mol KCl}}$$
$$= 45.5 \text{ g KCl}$$

CHECK
Check your answer. Are the units correct? Does the answer make physical sense?

The units (g KCl) are correct. The magnitude of the answer makes sense because if each liter of solution contains 1.25 mol, then the given amount of solution (which is about 0.5 L) should contain a bit more than 0.5 mol, which would have a mass that is bit more than about 37 g.

LO: Use the dilution equation in calculations (Section 13.7).

You are given the initial molarity and final molarity of a solution as well as the final volume. You are asked to find the initial volume.

Most solution dilution problems will use equation $M_1 V_1 = M_2 V_2$.

Solve the equation for the quantity you are trying to find (in this case, V_1) and substitute in the correct values to calculate it.

EXAMPLE **13.15**   **Solution Dilution**

How much of an 8.0 M HCl solution should you use to make 0.400 L of a 2.7 M HCl solution?

GIVEN: $M_1 = 8.0$ M
$M_2 = 2.7$ M
$V_2 = 0.400$ L

FIND: V_1

SOLUTION

$$M_1 V_1 = M_2 V_2$$
$$V_1 = \frac{M_2 V_2}{M_1}$$
$$= \frac{2.7 \frac{\text{mol}}{\text{L}} \times 0.400 \text{ L}}{8.0 \frac{\text{mol}}{\text{L}}}$$
$$= 0.14 \text{ L}$$

LO: Use volume and concentration to calculate the number of moles of reactants or products and then use stoichiometric coefficients to convert to other quantities in a reaction (Section 13.8).

SORT
You are given the volume and concentration of a hydrochloric acid solution as well as the concentration of a sodium hydroxide solution with which it reacts. You are asked to find the volume of the sodium hydroxide solution that will completely react with the hydrochloric acid.

EXAMPLE **13.16**   **Solution Stoichiometry**

Consider the reaction:

$$HCl(aq) + NaOH(aq) \longrightarrow NaCl(aq) + H_2O(l)$$

How much 0.113 M NaOH solution will completely neutralize 1.25 L of 0.228 M HCl solution?

GIVEN: 1.25 L HCl solution
0.228 M HCl
0.113 M NaOH

FIND: L NaOH solution

STRATEGIZE

Draw a solution map. Use the volume and molarity of HCl to get to mol HCl. Then use the stoichiometric coefficients to convert to mole NaOH. Finally, convert back to volume of NaOH using the molarity of NaOH.

SOLUTION MAP

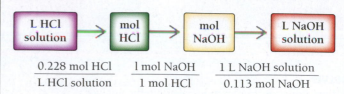

$$\frac{0.228 \text{ mol HCl}}{\text{L HCl solution}} \qquad \frac{1 \text{ mol NaOH}}{1 \text{ mol HCl}} \qquad \frac{1 \text{ L NaOH solution}}{0.113 \text{ mol NaOH}}$$

RELATIONSHIPS USED

$$M\,(\text{HCl}) = \frac{0.228 \text{ mol HCl}}{\text{L HCl solution}} \text{ (given concentration of HCl solution, written as a fraction)}$$

$$M\,(\text{NaOH}) = \frac{0.113 \text{ mol NaOH}}{\text{L NaOH solution}} \text{ (given concentration of NaOH solution, written as a fraction)}$$

1 mol HCl:1 mol NaOH (stoichiometric relationship between HCl and NaOH, from balanced equation)

SOLVE

Follow the solution map to calculate the answer.

SOLUTION

$$1.25 \text{ L HCl solution} \times \frac{0.228 \text{ mol HCl}}{\text{L HCl solution}}$$

$$\times \frac{1 \text{ mol NaOH}}{1 \text{ mol HCl}} \times \frac{\text{L NaOH solution}}{0.113 \text{ mol NaOH}}$$

$$= 2.52 \text{ L NaOH solution}$$

CHECK

Check your answer. Are the units correct? Does the answer make physical sense?

The units (L NaOH solution) are correct. The magnitude of the answer makes sense because the sodium hydroxide solution is about half as concentrated as the hydrochloric acid. Since the reaction stoichiometry is 1:1, the volume of the sodium hydroxide solution should be about twice the volume of the hydrochloric acid.

LO: Calculate molality (Section 13.9).

You are given the number of moles of sucrose and the mass of water into which it is dissolved. You are asked to find the molality of the resulting solution.

EXAMPLE **13.17**   **Calculating Molality**

Calculate the molality of a solution containing 0.183 mol of sucrose dissolved in 1.10 kg of water.

GIVEN: 0.183 mol of sucrose
1.10 kg H_2O

FIND: molality (m)

SOLUTION

Substitute the correct values into the definition of molality and calculate the answer. If any of the quantities are not in the correct units, convert them into the correct units before substituting into the equation.

$$\text{molality}\,(m) = \frac{\text{moles solute}}{\text{kilograms solvent}}$$

$$\text{molality}\,(m) = \frac{0.183 \text{ mol sucrose}}{1.10 \text{ kg } H_2O}$$

$$= 0.166 \ m$$

LO: Calculate freezing points and boiling points for solutions (Section 13.9).

You are given the molality of the solution and asked to find its freezing point.

To find ΔT_f or ΔT_b, simply substitute the values into the equation and calculate the answer.

The freezing point is the freezing point of pure water $(0.00\,°C) - \Delta T_f$.

EXAMPLE 13.18   Freezing Point Depression

Calculate the freezing point of a 2.5 *m* aqueous sucrose solution.

GIVEN: 2.5 *m* solution

FIND: ΔT_f

SOLUTION

$$\Delta T_f = m \times K_f$$

$$= 2.5\,\frac{\text{mol solute}}{\text{kg solvent}} \times 1.86\,\frac{°C\ \text{kg solvent}}{\text{mol solute}}$$

$$= 4.7\,°C$$

$$\text{freezing point} = 0.00\,°C - 4.7\,°C$$

$$= -4.7\,°C$$

Key Terms

boiling point elevation [13.9]
colligative properties [13.9]
concentrated solution [13.5]
dilute solution [13.5]
electrolyte solution [13.3]
freezing point depression [13.9]

Henry's law [13.4]
mass percent composition (or mass percent) [13.5]
membrane [13.10]
molality (*m*) [13.9]
molarity (M) [13.6]
nonelectrolyte solution [13.3]

osmosis [13.10]
osmotic pressure [13.10]
recrystallization [13.3]
saturated solution [13.3]
semipermeable membrane [13.10]
solubility [13.3]

solute [13.2]
solution [13.2]
solvent [13.2]
stock solution [13.7]
supersaturated solution [13.3]
unsaturated solution [13.3]

Exercises

Questions

1. What is a solution? List some examples.
2. What is an aqueous solution?
3. In a solution, what is the solvent? What is the solute? List some examples.
4. Explain what "like dissolves like" means.
5. What is solubility?
6. Describe what happens when additional solute is added to:
   (a) a saturated solution
   (b) an unsaturated solution
   (c) a supersaturated solution
7. Explain the difference between a strong electrolyte solution and a nonelectrolyte solution. What kinds of solutes form strong electrolyte solutions?
8. How does gas solubility depend on temperature?
9. Explain recrystallization.
10. How is rock candy made?
11. When you heat water on a stove, bubbles form on the bottom of the pot *before* the water boils. What are these bubbles? Why do they form?
12. Explain why warm soda pop goes flat faster than cold soda pop.

13. How does gas solubility depend on pressure? How does this relationship explain why a can of soda pop fizzes when opened?
14. What is the difference between a dilute solution and a concentrated solution?
15. Define the concentration units mass percent and molarity.
16. What is a stock solution?
17. How does the presence of a nonvolatile solute affect the boiling point and melting point of a solution relative to the boiling point and melting point of the pure solvent?
18. What are colligative properties?
19. Define molality.
20. What is osmosis?
21. Two shipwreck survivors were rescued from a life raft. One had drunk seawater while the other had not. The one who had drunk the seawater was more severely dehydrated than the one who did not. Explain.
22. Why are intravenous fluids always isoosmotic saline solutions? What would happen if pure water were administered intravenously?

Problems

SOLUTIONS

23. Determine whether or not each mixture is a solution.
  (a) sand and water mixture
  (b) oil and water mixture
  (c) salt and water mixture
  (d) sterling silver cup

24. Determine whether or not each mixture is a solution.
  (a) air
  (b) carbon dioxide and water mixture
  (c) a blueberry muffin
  (d) a brass buckle

25. Identify the solute and solvent in each solution.
  (a) saltwater
  (b) sugar water
  (c) soda water

26. Identify the solute and solvent in each solution.
  (a) 80-proof vodka (40% ethyl alcohol)
  (b) oxygenated water
  (c) antifreeze (ethylene glycol and water)

27. Pick an appropriate solvent from Table 13.2 to dissolve:
  (a) motor oil (nonpolar)
  (b) sugar (polar)
  (c) lard (nonpolar)
  (d) potassium chloride (ionic)

28. Pick an appropriate solvent from Table 13.2 to dissolve:
  (a) glucose (polar)
  (b) salt (ionic)
  (c) vegetable oil (nonpolar)
  (d) sodium nitrate (ionic)

SOLIDS DISSOLVED IN WATER

29. What are the dissolved particles in a solution containing an ionic solute? What is the name for this kind of solution?

30. What are the dissolved particles in a solution containing a molecular solute? What is the name for this kind of solution?

31. A solution contains 35 g of NaCl per 100 g of water at 25 °C. Is the solution unsaturated, saturated, or supersaturated? (See Figure 13.4.)

32. A solution contains 28 g of KNO_3 per 100 g of water at 25 °C. Is the solution unsaturated, saturated, or supersaturated? (See Figure 13.4.)

33. A KNO_3 solution containing 45 g of KNO_3 per 100 g of water is cooled from 40 °C to 0 °C. What happens during cooling? (See Figure 13.4.)

34. A KCl solution containing 42 g of KCl per 100 g of water is cooled from 60 °C to 0 °C. What happens during cooling? (See Figure 13.4.)

35. Refer to Figure 13.4 to determine whether each of the given amounts of solid will completely dissolve in the given amount of water at the indicated temperature.
  (a) 30.0 g $KClO_3$ in 85.0 g of water at 35 °C
  (b) 65.0 g $NaNO_3$ in 125 g of water at 15 °C
  (c) 32.0 g KCl in 70.0 g of water at 82 °C

36. Refer to Figure 13.4 to determine whether each of the given amounts of solid will completely dissolve in the given amount of water at the indicated temperature.
  (a) 45.0 g $CaCl_2$ in 105 g of water at 5 °C
  (b) 15.0 g $KClO_3$ in 115 g of water at 25 °C
  (c) 50.0 g $Pb(NO_3)_2$ in 95.0 g of water at 10 °C

GASES DISSOLVED IN WATER

37. Some laboratory procedures involving oxygen-sensitive reactants or products call for using preboiled (and then cooled) water. Explain why this is so.

38. A person preparing a fish tank uses preboiled (and then cooled) water to fill it. When the fish is put into the tank, it dies. Explain.

39. Scuba divers breathing air at increased pressure can suffer from nitrogen narcosis—a condition resembling drunkenness—when the partial pressure of nitrogen exceeds about 4 atm. What property of gas/water solutions causes this to happen? How could the diver reverse this effect?

40. Scuba divers breathing air at increased pressure can suffer from oxygen toxicity—too much oxygen in the bloodstream—when the partial pressure of oxygen exceeds about 1.4 atm. What happens to the amount of oxygen in a diver's bloodstream when he or she breathes oxygen at elevated pressures? How can this be reversed?

MASS PERCENT

41. Calculate the concentration of each solution in mass percent.
  (a) 41.2 g $C_{12}H_{22}O_{11}$ in 498 g H_2O
  (b) 178 mg $C_6H_{12}O_6$ in 4.91 g H_2O
  (c) 7.55 g NaCl in 155 g H_2O

42. Calculate the concentration of each solution in mass percent.
  (a) 132 g KCl in 598 g H_2O
  (b) 22.3 mg KNO_3 in 2.84 g H_2O
  (c) 8.72 g C_2H_6O in 76.1 g H_2O

43. A soft drink contains 42 g of sugar in 311 g of H_2O. What is the concentration of sugar in the soft drink in mass percent?

44. A soft drink contains 32 mg of sodium in 309 g of H_2O. What is the concentration of sodium in the soft drink in mass percent?

45. Complete the table.

Mass Solute	Mass Solvent	Mass Solution	Mass Percent
15.5 g	238.1 g	____	____
22.8 g	____	____	12.0%
____	183.3 g	212.1 g	____
____	315.2 g	____	15.3%

46. Complete the table.

Mass Solute	Mass Solvent	Mass Solution	Mass Percent
2.55 g	25.0 g	____	____
____	45.8 g	____	3.8%
1.38 g	____	27.2 g	____
23.7 g	____	____	5.8%

47. Ocean water contains 3.5% NaCl by mass. How much salt can be obtained from 254 g of seawater?

48. A saline solution contains 1.1% NaCl by mass. How much NaCl is present in 96.3 g of this solution?

49. Determine the amount of sucrose in each solution.
  (a) 48 g of a solution containing 3.7% sucrose by mass
  (b) 103 mg of a solution containing 10.2% sucrose by mass
  (c) 3.2 kg of a solution containing 14.3% sucrose by mass

50. Determine the amount of potassium chloride in each solution.
  (a) 19.7 g of a solution containing 1.08% KCl by mass
  (b) 23.2 kg of a solution containing 18.7% KCl by mass
  (c) 38 mg of a solution containing 12% KCl by mass

51. Determine the mass (in g) of each NaCl solution that contains 1.5 g of NaCl.
  (a) 0.058% NaCl by mass
  (b) 1.46% NaCl by mass
  (c) 8.44% NaCl by mass

52. Determine the mass (in g) of each sucrose solution that contains 12 g of sucrose.
  (a) 4.1% sucrose by mass
  (b) 3.2% sucrose by mass
  (c) 12.5% sucrose by mass

53. $AgNO_3$ solutions are often used to plate silver onto other metals. What is the maximum amount of silver in grams that can be plated out of 4.8 L of an $AgNO_3$ solution containing 3.4% Ag by mass? (Assume that the density of the solution is 1.01 g/mL.)

54. A dioxin-contaminated water source contains 0.085% dioxin by mass. How much dioxin is present in 2.5 L of this water? (Assume that the density of the solution is 1.01 g/mL.)

55. Ocean water contains 3.5% NaCl by mass. What mass of ocean water in grams contains 45.8 g of NaCl?

56. A hard-water sample contains 0.0085% Ca by mass (in the form of Ca^{2+} ions). What mass of water in grams contains 1.2 g of Ca? (1.2 g of Ca is the recommended daily allowance of calcium for 19- to 24-year-olds.)

57. Lead is a toxic metal that affects the central nervous system. A Pb-contaminated water sample contains 0.0011% Pb by mass. What volume of the water in milliliters contains 115 mg of Pb? (Assume that the density of the solution is 1.0 g/mL.)

58. Benzene is a carcinogenic (cancer-causing) compound. A benzene-contaminated water sample contains 0.000037% benzene by mass. What volume of the water in liters contains 175 mg of benzene? (Assume that the density of the solution is 1.0 g/mL.)

MOLARITY

59. Calculate the molarity of each solution.
  (a) 0.127 mol of sucrose in 655 mL of solution
  (b) 0.205 mol of KNO_3 in 0.875 L of solution
  (c) 1.1 mol of KCl in 2.7 L of solution

60. Calculate the molarity of each solution.
  (a) 1.54 mol of LiCl in 22.2 L of solution
  (b) 0.101 mol of $LiNO_3$ in 6.4 L of solution
  (c) 0.0323 mol of glucose in 76.2 mL of solution

61. Calculate the molarity of each solution.
(a) 22.6 g of $C_{12}H_{22}O_{11}$ in 0.442 L of solution
(b) 42.6 g of NaCl in 1.58 L of solution
(c) 315 mg of $C_6H_{12}O_6$ in 58.2 mL of solution

62. Calculate the molarity of each solution.
(a) 33.2 g of KCl in 0.895 L of solution
(b) 61.3 g of C_2H_6O in 3.4 L of solution
(c) 38.2 mg of KI in 112 mL of solution

63. A 205-mL sample of ocean water contains 6.8 g of NaCl. What is the molarity of the solution with respect to NaCl?

64. A 355-mL can of soda pop contains 41 g of sucrose $(C_{12}H_{22}O_{11})$. What is the molarity of the solution with respect to sucrose?

65. How many moles of NaCl are contained in each solution?
(a) 1.5 L of a 1.2 M NaCl solution
(b) 0.448 L of a 0.85 M NaCl solution
(c) 144 mL of a 1.65 M NaCl solution

66. How many moles of sucrose are contained in each solution?
(a) 3.4 L of a 0.100 M sucrose solution
(b) 0.952 L of a 1.88 M sucrose solution
(c) 21.5 mL of a 0.528 M sucrose solution

67. What volume of each solution contains 0.15 mol of KCl?
(a) 0.255 M KCl
(b) 1.8 M KCl
(c) 0.995 M KCl

68. What volume of each solution contains 0.325 mol of NaI?
(a) 0.152 M NaI
(b) 0.982 M NaI
(c) 1.76 M NaI

69. Complete the table.

Solute	Solute Mass	Mol Solute	Volume Solution	Molarity
KNO_3	22.5 g	____	125.0 mL	____
$NaHCO_3$	____	____	250.0 mL	0.100 M
$C_{12}H_{22}O_{11}$	55.38 g	____	____	0.150 M

70. Complete the table.

Solute	Solute Mass	Mol Solute	Volume Solution	Molarity
$MgSO_4$	0.588 g	____	25.0 mL	____
NaOH	____	____	0.100 mL	1.75 M
CH_3OH	12.5 g	____	____	0.500 M

71. Calculate the mass of NaCl in a 35-mL sample of a 1.3 M NaCl solution.

72. Calculate the mass of glucose $(C_6H_{12}O_6)$ in a 105-mL sample of a 1.02 M glucose solution.

73. A chemist wants to make 2.5 L of a 0.100 M KCl solution. How much KCl in grams should the chemist use?

74. A laboratory procedure calls for making 500.0 mL of a 1.4 M KNO_3 solution. How much KNO_3 in grams is needed?

75. How many liters of a 0.500 M sucrose $(C_{12}H_{22}O_{11})$ solution contain 1.5 kg of sucrose?

76. What volume of a 0.35 M $Mg(NO_3)_2$ solution contains 87 g of $Mg(NO_3)_2$?

77. Determine the concentration of Cl^- in each aqueous solution. (Assume complete dissociation of each compound.)
(a) 0.15 M NaCl
(b) 0.15 M $CuCl_2$
(c) 0.15 M $AlCl_3$

78. Determine the concentration of NO_3^- in each aqueous solution. (Assume complete dissociation of each compound.)
(a) 0.10 M KNO_3
(b) 0.10 M $Ca(NO_3)_2$
(c) 0.10 M $Cr(NO_3)_3$

79. Determine the concentration of the cation and anion in each aqueous solution. (Assume complete dissociation of each compound.)
(a) 0.12 M Na_2SO_4
(b) 0.25 M K_2CO_3
(c) 0.11 M RbBr

80. Determine the concentration of the cation and anion in each aqueous solution. (Assume complete dissociation of each compound.)
(a) 0.20 M $SrSO_4$
(b) 0.15 M $Cr_2(SO_4)_3$
(c) 0.12 M SrI_2

SOLUTION DILUTION

81. A 122-mL sample of a 1.2 M sucrose solution is diluted to 500.0 mL. What is the molarity of the diluted solution?

82. A 3.5-L sample of a 5.8 M NaCl solution is diluted to 55 L. What is the molarity of the diluted solution?

83. Describe how you would make 2.5 L of a 0.100 M KCl solution from a 5.5 M stock KCl solution.

84. Describe how you would make 500.0 mL of a 0.200 M NaOH solution from a 15.0 M stock NaOH solution.

85. To what volume should you dilute 25 mL of a 12 M stock HCl solution to obtain a 0.500 M HCl solution?

86. To what volume should you dilute 75 mL of a 10.0 M H_2SO_4 solution to obtain a 1.75 M H_2SO_4 solution?

87. How much of a 12.0 M HNO_3 solution should you use to make 850.0 mL of a 0.250 M HNO_3 solution?

88. How much of a 5.0 M sucrose solution should you use to make 85.0 mL of a 0.040 M solution?

SOLUTION STOICHIOMETRY

89. Determine the volume of 0.150 M NaOH solution required to neutralize each sample of hydrochloric acid. The neutralization reaction is:

$$NaOH(aq) + HCl(aq) \longrightarrow H_2O(l) + NaCl(aq)$$

(a) 25 mL of a 0.150 M HCl solution
(b) 55 mL of a 0.055 M HCl solution
(c) 175 mL of a 0.885 M HCl solution

90. Determine the volume of 0.225 M KOH solution required to neutralize each sample of sulfuric acid. The neutralization reaction is:

$$H_2SO_4(aq) + 2\,KOH(aq) \longrightarrow K_2SO_4(aq) + 2\,H_2O(l)$$

(a) 45 mL of 0.225 M H_2SO_4
(b) 185 mL of 0.125 M H_2SO_4
(c) 75 mL of 0.100 M H_2SO_4

91. Consider the reaction:

$$2\,K_3PO_4(aq) + 3\,NiCl_2(aq) \longrightarrow Ni_3(PO_4)_2(s) + 6\,KCl(aq)$$

What volume of 0.225 M K_3PO_4 solution is necessary to completely react with 134 mL of 0.0112 M $NiCl_2$?

92. Consider the reaction:

$$K_2S(aq) + Co(NO_3)_2(aq) \longrightarrow 2\,KNO_3(aq) + CoS(s)$$

What volume of 0.225 M K_2S solution is required to completely react with 175 mL of 0.115 M $Co(NO_3)_2$?

93. A 10.0-mL sample of an unknown H_3PO_4 solution requires 112 mL of 0.100 M KOH to completely react with the H_3PO_4. What was the concentration of the unknown H_3PO_4 solution?

$$H_3PO_4(aq) + 3\,KOH(aq) \longrightarrow 3\,H_2O(l) + K_3PO_4(aq)$$

94. A 25.0-mL sample of an unknown $HClO_4$ solution requires 45.3 mL of 0.101 M NaOH for complete neutralization. What was the concentration of the unknown $HClO_4$ solution? The neutralization reaction is:

$$HClO_4(aq) + NaOH(aq) \longrightarrow H_2O(l) + NaClO_4(aq)$$

95. What is the minimum amount of 6.0 M H_2SO_4 necessary to produce 15.0 g of $H_2(g)$ according to the reaction:

$$2\,Al(s) + 3\,H_2SO_4(aq) \longrightarrow Al_2(SO_4)_3(aq) + 3\,H_2(g)$$

96. What is the molarity of $ZnCl_2(aq)$ that forms when 15.0 g of zinc completely reacts with $CuCl_2(aq)$ according to the following reaction? (Assume a final volume of 175 mL.)

$$Zn(s) + CuCl_2(aq) \longrightarrow ZnCl_2 + Cu(s)$$

MOLALITY, FREEZING POINT DEPRESSION, AND BOILING POINT ELEVATION

97. Calculate the molality of each solution.
(a) 0.25 mol solute; 0.250 kg solvent
(b) 0.882 mol solute; 0.225 kg solvent
(c) 0.012 mol solute; 23.1 g solvent

98. Calculate the molality of each solution.
(a) 0.455 mol solute; 1.97 kg solvent
(b) 0.559 mol solute; 1.44 kg solvent
(c) 0.119 mol solute; 488 g solvent

99. Calculate the molality of a solution containing 12.5 g of ethylene glycol ($C_2H_6O_2$) dissolved in 135 g of water.

100. Calculate the molality of a solution containing 257 g glucose ($C_6H_{12}O_6$) dissolved in 1.62 L of water. (Assume a density of 1.00 g/mL for water.)

101. Calculate the freezing point of a water solution at each concentration.
(a) 0.85 m **(b)** 1.45 m **(c)** 4.8 m **(d)** 2.35 m

102. Calculate the freezing point of a water solution at each concentration.
(a) 0.100 m **(b)** 0.469 m **(c)** 1.44 m **(d)** 5.89 m

103. Calculate the boiling point of a water solution at each concentration.
(a) 0.118 m **(b)** 1.94 m **(c)** 3.88 m **(d)** 2.16 m

104. Calculate the boiling point of a water solution at each concentration.
(a) 0.225 m **(b)** 2.58 m **(c)** 4.33 m **(d)** 6.77 m

105. A glucose solution contains 55.8 g of glucose ($C_6H_{12}O_6$) in 455 g of water. Calculate the freezing point and boiling point of the solution. (Assume a density of 1.00 g/mL for water.)

106. An ethylene glycol solution contains 21.2 g of ethylene glycol ($C_2H_6O_2$) in 85.4 mL of water. Calculate the freezing point and boiling point of the solution. (Assume a density of 1.00 g/mL for water.)

Cumulative Problems

107. An NaCl solution is made using 133 g of NaCl and diluting to a total solution volume of 1.00 L. Calculate the molarity and mass percent of the solution. (Assume a density of 1.08 g/mL for the solution.)

108. A KNO$_3$ solution is made using 88.4 g of KNO$_3$ and diluting to a total solution volume of 1.50 L. Calculate the molarity and mass percent of the solution. (Assume a density of 1.05 g/mL for the solution.)

109. A 125-mL sample of an 8.5 M NaCl solution is diluted to 2.5 L. What volume of the diluted solution contains 10.8 g of NaCl? (*Hint:* Figure out the concentration of the diluted solution first.)

110. A 45.8-mL sample of a 5.8 M KNO$_3$ solution is diluted to 1.00 L. What volume of the diluted solution contains 15.0 g of KNO$_3$? (*Hint:* Figure out the concentration of the diluted solution first.)

111. To what final volume should you dilute 50.0 mL of a 5.00 M KI solution so that 25.0 mL of the diluted solution contains 3.25 g of KI?

112. To what volume should you dilute 125 mL of an 8.00 M CuCl$_2$ solution so that 50.0 mL of the diluted solution contains 5.9 g CuCl$_2$?

113. What is the molarity of an aqueous solution that is 5.88% NaCl by mass? (Assume a density of 1.02 g/mL for the solution.) (*Hint:* 5.88% NaCl by mass means 5.88 g NaCl/100.0 g solution.)

114. What is the molarity of an aqueous solution that is 6.75% glucose (C$_6$H$_{12}$O$_6$) by mass? (Assume a density of 1.03 g/mL for the solution.) (*Hint:* 6.75 % C$_6$H$_{12}$O$_6$ by mass means 6.75 g C$_6$H$_{12}$O$_6$/100.0 g solution.)

115. Consider the reaction:

$$2 \, Al(s) + 3 \, H_2SO_4(aq) \longrightarrow Al_2(SO_4)_3(aq) + 3 \, H_2(g)$$

What minimum volume of 4.0 M H$_2$SO$_4$ is required to produce 15.0 L of H$_2$ at STP?

116. Consider the reaction:

$$Mg(s) + 2 \, HCl(aq) \longrightarrow MgCl_2(aq) + H_2(g)$$

What minimum amount of 1.85 M HCl is necessary to produce 28.5 L of H$_2$ at STP?

117. How much of a 1.25 M sodium chloride solution in milliliters is required to completely precipitate all of the silver in 25.0 mL of a 0.45 M silver nitrate solution?

118. How much of a 1.50 M sodium sulfate solution in milliliters is required to completely precipitate all of the barium in 150.0 mL of a 0.250 M barium nitrate solution?

119. Nitric acid is usually purchased in concentrated form with a 70.3% HNO$_3$ concentration by mass and a density of 1.41 g/mL. How much of the concentrated stock solution in milliliters should you use to make 2.5 L of 0.500 M HNO$_3$?

120. Hydrochloric acid is usually purchased in concentrated form with a 37.0% HCl concentration by mass and a density of 1.20 g/mL. How much of the concentrated stock solution in milliliters should you use to make 2.5 L of 0.500 M HCl?

121. An ethylene glycol solution is made using 58.5 g of ethylene glycol (C$_2$H$_6$O$_2$) and diluting to a total volume of 500.0 mL. Calculate the freezing point and boiling point of the solution. (Assume a density of 1.09 g/mL for the solution.)

122. A sucrose solution is made using 144 g of sucrose (C$_{12}$H$_{22}$O$_{11}$) and diluting to a total volume of 1.00 L. Calculate the freezing point and boiling point of the solution. (Assume a density of 1.06 g/mL for the final solution.)

123. A 250.0-mL sample of a 5.00 M glucose (C$_6$H$_{12}$O$_6$) solution is diluted to 1.40 L. What are the freezing and boiling points of the final solution? (Assume a density of 1.06 g/mL for the final solution.)

124. A 135-mL sample of a 10.0 M ethylene glycol (C$_2$H$_6$O$_2$) solution is diluted to 1.50 L. What are the freezing and boiling points of the final solution? (Assume a density of 1.05 g/mL for the final solution.)

125. An aqueous solution containing 17.5 g of an unknown molecular (nonelectrolyte) compound in 100.0 g of water has a freezing point of −1.8 °C. Calculate the molar mass of the unknown compound.

126. An aqueous solution containing 35.9 g of an unknown molecular (nonelectrolyte) compound in 150.0 g of water has a freezing point of −1.3 °C. Calculate the molar mass of the unknown compound.

127. What is the boiling point of an aqueous solution that freezes at −6.7 °C? (*Hint:* Begin by calculating the molality of the solution.)

128. What is the freezing point of an aqueous solution that boils at 102.1 °C? (*Hint:* Begin by calculating the molality of the solution.)

129. A 125-g sample contains only glucose ($C_6H_{12}O_6$) and sucrose ($C_{12}H_{22}O_{11}$). When the sample is added to 0.500 kg of pure water, the resulting solution has a freezing point of $-1.75\,°C$. What were the masses of glucose and sucrose in the original sample?

130. A 13.03-g sample contains only ethylene glycol ($C_2H_6O_2$) and propylene glycol ($C_3H_8O_2$). When the sample is added to 100.0 g of pure water, the resulting solution has a freezing point of $-3.50\,°C$. What was the percent composition of ethylene glycol and propylene glycol in the original sample?

Highlight Problems

131. Consider the molecular views of osmosis cells. For each cell, determine the direction of water flow.

(a)

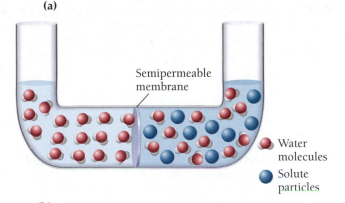

Semipermeable
membrane

Water
molecules

Solute
particles

(b)

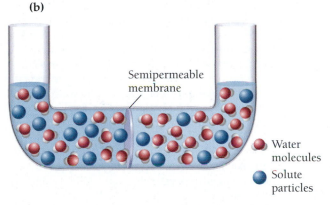

Semipermeable
membrane

Water
molecules

Solute
particles

(c)

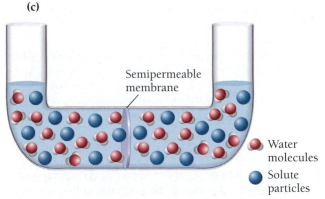

Semipermeable
membrane

Water
molecules

Solute
particles

132. What is wrong with this molecular view of a sodium chloride solution? What would make the picture correct?

133. The Safe Drinking Water Act (SDWA) sets a limit for mercury—a toxin to the central nervous system—at 0.002 mg/L. Water suppliers must periodically test their water to ensure that mercury levels do not exceed 0.002 mg/L. Suppose water is contaminated with mercury at twice the legal limit (0.004 mg/L). How much of this water would a person have to consume to ingest 0.100 g of mercury?

▲ Drinking water must be tested for the presence of various pollutants, including mercury compounds that can damage the nervous system.

134. Water softeners often replace calcium ions in hard water with sodium ions. Since sodium compounds are soluble, the presence of sodium ions in water does not result in the white, scaly residues caused by calcium ions. However, calcium is more beneficial to human health than sodium. Calcium is a necessary part of the human diet, while high levels of sodium intake are linked to increases in blood pressure. The Food and Drug Administration (FDA) recommends that adults ingest less than 2.4 g of sodium per day. How many liters of softened water, containing a sodium concentration of 0.050% sodium by mass, have to be consumed to exceed the FDA recommendation? (Assume a density of 1.0 g/mL for water.)

Questions for Group Work

Discuss these questions with the group and record your consensus answer.

135. 13.62 g (about one tablespoon) of table sugar (sucrose, $C_{12}H_{22}O_{11}$) is dissolved in 241.5 mL of water (density 0.997 g/mL). The final volume is 250.0 mL (about one cup). Have each group member calculate one of the following, and present their answer to the group.
   (a) mass percent
   (b) molarity
   (c) molality

136. Calculate the expected boiling and freezing points for the solution in Question 135. If you boil this syrup for a recipe, would you expect it to take more time to come to a boil than it takes to boil the same amount of pure water? Why or why not? Would the syrup freeze in a typical freezer ($-18\,°C$)? Why or why not?

137. 1 L of a 4 M solution of compound X is diluted to a final volume of 4 L. What is the final concentration? Draw a detailed diagram (including before and after) to illustrate the dilution process. In your diagram, indicate one mole of compound X with an "X" symbol.

138. It has been proposed that uranium be extracted from seawater to fuel nuclear power plants. If the concentration of uranium in seawater is 3.2 μg/L, how much seawater must be processed to generate one pound of uranium?

Data Interpretation and Analysis

139. Read CHEMISTRY IN THE ENVIRONMENT: **Water Pollution and the Flint River Water Crisis** at the end of Section 12.8. The table shown here features a set of data on lead levels in drinking water in Flint, Michigan, collected by the Virginia Tech team described in the box. The lead levels in water are expressed in units of parts per billion (ppb), which is a way of reporting solution concentration that is similar to mass percent. Mass percent is the number of grams of solute per 100 grams solution, while ppb is the number of grams of solute per 10^9 grams solution. In other words, 1 ppb Pb $= 1\,g\,Pb/10^9$ g solution. Examine the tabulated data and answer the questions that follow.

(a) Determine the average value of lead for first draw, 45-second flush, and 2-minute flush (round to three significant figures).

(b) Do the data support the idea that running the tap water before taking a sample made the lead levels in the water appear lower? Why might this be the case?

(c) The EPA requires water providers to monitor drinking water at customer taps. If lead concentrations exceed 15 ppb in 10% or more of the taps sampled, the water provider must notify the customer and take steps to control the corrosiveness of the water. If the water provider in Flint had used first-draw samples to monitor lead levels, would it have been required to take action by EPA requirements? If the Flint water provider used 2-minute flush samples, would it have had to take action? Which drawing technique do you think more closely mimics the way residents actually use their water?

(d) Using the highest value of lead from the first-draw data set, and assuming a resident drinks 2 L of water per day, calculate the mass of lead that the resident would consume over the course of 1 year. (Assume the water has a density of 1.0 g/mL.)

Lead Levels in Flint Tap Water

Sample #	Lead level first draw (ppb)	Lead level 45-sec flush (ppb)	Lead level 2-min flush (ppb)
1	0.344	0.226	0.145
2	8.133	10.77	2.761
3	1.111	0.11	0.123
4	8.007	7.446	3.384
5	1.951	0.048	0.035
6	7.2	1.4	0.2
7	40.63	9.726	6.132
8	1.1	2.5	0.1
9	10.6	1.038	1.294
10	6.2	4.2	2.3
11	4.358	0.822	0.147
12	24.37	8.796	4.347
13	6.609	5.752	1.433
14	4.062	1.099	1.085
15	29.59	3.258	1.843

Source: FlintWaterStudy.org (2015) "Lead Results from Tap Water Sampling in Flint, MI during the Flint Water Crisis"

Answers to Skillbuilder Exercises

Answers to Conceptual Checkpoints

13.1 (a) CH_3Cl and NH_3 are both polar compounds, and KF is ionic. All three would therefore interact more strongly with water molecules (which are polar) than CCl_4, which is nonpolar.

13.2 (b) Some potassium chloride precipitates out of solution. The solubility of most solids decreases with decreasing temperature. However, the solubility of gases increases with decreasing temperature. Therefore, the nitrogen becomes more soluble and will not bubble out of solution.

13.3 (c) The solution is 0.30 M in K^+ because the compound K_2SO_4 forms two moles of K^+ in solution for each mole of K_2SO_4 that dissolves.

13.4 (a) The original 100.0 mL solution contains 0.10 mol, which is present in 1.0 L after the dilution. Therefore, the molarity of the diluted solution is 0.10 M.

13.5 (d) You need twice as many moles of B as you have moles of A. But since the solution of A is twice as concentrated, you need four times the volume of B as you have of A.

13.6 (a) A 2.0 m solution would be made by adding 2 mol of solute to 1 kg of solvent. 1 kg of water has a volume of 1 L, but because of the dissolved solute, the final solution would have a volume of slightly *more than* 1 L. A 2.0 M solution, by contrast, would consist of 2 mol of solute in a solution of *exactly* 1 L. Therefore, a 2 M aqueous solution would be slightly more concentrated than a 2 m solution.

13.7 (d) Since boiling point elevation depends only on the *concentration* of the dissolved particles, and not on the *kind* of dissolved particles, all of these solutions have the same boiling point.

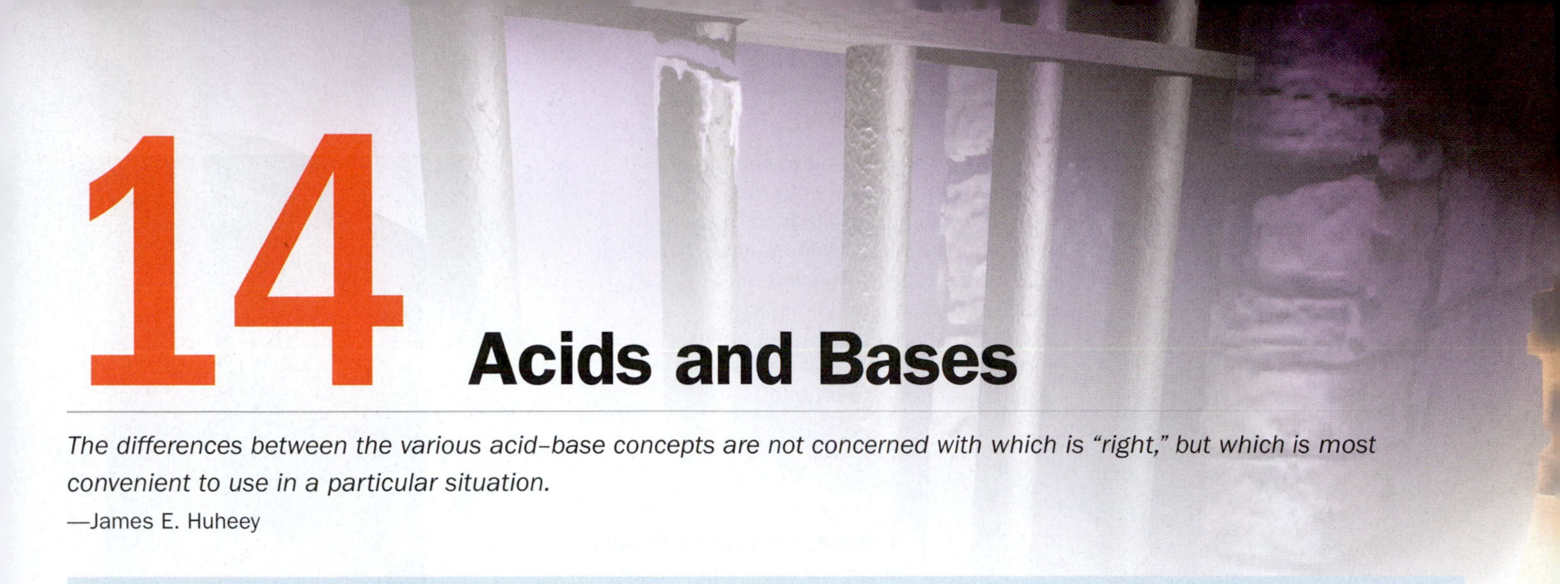

14

Acids and Bases

The differences between the various acid–base concepts are not concerned with which is "right," but which is most convenient to use in a particular situation.

—James E. Huheey

14.1 Sour Patch Kids and International Spy Movies

Gummy candies have a sweet taste and chewy texture that both children and adults can enjoy. From the original classic gummy bear to the gummy worm to the gummy just-about-any-shape-you-can-imagine, these candies are incredibly popular. A common variation is the *sour* gummy candy, whose best-known incarnation is the Sour Patch Kid. Sour Patch Kids are gummy candies shaped like children and coated with a white powder. When you first put a Sour Patch Kid in your mouth, it tastes incredibly sour. The taste is caused by the white powder coating, a mixture of citric acid and tartaric acid. Like all acids, citric and tartaric acid taste sour.

A number of other foods contain acids as well. Acids are responsible for the sour taste of lemons and limes, the bite of sourdough bread, and the tang of a tomato. Acids are substances that—by one definition that we elaborate on later in this chapter—produce H^+ ions in solution. When the citric and tartaric acids from a Sour Patch Kid combine with saliva in the mouth, they produce H^+ ions. Those H^+ ions react with protein molecules on the tongue. The protein molecules change shape, sending an electrical signal to the brain that we experience as a sour taste (▶ FIGURE 14.1, on the next page).

Acids are also famous from their appearances in spy movies. James Bond, for example, has been known to carry an acid-filled gold pen. When captured and imprisoned—as inevitably happens at least one time in each movie—Bond can squirt the acid out of his pen and onto the iron bars of his cell. The acid quickly dissolves the metal, allowing Bond to escape. Although acids do not dissolve iron bars with the ease depicted in movies, they do dissolve metals. A small piece of aluminum placed in hydrochloric acid, for example, dissolves in about 10 minutes (▶ FIGURE 14.2, on the next page). With enough acid, it would be possible to dissolve the iron bars of a prison cell, but it would take more acid than the amount that fits in a pen.

When we say that acids dissolve metals, we mean that acids react with metals in a way that causes the metals to go into solution as metal cations. Bond's pen is made of gold because gold is one of the few metals that is not dissolved by most acids (see Section 16.5).

◀ Acids have been used in spy movies and other thrillers to dissolve the metal bars of a prison cell.

◄ **FIGURE 14.1 Acids taste sour**
When a person eats a sour food, H^+ ions from the acid in the food react with protein molecules in the taste cells of the tongue. This interaction causes the protein molecules to change shape, triggering a nerve impulse to the brain that the person experiences as a sour taste.

▶ **FIGURE 14.2 Acids dissolve many metals** When aluminum is put into hydrochloric acid, the aluminum dissolves. **Question: What happens to the aluminum atoms? Where do they go?**

14.2 Acids: Properties and Examples

▶ Identify common acids and describe their key characteristics.

NEVER taste or touch laboratory chemicals.

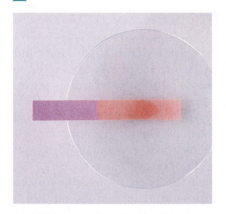

▲ **FIGURE 14.3 Acids turn blue litmus paper red**

For a review of naming acids, see Section 5.9.

Acids have the following properties:

• Acids have a sour taste.
• Acids dissolve many metals.
• Acids turn blue litmus paper red.

We just discussed examples of the sour taste of acids and their ability to dissolve metals. Acids also turn blue litmus paper red. Litmus paper contains a dye that turns red in acidic solutions (◄ **FIGURE 14.3**). In the laboratory, litmus paper is used routinely to test the acidity of solutions.

Table 14.1 lists some common acids. Hydrochloric acid is found in most chemistry laboratories. It is used in industry to clean metals, to prepare and process foods, and to refine metal ores.

HCl

Hydrochloric acid

Hydrochloric acid is also the main component of stomach acid. In the stomach, hydrochloric acid breaks down food and kills harmful bacteria that might enter the body through food. The sour taste sometimes associated with indigestion is caused by the stomach's hydrochloric acid refluxing up into the esophagus (the tube that joins the stomach and the mouth) and throat.

TABLE 14.1 Some Common Acids	
Name	**Uses**
hydrochloric acid (HCl)	metal cleaning; food preparation; ore refining; main component of stomach acid
sulfuric acid (H_2SO_4)	fertilizer and explosive manufacturing; dye and glue production; automobile batteries
nitric acid (HNO_3)	fertilizer and explosive manufacturing; dye and glue production
acetic acid ($HC_2H_3O_2$)	plastic and rubber manufacturing; food preservation; component of vinegar
carbonic acid (H_2CO_3)	component of carbonated beverages due to the reaction of carbon dioxide with water
hydrofluoric acid (HF)	metal cleaning; glass frosting and etching

Sulfuric acid—the most widely produced chemical in the United States—and nitric acid are commonly used in the laboratory. In addition, they are used in the manufacture of fertilizers, explosives, dyes, and glue. Sulfuric acid is contained in most automobile batteries.

Annual U.S. production of sulfuric acid exceeds 40 million tons.

H_2SO_4

O=S—O—H (sulfuric acid structure)

Sulfuric acid

HNO_3

O—N—O—H (nitric acid structure)

Nitric acid

Acetic acid is present in vinegar and is also produced in improperly stored wines. The word *vinegar* originates from the French *vin aigre*, which means "sour wine." The presence of vinegar in wines is considered a serious fault, making the wine taste like salad dressing.

$HC_2H_3O_2$

H—C—C—O—H (acetic acid structure)

Acetic acid

Acetic acid is an example of a **carboxylic acid**, an acid containing the grouping of atoms known as the carboxylic acid group.

— COOH   —C—O—H

Carboxylic acid group

▲ Vinegar is a solution of acetic acid and water.

We often find carboxylic acids (covered in more detail in Chapter 18, Section 18.15) in substances derived from living organisms. Other carboxylic acids include citric acid, the main acid in lemons and limes, and malic acid, an acid found in apples, grapes, and wine.

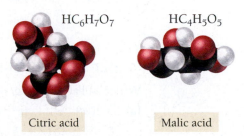

$HC_6H_7O_7$ $HC_4H_5O_5$

Citric acid   Malic acid

14.3   Bases: Properties and Examples

▶ Identify common bases and describe their key characteristics.

NEVER taste or touch laboratory chemicals.

Bases have the following properties:

- Bases have a bitter taste.
- Bases have a slippery feel.
- Bases turn red litmus paper blue.

Coffee is acidic overall, but bases present in coffee—such as caffeine—impart a bitter flavor.

▲ All these consumer products contain bases.

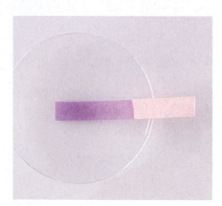

▲ **FIGURE 14.4 Bases turn red litmus paper blue**

Bases are less common in foods than acids because of their bitter taste. A Sour Patch Kid coated with a base would never sell. Our aversion to the taste of bases is probably an adaptation to protect us against **alkaloids**, organic bases found in plants (see the *Chemistry and Health* box, p. 510). Alkaloids are often poisonous—the toxic component of hemlock, for example, is the alkaloid coniine—and their bitter taste warns us against eating them. Nonetheless, some foods, such as coffee, contain small amounts of base (caffeine is a base). Many people enjoy the bitterness of coffee but only after acquiring the taste over time.

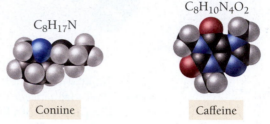

$C_8H_{17}N$ Coniine

$C_8H_{10}N_4O_2$ Caffeine

Bases feel slippery because they react with oils on our skin to form soaplike substances. Soap itself is basic, and its slippery feel is characteristic of bases. Some household cleaning solutions, such as ammonia, are also basic and have the typical slippery feel of a base. Bases turn red litmus paper blue (◀ FIGURE 14.4). In the laboratory, litmus paper is routinely used to test the basicity of solutions.

Table 14.2 lists some common bases. Sodium hydroxide and potassium hydroxide are found in most chemistry laboratories. They are also used in processing petroleum and cotton and manufacturing soap and plastic. Sodium hydroxide is the active ingredient in products such as Drano that work to unclog drains. Sodium bicarbonate can be found in most homes in the form of baking soda and is also an active ingredient in many antacids. When taken as an antacid, sodium bicarbonate neutralizes stomach acid (see Section 14.5), relieving heartburn and sour stomach.

TABLE 14.2	Some Common Bases
Name	**Uses**
sodium hydroxide (NaOH)	petroleum processing; soap and plastic manufacturing
potassium hydroxide (KOH)	cotton processing; electroplating; soap production
sodium bicarbonate (NaHCO₃)*	antacid; ingredient of baking soda; source of CO_2
ammonia (NH₃)	detergent; fertilizer and explosive manufacturing; synthetic fiber production

*Sodium bicarbonate is a salt whose anion $\left(HCO_3^-\right)$ is the conjugate base of a weak acid (see Section 14.4) and acts as a base.

PEARSON eText 2.0

CONCEPTUAL ✓ CHECKPOINT **14.1**

Which substance is most likely to have a sour taste?

(a) HCl(*aq*)   **(b)** NH₃(*aq*)   **(c)** KOH(*aq*)

14.4 Molecular Definitions of Acids and Bases

▶ Identify Arrhenius acids and bases.

▶ Identify Brønsted–Lowry acids and bases and their conjugates.

We have just seen some of the properties of acids and bases. In this section, we examine two different models that explain the molecular basis for acid and base behavior: the Arrhenius model and the Brønsted–Lowry model. The Arrhenius model, which was developed earlier, is more limited in its scope. The Brønsted–Lowry model was developed later and is more broadly applicable.

Key Concept Video
Definitions of Acids and Bases

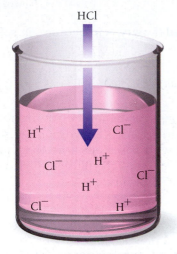

HCl

$$HCl(aq) \longrightarrow$$
$$H^+(aq) + Cl^-(aq)$$

▲ **FIGURE 14.5 Arrhenius acid** HCl is an Arrhenius acid: it produces H^+ ions in solution. The H^+ ions associate with H_2O to form H_3O^+ ions.

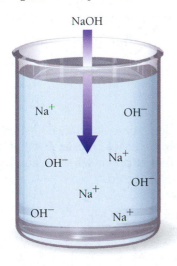

NaOH

$$NaOH(aq) \longrightarrow$$
$$Na^+(aq) + OH^-(aq)$$

▲ **FIGURE 14.6 Arrhenius base** NaOH is an Arrhenius base: it produces OH^- ions in solution.

Ionic compounds such as NaOH are composed of positive and negative ions. In solution, soluble ionic compounds dissociate into their component ions. Molecular compounds containing an OH group, such as methanol CH_3OH, do not dissociate and therefore do not act as bases.

The Arrhenius Definition

In the 1880s, the Swedish chemist Svante Arrhenius (1859–1927) proposed the following molecular definitions of acids and bases:

ARRHENIUS DEFINITION

Acid—An acid produces H^+ ions in aqueous solution.
Base—A base produces OH^- ions in aqueous solution.

For example, according to the **Arrhenius definition**, HCl is an **Arrhenius acid** because it produces H^+ ions in solution (◀ **FIGURE 14.5**):

$$HCl(aq) \longrightarrow H^+(aq) + Cl^-(aq)$$

HCl is a covalent compound and does not contain ions. However, in water it **ionizes** to form $H^+(aq)$ ions and $Cl^-(aq)$ ions. The H^+ ions are highly reactive. In aqueous solution, they bond to water molecules according to this reaction:

$$H^+ + :\ddot{O}:H \longrightarrow \left[H:\ddot{O}:H \right]^+$$

The H_3O^+ ion is the **hydronium ion**. In water, H^+ ions *always* associate with H_2O molecules. Chemists often use $H^+(aq)$ and $H_3O^+(aq)$ interchangeably, however, to refer to the same thing—a hydronium ion.

In the molecular formula for an acid, we often write the ionizable hydrogen first. For example, we write the formula for formic acid as follows:

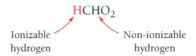

Ionizable hydrogen   Non-ionizable hydrogen

The structure of formic acid, however, is not indicated by the molecular formula in the preceding figure. We represent the *structure* of formic acid with its structural formula:

$$HC-OH \quad (O\,\|)$$

Ionizable hydrogen

Notice that the structural formula indicates how the atoms are bonded together; the molecular formula, by contrast, indicates only the number of each type of atom.

NaOH is an **Arrhenius base** because it produces OH^- ions in solution (◀ **FIGURE 14.6**).

$$NaOH(aq) \longrightarrow Na^+(aq) + OH^-(aq)$$

NaOH is an ionic compound and therefore contains Na^+ and OH^- ions. When NaOH is added to water, it **dissociates**, or breaks apart into its component ions.

Under the Arrhenius definition, acids and bases naturally combine to form water, neutralizing each other in the process.

$$H^+(aq) + OH^-(aq) \longrightarrow H_2O(l)$$

The Brønsted–Lowry Definition

Although the Arrhenius definition of acids and bases applies in many cases, it cannot easily explain why some substances act as bases even though they do not contain OH^-. The Arrhenius definition also does not apply to nonaqueous solvents. A second definition of acids and bases, called the Brønsted–Lowry definition, introduced in 1923, applies to a wider range of acid–base phenomena. This definition focuses on the *transfer* of H^+ ions in an acid–base reaction. Since an H^+ ion is a proton—a hydrogen atom with its electron taken away—this definition focuses on the idea of a proton donor and a proton acceptor.

BRØNSTED–LOWRY DEFINITION

Acid—An acid is a proton (H^+ ion) *donor.*

Base—A base is a proton (H^+ ion) *acceptor.*

According to this definition, HCl is a **Brønsted–Lowry acid** because, in solution, it donates a proton to water:

$$HCl(aq) + H_2O(l) \longrightarrow H_3O^+(aq) + Cl^-(aq)$$

This definition more clearly accounts for what happens to the H^+ ion from an acid: It associates with a water molecule to form H_3O^+ (a hydronium ion). The Brønsted–Lowry definition also works well with bases (such as NH_3) that do not contain OH^- ions but that still produce OH^- ions in solution. NH_3 is a **Brønsted–Lowry base** because it accepts a proton from water:

$$NH_3(aq) + H_2O(l) \rightleftharpoons NH_4^+(aq) + OH^-(aq)$$

According to the Brønsted–Lowry definition, acids (proton donors) and bases (proton acceptors) always occur together. In the reaction between HCl and H_2O, HCl is the proton donor (acid), and H_2O is the proton acceptor (base):

$$\underset{\substack{\text{Acid}\\\text{(Proton donor)}}}{HCl(aq)} + \underset{\substack{\text{Base}\\\text{(Proton acceptor)}}}{H_2O(l)} \longrightarrow H_3O^+(aq) + Cl^-(aq)$$

In the reaction between NH_3 and H_2O, H_2O is the proton donor (acid) and NH_3 is the proton acceptor (base):

$$\underset{\substack{\text{Base}\\\text{(Proton acceptor)}}}{NH_3(aq)} + \underset{\substack{\text{Acid}\\\text{(Proton donor)}}}{H_2O(l)} \rightleftharpoons NH_4^+(aq) + OH^-(aq)$$

Notice that under the Brønsted–Lowry definition, some substances—such as water in the previous two equations—can act as acids *or* bases. Substances that can act as acids or bases are **amphoteric**. Notice also what happens when an equation representing Brønsted–Lowry acid–base behavior is reversed:

$$\underset{\substack{\text{Acid}\\\text{(Proton donor)}}}{NH_4^+(aq)} + \underset{\substack{\text{Base}\\\text{(Proton acceptor)}}}{OH^-(aq)} \rightleftharpoons NH_3(aq) + H_2O(l)$$

In this reaction, NH_4^+ is the proton donor (acid) and OH^- is the proton acceptor (base). The substance that was the base (NH_3) becomes the acid (NH_4^+), and vice versa. NH_4^+ and NH_3 are often referred to as a **conjugate acid–base pair**, two substances related to each other by the transfer of a proton (◄ **FIGURE 14.7**). Going back to the original forward reaction, we can identify the conjugate acid–base pairs as follows:

$$\underset{\text{Base}}{NH_3(aq)} + \underset{\text{Acid}}{H_2O(l)} \rightleftharpoons \underset{\substack{\text{Conjugate}\\\text{acid}}}{NH_4^+(aq)} + \underset{\substack{\text{Conjugate}\\\text{base}}}{OH^-(aq)}$$

In an acid–base reaction, a base accepts a proton and becomes a conjugate acid. An acid donates a proton and becomes a conjugate base.

Johannes Brønsted (1879–1947), working in Denmark, and Thomas Lowry (1874–1936), working in England, developed the concept of proton transfer in acid–base behavior independently and simultaneously.

The double arrows in this equation indicate that the reaction does not go to completion. We discuss this concept in more detail in Section 14.7.

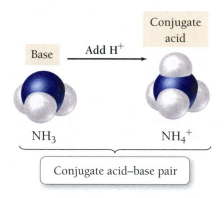

Base $\xrightarrow{\text{Add } H^+}$ Conjugate acid

NH_3 NH_4^+

Conjugate acid–base pair

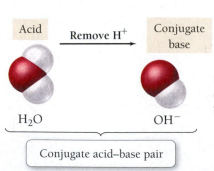

Acid $\xrightarrow{\text{Remove } H^+}$ Conjugate base

H_2O OH^-

Conjugate acid–base pair

▲ **FIGURE 14.7 A conjugate acid–base pair** Any two substances related to each other by the transfer of a proton can be considered a conjugate acid–base pair.

EXAMPLE **14.1** | **Identifying Brønsted–Lowry Acids and Bases and Their Conjugates**

In each reaction, identify the Brønsted–Lowry acid, the Brønsted–Lowry base, the conjugate acid, and the conjugate base.

(a) $H_2SO_4(aq) + H_2O(l) \longrightarrow H_3O^+(aq) + HSO_4^-(aq)$
(b) $HCO_3^-(aq) + H_2O(l) \rightleftharpoons H_2CO_3(aq) + OH^-(aq)$

SOLUTION

(a) Since H_2SO_4 donates a proton to H_2O in this reaction, it is the acid (the proton donor). After H_2SO_4 donates the proton, it becomes HSO_4^-, the conjugate base. Since H_2O accepts a proton, it is the base (the proton acceptor). After H_2O accepts the proton, it becomes H_3O^+, the conjugate acid.

$$H_2SO_4(aq) + H_2O(l) \longrightarrow HSO_4^-(aq) + H_3O^+(aq)$$

Acid   Base   Conjugate base   Conjugate acid

(b) Since H_2O donates a proton to HCO_3^- in this reaction, it is the acid (the proton donor). After H_2O donates the proton, it becomes OH^-, the conjugate base. Since HCO_3^- accepts a proton, it is the base (the proton acceptor). After HCO_3^- accepts the proton, it becomes H_2CO_3, the conjugate acid.

$$HCO_3^-(aq) + H_2O(l) \rightleftharpoons H_2CO_3(aq) + OH^-(aq)$$

Base   Acid   Conjugate acid   Conjugate base

▶ **SKILLBUILDER 14.1** | **Identifying Brønsted–Lowry Acids and Bases and Their Conjugates**

In each reaction, identify the Brønsted–Lowry acid, the Brønsted–Lowry base, the conjugate acid, and the conjugate base.

(a) $C_5H_5N(aq) + H_2O(l) \rightleftharpoons C_5H_5NH^+(aq) + OH^-(aq)$
(b) $HNO_3(aq) + H_2O(l) \longrightarrow NO_3^-(aq) + H_3O^+(aq)$

▶ **FOR MORE PRACTICE** Example 14.11; Problems 31, 32, 33, 34.

PEARSON
eText
2.0

CONCEPTUAL ✔ CHECKPOINT 14.2

Which species is the conjugate base of H_2SO_3?
(a) $H_3SO_3^+$ (b) HSO_3^- (c) SO_3^{2-}

14.5 Reactions of Acids and Bases

▶ Write equations for neutralization reactions.

▶ Write equations for the reactions of acids with metals and with metal oxides.

Acids and bases are typically reactive substances. In this section, we examine how they react with one another, as well as with other substances.

Neutralization Reactions

The reaction between HCl and KOH is also a double-displacement reaction (see Section 7.10).

One of the most important reactions of acids and bases is **neutralization**, first introduced in Chapter 7. When we mix an acid and a base, the $H^+(aq)$ from the acid combines with the $OH^-(aq)$ from the base to form $H_2O(l)$. For example, consider the reaction between hydrochloric acid and potassium hydroxide:

$$HCl(aq) + KOH(aq) \longrightarrow H_2O(l) + KCl(aq)$$

Acid   Base   Water   Salt

Neutralization reactions are covered in Section 7.8.

Acid–base reactions generally form water and a **salt**—an ionic compound—that usually remains dissolved in the solution. The salt contains the cation from the base and the anion from the acid.

Ionic compound that contains the cation from the base and the anion from the acid

$$\text{Acid} + \text{Base} \longrightarrow \text{Water} + \text{Salt}$$

Net ionic equations are explained in Section 7.7.

The net ionic equation for many neutralization reactions is:

$$H^+(aq) + OH^-(aq) \longrightarrow H_2O(l)$$

A slightly different but common type of neutralization reaction involves an acid reacting with carbonates or bicarbonates (compounds containing CO_3^{2-} or HCO_3^-). This type of neutralization reaction produces water, gaseous carbon dioxide, and a salt. As an example, consider the reaction of hydrochloric acid and sodium bicarbonate.

$$HCl(aq) + NaHCO_3(aq) \longrightarrow H_2O(l) + CO_2(g) + NaCl(aq)$$

Since this reaction produces gaseous CO_2, it is also called a *gas-evolution reaction* (Section 7.8).

$$HCl(aq) + NaHCO_3(aq) \longrightarrow$$

$$H_2O(l) + CO_2(g) + NaCl(aq)$$

▲ The reaction of carbonates or bicarbonates with acids produces water, gaseous carbon dioxide, and a salt.

EXAMPLE 14.2 Writing Equations for Neutralization Reactions

Write a molecular equation for the reaction between aqueous HCl and aqueous $Ca(OH)_2$.

SOLUTION	
First identify the acid and the base and write the skeletal reaction showing the production of water and the salt. The formulas for the ionic compounds in the equation must be charge-neutral (see Section 5.5).	$HCl(aq) + Ca(OH)_2(aq) \longrightarrow$ $H_2O(l) + CaCl_2(aq)$
Balance the equation. Notice that $Ca(OH)_2$ contains 2 mol of OH^- for every 1 mol of $Ca(OH)_2$ and therefore requires 2 mol of H^+ to neutralize it.	$2\,HCl(aq) + Ca(OH)_2(aq) \longrightarrow$ $2\,H_2O(l) + CaCl_2(aq)$

▶ **SKILLBUILDER 14.2** | **Writing Equations for Neutralization Reactions**

Write a molecular equation for the reaction that occurs between aqueous H_3PO_4 and aqueous NaOH. *Hint:* H_3PO_4 is a triprotic acid, meaning that 1 mol of H_3PO_4 requires 3 mol of OH^- to completely react with it.

▶ **FOR MORE PRACTICE** Example 14.12; Problems 39, 40.

$$2\,HCl(aq) + Mg(s) \longrightarrow$$

$$H_2(g) + MgCl_2(aq)$$

▲ The reaction between an acid and a metal usually produces hydrogen gas and a dissolved salt containing the metal ion.

Acid Reactions

Recall from Section 14.1 that acids dissolve metals, or more precisely, that acids react with metals in a way that causes metals to go into solution. The reaction between an acid and a metal usually produces hydrogen gas and a dissolved salt containing the metal ion as the cation. For example, hydrochloric acid reacts with magnesium metal to form hydrogen gas and magnesium chloride:

$$\underset{\text{Acid}}{2\,HCl(aq)} + \underset{\text{Metal}}{Mg(s)} \longrightarrow \underset{\substack{\text{Hydrogen}\\\text{gas}}}{H_2(g)} + \underset{\text{Salt}}{MgCl_2(aq)}$$

Similarly, sulfuric acid reacts with zinc to form hydrogen gas and zinc sulfate:

$$\underset{\text{Acid}}{H_2SO_4(aq)} + \underset{\text{Metal}}{Zn(s)} \longrightarrow \underset{\substack{\text{Hydrogen}\\\text{gas}}}{H_2(g)} + \underset{\text{Salt}}{ZnSO_4(aq)}$$

It is through reactions such as these that the acid from James Bond's pen (discussed in Section 14.1) dissolves the metal bars that imprison him. If the bars were made of iron and the acid in the pen were hydrochloric acid, the reaction would be:

$$\underset{\text{Acid}}{2\,HCl(aq)} + \underset{\text{Metal}}{Fe(s)} \longrightarrow \underset{\substack{\text{Hydrogen}\\\text{gas}}}{H_2(g)} + \underset{\text{Salt}}{FeCl_2(aq)}$$

Some metals, however, do not readily react with acids. If the bars that imprisoned James Bond were made of gold, for example, a pen filled with hydrochloric acid would not dissolve the bars. We will discuss the way to determine whether a particular metal dissolves in an acid in Section 16.5.

Acids also react with metal oxides to produce water and a dissolved salt. For example, hydrochloric acid reacts with potassium oxide to form water and potassium chloride:

$$\underset{\text{Acid}}{2\,HCl(aq)} + \underset{\text{Metal oxide}}{K_2O(s)} \longrightarrow \underset{\text{Water}}{H_2O(l)} + \underset{\text{Salt}}{2\,KCl(aq)}$$

Similarly, hydrobromic acid reacts with magnesium oxide to form water and magnesium bromide:

$$\underset{\text{Acid}}{2\,HBr(aq)} + \underset{\text{Metal oxide}}{MgO(s)} \longrightarrow \underset{\text{Water}}{H_2O(l)} + \underset{\text{Salt}}{MgBr_2(aq)}$$

EXAMPLE 14.3 | **Writing Equations for Acid Reactions**

Write an equation for each reaction.

(a) the reaction of hydroiodic acid with potassium metal
(b) the reaction of hydrobromic acid with sodium oxide

SOLUTION

(a) The reaction of hydroiodic acid with potassium metal forms hydrogen gas and a salt. The salt contains the ionized form of the metal (K^+) as the cation and the anion of the acid (I^-). Write the skeletal equation and then balance it.	$HI(aq) + K(s) \longrightarrow H_2(g) + KI(aq)$ $2\,HI(aq) + 2\,K(s) \longrightarrow H_2(g) + 2\,KI(aq)$
(b) The reaction of hydrobromic acid with sodium oxide forms water and a salt. The salt contains the cation from the metal oxide (Na^+) and the anion of the acid (Br^-). Write the skeletal equation and then balance it.	$HBr(aq) + Na_2O(s) \longrightarrow H_2O(l) + NaBr(aq)$ $2\,HBr(aq) + Na_2O(s) \longrightarrow H_2O(l) + 2\,NaBr(aq)$

▶ **SKILLBUILDER 14.3** | **Writing Equations for Acid Reactions**

Write an equation for each reaction.

(a) the reaction of hydrochloric acid with strontium metal
(b) the reaction of hydroiodic acid with barium oxide

▶ **FOR MORE PRACTICE** Example 14.13; Problems 41, 42, 43, 44.

EVERYDAY CHEMISTRY

What Is in My Antacid?

Heartburn, a burning sensation in the lower throat and above the stomach, is caused by the reflux or backflow of stomach acid into the esophagus (the tube that joins the stomach to the throat). In most individuals, this occurs only occasionally, typically after large meals. Physical activity—such as bending, stooping, or lifting—after meals also aggravates heartburn. In some people, the flap between the esophagus and the stomach that normally prevents acid

reflux becomes damaged, in which case heartburn becomes a regular occurrence.

Drugstores carry many products that either reduce the secretion of stomach acid or neutralize the acid that is produced. Antacids such as Mylanta or Phillips' Milk of Magnesia contain bases that neutralize the refluxed stomach acid, alleviating heartburn.

B14.1 CAN YOU ANSWER THIS? *Look at the label for Mylanta shown in the photograph. Can you identify the bases responsible for the antacid action? Write chemical equations showing the reactions of these bases with stomach acid (HCl).*

Base Reactions

The most important base reactions are those in which a base neutralizes an acid (see the beginning of this section). The only other kind of base reaction that we cover in this book is the reaction of sodium hydroxide with aluminum and water.

$$2\,NaOH(aq) + 2\,Al(s) + 6\,H_2O(l) \longrightarrow 2\,NaAl(OH)_4(aq) + 3\,H_2(g)$$

Aluminum is one of the few metals that dissolves in a base. Consequently, it is safe to use NaOH (the main ingredient in many drain-opening products) to unclog your drain as long as your pipes are not made of aluminum, which is generally the case as the use of aluminum pipe is forbidden by most building codes.

14.6 Acid–Base Titration: A Way to Quantify the Amount of Acid or Base in a Solution

▶ Use acid–base titration to determine the concentration of an unknown solution.

We can apply the principles we learned in Chapter 13 (Section 13.8) on solution stoichiometry to a common laboratory procedure called a titration. In a **titration**, we react a substance in a solution of known concentration with another substance in a solution of unknown concentration. For example, consider the acid–base reaction between hydrochloric acid and sodium hydroxide:

$$HCl(aq) + NaOH(aq) \longrightarrow H_2O(l) + NaCl(aq)$$

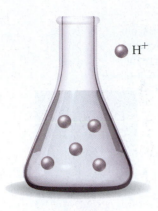

$\bullet$ H$^+$

The net ionic equation for this reaction is:

$$H^+(aq) + OH^-(aq) \longrightarrow H_2O(l)$$

Suppose we have an HCl solution represented by the molecular diagram at left. (The Cl$^-$ ions and the H$_2$O molecules not involved in the reaction have been omitted from this representation for clarity.)

In titrating this sample, we slowly add a solution of known OH$^-$ concentration. This process is represented by the following molecular diagrams:

The OH$^-$ solution also contains Na$^+$ cations that we do not show in this figure for clarity.

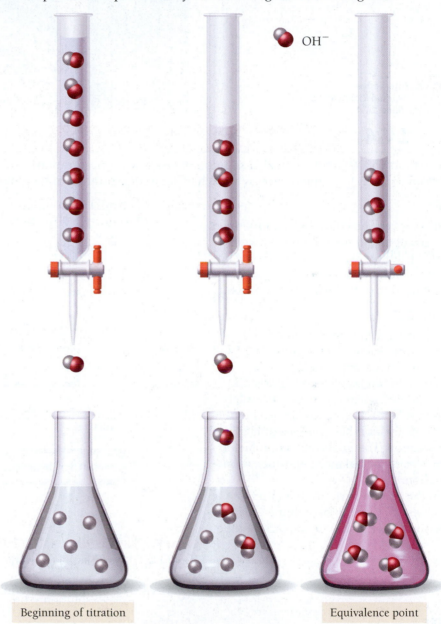

OH$^-$

Beginning of titration

Equivalence point

At the equivalence point, neither reactant is present in excess, and both are limiting. The number of moles of the reactants are related by the reaction stoichiometry (see Chapter 8).

As we add the OH$^-$, it reacts with and neutralizes the H$^+$, forming water. At the **equivalence point**—*the point in the titration when the number of moles of OH$^-$ added equals the number of moles of H$^+$ originally in solution*—the titration is complete. The equivalence point is usually signaled by an **indicator**, a dye whose color depends on the acidity of the solution (▶ **FIGURE 14.8**, on the next page).

In most laboratory titrations, the concentration of one of the reactant solutions is unknown, and the concentration of the other is precisely known. By carefully measuring the volume of each solution required to reach the equivalence point, we can determine the concentration of the unknown solution, as demonstrated in Example 14.4.

▶ FIGURE 14.8 Acid–base titration
In this titration, NaOH is added to an
HCl solution.

When the NaOH and HCl reach stoichiometric equivalence, the indicator changes
to pink, signaling the equivalence point.

PEARSON eText 2.0

Interactive
Worked Example
Video 14.4

EXAMPLE **14.4** **Acid–Base Titration**

The titration of 10.00 mL of an HCl solution of unknown concentration requires 12.54 mL of a 0.100 M NaOH
solution to reach the equivalence point. What is the concentration of the unknown HCl solution?

SORT

You are given the volume of an unknown
HCl solution and the volume of a known
NaOH solution required to titrate the unknown solution. You are asked to find the
concentration of the unknown solution.

GIVEN: 10.00 mL HCl solution
12.54 mL of a 0.100 M NaOH solution

FIND: concentration of HCl solution (mol/L)

STRATEGIZE

First write the balanced chemical equation for the reaction between the acid and
the base (see Example 14.2).

The solution map has two parts. In
the first part, use the volume of NaOH
required to reach the equivalence point to
calculate the number of moles of HCl in
the solution. The final conversion factor
comes from the balanced neutralization
equation.

In the second part, use the number
of moles of HCl and the volume of HCl
solution to determine the molarity of the
HCl solution.

SOLUTION MAP

$$HCl(aq) + NaOH(aq) \longrightarrow H_2O(l) + NaCl(aq)$$

| mL NaOH | → | L NaOH | → | mol NaOH | → | mol HCl |

$$\frac{1\ L}{1000\ mL} \qquad \frac{0.100\ mol\ NaOH}{1\ L\ NaOH} \qquad \frac{1\ mol\ HCl}{1\ mol\ NaOH}$$

| mol HCl, volume HCl solution | → | molarity |

$$M = \frac{mol}{L}$$

RELATIONSHIPS USED
1 mol HCl:1 mol NaOH (from balanced chemical equation)

$$\text{molarity (M)} = \frac{mol\ solute}{L\ solution} \text{ (definition of molarity, from Section 13.6)}$$

SOLVE

Calculate the moles of HCl in the unknown solution by following the first
part of the solution map.

To find the concentration of the solution, divide the number of moles of HCl
by the volume of the HCl solution in L.
(Note that 10.00 mL is equivalent to
0.01000 L.)

The unknown HCl solution therefore
has a concentration of 0.125 M.

SOLUTION

$$12.54\ \text{mL NaOH} \times \frac{1\ L}{1000\ mL} \times \frac{0.100\ mol\ NaOH}{1\ L\ NaOH} \times \frac{1\ mol\ HCl}{1\ mol\ NaOH}$$

$$= 1.25 \times 10^{-3}\ mol\ HCl$$

$$\text{molarity} = \frac{1.25 \times 10^{-3}\ mol\ HCl}{0.01000\ L} = 0.125\ M$$

CHECK	The units (M) are correct. The magnitude of the answer makes sense
Check your answer. Are the units correct? Does the answer make physical sense?	because the reaction has a one-to-one stoichiometry and the volumes of the two solutions are similar; therefore, their concentrations should also be similar.

▶ **SKILLBUILDER 14.4 | Acid–Base Titration**

The titration of a 20.0-mL sample of an H_2SO_4 solution of unknown concentration requires 22.87 mL of a 0.158 M KOH solution to reach the equivalence point. What is the concentration of the unknown H_2SO_4 solution?

▶ **FOR MORE PRACTICE** Example 14.14; Problems 47, 48, 49, 50, 51, 52.

CONCEPTUAL ✔ CHECKPOINT 14.3

The flask represents a sample of acid to be titrated.

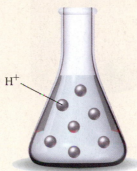

Which of the beakers contains the amount of OH^- required to reach the equivalence point in the titration?

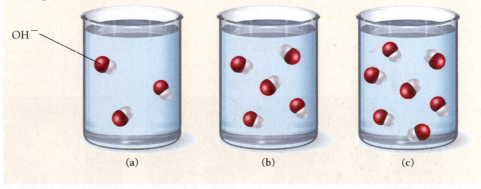

(a)   (b)   (c)

14.7 Strong and Weak Acids and Bases

▶ Identify strong and weak acids and strong and weak bases.

▶ Determine $[H_3O^+]$ in acid solutions.

▶ Determine $[OH^-]$ in base solutions.

We categorize acids and bases as strong or weak, depending on how much they ionize or dissociate in aqueous solution. In this section, we first look at strong and weak acids, and then turn to strong and weak bases.

Strong Acids

Hydrochloric acid (HCl) and hydrofluoric acid (HF) appear to be similar, but there is an important difference between these two acids. HCl is an example of a **strong acid**, one that completely ionizes in solution.

Single arrow indicates complete ionization

$$HCl(aq) + H_2O(l) \longrightarrow H_3O^+(aq) + Cl^-(aq)$$

[X] means "molar concentration of X."

We show the *complete* ionization of HCl with a single arrow pointing to the right in the equation. An HCl solution contains almost no intact HCl; virtually all the HCl has reacted with water to form $H_3O^+(aq)$ and $Cl^-(aq)$ (◀ FIGURE 14.9). A 1.0 M HCl solution therefore has an H_3O^+ concentration of 1.0 M. We often abbreviate the concentration of H_3O^+ as $[H_3O^+]$. Using this notation, a 1.0-M HCl solution has $[H_3O^+] = 1.0$ M.

A strong acid is also a **strong electrolyte** (first defined in Section 7.5), a substance whose aqueous solutions are good conductors of electricity (▼ FIGURE 14.10). Aqueous solutions require the presence of charged particles to conduct electricity. Strong acid solutions are also strong electrolyte solutions because each acid molecule ionizes into positive and negative ions. These mobile ions are good conductors of electricity. Pure water is not a good conductor of electricity because it has relatively few charged particles. The danger of using electrical devices—such as a hair dryer—while sitting in the bathtub is that water is seldom pure and often contains dissolved ions. If the device were to come in contact with the water, dangerously high levels of electricity could flow through the water and through your body.

Table 14.3 lists the six strong acids. The first five acids in the table are **monoprotic acids**, acids containing only one ionizable proton. Sulfuric acid is an example of a **diprotic acid**, an acid that contains two ionizable protons.

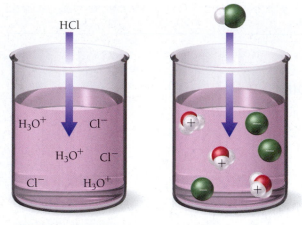

When HCl dissolves in water, it completely ionizes into H^+ and Cl^- ions.

HCl

H_3O^+ Cl^-

H_3O^+ Cl^-

Cl^- H_3O^+

▲ **FIGURE 14.9 A strong acid**

An ionizable proton is one that becomes an H^+ ion in solution.

TABLE 14.3 Strong Acids

hydrochloric acid (HCl)	nitric acid (HNO_3)
hydrobromic acid (HBr)	perchloric acid ($HClO_4$)
hydroiodic acid (HI)	sulfuric acid (H_2SO_4) (*diprotic*)

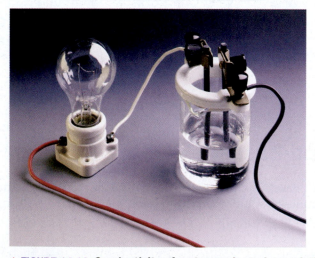

Pure water does not conduct electricity.

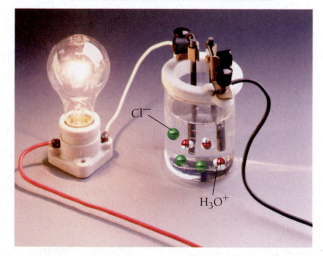

Strong electrolyte solution: Ions in an HCl solution conduct electricity, causing the light bulb to light.

Cl^-

H_3O^+

▲ **FIGURE 14.10 Conductivity of a strong electrolyte solution**

Weak Acids

It is a common mistake to confuse the terms *strong* and *weak acids* with the terms *concentrated* and *dilute acids*. Can you state the difference between these terms?

In contrast to HCl, HF is a **weak acid**, one that does not completely ionize in solution.

Double arrow indicates partial ionization

$$HF(aq) + H_2O(l) \rightleftharpoons H_3O^+(aq) + F^-(aq)$$

Calculating exact $[H_3O^+]$ for weak acids is beyond the scope of this text.

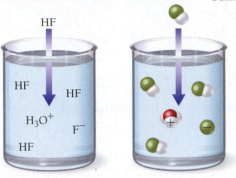

When HF dissolves in water, only a fraction of the dissolved molecules ionize into H^+ and F^- ions.

▲ **FIGURE 14.11 A weak acid**

To show that HF does not completely ionize in solution, the equation for its ionization has two opposing arrows, indicating that the reverse reaction occurs to some degree. An HF solution contains a lot of intact HF; it also contains some $H_3O^+(aq)$ and $F^-(aq)$ (◄ **FIGURE 14.11**). In other words, a 1.0 M HF solution has $[H_3O^+] < 1.0$ M because only some of the HF molecules ionize to form H_3O^+.

A weak acid is also a **weak electrolyte**, a substance whose aqueous solutions are poor conductors of electricity (▼ **FIGURE 14.12**). Weak acid solutions contain few charged particles because only a small fraction of the acid molecules ionize into positive and negative ions.

The degree to which an acid is strong or weak depends in part on the attraction between the anion of the acid (the conjugate base) and the hydrogen ion. Suppose HA is a generic formula for an acid. The degree to which the following reaction proceeds in the forward direction depends in part on the strength of the attraction between H^+ and A^-:

$$HA(aq) + H_2O(l) \longrightarrow H_3O^+(aq) + A^-(aq)$$
Acid   Conjugate base

Pure water does not conduct electricity.

Weak electrolyte solution: An HF solution contains some ions, but most of the HF is intact. The light bulb glows only dimly.

► **FIGURE 14.12 Conductivity of a weak electrolyte solution**

If the attraction between H^+ and A^- is *weak*, the reaction favors the forward direction and the acid is *strong* (◄ **FIGURE 14.13a**). If the attraction between H^+ and A^- is *strong*, the reaction favors the reverse direction and the acid is *weak* (◄ **FIGURE 14.13b**).

For example, in HCl, the conjugate base (Cl^-) has a relatively weak attraction to H^+, meaning that the reverse reaction does not occur to any significant extent. In HF, on the other hand, the conjugate base (F^-) has a greater attraction to H^+, meaning that the reverse reaction occurs to a significant degree. *In general, the stronger the acid, the weaker the conjugate base and vice versa.* This means that if the forward reaction (that of the acid) has a high tendency to occur, then the reverse reaction (that of the conjugate base) has a low tendency to occur. Table 14.4 lists some common weak acids.

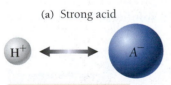

(a) Strong acid

Weak attraction
Complete ionization

(b) Weak acid

Strong attraction
Partial ionization

◄ **FIGURE 14.13 Strong and weak acids** **(a)** In a strong acid, the attraction between H^+ and A^- is low, resulting in complete ionization. **(b)** In a weak acid, the attraction between H^+ and A^- is high, resulting in partial ionization.

TABLE 14.4	Weak Acids
hydrofluoric acid (HF)	sulfurous acid (H_2SO_3) (diprotic)
acetic acid ($HC_2H_3O_2$)	carbonic acid (H_2CO_3) (diprotic)
formic acid ($HCHO_2$)	phosphoric acid (H_3PO_4) (triprotic)

Notice that two of the weak acids in Table 14.4 are diprotic (they have two ionizable protons) and one is triprotic (it has three ionizable protons). Let us return to sulfuric acid (see Table 14.3) for a moment. Sulfuric acid is a diprotic acid that is strong in its first ionizable proton:

$$H_2SO_4(aq) + H_2O(l) \longrightarrow H_3O^+(aq) + HSO_4^-(aq)$$

but weak in its second ionizable proton:

$$HSO_4^-(aq) + H_2O(l) \rightleftharpoons H_3O^+(aq) + SO_4^{2-}(aq)$$

Sulfurous acid and carbonic acid are weak in both of their ionizable protons, and phosphoric acid is weak in all three of its ionizable protons.

EXAMPLE **14.5**   **Determining [H₃O⁺] in Acid Solutions**

Determine the H_3O^+ concentration in each solution.

(a) 1.5 M HCl   **(b)** 3.0 M HC₂H₃O₂   **(c)** 2.5 M HNO₃

SOLUTION

(a) Since HCl is a strong acid, it completely ionizes. The concentration of H_3O^+ is 1.5 M.

$$[H_3O^+] = 1.5 \text{ M}$$

(b) Since HC₂H₃O₂ is a weak acid, it partially ionizes. The calculation of the exact concentration of H_3O^+ is beyond the scope of this text, but you know that it is less than 3.0 M.

$$[H_3O^+] < 3.0 \text{ M}$$

(c) Since HNO₃ is a strong acid, it completely ionizes. The concentration of H_3O^+ is 2.5 M.

$$[H_3O^+] = 2.5 \text{ M}$$

▶ **SKILLBUILDER 14.5** | **Determining [H₃O⁺] in Acid Solutions**

Determine the H_3O^+ concentration in each solution.

(a) 0.50 M HCHO₂   **(b)** 1.25 M HI   **(c)** 0.75 M HF

▶ **FOR MORE PRACTICE**   Example 14.15; Problems 55, 56.

CONCEPTUAL ✔ **CHECKPOINT 14.4**

Examine these molecular views of three different acid solutions. Which acid is a weak acid?

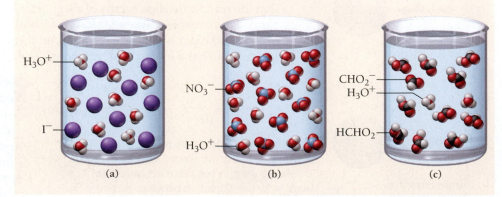

(a)   (b)   (c)

Strong Bases

By analogy to the definition of a strong acid, a **strong base** is one that completely dissociates in solution. NaOH, for example, is a strong base.

$$NaOH(aq) \longrightarrow Na^+(aq) + OH^-(aq)$$

An NaOH solution contains no intact NaOH—it has all dissociated to form $Na^+(aq)$ and $OH^-(aq)$ (▼ **FIGURE 14.14**), and a 1.0 M NaOH solution has $[OH^-] = 1.0$ M and $[Na^+] = 1.0$ M. Table 14.5 lists some common strong bases.

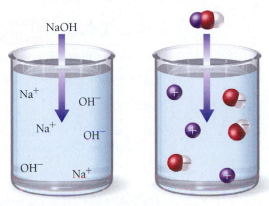

When NaOH dissolves in water, it completely dissociates into Na^+ and OH^-.

NaOH

Na^+
OH^-
Na^+
OH^-
OH^-
Na^+

▲ **FIGURE 14.14 A strong base**

TABLE 14.5 Strong Bases	
lithium hydroxide (LiOH)	strontium hydroxide (Sr(OH)$_2$)
sodium hydroxide (NaOH)	calcium hydroxide (Ca(OH)$_2$)
potassium hydroxide (KOH)	barium hydroxide (Ba(OH)$_2$)

> Unlike diprotic acids, which ionize in two steps, bases containing 2 OH⁻ ions dissociate in one step.

Some strong bases, such as $Sr(OH)_2$, contain two OH^- ions. These bases completely dissociate, producing 2 mol of OH^- per mole of base. For example, $Sr(OH)_2$ dissociates as follows:

$$Sr(OH)_2(aq) \longrightarrow Sr^{2+}(aq) + 2\,OH^-(aq)$$

Weak Bases

> Calculating exact [OH⁻] for weak bases is beyond the scope of this text.

A **weak base** is analogous to a weak acid. In contrast to strong bases that contain OH^- and dissociate in water, the most common weak bases produce OH^- by accepting a proton from water, ionizing water to form OH^-.

$$B(aq) + H_2O(l) \rightleftharpoons BH^+(aq) + OH^-(aq)$$

In this equation, B is generic for a weak base. Ammonia, for example, ionizes water according to the reaction:

$$NH_3(aq) + H_2O(l) \rightleftharpoons NH_4^+(aq) + OH^-(aq)$$

The double arrow indicates that the ionization is not complete. An NH_3 solution contains NH_3, NH_4^+, and OH^- (◄ **FIGURE 14.15**). A 1.0 M NH_3 solution has $[OH^-] < 1.0$ M. Table 14.6 lists some common weak bases.

> When NH_3 dissolves in water, it partially ionizes to form NH_4^+ and OH^-. However, only a fraction of the molecules ionize.

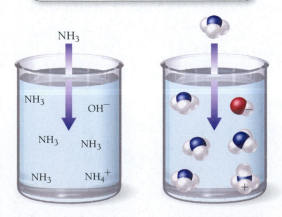

NH_3

NH_3
OH^-
NH_3
NH_3
NH_3
NH_4^+

◄ **FIGURE 14.15 A weak base**

TABLE 14.6   Some Weak Bases

Base	Ionization Reaction
ammonia (NH_3)	$NH_3(aq) + H_2O(l) \rightleftharpoons NH_4^+(aq) + OH^-(aq)$
pyridine (C_5H_5N)	$C_5H_5N(aq) + H_2O(l) \rightleftharpoons C_5H_5NH^+(aq) + OH^-(aq)$
methylamine (CH_3NH_2)	$CH_3NH_2(aq) + H_2O(l) \rightleftharpoons CH_3NH_3^+(aq) + OH^-(aq)$
ethylamine ($C_2H_5NH_2$)	$C_2H_5NH_2(aq) + H_2O(l) \rightleftharpoons C_2H_5NH_3^+(aq) + OH^-(aq)$
bicarbonate ion (HCO_3^-)*	$HCO_3^-(aq) + H_2O(l) \rightleftharpoons H_2CO_3(aq) + OH^-(aq)$

*The bicarbonate ion must occur with a positively charged ion such as Na^+ that serves to balance the charge but does not have any part in the ionization reaction. It is the bicarbonate ion that makes sodium bicarbonate ($NaHCO_3$) basic.

EXAMPLE **14.6**   **Determining [OH^-] in Base Solutions**

Determine the OH^- concentration in each solution.

(a) 2.25 M KOH   **(b)** 0.35 M CH_3NH_2 **(c)** 0.025 M $Sr(OH)_2$

SOLUTION

(a) Since KOH is a strong base, it completely dissociates into K^+ and OH^- in solution. The concentration of OH^- is 2.25 M.

$$[OH^-] = 2.25 \text{ M}$$

(b) Since CH_3NH_2 is a weak base, it only partially ionizes water. We cannot calculate the exact concentration of OH^-, but we know it is less than 0.35 M.

$$[OH^-] < 0.35 \text{ M}$$

(c) Since $Sr(OH)_2$ is a strong base, it completely dissociates into $Sr^{2+}(aq)$ and $2 OH^-(aq)$. $Sr(OH)_2$ forms 2 mol of OH^- for every 1 mol of $Sr(OH)_2$. Consequently, the concentration of OH^- is twice the concentration of $Sr(OH)_2$.

$$[OH^-] = 2(0.025 \text{ M}) = 0.050 \text{ M}$$

▶ **SKILLBUILDER 14.6** | **Determining [OH^-] in Base Solutions**

Determine the OH^- concentration in each solution.

(a) 0.055 M $Ba(OH)_2$ **(b)** 1.05 M C_5H_5N **(c)** 0.45 M NaOH

▶ **FOR MORE PRACTICE**   Example 14.16; Problems 59, 60.

14.8 Water: Acid and Base in One

▶ Calculate [H_3O^+] or [OH^-] from K_w.

Recall that water acts as a base when it reacts with HCl and as an acid when it reacts with NH_3.

Water acting as a base

$$HCl(aq) + H_2O(l) \longrightarrow H_3O^+(aq) + Cl^-(aq)$$

Acid (Proton donor)   Base (Proton acceptor)

Water acting as an acid

$$NH_3(aq) + H_2O(l) \rightleftharpoons NH_4^+(aq) + OH^-(aq)$$

Base (Proton acceptor)   Acid (Proton donor)

Water is *amphoteric*; it can act as either an acid or a base. Even in pure water, water acts as an acid and a base with itself, a process called self-ionization.

Water acting as both an *acid* and a *base*

$$H_2O(l) \quad + \quad H_2O(l) \quad \rightleftharpoons \quad H_3O^+(aq) \; + \; OH^-(aq)$$

Acid   Base

(Proton donor)   (Proton acceptor)

In pure water, at 25 °C, the preceding reaction occurs only to a very small extent, resulting in equal and small concentrations of H_3O^+ and OH^-.

$$[H_3O^+] = [OH^-] = 1.0 \times 10^{-7}\,M \quad \text{(in pure water at 25 °C)}$$

where $[H_3O^+]$ = the concentration of H_3O^+ in M and $[OH^-]$ = the concentration of OH^- in M. The *product* of the concentration of these two ions in aqueous solutions is the **ion product constant for water (K_w).**

$$K_w = [H_3O^+][OH^-]$$

| The units of K_w are normally dropped. |

We can find the value of K_w at 25 °C by multiplying the hydronium and hydroxide concentrations for pure water listed earlier.

$$\begin{aligned} K_w &= [H_3O^+][OH^-] \\ &= (1.0 \times 10^{-7})(1.0 \times 10^{-7}) \\ &= (1.0 \times 10^{-7})^2 \\ &= 1.0 \times 10^{-14} \end{aligned}$$

The preceding equation holds true for all aqueous solutions at 25 °C. The concentration of H_3O^+ times the concentration of OH^- is 1.0×10^{-14}. In pure water, since H_2O is the only source of these ions, there is one H_3O^+ ion for every OH^- ion. Consequently, the concentrations of H_3O^+ and OH^- are equal. Such a solution is a **neutral solution**.

| In a neutral solution, $[H_3O^+] = [OH^-]$. |

$$[H_3O^+] = [OH^-] = \sqrt{K_w} = 1.0 \times 10^{-7}\,M \text{ (in pure water)}$$

An **acidic solution** contains an acid that creates additional H_3O^+ ions, causing $[H_3O^+]$ to increase. However, the *ion product constant still applies*.

$$[H_3O^+][OH^-] = K_w = 1.0 \times 10^{-14}$$

| In an acidic solution, $[H_3O^+] > [OH^-]$. |

If $[H_3O^+]$ increases, then $[OH^-]$ must decrease for the ion product to remain 1.0×10^{-14}. For example, suppose $[H_3O^+] = 1.0 \times 10^{-3}\,M$; then we can find $[OH^-]$ by solving the ion product expression for $[OH^-]$.

$$(1.0 \times 10^{-3})[OH^-] = 1.0 \times 10^{-14}$$

$$[OH^-] = \frac{1.0 \times 10^{-14}}{1.0 \times 10^{-3}} = 1.0 \times 10^{-11}\,M$$

In an acidic solution, $[H_3O^+]$ is greater than $1.0 \times 10^{-7}\,M$, and $[OH^-]$ is less than $1.0 \times 10^{-7}\,M$.

| In a basic solution, $[H_3O^+] < [OH^-]$. |

A **basic solution** contains a base that creates additional OH^- ions, causing the $[OH^-]$ to increase and the $[H_3O^+]$ to decrease. For example, suppose

$[OH^-] = 1.0 \times 10^{-2}$ M; we can find $[H_3O^+]$ by solving the ion product expression for $[H_3O^+]$.

$$[H_3O^+](1.0 \times 10^{-2}) = 1.0 \times 10^{-14}$$

$$[H_3O^+] = \frac{1.0 \times 10^{-14}}{1.0 \times 10^{-2}} = 1.0 \times 10^{-12} \text{ M}$$

In a basic solution, $[OH^-]$ is greater than 1.0×10^{-7} M and $[H_3O^+]$ is less than 1.0×10^{-7} M.

▶ **FIGURE 14.16 Acidic and basic solutions**

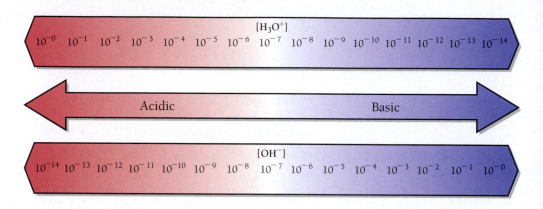

To summarize, at 25 °C (▲ FIGURE 14.16):

- In a neutral solution, $[H_3O^+] = [OH^-] = 1.0 \times 10^{-7}$ M
- In an acidic solution, $[H_3O^+] > 1.0 \times 10^{-7}$ M $[OH^-] < 1.0 \times 10^{-7}$ M
- In a basic solution, $[H_3O^+] < 1.0 \times 10^{-7}$ M $[OH^-] > 1.0 \times 10^{-7}$ M
- In all aqueous solutions, $[H_3O^+][OH^-] = K_w = 1.0 \times 10^{-14}$

EXAMPLE **14.7**   **Using K_w in Calculations**

Calculate $[OH^-]$ for each solution and determine whether the solution is acidic, basic, or neutral.

(a) $[H_3O^+] = 7.5 \times 10^{-5}$ M
(b) $[H_3O^+] = 1.5 \times 10^{-9}$ M
(c) $[H_3O^+] = 1.0 \times 10^{-7}$ M

To find $[OH^-]$ use the ion product constant, K_w. Substitute the given value for $[H_3O^+]$ and solve the equation for $[OH^-]$. Since $[H_3O^+] > 1.0 \times 10^{-7}$ M and $[OH^-] < 1.0 \times 10^{-7}$ M, the solution is acidic.	**SOLUTION** **(a)** $[H_3O^+][OH^-] = K_w = 1.0 \times 10^{-14}$ $[7.5 \times 10^{-5}][OH^-] = K_w = 1.0 \times 10^{-14}$ $[OH^-] = \dfrac{1.0 \times 10^{-14}}{7.5 \times 10^{-5}} = 1.3 \times 10^{-10}$ M acidic solution
Substitute the given value for $[H_3O^+]$ into the ion product constant equation and solve the equation for $[OH^-]$. Since $[H_3O^+] < 1.0 \times 10^{-7}$ M and $[OH^-] > 1.0 \times 10^{-7}$ M, the solution is basic.	**(b)** $[H_3O^+][OH^-] = K_w = 1.0 \times 10^{-14}$ $[1.5 \times 10^{-9}][OH^-] = 1.0 \times 10^{-14}$ $[OH^-] = \dfrac{1.0 \times 10^{-14}}{1.5 \times 10^{-9}} = 6.7 \times 10^{-6}$ M basic solution

Substitute the given value for $[H_3O^+]$ into the ion product constant equation and solve the equation for $[OH^-]$. Since $[H_3O^+] = 1.0 \times 10^{-7}$ M and $[OH^-] = 1.0 \times 10^{-7}$ M, the solution is neutral.	**(c)** $[H_3O^+][OH^-] = K_w = 1.0 \times 10^{-14}$ $[1.0 \times 10^{-7}][OH^-] = 1.0 \times 10^{-14}$ $[OH^-] = \dfrac{1.0 \times 10^{-14}}{1.0 \times 10^{-7}} = 1.0 \times 10^{-7}$ M neutral solution

▶ **SKILLBUILDER 14.7 | Using K_w in Calculations**

Calculate $[H_3O^+]$ in each solution and determine whether the solution is acidic, basic, or neutral.

(a) $[OH^-] = 1.5 \times 10^{-2}$ M
(b) $[OH^-] = 1.0 \times 10^{-7}$ M
(c) $[OH^-] = 8.2 \times 10^{-10}$ M

▶ **FOR MORE PRACTICE**   Example 14.17; Problems 63, 64, 65, 66.

CONCEPTUAL ✓ CHECKPOINT 14.5

Which substance is least likely to act as a base?

(a) H_2O **(b)** OH^- **(c)** NH_3 **(d)** NH_4^+

14.9 The pH and pOH Scales: Ways to Express Acidity and Basicity

▶ Calculate pH from $[H_3O^+]$.
▶ Calculate $[H_3O^+]$ from pH.
▶ Calculate $[OH^-]$ from pOH.
▶ Compare and contrast the pOH scale and the pH scale.

Chemists have devised a scale, called the **pH** scale, based on hydrogen ion concentration to express the acidity or basicity of solutions. At 25 °C according to the pH scale, a solution has these general characteristics:

- pH < 7 *acidic* solution
- pH > 7 *basic* solution
- pH = 7 *neutral* solution

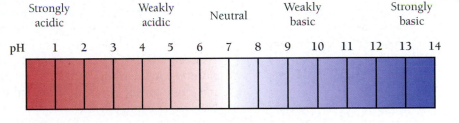

TABLE 14.7 The pH of Some Common Substances	
Substance	**pH**
gastric (human stomach) acid	1.0–3.0
limes	1.8–2.0
lemons	2.2–2.4
soft drinks	2.0–4.0
plums	2.8–3.0
wine	2.8–3.8
apples	2.9–3.3
peaches	3.4–3.6
cherries	3.2–4.0
beer	4.0–5.0
rainwater (unpolluted)	5.6
human blood	7.3–7.4
egg whites	7.6–8.0
milk of magnesia	10.5
household ammonia	10.5–11.5
4% NaOH solution	14

Table 14.7 lists the pH of some common substances. Notice that, as we discussed in Section 14.1, many foods, especially fruits, are acidic and therefore have low pH values. The foods with the lowest pH values are limes and lemons, and they are among the sourest. Relatively few foods, however, are basic.

The pH scale is a **logarithmic scale**; a change of 1 pH unit corresponds to a tenfold change in H_3O^+ concentration. For example, a lime with a pH of 2.0 is 10 times more acidic than a plum with a pH of 3.0 and 100 times more acidic than a cherry with a pH of 4.0. Each change of 1 in pH scale corresponds to a change of 10 in $[H_3O^+]$ (▶ **FIGURE 14.17**, on the next page).

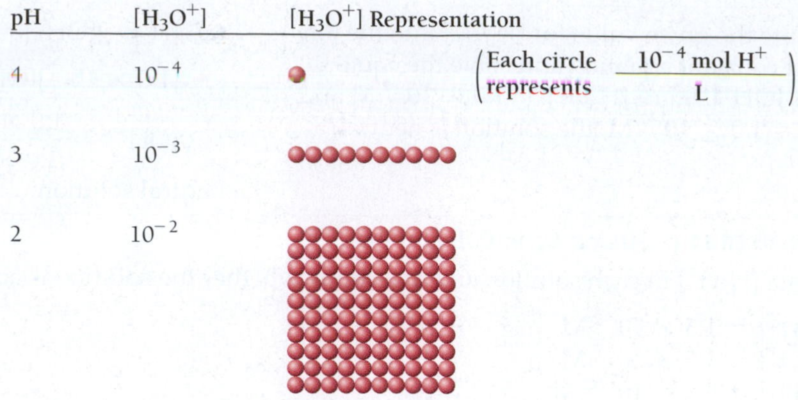

▶ FIGURE 14.17 The pH scale is a
logarithmic scale A *decrease* of
1 unit on the pH scale corresponds
to an *increase* in [H_3O^+] by a
factor of 10. Each circle stands for
10^{-4} mol H^+/L, or 6.022×10^{19} H^+
ions per liter. Question: How much of
an increase in H_3O^+ concentration corresponds to a decrease of 2 pH units?

pH	[H_3O^+]	[H_3O^+] Representation
4	10^{-4}	● $\left(\begin{matrix}\text{Each circle} \\ \text{represents}\end{matrix} \dfrac{10^{-4} \text{ mol } H^+}{L}\right)$
3	10^{-3}	●●●●●●●●●●
2	10^{-2}	(grid of circles)

Calculating pH from [H_3O^+]

Note that pH is defined using the log
function (base ten), which is different
from the natural log (abbreviated ln).

To calculate pH, we must use logarithms. Recall that the log of a number is the
exponent to which 10 must be raised to obtain that number, as shown in these
examples:

$$\log 10^1 = 1; \log 10^2 = 2; \log 10^3 = 3$$

$$\log 10^{-1} = -1; \log 10^{-2} = -2; \log 10^{-3} = -3$$

We define the pH of a solution as the negative of the logarithm of the hydronium
ion concentration:

$$pH = -\log[H_3O^+]$$

A solution having an [H_3O^+] = 1.5×10^{-7} M (acidic) has a pH of:

$$pH = -\log[H_3O^+]$$
$$= -\log(1.5 \times 10^{-7})$$
$$= -(-6.82)$$
$$= 6.82$$

Notice that the pH is reported to two decimal places here. This is because only the
numbers to the right of the decimal place are significant in a log. Because our original value for the concentration had two significant figures, the log of that number
has two decimal places:

2 decimal places

$$\log(1.0 \times 10^{-3}) = 3.00$$

When we take the log of a quantity, the
result should have the same number
of decimal places as the number
of significant figures in the original
quantity.

If the original number has three significant figures, we report the log to three decimal places:

3 decimal places

$$-\log(1.00 \times 10^{-3}) = 3.000$$

A solution having $[H_3O^+] = 1.0 \times 10^{-7}$ M (neutral) has a pH of:

$$pH = -\log[H_3O^+]$$
$$= -\log(1.0 \times 10^{-7})$$
$$= -(-7.00)$$
$$= 7.00$$

PEARSON
eText
2.0
Interactive
Worked Example
Video 14.8

EXAMPLE **14.8** **Calculating pH from [H₃O⁺]**

Calculate the pH of each solution and indicate whether the solution is acidic or basic.

(a) $[H_3O^+] = 1.8 \times 10^{-4}$ M
(b) $[H_3O^+] = 7.2 \times 10^{-9}$ M

SOLUTION	
To calculate pH, substitute the given $[H_3O^+]$ into the pH equation.	(a) $pH = -\log[H_3O^+]$ $= -\log(1.8 \times 10^{-4})$ $= -(-3.74)$ $= 3.74$
Since the pH $<$ 7, this solution is acidic.	
Again, substitute the given $[H_3O^+]$ into the pH equation.	(b) $pH = -\log[H_3O^+]$ $= -\log(7.2 \times 10^{-9})$ $= -(-8.14)$ $= 8.14$
Since the pH $>$ 7, this solution is basic.	

▶ SKILLBUILDER 14.8 | **Calculating pH from [H₃O⁺]**

Calculate the pH of each solution and indicate whether the solution is acidic or basic.

(a) $[H_3O^+] = 9.5 \times 10^{-9}$ M
(b) $[H_3O^+] = 6.1 \times 10^{-3}$ M

▶ SKILLBUILDER PLUS Calculate the pH of a solution with $[OH^-] = 1.3 \times 10^{-2}$ M and indicate whether the solution is acidic or basic. *Hint:* Begin by using K_w to find $[H_3O^+]$.

▶ FOR MORE PRACTICE Example 14.18; Problems 69, 70.

Calculating [H₃O⁺] from pH

> Ten raised to the log of a number is equal to that number: $10^{\log x} = x$.

To calculate $[H_3O^+]$ from a pH value, we *undo* the log. The log can be undone using the inverse log function (*Method 1*) on most calculators or using the 10^x key (*Method 2*). Both methods do the same thing; the one you use depends on the calculator you have.

Method 1: Inverse Log Function	Method 2: 10ˣ Function
$pH = -\log[H_3O^+]$	$pH = -\log[H_3O^+]$
$-pH = \log[H_3O^+]$	$-pH = \log[H_3O^+]$
$\text{invlog}(-pH) = \text{invlog}(\log[H_3O^+])$	$10^{-pH} = 10^{\log[H_3O^+]}$
$\text{invlog}(-pH) = [H_3O^+]$	$10^{-pH} = [H_3O^+]$

> The invlog function "undoes" log: $\text{invlog}(\log x) = x$.

> The inverse log is sometimes called the antilog.

So, to calculate $[H_3O^+]$ from a pH value, we take the inverse log of the negative of the pH value (*Method 1*) or raise 10 to the negative of the pH value (*Method 2*).

EXAMPLE **14.9** **Calculating [H₃O⁺] from pH**

Calculate the H_3O^+ concentration for a solution with a pH of 4.80.

SOLUTION

To find $[H_3O^+]$ from pH, undo the log function. Use either Method 1 or Method 2.

Method 1: Inverse Log Function	Method 2: 10ˣ Function
$pH = -\log[H_3O^+]$	$pH = -\log[H_3O^+]$
$4.80 = -\log[H_3O^+]$	$4.80 = -\log[H_3O^+]$
$-4.80 = \log[H_3O^+]$	$-4.80 = \log[H_3O^+]$
$\text{invlog}(-4.80) = \text{invlog}(\log[H_3O^+])$	$10^{-4.80} = 10^{\log[H_3O^+]}$
$\text{invlog}(-4.80) = [H_3O^+]$	$10^{-4.80} = [H_3O^+]$
$[H_3O^+] = 1.6 \times 10^{-5}$ M	$[H_3O^+] = 1.6 \times 10^{-5}$ M

The number of significant figures in the inverse log of a number is equal to the number of decimal places in the number.

▶ **SKILLBUILDER 14.9 | Calculating [H₃O⁺] from pH**

Calculate the H_3O^+ concentration for a solution with a pH of 8.37.

▶ **SKILLBUILDER PLUS**

Calculate the OH^- concentration for a solution with a pH of 3.66.

▶ **FOR MORE PRACTICE** Example 14.19; Problems 71, 72.

CONCEPTUAL ✔ CHECKPOINT 14.6

Solution A has a pH of 13. Solution B has a pH of 10. The concentration of H_3O^+ in solution B is _____ times that in solution A.

(a) 0.001

(b) $\frac{1}{3}$

(c) 3

(d) 1000

The pOH Scale

The **pOH** scale is analogous to the pH scale but is defined with respect to $[OH^-]$ instead of $[H_3O^+]$.

Notice that *p* is the mathematical function: $-\log$; thus, $pX = -\log X$.

$$pOH = -\log[OH^-]$$

A solution with an $[OH^-]$ of 1.0×10^{-3} M (basic) has a pOH of 3.00. On the pOH scale, a pOH less than 7 is basic and a pOH greater than 7 is acidic. A pOH of 7 is neutral. We can find $[OH^-]$ from the pOH just as we find $[H_3O^+]$ from the pH, as shown in Example 14.10.

EXAMPLE **14.10** **Calculating [OH⁻] from pOH**

Calculate $[OH^-]$ for a solution with a pOH of 8.55.

SOLUTION

To find $[OH^-]$ from pOH, undo the log function. Use either Method 1 or Method 2.

Method 1: Inverse Log Function	Method 2: 10^x Function
$pOH = -log[OH^-]$	$pOH = -log[OH^-]$
$8.55 = -log[OH^-]$	$8.55 = -log[HO^-]$
$-8.55 = log[OH^-]$	$-8.55 = log[OH^-]$
$invlog(-8.55) = invlog(log[OH^-])$	$10^{-8.55} = 10^{log\,[OH^-]}$
$invlog(-8.55) = [OH^-]$	$10^{-8.55} = [OH^-]$
$[OH^-] = 2.8 \times 10^{-9}$ M	$[OH^-] = 2.8 \times 10^{-9}$ M

▶ **SKILLBUILDER 14.10** | **Calculating [OH⁻] from pOH**

Calculate the OH^- concentration for a solution with a pOH of 4.25.

▶ **SKILLBUILDER PLUS**

Calculate the H_3O^+ concentration for a solution with a pOH of 5.68.

▶ **FOR MORE PRACTICE** Problems 79, 80, 81, 82.

We can derive a relationship between pH and pOH at 25 °C from the expression for K_w.

$$[H_3O^+][OH^-] = 1.0 \times 10^{-14}$$

$\boxed{log(AB) = log\,A + log\,B}$

Taking the log of both sides, we get:

$$log\{[H_3O^+][OH^-]\} = log\,(1.0 \times 10^{-14})$$

$$log[H_3O^+] + log[OH^-] = -14.00$$

$$-log[H_3O^+] - log[OH^-] = 14.00$$

$$pH + pOH = 14.00$$

The sum of pH and pOH is always equal to 14.00 at 25 °C. Therefore, a solution with a pH of 3 has a pOH of 11.

CONCEPTUAL ✔ **CHECKPOINT 14.7**

A solution has a pH of 5. What is the pOH of the solution?

(a) 5   (b) 10   (c) 14   (d) 9

14.10 Buffers: Solutions That Resist pH Change

▶ Describe how buffers resist pH change.

Key Concept Video
Buffers

Buffers can also be composed of a weak base and its conjugate acid.

Most solutions rapidly become more acidic (lower pH) upon addition of an acid or more basic (higher pH) upon addition of a base. A **buffer**, however, resists pH change by neutralizing added acid or added base. Human blood, for example, is a buffer. Acid or base that is added to blood gets neutralized by components within blood, resulting in a nearly constant pH. In healthy individuals, blood pH is between 7.36 and 7.40. If blood pH drops below 7.0 or rises above 7.8, death results.

How does blood maintain such a narrow pH range? Like all buffers, blood contains *significant* amounts of *both a weak acid and its conjugate base*. When additional base is added to blood, the weak acid reacts with the base, neutralizing it. When additional acid is added to blood, the conjugate base reacts with the acid, neutralizing it. In this way, blood maintains a constant pH.

CHEMISTRY AND HEALTH

Alkaloids

Alkaloids are organic bases that occur naturally in many plants (see Section 14.3) that often have medicinal qualities. Morphine, for example, is a powerful alkaloid drug that occurs in the opium poppy (▶ **FIGURE 14.18**) and is used to relieve severe pain. Morphine is an example of a *narcotic*, a drug that dulls the senses and induces sleep. It produces relief from and indifference to pain. Morphine can also produce feelings of euphoria and contentment, which lead to its abuse. Morphine is highly addictive, both psychologically and physically. A person who abuses morphine over long periods of time becomes physically dependent on the drug and suffers severe withdrawal symptoms upon termination of use.

▲ **FIGURE 14.18 Opium poppy** The opium poppy contains the alkaloids morphine and codeine.

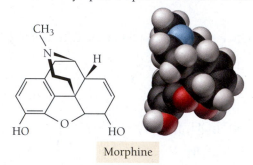

Morphine

Amphetamine is another powerful drug related to the alkaloid ephedrine. Whereas morphine slows down nerve signal transmissions, amphetamine enhances them. Amphetamine is an example of a *stimulant*, a drug that increases alertness and wakefulness. Amphetamine is widely used to treat attention-deficit hyperactivity disorder (ADHD) and is prescribed under the trade name Adderall. Patients suffering from ADHD find that amphetamine helps them to focus and concentrate more effectively. However, because amphetamine produces alertness and increased stamina, it, too, is often abused.

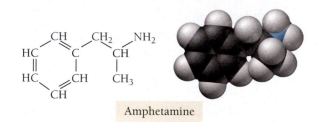

Amphetamine

Other common alkaloids include caffeine and nicotine, both of which are stimulants. Caffeine is found in the coffee bean, and nicotine is found in tobacco. Although both have some addictive qualities, nicotine is by far the more addictive. A nicotine addiction is among the most difficult to break, as any smoker can attest.

B14.2 CAN YOU ANSWER THIS? *What part of the amphetamine and morphine molecules makes them bases?*

We can make a simple buffer by mixing acetic acid ($HC_2H_3O_2$) and its conjugate base, sodium acetate ($NaC_2H_3O_2$), in water (▶ **FIGURE 14.19**). (The sodium in sodium acetate is just a spectator ion and does not contribute to buffering action.) Because $HC_2H_3O_2$ is a weak acid and $C_2H_3O_2^-$ is its conjugate base, a solution containing both of these is a buffer. Note that a weak acid by itself, even though it partially ionizes to form some of its conjugate base, does not contain sufficient base to be a buffer. A buffer must contain *significant* amounts of *both* a weak acid and its conjugate base. If we add more base, in the form of NaOH, to the buffer solution containing acetic acid and sodium acetate, the acetic acid neutralizes the base according to the reaction:

$$NaOH(aq) + HC_2H_3O_2(aq) \longrightarrow H_2O(l) + NaC_2H_3O_2(aq)$$

Base   Acid

As long as the amount of NaOH that we add is less than the amount of $HC_2H_3O_2$ in solution, the solution neutralizes the NaOH, and the resulting pH change is small.

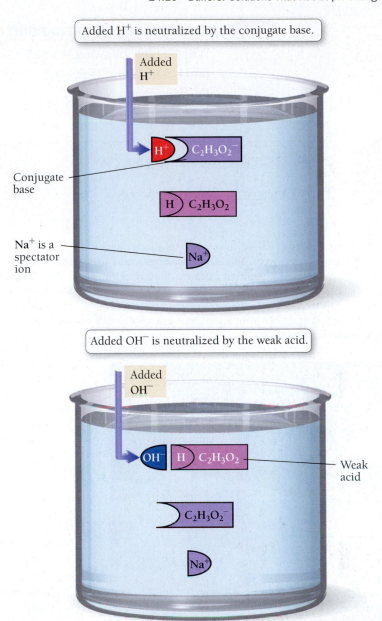

▲ **FIGURE 14.19 How buffers resist pH change** A buffer contains significant amounts of a weak acid and its conjugate base. The acid consumes any added base, and the base consumes any added acid. In this way, a buffer resists pH change.

Suppose, on the other hand, that we add more acid, in the form of HCl, to the solution. In this case, the conjugate base, $NaC_2H_3O_2$, neutralizes the added HCl according to the reaction:

$$\underset{\text{Acid}}{HCl(aq)} + \underset{\text{Base}}{NaC_2H_3O_2(aq)} \longrightarrow HC_2H_3O_2(aq) + NaCl(aq)$$

As long as the amount of HCl that we add is less than the amount of $NaC_2H_3O_2$ in solution, the solution neutralizes the HCl and the resulting pH change is small.

To summarize:

- Buffers resist pH change.
- Buffers contain significant amounts of both a weak acid and its conjugate base.
- The weak acid in a buffer neutralizes added base.
- The conjugate base in a buffer neutralizes added acid.

CONCEPTUAL ✔ CHECKPOINT 14.8

Which of the following is a buffer solution?

(a) $H_2SO_4(aq)$ and $H_2SO_3(aq)$

(b) $HF(aq)$ and $NaF(aq)$

(c) $HCl(aq)$ and $NaCl(aq)$

(d) $NaCl(aq)$ and $NaOH(aq)$

CHEMISTRY AND HEALTH

The Danger of Antifreeze

Most types of antifreeze used in cars are solutions of ethylene glycol. Every year, thousands of dogs and cats die from ethylene glycol poisoning because they consume improperly stored antifreeze or antifreeze that has leaked out of a radiator. Antifreeze has a somewhat sweet taste, which attracts a curious dog or cat. Young children are also at risk for ethylene glycol poisoning.

▲ Antifreeze contains ethylene glycol, which metabolizes in the liver to form glycolic acid.

The first stage of ethylene glycol poisoning is a drunken state. Ethylene glycol is an alcohol, and it affects the brain of a dog or cat much as an alcoholic beverage would. Once ethylene glycol begins to metabolize, however, the second and more deadly stage begins. Ethylene glycol is metabolized in the liver into glycolic acid ($HC_2H_3O_3$), which enters the bloodstream. If the original quantities of consumed antifreeze are significant, the glycolic acid overwhelms the blood's natural buffering system, causing blood pH to drop to dangerously low levels. At this point, the cat or dog may begin hyperventilating in an effort to overcome the acidic blood's reduced ability to carry oxygen. If no treatment is administered, the animal will eventually go into a coma and die.

One treatment for ethylene glycol poisoning is the administration of ethyl alcohol (the alcohol found in alcoholic beverages). The liver enzyme that metabolizes ethylene glycol is the same one that metabolizes ethyl alcohol, but it has a higher affinity for ethyl alcohol than for ethylene glycol. Consequently, the enzyme preferentially metabolizes ethyl alcohol, allowing the unmetabolized ethylene glycol to escape through the urine. If administered early, this treatment can save the life of a dog or cat that has consumed ethylene glycol.

B14.3 CAN YOU ANSWER THIS? *One of the main buffering systems found in blood consists of carbonic acid (H_2CO_3) and bicarbonate ion (HCO_3^-). Write an equation showing how this buffering system could neutralize glycolic acid ($HC_2H_3O_3$) that might enter the blood from ethylene glycol poisoning. Suppose a cat has 0.15 mol of HCO_3^- and 0.15 mol of H_2CO_3 in its bloodstream. How many grams of $HC_2H_3O_3$ could be neutralized before the buffering system in the cat's blood is overwhelmed?*

Chapter in 14 Review

MasteringChemistry™ provides end-of-chapter exercises, feedback-enriched tutorial problems, animations, and interactive activities to encourage problem solving practice and deeper understanding of key concepts and topics.

Self-Assessment Quiz

Q1. Which substance is most likely to have a bitter taste?

(a) $HCl(aq)$

(b) $NaCl(aq)$

(c) $NaOH(aq)$

(d) none of the above

Q2. Identify the Brønsted–Lowry base in the reaction.

$$HClO_2(aq) + H_2O(aq) \rightleftharpoons H_3O^+(aq) + ClO_2^-(aq)$$

(a) $HClO_2$

(b) H_2O

(c) H_3O^+

(d) ClO_2^-

Q3. What is the conjugate base of the acid $HClO_4$?
  (a) HCl
  (b) HO_3^+
  (c) ClO_3^-
  (d) ClO_4^-

Q4. What are the products of the reaction between HBr(*aq*) and KOH(*aq*)?
  (a) HK(*aq*) and BrOH(*aq*)
  (b) H_2(*g*) and KBr(*aq*)
  (c) H_2O(*l*) and KBr(*aq*)
  (d) H_2O(*l*) and BrOH(*aq*)

Q5. What are the products of the reaction between HCl and Sn?
  (a) $SnCl_2$(*aq*) and H_2(*g*)
  (b) SnH(*aq*) and Cl_2(*g*)
  (c) H_2O(*l*) and SnCl(*aq*)
  (d) H_2(*g*),Cl_2(*g*), and Sn(*s*)

Q6. A 25.00-mL sample of an HNO_3 solution is titrated with 0.102 M NaOH. The titration requires 28.52 mL to reach the equivalence point. What is the concentration of the HNO_3 solution?
  (a) 0.116 M
  (b) 8.89 M
  (c) 0.0894 M
  (d) 0.102 M

Q7. In which solution is $[H_3O^+]$ less than 0.100 M?
  (a) 0.100 M H_2SO_4(*aq*)
  (b) 0.100 M $HCHO_2$(*aq*)
  (c) 0.100 M $HClO_4$(*aq*)
  (d) none of the above (all have $[H_3O^+]$ less than 0.100 M)

Q8. In which solution is $[OH^-]$ equal to 0.100 M?
  (a) 0.100 M NH_3(*aq*)
  (b) 0.100 M $Ba(OH)_2$(*aq*)
  (c) 0.100 M NaOH(*aq*)
  (d) all of the above (all have $[OH^-]$ equal to 0.100 M)

Q9. What is $[H_3O^+]$ in a solution with $[OH^-] = 2.5 \times 10^{-4}$ M?
  (a) 2.5×10^{10} M
  (b) 4.0×10^{-11} M
  (c) 1.0×0^{-7} M
  (d) 1.0×10^{-14} M

Q10. What is the pH of a solution with $[H_3O^+] = 2.8 \times 10^{-5}$ M?
  (a) −4.55
  (b) 4.55
  (c) 10.48
  (d) 1.00

Q11. What is $[OH^-]$ in a solution with a pH of 9.55?
  (a) 2.82×10^{-10} M
  (b) 4.45 M
  (c) 2.82×10^4 M
  (d) 3.55×10^{-5} M

Q12. A buffer contains $HCHO_2$(*aq*) and $KCHO_2$(*aq*). Which statement correctly summarizes the action of this buffer?
  (a) $HCHO_2$(*aq*) neutralizes added acid, and $KCHO_2$(*aq*) neutralizes added base.
  (b) Both $HCHO_2$(*aq*) and $KCHO_2$(*aq*) neutralize added acid.
  (c) Both $HCHO_2$(*aq*) and $KCHO_2$(*aq*) neutralize added base.
  (d) $HCHO_2$(*aq*) neutralizes added base, and $KCHO_2$(*aq*) neutralizes added acid.

Answers: 1:c, 2:b, 3:d, 4:c, 5:a, 6:a, 7:b, 8:c, 9:b, 10:b, 11:d, 12:d

---

Chemical Principles

Relevance

Acid Properties

- Acids have a sour taste.
- Acids dissolve many metals.
- Acids turn blue litmus paper red.

Acids are responsible for the sour taste in foods such as lemons, limes, and vinegar. Acids are also frequently used in the laboratory and in industry.

Base Properties

- Bases have a bitter taste.
- Bases have a slippery feel.
- Bases turn red litmus paper blue.

Bases are less common in foods, but their presence in some foods—such as coffee and beer—is enjoyed by many as an acquired taste. Bases are also frequently used in the laboratory and in industry.

Molecular Definitions of Acids and Bases: Arrhenius definition

Acid—substance that produces H^+ ions in solution
Base—substance that produces OH^- ions in solution

The Arrhenius definition is simpler and easier to use. It also shows how an acid and a base neutralize each other to form water ($H^+ + OH^- \longrightarrow H_2O$).

Brønsted–Lowry definition

Acid—proton donor
Base—proton acceptor

The more generally applicable Brønsted–Lowry definition helps us see that, in water, H^+ ions usually associate with water molecules to form H_3O^+. It also shows how bases that do not contain OH^- ions can still act as bases by accepting a proton from water.

Reactions of Acids and Bases: Neutralization Reactions

An acid and a base react to form water and a salt.

$$\text{HCl}(aq) + \text{KOH}(aq) \longrightarrow \text{H}_2\text{O}(l) + \text{KCl}(aq)$$

Acid   Base   Water   Salt

Neutralization reactions are common in our everyday lives. Antacids, for example, are bases that react with acids from the stomach to alleviate heartburn and sour stomach.

Acid–Metal Reactions

Acids react with many metals to form hydrogen gas and a salt.

$$2\,\text{HCl}(aq) + \text{Mg}(s) \longrightarrow \text{H}_2(g) + \text{MgCl}_2(aq)$$

Acid   Metal   Hydrogen   Salt
gas

Acid–metal and acid–metal oxide reactions show the corrosive nature of acids. In both of these reactions, the acid dissolves the metal or the metal oxide. Some of the effects of these kinds of reactions can be seen in the damage to building materials caused by acid rain. Acids dissolve metals and metal oxides, so building materials composed of these substances are susceptible to acid rain.

Acid–Metal Oxide Reactions

Acids react with many metal oxides to form water and a salt.

$$2\,\text{HCl}(aq) + \text{K}_2\text{O}(s) \longrightarrow \text{H}_2\text{O}(l) + 2\,\text{KCl}(aq)$$

Acid   Metal oxide   Water   Salt

Acid–Base Titration

In an acid–base titration, we add an acid (or base) of known concentration to a base (or acid) of unknown concentration. We combine the two reactants until they are in exact stoichiometric proportions (moles of H^+ = moles of OH^-), which marks the equivalence point of the titration. In titration, since we know the moles of H^+ (or OH^-) that we added, we can determine the moles of OH^- (or H^+) in the unknown solution.

An acid–base titration is a laboratory procedure often used to determine the unknown concentration of an acid or a base.

Strong and Weak Acids and Bases

Strong acids completely ionize, and strong bases completely dissociate in aqueous solutions. For example:

$$\text{HCl}(aq) + \text{H}_2\text{O}(l) \longrightarrow \text{H}_3\text{O}^+(aq) + \text{Cl}^-(aq)$$

$$\text{NaOH}(aq) \longrightarrow \text{Na}^+(aq) + \text{OH}^-(aq)$$

A 1 M HCl solution has $[\text{H}_3\text{O}^+] = 1$ M and a 1 M NaOH solution has $[\text{OH}^-] = 1$ M.

Weak acids only partially ionize in solution. Most weak bases partially ionize water in solution. For example:

$$\text{HF}(aq) + \text{H}_2\text{O}(l) \rightleftharpoons \text{H}_3\text{O}^+(aq) + \text{F}^-(aq)$$

$$\text{NH}_3(aq) + \text{H}_2\text{O}(l) \rightleftharpoons \text{NH}_4^+(aq) + \text{OH}^-(aq)$$

A 1 M HF solution has $[\text{H}_3\text{O}^+] < 1$ M, and a 1 M NH_3 solution has $[\text{OH}^-] < 1$ M.

Whether an acid is strong or weak depends on the conjugate base: The stronger the conjugate base, the weaker the acid. Since the acidity or basicity of a solution depends on $[\text{H}_3\text{O}^+]$ and $[\text{OH}^-]$, we must know whether an acid is strong or weak to know the degree of acidity or basicity.

Self-Ionization of Water

Water can act as both an acid and a base with itself.

$$\text{H}_2\text{O}(l) + \text{H}_2\text{O}(l) \rightleftharpoons \text{H}_3\text{O}^+(aq) + \text{OH}^-(aq)$$

Acid   Base

The product of $[\text{H}_3\text{O}^+]$ and $[\text{OH}^-]$ in aqueous solutions is always equal to the ion product constant, $K_w(10^{-14})$.

$$[\text{H}_3\text{O}^+][\text{OH}^-] = K_w = 1.0 \times 10^{-14}$$

The self-ionization of water occurs because aqueous solutions always contain some H_3O^+ and some OH^-. In a neutral solution, the concentrations of these are equal (1.0×10^{-7} M). When an acid is added to water, $[\text{H}_3\text{O}^+]$ increases and $[\text{OH}^-]$ decreases. When a base is added to water, the opposite happens. The ion product constant, however, still equals 1.0×10^{-14}, allowing us to calculate $[\text{H}_3\text{O}^+]$ given $[\text{OH}^-]$ and vice versa.

pH and pOH Scales

$$\text{pH} = -\log[\text{H}_3\text{O}^+]$$
$$\text{pH} > 7 \text{ (basic)}$$
$$\text{pH} < 7 \text{ (acidic)}$$
$$\text{pH} = 7 \text{ (neutral)}$$
$$\text{pOH} = -\log[\text{OH}^-]$$

pH is a convenient way to specify acidity or basicity. Since the pH scale is logarithmic, a change of one on the pH scale corresponds to a tenfold change in the $[\text{H}_3\text{O}^+]$. The pOH scale, defined with respect to $[\text{OH}^-]$ instead of $[\text{H}_3\text{O}^+]$, is less commonly used.

Buffers

Buffers are solutions containing significant amounts of both a weak acid and its conjugate base. Buffers resist pH change by neutralizing added acid or base.

Buffers are important in blood chemistry because blood must stay within a narrow pH range in order to carry oxygen.

Chemical Skills

Examples

LO: Identify Brønsted–Lowry acids and bases and their conjugates (Section 14.4).

The substance that donates the proton is the acid (proton donor) and becomes the conjugate base (as a product). The substance that accepts the proton (proton acceptor) is the base and becomes the conjugate acid (as a product).

EXAMPLE 14.11 | **Identifying Brønsted–Lowry Acids and Bases and Their Conjugates**

Identify the Brønsted–Lowry acid, the Brønsted–Lowry base, the conjugate acid, and the conjugate base in this reaction:

$$HNO_3(aq) + H_2O(l) \longrightarrow H_3O^+(aq) + NO_3^-(aq)$$

SOLUTION

$$HNO_3(aq) + H_2O(l) \longrightarrow H_3O^+(aq) + NO_3^-(aq)$$
Acid   Base   Conjugate acid   Conjugate base

LO: Write equations for neutralization reactions (Section 14.5).

In a neutralization reaction, an acid and a base usually react to form water and a salt (ionic compound).

$$Acid + Base \longrightarrow Water + Salt$$

Write the skeletal equation first, making sure to write the formula of the salt so that it is charge-neutral. Then balance the equation.

EXAMPLE 14.12 | **Writing Equations for Neutralization Reactions**

Write a molecular equation for the reaction between aqueous HBr and aqueous $Ca(OH)_2$.

SOLUTION
Skeletal equation:

$$HBr(aq) + Ca(OH)_2(aq) \longrightarrow H_2O(l) + CaBr_2(aq)$$

Balanced equation:

$$2\,HBr(aq) + Ca(OH)_2(aq) \longrightarrow 2\,H_2O(l) + CaBr_2(aq)$$

LO: Write equations for the reactions of acids with metals and with metal oxides (Section 14.5).

Acids react with many metals to form hydrogen gas and a salt.

$$Acid + Metal \longrightarrow Hydrogen\ gas + Salt$$

Write the skeletal equation first, making sure to write the formula of the salt so that it is charge-neutral. Then balance the equation.

Acids react with many metal oxides to form water and a salt.

$$Acid + Metal\ oxide \longrightarrow Water + Salt$$

Write the skeletal equation first, making sure to write the formula of the salt so that it is charge-neutral. Then balance the equation.

EXAMPLE 14.13 | **Writing Equations for the Reactions of Acids with Metals and with Metal Oxides**

Write equations for the reaction of hydrobromic acid with calcium metal and for the reaction of hydrobromic acid with calcium oxide.

SOLUTION
Skeletal equation:

$$HBr(aq) + Ca(s) \longrightarrow H_2(g) + CaBr_2(aq)$$

Balanced equation:

$$2\,HBr(aq) + Ca(s) \longrightarrow H_2(g) + CaBr_2(aq)$$

Skeletal equation:

$$HBr(aq) + CaO(s) \longrightarrow H_2O(l) + CaBr_2(aq)$$

Balanced equation:

$$2\,HBr(aq) + CaO(s) \longrightarrow H_2O(l) + CaBr_2(aq)$$

LO: Use acid–base titration to determine the concentration of an unknown solution (Section 14.6).

EXAMPLE **14.14** **Acid–Base Titrations**

A 15.00-mL sample of a NaOH solution of unknown concentration requires 17.88 mL of a 0.1053 M H_2SO_4 solution to reach the equivalence point in a titration. What is the concentration of the NaOH solution?

SORT
You are given the volume of a sodium hydroxide solution and the volume and concentration of the sulfuric acid solution required for its titration. You are asked to find the concentration of the sodium hydroxide solution.

GIVEN: 15.00 mL NaOH
17.88 mL of a 0.1053 M H_2SO_4 solution

FIND: concentration of NaOH solution mol/L

STRATEGIZE
Begin by writing the balanced equation for the neutralization reaction (see Example 14.12).

Next draw a solution map. Use the volume and concentration of the known reactant to determine moles of the known reactant. (You have to convert from milliliters to liters first.) Then use the stoichiometric ratio from the balanced equation to get moles of the unknown reactant.

Then add a second part to the solution map indicating how to use moles and volume to determine molarity.

SOLUTION MAP

$$H_2SO_4(aq) + 2\,NaOH(aq) \longrightarrow 2\,H_2O(l) + Na_2SO_4(aq)$$

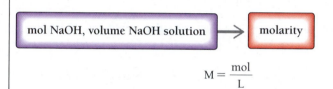

$$\frac{1\,L}{1000\,mL} \qquad \frac{0.1053\,mol\,H_2SO_4}{1\,L\,H_2SO_4} \qquad \frac{2\,mol\,NaOH}{1\,mol\,H_2SO_4}$$

| mol NaOH, volume NaOH solution | → | molarity |

$$M = \frac{mol}{L}$$

RELATIONSHIPS USED

2 mol NaOH : 1 mol H_2SO_4 (from balanced equation)

$$molarity(M) = \frac{mol\,solute}{L\,solution}$$

SOLVE
Follow the solution map to solve the problem. The first part of the solution gives you moles of the unknown reactant. In the second part of the solution, divide the moles from the first part by the volume to obtain molarity.

SOLUTION

$$17.88\,\cancel{mL\,H_2SO_4} \times \frac{1\,\cancel{L}}{1000\,\cancel{mL}} \times \frac{0.1053\,\cancel{mol\,H_2SO_4}}{\cancel{L\,H_2SO_4}}$$

$$\times \frac{2\,mol\,NaOH}{1\,\cancel{mol\,H_2SO_4}} = 3.7655 \times 10^{-3}\,mol\,NaOH$$

$$M = \frac{mol}{L} = \frac{3.7655 \times 10^{-3}\,mol\,NaOH}{0.01500\,L\,NaOH} = 0.2510\,M$$

The unknown NaOH solution has a concentration of 0.2510 M.

CHECK
Check your answer. Are the units correct? Does the answer make physical sense?

The units (M) are correct. The magnitude of the answer makes sense because the reaction has a two-to-one stoichiometry and the volumes of the two solutions are similar; therefore, the concentration of the NaOH solution must be approximately twice the concentration of the H_2SO_4 solution.

LO: Determine [H_3O^+] in acid solutions (Section 14.7).

In a strong acid, [H_3O^+] is equal to the concentration of the acid. In a weak acid, [H_3O^+] is less than the concentration of the acid.

EXAMPLE **14.15** **Determining [H_3O^+] in Acid Solutions**

What is the H_3O^+ concentration in a 0.25 M HCl solution and in a 0.25 M HF solution?

SOLUTION
In the 0.25 M HCl solution (strong acid), [H_3O^+] = 0.25 M.
In the 0.25 M HF solution (weak acid), [H_3O^+] < 0.25 M.

LO: Determine [OH$^-$] in base solutions (Section 14.7).

In a strong base, [OH$^-$] is equal to the concentration of the base times the number of hydroxide ions in the base. In a weak base, [OH$^-$] is less than the concentration of the base.

| EXAMPLE **14.16** | Determining [OH$^-$] in Base Solutions |

What is the OH$^-$ concentration in a 0.25 M NaOH solution, in a 0.25 M Sr(OH)$_2$ solution, and in a 0.25 M NH$_3$ solution?

SOLUTION

In the 0.25 M NaOH solution (strong base), [OH$^-$] = 0.25 M. In the 0.25 M Sr(OH)$_2$ solution (strong base), [OH$^-$] = 0.50 M. In the 0.25 M NH$_3$ solution (weak base), [OH$^-$] < 0.25 M.

LO: Calculate [H$_3$O$^+$] or [OH$^-$] from K_w (Section 14.8).

To find [H$_3$O$^+$] or [OH$^-$], use the ion product constant expression.

$$[H_3O^+][OH^-] = 1.0 \times 10^{-14}$$

Substitute the known quantity into the equation ([H$_3$O$^+$] or [OH$^-$]) and solve for the unknown quantity.

| EXAMPLE **14.17** | Finding [H$_3$O$^+$] or [OH$^-$] from K_w |

Calculate [OH$^-$] in a solution with
$$[H_3O^+] = 1.5 \times 10^{-4} \text{ M}.$$

SOLUTION

$$[H_3O^+][OH^-] = 1.0 \times 10^{-14}$$
$$[1.5 \times 10^{-4}][OH^-] = 1.0 \times 10^{-14}$$
$$[OH^-] = \frac{1.0 \times 10^{-14}}{1.5 \times 10^{-4}} = 6.7 \times 10^{-11} \text{ M}$$

LO: Calculate pH from [H$_3$O$^+$] (Section 14.9).

To calculate the pH of a solution from [H$_3$O$^+$], take the negative log of [H$_3$O$^+$].

$$pH = -\log[H_3O^+]$$

| EXAMPLE **14.18** | Calculating pH from [H$_3$O$^+$] |

Calculate the pH of a solution with [H$_3$O$^+$] = 2.4 $\times$ 10^{-5} M.

SOLUTION

$$pH = -\log[H_3O^+]$$
$$= -\log(2.4 \times 10^{-5})$$
$$= -(-4.62)$$
$$= 4.62$$

LO: Calculate [H$_3$O$^+$] from pH (Section 14.9).

Calculate [H$_3$O$^+$] from pH by taking the inverse log of the negative of the pH value (Method 1):

$$[H_3O^+] = \text{invlog}(-pH)$$

You can also calculate [H$_3$O$^+$] from pH by raising 10 to the negative of the pH (Method 2):

$$[H_3O^+] = 10^{-pH}$$

| EXAMPLE **14.19** | Calculating [H$_3$O$^+$] from pH |

Calculate [H$_3$O$^+$] for a solution with a pH of 6.22.

SOLUTION

Method 1: Inverse Log Function
$$[H_3O^+] = \text{invlog}(-pH)$$
$$= \text{invlog}(-6.22)$$
$$= 6.0 \times 10^{-7}$$

Method 2: 10^x Function
$$[H_3O^+] = 10^{-pH}$$
$$= 10^{-6.22}$$
$$= 6.0 \times 10^{-7}$$

Key Terms

acid [14.2]
acidic solution [14.8]
alkaloid [14.3]
amphoteric [14.4]
Arrhenius acid [14.4]
Arrhenius base [14.4]
Arrhenius definitions
   (of acid bases) [14.4]
base [14.3]
basic solution [14.8]

Brønsted–Lowry acid [14.4]
Brønsted–Lowry base [14.4]
Brønsted–Lowry definition
   [14.4]
buffer [14.10]
carboxylic acid [14.2]
conjugate acid–base pair
   [14.4]
diprotic acid [14.7]
dissociation [14.4]

equivalence point [14.6]
hydronium ion [14.4]
indicator [14.6]
ion product constant for
   water (K_w) [14.8]
ionize [14.4]
logarithmic scale [14.9]
monoprotic acid [14.7]
neutral solution [14.8]
neutralization [14.5]

pH [14.9]
pOH [14.9]
salt [14.5]
strong acid [14.7]
strong base [14.7]
strong electrolyte [14.7]
titration [14.6]
weak acid [14.7]
weak base [14.7]
weak electrolyte [14.7]

Exercises

Questions

1. What makes tart gummy candies, such as Sour Patch Kids, sour?
2. What are the properties of acids? List some foods that contain acids.
3. What is the main component of stomach acid? Why do we have stomach acid?
4. What are organic acids? List two examples of organic acids.
5. What are the properties of bases? Provide some examples of common substances that contain bases.
6. What are alkaloids?
7. Restate the Arrhenius definition of an acid and demonstrate the definition with a chemical equation.
8. Restate the Arrhenius definition of a base and demonstrate the definition with a chemical equation.
9. Restate the Brønsted–Lowry definitions of acids and bases and demonstrate the definitions with a chemical equation.
10. According to the Brønsted–Lowry definition of acids and bases, what is a conjugate acid–base pair? Provide an example.
11. What is an acid–base neutralization reaction? Provide an example.
12. Provide an example of a reaction between an acid and a metal.
13. List an example of a reaction between an acid and a metal oxide.
14. Name a metal that a base can dissolve and write an equation for the reaction.

15. What is a titration? What is the equivalence point?
16. If a solution contains 0.85 mol of OH^-, how many moles of H^+ are required to reach the equivalence point in a titration?
17. What is the difference between a strong acid and a weak acid?
18. How is the strength of an acid related to the strength of its conjugate base?
19. What are monoprotic and diprotic acids?
20. What is the difference between a strong base and a weak base?
21. Does pure water contain any H_3O^+ ions? Explain your answer.
22. What happens to $[OH^-]$ in an aqueous solution when $[H_3O^+]$ increases?
23. Give a possible value of $[OH^-]$ and $[H_3O^+]$ in a solution that is:
   (a) acidic   **(b)** basic   **(c)** neutral
24. How is pH defined? A change of 1.0 pH unit corresponds to how much of a change in $[H_3O^+]$?
25. How is pOH defined? A change of 2.0 pOH units corresponds to how much of a change in $[OH^-]$?
26. In any aqueous solution at 25 °C, the sum of pH and pOH is 14.0. Explain why this is so.
27. What is a buffer?
28. What are the main components in a buffer?

Problems

ACID AND BASE DEFINITIONS

29. Identify each substance as an acid or a base and write a chemical equation showing how it is an acid or a base according to the Arrhenius definition.
   (a) $H_2SO_4(aq)$
   (b) $Sr(OH)_2(aq)$
   (c) $HBr(aq)$
   (d) $NaOH(aq)$

30. Identify each substance as an acid or a base and write a chemical equation showing how it is an acid or a base according to the Arrhenius definition.
   (a) $Ca(OH)_2(aq)$
   (b) $HC_2H_3O_2(aq)$
   (c) $KOH(aq)$
   (d) $HNO_3(aq)$

31. For each reaction, identify the Brønsted–Lowry acid, the Brønsted–Lowry base, the conjugate acid, and the conjugate base.
(a) $HBr(aq) + H_2O(l) \longrightarrow H_3O^+(aq) + Br^-(aq)$
(b) $NH_3(aq) + H_2O(l) \rightleftharpoons NH_4^+(aq) + OH^-(aq)$
(c) $HNO_3(aq) + H_2O(l) \longrightarrow H_3O^+(aq) + NO_3^-(aq)$
(d) $C_2H_5N(aq) + H_2O(l) \rightleftharpoons C_2H_5NH^+(aq) + OH^-(aq)$

32. For each reaction, identify the Brønsted–Lowry acid, the Brønsted–Lowry base, the conjugate acid, and the conjugate base.
(a) $HI(aq) + H_2O(l) \longrightarrow H_3O^+(aq) + I^-(aq)$
(b) $CH_3NH_2(aq) + H_2O(l) \rightleftharpoons CH_3NH_3^+(aq) + OH^-(aq)$
(c) $CO_3^{2-}(aq) + H_2O(l) \rightleftharpoons HCO_3^-(aq) + OH^-(aq)$
(d) $H_2CO_3(aq) + H_2O(l) \rightleftharpoons H_3O^+(aq) + HCO_3^-(aq)$

33. Determine whether each pair is a conjugate acid–base pair.
(a) NH_3, NH_4^+ (b) HCl, HBr
(c) $C_2H_3O_2^-$, $HC_2H_3O_2$ (d) HCO_3^-, NO_3^-

34. Determine whether each pair is a conjugate acid–base pair.
(a) HI, I^- (b) $HCHO_2$, SO_4^{2-}
(c) PO_4^{3-}, HPO_4^{2-} (d) CO_3^{2-}, HCl

35. Write the formula for the conjugate base of each acid.
(a) HCl (b) H_2SO_3
(c) $HCHO_2$ (d) HF

36. Write the formula for the conjugate base of each acid.
(a) HBr (b) H_2CO_3
(c) $HClO_4$ (d) $HC_2H_3O_2$

37. Write the formula for the conjugate acid of each base.
(a) NH_3 (b) ClO_4^-
(c) HSO_4^- (d) CO_3^{2-}

38. Write the formula for the conjugate acid of each base.
(a) CH_3NH_2 (b) C_5H_5N
(c) Cl^- (d) F^-

ACID–BASE REACTIONS

39. Write a neutralization reaction for each acid and base pair.
(a) $HI(aq)$ and $NaOH(aq)$
(b) $HBr(aq)$ and $KOH(aq)$
(c) $HNO_3(aq)$ and $Ba(OH)_2(aq)$
(d) $HClO_4(aq)$ and $Sr(OH)_2(aq)$

40. Write a neutralization reaction for each acid and base pair.
(a) $HF(aq)$ and $Ba(OH)_2(aq)$
(b) $HClO_4(aq)$ and $NaOH(aq)$
(c) $HBr(aq)$ and $Ca(OH)_2(aq)$
(d) $HCl(aq)$ and $KOH(aq)$

41. Write a balanced chemical equation showing how each metal reacts with HBr.
(a) Rb (b) Mg (c) Ba (d) Al

42. Write a balanced chemical equation showing how each metal reacts with HCl.
(a) K (b) Ca (c) Na (d) Sr

43. Write a balanced chemical equation showing how each metal oxide reacts with HI.
(a) MgO (b) K_2O (c) Rb_2O (d) CaO

44. Write a balanced chemical equation showing how each metal oxide reacts with HCl.
(a) SrO (b) Na_2O (c) Li_2O (d) BaO

45. Predict the products of each reaction.
(a) $HClO_4(aq) + Fe_2O_3(s) \longrightarrow$
(b) $H_2SO_4(aq) + Sr(s) \longrightarrow$
(c) $H_3PO_4(aq) + KOH(aq) \longrightarrow$

46. Predict the products of each reaction.
(a) $HI(aq) + Al(s) \longrightarrow$
(b) $H_2SO_4(aq) + TiO_2(s) \longrightarrow$
(c) $H_2CO_3(aq) + LiOH(aq) \longrightarrow$

ACID–BASE TITRATIONS

47. Four solutions of unknown HCl concentration are titrated with solutions of NaOH. The following table lists the volume of each unknown HCl solution, the volume of NaOH solution required to reach the equivalence point, and the concentration of each NaOH solution. Calculate the concentration (in M) of the unknown HCl solution in each case.

HCl Volume (mL)	NaOH Volume (mL)	[NaOH] (M)
a. 25.00 mL	28.44 mL	0.1231 M
b. 15.00 mL	21.22 mL	0.0972 M
c. 20.00 mL	14.88 mL	0.1178 M
d. 5.00 mL	6.88 mL	0.1325 M

48. Four solutions of unknown NaOH concentration are titrated with solutions of HCl. The following table lists the volume of each unknown NaOH solution, the volume of HCl solution required to reach the equivalence point, and the concentration of each HCl solution. Calculate the concentration (in M) of the unknown NaOH solution in each case.

NaOH Volume (mL)	HCl Volume (mL)	[HCl] (M)
a. 5.00 mL	9.77 mL	0.1599 M
b. 15.00 mL	11.34 mL	0.1311 M
c. 10.00 mL	10.55 mL	0.0889 M
d. 30.00 mL	36.18 mL	0.1021 M

49. A 25.00-mL sample of an H_2SO_4 solution of unknown concentration is titrated with a 0.1322 M KOH solution. A volume of 41.22 mL of KOH is required to reach the equivalence point. What is the concentration of the unknown H_2SO_4 solution?

50. A 5.00-mL sample of an H_3PO_4 solution of unknown concentration is titrated with a 0.1090 M NaOH solution. A volume of 7.12 mL of the NaOH solution is required to reach the equivalence point. What is the concentration of the unknown H_3PO_4 solution?

51. What volume in milliliters of a 0.121 M sodium hydroxide solution is required to reach the equivalence point in the complete titration of a 10.0-mL sample of 0.102 M sulfuric acid?

52. What volume in milliliters of 0.0985 M sodium hydroxide solution is required to reach the equivalence point in the complete titration of a 15.0-mL sample of 0.124 M phosphoric acid?

STRONG AND WEAK ACIDS AND BASES

53. Classify each acid as strong or weak.
(a) HCl
(b) HF
(c) HBr
(d) H_2SO_3

54. Classify each acid as strong or weak.
(a) $HCHO_2$
(b) H_2SO_4
(c) HNO_3
(d) H_2CO_3

55. Determine $[H_3O^+]$ in each acid solution. If the acid is weak, indicate the value that $[H_3O^+]$ is less than.
(a) 1.7 M HBr
(b) 1.5 M HNO_3
(c) 0.38 M H_2CO_3
(d) 1.75 M $HCHO_2$

56. Determine $[H_3O^+]$ in each acid solution. If the acid is weak, indicate the value that $[H_3O^+]$ is less than.
(a) 0.125 M $HClO_2$
(b) 1.25 M H_3PO_4
(c) 2.77 M HCl
(d) 0.95 M H_2SO_3

57. Classify each base as strong or weak.
(a) LiOH
(b) NH_4OH
(c) $Ca(OH)_2$
(d) NH_3

58. Classify each base as strong or weak.
(a) C_5H_5N
(b) NaOH
(c) $Ba(OH)_2$
(d) KOH

59. Determine $[OH^-]$ in each base solution. If the acid is weak, indicate the value that $[OH^-]$ is less than.
(a) 0.25 M NaOH
(b) 0.25 M NH_3
(c) 0.25 M $Sr(OH)_2$
(d) 1.25 M KOH

60. Determine $[OH^-]$ in each base solution. If the acid is weak, indicate the value that $[OH^-]$ is less than.
(a) 2.5 M KOH
(b) 1.95 M NH_3
(c) 0.225 M $Ba(OH)_2$
(d) 1.8 M C_5H_5N

ACIDITY, BASICITY, AND K_w

61. Determine if each solution is acidic, basic, or neutral.
(a) $[H_3O^+] = 1 \times 10^{-5}$ M; $[OH^-] = 1 \times 10^{-9}$ M
(b) $[H_3O^+] = 1 \times 10^{-6}$ M; $[OH^-] = 1 \times 10^{-8}$ M
(c) $[H_3O^+] = 1 \times 10^{-7}$ M; $[OH^-] = 1 \times 10^{-7}$ M
(d) $[H_3O^+] = 1 \times 10^{-8}$ M; $[OH^-] = 1 \times 10^{-6}$ M

62. Determine if each solution is acidic, basic, or neutral.
(a) $[H_3O^+] = 1 \times 10^{-9}$ M; $[OH^-] = 1 \times 10^{-5}$ M
(b) $[H_3O^+] = 1 \times 10^{-10}$ M; $[OH^-] = 1 \times 10^{-4}$ M
(c) $[H_3O^+] = 1 \times 10^{-2}$ M; $[OH^-] = 1 \times 10^{-12}$ M
(d) $[H_3O^+] = 1 \times 10^{-13}$ M; $[OH^-] = 1 \times 10^{-1}$ M

63. Calculate $[OH^-]$ given $[H_3O^+]$ in each aqueous solution and classify the solution as acidic or basic.
(a) $[H_3O^+] = 1.5 \times 10^{-9}$ M
(b) $[H_3O^+] = 9.3 \times 10^{-9}$ M
(c) $[H_3O^+] = 2.2 \times 10^{-6}$ M
(d) $[H_3O^+] = 7.4 \times 10^{-4}$ M

64. Calculate $[OH^-]$ given $[H_3O^+]$ in each aqueous solution and classify the solution as acidic or basic.
(a) $[H_3O^+] = 1.3 \times 10^{-3}$ M
(b) $[H_3O^+] = 9.1 \times 10^{-12}$ M
(c) $[H_3O^+] = 5.2 \times 10^{-4}$ M
(d) $[H_3O^+] = 6.1 \times 10^{-9}$ M

65. Calculate $[H_3O^+]$ given $[OH^-]$ in each aqueous solution and classify each solution as acidic or basic.
   (a) $[OH^-] = 2.7 \times 10^{-12}$ M
   (b) $[OH^-] = 2.5 \times 10^{-2}$ M
   (c) $[OH^-] = 1.1 \times 10^{-10}$ M
   (d) $[OH^-] = 3.3 \times 10^{-4}$ M

66. Calculate $[H_3O^+]$ given $[OH^-]$ in each aqueous solution and classify each solution as acidic or basic.
   (a) $[OH^-] = 2.1 \times 10^{-11}$ M
   (b) $[OH^-] = 7.5 \times 10^{-9}$ M
   (c) $[OH^-] = 2.1 \times 10^{-4}$ M
   (d) $[OH^-] = 1.0 \times 10^{-2}$ M

pH

67. Classify each solution as acidic, basic, or neutral according to its pH value.
   (a) pH = 8.0
   (b) pH = 7.0
   (c) pH = 3.5
   (d) pH = 6.1

68. Classify each solution as acidic, basic, or neutral according to its pH value.
   (a) pH = 4.0
   (b) pH = 3.5
   (c) pH = 13.0
   (d) pH = 0.85

69. Calculate the pH of each solution.
   (a) $[H_3O^+] = 1.7 \times 10^{-8}$ M
   (b) $[H_3O^+] = 1.0 \times 10^{-7}$ M
   (c) $[H_3O^+] = 2.2 \times 10^{-6}$ M
   (d) $[H_3O^+] = 7.4 \times 10^{-4}$ M

70. Calculate the pH of each solution.
   (a) $[H_3O^+] = 2.4 \times 10^{-10}$ M
   (b) $[H_3O^+] = 7.6 \times 10^{-2}$ M
   (c) $[H_3O^+] = 9.2 \times 10^{-13}$ M
   (d) $[H_3O^+] = 3.4 \times 10^{-5}$ M

71. Calculate $[H_3O^+]$ for each solution.
   (a) pH = 8.55
   (b) pH = 11.23
   (c) pH = 2.87
   (d) pH = 1.22

72. Calculate $[H_3O^+]$ for each solution.
   (a) pH = 1.76
   (b) pH = 3.88
   (c) pH = 8.43
   (d) pH = 12.32

73. Calculate the pH of each solution.
   (a) $[OH^-] = 1.9 \times 10^{-7}$ M
   (b) $[OH^-] = 2.6 \times 10^{-8}$ M
   (c) $[OH^-] = 7.2 \times 10^{-11}$ M
   (d) $[OH^-] = 9.5 \times 10^{-2}$ M

74. Calculate the pH of each solution.
   (a) $[OH^-] = 2.8 \times 10^{-11}$ M
   (b) $[OH^-] = 9.6 \times 10^{-3}$ M
   (c) $[OH^-] = 3.8 \times 10^{-12}$ M
   (d) $[OH^-] = 6.4 \times 10^{-4}$ M

75. Calculate $[OH^-]$ for each solution.
   (a) pH = 4.25
   (b) pH = 12.53
   (c) pH = 1.50
   (d) pH = 8.25

76. Calculate $[OH^-]$ for each solution.
   (a) pH = 1.82
   (b) pH = 13.28
   (c) pH = 8.29
   (d) pH = 2.32

77. Calculate the pH of each solution:
   (a) 0.0155 M HBr
   (b) 1.28×10^{-3} M KOH
   (c) 1.89×10^{-3} M HNO_3
   (d) 1.54×10^{-4} M $Sr(OH)_2$

78. Calculate the pH of each solution:
   (a) 1.34×10^{-3} M $HClO_4$
   (b) 0.0211 M NaOH
   (c) 0.0109 M HBr
   (d) 7.02×10^{-5} M $Ba(OH)_2$

pOH

79. Detemine the pOH of each solution and classify it as acidic, basic, or neutral.
   (a) $[OH^-] = 1.5 \times 10^{-9}$ M
   (b) $[OH^-] = 7.0 \times 10^{-5}$ M
   (c) $[OH^-] = 1.0 \times 10^{-7}$ M
   (d) $[OH^-] = 8.8 \times 10^{-3}$ M

80. Detemine the pOH of each solution and classify it as acidic, basic, or neutral.
   (a) $[OH^-] = 4.5 \times 10^{-2}$ M
   (b) $[OH^-] = 3.1 \times 10^{-12}$ M
   (c) $[OH^-] = 5.4 \times 10^{-5}$ M
   (d) $[OH^-] = 1.2 \times 10^{-2}$ M

81. Determine the pOH of each solution.
(a) $[H_3O^+] = 1.2 \times 10^{-8}$ M
(b) $[H_3O^+] = 5.5 \times 10^{-2}$ M
(c) $[H_3O^+] = 3.9 \times 10^{-9}$ M
(d) $[H_3O^+] = 1.88 \times 10^{-13}$ M

82. Determine the pOH of each solution.
(a) $[H_3O^+] = 8.3 \times 10^{-10}$ M
(b) $[H_3O^+] = 1.6 \times 10^{-7}$ M
(c) $[H_3O^+] = 7.3 \times 10^{-2}$ M
(d) $[OH^-] = 4.32 \times 10^{-4}$ M

83. Determine the pH of each solution and classify it as acidic, basic, or neutral.
(a) pOH = 8.5
(b) pOH = 4.2
(c) pOH = 1.7
(d) pOH = 7.0

84. Determine the pH of each solution and classify it as acidic, basic, or neutral.
(a) pOH = 12.5
(b) pOH = 5.5
(c) pOH = 0.55
(d) pOH = 7.98

BUFFERS AND ACID RAIN

85. Determine whether or not each mixture is a buffer.
(a) HCl and HF
(b) NaOH and NH_3
(c) HF and NaF
(d) $HC_2H_3O_2$ and $KC_2H_3O_2$

86. Determine whether or not each mixture is a buffer.
(a) HBr and NaCl
(b) $HCHO_2$ and $NaCHO_2$
(c) HCl and HBr
(d) KOH and NH_3

87. Write reactions showing how each of the buffers in Problem 85 would neutralize added HCl.

88. Write reactions showing how each of the buffers in Problem 86 would neutralize added NaOH.

89. Which substance could you add to each solution to make it a buffer solution?
(a) 0.100 M $NaC_2H_3O_2$
(b) 0.500 M H_3PO_4
(c) 0.200 M $HCHO_2$

90. Which substance could you add to each solution to make it a buffer solution?
(a) 0.050 M $NaHSO_3$
(b) 0.150 M HF
(c) 0.200 M $KCHO_2$

Cumulative Problems

91. How much 0.100 M HCl is required to completely neutralize 20.0 mL of 0.250 M NaOH?

92. How much 0.200 M KOH is required to completely neutralize 25.0 mL of 0.150 M $HClO_4$?

93. What is the minimum volume of 5.0 M HCl required to completely dissolve 10.0 g of magnesium metal?

94. What is the minimum volume of 3.0 M HBr required to completely dissolve 15.0 g of potassium metal?

95. When 18.5 g of $K_2O(s)$ is completely dissolved by HI(aq), how many grams of KI(aq) form in solution?

96. When 5.88 g of CaO(s) is completely dissolved by HBr(aq), how many grams of $CaBr_2(aq)$ form in solution?

97. A 0.125-g sample of a monoprotic acid of unknown molar mass is dissolved in water and titrated with 0.1003 M NaOH. The equivalence point is reached after adding 20.77 mL of base. What is the molar mass of the unknown acid?

98. A 0.105-g sample of a diprotic acid of unknown molar mass is dissolved in water and titrated with 0.1288 M NaOH. The equivalence point is reached after adding 15.2 mL of base. What is the molar mass of the unknown acid?

99. People take antacids, such as milk of magnesia, to reduce the discomfort of acid stomach or heartburn. The recommended dose of milk of magnesia is 1 teaspoon, which contains 400 mg of $Mg(OH)_2$. What volume of HCl solution with a pH of 1.1 can be neutralized by 1 dose of milk of magnesia? (Assume two significant figures in your calculations.)

100. An antacid tablet requires 25.82 mL of 0.200 M HCl to titrate to its equivalence point. What volume in milliliters of stomach acid can be neutralized by the antacid tablet? Assume that stomach acid has a pH of 1.1. (Assume two significant figures in your calculations.)

101. For each $[H_3O^+]$, determine the pH and state whether the solution is acidic or basic.
   (a) $[H_3O^+] = 0.0025$ M
   (b) $[H_3O^+] = 1.8 \times 10^{-12}$ M
   (c) $[H_3O^+] = 9.6 \times 10^{-9}$ M
   (d) $[H_3O^+] = 0.0195$ M

102. For each $[OH^-]$, determine the pH and state whether the solution is acidic or basic.
   (a) $[OH^-] = 1.8 \times 10^{-5}$ M
   (b) $[OH^-] = 8.9 \times 10^{-12}$ M
   (c) $[OH^-] = 3.1 \times 10^{-2}$ M
   (d) $[OH^-] = 1.96 \times 10^{-9}$ M

103. Complete the table. (The first row is completed for you.)

$[H_3O^+]$	$[OH^-]$	pOH	pH	Acidic or Basic
1.0×10^{-4}	1.0×10^{-10}	10.00	4.00	acidic
5.5×10^{-3}	___	___	___	___
___	3.2×10^{-6}	___	___	___
4.8×10^{-9}	___	___	___	___
___	___	___	7.55	___

104. Complete the table. (The first row is completed for you.)

$[H_3O^+]$	$[OH^-]$	pOH	pH	Acidic or Basic
1.0×10^{-8}	1.0×10^{-6}	6.00	8.00	basic
___	___	___	3.55	___
1.7×10^{-9}	___	___	___	___
___	___	___	13.5	___
___	8.6×10^{-11}	___	___	___

105. For each strong acid solution, determine $[H_3O^+]$, $[OH^-]$, and pH.
   (a) 0.0088 M $HClO_4$
   (b) 1.5×10^{-3} M HBr
   (c) 9.77×10^{-4} M HI
   (d) 0.0878 M HNO_3

106. For each strong acid solution, determine $[H_3O^+]$, $[OH^-]$, and pH.
   (a) 0.0150 M HCl
   (b) 1.9×10^{-4} M HI
   (c) 0.0226 M HBr
   (d) 1.7×10^{-3} M HNO_3

107. For each strong base solution, determine $[OH^-]$, $[H_3O^+]$, pH, and pOH.
   (a) 0.15 M NaOH
   (b) 1.5×10^{-3} M $Ca(OH)_2$
   (c) 4.8×10^{-4} M $Sr(OH)_2$
   (d) 8.7×10^{-5} M KOH

108. For each strong base solution, determine $[OH^-]$, $[H_3O^+]$, pH, and pOH.
   (a) 8.77×10^{-3} M LiOH
   (b) 0.0112 M $Ba(OH)_2$
   (c) 1.9×10^{-4} M KOH
   (d) 5.0×10^{-4} M $Ca(OH)_2$

109. As described in Section 14.1, jailed spies on the big screen have been known to use acid stored in a pen to dissolve jail bars and escape. What minimum volume of 12.0 M hydrochloric acid would be required to completely dissolve a 500.0-g iron bar? Would this amount of acid fit into a pen?

110. A popular classroom demonstration consists of filing notches into a new penny and soaking the penny in hydrochloric acid overnight. Because new pennies are made of zinc coated with copper, and hydrochloric acid dissolves zinc and not copper, the inside of the penny is dissolved by the acid, while the outer copper shell remains. Suppose a penny contains 2.5 g of zinc and is soaked in 20.0 mL of 6.0 M HCl. Calculate the concentration of the HCl solution after all of the zinc has dissolved. *Hint:* The Zn from the penny is oxidized to Zn^{2+}.

111. What is the pH of a solution formed by mixing 125.0 mL of 0.0250 M HCl with 75.0 mL of 0.0500 M NaOH?

112. What is the pH of a solution formed by mixing 175.0 mL of 0.0880 M HI with 125.0 mL of 0.0570 M KOH?

113. How many H^+ (or H_3O^+) ions are present in one drop (0.050 mL) of pure water at 25 °C?

114. Calculate the number of H^+ (or H_3O^+) ions and OH^- ions in 1.0 mL of 0.100 M HCl.

115. A 4.00-L base solution contains 0.100 mol total of NaOH and $Sr(OH)_2$. The pOH of the solution is 1.51. Determine the amounts (in moles) of NaOH and $Sr(OH)_2$ in the solution.

116. A 1.50-L acid solution contains 0.35 g total of HCl and HBr. The pH of the solution is 2.40. What are the masses of HCl and HBr in the solution?

Highlight Problems

117. Based on the molecular view of each acid solution, determine whether the acid is weak or strong.

(a)

(b)
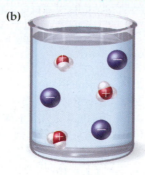

(c)
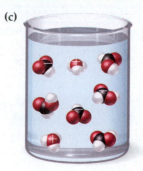

(d)

118. Lakes that have been acidified by acid rain can be neutralized by liming, the addition of limestone ($CaCO_3$). How much limestone in kilograms is required to completely neutralize a 3.8×10^9 L lake with a pH of 5.5?

119. Acid rain over the Great Lakes has a pH of about 4.5. Calculate the $[H_3O^+]$ of this rain and compare that value to the $[H_3O^+]$ of rain over the West Coast that has a pH of 5.4. How many times more concentrated is the acid in rain over the Great Lakes?

Questions for Group Work

Discuss these questions with the group and record your consensus answer.

120. Choose an example of a reaction featuring a chemical acting as an Arrhenius base. With group members representing atoms or ions, act out the reaction. Now repeat the process for a chemical acting as a Brønsted–Lowry acid. Write a script for a narrator to describe the processes that occur.

121. Divide your group in two. Have each half of your group write a quiz related to an acid–base titration. Make an answer key for your problem. Trade problems with the other half of your group and solve each other's problems.

122. Each group member needs to have the calculator that they will use for homework and exams in this class. For each member of your group, write out the specific instructions (the order in which you press the buttons) for using his or her calculator to verify the following:

(a) a solution with $[H_3O^+] = 2.8 \times 10^{-4}$ M has a pH of 3.55

(b) a solution with pOH = 6.83 has $[OH^-] = 1.47 \times 10^{-7}$ M

123. With group members acting as atoms or ions, act out the reaction that occurs when HCl is added to a buffer solution composed of $HC_2H_3O_2$ and $NaC_2H_3O_2$. Write a script for a narrator to describe the processes that occur, including how the buffer keeps the pH approximately the same, even though a strong acid is added.

Data Interpretation and Analysis

124. The progress of an acid–base titration can be monitored by measuring the pH of the solution being titrated while adding base. The resulting graph, called a *titration curve*, is a plot of the pH versus the volume of added base. The graph shown here is the titration curve for an unknown monoprotic acid (HX) titrated with 0.100 M NaOH. Examine the graph and answer the questions that follow.

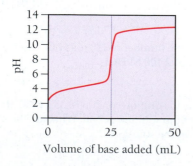

Volume of base added (mL)

(a) Write an equation for the neutralization reaction that occurs during the titration.

(b) Why does the pH of the solution increase during the course of the titration?

(c) The equivalence point of the titration occurs at the point where the titration curve is the steepest. What is the pH at the equivalence point? What is the pH halfway to the equivalence point?

(d) Determine the number of moles of the acid HX that were present in the solution that was titrated.

Answers to Skillbuilder Exercises

Skillbuilder 14.1

   (a) $C_5H_5N(aq) + H_2O(l) \rightleftharpoons C_5H_5NH^+(aq) + OH^-(aq)$
        Base   Acid   Conjugate acid   Conjugate base

   (b) $HNO_3(aq) + H_2O(l) \longrightarrow H_3O^+(aq) + NO_3^-(aq)$
        Acid   Base   Conjugate acid   Conjugate base

Skillbuilder 14.2

   $H_3PO_4(aq) + 3\,NaOH(aq) \longrightarrow 3\,H_2O(l) + Na_3PO_4(aq)$

Skillbuilder 14.3

   (a) $2\,HCl(aq) + Sr(s) \longrightarrow H_2(g) + SrCl_2(aq)$
   (b) $2\,HI(aq) + BaO(s) \longrightarrow H_2O(l) + BaI_2(aq)$

Skillbuilder 14.4 .... 9.03×10^{-2} M H_2SO_4

Skillbuilder 14.5 .... **(a)** $[H_3O^+] < 0.50$ M
                      (b) $[H_3O^+] = 1.25$ M
                      (c) $[H_3O^+] < 0.75$ M

Skillbuilder 14.6 .... **(a)** $[OH^-] = 0.11$ M
                      (b) $[OH^-] < 1.05$ M
                      (c) $[OH^-] = 0.45$ M

Skillbuilder 14.7 .... **(a)** $[H_3O^+] = 6.7 \times 10^{-13}$ M; basic
                      (b) $[H_3O^+] = 1.0 \times 10^{-7}$ M; neutral
                      (c) $[H_3O^+] = 1.2 \times 10^{-5}$ M; acidic

Skillbuilder 14.8 .... **(a)** pH = 8.02; basic
                      (b) pH = 2.21; acidic

Skillbuilder Plus, p. 507 ... pH = 12.11; basic

Skillbuilder 14.9 .... 4.3×10^{-9} M

Skillbuilder Plus, p. 508 ... 4.6×10^{-11} M

Skillbuilder 14.10 .... 5.6×10^{-5} M

Skillbuilder Plus, p. 509 ... 4.8×10^{-9} M

Answers to Conceptual Checkpoints

14.1 (a) HCl is an acid. Acids have a sour taste. The other two compounds are bases.

14.2 (b) The conjugate base of an acid always has one fewer proton and is one charge unit lower (more negative) than the acid.

14.3 (c) The acid solution contains 7 H^+ ions; therefore, 7 OH^- ions are required to reach the equivalence point.

14.4 (c) Both **(a)** and **(b)** show complete ionization and are therefore strong acids. Only the acid depicted in **(c)** undergoes partial ionization and is therefore a weak acid.

14.5 (d) Each of the others can accept a proton and thus acts as a base. NH_4^+, however, is the conjugate acid of NH_3 and therefore acts as an acid and not as a base.

14.6 (d) Because pH is the *negative* log of the H_3O^+ concentration, a higher pH corresponds to a lower $[H_3O^+]$, and each unit of pH represents a tenfold change in concentration.

14.7 (d) Since the pH is 5, the pOH = 14 − 5 = 9.

14.8 (b) A buffer solution consists of a weak acid and its conjugate base. Of the compounds listed, HF is the only weak acid, and F^- (from NaF in solution) is its conjugate base.

15 Chemical Equilibrium

A system is in equilibrium when the forces constituting it are arranged in such a way as to compensate each other, like the two weights pulling at the arms of a pair of scales.

—Rudolf Arnheim (1904–2007)

15.1 Life: Controlled Disequilibrium

Have you ever tried to define life? If you have, you know that it is not easily defined. How are living things different from nonliving things? You may try to define living things as those things that can move. But of course many living things do not move—many plants, for example, do not move very much—and some nonliving things, such as glaciers and Earth itself, do move. So motion is neither unique to nor definitive of life. You may try to define living things as those things that can reproduce. But again, many living things, such as mules or sterile humans, cannot reproduce; yet they are alive. In addition, some nonliving things—such as crystals—reproduce (in some sense). So what is unique about living things?

The concept of equilibrium underlies one definition of life. We define *chemical equilibrium* more carefully soon; for now, we can think generally of equilibrium as *sameness and constancy*. When an object is in equilibrium with its surroundings, some property of the object has reached sameness with the surroundings and is no longer changing. For example, a cup of hot water is not in equilibrium with its surroundings with respect to temperature. However, if left undisturbed, the cup of hot water will slowly cool until it reaches equilibrium with its surroundings. At that point, the temperature of the water is the *same as* that of the surroundings (sameness) and *no longer changes* (constancy).

So equilibrium involves sameness and constancy. Part of a definition for living things, then, is that living things *are not* in equilibrium with their surroundings. Human body temperature, for example, is not the same as the temperature of our surroundings. When we jump into a swimming pool, the pH of our blood

◀ Dynamic equilibrium involves two opposing processes that occur at the same rate. This image draws an analogy between a chemical equilibrium ($N_2O_4 \rightleftharpoons 2\ NO_2$), in which the two opposing reactions occur at the same rate, and a freeway with traffic moving in opposing directions at the same rate.

Even cold-blooded animals such as reptiles maintain a level of disequilibrium in their body temperature by moving into and out of the sun.

does not become the same as the pH of the surrounding water. Living things, even the simplest ones, maintain some measure of *disequilibrium* with their environment.

We must add one more concept, however, to complete our definition of life with respect to equilibrium. A cup of hot water is in disequilibrium with its environment, yet it is not alive. However, the cup of hot water has no control over its disequilibrium and will slowly come to equilibrium with its environment. In contrast, living things—as long as they are alive—maintain and *control* their disequilibrium. Your body temperature, for example, is not only in disequilibrium with your surroundings—it is in controlled disequilibrium. Your body maintains your temperature within a specific range that is not in equilibrium with the surrounding temperature.

So one definition of life is that living things are in *controlled disequilibrium* with their environment. A living thing comes into equilibrium with its surroundings only after it dies. In this chapter, we examine the concept of equilibrium, especially chemical equilibrium—the state that involves sameness and constancy.

15.2 The Rate of a Chemical Reaction

▶ Identify and explain the relationship between concentration and temperature and the rate of a chemical reaction.

Before we probe more deeply into the concept of chemical equilibrium, we must first understand something about the rates of chemical reactions. The **rate of a chemical reaction**—a measure of how fast the reaction proceeds—is defined as the amount of reactant that changes to product in a given period of time. A reaction with a fast rate proceeds quickly; a large amount of reactant is converted to product in a certain period of time (▶ FIGURE 15.1a). A reaction with a slow rate proceeds slowly; only a small amount of reactant is converted to product in the same period of time (▶ FIGURE 15.1b).

Reaction rates are related to chemical equilibrium because, as we will see in Section 15.3, a chemical system is at equilibrium when the rate of the forward reaction equals the rate of the reverse reaction.

Chemists seek to control reaction rates for many chemical reactions. For example, rockets can be propelled by the reaction of hydrogen and oxygen to form water. If the reaction proceeds too slowly, the rocket will not lift off the ground. If, however, the reaction proceeds too quickly, the rocket can explode. Reaction rates can be controlled if we understand the factors that influence them.

A reaction rate can also be defined as the amount of a product that forms in a given period of time.

Collision Theory

According to **collision theory**, chemical reactions occur through collisions between molecules or atoms. For example, consider the gas-phase chemical reaction between $H_2(g)$ and $I_2(g)$ to form $HI(g)$.

$$H_2(g) + I_2(g) \longrightarrow 2\,HI(g)$$

The gas-phase reaction between hydrogen and iodine can proceed by other mechanisms, but the mechanism here is valid for the low-temperature thermal reaction.

The reaction begins when an H_2 molecule collides with an I_2 molecule. If the collision occurs with enough energy—that is, if the colliding molecules are moving fast enough—the reaction can proceed to form the products. If the collision occurs with insufficient energy, the reactant molecules (H_2 and I_2) simply bounce off of one another. Gas-phase molecules have a wide distribution of velocities, so collisions occur with a wide distribution of energies. High-energy collisions lead to products, and low-energy collisions do not.

Whether a collision leads to a reaction also depends on the *orientation* of the colliding molecules, but this topic is beyond the scope of this text.

Higher-energy collisions are more likely to lead to products because most chemical reactions have an *activation energy* (or an activation barrier). We discuss the activation energy for chemical reactions in more detail in Section 15.12. For now, think of the activation energy as an energy barrier that must be overcome for the reaction to proceed. For example, in the case of H_2 reacting with I_2 to form HI, the product (HI) can begin to form only after the H—H bond and the I—I bond each begin to break. The activation energy is the energy required to begin to break these bonds.

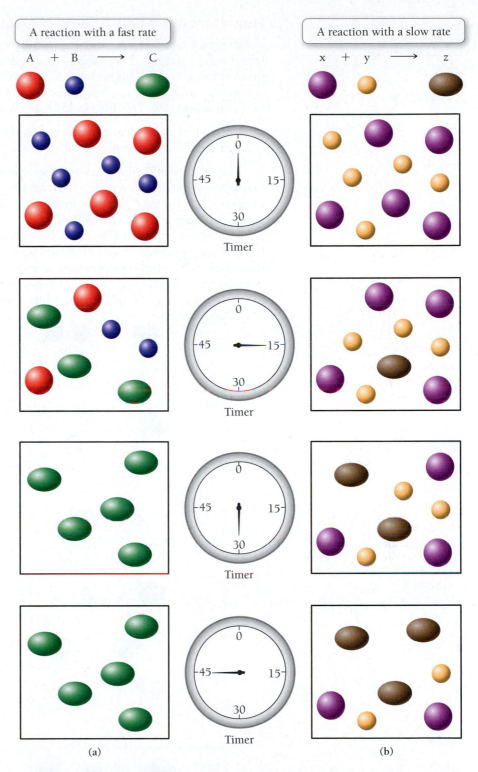

▲ **FIGURE 15.1 Reaction rates   (a)** In a reaction with a fast rate, the reactants react to form products in a short period of time. **(b)** In a reaction with a slow rate, the reactants react to form products over a long period of time.

If molecules react via high-energy collisions, then we know that the factors that influence the rate of a reaction must be the same factors that affect the number of high-energy collisions that occur per unit time. Here, we focus on the two most important factors that affect collisions: the *concentration* of the reacting molecules and the *temperature* of the reaction mixture.

How Concentration Affects the Rate of a Reaction

▼ FIGURE 15.2 shows various mixtures of H_2 and I_2 at the same temperature but different concentrations. If H_2 and I_2 react via collisions to form HI, which mixture do you think has the highest reaction rate? Since ▼ FIGURE 15.2c has the highest concentration of H_2 and I_2, it has the most collisions per unit time and therefore the fastest reaction rate. This idea holds true for most chemical reactions:

The rate of a chemical reaction generally increases with increasing concentration of the reactants.

The exact relationship between increase in concentration and increase in reaction rate varies for different reactions and is beyond the scope of this text. For our purposes, we just need to know that for most reactions, the reaction rate increases with increasing reactant concentration.

Armed with this knowledge, what can we say about the rate of a reaction as the reaction proceeds? Since reactants turn into products in the course of a reaction, their concentration decreases. Consequently, the reaction rate decreases as well. In other words, as a reaction proceeds, there are fewer reactant molecules (because they have turned into products), and the reaction slows down.

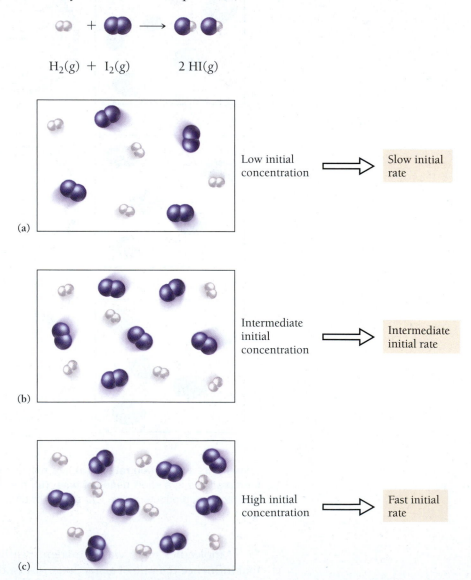

$$H_2(g) + I_2(g) \longrightarrow 2\,HI(g)$$

(a)   Low initial concentration ⟹ Slow initial rate

(b)   Intermediate initial concentration ⟹ Intermediate initial rate

(c)   High initial concentration ⟹ Fast initial rate

▲ **FIGURE 15.2 Effect of concentration on reaction rate** Question: Which reaction mixture will have the fastest initial rate? The mixture in (c) is fastest because it has the highest concentration of reactants and therefore the highest rate of collisions.

How Temperature Affects the Rate of a Reaction

Reaction rates also depend on temperature. ▼ FIGURE 15.3 shows various mixtures of H_2 and I_2 at the same concentration but different temperatures. Which will have the fastest rate? Recall that raising the temperature makes the molecules move faster (Section 3.10). They therefore experience more collisions per unit time, resulting in a faster reaction rate. In addition, a higher temperature results in more collisions that are (on average) of higher energy. Because the high-energy collisions are the ones that result in products, this also produces a faster rate. Consequently, ▼ FIGURE 15.3c (which has the highest temperature) has the fastest reaction rate. This relationship holds true for most chemical reactions.

▲ Cold-blooded animals become sluggish at low temperatures because the reactions that power their metabolism slow down.

> The rate of a chemical reaction generally increases with increasing temperature of the reaction mixture.

The temperature dependence of reaction rates is the reason that cold-blooded animals become more sluggish at lower temperatures. The reactions required for them to think and move become slower, resulting in the sluggish behavior.

$$H_2(g) \ + \ I_2(g) \qquad 2\,HI(g)$$

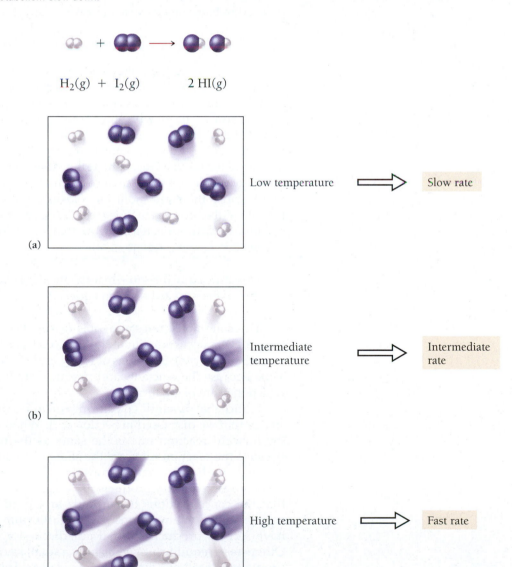

(a)   Low temperature ⇒ Slow rate

(b)   Intermediate temperature ⇒ Intermediate rate

(c)   High temperature ⇒ Fast rate

▶ FIGURE 15.3 **Effect of temperature on reaction rate**   Question: Which reaction mixture will have the fastest initial rate? The mixture in (c) is fastest because it has the highest temperature.

To summarize:

- Reaction rates generally increase with increasing reactant concentration.
- Reaction rates generally increase with increasing temperature.
- Reaction rates generally decrease as a reaction proceeds.

CONCEPTUAL ✔ **CHECKPOINT 15.1**

In a chemical reaction between two gases, what would you expect to be the result of increasing the pressure of the gases?

(a) an increase in reaction rate

(b) a decrease in reaction rate

(c) no effect on reaction rate

15.3 The Idea of Dynamic Chemical Equilibrium

▶ Define dynamic equilibrium.

What would happen if our reaction between H_2 and I_2 to form HI were able to proceed in both the forward and reverse directions?

$$H_2(g) + I_2(g) \rightleftharpoons 2\,HI(g)$$

In this case, H_2 and I_2 would collide and react to form 2 HI molecules, but the 2 HI molecules also collide and react to re-form H_2 and I_2. A reaction that can proceed in both the forward and reverse directions is a **reversible reaction**.

Suppose we begin with only H_2 and I_2 in a container (▶ **FIGURE 15.4a**). What happens initially? The H_2 and I_2 molecules begin to react to form HI (▶ **FIGURE 15.4b**). However, as H_2 and I_2 react, their concentration decreases, which in turn decreases the rate of the forward reaction. At the same time, HI begins to form. As the concentration of HI increases, the reverse reaction begins to occur at an increasingly faster rate because there are more HI collisions with other HI molecules. Eventually, the rate of the reverse reaction (which is increasing) equals the rate of the forward reaction (which is decreasing). At that point, **dynamic equilibrium** is reached (▶ **FIGURE 15.4c** and ▶ **FIGURE 15.4d**).

> **Dynamic equilibrium**—In a chemical reaction, the condition in which the rate of the forward reaction equals the rate of the reverse reaction.

This condition is not static—it is *dynamic* because the forward and reverse reactions are still occurring but at the same constant rate. When dynamic equilibrium is reached, the concentrations of H_2, I_2, and HI no longer change. They remain the same because the reactants and products are being depleted at the same rate at which they are being formed.

Notice that dynamic equilibrium includes the concepts of sameness and constancy that we discussed in Section 15.1. When dynamic equilibrium is reached, the forward reaction rate is the same as the reverse reaction rate (sameness). Because the reaction rates are the same, the concentrations of the reactants and products no longer change (constancy). However, just because the concentrations of reactants and products no longer change at equilibrium does *not* mean that the concentrations of reactants and products are *equal* to one another at equilibrium. Some reactions reach equilibrium only after most of the reactants have formed products. (Recall our discussion of strong acids in Chapter 14.) Others reach equilibrium when only a small fraction of the reactants have formed products. (Recall our discussion of weak acids in Chapter 14.) It depends on the reaction.

A reversible reaction

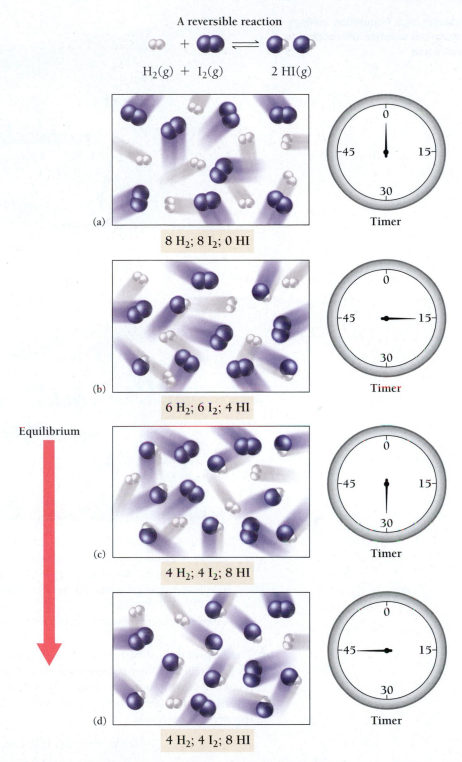

$$H_2(g) + I_2(g) \rightleftharpoons 2\,HI(g)$$

(a) 8 H$_2$; 8 I$_2$; 0 HI

(b) 6 H$_2$; 6 I$_2$; 4 HI

Equilibrium

(c) 4 H$_2$; 4 I$_2$; 8 HI

(d) 4 H$_2$; 4 I$_2$; 8 HI

▲ **FIGURE 15.4 Equilibrium** When the concentrations of the reactants and products no longer change, equilibrium has been reached.

We can better understand dynamic equilibrium with a simple analogy. Imagine that Narnia and Middle Earth are two neighboring kingdoms (▶ **FIGURE 15.5**, on the next page). Narnia is overpopulated, and Middle Earth is underpopulated. One day, however, the border between the two kingdoms opens, and people immediately begin to leave Narnia for Middle Earth (call this the forward reaction).

Narnia is the fictitious world featured in C.S. Lewis's *The Chronicles of Narnia*, and Middle Earth is the fictitious world featured in J.R.R. Tolkien's *The Lord of the Rings.*

$$\text{Narnia} \longrightarrow \text{Middle Earth} \quad \text{(forward reaction)}$$

▶ **FIGURE 15.5 Population analogy for a chemical reaction proceeding to equilibrium**

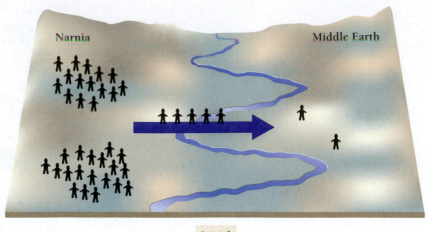

Initial

♦ Represents population

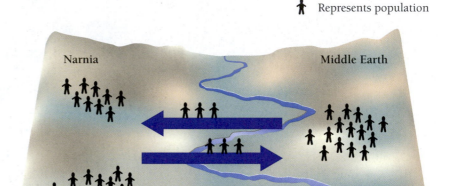

Equilibrium

> When the two kingdoms reach dynamic equilibrium, their populations no longer change because the number of people moving out equals the number of people moving in.

The population of Narnia decreases as the population of Middle Earth increases. As people leave Narnia, however, the *rate* at which they leave begins to slow down (because Narnia becomes less crowded). On the other hand, as people move into Middle Earth, some decide it was not for them and begin to move back (call this the reverse reaction).

Narnia ⟵ Middle Earth (reverse reaction)

As Middle Earth fills, the rate of people moving back to Narnia accelerates. Eventually, the *rate* of people moving out of Narnia (which has been slowing down as people leave) equals the *rate* of people moving back to Narnia (which has been increasing as Middle Earth gets more crowded). Dynamic equilibrium has been reached.

Narnia ⇌ Middle Earth

Notice that when the two kingdoms reach dynamic equilibrium, their populations no longer change because the number of people moving out equals the number of people moving in. However, one kingdom—because of its charm, or the character of its leader, or a lower tax rate, or whatever other reason—may have a higher population than the other kingdom, even when dynamic equilibrium is reached.

Similarly, when a chemical reaction reaches dynamic equilibrium, the rate of the forward reaction (analogous to people moving out of Narnia) equals the rate of the reverse reaction (analogous to people moving back into Narnia), and the relative concentrations of reactants and products (analogous to the relative populations of the two kingdoms) become constant. Also, like our two kingdoms, the concentrations of reactants and products are not necessarily equal at equilibrium, just as the populations of the two kingdoms are not equal at equilibrium.

15.4 The Equilibrium Constant: A Measure of How Far a Reaction Goes

▶ Write equilibrium constant expressions for chemical reactions.

We have just learned that the *concentrations* of reactants and products are not equal at equilibrium; rather, it is the *rates* of the forward and reverse reactions that are equal. But what about the concentrations? What can we know about them? The equilibrium constant (K_{eq}) is a way to quantify the relative concentrations of the reactants and products at equilibrium. Consider the generic chemical reaction:

$$aA + bB \rightleftharpoons cC + dD$$

where A and B are reactants, C and D are products, and a, b, c, and d are the respective stoichiometric coefficients in the chemical equation. The **equilibrium constant (K_{eq})** for the reaction is defined as the ratio—at equilibrium—of the concentrations of the products raised to their stoichiometric coefficients divided by the concentrations of the reactants raised to their stoichiometric coefficients.

$$K_{eq} = \frac{[C]^c\,[D]^d}{[A]^a\,[B]^b}$$

Products

Reactants

[A] means the molar concentration of A.

Notice that the equilibrium constant is a measure of the relative concentrations of reactants and products at equilibrium; the larger the equilibrium constant, the greater the concentration of products relative to reactants at equilibrium.

Writing Equilibrium Constant Expressions for Chemical Reactions

To write an equilibrium constant expression for a chemical reaction, we examine the chemical equation and follow the definition for the equilibrium constant. For example, suppose we want to write an equilibrium expression for this reaction:

$$2\,N_2O_5(g) \rightleftharpoons 4\,NO_2(g) + O_2(g)$$

The equilibrium constant is $[NO_2]$ raised to the fourth power multiplied by $[O_2]$ raised to the first power divided by $[N_2O_5]$ raised to the second power.

$$K_{eq} = \frac{[NO_2]^4[O_2]}{[N_2O_5]^2}$$

Notice that the *coefficients* in the chemical equation become the *exponents* in the equilibrium expression.

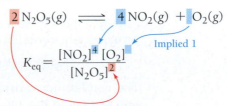

EXAMPLE **15.1** **Writing Equilibrium Constant Expressions for Chemical Reactions**

Write an equilibrium expression for the chemical equation.

$$CO(g) + 2\,H_2(g) \rightleftharpoons CH_3OH(g)$$

SOLUTION

The equilibrium expression is the concentration of the products raised to their stoichiometric coefficients divided by the concentration of the reactants raised to their stoichiometric coefficients. Notice that the expression is a ratio of products over reactants. Notice also that the coefficients in the chemical equation are the exponents in the equilibrium expression.

$$K_{eq} = \frac{[CH_3OH]}{[CO][H_2]^2}$$

Product

Reactants

▶ **SKILLBUILDER 15.1** | **Writing Equilibrium Expressions for Chemical Reactions**

Write an equilibrium expression for the chemical equation.

$$H_2(g) + F_2(g) \rightleftharpoons 2\,HF(g)$$

▶ **FOR MORE PRACTICE**   Example 15.10; Problems 43, 44.

The Significance of the Equilibrium Constant

> The symbol ≫ means *much greater than.*

What does an equilibrium constant tell us? For instance, what does a large equilibrium constant ($K_{eq} \gg 1$) imply about a reaction? It indicates that the forward reaction is largely favored and that there will be more products than reactants when equilibrium is reached. For example, consider the reaction:

$$H_2(g) + Br_2(g) \rightleftharpoons 2\,HBr(g) \quad K_{eq} = 1.9 \times 10^{19} \text{ at } 25\,°C$$

The equilibrium constant for this reaction is large, meaning that at equilibrium the reaction lies far to the right. In other words, at equilibrium there are high concentrations of products, tiny concentrations of reactants (▼ **FIGURE 15.6**).

$$H_2(g) + Br_2(g) \rightleftharpoons 2\,HBr(g)$$

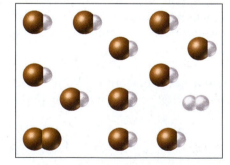

$$K_{eq} = \frac{[HBr]^2}{[H_2][Br_2]} = \text{Large Number}$$

A large equilibrium constant means a high concentration of products and a low concentration of reactants at equilibrium.

▶ **FIGURE 15.6 The meaning of a large equilibrium constant**

> The symbol ≪ means *much less than.*

Conversely, what does a *small* equilibrium constant ($K_{eq} \ll 1$) mean? It indicates that the reverse reaction is favored and that there will be more reactants than products when equilibrium is reached. For example, consider the reaction:

$$N_2(g) + O_2(g) \rightleftharpoons 2\,NO(g) \quad K_{eq} = 4.1 \times 10^{-31} \text{ at } 25\,°C$$

The equilibrium constant is very small, meaning that at equilibrium the reaction lies far to the left—high concentrations of reactants, low concentrations of

$$N_2(g) + O_2(g) \rightleftharpoons 2\,NO(g)$$

$$K_{eq} = \frac{[NO]^2}{[N_2][O_2]} = \text{Small Number}$$

A small equilibrium constant means a high concentration of reactants and a low concentration of products at equilibrium.

▲ **FIGURE 15.7 The meaning of a small equilibrium constant**

products (▲ **FIGURE 15.7**). This is fortunate because N_2 and O_2 are the main components of air. If this equilibrium constant were large, much of the N_2 and O_2 in air would react to form NO, a toxic gas.

To summarize:

- $K_{eq} \ll 1$ Reverse reaction is favored; forward reaction does not proceed very far.
- $K_{eq} \approx 1$ Neither direction is favored; forward reaction proceeds about halfway (significant amounts of both reactants and products are present at equilibrium).
- $K_{eq} \gg 1$ Forward reaction is favored; forward reaction proceeds virtually to completion.

The symbol ≈ means "approximately equal to."

CONCEPTUAL ✔ CHECKPOINT 15.2

Consider a generic chemical reaction in which the reactant, A, turns directly into the product, B.

$$A(g) \rightleftharpoons B(g)$$

The reaction is allowed to come to equilibrium at three different temperatures as represented in the three circles. At which temperature is the equilibrium constant the largest?

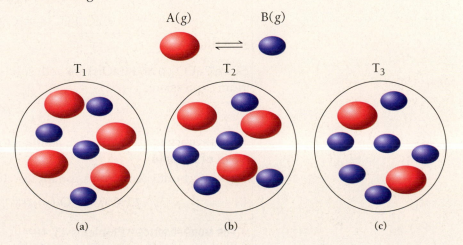

A(g)   B(g)

T_1 T_2 T_3

(a)   (b)   (c)

15.5 Heterogeneous Equilibria: The Equilibrium Expression for Reactions Involving a Solid or a Liquid

▶ Write equilibrium expressions for chemical reactions involving a solid or a liquid.

Consider the chemical reaction shown here:

$$2\,CO(g) \rightleftharpoons CO_2(g) + C(s)$$

We might expect the expression for the equilibrium constant to be:

$$K_{eq} = \frac{[CO_2][C]}{[CO]^2} \quad \text{(incorrect)}$$

However, since carbon is a solid, its concentration is constant—it does not change. Adding more or less carbon to the reaction mixture does not change the concentration of carbon. The concentration of a solid does not change because a solid does not expand to fill its container. The concentration of a solid, therefore, depends only on its density, which (except for slight variations due to temperature) is constant as long as *some solid is present*. Consequently, we do not include pure solids—those reactants or products labeled in the chemical equation with an (s)—in the equilibrium expression. The correct equilibrium expression is:

We exclude the concentrations of pure solids and pure liquids from equilibrium expressions because they are constant.

$$K_{eq} = \frac{[CO_2]}{[CO]^2} \quad \text{(correct)}$$

Similarly, the concentration of a pure liquid does not change. Consequently, we also exclude pure liquids—those reactants or products labeled in the chemical equation with an (l)— from the equilibrium expression. For example, what is the equilibrium expression for the following reaction?

$$CO_2(g) + H_2O(l) \rightleftharpoons H^+(aq) + HCO_3^-(aq)$$

Because $H_2O(l)$ is pure liquid, it is omitted from the equilibrium expression:

$$K_{eq} = \frac{[H^+][HCO_3^-]}{[CO_2]}$$

EXAMPLE **15.2** | **Writing Equilibrium Expressions for Reactions Involving a Solid or a Liquid**

Write an equilibrium expression for the chemical equation.

$$CaCO_3(s) \rightleftharpoons CaO(s) + CO_2(g)$$

SOLUTION

Since $CaCO_3(s)$ and $CaO(s)$ are both solids, you omit them from the equilibrium expression:

$$K_{eq} = [CO_2]$$

▶ **SKILLBUILDER 15.2** | **Writing Equilibrium Expressions for Reactions Involving a Solid or a Liquid**

Write an equilibrium expression for the chemical equation.

$$4\,HCl(g) + O_2(g) \rightleftharpoons 2\,H_2O(l) + 2\,Cl_2(g)$$

▶ **FOR MORE PRACTICE** Problems 45, 46.

15.6 Calculating and Using Equilibrium Constants

▶ Calculate equilibrium constants.

▶ Use the equilibrium constant to find the concentration of a reactant or product at equilibrium.

The equilibrium constant expression is a quantitative relationship between the constant itself and the amounts of reactants and products at equilibrium. Therefore, we can use measurements of the amounts of reactants and products at equilibrium to calculate the equilibrium constant. We can also use the equilibrium constant to determine amounts of reactants and products at equilibrium.

Calculating Equilibrium Constants

The most direct way to obtain a value for the equilibrium constant of a reaction is to measure the concentrations of the reactants and products in a reaction mixture at equilibrium. For example, consider the reaction:

$$H_2(g) + I_2(g) \rightleftharpoons 2\,HI(g)$$

Suppose a mixture of H_2 and I_2 comes to equilibrium at 445 °C. The measured equilibrium concentrations are $[H_2] = 0.11$ M, $[I_2] = 0.11$ M, and $[HI] = 0.78$ M. What is the value of the equilibrium constant? We begin by sorting the information in the problem statement.

> Equilibrium constants depend on temperature, so temperatures are often included with equilibrium data. However, the temperature is not part of the equilibrium expression.

GIVEN: $[H_2] = 0.11$ M
$[I_2] = 0.11$ M
$[HI] = 0.78$ M

FIND: K_{eq}

SOLUTION

We can write the expression for K_{eq} from the balanced equation.

$$K_{eq} = \frac{[HI]^2}{[H_2][I_2]}$$

To calculate the value of K_{eq}, we substitute the correct equilibrium concentrations into the expression for K_{eq}.

$$K_{eq} = \frac{[HI]^2}{[H_2][I_2]}$$

$$= \frac{[0.78]^2}{[0.11][0.11]}$$

$$= 5.0 \times 10^1$$

> The concentrations in an equilibrium expression should always be in units of molarity (M), but the units themselves are normally dropped.

We must always write the concentrations within K_{eq} in moles per liter (M); however, we normally drop the units in expressing the equilibrium constant.

The *particular concentrations* of reactants and products for a reaction at equilibrium are *not* always the same for a given reaction; they depend on the initial concentrations. However, the *equilibrium constant* is always the same at a given temperature, regardless of the initial concentrations. For example, Table 15.1 (on the next page) shows several different equilibrium concentrations of H_2, I_2, and HI, each from a different set of initial concentrations. Notice that the equilibrium constant is always the same, regardless of the initial concentrations. In other words, no matter what the initial concentrations are, the reaction always goes in a direction so that the equilibrium concentrations—when substituted into the equilibrium expression—give the same constant, K_{eq}.

> A reaction can approach equilibrium from either direction, depending on the initial concentrations, but its K_{eq} at a given temperature is always the same.

TABLE 15.1 Initial and Equilibrium Concentrations at 445 °C for the Reaction
$H_2(g) + I_2(g) \rightleftharpoons 2\,HI(g)$

Initial			Equilibrium			Equilibrium Constant
$[H_2]$	$[I_2]$	$[HI]$	$[H_2]$	$[I_2]$	$[HI]$	$\dfrac{[HI]^2}{[H_2][I_2]} = K_{eq}$
0.50	0.50	0.0	0.11	0.11	0.78	$\dfrac{[0.78]^2}{[0.11][0.11]} = 50$
0.0	0.0	0.50	0.055	0.055	0.39	$\dfrac{[0.39]^2}{[0.055][0.055]} = 50$
0.50	0.50	0.50	0.165	0.165	1.17	$\dfrac{[1.17]^2}{[0.165][0.165]} = 50$
1.0	0.5	0.0	0.53	0.033	0.934	$\dfrac{[0.934]^2}{[0.53][0.033]} = 50$

PEARSON
eText 2.0

Interactive
Worked Example
Video 15.3

EXAMPLE **15.3** | **Calculating Equilibrium Constants**

Consider the reaction:

$$2\,CH_4(g) \rightleftharpoons C_2H_2(g) + 3\,H_2(g)$$

A mixture of CH_4, C_2H_2, and H_2 comes to equilibrium at 1700 °C. The measured equilibrium concentrations are $[CH_4] = 0.0203$ M, $[C_2H_2] = 0.0451$ M, and $[H_2] = 0.112$ M. What is the value of the equilibrium constant at this temperature?

You are given the concentrations of the reactants and products of a reaction at equilibrium. You are asked to find the equilibrium constant.	**GIVEN:** $[CH_4] = 0.0203$ M $[C_2H_2] = 0.0451$ M $[H_2] = 0.112$ M **FIND:** K_{eq}
Write the expression for K_{eq} from the balanced equation. To calculate the value of K_{eq}, substitute the correct equilibrium concentrations into the expression for K_{eq}.	**SOLUTION** $K_{eq} = \dfrac{[C_2H_2][H_2]^3}{[CH_4]^2}$ $K_{eq} = \dfrac{[0.0451][0.112]^3}{[0.0203]^2}$ $= 0.154$

▶ **SKILLBUILDER 15.3** | **Calculating Equilibrium Constants**

Consider the reaction:

$$CO(g) + 2\,H_2(g) \rightleftharpoons CH_3OH(g)$$

A mixture of CO, H_2, and CH_3OH comes to equilibrium at 225 °C. The measured equilibrium concentrations are $[CO] = 0.489$ M, $[H_2] = 0.146$ M, and $[CH_3OH] = 0.151$ M. What is the value of the equilibrium constant at this temperature?

▶ **SKILLBUILDER PLUS**

Suppose that the preceding reaction is carried out at a different temperature and that the initial concentrations of the reactants are $[CO] = 0.500$ M and $[H_2] = 1.00$ M. Assuming there is no product at the beginning of the reaction and at equilibrium $[CO] = 0.15$ M, find the equilibrium constant at this new temperature. *Hint:* Use the stoichiometric relationships from the balanced equation to find the equilibrium concentrations of H_2 and CH_3OH.

▶ **FOR MORE PRACTICE** Example 15.11; Problems 51, 52, 53, 54, 55, 56.

Using Equilibrium Constants in Calculations

We can also use the equilibrium constant to calculate the equilibrium concentration for one of the reactants or products, given the equilibrium concentrations of the others. For example, consider the reaction:

$$2\,COF_2(g) \rightleftharpoons CO_2(g) + CF_4(g) \qquad K_{eq} = 2.00 \text{ at } 1000\,°C$$

In an equilibrium mixture, the concentration of COF_2 is 0.255 M and the concentration of CF_4 is 0.118 M. What is the equilibrium concentration of CO_2? We begin by sorting the information in the problem statement.

GIVEN: $[COF_2] = 0.255$ M
$[CF_4] = 0.118$ M
$K_{eq} = 2.00$

FIND: $[CO_2]$

SOLUTION MAP

We then draw a solution map showing how the expression for the equilibrium constant provides the equation that gets us from the given quantities to the quantity we are trying to find.

$$\boxed{[COF_2], [CF_4], K_{eq}} \longrightarrow \boxed{[CO_2]}$$

$$K_{eq} = \frac{[CO_2][CF_4]}{[COF_2]^2}$$

SOLUTION

We write the equilibrium expression for the reaction, and then we solve it for the quantity we are trying to find ($[CO_2]$).

$$K_{eq} = \frac{[CO_2][CF_4]}{[COF_2]^2}$$

$$[CO_2] = K_{eq}\frac{[COF_2]^2}{[CF_4]}$$

We substitute the appropriate values and calculate $[CO_2]$.

$$[CO_2] = 2.00\frac{[0.255]^2}{[0.118]}$$

$$= 1.10 \text{ M}$$

PEARSON
eText 2.0
Interactive
Worked Example
Video 15.4

EXAMPLE **15.4**   **Using Equilibrium Constants in Calculations**

Consider the reaction:

$$H_2(g) + I_2(g) \rightleftharpoons 2\,HI(g) \qquad K_{eq} = 69 \text{ at } 340\,°C$$

In an equilibrium mixture, the concentrations of H_2 and I_2 are both 0.020 M. What is the equilibrium concentration of HI?

SORT	**GIVEN:** $[H_2] = [I_2] = 0.020$ M
You are given the equilibrium concentrations of the reactants in a chemical reaction and also the value of the equilibrium constant. You are asked to find the concentration of the product.	$K_{eq} = 69$ **FIND:** $[HI]$

STRATEGIZE	**SOLUTION MAP**
Draw a solution map showing how the equilibrium constant expression gives the relationship between the given concentrations and the concentration you are asked to find.	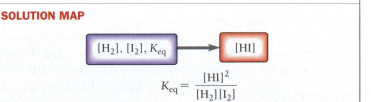

continued on page 542 ▶

continued from page 541

SOLVE	SOLUTION
Solve the equilibrium expression for [HI] and substitute in the appropriate values to calculate it. Because the value of [HI] is squared, you must take the square root of both sides of the equation to solve for [HI] because $\sqrt{[HI]^2} = [HI]$.	$$K_{eq} = \frac{[HI]^2}{[H_2][I_2]}$$ $$[HI]^2 = K_{eq}[H_2][I_2]$$ $$\sqrt{[HI]^2} = \sqrt{K_{eq}[H_2][I_2]}$$ $$[HI] = \sqrt{69(0.020)(0.020)}$$ $$= 0.17 \text{ M}$$
CHECK You can check your answer by substituting it back into the expression for K_{eq}.	$$K_{eq} = \frac{[HI]^2}{[H_2][I_2]}$$ $$= \frac{(0.17)^2}{(0.020)(0.020)}$$ $$= 72$$ The calculated value of K_{eq} is about equal to the given value of K_{eq} (which was 69), indicating that your answer is correct. The slight difference is due to rounding error, which is common in problems like these.

▶ **SKILLBUILDER 15.4** | **Using Equilibrium Constants in Calculations**

Diatomic iodine (I_2) decomposes at high temperature to form I atoms according to the reaction:

$$I_2(g) \rightleftharpoons 2\,I(g) \quad K_{eq} = 0.011 \text{ at } 1200\,°C$$

In an equilibrium mixture, the concentration of I_2 is 0.10 M. What is the equilibrium concentration of I?

▶ **FOR MORE PRACTICE** Example 15.12; Problems 57, 58, 59, 60.

PEARSON eText 2.0

CONCEPTUAL ✔ **CHECKPOINT 15.3**

When the reaction A(*aq*) $\rightleftharpoons$ B(*aq*) + C(*aq*) is at equilibrium, each of the three compounds has a concentration of 2 M. What is the equilibrium constant for this reaction?

(a) 4   (b) 2   (c) 1   (d) 1/2

15.7 Disturbing a Reaction at Equilibrium: Le Châtelier's Principle

▶ Restate Le Châtelier's principle.

Pronounced "le-sha-te-lyay."

Key Concept Video
Le Châtelier's Principle

We have seen that a chemical system not in equilibrium tends to go toward equilibrium and that the concentrations of the reactants and products at equilibrium correspond to the equilibrium constant, K_{eq}. What happens, in contrast, when a chemical system already at equilibrium is disturbed? **Le Châtelier's principle** states that the chemical system will respond to minimize the disturbance.

Le Châtelier's principle—When a chemical system at equilibrium is disturbed, the system shifts in a direction that minimizes the disturbance.

In other words, a system at equilibrium tries to maintain that equilibrium—it fights back when disturbed.

We can understand Le Châtelier's principle by returning to our Narnia and Middle Earth analogy. Suppose the populations of Narnia and Middle Earth are at equilibrium. This means that the rate of people moving out of Narnia (and

▶ **FIGURE 15.8 Population analogy for Le Châtelier's principle** When a system at equilibrium is disturbed, it shifts to minimize the disturbance. In this case, adding population to Middle Earth (the disturbance) causes population to move out of Middle Earth (minimizing the disturbance). Question: What would happen if you disturbed the equilibrium by taking population out of Middle Earth? In which direction would the population move to minimize the disturbance?

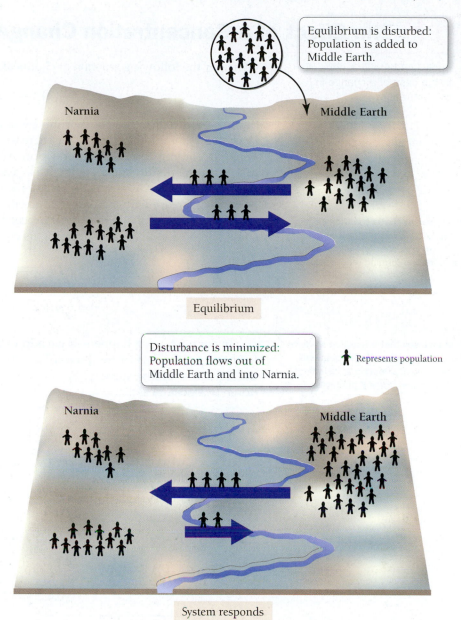

Equilibrium is disturbed: Population is added to Middle Earth.

Narnia   Middle Earth

Equilibrium

Disturbance is minimized: Population flows out of Middle Earth and into Narnia.

† Represents population

Narnia   Middle Earth

System responds

into Middle Earth) is equal to the rate of people moving into Narnia (and out of Middle Earth). It also means that the populations of the two kingdoms are stable. Now imagine disturbing that balance (▲ **FIGURE 15.8**). Suppose we add extra people to Middle Earth. What happens? Since Middle Earth suddenly becomes more crowded, the rate of people leaving Middle Earth increases. The net flow of people is *out of Middle Earth and into Narnia*. Notice what happened. We disturbed the equilibrium by adding more people to Middle Earth. The system responded by moving people out of Middle Earth—it shifted in the direction that minimized the disturbance.

On the other hand, what happens if we add extra people to Narnia? Since Narnia suddenly gets more crowded, the rate of people leaving Narnia increases. The net flow of people is out of Narnia and into Middle Earth. We added people to Narnia, and the system responded by moving people out of Narnia. When systems at equilibrium are disturbed, they react to counter the disturbance. Chemical systems behave similarly. There are several ways to disturb a system in chemical equilibrium. We consider each of these separately in the next three sections of the chapter.

15.8 The Effect of a Concentration Change on Equilibrium

▶ Apply Le Châtelier's principle in the case of a change in concentration.

Consider the following reaction at chemical equilibrium:

$$N_2O_4(g) \rightleftharpoons 2\,NO_2(g)$$

Suppose we disturb the equilibrium by adding NO_2 to the equilibrium mixture (▼ FIGURE 15.9). In other words, we increase the concentration of NO_2. What happens? According to Le Châtelier's principle, the system shifts in a direction to minimize the disturbance. The shift is caused by the increased concentration of NO_2, which in turn increases the rate of the reverse reaction because reaction rates generally increase with increasing concentration (as we discussed in Section 15.2).

The reaction shifts to the left (it proceeds in the reverse direction), consuming some of the added NO_2 and bringing its concentration back down:

> When we say that a reaction *shifts to the left*, we mean that it proceeds in the reverse direction, consuming products and forming reactants.

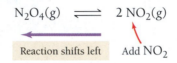

$$N_2O_4(g) \rightleftharpoons 2\,NO_2(g)$$

Reaction shifts left   Add NO_2

In contrast, what happens if we add extra N_2O_4, increasing its concentration? In this case, the rate of the forward reaction increases and the reaction shifts to the right, consuming some of the added N_2O_4 and bringing *its* concentration back down (▶ FIGURE 15.10):

> When we say that a reaction *shifts to the right*, we mean that it proceeds in the forward direction, consuming reactants and forming products.

$$N_2O_4(g) \rightleftharpoons 2\,NO_2(g)$$

Add N_2O_4 Reaction shifts right

In each case, the system shifts in a direction that minimizes the disturbance.

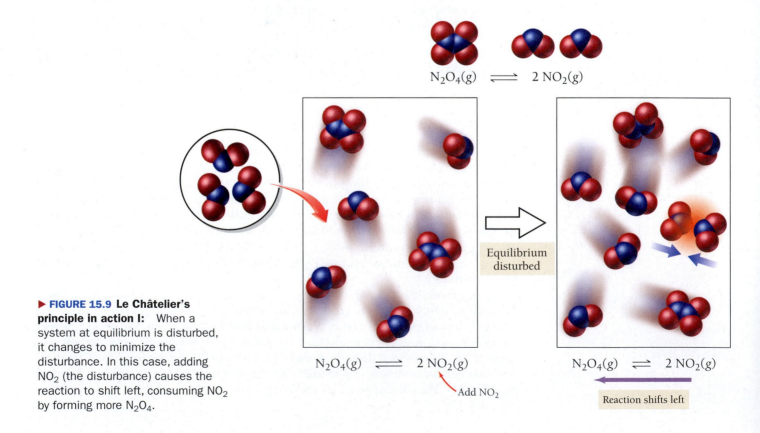

$$N_2O_4(g) \rightleftharpoons 2\,NO_2(g)$$

Equilibrium disturbed

$$N_2O_4(g) \rightleftharpoons 2\,NO_2(g)$$

Add NO_2

$$N_2O_4(g) \rightleftharpoons 2\,NO_2(g)$$

Reaction shifts left

▶ FIGURE 15.9 **Le Châtelier's principle in action I:** When a system at equilibrium is disturbed, it changes to minimize the disturbance. In this case, adding NO_2 (the disturbance) causes the reaction to shift left, consuming NO_2 by forming more N_2O_4.

▶ **FIGURE 15.10 Le Châtelier's principle in action II:**   When a system at equilibrium is disturbed, it changes to minimize the disturbance. In this case, adding N_2O_4 (the disturbance) causes the reaction to shift right, consuming N_2O_4 by producing more NO_2.

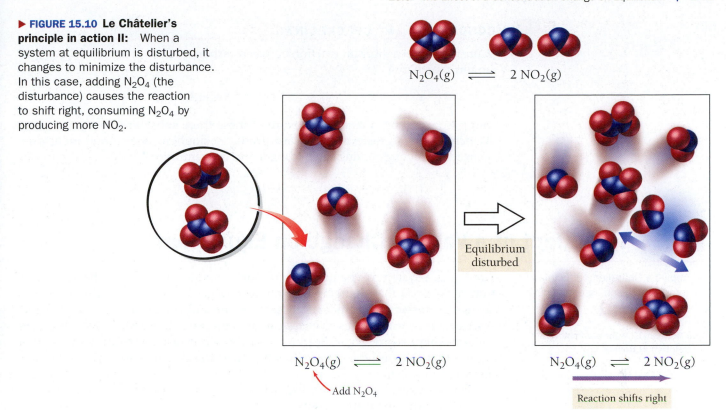

$$N_2O_4(g) \rightleftharpoons 2\,NO_2(g)$$

Equilibrium disturbed

$$N_2O_4(g) \rightleftharpoons 2\,NO_2(g)$$

Add N_2O_4

$$N_2O_4(g) \rightleftharpoons 2\,NO_2(g)$$

Reaction shifts right

To summarize, if a chemical system is at equilibrium:

- Increasing the concentration of one or more of the reactants causes the reaction to shift to the right (in the direction of the products).
- Increasing the concentration of one or more of the products causes the reaction to shift to the left (in the direction of the reactants).

EXAMPLE **15.5**   **The Effect of a Concentration Change on Equilibrium**

Consider the following reaction at equilibrium:

$$CaCO_3(s) \rightleftharpoons CaO(s) + CO_2(g)$$

What is the effect of adding more CO_2 to the reaction mixture? What is the effect of adding more $CaCO_3$?

SOLUTION

Adding more CO_2 increases the concentration of CO_2 and causes the reaction to shift to the left. Adding more $CaCO_3$ does not increase the concentration of $CaCO_3$ because $CaCO_3$ is a solid and thus has a constant concentration. It is therefore not included in the equilibrium expression and has no effect on the position of the equilibrium.

▶ **SKILLBUILDER 15.5 | The Effect of a Concentration Change on Equilibrium**

Consider the following reaction in chemical equilibrium:

$$2\,BrNO(g) \rightleftharpoons 2\,NO(g) + Br_2(g)$$

What is the effect of adding more Br_2 to the reaction mixture? What is the effect of adding more BrNO?

▶ **SKILLBUILDER PLUS**

What is the effect of removing some Br_2 from the preceding reaction mixture?

▶ **FOR MORE PRACTICE**   Example 15.13a, b; Problems 63, 64, 65, 66.

CONCEPTUAL ✔ CHECKPOINT 15.4

Consider the equilibrium reaction between carbon monoxide and hydrogen gas to form methanol.

$$CO(g) + 2\,H_2(g) \rightleftharpoons CH_3OH(g)$$

Suppose you have a reaction mixture of these three substances at equilibrium. Which causes a greater shift toward products: doubling the carbon monoxide concentration or doubling the hydrogen gas concentration?

15.9 The Effect of a Volume Change on Equilibrium

▶ Apply Le Châtelier's principle in the case of a change in volume.

See Section 11.4 for a complete description of Boyle's law.

How does a system in chemical equilibrium respond to a volume change? Recall from Chapter 11 that changing the volume of a gas (or a gas mixture) results in a change in pressure. Remember also that pressure and volume are inversely related: A *decrease* in volume causes an *increase* in pressure, and an *increase* in volume causes a *decrease* in pressure. So, if the volume of a gaseous reaction mixture at chemical equilibrium changes, the pressure changes and the system shifts in a direction to minimize that change.

For example, consider the following reaction at equilibrium in a cylinder equipped with a moveable piston:

$$N_2(g) + 3\,H_2(g) \rightleftharpoons 2\,NH_3(g)$$

From the ideal gas law ($PV = nRT$), we can see that lowering the number of moles of a gas (*n*) results in a lower pressure (*P*) at constant temperature and volume.

What happens if we push down on the piston, lowering the volume and raising the pressure (▶ FIGURE 15.11)? How can the chemical system bring the pressure back down? Look carefully at the reaction coefficients in the balanced equation. If the reaction shifts to the right, 4 mol of gas particles (1 mol of N_2 and 3 mol of H_2) are converted to 2 mol of gas particles (2 mol of NH_3). So, as the reaction shifts toward products, the pressure is lowered (because the reaction mixture contains fewer gas particles). Therefore, the system then shifts to the right, bringing the pressure back down and minimizing the disturbance.

Consider the same reaction mixture at equilibrium again. What happens if, this time, we pull *up* on the piston, *increasing* the volume (▶ FIGURE 15.12)? The higher volume results in a lower pressure, and the system responds to bring the pressure back up. It can do this by shifting to the left, converting 2 mol of gas particles into 4 mol of gas particles. As the reaction shifts toward reactants, the pressure increases again (because the reaction mixture contains more gas particles), minimizing the disturbance.

To summarize, if a chemical system is at equilibrium:

- Decreasing the volume causes the reaction to shift in the direction that has fewer moles of gas particles.
- Increasing the volume causes the reaction to shift in the direction that has more moles of gas particles.

Notice that if a chemical reaction has an equal number of moles of gas particles on both sides of the chemical equation, a change in volume has no effect. For example, consider the following reaction:

$$H_2(g) + I_2(g) \rightleftharpoons 2\,HI(g)$$

Both the left and the right sides of the equation contain 2 mol of gas particles, so a change in volume has no effect on this reaction. In addition, a change in volume has no effect on a reaction that has no gaseous reactants or products.

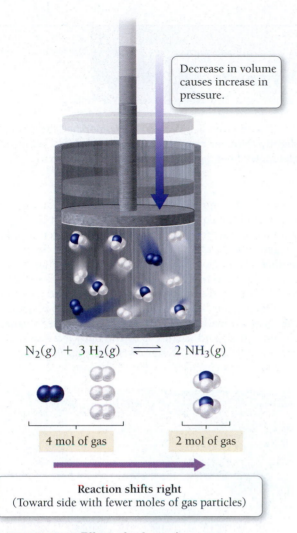

N₂(g) + 3 H₂(g) ⇌ 2 NH₃(g)

4 mol of gas   2 mol of gas

Reaction shifts right
(Toward side with fewer moles of gas particles)

▲ **FIGURE 15.11 Effect of volume decrease on equilibrium** When the volume of an equilibrium mixture decreases, the pressure increases. The system responds (to bring the pressure back down) by shifting to the right, the side of the reaction with fewer moles of gas particles.

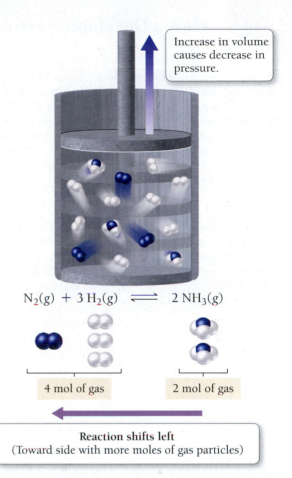

N₂(g) + 3 H₂(g) ⇌ 2 NH₃(g)

4 mol of gas   2 mol of gas

Reaction shifts left
(Toward side with more moles of gas particles)

▲ **FIGURE 15.12 Effect of volume increase on equilibrium** When the volume of an equilibrium mixture increases, the pressure decreases. The system responds (to raise the pressure) by shifting to the left, the side of the reaction with more moles of gas particles.

EXAMPLE **15.6** | **The Effect of a Volume Change on Equilibrium**

Consider the reaction at chemical equilibrium:

$$2 \text{ KClO}_3(s) \rightleftharpoons 2 \text{ KCl}(s) + 3 \text{ O}_2(g)$$

What is the effect of decreasing the volume of the reaction mixture? Increasing the volume of the reaction mixture?

SOLUTION

The chemical equation has 3 mol of gas on the right and 0 mol of gas on the left. Decreasing the volume of the reaction mixture increases the pressure and causes the reaction to shift to the left (toward the side with fewer moles of gas particles). Increasing the volume of the reaction mixture decreases the pressure and causes the reaction to shift to the right (toward the side with more moles of gas particles).

▶ **SKILLBUILDER 15.6 | The Effect of a Volume Change on Equilibrium**

Consider the reaction at chemical equilibrium:

$$2 \text{ SO}_2(g) + \text{O}_2(g) \rightleftharpoons 2 \text{ SO}_3(g)$$

What is the effect of decreasing the volume of the reaction mixture? Increasing the volume of the reaction mixture?

▶ **FOR MORE PRACTICE** Example 15.13c; Problems 67, 68, 69, 70.

CHEMISTRY AND HEALTH

How a Developing Fetus Gets Oxygen from Its Mother

Have you ever wondered how a fetus in the womb gets oxygen? Unlike you and me, a fetus cannot breathe. Yet like you and me, a fetus needs oxygen. Where does that oxygen come from? In adults, oxygen is absorbed in the lungs and carried in the blood by a protein molecule called hemoglobin, which is abundantly present in red blood cells. Hemoglobin (Hb) reacts with oxygen according to the equilibrium equation:

$$Hb + O_2 \rightleftharpoons HbO_2$$

The equilibrium constant for this reaction is neither large nor small but intermediate. Consequently, the reaction shifts toward the right or the left, depending on the concentration of oxygen. As blood flows through the lungs, where oxygen concentrations are high, the equilibrium shifts to the right—hemoglobin loads oxygen.

Lung [O$_2$] high

$$Hb + O_2 \rightleftharpoons HbO_2$$

Reaction shifts right

As blood flows through muscles and organs that are using oxygen (where oxygen concentrations have been depleted), the equilibrium shifts to the left—hemoglobin unloads oxygen.

Muscle [O$_2$] low

$$Hb + O_2 \rightleftharpoons HbO_2$$

Reaction shifts left

A fetus has its own blood circulatory system. The mother's blood never flows into the fetus's body, and the fetus cannot get any air in the womb. So how does the fetus get oxygen?

The answer lies in fetal hemoglobin (HbF), which is slightly different from adult hemoglobin. Like adult hemoglobin, fetal hemoglobin is in equilibrium with oxygen.

$$HbF + O_2 \rightleftharpoons HbFO_2$$

However, the equilibrium constant for fetal hemoglobin is larger than the equilibrium constant for adult hemoglobin. In other words, fetal hemoglobin loads oxygen at a lower oxygen concentration than adult hemoglobin. So, when the mother's hemoglobin flows through the placenta, it unloads oxygen into the placenta. The baby's blood also flows into the placenta, and even though the baby's blood never mixes with the mother's blood, the fetal hemoglobin within the baby's blood loads the oxygen (that the mother's hemoglobin unloaded) and carries it to the baby. Nature has thus engineered a chemical system where the mother's hemoglobin can in effect *hand off* oxygen to the baby's hemoglobin.

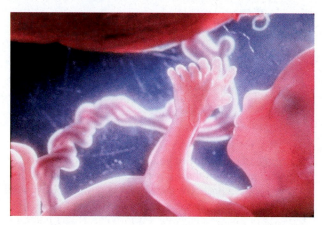

▲ A human fetus. Question: How does the fetus get oxygen?

B15.1 CAN YOU ANSWER THIS? *What would happen if fetal hemoglobin had the same equilibrium constant for the reaction with oxygen as adult hemoglobin?*

PEARSON
eText
2.0

CONCEPTUAL ✔ CHECKPOINT 15.5

Consider the reaction:

$$H_2(g) + I_2(g) \rightleftharpoons 2\,HI(g)$$

Which change would cause the reaction to shift to the right (toward products)?

(a) decreasing the volume

(b) increasing the volume

(c) increasing the concentration of hydrogen gas

(d) decreasing the concentration of hydrogen gas

15.10 The Effect of a Temperature Change on Equilibrium

▶ Apply Le Châtelier's principle in the case of a change in temperature.

According to Le Châtelier's principle, if the temperature of a system at equilibrium is changed, the system should shift in a direction to counter that change. So if the temperature is increased, the reaction should shift in the direction that attempts to lower the temperature and vice versa. Recall from Section 3.9 that energy changes are often associated with chemical reactions. If we want to predict the direction in which a reaction will shift upon a temperature change, we must understand how a shift in the reaction affects the temperature.

In Section 3.9, we classify chemical reactions according to whether they absorb or emit heat energy in the course of the reaction. Recall that an *exothermic reaction* (one with a negative ΔH_{rxn}) emits heat.

Exothermic reaction: $A + B \rightleftharpoons C + D + Heat$

In an exothermic reaction, we can think of heat as a product. Consequently, raising the temperature of an exothermic reaction—think of this as adding heat—causes the reaction to shift left.

For example, the reaction of nitrogen with hydrogen to form ammonia is exothermic:

$$N_2(g) + 3\,H_2(g) \rightleftharpoons 2\,NH_3(g) + Heat$$

2) Reaction shifts left   1) Add heat

Raising the temperature of an equilibrium mixture of these three gases causes the reaction to shift left, absorbing some of the added heat. Conversely, lowering the temperature of an equilibrium mixture of these three gases causes the reaction to shift right, releasing heat.

$$N_2(g) + 3\,H_2(g) \rightleftharpoons 2\,NH_3(g) + Heat$$

2) Reaction shifts right   1) Remove heat

In contrast, an *endothermic reaction* (one with a positive ΔH_{rxn}) absorbs heat.

Endothermic reaction: $A + B + Heat \rightleftharpoons C + D$

In an endothermic reaction, we can think of heat as a reactant. Consequently, raising the temperature (or adding heat) causes an endothermic reaction to shift right.

For example, the following reaction is endothermic:

Colorless   Brown
$$N_2O_4(g) + Heat \rightleftharpoons 2\,NO_2(g)$$

1) Add heat   2) Reaction shifts right

Raising the temperature of an equilibrium mixture of these two gases causes the reaction to shift right, absorbing some of the added heat. Because N_2O_4 is colorless and NO_2 is brown, we can see the effects of changing the temperature of this reaction (▶ FIGURE 15.13, on the next page).

On the other hand, lowering the temperature of a reaction mixture of these two gases causes the reaction to shift left, releasing heat:

Colorless   Brown
$$N_2O_4(g) + Heat \rightleftharpoons 2\,NO_2(g)$$

1) Remove heat   2) Reaction shifts left

▶ FIGURE 15.13 **Equilibrium as a function of temperature**

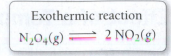

Exothermic reaction

$N_2O_4(g) \rightleftharpoons 2 NO_2(g)$

Cool temperatures cause a shift to the left, to colorless N_2O_4.

Warm temperatures cause a shift to the right, toward the production of brown NO_2.

To summarize:

In an exothermic chemical reaction, heat is a product and:

- Increasing the temperature causes the reaction to shift left (in the direction of the reactants).
- Decreasing the temperature causes the reaction to shift right (in the direction of the products).

In an endothermic chemical reaction, heat is a reactant and:

- Increasing the temperature causes the reaction to shift right (in the direction of the products).
- Decreasing the temperature causes the reaction to shift left (in the direction of the reactants).

EXAMPLE **15.7**   **The Effect of a Temperature Change on Equilibrium**

The following reaction is endothermic:

$$CaCO_3(s) \rightleftharpoons CaO(s) + CO_2(g)$$

What is the effect of increasing the temperature of the reaction mixture? Decreasing the temperature?

SOLUTION

Because the reaction is endothermic, we can think of heat as a reactant.

$$\text{Heat} + CaCO_3(s) \rightleftharpoons CaO(s) + CO_2(g)$$

Raising the temperature is adding heat, causing the reaction to shift to the right. Lowering the temperature is removing heat, causing the reaction to shift to the left.

▶ SKILLBUILDER 15.7 | **The Effect of a Temperature Change on Equilibrium**

The following reaction is exothermic.

$$2 SO_2(g) + O_2(g) \rightleftharpoons 2 SO_3(g)$$

What is the effect of increasing the temperature of the reaction mixture? Decreasing the temperature?

▶ **FOR MORE PRACTICE**   Example 15.13d; Problems 71, 72, 73, 74.

CONCEPTUAL ✓ CHECKPOINT 15.6

Consider the endothermic reaction.

$$Cl_2(g) \rightleftharpoons 2\,Cl(g)$$

If the reaction mixture is at equilibrium, which disturbances increase the amount of product the most?

(a) increasing the temperature and increasing the volume

(b) increasing the temperature and decreasing the volume

(c) decreasing the temperature and increasing the volume

(d) decreasing the temperature and decreasing the volume

15.11   The Solubility-Product Constant

▶ Use K_{sp} to determine molar solubility.

▶ Write an expression for the solubility-product constant.

Recall from Section 7.5 that a compound is considered soluble if it dissolves in water and insoluble if it does not. Recall also that, by applying the *solubility rules* (Table 7.2), we can classify many ionic compounds as soluble or insoluble. We can better understand the solubility of an ionic compound with the concept of equilibrium. The process by which an ionic compound dissolves is an equilibrium process. For example, we can represent the dissolving of calcium fluoride in water with the following chemical equation:

$$CaF_2(s) \rightleftharpoons Ca^{2+}(aq) + 2\,F^-(aq)$$

The equilibrium expression for a chemical equation that represents the dissolving of an ionic compound is the **solubility-product constant (K_{sp})**. For CaF₂, the solubility-product constant is:

$$K_{sp} = [Ca^{2+}][F^-]^2$$

Notice that, as we discussed in Section 15.5, solids are omitted from the equilibrium expression.

> K_{sp} values are normally used for slightly soluble or insoluble compounds only.

The K_{sp} value is therefore a measure of the solubility of a compound. A large K_{sp} (forward reaction favored) means that the compound is very soluble. A small K_{sp} (reverse reaction favored) means that the compound is not very soluble. Table 15.2 lists the value of K_{sp} for a number of ionic compounds.

TABLE 15.2   Selected Solubility-Product Constants (K_{sp})

Compound	Formula	K_{sp}
barium sulfate	$BaSO_4$	1.07×10^{-10}
calcium carbonate	$CaCO_3$	4.96×10^{-9}
calcium fluoride	CaF_2	1.46×10^{-10}
calcium hydroxide	$Ca(OH)_2$	4.68×10^{-6}
calcium sulfate	$CaSO_4$	7.10×10^{-5}
copper(II) sulfide	CuS	1.27×10^{-36}
iron(II) carbonate	$FeCO_3$	3.07×10^{-11}
iron(II) hydroxide	$Fe(OH)_2$	4.87×10^{-17}
lead(II) chloride	$PbCl_2$	1.17×10^{-5}
lead(II) sulfate	$PbSO_4$	1.82×10^{-8}
lead(II) sulfide	PbS	9.04×10^{-29}
magnesium carbonate	$MgCO_3$	6.82×10^{-6}
magnesium hydroxide	$Mg(OH)_2$	2.06×10^{-13}
silver chloride	$AgCl$	1.77×10^{-10}
silver chromate	Ag_2CrO_4	1.12×10^{-12}
silver iodide	AgI	8.51×10^{-17}

EXAMPLE **15.8** | **Writing Expressions for K_{sp}**

Write expressions for K_{sp} for each ionic compound.

(a) $BaSO_4$ (b) $Mn(OH)_2$ (c) Ag_2CrO_4

SOLUTION

To write the expression for K_{sp}, first write the chemical reaction showing the solid compound in equilibrium with its dissolved aqueous ions. Then write the equilibrium expression based on this equation.

(a) $BaSO_4(s) \rightleftharpoons Ba^{2+}(aq) + SO_4^{2-}(aq)$

$K_{sp} = [Ba^{2+}][SO_4^{2-}]$

(b) $Mn(OH)_2(s) \rightleftharpoons Mn^{2+}(aq) + 2\,OH^-(aq)$

$K_{sp} = [Mn^{2+}][OH^-]^2$

(c) $Ag_2CrO_4(s) \rightleftharpoons 2\,Ag^+(aq) + CrO_4^{2-}(aq)$

$K_{sp} = [Ag^+]^2[CrO_4^{2-}]$

▶ **SKILLBUILDER 15.8 | Writing Expressions for K_{sp}**

Write expressions for K_{sp} for each ionic compound.

(a) AgI (b) $Ca(OH)_2$

▶ **FOR MORE PRACTICE** Example 15.14; Problems 77, 78.

Using K_{sp} to Determine Molar Solubility

Recall from Section 13.3 that the solubility of a compound is the amount of the compound that dissolves in a certain amount of liquid. The **molar solubility** is the solubility in units of moles per liter. We can calculate the molar solubility of a compound directly from K_{sp}.

For example, consider silver chloride:

$$AgCl(s) \rightleftharpoons Ag^+(aq) + Cl^-(aq) \quad K_{sp} = 1.77 \times 10^{-10}$$

First, notice that K_{sp} is *not* the molar solubility; rather, it is the solubility-product constant. Second, notice that the concentration of either Ag^+ or Cl^- at equilibrium will be equal to the amount of AgCl that dissolved. We know this from the relationship of the stoichiometric coefficients in the balanced equation:

$$1\text{ mol AgCl:}1\text{ mol }Ag^+\text{:}1\text{ mol }Cl^-$$

Consequently, to find the solubility, we need to find $[Ag^+]$ or $[Cl^-]$ at equilibrium. We can do this by writing the expression for the solubility-product constant:

$$K_{sp} = [Ag^+][Cl^-]$$

Because both Ag^+ and Cl^- come from AgCl, their concentrations must be equal. Since the solubility of AgCl is equal to the equilibrium concentration of either dissolved ion, we write:

$$\text{Solubility} = S = [Ag^+] = [Cl^-]$$

Substituting this into the expression for the solubility constant, we get:

$$K_{sp} = [Ag^+][Cl^-]$$
$$= S \times S$$
$$= S^2$$

Therefore,

$$S = \sqrt{K_{sp}}$$
$$= \sqrt{1.77 \times 10^{-10}}$$
$$= 1.33 \times 10^{-5}\,M$$

So the molar solubility of AgCl is $1.33 \times 10^{-5}\,mol/L$.

In this text, we limit the calculation of molar solubility to ionic compounds whose chemical formulas have one cation and one anion.

CONCEPTUAL ✔ CHECKPOINT 15.7

Which substance is more soluble?

$CaSO_4$	$K_{sp} = 7.10 \times 10^{-5}$
AgCl	$K_{sp} = 1.77 \times 10^{-10}$

PEARSON eText 2.0 Interactive Worked Example Video 15.9

EXAMPLE **15.9** **Calculating Molar Solubility from K_{sp}**

Calculate the molar solubility of $BaSO_4$.

SOLUTION

Begin by writing the reaction by which solid $BaSO_4$ dissolves into its constituent aqueous ions.	$BaSO_4(s) \rightleftharpoons Ba^{2+}(aq) + SO_4^{2-}(aq)$
Next, write the expression for K_{sp}.	$K_{sp} = [Ba^{2+}][SO_4^{2-}]$
Define the molar solubility (S) as $[Ba^{2+}]$ or $[SO_4^{2-}]$ at equilibrium.	$S = [Ba^{2+}] = [SO_4^{2-}]$
Substitute S into the equilibrium expression and solve for it.	$K_{sp} = [Ba^{2+}][SO_4^{2-}]$ $= S \times S$ $= S^2$ Therefore $S = \sqrt{K_{sp}}$
Finally, look up the value of K_{sp} in Table 15.2 and calculate S. The molar solubility of $BaSO_4$ is $1.03 \times 10^{-5}\,mol/L$.	$S = \sqrt{K_{sp}}$ $= \sqrt{1.07 \times 10^{-10}}$ $= 1.03 \times 10^{-5}\,M$

▶ **SKILLBUILDER 15.9 | Calculating Molar Solubility from K_{sp}**

Calculate the molar solubility of $CaSO_4$.

▶ **FOR MORE PRACTICE** Example 15.15; Problems 85, 86, 87, 88.

15.12 The Path of a Reaction and the Effect of a Catalyst

▶ Describe the relationship between activation energy and reaction rates and the role catalysts play in reactions.

In this chapter, we have learned that the equilibrium constant describes the ultimate fate of a chemical reaction. Large equilibrium constants indicate that the reaction favors the products. Small equilibrium constants indicate that the reaction favors the reactants. But the equilibrium constant alone does not tell the whole story.

Warning: Hydrogen gas is explosive and should never be handled without proper training.

For example, consider the reaction between hydrogen gas and oxygen gas to form water:

$$2\,H_2(g) + O_2(g) \rightleftharpoons 2\,H_2O(g) \quad K_{eq} = 3.2 \times 10^{81} \text{ at } 25\,°C$$

The equilibrium constant for this reaction is huge, meaning that the forward reaction is heavily favored. Yet we can mix hydrogen and oxygen in a balloon at room temperature, and no reaction occurs. Hydrogen and oxygen peacefully coexist together inside of the balloon and form virtually no water. Why?

To answer this question, we revisit a topic from the beginning of this chapter—*the reaction rate*. At 25 °C, the reaction rate between hydrogen gas and oxygen gas is virtually zero. Even though the equilibrium constant is large, the reaction rate is small and no reaction occurs. The reaction rate between hydrogen and oxygen is slow because the reaction has a large *activation energy*. The **activation energy** (or activation barrier) for a reaction is the energy barrier that must be overcome in order for the reactants to be converted into products.

Activation energies exist for most chemical reactions because the original bonds must begin to break before new bonds begin to form, and this requires energy. For example, for H_2 and O_2 to react to form H_2O, the H—H and O=O bonds must begin to break before the new bonds can form. The initial weakening of H_2 and O_2 bonds takes energy—this is the activation energy of the reaction.

> The equilibrium constant describes *how far* a chemical reaction will go. The reaction rate describes *how fast* it will get there.

> The activation energy is sometimes called the *activation barrier*.

How Activation Energies Affect Reaction Rates

We can illustrate how activation energies affect reaction rates with a graph showing the energy progress of a reaction (▼ FIGURE 15.14). In the figure the products have less energy than the reactants, so we know the reaction is exothermic (the reaction releases energy when it occurs). However, before the reaction can take place, some energy must first be *added*—the energy of the reactants must be raised by an amount that we call the activation energy. The activation energy is a kind of "energy hump" that normally exists between the reactants and products.

We can explain this concept with a simple analogy—getting a chemical reaction to occur is much like trying to push a bunch of boulders over a hill (▶ FIGURE 15.15a). We can think of each collision that occurs between reactant molecules as an attempt to roll a boulder over the hill. A successful collision between two molecules (one that leads to product) is like a successful attempt to roll a boulder over the hill and down the other side.

The higher the hill is, the harder it is to get the boulders over the hill, and the fewer the number of boulders that make it over the hill in a given period of time. Similarly, for chemical reactions, the higher the activation energy, the fewer the number of reactant molecules that make it over the barrier, and the slower the reaction rate. In general:

At a given temperature, the higher the activation energy for a chemical reaction, the slower the reaction rate.

Are there any ways to speed up a slow reaction (one with a high activation barrier)? In Section 15.2, we discussed two ways to increase reaction rates. The first

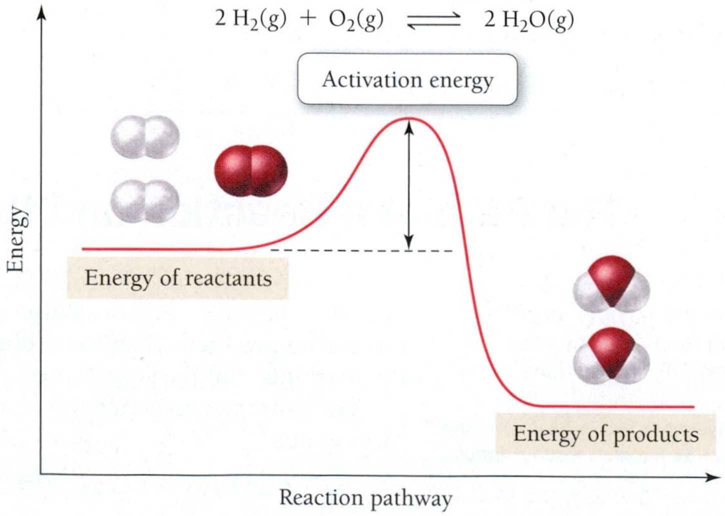

▶ FIGURE 15.14 **Activation energy** This plot represents the energy of the reactants and products along the reaction pathway (as the reaction occurs). Notice that the energy of the products is lower than the energy of the reactants, so this is an exothermic reaction. However, notice that the reactants must get over an energy hump—called the *activation energy*—to proceed from reactants to products.

$$2 H_2(g) + O_2(g) \rightleftharpoons 2 H_2O(g)$$

Activation energy

Energy

Energy of reactants

Energy of products

Reaction pathway

(a) Without catalyst

(b) With catalyst

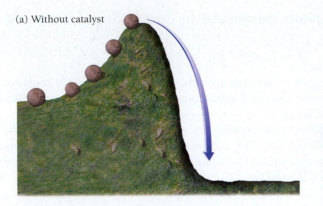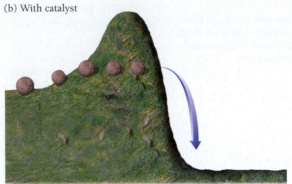

▲ **FIGURE 15.15 Hill analogy for activation energy** There are several ways to get these boulders over the hill as fast as possible. **(a)** One way is to push more boulders at the hill or simply push them harder—this is analogous to an increase in concentration or temperature (respectively) for a chemical reaction. **(b)** Another way is to find a path that goes *around* the hill—this is analogous to the role of a catalyst for a chemical reaction.

way is to increase the concentrations of the reactants, which results in more collisions per unit time. This is analogous to pushing more boulders toward the hill in a given period of time. The second way is to increase the temperature. This results in more collisions per unit time and in higher-energy collisions. Higher-energy collisions are analogous to pushing the boulders harder (with more force), which results in more boulders making it over the hill per unit time—a faster reaction rate. There is, however, a third way to speed up a slow chemical reaction: by using a *catalyst*.

PEARSON
eText
2.0

CONCEPTUAL ✔ **CHECKPOINT 15.8**

Reaction A has an activation barrier of 35 kJ/mol, and reaction B has an activation barrier of 55 kJ/mol. Which of the two reactions is likely to have the faster rate at room temperature?

Catalysts Lower the Activation Energy

> A catalyst does not change the *position* of equilibrium, only *how fast* equilibrium is reached.

A **catalyst** is a substance that increases the rate of a chemical reaction but is not consumed by the reaction. A catalyst works by lowering the activation energy for the reaction, making it easier for reactants to get over the energy barrier (▼ **FIGURE 15.16**). In our boulder analogy, a catalyst creates another path for the boulders to travel—a path with a smaller hill (see ▲ **FIGURE 15.15b**).

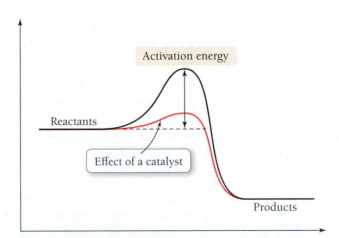

Activation energy

Reactants

Effect of a catalyst

Products

A catalyst provides an alternate pathway with a lower activation energy barrier for the reaction.

▶ **FIGURE 15.16 Function of a catalyst**

Upper-atmospheric ozone forms a shield against harmful ultraviolet light that would otherwise enter Earth's atmosphere.

For example, consider the noncatalytic destruction of ozone in the upper atmosphere:

$$O_3 + O \longrightarrow 2\,O_2$$

We have a protective ozone layer because this reaction has a fairly high activation barrier and therefore proceeds at a fairly slow rate. The ozone layer does not rapidly decompose into O_2.

However, the addition of Cl (from synthetic chlorofluorocarbons) to the upper atmosphere has resulted in another pathway by which O_3 can be destroyed. The first step in this pathway—called the *catalytic* destruction of ozone—is the reaction of Cl with O_3 to form ClO and O_2.

$$Cl + O_3 \longrightarrow ClO + O_2$$

This is followed by a second step in which ClO reacts with O, regenerating Cl.

$$ClO + O \longrightarrow Cl + O_2$$

Notice that, if we add the two reactions, the overall reaction is identical to the noncatalytic reaction.

$$\cancel{Cl} + O_3 \longrightarrow \cancel{ClO} + O_2$$
$$\cancel{ClO} + O \longrightarrow \cancel{Cl} + O_2$$
$$\overline{\quad O_3 + O \longrightarrow 2\,O_2 \quad}$$

However, the activation energies for the two reactions in this pathway are much smaller than for the first, uncatalyzed pathway, and therefore the reaction occurs at a much faster rate. The Cl is not consumed in the overall reaction; this is characteristic of a catalyst.

In the case of the catalytic destruction of ozone, the catalyst speeds up a reaction that we do *not* want to happen. Most of the time, however, we use catalysts to speed up reactions that we *do* want to happen. For example, most cars have a catalytic converter in their exhaust system. The catalytic converter contains a catalyst that converts exhaust pollutants (such as carbon monoxide) into less harmful substances (such as carbon dioxide). These reactions occur only with the help of a catalyst because they are too slow to occur otherwise.

The role of catalysis in chemistry cannot be overstated. Without catalysts, chemistry would be a different field. For many reactions, increasing the reaction rate in another way—such as raising the temperature—is simply not feasible. Many reactants are thermally sensitive—increasing the temperature can destroy them. The only way to carry out many reactions is to use catalysts.

A catalyst cannot change the value of K_{eq} for a reaction—it affects only the *rate* of the reaction.

Enzymes: Biological Catalysts

Perhaps the best example of chemical catalysis is found in living organisms. Most of the thousands of reactions that must occur for a living organism to survive would be too slow at normal temperatures. So living organisms use **enzymes**—biological catalysts that increase the rates of biochemical reactions.

For example, when we eat sucrose (table sugar), our bodies must break it into two smaller molecules: glucose and fructose. The equilibrium constant for this reaction is large, favoring the products. However, at room temperature, or even at body temperature, the sucrose does not break down into glucose and fructose because the activation energy is high, resulting in a slow reaction rate. In other words, table sugar remains table sugar at room temperature, even though the

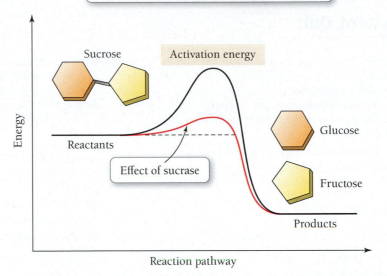

The enzyme sucrase creates a pathway with a lower activation energy for the conversion of sucrose to glucose and fructose.

▲ **FIGURE 15.17 An enzyme catalyst**

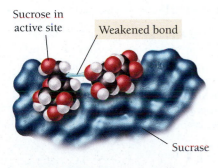

Sucrose in active site — Weakened bond — Sucrase

▲ **FIGURE 15.18 How an enzyme works**   Sucrase has a pocket called the active site where sucrose binds. When a molecule of sucrose enters the active site, the bond between glucose and fructose is weakened, lowering the activation energy for the reaction.

equilibrium constant for its reaction to glucose and fructose is relatively large (▲ FIGURE 15.17).

In the body, however, an enzyme called *sucrase* catalyzes the conversion of sucrose to glucose and fructose. Sucrase has a pocket—called the active site—into which sucrose snugly fits (like a key into a lock). When sucrose is in the active site, the bond between the glucose and fructose units weakens, lowering the activation energy for the reaction and increasing the reaction rate (◄ FIGURE 15.18). The reaction can then proceed toward equilibrium—which favors the products—at a much lower temperature.

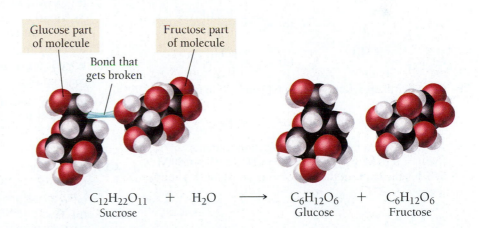

Glucose part of molecule   Fructose part of molecule

Bond that gets broken

$$C_{12}H_{22}O_{11} \;+\; H_2O \;\longrightarrow\; C_6H_{12}O_6 \;+\; C_6H_{12}O_6$$
Sucrose   Glucose   Fructose

Not only do enzymes allow otherwise slow reactions to occur at a reasonable rate, they also allow living organisms to have tremendous control over which reactions occur and when. Enzymes are extremely specific—each enzyme catalyzes only a single reaction. To turn a particular reaction on, a living organism simply needs to produce or activate the correct enzyme to catalyze that reaction.

Chapter 15 in Review

MasteringChemistry™ provides end-of-chapter exercises, feedback-enriched tutorial problems, animations, and interactive activities to encourage problem solving practice and deeper understanding of key concepts and topics.

Self-Assessment Quiz

PEARSON eText 2.0

Q1. Which change is likely to increase the *rate* of reaction in a reaction mixture?
(a) decreasing the temperature
(b) increasing the concentration of the reactants
(c) increasing the volume of the reaction vessel
(d) none of the above

Q2. The equilibrium constants at a fixed temperature for several reactions are listed here. Which equilibrium constant indicates a reaction that is favored in the forward direction?
(a) $K_{eq} = 1.0 \times 10^5$
(b) $K_{eq} = 1.0 \times 10^{-5}$
(c) $K_{eq} = 1.0$
(d) none of the above

Q3. The concentrations of A, B, and C for the reaction $A(g) \rightleftharpoons B(g) + C(g)$ are measured at several different temperatures. At which temperature is the equilibrium constant the largest?
(a) T_1: [A] = 0.10 M, [B] = 0.30 M, [C] = 0.30 M
(b) T_2: [A] = 0.20 M, [B] = 0.20 M, [C] = 0.20 M
(c) T_3: [A] = 0.30 M, [B] = 0.10 M, [C] = 0.10 M
(d) none of the above (the equilibrium constant is the same at all three temperatures)

Q4. What is the correct expression for the equilibrium constant (K_{eq}) for the reaction between carbon and hydrogen gas to form methane?

$$C(s) + 2\,H_2(g) \rightleftharpoons CH_4(g)$$

(a) $K_{eq} = \dfrac{[CH_4]}{[H_2]}$
(b) $K_{eq} = \dfrac{[CH_4]}{[C][H_2]}$
(c) $K_{eq} = \dfrac{[CH_4]}{[C][H_2]^2}$
(d) $K_{eq} = \dfrac{[CH_4]}{[H_2]^2}$

Q5. Consider the reaction between NO and Cl_2 to form NOCl.

$$2\,NO(g) + Cl_2(g) \rightleftharpoons 2\,NOCl(g)$$

An equilibrium mixture of this reaction contains [NO] = 0.20 M, [Cl_2] = 0.35 M, and [NOCl] = 0.30 M. What is the value of the equilibrium constant (K_c) at this temperature?
(a) 11
(b) 4.3
(c) 6.4
(d) 0.22

Q6. The equilibrium constant for this reaction is $K_{eq} = 1.0 \times 10^3$. The reaction mixture at equilibrium contains [A] = 1.0×10^{-3} M. What is the concentration of B in the mixture?

$$A(g) \rightleftharpoons B(g)$$

(a) 1.0×10^{-3} M
(b) 1.0 M
(c) 2.0 M
(d) 1.0×10^3 M

Q7. What is the effect of adding chlorine gas (at constant volume and temperature) to an equilibrium mixture of this reaction?

$$CO(g) + Cl_2(g) \rightleftharpoons COCl_2(g)$$

(a) The reaction shifts toward the products.
(b) The reaction shifts toward the reactants.
(c) The reaction does not shift in either direction.
(d) The reaction slows down.

Q8. The decomposition of NH_4HS is endothermic.

$$NH_4HS(s) \rightleftharpoons NH_3(g) + H_2S(g)$$

Which change to an equilibrium mixture of this reaction results in the formation of more H_2S?
(a) a decrease in the volume of the reaction vessel (at constant temperature)
(b) an increase in the amount of NH_3 in the reaction vessel
(c) an increase in temperature
(d) all of the above

Q9. What is the correct expression for the solubility-product constant (K_{sp}) of $Al(OH)_3$?
(a) $K_{sp} = [Al^{3+}][OH^-]$
(b) $K_{sp} = [Al^{3+}]\,3[OH^-]$
(c) $K_{sp} = [Al^{3+}]^3\,[OH^-]$
(d) $K_{sp} = [Al^{3+}]\,[OH^-]^3$

Q10. What is the molar solubility of CuBr? For CuBr, $K_{sp} = 6.27 \times 10^{-9}$.
(a) 6.27×10^{-9} M
(b) 3.93×10^{-17} M
(c) 7.92×10^{-5} M
(d) 1.25×10^{-8} M

Answers: 1:b, 2:a, 3:a, 4:d, 5:c, 6:b, 7:a, 8:c, 9:d, 10:c

Chemical Principles	*Relevance*

The Concept of Equilibrium

Equilibrium involves the ideas of sameness and constancy. When a system is in equilibrium, some property of the system remains the same and does not change.

The equilibrium concept explains many phenomena such as the human body's oxygen delivery system. Life itself can be defined as controlled disequilibrium with the environment.

Rates of Chemical Reactions

The rate of a chemical reaction is the amount of reactant(s) that goes to product(s) in a given period of time. In general, reaction rates increase with increasing reactant concentration and increasing temperature. Reaction rates depend on the concentration of reactants, and the concentration of reactants decreases as a reaction proceeds, so reaction rates usually slow down as a reaction proceeds.

The rate of a chemical reaction determines how fast a reaction will reach its equilibrium. Chemists want to understand the factors that influence reaction rates so that they can control them.

Dynamic Chemical Equilibrium

Dynamic chemical equilibrium occurs when the rate of the forward reaction equals the rate of the reverse reaction.

When dynamic chemical equilibrium is reached, the concentrations of the reactants and products become constant.

The Equilibrium Constant

For the generic reaction

$$aA + bB \rightleftharpoons cC + dD$$

we define the equilibrium constant (K_{eq}) as:

$$K_{eq} = \frac{[C]^c[D]^d}{[A]^a[B]^b}$$

Only the concentrations of gaseous or aqueous reactants and products are included in the equilibrium constant—the concentrations of solid or liquid reactants or products are omitted.

The equilibrium constant is a measure of how far a reaction will proceed. A large K_{eq} indicates that the forward reaction is favored (lots of products at equilibrium). A small K_{eq} indicates that the reverse reaction is favored (lots of reactants at equilibrium). An intermediate K_{eq} indicates that there will be significant amounts of both reactants and products at equilibrium.

Le Châtelier's Principle

Le Châtelier's principle states that when a chemical system at equilibrium is disturbed, the system shifts in a direction that minimizes the disturbance.

Le Châtelier's principle helps us predict what happens to a chemical system at equilibrium when conditions change. This allows chemists to modify the conditions of a chemical reaction to obtain a desired result.

Effect of a Concentration Change on Equilibrium

- Increasing the concentration of one or more of the *reactants* causes the reaction to shift to the *right*.
- Increasing the concentration of one or more of the *products* causes the reaction to shift to the *left*.

In many cases, a chemist may want to drive a reaction in one direction or another. For example, suppose a chemist is carrying out a reaction to make a desired compound. The reaction can be pushed to the right by continuously removing the product from the reaction mixture as it forms, maximizing the amount of product that can be made.

Effect of a Volume Change on Equilibrium

- Decreasing the volume causes the reaction to shift in the direction that has *fewer* moles of gas particles.
- Increasing the volume causes the reaction to shift in the direction that has *more* moles of gas particles.

Like the effect of concentration, the effect of pressure on equilibrium allows a chemist to choose the best conditions under which to carry out a chemical reaction. Some reactions are favored in the forward direction by high pressure (those with fewer moles of gas particles in the products), and others (those with fewer moles of gas particles in the reactants) are favored in the forward direction by low pressure.

Effect of a Temperature Change on Equilibrium

Exothermic chemical reaction (heat is a product):

- *Increasing* the temperature causes the reaction to shift *left*.
- *Decreasing* the temperature causes the reaction to shift *right*.

Endothermic chemical reaction (heat is a reactant):

- *Increasing* the temperature causes the reaction to shift *right*.
- *Decreasing* the temperature causes the reaction to shift *left*.

Again, the effect of temperature on a reaction allows chemists to choose conditions that favor desired reactions. Higher temperatures favor endothermic reactions, while lower temperatures favor exothermic reactions. Most reactions occur faster at higher temperature, so the effect of temperature on the rate, not just on the equilibrium constant, must be considered.

The Solubility-Product Constant, K_{sp}

The solubility-product constant of an ionic compound is the equilibrium constant for the chemical equation that describes the dissolving of the compound.

The solubility-product constant reflects the solubility of a compound. The greater the solubility-product constant, the greater the solubility of the compound.

Reaction Paths and Catalysts

Most chemical reactions must overcome an energy hump, called the *activation energy*, as they proceed from reactants to products. Increasing the temperature of a reaction mixture increases the fraction of reactant molecules that make it over the energy hump, therefore increasing the rate. A catalyst—a substance that increases the rate of the reaction but is not consumed by it—lowers the activation energy so that it is easier to get over the energy hump without increasing the temperature.

Catalysts are used in many chemical reactions to increase the rates. Without catalysts, many reactions occur too slowly to be of any value. Biological catalysts called *enzymes* control the thousands of reactions that occur in living organisms.

Chemical Skills

Examples

LO: Write equilibrium expressions for chemical reactions (Section 15.4).

To write the equilibrium expression for a reaction, write the concentrations of the products raised to their stoichiometric coefficients divided by the concentrations of the reactants raised to their stoichiometric coefficients. Remember that reactants or products that are liquids or solids are omitted from the equilibrium expression.

EXAMPLE **15.10**   **Writing Equilibrium Expressions for Chemical Reactions**

Write an equilibrium expression for the chemical equation:

$$2\,NO(g) + Br_2(g) \rightleftharpoons 2\,NOBr(g)$$

SOLUTION

$$K_{eq} = \frac{[NOBr]^2}{[NO]^2[Br_2]}$$

LO: Calculate equilibrium constants (Section 15.6).

EXAMPLE **15.11**   **Calculating Equilibrium Constants**

An equilibrium mixture of the following reaction has

$$[I] = 0.075\ M \text{ and } [I_2] = 0.88\ M.$$

What is the value of the equilibrium constant?

$$I_2(g) \rightleftharpoons 2\,I(g)$$

Begin by setting up the problem in the usual way.

GIVEN: $[I] = 0.075\ M$

$[I_2] = 0.88\ M$

FIND: K_{eq}

SOLUTION

Then write the expression for K_{eq} from the balanced equation. To calculate the value of K_{eq}, substitute the correct equilibrium concentrations into the expression for K_{eq}. Always write the concentrations within K_{eq} in moles per liter, M. Units are normally dropped in expressing the equilibrium constant so that K_{eq} is unitless.

$$K_{eq} = \frac{[I]^2}{[I_2]}$$

$$= \frac{[0.075]^2}{[0.88]}$$

$$= 0.0064$$

LO: **Use the equilibrium constant to find the concentration of a reactant or product at equilibrium (Section 15.6).**

EXAMPLE **15.12**

Consider the reaction:

$$N_2(g) + 3 H_2(g) \rightleftharpoons 2 NH_3(g)$$

$$K_{eq} = 152 \text{ at } 225 \degree C$$

In an equilibrium mixture, $[N_2] = 0.110$ M and $[H_2] = 0.0935$ M. What is the equilibrium concentration of NH_3?

SORT
You are given the initial concentrations of nitrogen and hydrogen as well as the equilibrium constant for their reaction to form ammonia. You are asked to find the equilibrium concentration of ammonia.

GIVEN: $[N_2] = 0.110$ M
$[H_2] = 0.0935$ M
$K_{eq} = 152$

FIND: $[NH_3]$

STRATEGIZE
Write a solution map that shows how you can use the given concentrations and the equilibrium constant to find the unknown concentration.

SOLUTION MAP

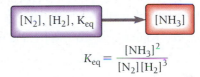

$$K_{eq} = \frac{[NH_3]^2}{[N_2][H_2]^3}$$

SOLVE
Solve the equilibrium expression for the quantity you are trying to find and substitute in the appropriate values to calculate the unknown quantity.

SOLUTION

$$K_{eq} = \frac{[NH_3]^2}{[N_2][H_2]^3}$$

$$[NH_3]^2 = K_{eq}[N_2][H_2]^3$$

$$\sqrt{[NH_3]^2} = \sqrt{K_{eq}[N_2][H_2]^3}$$

$$[NH_3] = \sqrt{(152)(0.110)(0.0935)^3}$$

$$= 0.117 \text{ M}$$

CHECK
Check your answer by substituting it back into the expression for K_{eq}.

$$K_{eq} = \frac{[NH_3]^2}{[N_2][H_2]^3} = \frac{(0.117)^2}{(0.110)(0.0935)^3}$$

$$= 141$$

The calculated value of K_{eq} is about equal to the given value of K_{eq} (which was 152), indicating that your answer is correct. The slight difference is due to rounding, which is common in problems like these.

LO: **Apply Le Châtelier's principle (Sections 15.8, 15.9, 15.10).**

EXAMPLE **15.13** **Using Le Châtelier's Principle**

Consider the *endothermic* chemical reaction:

$$C(s) + H_2O(g) \rightleftharpoons CO(g) + H_2(g)$$

To apply Le Châtelier's principle, review the effects of concentration, volume, and temperature in the end-of-chapter Chemical Principles section (pages 559–560). For each disturbance, predict how the reaction changes to counter the disturbance.

Predict the effect of:
(a) increasing [CO]
(b) increasing [H₂O]
(c) increasing the reaction volume
(d) increasing the temperature

SOLUTION
(a) shift left
(b) shift right
(c) shift right (more moles of gas on right)
(d) shift right (heat is a reactant)

LO: Write an expression for the solubility-product constant (Section 15.11).

To write the expression for K_{sp}, first write the chemical reaction showing the solid compound in equilibrium with its dissolved aqueous ions. Then write the equilibrium expression as the product of the concentrations of the aqueous ions raised to their stoichiometric coefficients.

EXAMPLE 15.14 Writing an Expression for the Solubility-Product Constant

Write an expression for K_{sp} for $PbCl_2$.

SOLUTION

$$PbCl_2(s) \rightleftharpoons Pb^{2+}(aq) + 2\,Cl^-(aq)$$

$$K_{sp} = [Pb^{2+}][Cl^-]^2$$

LO: Use K_{sp} to determine molar solubility (Section 15.11).

Begin by writing the reaction by which the solid dissolves into its constituent aqueous ions. Next, write the expression for K_{sp}.

For the problems assigned in this book, the concentration of individual aqueous ions is equal to the solubility, S.

Substitute S into the equilibrium expression and solve the expression for S.

Look up the value of K_{sp} in Table 15.2 and calculate S.

EXAMPLE 15.15 Using K_{sp} to Determine Molar Solubility

Calculate the molar solubility of AgI.

SOLUTION

$$AgI(s) \rightleftharpoons Ag^+(aq) + I^-(aq)$$

$$K_{sp} = [Ag^+][I^-]$$
$$S = [Ag^+] = [I^-]$$
$$K_{sp} = [Ag^+][I^-]$$
$$= S \times S$$
$$= S^2$$
$$S = \sqrt{K_{sp}}$$
$$S = \sqrt{8.51 \times 10^{-17}}$$
$$= 9.22 \times 10^{-9}\,M$$

Key Terms

activation energy [15.12]
catalyst [15.12]
collision theory [15.2]
dynamic equilibrium [15.3]

enzyme [15.12]
equilibrium constant (K_{eq}) [15.4]
Le Châtelier's principle [15.7]

molar solubility [15.11]
rate of a chemical reaction (reaction rate) [15.2]

reversible reaction [15.3]
solubility-product constant (K_{sp}) [15.11]

Exercises

Questions

1. What are the two *general* concepts involved in equilibrium?
2. What is the rate of a chemical reaction? What is the difference between a chemical reaction with a fast rate and one with a slow rate?
3. Why do chemists seek to control reaction rates?
4. How do most chemical reactions occur?
5. What factors influence reaction rates? How?
6. What normally happens to the rate of the forward reaction as a reaction proceeds?
7. What is dynamic chemical equilibrium?
8. Explain how dynamic chemical equilibrium involves the concepts of sameness and constancy.
9. Explain why the concentrations of reactants and products are not necessarily the same at equilibrium.
10. Devise your own analogy—like the Narnia and Middle Earth analogy in the chapter—to explain chemical equilibrium.
11. What is the equilibrium constant? Why is it significant?
12. Write the expression for the equilibrium constant for the following generic chemical equation.

$$aA + bB \rightleftharpoons cC + dD$$

13. What does a small equilibrium constant tell you about a reaction? A large equilibrium constant?
14. Why are solids and liquids omitted from the equilibrium expression?

15. Will the concentrations of reactants and products always be the same in every equilibrium mixture of a particular reaction at a given temperature? Explain.

16. What is Le Châtelier's principle?

17. Apply Le Châtelier's principle to your analogy from Question 10.

18. What is the effect of *increasing* the concentration of a reactant in a reaction mixture at equilibrium?

19. What is the effect of *decreasing* the concentration of a reactant in a reaction mixture at equilibrium?

20. What is the effect of *increasing* the concentration of a product in a reaction mixture at equilibrium?

21. What is the effect of *decreasing* the concentration of a product in a reaction mixture at equilibrium?

22. What is the effect of *increasing* the pressure of a reaction mixture at equilibrium if the reactant side has fewer moles of gas particles than the product side?

23. What is the effect of *increasing* the pressure of a reaction mixture at equilibrium if the product side has fewer moles of gas particles than the reactant side?

24. What is the effect of *decreasing* the pressure of a reaction mixture at equilibrium if the reactant side has fewer moles of gas particles than the product side?

25. What is the effect of *decreasing* the pressure of a reaction mixture at equilibrium if the product side has fewer moles of gas particles than the reactant side?

26. What is the effect of *increasing* the temperature of an endothermic reaction mixture at equilibrium? Of *decreasing* the temperature?

27. What is the effect of *increasing* the temperature of an exothermic reaction mixture at equilibrium? Of *decreasing* the temperature?

28. What is the solubility-product constant? What does it signify?

29. Write an expression for the solubility-product constant of $AB_2(s)$. Assume that an ion of B has a charge of $1-$ (that is, B^-).

30. Write an expression for the solubility-product constant of $A_2B(s)$. Assume that an ion of B has a charge of $2-$ (that is, B^{2-}).

31. What are solubility and molar solubility?

32. What is activation energy for a chemical reaction?

33. Explain why two reactants with a large K_{eq} for a particular reaction might not react immediately when combined.

34. What is the effect of a catalyst on a reaction? Why are catalysts so important to chemistry?

35. Does a catalyst affect the value of the equilibrium constant?

36. What are enzymes?

Problems

THE RATE OF REACTION

37. Two gaseous reactants are allowed to react in a 1-L flask, and the reaction rate is measured. The experiment is repeated with the same amount of each reactant and at the same temperature in a 2-L flask (so the concentration of each reactant is less). What is likely to happen to the measured reaction rate in the second experiment compared to the first?

38. The rate of phosphorus pentachloride decomposition is measured at a PCl_5 pressure of 0.015 atm and then again at a PCl_5 pressure of 0.30 atm. The temperature is identical in both measurements. Which rate is likely to be faster?

39. The body temperature of cold-blooded animals varies with the ambient temperature. From the point of view of reaction rates, explain why cold-blooded animals are more sluggish at cold temperatures.

40. The rate of a particular reaction doubles when the temperature increases from 25 °C to 35 °C. Explain why this happens.

41. The initial rate of a chemical reaction was measured, and one of the reactants was found to be reacting at a rate of 0.0011 mol/L s. The reaction was allowed to proceed for 15 minutes, and the rate was measured again. What would you predict about the second measured rate relative to the first?

42. When vinegar is added to a solution of sodium bicarbonate, the mixture immediately begins to bubble furiously. As time passes, however, there is less and less bubbling. Explain why this happens.

THE EQUILIBRIUM CONSTANT

43. Write an equilibrium expression for each chemical equation.
  (a) $2\,NO_2(g) \rightleftharpoons N_2O_4(g)$
  (b) $2\,BrNO(g) \rightleftharpoons 2\,NO(g) + Br_2(g)$
  (c) $H_2O(g) + CO(g) \rightleftharpoons H_2(g) + CO_2(g)$
  (d) $CH_4(g) + 2\,H_2S(g) \rightleftharpoons CS_2(g) + 4\,H_2(g)$

44. Write an equilibrium expression for each chemical equation.
  (a) $2\,CO(g) + O_2(g) \rightleftharpoons 2\,CO_2(g)$
  (b) $N_2(g) + O_2(g) \rightleftharpoons 2\,NO(g)$
  (c) $SbCl_5(g) \rightleftharpoons SbCl_3(g) + Cl_2(g)$
  (d) $CO(g) + Cl_2(g) \rightleftharpoons COCl_2(g)$

45. Write an equilibrium expression for each chemical equation involving one or more solid or liquid reactants or products.
(a) $PCl_5(g) \rightleftharpoons PCl_3(l) + Cl_2(g)$
(b) $2\,KClO_3(s) \rightleftharpoons 2\,KCl(s) + 3\,O_2(g)$
(c) $HF(aq) + H_2O(l) \rightleftharpoons H_3O^+(aq) + F^-(aq)$
(d) $NH_3(aq) + H_2O(l) \rightleftharpoons NH_4^+(aq) + OH^-(aq)$

46. Write an equilibrium expression for each chemical equation involving one or more solid or liquid reactants or products.
(a) $HCHO_2(aq) + H_2O(l) \rightleftharpoons H_3O^+(aq) + CHO_2^-(aq)$
(b) $CO_3^{2-}(aq) + H_2O(l) \rightleftharpoons HCO_3^-(aq) + OH^-(aq)$
(c) $2\,C(s) + O_2(g) \rightleftharpoons 2\,CO(g)$
(d) $C(s) + CO_2(g) \rightleftharpoons 2\,CO(g)$

47. Consider the reaction.

$$2\,H_2S(g) \rightleftharpoons 2\,H_2(g) + S_2(g)$$

Find the mistakes in the equilibrium expression and fix them.

$$K_{eq} = \frac{[H_2][S_2]}{[H_2S]}$$

48. Consider the reaction.

$$CO(g) + Cl_2(g) \rightleftharpoons COCl_2(g)$$

Find the mistake in the equilibrium expression and fix it.

$$K_{eq} = \frac{[CO][Cl_2]}{[COCl_2]}$$

49. For each equilibrium constant, indicate if you would expect an equilibrium reaction mixture to be dominated by reactants or by products, or to contain significant amounts of both.
(a) $K_{eq} = 5.2 \times 10^{17}$
(b) $K_{eq} = 1.24$
(c) $K_{eq} = 3.22 \times 10^{-21}$
(d) $K_{eq} = 0.47$

50. For each equilibrium constant, indicate if you would expect an equilibrium reaction mixture to be dominated by reactants or by products, or to contain significant amounts of both.
(a) $K_{eq} = 0.75$
(b) $K_{eq} = 8.5 \times 10^{-7}$
(c) $K_{eq} = 1.4 \times 10^{19}$
(d) $K_{eq} = 4.7 \times 10^{-9}$

CALCULATING AND USING EQUILIBRIUM CONSTANTS

51. Consider the reaction.

$$COCl_2(g) \rightleftharpoons CO(g) + Cl_2(g)$$

An equilibrium mixture of this reaction at a certain temperature has $[COCl_2] = 0.225\,M$, $[CO] = 0.105\,M$, and $[Cl_2] = 0.0844\,M$. What is the value of the equilibrium constant at this temperature?

52. Consider the reaction.

$$CO(g) + 2\,H_2(g) \rightleftharpoons CH_3OH(g)$$

An equilibrium mixture of this reaction at a certain temperature has $[CO] = 0.105\,M$, $[H_2] = 0.114\,M$, and $[CH_3OH] = 0.185\,M$. What is the value of the equilibrium constant at this temperature?

53. Consider the reaction.

$$2\,H_2S(g) \rightleftharpoons 2\,H_2(g) + S_2(g)$$

An equilibrium mixture of this reaction at a certain temperature has $[H_2S] = 0.562\,M$, $[H_2] = 2.74 \times 10^{-2}\,M$, and $[S_2] = 7.54 \times 10^{-3}\,M$. What is the value of the equilibrium constant at this temperature?

54. Consider the reaction.

$$CO(g) + H_2O(g) \rightleftharpoons CO_2(g) + H_2(g)$$

An equilibrium mixture of this reaction at a certain temperature has $[CO] = 0.0233\,M$, $[H_2O] = 0.0115\,M$, $[CO_2] = 0.175\,M$, and $[H_2] = 0.0274\,M$. What is the value of the equilibrium constant at this temperature?

55. Consider the reaction.

$$NH_4HS(s) \rightleftharpoons NH_3(g) + H_2S(g)$$

An equilibrium mixture of this reaction at a certain temperature has $[NH_3] = 0.278\,M$ and $[H_2S] = 0.355\,M$. What is the value of the equilibrium constant at this temperature?

56. Consider the reaction.

$$CaCO_3(s) \rightleftharpoons CaO(s) + CO_2(g)$$

An equilibrium mixture of this reaction at a certain temperature has $[CO_2] = 0.548\,M$. What is the value of the equilibrium constant at this temperature?

57. An equilibrium mixture of the following reaction is found to have $[SbCl_3] = 0.0255\,M$ and $[Cl_2] = 0.135\,M$ at 248 °C. What is the concentration of $SbCl_5$?

$$SbCl_5(g) \rightleftharpoons SbCl_3(g) + Cl_2(g)$$

$$K_{eq} = 4.9 \times 10^{-4} \text{ at } 248 \text{ °C}$$

58. An equilibrium mixture of the following reaction has $[I_2] = 0.0205\,M$ at 1200 °C. What is the concentration of I?

$$I_2(g) \rightleftharpoons 2\,I(g)$$

$$K_{eq} = 1.1 \times 10^{-2} \text{ at } 1200 \text{ °C}$$

59. An equilibrium mixture of the following reaction has $[I_2] = 0.0112$ M and $[Cl_2] = 0.0155$ M at 25 °C. What is the concentration of ICl?

$$I_2(g) + Cl_2(g) \rightleftharpoons 2\,ICl(g)$$

$$K_{eq} = 81.9 \text{ at } 25\,°C$$

60. An equilibrium mixture of the following reaction has $[SO_3] = 0.391$ M and $[O_2] = 0.125$ M at 600 °C. What is the concentration of SO_2?

$$2\,SO_2(g) + O_2(g) \rightleftharpoons 2\,SO_3(g)$$

$$K_{eq} = 4.34 \text{ at } 600\,°C$$

61. Consider the reaction:

$$N_2(g) + 3\,H_2(g) \rightleftharpoons 2\,NH_3(g)$$

Complete the table. Assume that all concentrations are equilibrium concentrations in moles per liter, M.

T (K)	$[N_2]$	$[H_2]$	$[NH_3]$	K_{eq}
500	0.115	0.105	0.439	____
575	0.110	____	0.128	9.6
775	0.120	0.140	____	0.0584

62. Consider the reaction:

$$H_2(g) + I_2(g) \rightleftharpoons 2\,HI(g)$$

Complete the table. Assume that all concentrations are equilibrium concentrations in moles per liter, M.

T (°C)	$[H_2]$	$[I_2]$	$[HI]$	K_{eq}
25	0.355	0.388	0.0922	____
340	____	0.0455	0.387	9.6
445	0.0485	0.0468	____	50.2

LE CHÂTELIER'S PRINCIPLE

63. Consider this reaction at equilibrium.

$$CO(g) + Cl_2(g) \rightleftharpoons COCl_2(g)$$

Predict the effect (shift right, shift left, or no effect) of these changes.
(a) adding Cl_2 to the reaction mixture
(b) adding $COCl_2$ to the reaction mixture
(c) adding CO to the reaction mixture

64. Consider this reaction at equilibrium.

$$2\,BrNO(g) \rightleftharpoons 2\,NO(g) + Br_2(g)$$

Predict the effect (shift right, shift left, or no effect) of these changes.
(a) adding BrNO to the reaction mixture
(b) adding NO to the reaction mixture
(c) adding Br_2 to the reaction mixture

65. Consider this reaction at equilibrium.

$$C(s) + H_2O(g) \rightleftharpoons CO(g) + H_2(g)$$

Predict the effect (shift right, shift left, or no effect) of these changes.
(a) adding C to the reaction mixture
(b) condensing H_2O and removing it from the reaction mixture
(c) adding CO to the reaction mixture
(d) removing H_2 from the reaction mixture

66. Consider this reaction at equilibrium.

$$2\,KClO_3(s) \rightleftharpoons 2\,KCl(s) + 3\,O_2(g)$$

Predict the effect (shift right, shift left, or no effect) of these changes.
(a) adding KCl to the reaction mixture
(b) adding $KClO_3$ to the reaction mixture
(c) adding O_2 to the reaction mixture
(d) removing O_2 from the reaction mixture

67. Consider the effect of a volume change on this reaction at equilibrium.

$$I_2(g) \rightleftharpoons 2\,I(g)$$

Predict the effect (shift right, shift left, or no effect) of these changes.
(a) increasing the reaction volume
(b) decreasing the reaction volume

68. Consider the effect of a volume change on this reaction at equilibrium.

$$2\,H_2S(g) \rightleftharpoons 2\,H_2(g) + S_2(g)$$

Predict the effect (shift right, shift left, or no effect) of these changes.
(a) increasing the reaction volume
(b) decreasing the reaction volume

69. Consider the effect of a volume change on this reaction at equilibrium.

$$I_2(g) + Cl_2(g) \rightleftharpoons 2\,ICl(g)$$

Predict the effect (shift right, shift left, or no effect) of these changes.
(a) increasing the reaction volume
(b) decreasing the reaction volume

70. Consider the effect of a volume change on this reaction at equilibrium.

$$CO(g) + H_2O(g) \rightleftharpoons CO_2(g) + H_2(g)$$

Predict the effect (shift right, shift left, or no effect) of these changes.
(a) increasing the reaction volume
(b) decreasing the reaction volume

71. This reaction is endothermic.

$$C(s) + CO_2(g) \rightleftharpoons 2\,CO(g)$$

Predict the effect (shift right, shift left, or no effect) of these changes.
(a) increasing the reaction temperature
(b) decreasing the reaction temperature

72. This reaction is endothermic.

$$I_2(g) \rightleftharpoons 2\,I(g)$$

Predict the effect (shift right, shift left, or no effect) of these changes.
(a) increasing the reaction temperature
(b) decreasing the reaction temperature

73. This reaction is exothermic.

$$C_6H_{12}O_6(s) + 6\,O_2(g) \rightleftharpoons 6\,CO_2(g) + 6\,H_2O(g)$$

Predict the effect (shift right, shift left, or no effect) of these changes.
(a) increasing the reaction temperature
(b) decreasing the reaction temperature

74. The following reaction is exothermic.

$$C_2H_4(g) + Br_2(g) \rightleftharpoons C_2H_4Br_2(g)$$

Predict the effect (shift right, shift left, or no effect) of these changes.
(a) increasing the reaction temperature
(b) decreasing the reaction temperature

75. Coal, which is primarily carbon, can be converted to natural gas, primarily CH_4, by this exothermic reaction.

$$C(s) + 2\,H_2(g) \rightleftharpoons CH_4(g)$$

If this reaction mixture is at equilibrium, predict the effect (shift right, shift left, or no effect) of these changes.
(a) adding more C to the reaction mixture
(b) adding more H_2 to the reaction mixture
(c) raising the temperature of the reaction mixture
(d) lowering the volume of the reaction mixture
(e) adding a catalyst to the reaction mixture

76. Coal can be used to generate hydrogen gas (a potential fuel) by this endothermic reaction.

$$C(s) + H_2O(g) \rightleftharpoons CO(g) + H_2(g)$$

If this reaction mixture is at equilibrium, predict the effect (shift right, shift left, or no effect) of these changes.
(a) adding more C to the reaction mixture
(b) adding more $H_2O(g)$ to the reaction mixture
(c) raising the temperature of the reaction mixture
(d) increasing the volume of the reaction mixture
(e) adding a catalyst to the reaction mixture

THE SOLUBILITY-PRODUCT CONSTANT

77. For each compound, write an equation showing how the compound dissolves in water and write an expression for K_{sp}.
(a) $CaSO_4$
(b) $AgCl$
(c) CuS
(d) $FeCO_3$

78. For each compound, write an equation showing how the compound dissolves in water and write an expression for K_{sp}.
(a) $Mg(OH)_2$
(b) $FeCO_3$
(c) PbS
(d) $PbSO_4$

79. Determine what is wrong with the K_{sp} expression for $Fe(OH)_2$ and correct it.

$$K_{sp} = [Fe^{2+}][OH^-]$$

80. Determine what is wrong with the K_{sp} expression for $Ba(OH)_2$ and correct it.

$$K_{sp} = \frac{[Ba(OH)_2]}{[Ba^{2+}][OH^-]^2}$$

81. A saturated solution of MgF_2 has $[Mg^{2+}] = 2.6 \times 10^{-4}$ M and $[F^-] = 5.2 \times 10^{-4}$ M. What is the value of K_{sp} for MgF_2?

82. A saturated solution of AgI has $[Ag^+] = 9.2 \times 10^{-9}$ M and $[I^-] = 9.2 \times 10^{-9}$ M. What is the value of K_{sp} for AgI?

83. A saturated solution of $PbSO_4$ has $[Pb^{2+}] = 1.35 \times 10^{-4}$ M. What is the concentration of SO_4^{2-}?

84. A saturated solution of $PbCl_2$ has $[Cl^-] = 2.86 \times 10^{-2}$ M. What is the concentration of Pb^{2+}?

85. Calculate the molar solubility of $CaCO_3$.

86. Calculate the molar solubility of PbS.

87. Calculate the molar solubility of $MgCO_3$.

88. Calculate the molar solubility of CuI ($K_{sp} = 1.27 \times 10^{-12}$).

89. Complete the table. Assume that all concentrations are equilibrium concentrations in moles per liter, M.

Compound	[Cation]	[Anion]	K_{eq}
$SrCO_3$	2.4×10^{-5}	2.4×10^{-5}	——
SrF_2	1.0×10^{-3}	——	4.0×10^{-9}
Ag_2CO_3	——	1.3×10^{-4}	8.8×10^{-12}

90. Complete the table. Assume that all concentrations are equilibrium concentrations in moles per liter, M.

Compound	[Cation]	[Anion]	K_{eq}
CdS	3.7×10^{-15}	3.7×10^{-15}	——
BaF_2	——	7.2×10^{-3}	1.9×10^{-7}
Ag_2SO_4	2.8×10^{-2}	——	1.1×10^{-5}

Cumulative Problems

91. Consider the reaction.

$$Fe^{3+}(aq) + SCN^-(aq) \rightleftharpoons FeSCN^{2+}(aq)$$

A solution is made containing initial $[Fe^{3+}] = 1.0 \times 10^{-3}$ M and initial $[SCN^-] = 8.0 \times 10^{-4}$ M. At equilibrium, $[FeSCN^{2+}] = 1.7 \times 10^{-4}$ M. Calculate the value of the equilibrium constant. *Hint:* Use the chemical reaction stoichiometry to calculate the equilibrium concentrations of Fe^{3+} and SCN^-.

92. Consider the reaction.

$$SO_2Cl_2(g) \rightleftharpoons SO_2(g) + Cl_2(g)$$

A solution is made containing initial $[SO_2Cl_2] = 0.020$ M. At equilibrium, $[Cl_2] = 1.2 \times 10^{-2}$ M. Calculate the value of the equilibrium constant. *Hint:* Use the chemical reaction stoichiometry to calculate the equilibrium concentrations of SO_2Cl_2 and SO_2.

93. Consider the reaction.

$$CO(g) + Cl_2(g) \rightleftharpoons COCl_2(g)$$

$$K_{eq} = 6.17 \times 10^{-2} \text{ at } 25\ °C$$

A 3.67-L flask containing an equilibrium reaction mixture has $[H_2] = 0.104$ M and $[I_2] = 0.0202$ M. What mass of HI in grams is in the equilibrium mixture?

94. Consider the reaction.

$$H_2(g) + I_2(g) \rightleftharpoons 2\ HI(g)$$

$$K_{eq} = 2.9 \times 10^{10} \text{ at } 25\ °C$$

A 5.19-L flask containing an equilibrium reaction mixture has $[CO] = 1.8 \times 10^{-6}$ M and $[Cl_2] = 7.3 \times 10^{-7}$ M. What mass of $COCl_2$ in grams is in the equilibrium mixture?

95. This reaction is exothermic.

$$C_2H_4(g) + Cl_2(g) \rightleftharpoons C_2H_4Cl_2(g)$$

If you were a chemist trying to maximize the amount of $C_2H_4Cl_2$ produced, which of the following might you try? Assume that the reaction mixture reaches equilibrium.
(a) increasing the reaction volume
(b) removing $C_2H_4Cl_2$ from the reaction mixture as it forms
(c) lowering the reaction temperature
(d) adding Cl_2

96. This reaction is endothermic.

$$C_2H_4(g) + I_2(g) \rightleftharpoons C_2H_4I_2(g)$$

If you were a chemist trying to maximize the amount of $C_2H_4I_2$ produced, which of the following might you try? Assume that the reaction mixture reaches equilibrium.
(a) decreasing the reaction volume
(b) removing I_2 from the reaction mixture
(c) raising the reaction temperature
(d) adding C_2H_4 to the reaction mixture

97. Calculate the molar solubility of CuS. How many grams of CuS are present in 15.0 L of a saturated CuS solution?

98. Calculate the molar solubility of $FeCO_3$. How many grams of $FeCO_3$ are present in 15.0 L of a saturated $FeCO_3$ solution?

99. A sample of tap water is found to be 0.025 M in Ca^{2+}. If 105 mg of Na_2SO_4 is added to 100.0 mL of the tap water, will any $CaSO_4$ precipitate out of solution?

100. If 50.0 mg of Na_2CO_3 is added to 150.0 mL of a solution that is 1.5×10^{-3} M in Mg^{2+}, will any $MgCO_3$ precipitate from the solution?

101. The solubility of $CaCrO_4$ at 25 °C is 4.15 g/L. Calculate K_{sp} for $CaCrO_4$.

102. The solubility of nickel(II) carbonate at 25 °C is 0.042 g/L. Calculate K_{sp} for nickel(II) carbonate.

103. Consider the reaction:

$$CaCO_3 \rightleftharpoons CaO(s) + CO_2(g)$$

A sample of $CaCO_3$ is placed into a sealed 0.500-L container and heated to 550 K at which the equilibrium constant is 4.1×10^{-4}. When the reaction has come to equilibrium, what mass of solid CaO is in the container? (Assume that the sample of $CaCO_3$ was large enough that equilibrium could be achieved.)

104. Consider the reaction:

$$NH_4HS(s) \rightleftharpoons NH_3(g) + H_2S(g)$$

A sample of pure NH_4HS is placed in a sealed 2.0-L container and heated to 550 K at which the equilibrium constant is 3.5×10^{-3}. Once the reaction reaches equilibrium, what mass of NH_3 is in the container? (Assume that the sample of NH_4HS was large enough that equilibrium could be achieved.)

105. A 2.55-L solution is 0.115 M in Mg^{2+}. If K_2CO_3 is added to the solution in order to precipitate the magnesium, what minimum mass of K_2CO_3 is required to get a precipitate?

106. A 75.0-L solution is 0.0251 M in Ca^{2+}. If Na_2SO_4 is added to the solution in order to precipitate the calcium, what minimum mass of Na_2SO_4 is required to get a precipitate?

Highlight Problems

107. H_2 and I_2 are combined in a flask and allowed to react according to the reaction:

$$H_2(g) + I_2(g) \rightleftharpoons 2\,HI(g)$$

Examine the figures (sequential in time) and determine which figure represents the point where equilibrium is reached.

(a)

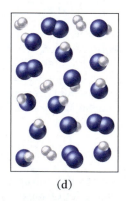

(b)

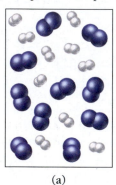

(c)

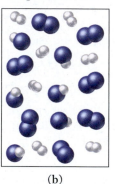

(d)

(e)

(f)

108. Ethene (C_2H_4) can be halogenated by the reaction:

$$C_2H_4(g) + X_2(g) \rightleftharpoons C_2H_4X_2(g)$$

where X_2 can be Cl_2, Br_2, or I_2. Examine the figures representing equilibrium concentrations of this reaction at the same temperature for the three different halogens. Rank the equilibrium constants for these three reactions from largest to smallest.

$$C_2H_4 + Cl_2 \rightleftharpoons C_2H_4Cl_2$$

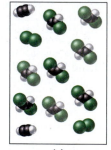

(a)

$$C_2H_4 + Br_2 \rightleftharpoons C_2H_4Br_2$$

(b)

$$C_2H_4 + I_2 \rightleftharpoons C_2H_4I_2$$

(c)

109. One of the main components of hard water is $CaCO_3$. When hard water evaporates, some of the $CaCO_3$ is left behind as a white mineral deposit. Plumbing fixtures in homes with hard water often acquire these deposits over time. Toilets, for example, may develop deposits at the water line as the water in the toilet slowly evaporates away. If water is saturated with $CaCO_3$, how much of it has to evaporate to deposit 0.250 g of $CaCO_3$? *Hint:* Begin by using K_{sp} for $CaCO_3$ to determine its solubility.

110. Consider the following generic equilibrium in which a solid reactant is in equilibrium with a gaseous product:

$$A(s) \rightleftharpoons B(g)$$

These diagrams represent the reaction mixture at the following points: (a) initially; (b) after a short period of time has passed; and (c) at equilibrium.

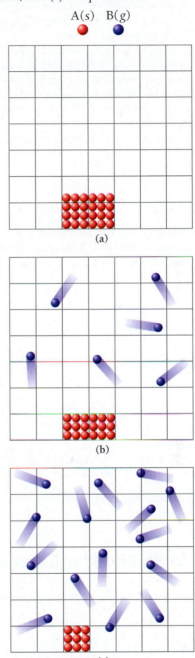

For each diagram, calculate the concentrations of the spheres representing A(s) and B(g). Assume that each block in the grid has an area of 1 cm² and report your answer in units of spheres/cm². (Since the spheres in the solid are not free to move, the solid only occupies the area that it covers. The spheres in the solid, however, are free to move and therefore occupy the entire grid.) What do you notice about the concentrations of A(s) and B(g) in these representations? Write an equilibrium expression for the generic reaction and use the results of your calculations to explain why A(s) is not included in the expression.

Questions for Group Work

Discuss these questions with the group and record your consensus answer.

111. A certain town gets its water from an underground aqui-
fer that contains water in equilibrium with calcium carbonate limestone.
  (a) What is the symbol for the equilibrium constant that
    describes calcium carbonate dissolving in water?
    What is the numerical value?
  (b) Calculate the molar solubility of calcium carbonate.
  (c) If an entire coffee cup full of water (about 200 mL)
    evaporated on your desk over spring break, how many
    grams of calcium carbonate would be left behind?

112. The reaction A(g) + B(g) ⟶ 2 C(g) has an equilibrium
constant of 16.
  (a) Without doing any calculations, at equilibrium will
    there be more A, B, or C around? Explain your answer.

  (b) A certain chemist mixes up 3 mol of A and 3 mol of
    B in a liter of water. Draw a picture of the contents of
    the beaker before any reaction occurs. Let a circle rep-
    resent 1 mol of A, a square represent 1 mol of B, and a
    triangle represent 1 mol of C.
  (c) At equilibrium, how many As, Bs, and Cs are pres-
    ent? Draw this on a new diagram.
  (d) If the chemist removes some of the C that has been
    produced, what will happen?

113. Describe three ways a reaction at equilibrium can be
changed such that it is no longer in equilibrium. For each
case, indicate which way the reaction will shift to return to
equilibrium.

Data Interpretation and Analysis

114. Solid $CaCO_3$ decomposes into solid CaO and gaseous
CO_2. The concentration of CO_2 in an equilibrium mixture
of this reaction is measured as a function of temperature,

and the results are shown. Study the graph and answer
the questions.

(a) Write a balanced chemical equation for the reaction.
(b) Write an equilbrium constant expression for the
  reaction.
(c) What is the value of the equilibrium constant at
  950 K? At 1050 K?
(d) What mass of carbon dioxide is present in a 1.00-L
  reaction vessel at equilibrium at 1000 K?
(e) What minimum mass of $CaCO_3$ is required to produce the mass of carbon dioxide in part d?

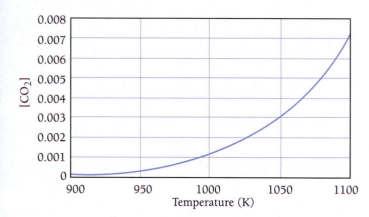

▲ Carbon dioxide concentration (M) in equilibrium with calcium carbonate
and calcium oxide

Answers to Skillbuilder Exercises

Skillbuilder 15.1 .... $K_{eq} = \dfrac{[HF]^2}{[H_2][F_2]}$

Skillbuilder 15.2 .... $K_{eq} = \dfrac{[Cl_2]^2}{[HCl]^4[O_2]}$

Skillbuilder 15.3 .... $K_{eq} = 14.5$

Skillbuilder Plus, p. 540 ... $K_{eq} = 26$

Skillbuilder 15.4 .... 0.033 M

Skillbuilder 15.5 .... Adding Br_2 causes a shift to the left;
adding BrNO causes a shift to the
right.

Skillbuilder Plus, p. 545 ... Removing Br_2 causes a shift to the
right.

Skillbuilder 15.6 .... Decreasing volume causes a shift to
the right; increasing volume causes a
shift to the left.

Skillbuilder 15.7 .... Increasing the temperature shifts the
reaction to the left; decreasing the tem-
perature shifts the reaction to the right.

Skillbuilder 15.8 .... **(a)** $K_{sp} = [Ag^+][I^-]$
(b) $K_{sp} = [Ca^{2+}][OH^-]^2$

Skillbuilder 15.9 .... 8.43×10^{-3} M

Answers to Conceptual Checkpoints

15.1 (a) In accordance with the gas laws (Chapter 11), increasing the pressure would increase the temperature, decrease the volume, or both. Increasing the temperature would increase the reaction rate. Decreasing the volume would increase the concentration of the reactants, which would also increase the reaction rate. Therefore, we would expect that increasing the pressure would speed up the reaction.

15.2 (c) Since the image in (c) has the greatest amount of product, the equilibrium constant must be largest at T_3.

15.3 (b) For this reaction, $K_{eq} = [B][C]/[A] = (2 \times 2)/2 = 2$.

15.4 Doubling the hydrogen concentration leads to a larger shift toward products because the concentration of hydrogen in the denominator is squared.

15.5 (c) Increasing the concentration of a reactant causes the reaction to shift right. For this reaction, changing the volume has no effect because the number of gas particles on both sides of the equation is equal.

15.6 (a) Since the reaction is endothermic, increasing the temperature drives it to the right (toward the product). Since the reaction has 1 mol of particles on the left and 2 mol of particles on the right, increasing the volume drives it to the right. Therefore, increasing the temperature and increasing the volume will create the greatest amount of product.

15.7 Calcium sulfate is more soluble because it has the larger K_{sp}.

15.8 Reaction A is likely to have the faster rate. In general, the lower the activation barrier, the faster the rate at a given temperature.

16 Oxidation and Reduction

We have to learn to understand nature and not merely to observe it and endure what it imposes on us.

—John Desmond Bernal (1901–1971)

16.1 The End of the Internal Combustion Engine?

▲ **FIGURE 16.1 A fuel-cell car** The Toyota Mirai is the first mass-produced fuel-cell vehicle. It is powered by hydrogen and its only emission is water.

It is possible, even likely, that you will see the end of the internal combustion engine within your lifetime. Although it has served us well—powering our airplanes, automobiles, and trains—its time is running out. What will replace it? If our cars don't run on gasoline, what will fuel them? The answers to these questions are not completely settled, but new and better technologies are here. Among the most promising are **fuel cells** that power electric vehicles. Such whisper-quiet, environmentally friendly supercars are now commercially available in select areas where hydrogen refueling stations have been built, primarily California.

In 2015, Toyota started selling its fuel-cell vehicle—called the Mirai (Japanese for "future")—in the United States (◄ **FIGURE 16.1**). The four-passenger car has 151 horsepower and can travel up to 312 miles on one tank of fuel. Unlike electric vehicles, such as those produced by Tesla Motors, which take hours to recharge, the Mirai refuels in 5 minutes. The electric motor is powered by hydrogen, stored as a compressed gas, and its only emission is water—which is so clean you can drink it. Other automakers have similar models in development. In addition, several cities around the world—including Palm Springs, California; Washington, D.C.; Vancouver, British Columbia; Whistler, British Columbia; and São Paulo, Brazil—currently run buses powered by fuel cells.

Fuel-cell technology is possible because of the tendency of some elements to gain electrons from other elements. The most common type of fuel cell—called the *hydrogen–oxygen fuel cell*—is based on the reaction between hydrogen and oxygen.

$$2\,H_2(g) + O_2(g) \longrightarrow 2\,H_2O(g)$$

In this reaction, hydrogen and oxygen form covalent bonds with one another. Recall from Section 10.2 that a single covalent bond is a shared electron pair. However,

◄ Fuel-cell vehicles (FCVs), such as the one shown here, may someday replace vehicles powered by internal combustion engines. FCVs produce only water as exhaust.

because oxygen is more electronegative than hydrogen (see Section 10.8), the electron pair in a hydrogen–oxygen bond is *unequally* shared, with oxygen getting the larger portion. In effect, oxygen has more electrons in H_2O than in elemental O_2—oxygen has *gained electrons* in the reaction.

In a typical reaction between hydrogen and oxygen, oxygen atoms gain the electrons directly from hydrogen atoms as the reaction proceeds. In a hydrogen–oxygen fuel cell, the same reaction occurs, but the hydrogen and oxygen are separated, forcing the electrons to move through an external wire to get from hydrogen to oxygen. These moving electrons constitute an electrical current, which is used to power the electric motor of a fuel-cell vehicle. In effect, fuel cells use the electron-gaining tendency of oxygen and the electron-losing tendency of hydrogen to force electrons to move through a wire, creating the electricity that powers the car.

Reactions involving the transfer of electrons are **oxidation–reduction reactions** or **redox reactions**. In addition to their application to fuel-cell vehicles, redox reactions are prevalent in nature, in industry, and in many everyday processes. For example, the rusting of iron, the bleaching of hair, and the reactions occurring in batteries all involve redox reactions. Redox reactions are also responsible for providing the energy our bodies need to move, think, and stay alive.

16.2 Oxidation and Reduction: Some Definitions

▶ Define and identify oxidation and reduction.

▶ Identify oxidizing agents and reducing agents.

Consider the following redox reactions:

$$2\,H_2(g) + O_2(g) \longrightarrow 2\,H_2O(g) \text{ (hydrogen–oxygen fuel-cell reaction)}$$

$$4\,Fe(s) + 3\,O_2(g) \longrightarrow 2\,Fe_2O_3(s) \text{ (rusting of iron)}$$

$$CH_4(g) + 2\,O_2(g) \longrightarrow CO_2(g) + 2\,H_2O(g) \text{ (combustion of methane)}$$

What do they all have in common? Each of these reactions involves one or more elements gaining oxygen. In the hydrogen–oxygen fuel-cell reaction, *hydrogen* gains oxygen as it turns into water. In the rusting of iron, *iron* gains oxygen as it turns into iron oxide, the familiar orange substance we call rust (▼ FIGURE 16.2). In the combustion of methane, *carbon* gains oxygen to form carbon dioxide, producing the brilliant blue flame we see on gas stoves (▼ FIGURE 16.3). In each case,

▲ FIGURE 16.2 **Slow oxidation** Rust is produced by the oxidation of iron to form iron(III) oxide.

▲ FIGURE 16.3 **Rapid oxidation** The flame on a gas stove results from the oxidation of carbon in natural gas.

the substance that gains oxygen is oxidized in the reaction. One definition of *oxidation*—though not the most fundamental one—is the *gaining of oxygen*.

Now consider these same three reactions in reverse:

$$2\,H_2O(g) \longrightarrow 2\,H_2(g) + O_2(g)$$

$$2\,Fe_2O_3(s) \longrightarrow 4\,Fe(s) + 3\,O_2(g)$$

$$CO_2(g) + 2\,H_2O(g) \longrightarrow CH_4(g) + 2\,O_2(g)$$

Each of these reactions involves loss of oxygen. In the first reaction, hydrogen loses oxygen; in the second reaction, iron loses oxygen; and in the third reaction, carbon loses oxygen. In each case, the substance that loses oxygen is reduced in the reaction. One definition of *reduction* is the *loss of oxygen*.

Redox reactions need not involve oxygen, however. Consider, for example, the similarities between the following two reactions:

$$4\,Li(s) + O_2(g) \longrightarrow 2\,Li_2O(s)$$

$$2\,Li(s) + Cl_2(g) \longrightarrow 2\,LiCl(s)$$

In both cases, lithium (a metal with a strong tendency to lose electrons) reacts with an electronegative nonmetal (which has a tendency to gain electrons). In both cases, lithium atoms lose electrons to become positive ions—lithium is oxidized.

$$Li \longrightarrow Li^+ + e^-$$

The electrons lost by lithium are gained by the nonmetals, which become negative ions—the nonmetals are reduced.

$$O_2 + 4e^- \longrightarrow 2\,O^{2-}$$

$$Cl_2 + 2e^- \longrightarrow 2\,Cl^-$$

A more fundamental definition of **oxidation**, then, is the *loss of electrons*, and a more fundamental definition of **reduction** is the *gain of electrons*.

Notice that *oxidation and reduction must occur together*. If one substance loses electrons (oxidation), then another substance must gain electrons (reduction) (◀ **FIGURE 16.4**). The substance that is oxidized is the **reducing agent** because it causes the reduction of the other substance. Similarly, the substance that is reduced is the **oxidizing agent** because it causes the oxidation of the other substance.

For example, consider our hydrogen–oxygen fuel-cell reaction:

$$\underset{\text{Reducing agent}}{2\,H_2(g)} \quad + \quad \underset{\text{Oxidizing agent}}{O_2(g)} \longrightarrow 2\,H_2O(g)$$

In this reaction, hydrogen is oxidized, making it the reducing agent. Oxygen is reduced, making it the oxidizing agent. Substances such as oxygen, which have a strong tendency to attract electrons, are good oxidizing agents—they tend to cause the oxidation of other substances. Substances such as hydrogen, which have a strong tendency to give up electrons, are good reducing agents—they tend to cause the reduction of other substances.

To summarize:

- Oxidation—the loss of electrons
- Reduction—the gain of electrons
- Oxidizing agent—the substance being reduced
- Reducing agent—the substance being oxidized

These definitions of oxidation and reduction are useful because they show the origin of the term *oxidation*, and they allow us to quickly identify reactions involving elemental oxygen as oxidation and reduction reactions. However, as you will see, these definitions are *not* the most fundamental.

In redox reactions between a metal and a nonmetal, the metal is oxidized and the nonmetal is reduced.

Helpful mnemonics:
OIL RIG—**O**xidation **I**s **L**oss (of electrons); **R**eduction **I**s **G**ain (of electrons).
LEO the lion says GER—**L**ose **E**lectrons **O**xidation; **G**ain **E**lectrons **R**eduction.

In a redox reaction, one substance loses electrons and another substance gains electrons.

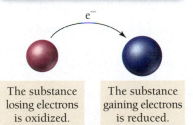

| The substance losing electrons is oxidized. | The substance gaining electrons is reduced. |

▲ **FIGURE 16.4 Oxidation and reduction**

The oxidizing agent may be an element that is itself reduced, or a compound or ion containing an element that is reduced. The reducing agent may be an element that is itself oxidized, or a compound or ion containing an element that is oxidized.

EXAMPLE **16.1** | **Identifying Oxidation and Reduction**

Identify the substance being oxidized and the substance being reduced in each reaction.

(a) $2 Mg(s) + O_2(g) \longrightarrow 2 MgO(s)$
(b) $Fe(s) + Cl_2(g) \longrightarrow FeCl_2(s)$
(c) $Zn(s) + Fe^{2+}(aq) \longrightarrow Zn^{2+}(aq) + Fe(s)$

SOLUTION

(a) Magnesium is gaining oxygen and losing electrons to oxygen. Mg is therefore oxidized, and O_2 is reduced.
(b) A metal (Fe) is reacting with an electronegative nonmetal (Cl_2). Fe loses electrons and is therefore oxidized, while Cl_2 gains electrons and is therefore reduced.
(c) Electrons are transferred from the Zn to the Fe^{2+}. Zn loses electrons and is oxidized. Fe^{2+} gains electrons and is reduced.

▶ **SKILLBUILDER 16.1** | **Identifying Oxidation and Reduction**

Identify the substance being oxidized and the substance being reduced in each reaction.

(a) $2 K(s) + Cl_2(g) \longrightarrow 2 KCl(s)$
(b) $2 Al(s) + 3 Sn^{2+}(aq) \longrightarrow 2 Al^{3+}(aq) + 3 Sn(s)$
(c) $C(s) + O_2(g) \longrightarrow CO_2(g)$

▶ **FOR MORE PRACTICE** Example 16.11; Problems 33, 34, 35, 36.

EXAMPLE **16.2** | **Identifying Oxidizing and Reducing Agents**

Identify the oxidizing agent and the reducing agent in each reaction.

(a) $2 Mg(s) + O_2(g) \longrightarrow 2 MgO(s)$
(b) $Fe(s) + Cl_2(g) \longrightarrow FeCl_2(s)$
(c) $Zn(s) + Fe^{2+}(aq) \longrightarrow Zn^{2+}(aq) + Fe(s)$

SOLUTION

In the previous example, we identified the substances being oxidized and reduced for these reactions. Recall that the substance being oxidized is the reducing agent, and the substance being reduced is the oxidizing agent.

(a) Mg is oxidized and is therefore the reducing agent; O_2 is reduced and is therefore the oxidizing agent.
(b) Fe is oxidized and is therefore the reducing agent; Cl_2 is reduced and is therefore the oxidizing agent.
(c) Zn is oxidized and is therefore the reducing agent; Fe^{2+} is reduced and is therefore the oxidizing agent.

▶ **SKILLBUILDER 16.2** | **Identifying Oxidizing and Reducing Agents**

Identify the oxidizing agent and the reducing agent in each reaction.

(a) $2 K(s) + Cl_2(g) \longrightarrow 2 KCl(s)$
(b) $2 Al(s) + 3 Sn^{2+}(aq) \longrightarrow 2 Al^{3+}(aq) + 3 Sn(s)$
(c) $C(s) + O_2(g) \longrightarrow CO_2(g)$

▶ **FOR MORE PRACTICE** Example 16.12; Problems 37, 38.

CONCEPTUAL ✓ CHECKPOINT 16.1

An oxidizing agent:

(a) is always oxidized.

(b) is always reduced.

(c) can either be oxidized or reduced, depending on the reaction.

16.3 Oxidation States: Electron Bookkeeping

▶ Assign oxidation states.

▶ Use oxidation states to identify oxidation and reduction.

For many redox reactions, such as those involving oxygen or other highly electronegative elements, we can readily identify the substances being oxidized and reduced by inspection. For other redox reactions, identification is more difficult.

For example, consider the redox reaction between carbon and sulfur:

$$C + 2\,S \longrightarrow CS_2$$

> Do not confuse oxidation state with ionic charge. A substance need not be ionic to have an assigned oxidation state.

What is oxidized here? What is reduced? In order to readily identify oxidation and reduction, chemists have devised a scheme to track electrons and where they go in chemical reactions. In this scheme—which is like bookkeeping for electrons—we assign all shared electrons to the most electronegative element. Then we calculate a number—called the **oxidation state** or **oxidation number**—for each element based on the number of electrons assigned to it.

The procedure just described is a bit cumbersome in practice. However, we can summarize its main results in a series of rules. The easiest way to assign oxidation states is to follow these rules.

> These rules are hierarchical. If any two rules conflict, follow the rule that is higher on the list.

Nonmetal	Oxidation State	Example
fluorine	−1	MgF_2 −1 ox state
hydrogen	+1	H_2O +1 ox state
oxygen	−2	CO_2 −2 ox state
Group 7A	−1	CCl_4 −1 ox state
Group 6A	−2	H_2S −2 ox state
Group 5A	−3	NH_3 −3 ox state

Rules for Assigning Oxidation States

Examples

1. The oxidation state of an atom in a free element is 0.

    Cu Cl_2
    0 ox state   0 ox state

2. The oxidation state of a monoatomic ion is equal to its charge.

    Ca^{2+} Cl^-
    +2 ox state   −1 ox state

3. The sum of the oxidation states of all atoms in:
    - a neutral molecule or formula unit is 0.

      H_2O
      2(H ox state) + 1(O ox state) = 0
    - an ion is equal to the charge of the ion.

      NO_3^-
      1(N ox state) + 3(O ox state) = −1

4. In their compounds,
    - Group I metals have an oxidation state of +1.

      NaCl
      +1 ox state
    - Group II metals have an oxidation state of +2.

      CaF_2
      +2 ox state

5. In their compounds, we assign nonmetals oxidation states according to the hierarchical table shown at left. Entries at the top of the table have priority over entries at the bottom.

EXAMPLE **16.3** | **Assigning Oxidation States**

Assign an oxidation state to each atom in each species.

(a) Br_2 (b) K^+ (c) LiF   (d) CO_2 (e) SO_4^{2-} (f) Na_2O_2

Br_2 is a free element, so the oxidation state of both Br atoms is 0 (Rule 1).	**SOLUTION**
	
	 (a) Br_2
	
	     Br Br
	     0 0

K^+ is a monoatomic ion, so the oxidation state of the K^+ ion is +1 (Rule 2).	(b) K^+ K^+ +1
The oxidation state of Li is +1 (Rule 4). The oxidation state of F is −1 (Rule 5). This is a neutral compound, so the sum of the oxidation states is 0 (Rule 3).	(c) LiF Li F +1 −1 sum: +1 − 1 = 0
The oxidation state of oxygen is −2 (Rule 5). You deduce the oxidation state of carbon from Rule 3, which states that the sum of the oxidation states of all the atoms must be 0. Since there are two oxygen atoms, you multiply the oxidation state of oxygen by 2 when calculating the sum.	(d) CO_2 (C ox state) + 2(O ox state) = 0 (C ox state) + 2(−2) = 0 (C ox state) − 4 = 0 C ox state = +4 C O_2 +4 −2 sum: +4 + 2(−2) = 0
The oxidation state of oxygen is −2 (Rule 5). The oxidation state of S is expected to be −2 (Rule 5). However, if that were the case, the sum of the oxidation states would not equal the charge of the ion. Since O is higher on the list, it takes priority, and you calculate the oxidation state of sulfur by setting the sum of all the oxidation states equal to −2 (the charge of the ion).	(e) SO_4^{2-} (S ox state) + 4(O ox state) = −2 (S ox state) + 4(−2) = −2 (Oxygen takes priority over sulfur.) (S ox state) − 8 = −2 S ox state = −2 + 8 S ox state = +6 S O_4^{2-} +6 −2 sum: +6 + 4(−2) = −2
The oxidation state of sodium is +1 (Rule 4). The oxidation state of oxygen is expected to be −2 (Rule 5). However, Na takes priority, and you deduce the oxidation state of oxygen by setting the sum of all the oxidation states equal to 0.	(f) Na_2O_2 2(Na ox state) + 2(O ox state) = 0 2(+1) + 2(O ox state) = 0 (Sodium takes priority over oxygen.) +2 + 2(O ox state) = 0 O ox state = −1 Na_2 O_2 +1 −1 sum: 2(+1) + 2(−1) = 0

▶ **SKILLBUILDER 16.3** | **Assigning Oxidation States**

Assign an oxidation state to each atom in each substance.

(a) Zn   (b) Cu^{2+} (c) $CaCl_2$ (d) CF_4 (e) NO_2^- (f) SO_3

▶ **FOR MORE PRACTICE** Example 16.13; Problems 45, 46, 47, 48, 49, 50, 51, 52, 53, 54.

EVERYDAY CHEMISTRY

The Bleaching of Hair

Students with bleached hair are a common sight on most campuses. Many students bleach their own hair with home-bleaching kits available at drugstores and supermarkets. These kits normally contain hydrogen peroxide (H_2O_2), an excellent oxidizing agent. When applied to hair, hydrogen peroxide oxidizes melanin, the dark pigment that gives hair color. Once melanin is oxidized, it no longer imparts a dark color to hair, leaving the hair with the familiar bleached look.

Hydrogen peroxide also oxidizes other components of hair. For example, the protein molecules in hair contain SH groups called *thiols*. Thiols are normally slippery (they slide across each other). Hydrogen peroxide oxidizes these thiol groups to sulfonic acid groups (—SO_3H). Sulfonic acid groups are stickier, causing hair to tangle more easily. Consequently, people with heavily bleached hair often use conditioners. Conditioners contain compounds that form thin, lubricating coatings on individual hair shafts. These coatings prevent tangling and make hair softer and more manageable.

B16.1 CAN YOU ANSWER THIS? *Assign oxidation states to the atoms of H_2O_2. Which atoms in H_2O_2 do you think change oxidation state when H_2O_2 oxidizes hair?*

▲ Hydrogen peroxide, a good oxidizing agent, bleaches hair.

Now let's return to our original question. What is being oxidized, and what is being reduced in the following reaction?

$$C + 2\,S \longrightarrow CS_2$$

We use the oxidation state rules to assign oxidation states to all elements on both sides of the equation:

$$\underset{0}{C} + \underset{0}{2\,S} \longrightarrow \underset{+4\ \ -2}{CS_2}$$

The oxidation state of carbon changed from 0 to +4. In terms of our electron bookkeeping scheme (the assigned oxidation state), carbon *lost electrons* and was *oxidized*. The oxidation state of sulfur changed from 0 to −2. In terms of our electron bookkeeping scheme, sulfur *gained electrons* and was *reduced*.

$$\underset{0}{C} + \underset{0}{2\,S} \longrightarrow \underset{+4\ \ -2}{CS_2}$$

Reduction
Oxidation

In terms of oxidation states, we define oxidation and reduction as follows:

Oxidation—an increase in oxidation state
Reduction—a decrease in oxidation state

EXAMPLE **16.4**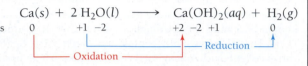

Use oxidation states to identify the element that is being oxidized and the element that is being reduced in the redox reaction.

$$Ca(s) + 2\,H_2O(l) \longrightarrow Ca(OH)_2(aq) + H_2(g)$$

Assign an oxidation state to each atom in the reaction. Ca increased in oxidation state; it was oxidized. H decreased in oxidation state; it was reduced. (Note that oxygen has the same oxidation state on both sides of the equation and was therefore neither oxidized nor reduced.)	**SOLUTION**

$$Ca(s) + 2\,H_2O(l) \longrightarrow Ca(OH)_2(aq) + H_2(g)$$

Oxidation states   0   +1 −2   +2 −2 +1   0

⎣—————— Oxidation ——————⎦   Reduction

▶ **SKILLBUILDER 16.4 | Using Oxidation States to Identify Oxidation and Reduction**

Use oxidation states to identify the element that is being oxidized and the element that is being reduced in the redox reaction.

$$Sn(s) + 4\,HNO_3(aq) \longrightarrow SnO_2(s) + 4\,NO_2(g) + 2\,H_2O(g)$$

▶ **FOR MORE PRACTICE**   Problems 57, 58, 59, 60.

CONCEPTUAL ✓ CHECKPOINT 16.2

In which substance does nitrogen have the *lowest* oxidation state?

(a) N_2

(b) NO

(c) NO_2

(d) NH_3

16.4 Balancing Redox Equations

▶ Balance redox reactions.

In Chapter 7, we learned how to balance chemical equations by inspection. We can balance some redox reactions in this way. However, redox reactions occurring in aqueous solutions are usually difficult to balance by inspection and require a special procedure called the *half-reaction method of balancing*. In this procedure, we break down the overall equation into two **half-reactions**: one for oxidation and one for reduction. We balance the half-reactions individually and then add them together.

For example, consider the redox reaction:

$$Al(s) + Ag^+(aq) \longrightarrow Al^{3+}(aq) + Ag(s)$$

We assign oxidation numbers to all atoms to determine what is being oxidized and what is being reduced.

$$Al(s) + Ag^+(aq) \longrightarrow Al^{3+}(aq) + Ag(s)$$

Oxidation states   0   +1   +3   0

Reduction

Oxidation

We then divide the reaction into two half-reactions, one for oxidation and one for reduction.

Oxidation: $Al(s) \longrightarrow Al^{3+}(aq)$

Reduction: $Ag^+(aq) \longrightarrow Ag(s)$

Next we balance the two half-reactions individually. In this case, the half-reactions are already balanced with respect to mass—the number of each type of atom on both sides of each half-reaction is the same. However, the equations are not balanced with respect to charge—in the oxidation half-reaction, the left side of the equation has 0 charge while the right side has +3 charge, and in the reduction half-reaction, the left side has +1 charge and the right side has 0 charge. We balance the charge of each half-reaction individually by adding the appropriate number of electrons to make the charges on both sides equal.

$$Al(s) \longrightarrow Al^{3+}(aq) + 3e^- \quad \text{(zero charge on both sides)}$$
$$1e^- + Ag^+(aq) \longrightarrow Ag(s) \quad \text{(zero charge on both sides)}$$

Since these half-reactions must occur together, the number of electrons lost in the oxidation half-reaction must equal the number gained in the reduction half-reaction. We equalize these by multiplying one or both half-reactions by appropriate whole numbers to balance the electrons lost and gained. In this case, we multiply the reduction half-reaction by 3:

$$Al(s) \longrightarrow Al^{3+}(aq) + 3e^-$$
$$3 \times [1e^- + Ag^+(aq) \longrightarrow Ag(s)]$$

We then add the half-reactions together, canceling electrons and other species as necessary.

$$Al(s) \longrightarrow Al^{3+}(aq) + 3e^-$$
$$3e^- + 3\,Ag^+(aq) \longrightarrow 3\,Ag(s)$$
$$\overline{Al(s) + 3\,Ag^+(aq) \longrightarrow Al^{3+}(aq) + 3\,Ag(s)}$$

Reactants	Products
1 Al	1 Al
3 Ag	3 Ag
+3 charge	+3 charge

Lastly, we verify that the equation is balanced, with respect to both mass and charge as shown in the margin. Notice that the charge need not be zero on both sides of the equation—it just has to be *equal* on both sides. The equation is balanced.

A general procedure for balancing redox reactions is demonstrated in the following examples. Because aqueous solutions are often acidic or basic, the procedure must account for the presence of H^+ ions or OH^- ions. We cover acidic solutions in Examples 16.5 through 16.7 and demonstrate how to balance redox reactions in basic solutions in Example 16.8.

Balancing Redox Equations Using the Half-Reaction Method	EXAMPLE **16.5** Balance the redox reaction.	EXAMPLE **16.6** Balance the redox reaction.
1. **Assign oxidation states** to all atoms and identify the substances being oxidized and reduced.	**SOLUTION** $Al(s) + Cu^{2+}(aq) \longrightarrow Al^{3+}(aq) + Cu(s)$ (0) (+2) (+3) (0) Oxidation / Reduction	**SOLUTION** $Fe^{2+}(aq) + MnO_4^-(aq) \longrightarrow Fe^{3+}(aq) + Mn^{2+}(aq)$ (+2) (+7 −2) (+3) (+2) Oxidation / Reduction
2. **Separate the overall reaction into two half-reactions**, one for oxidation and one for reduction.	**OXIDATION** $Al(s) \longrightarrow Al^{3+}(aq)$ **REDUCTION** $Cu^{2+}(aq) \longrightarrow Cu(s)$	**OXIDATION** $Fe^{2+}(aq) \longrightarrow Fe^{3+}(aq)$ **REDUCTION** $MnO_4^-(aq) \longrightarrow Mn^{2+}(aq)$

continued on page 582 ▶

continued from page 581

3. **Balance each half-reaction with respect to mass** in the following order: • Balance all elements other than H and O. • Balance O by adding H_2O. • Balance H by adding H^+.	All elements other than hydrogen and oxygen are balanced, so you can proceed to the next step. No oxygen; proceed to the next step. No hydrogen; proceed to the next step.	All elements other than hydrogen and oxygen are balanced, so you can proceed to the next step. $Fe^{2+}(aq) \longrightarrow Fe^{3+}(aq)$ $MnO_4^-(aq) \longrightarrow$ $\qquad Mn^{2+}(aq) + \textbf{4 H}_2\textbf{O}(l)$ $Fe^{2+}(aq) \longrightarrow Fe^{3+}(aq)$ $\textbf{8 H}^+\textbf{(aq)} + MnO_4^-(aq) \longrightarrow$ $\qquad Mn^{2+}(aq) + 4 H_2O(l)$
4. **Balance each half-reaction with respect to charge** by adding electrons to the right side of the oxidation half-reaction and the left side of the reduction half-reaction. (The sum of the charges on both sides of each equation should be equal.)	$Al(s) \longrightarrow Al^{3+}(aq) + \textbf{3e}^-$ $\textbf{2e}^- + Cu^{2+}(aq) \longrightarrow Cu(s)$	$Fe^{2+}(aq) \longrightarrow Fe^{3+}(aq) + \textbf{1e}^-$ $\textbf{5e}^- + 8 H^+(aq) + MnO_4^-(aq) \longrightarrow$ $\qquad Mn^{2+}(aq) + 4 H_2O(l)$
5. **Make the number of electrons in both half-reactions equal** by multiplying one or both half-reactions by a small whole number.	$\textbf{2} \times [Al(s) \longrightarrow Al^{3+}(aq) + 3e^-]$ $\textbf{3} \times [2e^- + Cu^{2+}(aq) \longrightarrow Cu(s)]$	$\textbf{5} \times [Fe^{2+}(aq) \longrightarrow Fe^{3+}(aq) + 1e^-]$ $5e^- + 8 H^+(aq) + MnO_4^-(aq) \longrightarrow$ $\qquad Mn^{2+}(aq) + 4 H_2O(l)$
6. **Add the two half-reactions together**, canceling electrons and other species as necessary.	$2 Al(s) \longrightarrow 2 Al^{3+}(aq) + \cancel{6e^-}$ $\cancel{6e^-} + 3 Cu^{2+}(aq) \longrightarrow 3 Cu(s)$ ───────────────── $2 Al(s) + 3 Cu^{2+}(aq) \longrightarrow$ $\qquad 2 Al^{3+}(aq) + 3 Cu(s)$	$5 Fe^{2+}(aq) \longrightarrow 5 Fe^{3+}(aq) + \cancel{5e^-}$ $\cancel{5e^-} + 8 H^+(aq) + MnO_4^-(aq) \longrightarrow$ $\qquad Mn^{2+}(aq) + 4 H_2O(l)$ ───────────────── $5 Fe^{2+}(aq) + 8 H^+(aq) + MnO_4^-(aq)$ $\longrightarrow 5 Fe^{3+}(aq) + Mn^{2+}(aq) + 4 H_2O(l)$
7. **Verify that the reaction is balanced** with respect to both mass and charge.	 **Reactants** — **Products** 2 Al — 2 Al 3 Cu — 3 Cu +6 charge — +6 charge	 **Reactants** — **Products** 5 Fe — 5 Fe 8 H — 8 H 1 Mn — 1 Mn 4 O — 4 O +17 charge — +17 charge
	▶ **SKILLBUILDER 16.5** | Balance the redox reaction occurring in acidic solution. $H^+(aq) + Cr(s) \longrightarrow$ $\qquad H_2(g) + Cr^{3+}(aq)$	▶ **SKILLBUILDER 16.6** | Balance the redox reaction occurring in acidic solution. $Cu(s) + NO_3^-(aq) \longrightarrow$ $\qquad Cu^{2+}(aq) + NO_2(g)$ ▶ **FOR MORE PRACTICE** Example 16.14; Problems 61, 62, 63, 64.

Interactive
Worked Example
Video 16.7

EXAMPLE **16.7** | **Balancing Redox Reactions**

Balance the following redox reaction occurring in acidic solution.

$$I^-(aq) + Cr_2O_7^{2-}(aq) \longrightarrow Cr^{3+}(aq) + I_2(s)$$

1. Follow the half-reaction method for balancing redox reactions. Begin by assigning oxidation states.	**SOLUTION** $I^-(aq) + Cr_2O_7^{2-}(aq) \longrightarrow Cr^{3+}(aq) + I_2(s)$ $-1 \qquad +6 \; -2 \qquad\qquad +3 \qquad 0$ Reduction Oxidation
2. Separate the overall reaction into two half-reactions.	**OXIDATION** $\quad I^-(aq) \longrightarrow I_2(s)$ **REDUCTION** $\quad Cr_2O_7^{2-}(aq) \longrightarrow Cr^{3+}(aq)$
3. Balance each half-reaction with respect to mass.	
• Balance all elements other than H and O.	$2\,I^-(aq) \longrightarrow I_2(aq)$ $Cr_2O_7^{2-}(aq) \longrightarrow 2\,Cr^{3+}(s)$
• Balance O by adding H_2O.	$2\,I^-(aq) \longrightarrow I_2(s)$ $Cr_2O_7^{2-}(aq) \longrightarrow 2\,Cr^{3+}(aq) + 7\,H_2O(l)$
• Balance H by adding H^+.	$2\,I^-(aq) \longrightarrow I_2(s)$ $14\,H^+(aq) + Cr_2O_7^{2-}(aq) \longrightarrow 2\,Cr^{3+}(aq) + 7\,H_2O(l)$
4. Balance each half-reaction with respect to charge.	$2\,I^-(aq) \longrightarrow I_2(s) + 2e^-$ $6e^- + 14\,H^+(aq) + Cr_2O_7^{2-}(aq) \longrightarrow 2\,Cr^{3+}(aq) + 7\,H_2O(l)$
5. Make the number of electrons in both half-reactions equal.	$3 \times [2\,I^-(aq) \longrightarrow I_2(s) + 2e^-]$ $6e^- + 14\,H^+(aq) + Cr_2O_7^{2-}(aq) \longrightarrow 2\,Cr^{3+}(aq) + 7\,H_2O(l)$
6. Add the half-reactions together.	$6\,I^-(aq) \longrightarrow 3\,I_2(s) + 6e^-$ $6e^- + 14\,H^+(aq) + Cr_2O_7^{2-}(aq) \longrightarrow 2\,Cr^{3+}(aq) + 7\,H_2O(l)$ ─────────────────────── $6\,I^-(aq) + 14\,H^+(aq) + Cr_2O_7^{2-}(aq) \longrightarrow 3\,I_2(s) + 2\,Cr^{3+}(aq) + 7\,H_2O(l)$
7. Verify that the reaction is balanced.	**Reactants** **Products** 6 I 6 I 14 H 14 H 2 Cr 2 Cr 7 O 7 O +6 charge +6 charge

▶ **SKILLBUILDER 16.7** | **Balancing Redox Reactions**

Balance the redox reaction occurring in acidic solution.

$$Sn(s) + MnO_4^-(aq) \longrightarrow Sn^{2+}(aq) + Mn^{2+}(aq)$$

▶ **FOR MORE PRACTICE**   Problems 65, 66, 67, 68.

When a redox reaction occurs in basic solution, we follow the same general procedure with one addition: the neutralization of H^+ with OH^-. The H^+ and OH^- combine to form water, as shown in Example 16.8.

EXAMPLE **16.8** | **Balancing Redox Reactions Occurring in Basic Solution**

Balance the redox reaction occurring in basic solution.

$$CN^-(aq) + MnO_4^-(aq) \longrightarrow CNO^-(aq) + MnO_2(s) \quad \text{(basic solution)}$$

1. Follow the half-reaction method for balancing redox reactions. Begin by assigning oxidation states.	**SOLUTION** $$CN^-(aq) + MnO_4^-(aq) \longrightarrow CNO^-(aq) + MnO_2(s)$$ $$\underset{+2\ -3}{}\quad\underset{+7\ -2}{}\qquad\qquad\underset{+4\ -3\ -2}{}\quad\underset{+4\ -2}{}$$ Reduction — Oxidation
2. Separate the overall reaction into two half-reactions.	**OXIDATION** $CN^-(aq) \longrightarrow CNO^-(aq)$ **REDUCTION** $MnO_4^-(aq) \longrightarrow MnO_2(s)$
3. Balance each half-reaction with respect to mass. • Balance all elements other than H and O. • Balance O by adding H_2O.	All elements other than H and O are already balanced. $$CN^-(aq) + \mathbf{H_2O}(l) \longrightarrow CNO^-(aq)$$ $$MnO_4^-(aq) \longrightarrow MnO_2(s) + \mathbf{2\,H_2O}(l)$$
• Balance H by adding H^+.	$$CN^-(aq) + H_2O(l) \longrightarrow CNO^-(aq) + \mathbf{2\,H^+}(aq)$$ $$MnO_4^-(aq) + \mathbf{4\,H^+}(aq) \longrightarrow MnO_2(s) + 2\,H_2O(l)$$
• Neutralize H^+ by adding OH^-. Add the same number of OH^- to each side of the equation (to preserve mass balance).	$$CN^-(aq) + H_2O(l) + \mathbf{2\,OH^-}(aq) \longrightarrow CNO^-(aq) + \underbrace{2\,H^+(aq) + \mathbf{2\,OH^-}(aq)}_{\mathbf{2\,H_2O}(l)}$$ $$MnO_4^-(aq) + \underbrace{4\,H^+(aq) + \mathbf{4\,OH^-}(aq)}_{\mathbf{4\,H_2O}(l)} \longrightarrow MnO_2(s) + 2\,H_2O(l) + \mathbf{4\,OH^-}(aq)$$
• Cancel any water molecules that occur on both sides of the half-reaction.	$$CN^-(aq) + \cancel{H_2O}(l) + 2\,OH^-(aq) \longrightarrow CNO^-(aq) + \overset{1}{\cancel{2}}\,H_2O(l)$$ $$MnO_4^-(aq) + \overset{2}{\cancel{4}}\,H_2O(l) \longrightarrow MnO_2(s) + \cancel{2\,H_2O}(l) + 4\,OH^-(aq)$$
4. Balance each half-reaction with respect to charge.	$$CN^-(aq) + 2\,OH^-(aq) \longrightarrow CNO^-(aq) + H_2O(l) + \mathbf{2e^-}$$ $$\mathbf{3e^-} + MnO_4^-(aq) + 2\,H_2O(l) \longrightarrow MnO_2(s) + 4\,OH^-(aq)$$
5. Make the number of electrons in both half-reactions equal.	$\mathbf{3} \times [CN^-(aq) + 2\,OH^-(aq) \longrightarrow CNO^-(aq) + H_2O(l) + 2e^-]$ $\mathbf{2} \times [3e^- + MnO_4^-(aq) + 2\,H_2O(l) \longrightarrow MnO_2(s) + 4\,OH^-(aq)]$
6. Add the half-reactions together and cancel.	$$3\,CN^-(aq) + \overset{2}{\cancel{6}}\,OH^-(aq) \longrightarrow 3\,CNO^-(aq) + 3\,\cancel{H_2O}(l) + \cancel{6e^-}$$ $$\cancel{6e^-} + 2\,MnO_4^-(aq) + \overset{1}{\cancel{4}}\,H_2O(l) \longrightarrow 2\,MnO_2(s) + \overset{8}{\cancel{2}}\,OH^-(aq)$$ $$\overline{3\,CN^-(aq) + 2\,MnO_4^-(aq) + H_2O(l) \longrightarrow 3\,CNO^-(aq) + 2\,MnO_2(s) + 2\,OH^-(aq)}$$
7. Verify that the reaction is balanced.	

Reactants	Products
3 C	3 C
3 N	3 N
2 Mn	2 Mn
9 O	9 O
2 H	2 H
−5 charge	−5 charge

▶ **SKILLBUILDER 16.8** | **Balancing Redox Reactions**

Balance the redox reaction occurring in basic solution.

$$H_2O_2(aq) + ClO_2(aq) \longrightarrow ClO_2^-(aq) + O_2(g)$$

▶ **FOR MORE PRACTICE** Problems 69, 70.

CHEMISTRY IN THE ENVIRONMENT

Photosynthesis and Respiration: Energy for Life

All living things require energy, and most of that energy comes from the sun. Solar energy reaches Earth in the form of electromagnetic radiation (Chapter 9). This radiation keeps our planet at a temperature that allows life as we know it to flourish. But the wavelengths that make up visible light have an additional and very crucial role to play in the maintenance of life. Plants capture this light and use it to make energy-rich organic molecules such as carbohydrates. (These compounds are discussed more fully in Chapter 19.) Animals get their energy by eating plants or by eating other animals that have eaten plants. So ultimately, virtually all of the energy for life comes from sunlight.

But in chemical terms, how is this energy captured, transferred from organism to organism, and used? *The key reactions in these processes all involve oxidation and reduction.*

Most living things use chemical energy through a process known as *respiration.* In respiration, energy-rich molecules, typified by the sugar glucose, are "burned" in a reaction that we summarize as follows:

$$C_6H_{12}O_6 + 6\,O_2 \longrightarrow 6\,H_2O + 6\,CO_2 + energy$$
$$\text{Glucose} \quad \text{Oxygen} \quad\quad \text{Water} \quad \text{Carbon} \atop \text{dioxide}$$

We can readily see that respiration is a redox reaction. On the simplest level, it is clear that some of the atoms in glucose are gaining oxygen. More precisely, we can use the rules for assigning oxidation states to show that carbon is oxidized from an oxidation number of 0 in glucose to +4 in carbon dioxide.

Respiration is also an exothermic reaction—it releases energy. If we burn glucose in a test tube, the energy is lost as heat. Living things, however, have devised ways to capture the energy released and use it to power their life processes, such as movement, growth, and the synthesis of other life-sustaining molecules.

Respiration is one half of a larger cycle; the other half is *photosynthesis.* Photosynthesis is the series of reactions by which green plants capture the energy of sunlight and store it as chemical energy in compounds such as glucose. We summarize photosynthesis as follows:

$$6\,CO_2 + 6\,H_2O + energy\ (sunlight) \longrightarrow C_6H_{12}O_6 + 6\,O_2$$
$$\text{Carbon} \quad \text{Water} \quad\quad\quad\quad\quad\quad\quad\quad \text{Glucose} \quad \text{Oxygen} \atop \text{dioxide}$$

This reaction—the exact reverse of respiration—is the ultimate source of the molecules that are oxidized in respiration. And just as the key process in respiration is the oxidation of carbon, the key process in photosynthesis is the reduction of carbon. This reduction is driven by solar energy—energy that is then stored in the resulting glucose molecule. Living things harvest that energy when they "burn" glucose in the respiration half of the cycle.

Thus, oxidation and reduction reactions are at the very center of all life on Earth.

B16.2 CAN YOU ANSWER THIS?   *What is the oxidation state of the oxygen atoms in CO_2, H_2O, and O_2? What does this information tell you about the reactions of photosynthesis and respiration?*

▲ Sunlight, captured by plants in photosynthesis, is the ultimate source of the chemical energy for nearly all living things on Earth.

16.5 The Activity Series: Predicting Spontaneous Redox Reactions

▶ Predict spontaneous redox reactions.

▶ Predict whether a metal will dissolve in acid.

As we have seen, redox reactions depend on one substance gaining electrons and another losing them. Is there a way to predict whether a particular redox reaction will spontaneously occur? Suppose we knew that substance A has a greater tendency to lose electrons than substance B (A is more easily oxidized than B). Then we could predict that if we mix A with cations of B, a redox reaction would occur in which A loses its electrons (A is oxidized) to the cations of B (B cations are reduced).

▶ **FIGURE 16.5 Cu^{2+} oxidizes magnesium** When we put a magnesium strip into a Cu^{2+} solution, the magnesium is oxidized to Mg^{2+} and the copper(II) ion is reduced to Cu(s). Notice the fading of the blue color (due to Cu^{2+} ions) in solution and the appearance of solid copper on the magnesium strip.

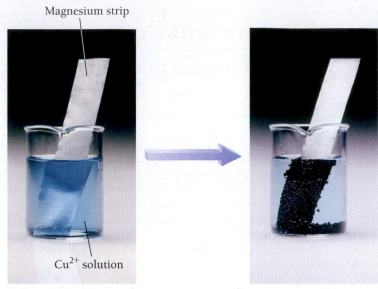

Magnesium strip

Cu^{2+} solution

$$Mg(s) + Cu^{2+}(aq) \longrightarrow Mg^{2+}(aq) + Cu(s)$$

For example, Mg has a greater tendency to lose electrons than Cu. Consequently, if we put solid Mg into a solution containing Cu^{2+} ions, Mg is oxidized and Cu^{2+} is reduced.

$$Mg(s) + Cu^{2+}(aq) \longrightarrow Mg^{2+}(aq) + Cu(s)$$

We see this as the fading of blue (the color of the Cu^{2+} ions in solution), the dissolving of the solid magnesium, and the appearance of solid copper on the remaining magnesium surface (▲ FIGURE 16.5). This reaction is spontaneous—it occurs on its own when Mg(s) and $Cu^{2+}(aq)$ come into contact.

In contrast, if we put Cu(s) in a solution containing $Mg^{2+}(aq)$ ions, no reaction occurs (◀ FIGURE 16.6).

$$Cu(s) + Mg^{2+}(aq) \longrightarrow NO\ REACTION$$

No reaction occurs because, as we said previously, Mg atoms have a greater tendency to lose electrons than do Cu atoms; Cu atoms will therefore not lose electrons to Mg^{2+} ions.

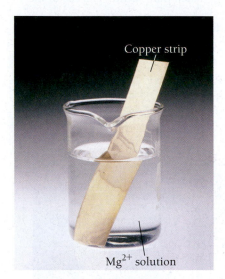

▲ **FIGURE 16.6 Mg^{2+} does not oxidize copper** When we place solid copper in a solution containing Mg^{2+} ions, no reaction occurs.
Question: Why?

Copper strip

Mg^{2+} solution

The Activity Series of Metals

Table 16.1 is the **activity series of metals**. This table lists metals in order of decreasing tendency to lose electrons. The metals at the top of the list have the greatest tendency to lose electrons—they are most easily oxidized and therefore the most reactive. The metals at the bottom of the list have the lowest tendency to lose electrons—they are the most difficult to oxidize and therefore the least reactive. It is not a coincidence that the metals used for jewelry, such as silver and gold, are near the bottom of the list. They are among the least reactive metals and therefore do not form compounds easily. Instead, they tend to remain as solid silver and solid gold rather than being oxidized to silver and gold cations by elements in the environment (such as the oxygen in the air).

Each reaction in the activity series is an oxidation half-reaction. The half-reactions at the top are most likely to occur in the *forward* direction, and the half-reactions at the bottom are most likely to occur in the *reverse* direction. Consequently, if we pair a half-reaction from the top of the list with the reverse of a half-reaction from the bottom of the list, we get a spontaneous reaction. More specifically,

Any half-reaction on the list is spontaneous when paired with the reverse of any half-reaction below it on the list.

For example, consider the two half-reactions:

$$Mn(s) \longrightarrow Mn^{2+}(aq) + 2e^-$$

$$Ni(s) \longrightarrow Ni^{2+}(aq) + 2e^-$$

▲ Gold is very low on the activity series. Because it is so difficult to oxidize, it resists the tarnishing and corrosion that more active metals undergo.

TABLE 16.1 Activity Series of Metals	
$Li(s) \longrightarrow Li^+(aq) + e^-$	Most reactive
$K(s) \longrightarrow K^+(aq) + e^-$	Most easily oxidized
$Ca(s) \longrightarrow Ca^{2+}(aq) + 2e^-$	Strongest tendency to lose electrons
$Na(s) \longrightarrow Na^+(aq) + e^-$	
$Mg(s) \longrightarrow Mg^{2+}(aq) + 2e^-$	
$Al(s) \longrightarrow Al^{3+}(aq) + 3e^-$	
$Mn(s) \longrightarrow Mn^{2+}(aq) + 2e^-$	
$Zn(s) \longrightarrow Zn^{2+}(aq) + 2e^-$	
$Cr(s) \longrightarrow Cr^{3+}(aq) + 3e^-$	
$Fe(s) \longrightarrow Fe^{2+}(aq) + 2e^-$	
$Ni(s) \longrightarrow Ni^{2+}(aq) + 2e^-$	
$Sn(s) \longrightarrow Sn^{2+}(aq) + 2e^-$	
$Pb(s) \longrightarrow Pb^{2+}(aq) + 2e^-$	
$\mathbf{H_2(g) \longrightarrow 2\,H^+(aq) + 2\,e^-}$	
$Cu(s) \longrightarrow Cu^{2+}(aq) + 2e^-$	Least reactive
$Ag(s) \longrightarrow Ag^+(aq) + e^-$	Most difficult to oxidize
$Au(s) \longrightarrow Au^{3+}(aq) + 3e^-$	Least tendency to lose electrons

H_2 is in bold in Table 16.1 because it is the only non-metal in the table and a common reference point in the activity series.

The oxidation of Mn is spontaneous when paired with the reduction of Ni^{2+}.

$$Mn(s) \longrightarrow Mn^{2+}(aq) + 2e^-$$
$$Ni^{2+}(aq) + 2e^- \longrightarrow Ni(s)$$
$$Mn(s) + Ni^{2+}(aq) \longrightarrow Mn^{2+}(aq) + Ni(s) \quad \text{(spontaneous reaction)}$$

In contrast, if we pair a half-reaction on the list with the reverse of a half-reaction above it, there is no reaction.

$$Mn(s) \longrightarrow Mn^{2+}(aq) + 2e^-$$
$$Mg^{2+}(aq) + 2e^- \longrightarrow Mg(s)$$
$$Mn(s) + Mg^{2+}(aq) \longrightarrow \text{NO REACTION}$$

No reaction occurs because Mg has a greater tendency to be oxidized than Mn. Since it is already oxidized in this reaction, nothing else happens.

EXAMPLE 16.9   **Predicting Spontaneous Redox Reactions**

Determine whether each redox reaction is spontaneous.

(a) $Fe(s) + Mg^{2+}(aq) \longrightarrow Fe^{2+}(aq) + Mg(s)$

(b) $Fe(s) + Pb^{2+}(aq) \longrightarrow Fe^{2+}(aq) + Pb(s)$

SOLUTION

(a) $Fe(s) + Mg^{2+}(aq) \longrightarrow Fe^{2+}(aq) + Mg(s)$

This reaction involves the oxidation of Fe:

$$Fe(s) \longrightarrow Fe^{2+}(aq) + 2e^-$$

with the reverse of a half-reaction *above it* in the activity series:

$$Mg^{2+}(aq) + 2e^- \longrightarrow Mg(s)$$

Therefore, the reaction *is not* spontaneous.

(b) $Fe(s) + Pb^{2+}(aq) \longrightarrow Fe^{2+}(aq) + Pb(s)$

This reaction involves the oxidation of Fe:

$$Fe(s) \longrightarrow Fe^{2+}(aq) + 2e^-$$

with the reverse of a half-reaction *below it* in the activity series:

$$Pb^{2+}(aq) + 2e^- \longrightarrow Pb(s)$$

Therefore, the reaction *is* spontaneous.

continued on page 588 ▶

continued from page 587

▶ **SKILLBUILDER 16.9** | **Predicting Spontaneous Redox Reactions**

Determine whether each redox reaction is spontaneous.

(a) $Zn(s) + Ni^{2+}(aq) \longrightarrow Zn^{2+}(aq) + Ni(s)$
(b) $Zn(s) + Ca^{2+}(aq) \longrightarrow Zn^{2+}(aq) + Ca(s)$

▶ **FOR MORE PRACTICE** Example 16.15; Problems 77, 78.

CONCEPTUAL ✓ CHECKPOINT 16.3

Which metal is most easily oxidized?

(a) Na   **(b)** Cr   **(c)** Au

$Zn(s) + 2 H^+(aq) \longrightarrow$

$Zn^{2+}(aq) + H_2(g)$

▲ **FIGURE 16.7 Zinc dissolves in hydrochloric acid** The zinc metal is oxidized to Zn^{2+} ions, and the H^+ ions are reduced to hydrogen gas.

Predicting Whether a Metal Will Dissolve in Acid

In Chapter 14, we learned that acids dissolve metals. Most acids dissolve metals by the reduction of H^+ ions to hydrogen gas and the corresponding oxidation of the metal to its ion. For example, if we drop solid Zn into hydrochloric acid, the following reaction occurs:

$$Zn(s) \longrightarrow Zn^{2+}(aq) + 2e^-$$
$$2 H^+(aq) + 2e^- \longrightarrow H_2(g)$$
$$Zn(s) + 2 H^+(aq) \longrightarrow Zn^{2+}(aq) + H_2(g)$$

We observe the reaction as the dissolving of the zinc and the bubbling of hydrogen gas (◀ **FIGURE 16.7**). The zinc is oxidized and the H^+ ions are reduced, dissolving the zinc. Notice that this reaction involves pairing the oxidation half-reaction of Zn with the reverse of a half-reaction below zinc on the activity series (the reduction of H^+). Therefore, this reaction is spontaneous.

What happens, in contrast, if we pair the oxidation of Cu with the reduction of H^+? The reaction is not spontaneous because it involves pairing the oxidation of copper with the reverse of a half-reaction *above it* in the activity series. Consequently, copper does not react with H^+ and does not dissolve in acids such as HCl. In general,

Metals above H_2 on the activity series dissolve in acids, while metals below H_2 do not dissolve in acids.

An important exception to this rule is nitric acid (HNO_3), which through a different reduction half-reaction dissolves some of the metals below H_2 in the activity series.

EXAMPLE 16.10 | **Predicting Whether a Metal Will Dissolve in Acid**

Does Cr dissolve in hydrochloric acid?

SOLUTION

Yes. Since Cr is above H_2 in the activity series, it dissolves in HCl.

▶ **SKILLBUILDER 16.10** | **Predicting Whether a Metal Will Dissolve in Acid**

Does Ag dissolve in hydrobromic acid?

▶ **FOR MORE PRACTICE** Problems 83, 84.

CONCEPTUAL ✔ CHECKPOINT 16.4

It has been suggested that one cause of the decline of the Roman Empire was widespread chronic poisoning. The suspected source was a metal present in the vessels commonly used to store and serve acidic substances such as wine. Which metal would you expect to pose such a danger?

(a) silver   **(b)** gold   **(c)** lead   **(d)** copper

16.6 Batteries: Using Chemistry to Generate Electricity

▶ Describe how a voltaic cell functions.

▶ Compare and contrast the various types of batteries.

▲ **FIGURE 16.8 An electrical current** Electrical current is the flow of electrical charge. In this figure, electrons are flowing through a wire.

> The fuel cell discussed in Section 16.1 is a type of electrochemical cell.

> The salt bridge completes the circuit— it allows the flow of ions between the two half-cells.

Electrical current is the flow of electric charge (◀ FIGURE 16.8). Electrons flowing through a wire or ions flowing through a solution are both examples of electrical current. Since redox reactions involve the transfer of electrons from one species to another, they can create electrical current.

For example, consider the following spontaneous redox reaction:

$$Zn(s) + Cu^{2+}(aq) \longrightarrow Zn^{2+}(aq) + Cu(s)$$

When we place Zn metal into a Cu^{2+} solution, Zn is oxidized and Cu^{2+} is reduced—electrons are transferred directly from the Zn to the Cu^{2+}. Suppose we separate the reactants and force the electrons to travel through a wire to get from the Zn to the Cu^{2+}. The flowing electrons constitute an electrical current and can be used to do electrical work. This process is normally carried out in an **electrochemical cell**, a device that creates electrical current from a spontaneous redox reaction (or that uses electrical current to drive a nonspontaneous redox reaction). Electrochemical cells that create electrical current from spontaneous reactions are **voltaic cells** or **galvanic cells**. A *battery* is a voltaic cell that (usually) has been designed for portability.

Consider the voltaic cell in ▶ FIGURE 16.9, on the next page. In this cell, a solid strip of Zn is placed into a $Zn(NO_3)_2$ solution to form a **half-cell**. Similarly, a solid strip of Cu is placed into a $Cu(NO_3)_2$ solution to form a second half-cell. The two half-cells are connected with a wire from the zinc, through a light bulb or other electrical device, to the copper. The natural tendency of Zn to oxidize and Cu^{2+} to reduce results in a flow of electrons through the wire. The flowing electrons constitute an electrical current that lights the bulb.

In a voltaic cell, the metal strip where oxidation occurs is the **anode** and is labeled with a negative (−) sign. The metal strip where reduction occurs is the **cathode** and is labeled with a (+) sign. Electrons flow from the anode to the cathode (away from negative and toward positive).

As electrons flow out of the anode, positive ions form in the oxidation half-cell (Zn^{2+} forms in the preceding example). As electrons flow into the cathode, positive ions deposit as charge-neutral atoms at the reduction half-cell (Cu^{2+} deposits as $Cu(s)$ in the preceding example). However, if this were the only flow of charge, the flow would soon stop as positive charge accumulated at the anode and negative charge at the cathode. The circuit must be completed with a **salt bridge**, an inverted U-shaped tube that joins the two half-cells and contains a strong electrolyte such as KNO_3. The salt bridge allows for the flow of ions that neutralizes the charge imbalance. The negative ions within the salt bridge flow to neutralize the accumulation of positive charge at the anode, and the positive ions flow to neutralize the accumulation of negative charge at the cathode.

In a voltaic cell, electrical voltage is the driving force that causes electrons to flow. A high voltage corresponds to a high driving force, while a low voltage corresponds to a low driving force. We can understand electrical voltage with an analogy. Electrons

▶ **FIGURE 16.9 A voltaic cell**
Question: Why do electrons flow from
left to right in this figure?

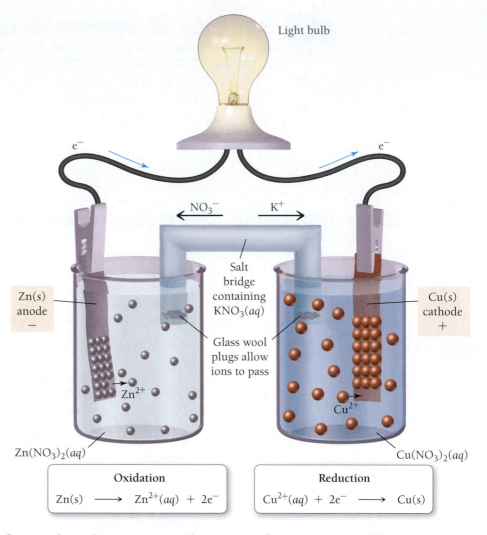

Light bulb

e^-

e^-

NO_3^- K^+

Salt
bridge
containing
$KNO_3(aq)$

Zn(s)
anode
−

Cu(s)
cathode
+

Glass wool
plugs allow
ions to pass

Zn^{2+}

Cu^{2+}

$Zn(NO_3)_2(aq)$

$Cu(NO_3)_2(aq)$

Oxidation
$Zn(s) \longrightarrow Zn^{2+}(aq) + 2e^-$

Reduction
$Cu^{2+}(aq) + 2e^- \longrightarrow Cu(s)$

flowing through a wire are similar to water flowing in a river (▼ **FIGURE 16.10**). The
quantity of electrons that flows through the wire (electrical current) is analogous to the
amount of water that flows through the river (the river's current). The driving force
that causes the electrons to flow through a wire—*potential difference* or **voltage**—is
analogous to the force of gravity that causes water to flow in a river. A high voltage is
analogous to a steeply descending streambed (▼ **FIGURE 16.11**).

The flow of electrons through a wire is
analogous to the flow of water in a river.

A high voltage for electricity is analogous
to a steep descent for a river.

e^- e^- e^-

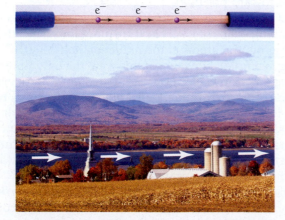

e^- e^- e^- e^- e^-

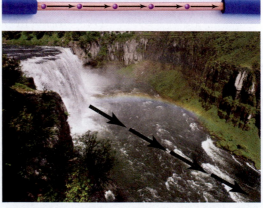

▲ **FIGURE 16.10 River analogy for electrical current**

▲ **FIGURE 16.11 River analogy for voltage**

▶ **FIGURE 16.12 Used voltaic cell**
A voltaic cell dies with extended use because the reactants [in this case $Zn(s)$ and $Cu^{2+}(aq)$] become depleted while the products [in this case $Zn^{2+}(aq)$ and $Cu(s)$] accumulate.

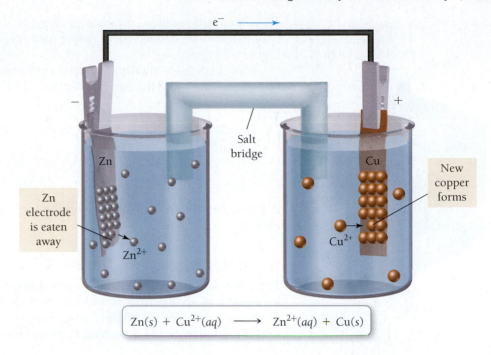

$$Zn(s) + Cu^{2+}(aq) \longrightarrow Zn^{2+}(aq) + Cu(s)$$

The voltage of a voltaic cell depends on the relative tendencies of the reactants to undergo oxidation and reduction. Combining the oxidation of a metal high on the activity series with the reduction of a metal ion low on the activity series produces a battery with a relatively high voltage. For example, the oxidation of $Li(s)$ combined with the reduction of $Cu^{2+}(aq)$ results in a relatively high voltage. On the other hand, combining the oxidation of a metal on the activity series with the reduction of a metal ion just below it results in a voltaic cell with a relatively low voltage. Combining the oxidation of a metal on the activity series with the reduction of a metal ion above it on the activity series does not produce a voltaic cell at all. For example, we cannot make a voltaic cell by trying to oxidize $Cu(s)$ and reduce $Li^+(aq)$. Such a reaction is not spontaneous and does not produce electrical current.

Why do voltaic cells (and batteries) go dead after extended use? As the simple voltaic cell we have just described is used, the zinc electrode dissolves away as zinc is oxidized to zinc ions. Similarly, the Cu^{2+} solution is depleted of Cu^{2+} ions as they deposit as solid Cu (▲ **FIGURE 16.12**). Once the zinc electrode is dissolved and the Cu^{2+} ions are depleted, the cell is dead. Some voltaic cells, such as those employed in rechargeable batteries, can be recharged by running electrical current—from an external source—in the opposite direction. This causes the regeneration of the reactants, allowing repeated use of the battery.

Dry-Cell Batteries

Flashlight batteries are called **dry cells** because they do not contain large amounts of liquid water. There are several common types of dry-cell batteries. The most inexpensive type of dry cell is composed of a zinc case that acts as the anode (◀ **FIGURE 16.13**). The zinc is oxidized according to the following reaction:

$$\text{Anode reaction: } Zn(s) \longrightarrow Zn^{2+}(aq) + 2e^- \quad \text{(oxidation)}$$

The cathode is a carbon rod immersed in a moist paste of MnO_2 that also contains NH_4Cl. The MnO_2 is reduced to Mn_2O_3 according to the following reaction:

$$\text{Cathode reaction: } 2\,MnO_2(s) + 2\,NH_4^+(aq) + 2e^- \longrightarrow$$

$$Mn_2O_3(s) + 2\,NH_3(g) + H_2O(l) \quad \text{(reduction)}$$

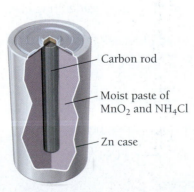

▲ **FIGURE 16.13 Common dry-cell battery**

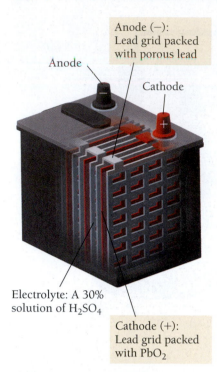

Anode

Anode (−):
Lead grid packed
with porous lead

Cathode

Electrolyte: A 30%
solution of H_2SO_4

Cathode (+):
Lead grid packed
with PbO_2

▲ **FIGURE 16.14 Lead-acid storage battery** Question: Why do batteries like this become depleted? How are they recharged?

The porosity of the lead anode in a lead-acid storage battery increases the surface area where electrons can be transferred from the solid lead to the solution.

These two half-reactions produce a voltage of about 1.5 volts. Two or more of these batteries can be connected in series (cathode-to-anode connection) to produce higher voltages.

More expensive **alkaline batteries** employ slightly different half-reactions that use a base (therefore the name *alkaline*). In an alkaline battery, the reactions are as follows:

Anode reaction: $Zn(s) + 2\,OH^-(aq) \longrightarrow Zn(OH)_2(s) + 2e^-$ (oxidation)

Cathode reaction: $2\,MnO_2(s) + 2\,H_2O(l) + 2e^- \longrightarrow$
$2\,MnO(OH)(s) + 2\,OH^-(aq)$ (reduction)

Alkaline batteries have a longer working life and a longer shelf life than their non-alkaline counterparts.

Lead-Acid Storage Batteries

The batteries in most automobiles are **lead-acid storage batteries**. These batteries consist of six electrochemical cells wired in series (◀ FIGURE 16.14). Each cell produces 2 volts for a total of 12 volts. The cells each contain a porous lead anode where oxidation occurs according to the following reaction:

Anode reaction: $Pb(s) + SO_4^{2-}(aq) \longrightarrow PbSO_4(s) + 2e^-$ (oxidation)

Each cell also contains a lead(IV) oxide cathode where reduction occurs according to the following reaction:

Cathode reaction: $PbO_2(s) + 4\,H^+(aq) + SO_4^{2-}(aq) + 2e^- \longrightarrow$
$PbSO_4(s) + 2\,H_2O(l)$ (reduction)

Both the anode and the cathode are immersed in sulfuric acid (H_2SO_4). As electrical current is drawn from the battery, both the anode and the cathode become coated with $PbSO_4(s)$. If the battery is run for a long time without recharging, too much $PbSO_4(s)$ develops and the battery goes dead. The lead-acid storage battery can be recharged, however, by running electrical current through it in reverse. The electrical current has to come from an external source, such as an alternator in a car. This causes the preceding reactions to occur in reverse, converting the $PbSO_4(s)$ back to $Pb(s)$ and $PbO_2(s)$, recharging the battery.

Fuel Cells

As suggested in Section 16.1, electric vehicles powered by fuel cells may one day replace internal combustion vehicles. Fuel cells are like batteries, but the reactants are constantly replenished. Normal batteries lose their voltage with use because the reactants are depleted as electrical current is drawn from the battery. In a fuel cell, the reactants—the fuel—constantly flow through the battery, generating electrical current as they undergo a redox reaction.

The most common fuel cell is the hydrogen–oxygen fuel cell (▶ FIGURE 16.15). In this cell, hydrogen gas flows past the anode (a screen coated with platinum catalyst) and undergoes oxidation.

Anode reaction: $2\,H_2(g) + 4\,OH^-(aq) \longrightarrow 4\,H_2O(l) + 4e^-$

Oxygen gas flows past the cathode (a similar screen) and undergoes reduction.

Cathode reaction: $O_2(g) + 2\,H_2O(l) + 4e^- \longrightarrow 4\,OH^-(aq)$

The half-reactions sum to the following overall reaction:

Overall reaction: $2\,H_2(g) + O_2(g) \longrightarrow 2\,H_2O(l)$

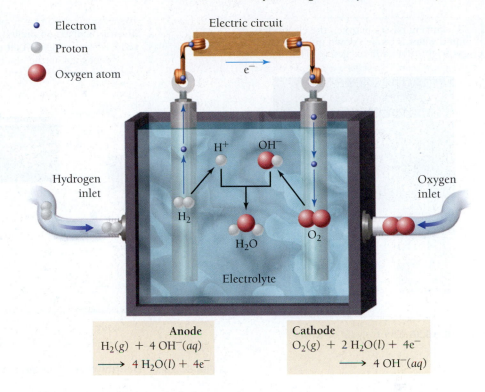

◀ **FIGURE 16.15 Hydrogen–oxygen fuel cell**

Anode	Cathode
$H_2(g) + 4 OH^-(aq)$	$O_2(g) + 2 H_2O(l) + 4 e^-$
$\longrightarrow 4 H_2O(l) + 4 e^-$	$\longrightarrow 4 OH^-(aq)$

Notice that the only product is water. In the space shuttle program, hydrogen–oxygen fuel cells provide electricity and astronauts drink the resulting water.

CONCEPTUAL ✔ CHECKPOINT 16.5

Suppose you are making a battery composed of a carbon rod inserted into a moist paste of lead(II) ions that acts as the cathode. You want the metal casing that encloses the battery to act as the anode. Which metal should you use for the casing to achieve the battery with the highest voltage?

(a) Mg
(b) Zn
(c) Ni

16.7 Electrolysis: Using Electricity to Do Chemistry

▶ Describe the process of electrolysis and how an electrolytic cell functions.

In a voltaic cell, a spontaneous redox reaction produces electrical current. In **electrolysis**, electrical current is used to drive an otherwise nonspontaneous redox reaction. An electrochemical cell used for electrolysis is an **electrolytic cell**. We discussed how the reaction of hydrogen with oxygen to form water is spontaneous and can be used to produce an electrical current in a fuel cell. By providing electrical current, we can cause the reverse reaction to occur, breaking water into hydrogen and oxygen (▶ **FIGURE 16.16**, on the next page).

$2 H_2(g) + O_2(g) \longrightarrow 2 H_2O(l)$ (spontaneous—produces electrical current; occurs in a voltaic cell)

$2 H_2O(l) \longrightarrow 2 H_2(g) + O_2(g)$ (nonspontaneous—consumes electrical current; occurs in an electrolytic cell)

As a current passes between the electrodes, liquid water is broken down into hydrogen gas (right tube) and oxygen gas (left tube).

Silver is oxidized on the left side of the cell and reduced on the right. As it is reduced, silver is deposited on the object to be plated.

▲ **FIGURE 16.16 Electrolysis of water**

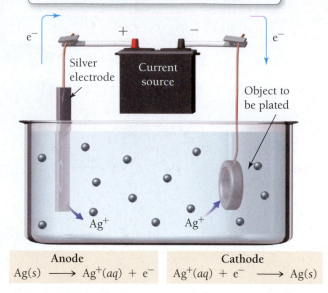

Anode	Cathode
$Ag(s) \longrightarrow Ag^+(aq) + e^-$	$Ag^+(aq) + e^- \longrightarrow Ag(s)$

▲ **FIGURE 16.17 Electrolytic cell for silver plating**

One problem associated with the widespread adoption of fuel cells is the scarcity of hydrogen. Where is the hydrogen to power these fuel cells going to come from? One possible answer is that the hydrogen can come from water through solar- or wind-powered electrolysis. In other words, a solar- or wind-powered electrolytic cell can make hydrogen from water when the sun is shining or when the wind is blowing. The hydrogen can then be converted back to water to generate electricity when needed. Hydrogen made in this way could also be used to power fuel-cell vehicles.

Electrolysis also has numerous other applications. For example, most metals are found in Earth's crust as metal oxides. Converting them to pure metals requires the reduction of the metal, a nonspontaneous process. Electrolysis can be used to produce these metals. Electrolysis can also be used to plate metals onto other metals. For example, we can plate silver onto another, less expensive metal using the electrolytic cell shown in ▲ **FIGURE 16.17**. In this cell, we place a silver electrode in a solution containing silver ions. An electrical current then causes the oxidation of silver at the anode (replenishing the silver ions in solution) and the reduction of silver ions at the cathode (coating the ordinary metal with solid silver).

$$\text{Anode reaction:} \quad Ag(s) \longrightarrow Ag^+(aq) + e^-$$

$$\text{Cathode reaction:} \quad Ag^+(aq) + e^- \longrightarrow Ag(s)$$

16.8 Corrosion: Undesirable Redox Reactions

▶ Describe the process of corrosion and the various methods used to prevent rust.

Corrosion is the oxidation of metals. The most common kind of corrosion is the rusting of iron. A significant part of the iron produced each year goes to replace rusted iron. Rusting is a redox reaction in which iron is oxidized and oxygen is reduced.

Oxidation: $\quad 2\,Fe(s) \longrightarrow 2\,Fe^{2+}(aq) + 4e^-$

Reduction: $\quad O_2(g) + 2\,H_2O(l) + 4e^- \longrightarrow 4\,OH^-(aq)$

Overall: $\quad \dfrac{}{2\,Fe(s) + O_2(g) + 2\,H_2O(l) \longrightarrow 2\,Fe(OH)_2(s)}$

The $Fe(OH)_2$ formed in the overall reaction undergoes several additional reactions to form Fe_2O_3, the familiar orange substance that we call rust. One of the

EVERYDAY CHEMISTRY

The Fuel-Cell Breathalyzer

Police use a device called a breathalyzer to measure the amount of ethyl alcohol (C_2H_5OH) in the bloodstream of a person suspected of driving under the influence of alcohol.

Breathalyzers work because the amount of ethyl alcohol in the breath is proportional to the amount of ethyl alcohol in the bloodstream. One type of breathalyzer employs a fuel cell to measure the amount of alcohol in the breath. The fuel cell consists of two platinum electrodes (▼ FIGURE 16.18). When a suspect blows into the breathalyzer, any ethyl alcohol in the breath is oxidized to acetic acid at the anode.

$$\text{Anode: } \underset{\text{Ethyl alcohol}}{C_2H_5OH} + 4\,OH^-(aq) \longrightarrow$$

$$\underset{\text{Acetic acid}}{CH_3COOH(aq)} + 3\,H_2O + 4e^-$$

▶ FIGURE 16.18 Schematic diagram of a fuel-cell breathalyzer

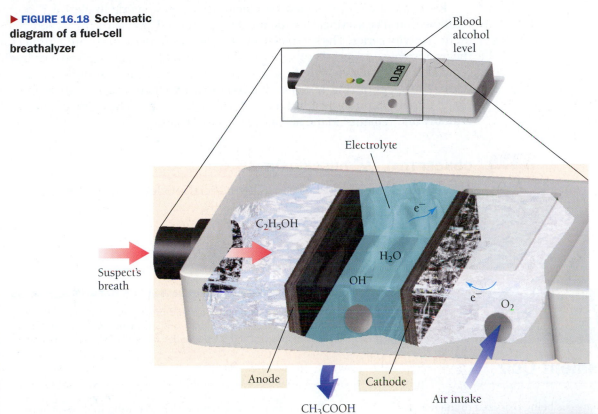

At the cathode, oxygen is reduced.

$$\text{Cathode: } O_2(g) + 2\,H_2O(l) + 4e^- \longrightarrow 4\,OH^-(aq)$$

The overall reaction is the oxidation of ethyl alcohol to acetic acid and water.

$$\text{Overall: } C_2H_5OH(g) + O_2(g) \longrightarrow CH_3COOH(g) + H_2O$$

The amount of electrical current produced depends on the amount of alcohol in the breath. A higher current reveals a higher blood alcohol level. When calibrated correctly, the fuel-cell breathalyzer can precisely measure the blood alcohol level of a suspected drunk driver.

B16.3 CAN YOU ANSWER THIS? *Assign oxidation states to each element in the reactants and products in the overall equation for the fuel-cell breathalyzer. What element is oxidized, and what element is reduced in the reaction?*

▲ **Fuel-cell breathalyzer** A suspect blows into the top of this device, and blood alcohol level is determined.

▲ Paint can prevent underlying iron from rusting. However, if the paint becomes scratched, the iron will rust at the point of the chip. **Question: Why?**

main problems with Fe_2O_3 is that it crumbles off the solid iron below it, exposing more iron to further rusting. Under the right conditions, an entire piece of iron can rust away.

Iron is not the only metal that undergoes oxidation. Most other metals, such as copper and aluminum, also undergo oxidation. However, the oxides of copper and aluminum do not flake off as iron oxide does. When aluminum oxidizes, the aluminum oxide actually forms a tough clear coating on the underlying metal. This coating protects the underlying metal from further oxidation.

Preventing iron from rusting is a major industry. The most obvious way to prevent rust is to keep iron dry. Without water, the redox reaction cannot occur. Another way to prevent rust is to coat the iron with a substance that is impervious to water. Cars, for example, are painted and sealed to prevent rust. A scratch in the paint, however, can lead to rusting of the underlying iron.

Rust can also be prevented by placing a *sacrificial anode* in electrical contact with the iron. The sacrificial anode must be composed of a metal that is above iron on the activity series. The sacrificial anode oxidizes in place of the iron, protecting the iron from oxidation. Another way to protect iron from rusting is to coat it with a metal above it in the activity series. Galvanized nails, for example, are coated with a thin layer of zinc. Because zinc is more active than iron, it oxidizes in place of the underlying iron (just like a sacrificial anode does). The oxide of zinc is not crumbly and remains on the nail as a protective coating.

CONCEPTUAL ✓ **CHECKPOINT 16.6**

Which metal could NOT be used as a sacrificial anode to prevent rusting?

  (a) Mg   **(b)** Mn   **(c)** Zn   **(d)** Sn

Chapter 16 in Review

MasteringChemistry™ provides end-of-chapter exercises, feedback-enriched tutorial problems, animations, and interactive activities to encourage problem solving practice and deeper understanding of key concepts and topics.

Self-Assessment Quiz

Q1. Which substance is being oxidized in the reaction?

$$2\,Na(s)\ +\ Br_2(g)\ \longrightarrow\ 2\,NaBr(s)$$

  (a) Na   **(b)** Br_2
  (c) NaBr   **(d)** none of the above

Q2. What always happens to an oxidizing agent during a redox reaction?
  (a) It is oxidized.   **(b)** It is reduced.
  (c) It becomes a liquid.   **(d)** It becomes a gas.

Q3. What is the oxidation state of carbon in CO_3^{2-}?
  (a) -3 **(b)** -2 **(c)** $+3$ **(d)** $+4$

Q4. In which compound does phosphorus have the lowest oxidation state?
  (a) P_2S_5
  (b) P_2O_3
  (c) K_3PO_4
  (d) none of the above (the oxidation state of phosphorus is the same in all three compounds)

Q5. Sodium reacts with water according to the reaction:

$$2\,Na(s) + 2\,H_2O(l)\ \longrightarrow\ 2\,NaOH(aq) + H_2(g)$$

Identify the element that is reduced.
  (a) Na   **(b)** O
  (c) H   **(d)** none of the above

Q6. How many electrons are exchanged when this reaction is balanced?

$$Mn(s) + Cr^{3+}(aq)\ \longrightarrow\ Mn^{2+}(aq) + Cr(s)$$

  (a) 2   **(b)** 3   **(c)** 4   **(d)** 6

Q7. Balance the redox reaction equation (occurring in acidic solution) and choose the correct coefficients for each reactant and product.

$$\_VO_2^{+}(aq) + \_Sn(s) + \_H^{+}(aq) \longrightarrow$$
$$\_VO^{2+}(aq) + \_Sn^{2+}(aq) + \_H_2O(l)$$

  (a) $1,1,2 \longrightarrow 1,1,1$ **(b)** $2,1,2 \longrightarrow 2,1,1$
  (c) $2,1,4 \longrightarrow 2,1,2$ **(d)** $2,1,2 \longrightarrow 2,1,2$

Q8. Which metal is least reactive (based on the activity series)?
(a) Ca   (b) Mn   (c) Cr   (d) Cu

Q9. Which redox reaction is spontaneous?
(a) $Pb^{2+}(aq) + Mg(s) \longrightarrow Mg^{2+}(aq) + Pb(s)$
(b) $Zn^{2+}(aq) + Sn(s) \longrightarrow Sn^{2+}(aq) + Zn(s)$
(c) $Ca^{2+}(aq) + Al(s) \longrightarrow Al^{3+}(aq) + Ca(s)$
(d) none of the above

Q10. Which metal does *not* dissolve in hydrochloric acid?
(a) Mn
(b) Ni
(c) Sn
(d) Ag

Answers: 1:a, 2:b, 3:d, 4:b, 5:c, 6:d, 7:c, 8:d, 9:a, 10:d

Chemical Principles

Oxidation and Reduction

Oxidation is:

- the loss of electrons.
- an increase in oxidation state.

Reduction is:

- the gain of electrons.
- a decrease in oxidation state.

Oxidation and reduction reactions always occur together and are sometimes called redox reactions. The substance that is oxidized is the reducing agent, and the substance that is reduced is the oxidizing agent.

Oxidation States

The oxidation state is a fictitious charge assigned to each atom in a compound. It is calculated by assigning all bonding electrons in a compound to the most electronegative element.

The Activity Series

The activity series is a listing of metals from those that are easiest to oxidize to those that are most difficult to oxidize. Any half-reaction in the activity series is spontaneous when paired with the reverse of a half-reaction below it on the list.

Batteries

In a battery, the reactants of a spontaneous redox reaction are physically separated. As the redox reaction occurs, the transferred electrons are forced to travel through a wire or other external circuit, creating an electrical current that can be used to do electrical work.

Electrolysis

In a battery, a spontaneous redox reaction is used to generate an electrical current. In electrolysis, an electrical current is used to drive a nonspontaneous redox reaction.

Corrosion

Corrosion is the oxidation of iron and other metals by atmospheric oxygen. Corrosion can be prevented by keeping the metal dry, sealing it with a protective coating, or depositing a more active metal onto the surface of the metal to be protected.

Relevance

Redox reactions are common in nature, in industry, and in many everyday processes. Batteries use redox reactions to generate electrical current. Our bodies use redox reactions to obtain energy from glucose. In addition, the bleaching of hair, the rusting of iron, and the electroplating of metals all involve redox reactions.

Good oxidizing agents, such as oxygen, hydrogen peroxide, and chlorine, have a strong tendency to gain electrons. Good reducing agents, such as sodium and hydrogen, have a strong tendency to lose electrons.

Oxidation states help us more easily identify substances being oxidized and reduced in a chemical reaction.

The activity series allows us to predict whether a redox reaction (involving half-reactions from the series) is spontaneous.

Batteries are common as portable sources of electrical current. They are used in flashlights, watches, automobiles, and other electrical devices.

Electrolysis has many applications. For example, electrolysis is used to reduce metal oxides found in Earth's crust to their metals and to plate metals onto other metals.

The most common form of corrosion is the rusting of iron. Since a significant fraction of all iron produced is used to replace rusted iron, the prevention of rust is a major industry.

Chemical Skills

Examples

Oxidation can be identified as the gain of oxygen, the loss of electrons, or an increase in oxidation state. Reduction can be identified as the loss of oxygen, the gain of electrons, or a decrease in oxidation state.

When a substance gains oxygen, the substance is oxidized and the oxygen is reduced.

When a metal reacts with an electronegative element, the metal is oxidized and the electronegative element is reduced.

When a metal transfers electrons to a metal ion, the metal is oxidized and the metal ion is reduced.

EXAMPLE 16.11 **Identifying Oxidation and Reduction**

Determine the substance being oxidized and the substance being reduced in each redox reaction.

(a) $Sn(s) + O_2(g) \longrightarrow SnO_2(s)$
(b) $2\,Na(s) + F_2(g) \longrightarrow 2\,NaF(s)$
(c) $Mg(s) + Cu^{2+}(aq) \longrightarrow Mg^{2+}(aq) + Cu(s)$

SOLUTION

(a) Sn oxidized; O_2 reduced
(b) Na oxidized; F_2 reduced
(c) Mg oxidized; Cu^{2+} reduced

LO: Identify oxidizing agents and reducing agents (Section 16.2).

The reducing agent is the substance that is oxidized. The oxidizing agent is the substance that is reduced.

EXAMPLE 16.12 **Identifying Oxidizing Agents and Reducing Agents**

Identify the oxidizing and reducing agents in each redox reaction.

(a) $Sn(s) + O_2(g) \longrightarrow SnO_2(s)$
(b) $2\,Na(s) + F_2(g) \longrightarrow 2\,NaF(s)$
(c) $Mg(s) + Cu^{2+}(aq) \longrightarrow Mg^{2+}(aq) + Cu(s)$

SOLUTION

(a) Sn is the reducing agent; O_2 is the oxidizing agent.
(b) Na is the reducing agent; F_2 is the oxidizing agent.
(c) Mg is the reducing agent; Cu^{2+} is the oxidizing agent.

LO: Assign oxidation states (Section 16.3).

Rules for Assigning Oxidation States
(These rules are hierarchical. If two rules conflict, follow the rule higher on the list.)

1. The oxidation state of an atom in a free element is 0.

2. The oxidation state of a monoatomic ion is equal to its charge.

3. The sum of the oxidation states of all atoms in:
   - a neutral molecule or formula unit is 0.
   - an ion is equal to the charge of the ion.

4. In their compounds,
   - Group I metals have an oxidation state of +1.
   - Group II metals have an oxidation state of +2.

5. In their compounds, nonmetals are assigned oxidation states according to the following hierarchical table.

Fluorine	−1
Hydrogen	+1
Oxygen	−2
Group 7A	−1
Group 6A	−2
Group 5A	−3

EXAMPLE 16.13 **Assigning Oxidation States**

Assign an oxidation state to each atom in each species.

(a) Al
(b) Al^{3+}
(c) N_2O
(d) CO_3^{2-}

SOLUTION

(a) $\underset{0}{Al}$ (Rule 1)

(b) $\underset{+3}{Al^{3+}}$ (Rule 2)

(c) $\underset{+1\ -2}{N_2O}$ (Rule 5, O takes priority over N)

(d) $\underset{+4\ -2}{CO_3^{2-}}$ (Rules 5, 3)

LO: Balance redox reactions (Section 16.4).

To balance redox reactions in aqueous acidic solutions, follow this procedure (brief version).

1. Assign oxidation states.

2. Separate the overall reaction into two half-reactions.

3. Balance each half-reaction with respect to mass.
   - Balance all elements other than H and O.

   - Balance O by adding H_2O.

   - Balance H by adding H^+.

4. Balance each half-reaction with respect to charge by adding electrons.

5. Make the number of electrons in both half-reactions equal.

6. Add the two half-reactions together.

7. Verify that the reaction is balanced.

EXAMPLE 16.14 Balancing Redox Reactions

Balance the reaction occurring in acidic solution.

$$IO_3^-(aq) + Fe^{2+}(aq) \longrightarrow I_2(s) + Fe^{3+}(aq)$$

SOLUTION

$$IO_3^-(aq) + Fe^{2+}(aq) \longrightarrow I_2(s) + Fe^{3+}(aq)$$
$$\substack{+5 \ -2} \qquad \substack{+2} \qquad \substack{0} \qquad \substack{+3}$$

Reduction — Oxidation

OXIDATION $Fe^{2+}(aq) \longrightarrow Fe^{3+}(aq)$

REDUCTION $IO_3^-(aq) \longrightarrow I_2(s)$

$Fe^{2+}(aq) \longrightarrow Fe^{3+}(aq)$
$\mathbf{2}\,IO_3^-(aq) \longrightarrow I_2(s)$

$Fe^{2+}(aq) \longrightarrow Fe^{3+}(aq)$
$2\,IO_3^-(aq) \longrightarrow I_2(s) + \mathbf{6\,H_2O}(l)$

$Fe^{2+}(aq) \longrightarrow Fe^{3+}(aq)$
$\mathbf{12\,H^+}(aq) + 2\,IO_3^-(aq) \longrightarrow I_2(s) + 6\,H_2O(l)$

$Fe^{2+}(aq) \longrightarrow Fe^{3+}(aq) + \mathbf{e^-}$
$\mathbf{10e^-} + 12\,H^+(aq) + 2\,IO_3^-(aq) \longrightarrow I_2(s) + 6\,H_2O(l)$

$\mathbf{10 \times}[Fe^{2+}(aq) \longrightarrow Fe^{3+}(aq) + e^-]$
$10e^- + 12\,H^+(aq) + 2\,IO_3^-(aq) \longrightarrow I_2(s) + 6\,H_2O$

$10\,Fe^{2+}(aq) \longrightarrow 10\,Fe^{3+}(aq) + 10e^-$
$\underline{10e^- + 12\,H^+(aq) + 2\,IO_3^-(aq) \longrightarrow I_2(s) + 6\,H_2O}$
$10\,Fe^{2+}(aq) + 12\,H^+(aq) + 2\,IO_3^-(aq) \longrightarrow$
$\qquad\qquad 10\,Fe^{3+}(aq) + I_2(s) + 6\,H_2O$

Reactants	Products
10 Fe	10 Fe
12 H	12 H
2 I	2 I
6 O	6 O
+30 charge	+30 charge

LO: Predict spontaneous redox reactions (Section 16.5).

Any half-reaction in the activity series is spontaneous when paired with the reverse of any half-reaction below it.

EXAMPLE 16.15 Predicting Spontaneous Redox Reactions

Predict whether each redox reaction is spontaneous.

(a) $Cr(s) + 3\,Ag^+(aq) \longrightarrow Cr^{3+}(aq) + 3\,Ag(s)$
(b) $Mn^{2+}(aq) + Fe(s) \longrightarrow Mn(s) + Fe^{2+}(aq)$

SOLUTION

(a) spontaneous
(b) nonspontaneous

Key Terms

activity series of metals [16.5]
alkaline battery [16.6]
anode [16.6]
cathode [16.6]
corrosion [16.8]
dry cell [16.6]
electrical current [16.6]

electrochemical cell [16.6]
electrolysis [16.7]
electrolytic cell [16.7]
fuel cell [16.1]
half-cell [16.6]
half-reaction [16.4]

lead-acid storage battery
[16.6]
oxidation [16.2]
oxidation state (oxidation
number) [16.3]
oxidizing agent [16.2]

redox (oxidation–reduction)
reaction [16.1]
reducing agent [16.2]
reduction [16.2]
salt bridge [16.6]
voltage [16.6]
voltaic (galvanic) cell [16.6]

Exercises

Questions

1. What is a fuel-cell electric vehicle?
2. What is an oxidation–reduction or redox reaction?
3. Define oxidation and reduction with respect to:
   (a) oxygen
   (b) electrons
   (c) oxidation state
4. What is an oxidizing agent? What is a reducing agent?
5. Good oxidizing agents have a strong tendency to _____ electrons in reactions.
6. Good reducing agents have a strong tendency to _____ electrons in reactions.
7. What is the oxidation state of a free element? Of a monoatomic ion?
8. For a neutral molecule, the oxidation states of the individual atoms must add up to _____.
9. For an ion, the oxidation states of the individual atoms must add up to _____.
10. In their compounds, elements have oxidation states equal to _____. Are there exceptions to this rule? Explain.
11. In a redox reaction, an atom that undergoes an increase in oxidation state is _____. An atom that undergoes a decrease in oxidation state is _____.
12. How does hydrogen peroxide change hair color?
13. When balancing redox equations, the number of electrons lost in the oxidation half-reaction must _____ the number of electrons gained in the reduction half-reaction.
14. When balancing aqueous redox reactions, oxygen is balanced using _____, and hydrogen is balanced using _____.
15. When balancing aqueous redox reactions, charge is balanced using _____.

16. When balancing aqueous redox reactions in basic media, hydrogen ions are neutralized using _____.
17. Are metals at the top of the activity series the most reactive or least reactive?
18. Are metals at the top of the activity series the easiest or hardest to oxidize?
19. Are metals at the bottom of the activity series most likely or least likely to lose electrons?
20. Any half-reaction in the activity series will be spontaneous when paired with the reverse of any half-reaction _____ it.
21. How can you use the activity series to determine whether a metal will dissolve in acids such as HCl or HBr?
22. What is electrical current? Explain how a simple battery creates electrical current.
23. Oxidation occurs at the _____ of an electrochemical cell.
24. Reduction occurs at the _____ of an electrochemical cell.
25. Explain the role of a salt bridge in an electrochemical cell.
26. A high voltage in an electrochemical cell is analogous to _____ in a river.
27. Describe a common dry-cell battery. Include equations for the anode and cathode reactions.
28. Describe a lead-acid storage battery. Include equations for the anode and cathode reactions.
29. Describe a fuel cell. Include equations for the anode and cathode reactions of the hydrogen–oxygen fuel cell.
30. What is electrolysis? Why is it useful?
31. What is corrosion? List reactions for the corrosion of iron.
32. How can rust be prevented?

Problems

OXIDATION AND REDUCTION

33. Which substance is oxidized in each reaction?
    (a) $2 H_2(g) + O_2(g) \longrightarrow 2 H_2O(l)$
    (b) $4 Al(s) + 3 O_2(g) \longrightarrow 2 Al_2O_3(s)$
    (c) $2 Al(s) + 3 Cl_2(g) \longrightarrow 2 AlCl_3(s)$

34. Which substance is oxidized in each reaction?
    (a) $2 Zn(s) + O_2(g) \longrightarrow 2 ZnO(s)$
    (b) $CH_4(g) + 2 O_2(g) \longrightarrow CO_2(g) + 2 H_2O(g)$
    (c) $Sr(s) + F_2(g) \longrightarrow SrF_2(s)$

35. For each reaction, identify the substance being oxidized and the substance being reduced.
    (a) $2 Sr(s) + O_2(g) \longrightarrow 2 SrO(s)$
    (b) $Ca(s) + Cl_2(g) \longrightarrow CaCl_2(s)$
    (c) $Ni^{2+}(aq) + Mg(s) \longrightarrow Mg^{2+}(aq) + Ni(s)$

36. For each reaction, identify the substance being oxidized and the substance being reduced.
    (a) $Mg(s) + Br_2(g) \longrightarrow MgBr_2(s)$
    (b) $2 Cr^{3+}(aq) + 3 Mn(s) \longrightarrow 2 Cr(s) + 3 Mn^{2+}(aq)$
    (c) $2 H^+(aq) + Ni(s) \longrightarrow H_2(g) + Ni^{2+}(aq)$

37. For each of the reactions in Problem 35, identify the oxidizing agent and the reducing agent.

38. For each of the reactions in Problem 36, identify the oxidizing agent and the reducing agent.

39. Based on periodic trends, which elements would you expect to be good oxidizing agents?
  (a) potassium
  (b) fluorine
  (c) iron
  (d) chlorine

40. Based on periodic trends, which elements would you expect to be good oxidizing agents?
  (a) oxygen
  (b) bromine
  (c) lithium
  (d) sodium

41. Based on periodic trends, which elements in Problem 39 (in their elemental form) would you expect to be good reducing agents?

42. Based on periodic trends, which elements in Problem 40 (in their elemental form) would you expect to be good reducing agents?

43. For each redox reaction, identify the substance being oxidized, the substance being reduced, the oxidizing agent, and the reducing agent.
  (a) $N_2(g) + O_2(g) \longrightarrow 2\ NO(g)$
  (b) $2\ CO(g) + O_2(g) \longrightarrow 2\ CO_2(g)$
  (c) $SbCl_3(g) + Cl_2(g) \longrightarrow SbCl_5(g)$

44. For each redox reaction, identify the substance being oxidized, the substance being reduced, the oxidizing agent, and the reducing agent.
  (a) $H_2(g) + I_2(g) \longrightarrow 2\ HI(g)$
  (b) $CO(g) + H_2(g) \longrightarrow C(s) + H_2O(g)$
  (c) $2\ Al(s) + 6\ H^+(aq) \longrightarrow 2\ Al^{3+}(aq) + 3\ H_2(g)$

OXIDATION STATES

45. Assign an oxidation state to each element or ion.
  (a) V
  (b) Mg^{2+}
  (c) Cr^{3+}
  (d) O_2

46. Assign an oxidation state to each element or ion.
  (a) Ne
  (b) Br_2
  (c) Cu^+
  (d) Fe^{3+}

47. Assign an oxidation state to each atom in each compound.
  (a) NaCl
  (b) CaF_2
  (c) SO_2
  (d) H_2S

48. Assign an oxidation state to each atom in each compound.
  (a) CH_4
  (b) CH_2Cl_2
  (c) $CuCl_2$
  (d) HI

49. What is the oxidation state of nitrogen in each compound?
  (a) NO
  (b) NO_2
  (c) N_2O

50. What is the oxidation state of Cr in each compound?
  (a) CrO
  (b) CrO_3
  (c) Cr_2O_3

51. Assign an oxidation state to each atom in each polyatomic ion.
  (a) CO_3^{2-}
  (b) OH^-
  (c) NO_3^-
  (d) NO_2^-

52. Assign an oxidation state to each atom in each polyatomic ion.
  (a) CrO_4^{2-}
  (b) $Cr_2O_7^{2-}$
  (c) PO_4^{3-}
  (d) MnO_4^-

53. What is the oxidation state of Cl in each ion?
  (a) ClO^-
  (b) ClO_2^-
  (c) ClO_3^-
  (d) ClO_4^-

54. What is the oxidation state of S in each ion?
  (a) SO_4^{2-}
  (b) SO_3^{2-}
  (c) HSO_3^-
  (d) HSO_4^-

55. Assign an oxidation state to each element in each compound.
(a) $Cu(NO_3)_2$
(b) $Sr(OH)_2$
(c) $K_2Cr_2O_7$
(d) $NaHCO_3$

56. Assign an oxidation state to each element in each compound.
(a) Na_3PO_4
(b) Hg_2S
(c) $Fe(CN)_3$
(d) NH_4Cl

57. Assign an oxidation state to each element in each reaction and use the change in oxidation state to determine which element is being oxidized and which element is being reduced.
(a) $SbCl_5(g) \longrightarrow SbCl_3(g) + Cl_2(g)$
(b) $CO(g) + Cl_2(g) \longrightarrow COCl_2(g)$
(c) $2\,NO(g) + Br_2(g) \longrightarrow 2\,BrNO(g)$
(d) $H_2(g) + CO_2(g) \longrightarrow H_2O(g) + CO(g)$

58. Assign an oxidation state to each element in each reaction and use the change in oxidation state to determine which element is being oxidized and which element is being reduced.
(a) $CH_4(g) + 2\,H_2S(g) \longrightarrow CS_2(g) + 4\,H_2(g)$
(b) $2\,H_2S(g) \longrightarrow 2\,H_2(g) + S_2(g)$
(c) $C_6H_{12}O_6(s) + 6\,O_2(g) \longrightarrow 6\,CO_2(g) + 6\,H_2O(g)$
(d) $C_2H_4(g) + Cl_2(g) \longrightarrow C_2H_4Cl_2(g)$

59. Use oxidation states to identify the oxidizing agent and the reducing agent in the redox reaction.

$$2\,Na(s) + 2\,H_2O(l) \longrightarrow 2\,NaOH(aq) + H_2(g)$$

60. Use oxidation states to identify the oxidizing agent and the reducing agent in the redox reaction.

$$N_2(g) + 3\,H_2(g) \longrightarrow 2\,NH_3(g)$$

BALANCING REDOX REACTIONS

61. Balance each redox reaction using the half-reaction method.
(a) $K(s) + Cr^{3+}(aq) \longrightarrow Cr(s) + K^+(aq)$
(b) $Mg(s) + Ag^+(aq) \longrightarrow Mg^{2+}(aq) + Ag(s)$
(c) $Al(s) + Fe^{2+}(aq) \longrightarrow Al^{3+}(aq) + Fe(s)$

62. Balance each redox reaction using the half-reaction method.
(a) $Zn(s) + Sn^{2+}(aq) \longrightarrow Zn^{2+}(aq) + Sn(s)$
(b) $Mg(s) + Cr^{3+}(aq) \longrightarrow Mg^{2+}(aq) + Cr(s)$
(c) $Al(s) + Ag^+(aq) \longrightarrow Al^{3+}(aq) + Ag(s)$

63. Classify each half-reaction occurring in acidic aqueous solution as an oxidation or a reduction and balance the half-reaction.
(a) $MnO_4^-(aq) \longrightarrow Mn^{2+}(aq)$
(b) $Pb^{2+}(aq) \longrightarrow PbO_2(s)$
(c) $IO_3^-(aq) \longrightarrow I_2(s)$
(d) $SO_2(g) \longrightarrow SO_4^{2-}(aq)$

64. Classify each half-reaction occurring in acidic aqueous solution as an oxidation or a reduction and balance the half-reaction.
(a) $S(s) \longrightarrow H_2S(g)$
(b) $S_2O_8^{2-}(aq) \longrightarrow 2\,SO_4^{2-}(aq)$
(c) $Cr_2O_7^{2-}(aq) \longrightarrow Cr^{3+}(aq)$
(d) $NO(g) \longrightarrow NO_3^-(aq)$

65. Use the half-reaction method to balance each redox reaction occurring in acidic aqueous solution.
(a) $PbO_2(s) + I^-(aq) \longrightarrow Pb^{2+}(aq) + I_2(s)$
(b) $SO_3^{2-}(aq) + MnO_4^-(aq) \longrightarrow SO_4^{2-}(aq) + Mn^{2+}(aq)$
(c) $S_2O_3^{2-}(aq) + Cl_2(g) \longrightarrow SO_4^{2-}(aq) + Cl^-(aq)$

66. Use the half-reaction method to balance each redox reaction occurring in acidic aqueous solution.
(a) $I^-(aq) + NO_2^-(aq) \longrightarrow I_2(s) + NO(g)$
(b) $BrO_3^-(aq) + N_2H_4(g) \longrightarrow Br^-(aq) + N_2(g)$
(c) $NO_3^-(aq) + Sn^{2+}(aq) \longrightarrow Sn^{4+}(aq) + NO(g)$

67. Use the half-reaction method to balance each redox reaction occurring in acidic aqueous solution.
(a) $ClO_4^-(aq) + Cl^-(aq) \longrightarrow ClO_3^-(aq) + Cl_2(g)$
(b) $MnO_4^-(aq) + Al(s) \longrightarrow Mn^{2+}(aq) + Al^{3+}(aq)$
(c) $Br_2(aq) + Sn(s) \longrightarrow Sn^{2+}(aq) + Br^-(aq)$

68. Use the half-reaction method to balance each redox reaction occurring in acidic aqueous solution.
(a) $IO_3^-(aq) + SO_2(g) \longrightarrow I_2(s) + SO_4^{2-}(aq)$
(b) $Sn^{4+}(aq) + H_2(g) \longrightarrow Sn^{2+}(aq) + H^+(aq)$
(c) $Cr_2O_7^{2-}(aq) + Br^-(aq) \longrightarrow Cr^{3+}(aq) + Br_2(aq)$

69. Balance each redox reaction occurring in basic solution.
(a) $ClO^-(aq) + Cr(OH)_4^-(aq) \longrightarrow CrO_4^{2-}(aq) + Cl^-(aq)$
(b) $MnO_4^-(aq) + Br^-(aq) \longrightarrow MnO_2(s) + BrO_3^-(aq)$

70. Balance each redox reaction occurring in basic solution.
(a) $NO_2^-(aq) + Al(s) \longrightarrow NH_3(g) + AlO_2^-(aq)$
(b) $Al(s) + MnO_4^-(aq) \longrightarrow MnO_2(s) + Al(OH)_4^-(aq)$

THE ACTIVITY SERIES

71. Which metal has the least tendency to be oxidized?
(a) Ag
(b) Na
(c) Ni
(d) Pb

72. Which metal has the least tendency to be oxidized?
(a) Sn
(b) Mg
(c) Cu
(d) Fe

73. Which metal cation has the greatest tendency to be reduced?
(a) Mn^{2+}
(b) Cu^{2+}
(c) K^+
(d) Ni^{2+}

74. Which metal cation has the greatest tendency to be reduced?
(a) Pb^{2+}
(b) Cr^{3+}
(c) Fe^{2+}
(d) Sn^{2+}

75. Which metal is the best reducing agent?
(a) Mn
(b) Al
(c) Ni
(d) Cr

76. Which metal is the best reducing agent?
(a) Ag
(b) Mg
(c) Fe
(d) Pb

77. Determine whether each redox reaction occurs spontaneously in the forward direction.
(a) $Ni(s) + Zn^{2+}(aq) \longrightarrow Ni^{2+}(aq) + Zn(s)$
(b) $Ni(s) + Pb^{2+}(aq) \longrightarrow Ni^{2+}(aq) + Pb(s)$
(c) $Al(s) + 3 Ag^+(aq) \longrightarrow 3 Al^{3+}(aq) + Ag(s)$
(d) $Pb(s) + Mn^{2+}(aq) \longrightarrow Pb^{2+}(aq) + Mn(s)$

78. Determine whether each redox reaction occurs spontaneously in the forward direction.
(a) $Ca^{2+}(aq) + Zn(s) \longrightarrow Ca(s) + Zn^{2+}(aq)$
(b) $2 Ag^+(aq) + Ni(s) \longrightarrow 2 Ag(s) + Ni^{2+}(aq)$
(c) $Fe(s) + Mn^{2+}(aq) \longrightarrow Fe^{2+}(aq) + Mn(s)$
(d) $2 Al(s) + 3 Pb^{2+}(aq) \longrightarrow 2 Al^{3+}(aq) + 3 Pb(s)$

79. Suppose you wanted to cause Ni^{2+} ions to come out of solution as solid Ni. What metal could you use to accomplish this?

80. Suppose you wanted to cause Pb^{2+} ions to come out of solution as solid Pb. What metal could you use to accomplish this?

81. Which metal in the activity series reduces Al^{3+} ions but not Na^+ ions?

82. Which metal in the activity series is oxidized with a Ni^{2+} solution but not with a Cr^{3+} solution?

83. Which metals dissolve in HCl? For those metals that do dissolve, write a balanced redox reaction showing what happens when the metal dissolves.
(a) Ag
(b) Fe
(c) Cu
(d) Al

84. Which metals dissolve in HCl? For those metals that do dissolve, write a balanced redox reaction showing what happens when the metal dissolves.
(a) Cr
(b) Pb
(c) Au
(d) Zn

BATTERIES, ELECTROCHEMICAL CELLS, AND ELECTROLYSIS

85. Make a sketch of an electrochemical cell with the overall reaction shown here. Label the anode, the cathode, and the salt bridge. Indicate the direction of electron flow. *Hint:* When drawing electrochemical cells, the anode is usually drawn on the left side.

$$Mn(s) + Pb^{2+}(aq) \longrightarrow Mn^{2+}(aq) + Pb(s)$$

86. Make a sketch of an electrochemical cell with the overall reaction shown here. Label the anode, the cathode, and the salt bridge. Indicate the direction of electron flow. *Hint:* When drawing electrochemical cells, the anode is usually drawn on the left side.

$$Mg(s) + Ni^{2+}(aq) \longrightarrow Mg^{2+}(aq) + Ni(s)$$

87. The following reaction occurs at the anode of an electrochemical cell:

$$Zn(s) \longrightarrow Zn^{2+}(aq) + 2e^-$$

Which cathode reaction would produce a battery with the highest voltage?
(a) $Mg^{2+}(aq) + 2e^- \longrightarrow Mg(s)$
(b) $Pb^{2+}(aq) + 2e^- \longrightarrow Pb(s)$
(c) $Cr^{3+}(aq) + 3e^- \longrightarrow Cr(s)$
(d) $Cu^{2+}(aq) + 2e^- \longrightarrow Cu(s)$

88. The following reaction occurs at the cathode of an electrochemical cell:

$$Ni^{2+}(aq) + 2e^- \longrightarrow Ni(s)$$

Which anode reaction would produce a battery with the highest voltage?
(a) $Ag(s) \longrightarrow Ag^+(aq) + e^-$
(b) $Mg(s) \longrightarrow Mg^{2+}(aq) + 2e^-$
(c) $Cr(s) \longrightarrow Cr^{3+}(aq) + 3e^-$
(d) $Cu(s) \longrightarrow Cu^{2+}(aq) + 2e^-$

89. Use half-cell reactions to determine the overall reaction that occurs in an alkaline battery.

90. Use half-cell reactions to determine the overall reaction that occurs in a lead-acid storage battery.

91. Make a sketch of an electrolysis cell that could be used to electroplate copper onto other metal surfaces. Label the anode and the cathode and show the reactions that occur at each.

92. Make a sketch of an electrolysis cell that could be used to electroplate nickel onto other metal surfaces. Label the anode and the cathode and show the reactions that occur at each.

CORROSION

93. Which metal, if coated onto iron, would prevent the corrosion of iron?
(a) Zn
(b) Sn
(c) Mn

94. Which metal, if coated onto iron, would prevent the corrosion of iron?
(a) Mg
(b) Cr
(c) Cu

Cumulative Problems

95. Determine whether each reaction is a redox reaction. For those reactions that are redox reactions, identify the substance being oxidized and the substance being reduced.
(a) $Zn(s) + CoCl_2(aq) \longrightarrow ZnCl_2(aq) + Co(s)$
(b) $HI(aq) + NaOH(aq) \longrightarrow H_2O(l) + NaI(aq)$
(c) $AgNO_3(aq) + NaCl(aq) \longrightarrow AgCl(s) + NaNO_3(aq)$
(d) $2 K(s) + Br_2(l) \longrightarrow 2 KBr(s)$

96. Determine whether each reaction is a redox reaction. For those reactions that are redox reactions, identify the substance being oxidized and the substance being reduced.
(a) $Pb(NO_3)_2(aq) + 2 LiCl(aq) \longrightarrow$
$PbCl_2(s) + 2 LiNO_3(aq)$
(b) $2 HBr(aq) + Ca(OH)_2(aq) \longrightarrow 2 H_2O(l) + CaBr_2(aq)$
(c) $2 Al(s) + Fe_2O_3(s) \longrightarrow Al_2O_3(s) + 2 Fe(l)$
(d) $Na_2O(s) + H_2O(l) \longrightarrow 2 NaOH(aq)$

97. Consider the unbalanced redox reaction.

$$MnO_4^-(aq) + Zn(s) \longrightarrow Mn^{2+}(aq) + Zn^{2+}(aq)$$

Balance the equation in acidic solution and determine how much of a 0.500 M $KMnO_4$ solution is required to completely dissolve 2.85 g of Zn.

98. Consider the unbalanced redox reaction.

$$Cr_2O_7^{2-}(aq) + Cu(s) \longrightarrow Cr^{3+}(aq) + Cu^{2+}(aq)$$

Balance the equation in acidic solution and determine how much of a 0.850 M $K_2Cr_2O_7$ solution is required to completely dissolve 5.25 g of Cu.

99. If a strip of magnesium metal is dipped into a solution containing silver ions, does a spontaneous reaction occur? If so, write the two half-reactions and the balanced overall equation for the reaction.

100. If a strip of tin metal is dipped into a solution containing zinc ions, does a spontaneous reaction occur? If so, write the two half-reactions and the balanced overall equation for the reaction.

101. A 10.0-mL sample of a commercial hydrogen peroxide (H_2O_2) solution is titrated with 0.0998 M $KMnO_4$. The endpoint is reached at a volume of 34.81 mL. Find the mass percent of H_2O_2 in the commercial hydrogen peroxide solution. (Assume a density of 1.00 g/mL for the hydrogen peroxide solution.) The unbalanced redox reaction that occurs in acidic solution during the titration is:

$$H_2O_2(aq) + MnO_4^-(aq) \longrightarrow O_2(g) + Mn^{2+}(aq)$$

102. A 1.012-g sample of a salt containing Fe^{2+} is titrated with 0.1201 M $KMnO_4$. The endpoint of the titration is reached at 22.45 mL. Find the mass percent of Fe^{2+} in the sample. The unbalanced redox reaction that occurs in acidic solution during the titration is:

$$Fe^{2+}(aq) + MnO_4^-(aq) \longrightarrow Fe^{3+}(aq) + Mn^{2+}(aq)$$

103. Silver is electroplated at the cathode of an electrolysis cell by this half-reaction.

$$Ag^+(aq) + e^- \longrightarrow Ag(s)$$

How many moles of electrons are required to electroplate 5.8 g of Ag?

104. Gold is electroplated at the cathode of an electrolysis cell by this half-reaction.

$$Au^{3+}(aq) + 3e^- \longrightarrow Au(s)$$

How many moles of electrons are required to electroplate 1.40 g of Au?

105. Determine whether HI can dissolve each metal sample. If it can, write a balanced chemical reaction showing how the metal dissolves in HI and determine the minimum amount of 3.5 M HI required to completely dissolve the sample.
(a) 5.95 g Cr
(b) 2.15 g Al
(c) 4.85 g Cu
(d) 2.42 g Au

106. Determine whether HCl can dissolve each metal sample. If it can, write a balanced chemical reaction showing how the metal dissolves in HCl and determine the minimum amount of 6.0 M HCl required to completely dissolve the sample.
(a) 5.90 g Ag
(b) 2.55 g Pb
(c) 4.83 g Sn
(d) 1.25 g Mg

107. One drop (assume 0.050 mL) of 6.0 M HCl is placed onto the surface of 0.028-mm-thick aluminum foil. What is the maximum diameter of the hole that will result from the HCl dissolving the aluminum? (Density of aluminum = 2.7 g/cm³)

108. A graduated cylinder containing 1.00 mL of 12.0 M HCl is accidentally tipped over, and the contents spill onto manganese foil with a thickness of 0.055 mm. Calculate the maximum diameter of the hole that will be dissolved in the foil by the reaction between the manganese and hydrochloric acid. (Density of manganese = 7.47 g/cm³)

109. The electrolytic cell represented in Figure 16.17 can be used to plate silver onto other metal surfaces. The plating reaction is: $Ag^+(aq) + e^- \longrightarrow Ag(s)$. Notice from the reaction that 1 mol e^- plates out 1 mol Ag(s). Use this stoichiometric relationship to determine how much time is required with an electrical current 0.100 amp to plate out 1.0 g Ag. The amp is a unit of electrical current equivalent to 1 C/s. (*Hint:* Recall that the charge of an electron is 1.60×10^{-19} C.)

110. An electrolytic cell similar to the one represented in Figure 16.17 can be used to plate gold onto other metal surfaces. The plating reaction is: $Au^+(aq) + e^- \longrightarrow Au(s)$. Notice from the reaction that 1 mol e^- plates out 1 mol Au(s). Use this stoichiometric relationship to determine how much time is required with an electrical current of 0.200 amp to plate out 0.400 g Au. The amp is a unit of electrical current equivalent to 1 C/s. (*Hint:* Recall that the charge of an electron is 1.60×10^{-19} C.)

Highlight Problems

111. Consider the molecular views of an Al strip and Cu^{2+} solution. Draw a similar sketch showing what happens to the atoms and ions if the Al strip is submerged in the solution for a few minutes.

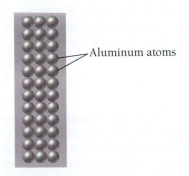
Aluminum atoms

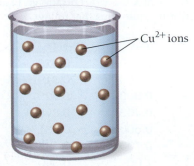
Cu^{2+} ions

112. Suppose a fuel-cell generator produces electricity for a house. If each H_2 molecule produces $2e^-$, how many kilograms of hydrogen are required to generate the electricity needed for a typical house? Assume the home uses about 850 kWh of electricity per month, which corresponds to approximately 2.65×10^4 mol of electrons at the voltage of a fuel cell.

113. Consider the molecular view of an electrochemical cell involving the overall reaction:

$$Zn(s) + Ni^{2+}(aq) \longrightarrow Zn^{2+}(aq) + Ni(s)$$

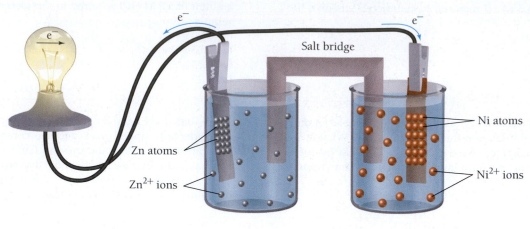

Anode: $Zn(s) \longrightarrow Zn^{2+}(aq) + 2e^-$ **Cathode:** $Ni^{2+}(aq) + 2e^- \longrightarrow Ni(s)$

Draw a similar sketch showing how the cell might appear after it has generated a substantial amount of electrical current.

Questions for Group Work

Discuss these questions with the group and record your consensus answer.

114. Design one electrochemical cell for each member in the group. Each cell should be made from two metals in the activity series. Make a diagram of each cell.
  (a) List the metal and solution on each side.
  (b) Label the anode, the cathode, and the salt bridge.
  (c) Write the reaction occurring on each side.
  (d) Indicate the direction of electron flow.
  Which of your group's cells do you think would produce the largest voltage? Why?

115. A promising technology based on a redox reaction is the *direct methanol fuel cell*. Instead of hydrogen, it uses liquid methanol, CH_3OH, as a fuel. The unbalanced reaction is $CH_3OH + O_2 \longrightarrow CO_2 + H_2O$.
  (a) Assign oxidation states to each atom in the reaction.
  (b) Determine what is being oxidized and what is being reduced.
  (c) Write and balance the separate half-reactions. (*Hint:* Methanol reacts to form carbon dioxide, and oxygen reacts to form water.)
  (d) Balance the overall reaction if it occurs in acidic solution.
  (e) Methanol fuel cells must be designed to allow H^+ to pass from one electrode to the other. Do they start at the electrode with the methanol or at the electrode with the oxygen? How do you know?

Data Interpretation and Analysis

116. We can use electrochemical cells to measure the concentrations of certain species in solution because the voltage of the cell is sensitive to the concentrations of the reactants and products in the redox reaction associated with the cell. For example, the voltage of an electrochemical cell based on the reaction $H_2(g) + Cu^{2+}(aq) \longrightarrow 2\,H^+(aq) + Cu(s)$ is sensitive to both the Cu^{2+} concentration and the H^+ concentration in solution. If the H^+ concentration is held constant, then the voltage only depends on the Cu^{2+} concentration, and we can use the cell to measure the Cu^{2+} concentration in an unknown solution.

When using an electrochemical cell for this purpose, chemists often make voltage measurements of several Cu^{2+} solutions of known concentrations and construct a *calibration curve*, a graph that shows the relationship between voltage and concentration. The calibration curve can be used to determine the concentration of an unknown solution based on the voltage associated with that solution. The tabulated data show the measured voltage in the hydrogen/copper electrochemical cell just discussed for several different Cu^{2+} concentrations. Examine the data and answer the questions that follow.

[Cu²⁺]	Voltage (V)
0.100	0.310
0.200	0.319
0.300	0.325
0.400	0.328
0.500	0.331
0.700	0.335
1.00	0.340

(a) Construct a graph of the measured voltage versus the log of the copper concentration. Is the graph linear?

(b) Determine the slope and y-intercept of the best fitting line to the points in your graph from part a.

(c) The voltage of two unknown solutions are measured and recorded. Use the slope and intercept from part b, together with the equation, voltage $=$ slope $\times$ log $[Cu^{2+}] + y$-intercept, to determine the Cu^{2+} concentrations of the unknown solutions.

Unknown Cu^{2+} Solution	Voltage (V)
i	0.303
ii	0.338

Answers to Skillbuilder Exercises

Skillbuilder 16.1 (a) K is oxidized; Cl_2 is reduced.

(b) Al is oxidized; Sn^{2+} is reduced.

(c) C is oxidized; O_2 is reduced.

Skillbuilder 16.2 (a) K is the reducing agent; Cl_2 is the oxidizing agent.

(b) Al is the reducing agent; Sn^{2+} is the oxidizing agent.

(c) C is the reducing agent; O_2 is the oxidizing agent.

Skillbuilder 16.3 (a) $\underset{0}{Zn}$

(b) $\underset{+2}{Cu^{2+}}$

(c) $\underset{+2\ -1}{CaCl_2}$

(d) $\underset{+4\ -1}{CF_4}$

(e) $\underset{+3\ -2}{NO_2^-}$

(f) $\underset{+6\ -2}{SO_3}$

Skillbuilder 16.4

Sn oxidized $(0 \longrightarrow +4)$; N reduced $(+5 \longrightarrow +4)$

Skillbuilder 16.5

$6\,H^+(aq) + 2\,Cr(s) \longrightarrow 3\,H_2(g) + 2\,Cr^{3+}(aq)$

Skillbuilder 16.6

$Cu(s) + 4\,H^+(aq) + 2\,NO_3^-(aq) \longrightarrow$
$\qquad\qquad Cu^{2+}(aq) + 2\,NO_2(g) + 2\,H_2O(l)$

Skillbuilder 16.7

$5\,Sn(s) + 16\,H^+(aq) + 2\,MnO_4^-(aq) \longrightarrow$
$\qquad\qquad 5\,Sn^{2+}(aq) + 2\,Mn^{2+}(aq) + 8\,H_2O(l)$

Skillbuilder 16.8

$H_2O_2(aq) + 2\,ClO_2(aq) + 2\,OH^-(aq) \longrightarrow$
$\qquad\qquad O_2(g) + 2\,ClO_2^-(aq) + 2\,H_2O(l)$

Skillbuilder 16.9 (a) Yes

(b) No

Skillbuilder 16.10 No

Answers to Conceptual Checkpoints

16.1 (b) The oxidizing agent oxidizes another species and is itself always reduced.

16.2 (d) From Rule 1, you know that the oxidation state of nitrogen in N_2 is 0. According to Rule 3, the sum of the oxidation states of all atoms in a compound $= 0$. Therefore, by applying Rule 5, you can determine that the oxidation state of nitrogen in NO is $+2$; in NO_2 it is $+4$; and in NH_3 it is -3.

16.3 (a) Na is highest on the activity series and therefore most easily oxidized.

16.4 (c) Lead is the only one of these metals that is above hydrogen in the activity series and therefore the only one that dissolves in an acidic solution.

16.5 (a) Magnesium would lead to the highest voltage because it is highest on the activity series. Of the metals listed, it is most easily oxidized and therefore produces the highest voltage when combined with the reduction of Pb^{2+} ions.

16.6 (d) Tin is the only metal in the list that is below iron in the activity series. Tin is therefore more difficult to oxidize than iron and cannot prevent the oxidation of the iron.

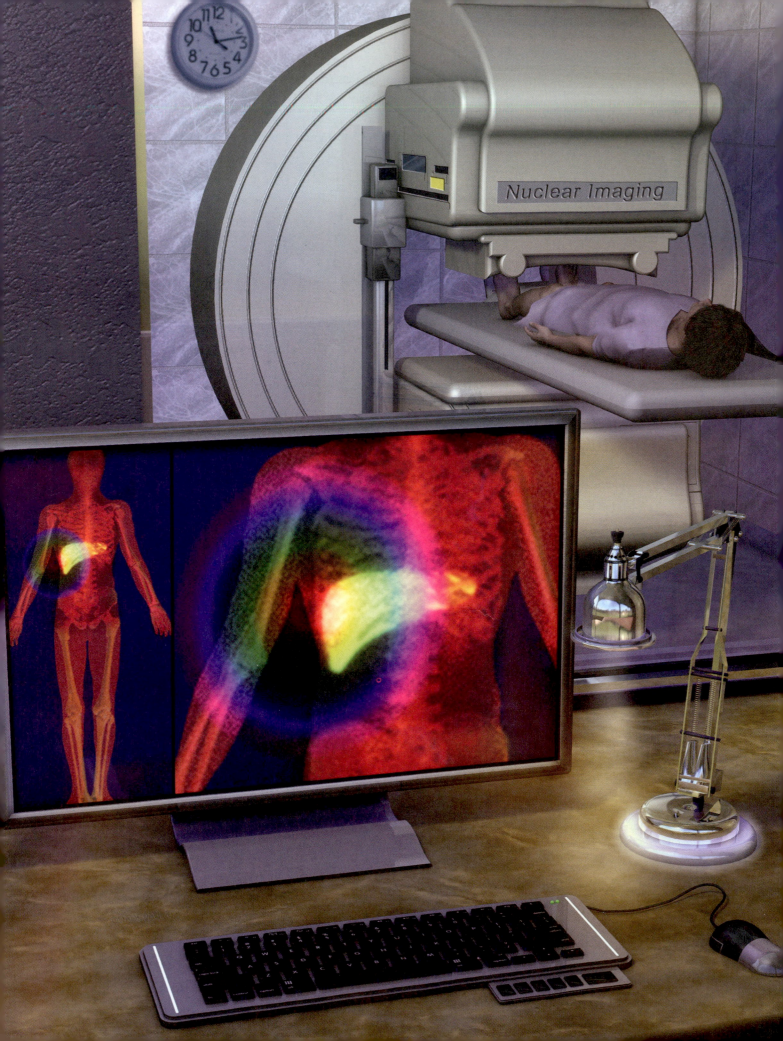

17 Radioactivity and Nuclear Chemistry

Nuclear energy is incomparably greater than the molecular energy which we use today . . . . What is lacking is the match to set the bonfire alight . . . . The scientists are looking for this.

—Winston Spencer Churchill (1874–1965), in 1931

17.1 Diagnosing Appendicitis

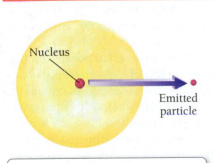

Radioactivity is the emission of particles by the nuclei of certain atoms.

▲ **FIGURE 17.1 Radioactivity**

Several years ago I awoke with a dull pain on the lower right side of my stomach. The pain worsened over several hours, so I went to the hospital emergency room for evaluation. I was examined by a doctor who said it might be appendicitis, an inflammation of the appendix. The appendix, which has no known function, is a small pouch that extends from the right side of the large intestine. Occasionally, it becomes infected and requires surgical removal.

Patients with appendicitis usually have a high number of white blood cells, so the hospital performed a blood test to determine my white blood cell count. The test was negative—I had a normal white blood cell count. Although my symptoms were consistent with appendicitis, the negative blood test clouded the diagnosis. The doctor gave me the choice of either having my appendix removed (even though there was a chance it was healthy) or performing an additional test to confirm appendicitis. I chose the additional test.

The additional test involved nuclear medicine, an area of medical practice that uses *radioactivity* to diagnose and treat disease. **Radioactivity** is the emission of tiny, invisible particles by the nuclei of certain atoms (◄ **FIGURE 17.1**). Many of these particles can pass right through matter. The atoms that emit these particles are said to be **radioactive**.

To perform the test, antibodies—naturally occurring molecules that fight infection—labeled with radioactive atoms were injected into my bloodstream. Since antibodies attack infection, they migrate to areas of the body where

◄ In nuclear medicine, radioactivity helps physicians obtain clear images of internal organs.

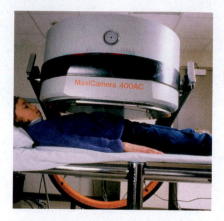

▲ **FIGURE 17.2 Nuclear medicine** In a test for appendicitis, radioactively tagged antibodies are given to the patient. If the patient has an infection in the appendix, the antibodies accumulate there and the emitted radiation is detected.

infection is present. If my appendix was infected, the antibodies would accumulate there. After waiting about an hour, I was taken to a room and laid on a table. A photographic film was inserted in a panel above me. Although radioactivity is invisible to the eye, it does expose photographic film. If my appendix was indeed infected, it would contain a high concentration of the radioactively tagged antibodies, and the film would show a bright spot at the location of my appendix (◄ FIGURE 17.2). In this test, I—or my appendix to be specific—was the radiation source that would expose film. The test, however, was negative. No radioactivity was emanating from my appendix. It was healthy. After several hours, the pain subsided and I went home, appendix and all. I never did find out what caused the pain.

Radioactivity is used to diagnose and treat many conditions, including cancer, thyroid disease, abnormal kidney and bladder function, and heart disease. These examples from medicine are just a few of the many applications of radioactivity. Naturally occurring radioactivity allows us to estimate the age of fossils and rocks as well. Radioactivity also led to the discovery of nuclear fission, used for electricity generation and nuclear weapons. In this chapter, we explore radioactivity—how it was discovered, what it is, and how it is used.

17.2 The Discovery of Radioactivity

▶ Explain how the experiments of Becquerel and Curie led to the discovery of radioactivity.

In 1896, a French scientist named Antoine-Henri Becquerel (1852–1908) discovered radioactivity. Becquerel was not looking for radioactivity at the time. Instead, he was interested in the newly discovered X-rays (see Section 9.3), which became the hot topic of physics research in his time. He hypothesized that X-rays were emitted in conjunction with **phosphorescence**. Phosphorescence is the long-lived *emission* of light that sometimes follows the *absorption* of light by some atoms and molecules. Phosphorescence is probably most familiar to you as the *glow* in glow-in-the-dark toys. After one of these toys is exposed to light, it reemits some of that light, usually at slightly longer wavelengths. If you turn off the lights or put the toy in the dark, you can see the greenish glow of the emitted light. Becquerel hypothesized that the visible greenish glow was associated with the emission of X-rays (which are invisible).

To test his hypothesis, Becquerel placed crystals—composed of potassium uranyl sulfate, a compound known to phosphoresce—on top of a photographic plate wrapped in black cloth (▼ FIGURE 17.3). He then placed the wrapped plate and the crystals outdoors to expose them to sunlight. He knew that the crystals phosphoresced because he could see the emitted light when he brought them back into the dark. If the crystals also emitted X-rays, the X-rays would pass through the black cloth and expose the underlying photographic plate. Becquerel performed the experiment several times and always got the same result: The photographic plate showed a bright exposure spot where the crystals had been. Becquerel believed his hypothesis was correct, and he presented the results—that phosphorescence and X-rays were linked—to the French Academy of Sciences.

▲ Marie Curie with her two daughters. Irene (right) became a distinguished nuclear physicist in her own right, winning a Nobel Prize in 1935. Eve (left) wrote a highly acclaimed biography of her mother.

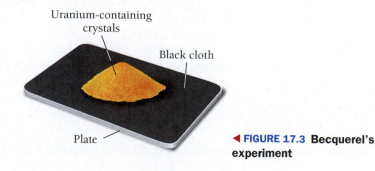

◄ **FIGURE 17.3 Becquerel's experiment**

▲ In the past, radium was added to some paints that were used on watch dials. The radium made the dial glow.

Element 96 (curium) is named in honor of Marie Curie and her contributions to our understanding of radioactivity.

Becquerel later retracted his results, however, when he discovered that a photographic plate with the same crystals showed a bright exposure spot even when the plate and the crystals were stored in a dark drawer and not exposed to sunlight. Becquerel realized that the crystals themselves were constantly emitting something (independent of whether or not they phosphoresced) that exposed the photographic plate. Becquerel concluded that the uranium within the crystals was the source of the emissions, and he called the emissions *uranic rays.*

Soon after Becquerel's discovery, a young graduate student named Marie Sklodowska Curie (1867–1934), one of the first women in France to attempt doctoral work, decided to pursue the study of uranic rays for her doctoral thesis. Her first task was to determine whether any other substances besides uranium (the heaviest known element at the time) emitted these rays. In her search, Curie discovered two new elements, both of which also emitted uranic rays. Curie named one of her newly discovered elements *polonium* after her home country of Poland. The other element she named *radium* because of the very high amount of radioactivity that it produced. Radium is so radioactive that it gently glows in the dark and emits significant amounts of heat. Since it was now clear that these rays were not unique to uranium, Curie changed the name of uranic rays to *radioactivity.* In 1903, Curie received the Nobel Prize in physics—which she shared with Becquerel and her husband, Pierre Curie—for the discovery of radioactivity. In 1911, Curie was awarded a second Nobel Prize, this time in chemistry, for her discovery of the two new elements.

17.3 Types of Radioactivity: Alpha, Beta, and Gamma Decay

▶ Write nuclear equations for alpha decay.

▶ Write nuclear equations for beta decay.

▶ Write nuclear equations for positron emission.

Key Concept Video
Types of Radioactivity

The atomic number equals the number of protons, and the mass number equals the number of protons and neutrons.

While Curie focused her work on discovering the different kinds of radioactive elements, Ernest Rutherford (1871–1937) and others focused on characterizing the radioactivity itself. These scientists found that the emissions were produced by the nuclei of radioactive atoms. These nuclei were unstable and would emit small pieces of themselves to gain stability. These were the particles that Becquerel and Curie detected. There are several different types of radioactive emissions: alpha (α) rays, beta (β) rays, gamma (γ) rays, and positrons.

In order to understand these different types of radioactivity, we must briefly review the notation to symbolize isotopes we first introduced in Section 4.8. Recall that we can represent any isotope with the notation:

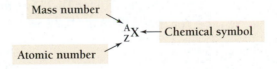

For example, the symbol

$$^{21}_{10}\text{Ne}$$

represents the neon isotope containing 10 protons and 11 neutrons. The symbol

$$^{20}_{10}\text{Ne}$$

represents the neon isotope containing 10 protons and 10 neutrons. Remember that many elements have several different isotopes.

We also represent the main subatomic particles—protons, neutrons, and electrons—with similar notation.

Proton symbol $^{1}_{1}\text{p}$ Neutron symbol $^{1}_{0}\text{n}$ Electron symbol $^{0}_{-1}\text{e}$

> Nuclei are unstable when they are too large or contain an unbalanced ratio of neutrons to protons. Small nuclei need about 1 neutron to every proton to be stable, while larger nuclei need about 1.5 neutrons to every proton.

> In nuclear chemistry, we are primarily interested in changes within the nucleus; therefore, the 2+ charge that we would normally write for a helium nucleus is omitted for an alpha particle.

> The term *nuclide* is used in nuclear chemistry to mean a specific isotope.

Alpha (α) Radiation

Alpha (α) radiation occurs when an unstable nucleus emits a small piece of itself consisting of 2 protons and 2 neutrons (▼ FIGURE 17.4). Because 2 protons and 2 neutrons are identical to a helium-4 nucleus, the symbol for an **alpha (α) particle** is identical to the symbol for helium-4.

$$\text{alpha (α) particle} \quad {}^{4}_{2}\text{He}$$

An α particle

When an atom emits an alpha particle, it becomes a lighter atom. We represent this process with a **nuclear equation**, an equation that represents the changes that occur during radioactivity and other nuclear processes. For example, the nuclear equation for the alpha decay of uranium-238 is:

Parent nuclide

Daughter nuclides

$$\text{alpha decay} \qquad {}^{238}_{92}\text{U} \longrightarrow {}^{234}_{90}\text{Th} + {}^{4}_{2}\text{He}$$

The original atom is the **parent nuclide**, and the products are the **daughter nuclides**. When an element emits an alpha particle, the number of protons in its nucleus changes, transforming it into a different element. In the example shown here, uranium-238 becomes thorium-234. Unlike a chemical reaction, in which elements retain their identity, a nuclear reaction often changes the identity of the elements involved. Like a chemical equation, however, nuclear equations must be balanced.

The sum of the atomic numbers on both sides of a nuclear equation must be equal, and the sum of the mass numbers on both sides must also be equal.

$$ {}^{238}_{92}\text{U} \longrightarrow {}^{234}_{90}\text{Th} + {}^{4}_{2}\text{He}$$

Left Side	Right Side
Sum of mass numbers = 238	Sum of mass numbers = 234 + 4 = 238
Sum of atomic numbers = 92	Sum of atomic numbers = 90 + 2 = 92

We can deduce the identity and symbol of the daughter nuclide of any alpha decay from the mass and atomic numbers of the parent nuclide. During alpha decay, the mass number decreases by 4 and the atomic number decreases by 2. For example, to write a nuclear equation for the alpha decay of Th-232, we begin with the symbol for Th-232 on the left side of the equation and the symbol for an alpha particle on the right side:

$$ {}^{232}_{90}\text{U} \longrightarrow {}^{x}_{y}\text{?} + {}^{4}_{2}\text{He}$$

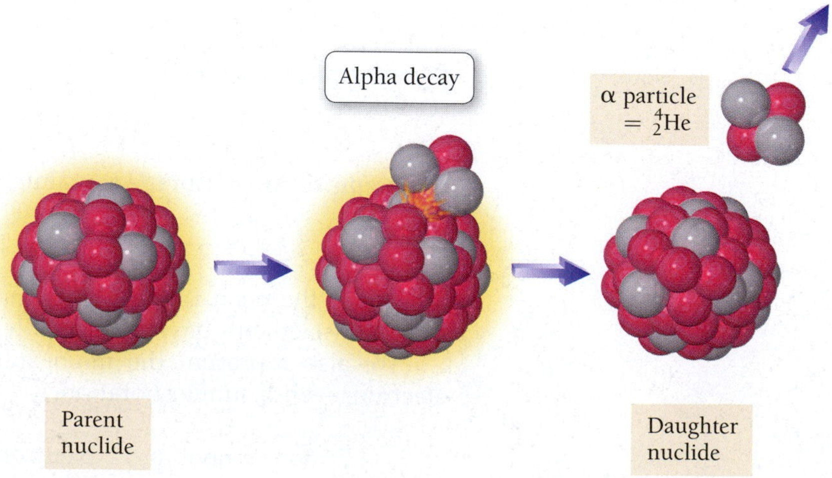

Alpha decay

α particle = ${}^{4}_{2}\text{He}$

Parent nuclide

Daughter nuclide

▶ **FIGURE 17.4 Alpha radiation** Alpha radiation occurs when an unstable nucleus emits a particle composed of 2 protons and 2 neutrons. Question: What happens to the atomic number of an element upon emission of an alpha particle?

We can deduce the mass number and atomic number of the unknown daughter nuclide because the equation must be balanced.

$$\overbrace{^{232}_{90}\text{Th} \longrightarrow {^{x}_{y}}? + {^{4}_{2}\text{He}}}$$

$x + 4 = 232; x = 228$

$y + 2 = 90; y = 88$

Therefore,

$$^{232}_{90}\text{Th} \longrightarrow {^{228}_{88}}? + {^{4}_{1}\text{He}}$$

The atomic number is 88 and the mass number is 228. Finally, we can deduce the identity of the daughter nuclide and its symbol from its atomic number. The atomic number is 88, so the daughter nuclide is radium (Ra).

$$^{232}_{90}\text{Th} \longrightarrow {^{228}_{88}\text{Ra}} + {^{4}_{2}\text{He}}$$

PEARSON
eText
2.0

Interactive Worked Example Video 17.1

EXAMPLE **17.1** | **Writing Nuclear Equations for Alpha (α) Decay**

Write a nuclear equation for the alpha decay of Ra-224.

SOLUTION	
Begin with the symbol for Ra-224 on the left side of the equation and the symbol for an alpha particle on the right side.	$^{224}_{88}\text{Ra} \longrightarrow {^{x}_{y}}? + {^{4}_{2}\text{He}}$
Equalize the sum of the mass numbers and the sum of the atomic numbers on both sides of the equation by writing the appropriate mass number and atomic number for the unknown daughter nuclide.	$^{224}_{88}\text{Ra} \longrightarrow {^{220}_{86}}? + {^{4}_{2}\text{He}}$
Deduce the identity of the unknown daughter nuclide from the atomic number. Since the atomic number is 86, the daughter nuclide is radon (Rn).	$^{224}_{88}\text{Ra} \longrightarrow {^{220}_{86}\text{Rn}} + {^{4}_{2}\text{He}}$

▶ **SKILLBUILDER 17.1** | **Writing Nuclear Equations for Alpha (α) Decay**

Write a nuclear equation for the alpha decay of Po-216.

▶ **FOR MORE PRACTICE** Example 17.6; Problems 59, 60.

Recall that Rutherford used alpha particles to probe the structure of the atom when he discovered the nucleus (Section 4.3).

To ionize means *to create ions* (charged particles).

Alpha radiation is the semi-truck of radioactivity. The alpha particle is by far the most massive of all particles emitted by radioactive nuclei. Consequently, alpha radiation has the most potential to interact with and damage other molecules, including biological ones. Radiation interacts with other molecules and atoms by ionizing them. If radiation ionizes molecules within the cells of living organisms, those molecules become damaged and the cell can die or begin to reproduce abnormally. The ability of radiation to ionize molecules and atoms is its **ionizing power**. Of all types of radioactivity, alpha radiation has the highest ionizing power.

Because of its large size, alpha radiation also has the lowest **penetrating power**—the ability to penetrate matter. (Imagine a semi-truck trying to get through a traffic jam.) In order for radiation to damage important molecules within living cells, it must penetrate the cell. Alpha radiation does not easily penetrate cells; it can be stopped by a sheet of paper, by clothing, or even by air. Consequently, a low-level alpha emitter kept outside the body is relatively safe. However, if an alpha emitter is ingested or inhaled, it becomes very dangerous because the alpha particles then have direct access to the molecules that compose organs and tissues.

To summarize:

- Alpha particles are composed of 2 protons and 2 neutrons.
- The symbol for an alpha particle is $_{2}^{4}\text{He}$.
- Alpha particles have a high ionizing power.
- Alpha particles have a low penetrating power.

Beta (β) Radiation

Beta (β) radiation occurs when an unstable nucleus emits an electron (▼ FIGURE 17.5). How does a nucleus, which contains only protons and neutrons, emit an electron? The electron results from the conversion of a neutron to a proton. In other words, in some unstable nuclei, a neutron changes into a proton and emits an electron in the process.

beta decay   Neutron $\rightarrow$ Proton $+$ Electron

The symbol for a **beta (β) particle** in a nuclear equation is:

beta (β) particle $_{-1}^{0}e$ ●

> Beta radiation is also called beta-minus (β^-) radiation because of its negative charge.

The 0 in the upper-left corner reflects the mass number of the electron. The -1 in the lower-left corner reflects the charge of the electron, which is equivalent to an atomic number of -1 in a nuclear equation. When an atom emits a beta particle, its atomic number increases by one because it now has an additional proton. For example, the nuclear equation for the beta decay of radium-228 is:

$$_{88}^{228}\text{Ra} \longrightarrow {}_{89}^{228}\text{Ac} + {}_{-1}^{0}e$$

> Remember that the mass number is the sum of the number of protons and neutrons. Since an electron has no protons or neutrons, its mass number is zero.

Notice that the nuclear equation is balanced—the sums of the mass numbers on both sides are equal, and the sums of the atomic numbers on both sides are equal.

Left Side		Right Side
$_{88}^{228}\text{Ra}$	$\longrightarrow$	$_{89}^{228}\text{Ac} + {}_{-1}^{0}e$
Sum of mass numbers = 228		Sum of mass numbers = 228 + 0 = 228
Sum of atomic numbers = 88		Sum of atomic numbers = 89 − 1 = 88

We can determine the identity and symbol of the daughter nuclide of any beta decay in a manner similar to the method we used for alpha decay, as demonstrated in Example 17.2.

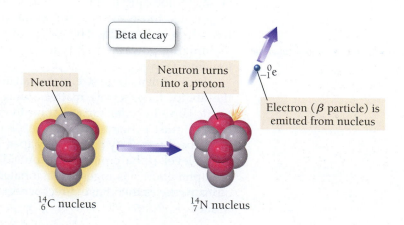

Beta decay

Neutron

Neutron turns into a proton

$_{-1}^{0}e$

Electron (β particle) is emitted from nucleus

$_{6}^{14}\text{C}$ nucleus $_{7}^{14}\text{N}$ nucleus

▶ **FIGURE 17.5 Beta radiation** Beta radiation occurs when an unstable nucleus emits an electron. As the emission occurs, a neutron turns into a proton. **Question: What happens to the atomic number of an element upon emission of a beta particle?**

EXAMPLE **17.2** | **Writing Nuclear Equations for Beta (β) Decay**

Write a nuclear equation for the beta decay of Bk-249.

SOLUTION	
Begin with the symbol for Bk-249 on the left side of the equation and the symbol for a beta particle on the right side.	$^{249}_{97}\text{Bk} \longrightarrow {}^{x}_{y}? + {}^{0}_{-1}\text{e}$
Equalize the sum of the mass numbers and the sum of the atomic numbers on both sides of the equation by writing the appropriate mass number and atomic number for the unknown daughter nuclide.	$^{249}_{97}\text{Bk} \longrightarrow {}^{249}_{98}? + {}^{0}_{-1}\text{e}$
Deduce the identity of the unknown daughter nuclide from the atomic number. Since the atomic number is 98, the daughter nuclide is californium (Cf).	$^{249}_{97}\text{Bk} \longrightarrow {}^{249}_{98}\text{Cf} + {}^{0}_{-1}\text{e}$

▶ **SKILLBUILDER 17.2** | **Writing Nuclear Equations for Beta (β) Decay**

Write a nuclear equation for the beta decay of Ac-228.

▶ **SKILLBUILDER PLUS**

Write three nuclear equations to represent the nuclear decay sequence that begins with the alpha decay of U-235 followed by a beta decay of the daughter nuclide and then another alpha decay.

▶ **FOR MORE PRACTICE** Example 17.7; Problems 61, 62.

Beta radiation is the midsized car of radioactivity. Beta particles are much less massive than alpha particles and consequently have a lower ionizing power. However, because of their smaller size, beta particles have greater penetrating power; a sheet of metal or a thick piece of wood is required to stop them. Consequently, a low-level beta emitter outside the body poses a higher risk than an alpha emitter. Inside the body, however, a beta emitter does less damage than an alpha emitter.

To summarize:

- Beta particles are electrons emitted from atomic nuclei when a neutron changes into a proton.
- The symbol for a beta particle is ${}^{0}_{-1}\text{e}$.
- Beta particles have intermediate ionizing power.
- Beta particles have intermediate penetrating power.

CONCEPTUAL ✓ **CHECKPOINT 17.1**

The element with atomic number 84 undergoes alpha decay followed by beta decay. What is the atomic number of the daughter nuclide following these two decays?

(a) 81   **(b)** 82   **(c)** 83   **(d)** 84

Gamma (γ) Radiation

See Section 9.3 for a review of electromagnetic radiation.

Gamma (γ) radiation is significantly different from alpha or beta radiation. Gamma radiation is not matter but *electromagnetic radiation*—gamma rays are high-energy (short-wavelength) photons. The symbol for a **gamma ray** is:

gamma (γ) ray ${}^{0}_{0}\gamma$

A gamma ray has no charge and no mass. When a gamma ray is emitted from a radioactive atom, it does not change the mass number or the atomic number of the element. Gamma rays are usually emitted in conjunction with other types of radiation. For example, the alpha emission of U-238 (discussed previously) is also accompanied by the emission of a gamma ray.

$$^{238}_{92}\text{U} \longrightarrow {}^{234}_{90}\text{Th} + {}^{4}_{2}\text{He} + {}^{0}_{0}\gamma$$

Gamma rays are the motorbikes of radioactivity. They have the lowest ionizing power but the highest penetrating power. (Imagine a motorbike zipping through a traffic jam.) Stopping gamma rays requires several inches of lead shielding or thick slabs of concrete.

To summarize:

- Gamma rays are electromagnetic radiation—high-energy short-wavelength photons.
- The symbol for a gamma ray is ${}^{0}_{0}\gamma$.
- Gamma rays have low ionizing power.
- Gamma rays have high penetrating power.

Positron Emission

> Positron emission can be thought of as a type of beta emission. It is sometimes referred to as beta-plus (β^{+}) emission.

Positron emission occurs when an unstable nucleus emits a positron (▼ FIGURE 17.6). A **positron** has the mass of an electron but carries a 1+ charge. In some unstable nuclei, a proton changes into a neutron and emits a positron in the process.

$$\textbf{position emission} \qquad \text{Proton} \longrightarrow \text{Neutron} + \text{Positron}$$

The symbol for a positron in a nuclear equation is:

$$\textbf{positron} \qquad {}^{0}_{+1}\text{e} \; \bullet$$

The 0 in the upper-left corner indicates that a positron has a mass number of 0. The +1 in the lower-left corner reflects the charge of the positron, which is equivalent to an atomic number of +1 in a nuclear equation. After an atom emits a positron, its atomic number decreases by 1 because it has 1 less proton. For example, the nuclear equation for the positron emission of phosphorus-30 is:

$$^{30}_{15}\text{P} \longrightarrow {}^{30}_{14}\text{Si} + {}^{0}_{+1}\text{e}$$

We can determine the identity and symbol of the daughter nuclide of any positron emission using a method similar to the one we use for alpha and beta decay, as demonstrated in Example 17.3. Positron emission is similar to beta emission in its ionizing and penetrating power. Table 17.1 summarizes the different kinds of radioactivity covered in this chapter.

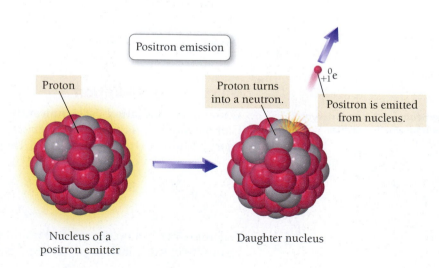

▶ **FIGURE 17.6 Positron emission** Positron emission occurs when an unstable nucleus emits a positron. As the emission occurs, a proton turns into a neutron. Question: What happens to the atomic number of an element upon positron emission?

TABLE 17.1	Selected Types of Radioactive Decay				
Decay Mode	**Process**		**Ionizing Power**	**Penetrating Power**	**Example**
α	Parent nuclide → Daughter nuclide + α particle (^4_2He)		High	Low	$^{238}_{92}\text{U} \longrightarrow \, ^{234}_{90}\text{Th} + \, ^4_2\text{He}$
β	Neutron — Parent nuclide → Neutron becomes a proton — Daughter nuclide + β particle ($^0_{-1}\text{e}$)		Moderate	Moderate	$^{228}_{88}\text{Ra} \longrightarrow \, ^{228}_{89}\text{Ac} + \, ^0_{-1}\text{e}$
γ	Excited nuclide → Stable nuclide + Photon ($^0_0\gamma$)		Low	High	$^{234}_{90}\text{Th} \longrightarrow \, ^{234}_{90}\text{Th} + \, ^0_0\gamma$
Positron emission	Proton — Parent nuclide → Proton becomes a neutron — Daughter nuclide + Positron ($^0_{+1}\text{e}$)		Moderate	Moderate	$^{30}_{15}\text{P} \longrightarrow \, ^{30}_{14}\text{Si} + \, ^0_{+1}\text{e}$

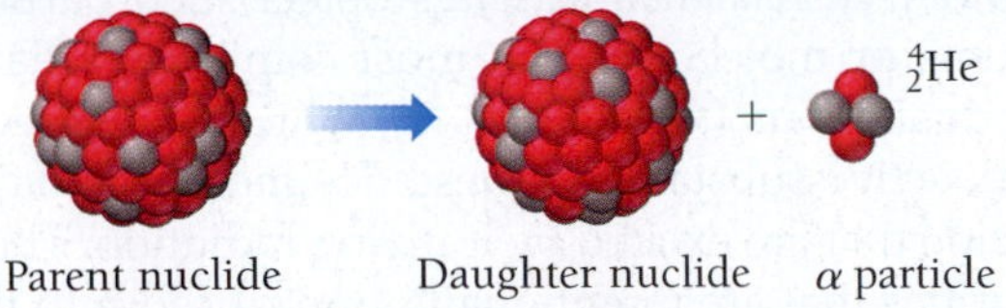

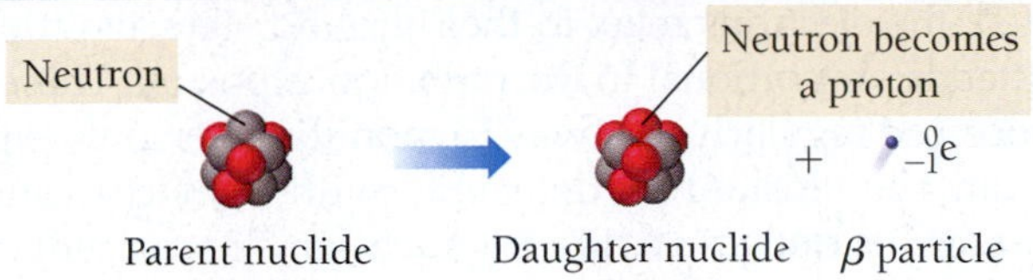

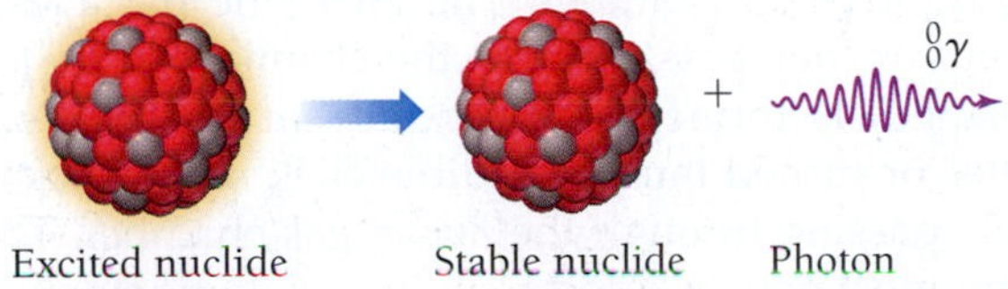

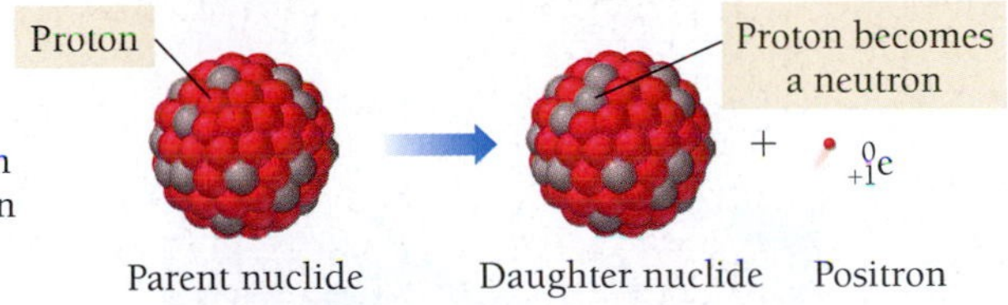

EXAMPLE **17.3**   Writing Nuclear Equations for Positron Emission

Write a nuclear equation for the positron emission of potassium-40.

SOLUTION	
Begin with the symbol for K-40 on the left side of the equation and the symbol for a positron on the right side.	$^{40}_{19}\text{K} \longrightarrow \, ^{x}_{y}? + \, ^0_{+1}\text{e}$
Equalize the sum of the mass numbers and the sum of the atomic numbers on both sides of the equation by writing the appropriate mass number and atomic number for the unknown daughter nuclide.	$^{40}_{19}\text{K} \longrightarrow \, ^{40}_{18}? + \, ^0_{+1}\text{e}$
Deduce the identity of the unknown daughter nuclide from the atomic number. Since the atomic number is 18, the daughter nuclide is argon (Ar).	$^{40}_{19}\text{K} \longrightarrow \, ^{40}_{18}\text{Ar} + \, ^0_{+1}\text{e}$

▶ **SKILLBUILDER 17.3 | Writing Nuclear Equations for Positron Emission**

Write a nuclear equation for the positron emission of sodium-22.

▶ **FOR MORE PRACTICE**   Example 17.8; Problems 63, 64.

CONCEPTUAL ✔ CHECKPOINT **17.2**

Which kind of radioactive decay changes the mass number of the parent element?

(a) alpha (α) decay   (b) beta (β) decay

(c) gamma (γ) decay   (d) positron decay

17.4 Detecting Radioactivity

▶ Describe and explain the methods used to detect radioactivity.

The particles emitted from radioactive nuclei contain a large amount of energy and therefore are easy to detect. *Radiation detectors* detect such particles through their interactions with atoms or molecules. The most common radiation detectors are **thermoluminescent dosimeters** (▼ FIGURE 17.7a), which are issued to people working with or near radioactive substances. These dosimeters contain crystals of salts such as calcium fluoride that are excited by ionizing radiation. The excited electrons are trapped by impurities that are intentionally introduced into the crystals. When the crystals are heated, the electrons relax to their ground state, emitting light. The amount of light emitted is proportional to the radiation exposure. These dosimeters are collected and processed regularly as a way to monitor a person's exposure.

Radioactivity can be instantly detected with devices such as the **Geiger-Müller counter** (▼ FIGURE 17.7b). In such an instrument (commonly referred to simply as a Geiger counter), particles emitted by radioactive nuclei pass through an argon-filled chamber. The energetic particles create a trail of ionized argon atoms as they pass through the chamber. If the applied voltage is high enough, these newly formed ions produce an electrical signal that can be detected on a meter or turned into an audible click. Each click corresponds to a radioactive particle passing through the argon gas chamber. This clicking is the stereotypical sound most people associate with a radiation detector.

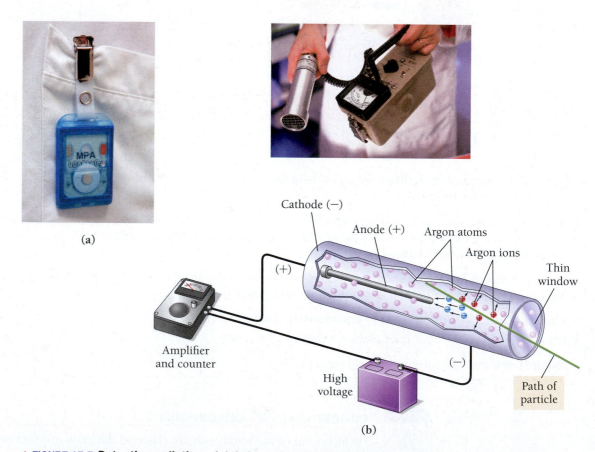

(a)

(b)

▲ **FIGURE 17.7 Detecting radiation** **(a)** A thermoluminescent dosimeter is issued to workers whose jobs entail radiation exposure risk. **(b)** A Geiger counter records the passage of individual energetic particles emitted by radioactive nuclei as they pass through a chamber filled with argon gas. When an argon atom is ionized, the resulting ion is attracted to the anode and the dislodged electron is attracted to the cathode, creating a tiny electrical current that can be recorded.

A second device commonly used to detect radiation instantly is a **scintillation counter**. In a scintillation counter, the radioactive emissions pass through a material (such as NaI or CsI) that emits ultraviolet or visible light in response to excitation by energetic particles. This light is detected and turned into an electrical signal that is read on a meter.

17.5 Natural Radioactivity and Half-Life

▶ Use half-life to relate radioactive sample amounts to elapsed time.

Radioactivity is a natural component of the environment. The ground beneath you most likely contains radioactive atoms that emit radiation into the air around you. The food you eat contains a residual amount of radioactive atoms that enter into your body fluids and tissues. Small amounts of radiation from space make it through the atmosphere and constantly bombard Earth. Humans and other living organisms have evolved in this environment and have adapted to survive in it. One reason for the radioactivity in our environment is the instability of all atomic nuclei beyond atomic number 83 (bismuth). In addition, some isotopes of elements with fewer than 83 protons are also unstable and radioactive.

Half-Life

A *decay event* is the emission of radiation by a single radioactive nuclide.

Different radioactive nuclides decay into their daughter nuclides at different rates. Some nuclides decay quickly while others decay slowly. The time it takes for half of the parent nuclides in a radioactive sample to decay to the daughter nuclides is the **half-life**. Nuclides that decay quickly have short half-lives and are very active (many decay events per unit time), while those that decay slowly have long half-lives and are less active (fewer decay events per unit time).

For example, Th-232 is an alpha emitter that decays according to the nuclear reaction:

$$^{232}_{90}\text{Th} \longrightarrow ^{228}_{88}\text{Ra} + ^{4}_{2}\text{He}$$

Th-232 has a half-life of 1.4×10^{10} years or 14 billion years, so it is not particularly active. If we start with a sample of Th-232 containing 1 million atoms, the sample would decay to half a million atoms in 14 billion years and to a quarter of a million in another 14 billion years and so on (▼ **FIGURE 17.8**).

$$\begin{array}{ccccc}
\text{1 million} & & \tfrac{1}{2}\,\text{million} & & \tfrac{1}{4}\,\text{million} \\
\text{Th-232 atoms} & \xrightarrow[\text{14 billion years}]{} & \text{Th-232 atoms} & \xrightarrow[\text{14 billion years}]{} & \text{Th-232 atoms}
\end{array}$$

Notice that a radioactive sample *does not* decay to zero atoms in two half-lives— *we can't add two half-lives together to get a "whole" life.* The amount that remains

▶ FIGURE 17.8 **The concept of half-life** A plot of the number of Th-232 atoms in a sample initially containing 1 million atoms as a function of time.

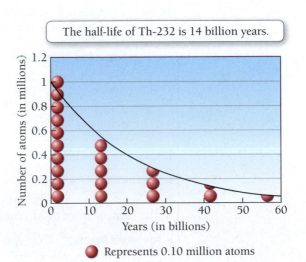

The half-life of Th-232 is 14 billion years.

Number of atoms (in millions)

Years (in billions)

● Represents 0.10 million atoms

> Each radioactive nuclide has a unique half-life, which is not affected by physical conditions or chemical environment.

TABLE 17.2 Selected Nuclides and Their Half-Lives

Nuclide	Half-Life	Type of Decay
$^{232}_{90}\text{Th}$	1.4×10^{10} yr	alpha
$^{238}_{92}\text{U}$	4.5×10^{9} yr	alpha
$^{14}_{6}\text{C}$	5715 yr	beta
$^{220}_{86}\text{Rn}$	55.6 s	alpha
$^{219}_{90}\text{Th}$	1.05×10^{-6} s	alpha

after one half-life is always half of what was present at the start. The amount that remains after two half-lives is a quarter of what was present at the start, and so on.

Some nuclides have short half-lives. Radon-220, for example, has a half-life of approximately 1 minute. If we had a sample of radon-220 that contained 1 million atoms, it would be diminished to $\frac{1}{4}$ million radon-220 atoms in just 2 minutes.

$$\underset{\text{Rn-220 atoms}}{1 \text{ million}} \xrightarrow[\text{1 minute}]{} \underset{\text{Rn-220 atoms}}{\tfrac{1}{2} \text{ million}} \xrightarrow[\text{1 minute}]{} \underset{\text{Rn-220 atoms}}{\tfrac{1}{4} \text{ million}}$$

Rn-220 is much more active than Th-232 because it undergoes many more decay events in a given period of time. Some nuclides have even shorter half-lives. Table 17.2 lists several nuclides and their half-lives.

CONCEPTUAL ✓ CHECKPOINT 17.3

The graph shows the number of moles of a radioactive nuclide as a function of time. What is the half-life of the nuclide?

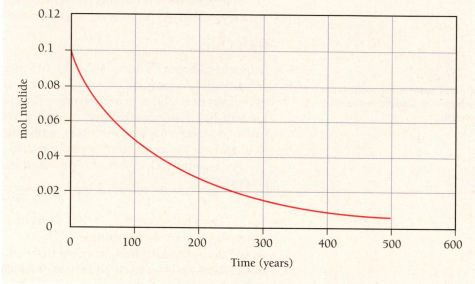

EXAMPLE 17.4   Half-Life

How long does it take for a 1.80-mol sample of Th-228 (which has a half-life of 1.9 years) to decay to 0.225 mol?

SOLUTION

Draw a table showing the amount of Th-228 as a function of number of half-lives. For each half-life, divide the amount of Th-228 by 2.

Amount of Th-228	Number of Half-Lives	Time (yrs)
1.80 mol	0	0
0.900 mol	1	1.9
0.450 mol	2	3.8
0.225 mol	3	5.7

It takes three half-lives or 5.7 years for the sample to decay to 0.225 mol.

▶ **SKILLBUILDER 17.4 | Half-Life**

A radium-226 sample initially contains 0.112 mol. How much radium-226 is left in the sample after 6400 years? The half-life of radium-226 is 1600 years.

▶ **FOR MORE PRACTICE**   Example 17.9; Problems 69, 70, 71, 72, 73, 74, 75, 76.

CHEMISTRY AND HEALTH

Environmental Radon

Radon—a radioactive gas—is one of the products of the radioactive decay series of uranium. Wherever there is uranium in the ground, there is likely to be radon seeping up into the air. Radon and its daughter nuclides (which attach to dust particles) can therefore be inhaled into the lungs, where they decay and increase lung cancer risk. The radioactive decay of radon is by far the single greatest source of human radiation exposure.

Homes built in areas with significant uranium deposits in the ground pose the greatest risk. These homes can accumulate radon levels that are above what the Environmental Protection Agency (EPA) considers safe. Simple test kits are available to test indoor air and determine radon levels. The higher the radon level is, the greater the risk. The health risk is even higher for smokers who live in these houses. Excessively high indoor radon levels require the installation of a ventilation system to purge radon from the house. Lower levels can be ventilated by keeping windows and doors open.

B17.1 CAN YOU ANSWER THIS? *Suppose that a house contains* 1.80×10^{-3} *mol of radon-222 (which has a half-life of 3.8 days). If no new radon entered the house, how long would it take for the radon to decay to* 4.50×10^{-4} *mol?*

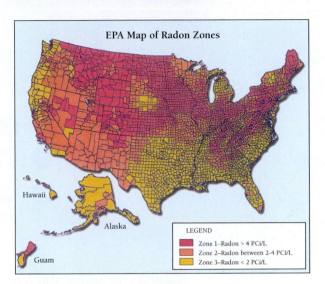

▲ Map of the United States showing radon levels. Zone 1 counties have the highest levels, and zone 3 counties have the lowest.

A Natural Radioactive Decay Series

The radioactive elements in our environment are all undergoing radioactive decay. They are always present in our environment because they either have very long half-lives (billions of years) or they are continuously being formed by some other process in the environment. In many cases, the daughter nuclide of a radioactive decay is itself radioactive and in turn produces another daughter nuclide that is radioactive and so on, resulting in a radioactive decay series.

For example, uranium (atomic number 92) is the heaviest naturally occurring element. U-238 is an alpha emitter that decays to Th-234 with a half-life of 4.47 billion years.

$$^{238}_{91}U \longrightarrow ^{234}_{90}Th + ^{4}_{2}He$$

The daughter nuclide, Th-234, is itself radioactive—it is a beta emitter that decays to Pa-234 with a half-life of 24.1 days.

$$^{234}_{90}Th \longrightarrow ^{234}_{91}Pa + ^{0}_{-1}e$$

Pa-234 is also radioactive, decaying to U-234 via beta emission with a half-life of 244,500 years. This process continues until it produces Pb-206, which is stable.

The entire uranium-238 decay series is shown in ▶ **FIGURE 17.9**, on the next page. All of the uranium-238 in the environment is slowly decaying away to lead. Since the half-life for the first step in the series is so long, however, there is still plenty of uranium-238 in the environment. All of the other nuclides in the decay series are also present in the environment in varying amounts, depending on their half-lives.

▶ **FIGURE 17.9 Uranium-238 decay series** The red arrows represent alpha decay, and the blue arrows represent beta decay.

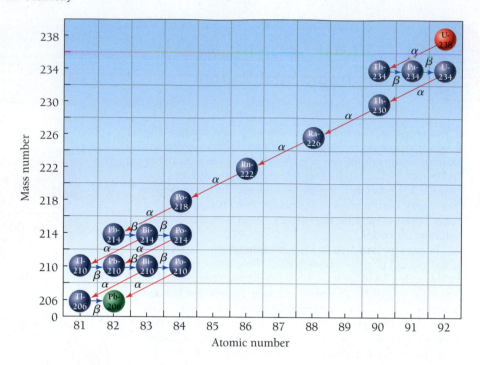

CONCEPTUAL ✔ CHECKPOINT 17.4

If you start with 1 million atoms of a particular radioactive isotope, how many half-lives are required to reduce the number of undecayed atoms to fewer than 1000?

(a) 10   **(b)** 100   **(c)** 1000   **(d)** 1001

17.6 Radiocarbon Dating: Using Radioactivity to Measure the Age of Fossils and Other Artifacts

▶ Use carbon-14 content to determine the age of fossils or artifacts.

Archaeologists, geologists, anthropologists, and other scientists take advantage of the presence of natural radioactivity in our environment to estimate the ages of fossils and artifacts with a technique called **radiocarbon dating**. For example, in 1947, young shepherds searching for a stray goat near the Dead Sea (east of Jerusalem) entered a cave and discovered ancient scrolls stuffed into jars. These scrolls—subsequently named the Dead Sea Scrolls—are 2000-year-old biblical manuscripts, predating other existing manuscripts by almost 1000 years.

The Dead Sea Scrolls—like other ancient artifacts—contain a radioactive signature that reveals their age. This signature results from the presence of carbon-14—which is radioactive—in the environment. Carbon-14 is constantly formed in the upper atmosphere by the neutron bombardment of nitrogen.

$$\,^{14}_{7}\text{N} + \,^{1}_{0}\text{n} \longrightarrow \,^{14}_{6}\text{C} + \,^{1}_{1}\text{H}$$

Carbon-14 decays back to nitrogen by beta emission, with a half-life of 5715 years.

$$\,^{14}_{6}\text{C} \longrightarrow \,^{14}_{7}\text{N} + \,^{0}_{-1}\text{e}$$

The continuous formation of carbon-14 in the atmosphere and its continuous decay back to nitrogen-14 produces a nearly constant equilibrium concentration of atmospheric carbon-14. That carbon-14 is oxidized to carbon dioxide and then incorporated into plants by photosynthesis. It is also incorporated into animals because animals ultimately depend on plants for food (they either eat plants or eat other animals that eat plants). Consequently, all living organisms contain a residual amount of carbon-14. When a living organism dies, it stops incorporating new

The concentration of carbon-14 in all *living* organisms is the same.

carbon-14 into its tissues. The carbon-14 present at the time of death decays with a half-life of 5715 years. Since many artifacts, such as the Dead Sea Scrolls, are made from materials that were once living—such as papyrus, wood, and other plant and animal derivatives—the amount of carbon-14 in these artifacts indicates their age.

For example, suppose an ancient artifact has a carbon-14 concentration that is 50% of that found in living organisms. How old is the artifact? Because it contains half as much carbon-14 as a living organism, it must be one half-life, or 5715 years old. If the artifact has a carbon-14 concentration that is 25% of that found in living organisms, its age is two half-lives or 11,430 years old. Table 17.3 lists the ages of objects based on carbon-14 content. Ages of less than one half-life, or intermediate between a whole number of half-lives, can also be calculated, but the method for doing so is beyond the scope of this book.

We know that carbon-14 dating is accurate because we check against objects whose ages are known from other methods. For example, old trees can be dated by counting the tree rings within their trunks and by carbon-14 dating. The two methods generally agree to within a few percent. However, carbon-14 dating is not dependable for objects that are more than 50,000 years old; the amount of carbon-14 becomes too low to measure.

TABLE 17.3 Age of Objects Based on Carbon-14 Content	
Concentration of C-14 (% Relative to Living Organisms)	**Age of Object (yrs)**
100.0	0
50.0	5,715
25.00	11,430
12.50	17,145
6.250	22,860
3.125	28,575
1.563	34,290

CHEMISTRY IN THE MEDIA

The Shroud of Turin

The shroud of Turin—kept in the Cathedral of Turin in Italy—is an old linen cloth that bears the image of a man who appears to have been crucified.

The image becomes clearer if the shroud is photographed and viewed as a negative. Many believe that the shroud is the original burial cloth of Jesus Christ, miraculously imprinted with his image. In 1988, the Roman Catholic Church chose three independent laboratories to perform radiocarbon dating on the shroud. The laboratories took samples from the shroud and measured the carbon-14 content. They all arrived at similar results—the shroud was made from linen originating in about A.D. 1325. Although some have disputed the results, and although no scientific test is 100% reliable, newspapers around the world quickly announced that the shroud could not have been the burial cloth of Jesus.

B17.2 CAN YOU ANSWER THIS? *An artifact is said to have originated in 3000 B.C. Examination of the C-14 content of the artifacts reveals that the concentration of C-14 is 55% of that found in living organisms. Is the artifact authentic?*

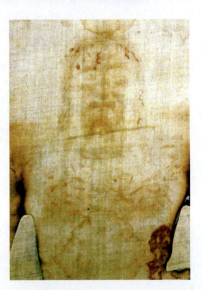

▲ The Shroud of Turin.

EXAMPLE **17.5** | **Radiocarbon Dating**

A skull believed to belong to an early human being is found to have a carbon-14 content of 3.125% of that found in living organisms. How old is the skull?

SOLUTION

Examine Table 17.3 to determine that a carbon-14 content of 3.125% of that found in living organisms corresponds to an age of 28,575 years.

▶ **SKILLBUILDER 17.5** | **Radiocarbon Dating**

An ancient scroll is claimed to have originated from Greek scholars in about 500 B.C. A measure of its carbon-14 content reveals it to contain 100.0% of that found in living organisms. Is the scroll authentic?

▶ **FOR MORE PRACTICE**   Example 17.10; Problems 79, 80, 81, 82.

17.7   The Discovery of Fission and the Atomic Bomb

▶ Explain how the experiments of Fermi, Meitner, Strassmann, and Hahn led to the discovery of nuclear fission.

In the mid-1930s Enrico Fermi (1901–1954), an Italian physicist, tried to synthesize a new element by bombarding uranium—the heaviest known element at that time—with neutrons. Fermi hypothesized that if a neutron were incorporated into the nucleus of a uranium atom, the nucleus might undergo beta decay, converting a neutron into a proton. If that happened, a new element, with atomic number 93, would be synthesized for the first time. The nuclear equation for the process is:

> The element with atomic number 100 is named *fermium*, in honor of Enrico Fermi.

$$^{238}_{92}\text{U} + {}^{1}_{0}\text{n} \longrightarrow {}^{239}_{92}\text{U} \longrightarrow {}^{239}_{93}\text{X} + {}^{0}_{-1}\text{e}$$

Neutron

Newly synthesized element

Fermi performed the experiment and detected the emission of beta particles. However, his results were inconclusive. Had he synthesized a new element? Fermi never chemically examined the products to determine their composition and therefore could not say with certainty whether he had.

Three researchers in Germany—Lise Meitner (1878–1968), Fritz Strassmann (1902–1980), and Otto Hahn (1879–1968)—repeated Fermi's experiments and then performed careful chemical analysis of the products. What they found in the products—several elements *lighter* than uranium—would change the world forever.

> The element with atomic number 109 is named *meitnerium*, in honor of Lise Meitner.

On January 6, 1939, Meitner, Strassmann, and Hahn reported that the neutron bombardment of uranium resulted in **nuclear fission**—the splitting of the atom. The nucleus of the neutron-bombarded uranium atom had broken apart into barium, krypton, and other smaller products. They also realized that the process emitted enormous amounts of energy. The following nuclear equation for a fission reaction shows how uranium breaks apart into the daughter nuclides:

$$^{235}_{92}\text{U} + {}^{1}_{0}\text{n} \longrightarrow {}^{142}_{56}\text{Ba} + {}^{91}_{36}\text{Kr} + 3{}^{1}_{0}\text{n} + \text{Energy}$$

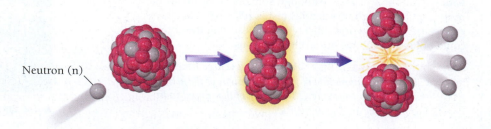

Neutron (n)

▲ Lise Meitner in Otto Hahn's Berlin laboratory.

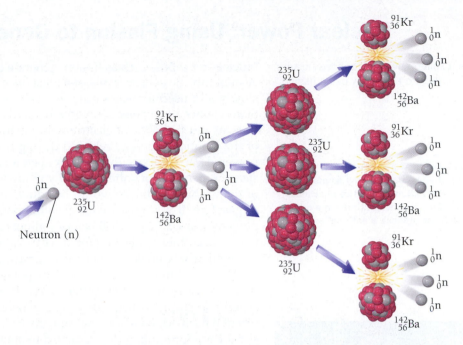

▲ **FIGURE 17.10** **Fission chain reaction**   The neutrons produced by the fission of one uranium nucleus induce fission in other uranium nuclei to produce a self-amplifying reaction. Question: Why must each fission event produce more than one neutron to sustain the chain reaction?

Notice that the initial uranium atom in the nuclear equation is the U-235 isotope, which makes up less than 1% of all naturally occurring uranium. The most abundant uranium isotope, which is U-238, does not undergo fission. Therefore, the uranium used for fuel in nuclear reactions must be *enriched* in U-235 (it must contain more than the naturally occurring percentage of U-235). Notice also that the process produces three neutrons, which have the potential to initiate fission in three other U-235 atoms.

U.S. scientists quickly realized that uranium enriched with U-235 could undergo a **chain reaction** in which neutrons produced by the fission of one uranium nucleus would induce fission in other uranium nuclei (▲ **FIGURE 17.10**).

The result is a self-amplifying reaction capable of producing an enormous amount of energy—an atomic bomb. However, to make a bomb, a **critical mass** of U-235—enough U-235 to produce a self-sustaining reaction—is necessary. Fearing that Nazi Germany would develop such a bomb, several U.S. scientists persuaded Albert Einstein (1879–1955), the most well-known scientist of the time, to write a letter to President Franklin Roosevelt in 1939 warning of this possibility. Einstein wrote, "and it is conceivable—though much less certain—that extremely powerful bombs of a new type may thus be constructed. A single bomb of this type, carried by boat and exploded in a port, might very well destroy the whole port together with some of the surrounding territory."

Roosevelt was convinced by Einstein's letter, and in 1941 he assembled the resources to begin the costliest scientific project ever attempted. The top-secret endeavor was called the *Manhattan Project*, and its main goal was to build an atomic bomb before the Germans did. The project was led by physicist J. R. Oppenheimer (1904–1967) and was headquartered at a high-security research facility in Los Alamos, New Mexico.

Four years later, on July 16, 1945, the world's first nuclear weapon was successfully detonated at a test site in New Mexico. The first atomic bomb exploded with a force equivalent to 18,000 tons of dynamite. The Germans— who had *not* made a successful nuclear bomb—had already been defeated by this time. Instead, the atomic bomb was used on Japan. One bomb was dropped on Hiroshima, and a second bomb was dropped on Nagasaki. Together, the bombs killed approximately 200,000 people and forced Japan to surrender. World War II was over. The atomic age had begun.

▲ The testing of the world's first nuclear bomb at Alamogordo, New Mexico, in 1945.

17.8 Nuclear Power: Using Fission to Generate Electricity

▶ Explain how nuclear power plants generate electricity using fission.

A nuclear-powered car really is hypothetical because the amount of uranium-235 in a pencil-sized cylinder would not be enough to reach a critical mass and produce a self-sustaining reaction.

▲ Technicians inspect the core of a nuclear reactor, which houses the fuel rods and control rods.

Nuclear reactions, such as fission, generate enormous amounts of energy. In a nuclear bomb, the energy is released all at once. The energy can also be released more slowly and used for other purposes such as electricity generation. In the United States, about 20% of electricity is generated by nuclear fission. In some other countries, as much as 70% of electricity is generated by nuclear fission. To get an idea of the amount of energy released during fission, imagine a hypothetical nuclear-powered car. Suppose the fuel for such a car was a uranium cylinder about the size of a pencil. How often would you have to refuel the car? The energy content of the uranium cylinder would be equivalent to about 1000 twenty-gallon tanks of gasoline. If you refuel your gasoline-powered car once a week, your nuclear-powered car could go 1000 weeks—almost 20 years—before refueling. Imagine a pencil-sized fuel rod lasting for 20 years of driving!

Similarly, a nuclear-powered electrical plant can produce a lot of electricity with a small amount of fuel. Nuclear power plants generate electricity by using fission to generate heat (▼ FIGURE 17.11). The heat is used to boil water and create steam, which turns the turbine on a generator to produce electricity. The fission reaction itself occurs in the nuclear core of the power plant, or *reactor*. The core consists of uranium fuel rods—enriched to about 3.5% U-235—interspersed between retractable neutron-absorbing control rods. When the control rods are fully retracted from the fuel rod assembly, the chain reaction occurs unabated. However, when the control rods are fully inserted into the fuel assembly, they absorb the neutrons that would otherwise induce fission, shutting down the chain reaction.

By inserting or retracting the control rods, the operator controls the rate of fission. If more heat is needed, the control rods are retracted slightly. If the fission reaction begins to get too hot, the control rods are inserted a little more. In this way, the fission reaction is controlled to produce the right amount of heat needed to generate electricity. In case of a power failure, the fuel rods automatically drop into the fuel rod assembly, shutting down the fission reaction.

▶ FIGURE 17.11 Nuclear power In a nuclear power plant, fission generates heat that is used to boil water and create steam. The steam turns a turbine on a generator to produce electricity. Note that the superheated water carrying heat from the reactor core is contained within separate pipes and does not come into direct contact with the steam that drives the turbines.

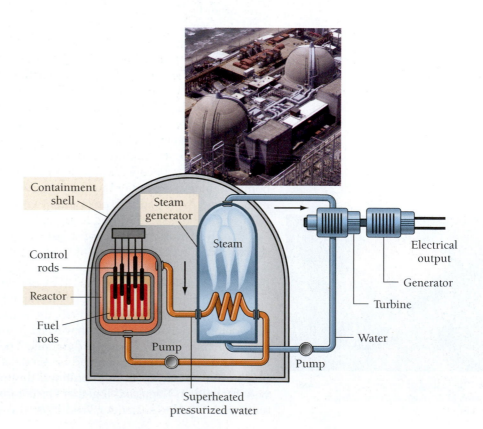

A typical nuclear power plant generates enough electricity for a city of about 1 million people and uses about 50 kg of fuel per day. In contrast, a coal-burning power plant uses about 2 million kg of fuel to generate the electricity for that same city. Furthermore, a nuclear power plant generates no air pollution and no greenhouse gases. A coal-burning power plant, on the other hand, emits pollutants such as carbon monoxide, nitrogen oxides, and sulfur oxides. Coal-burning power plants also emit carbon dioxide, a greenhouse gas.

Nuclear power generation, however, is not without potential risks. Foremost among them is the danger of nuclear accidents. In spite of safety precautions, the fission reaction occurring in a nuclear power plant can overheat. The most famous examples of this type of accident occurred in Chernobyl in the former Soviet Union on April 26, 1986, and at the Fukushima Daiichi Nuclear Power Plant in Japan in March of 2011.

> Reactor cores in the United States are not made of graphite and could not burn in the way that the Chernobyl core did.

In the Chernobyl incident, operators of the plant were performing an experiment designed to reduce maintenance costs. In order to perform the experiment, many of the safety features of the reactor core were disabled. The experiment failed, with disastrous results. The nuclear core, composed partly of graphite, overheated and began to burn. The accident directly caused 31 deaths and produced a fire that scattered radioactive debris into the atmosphere, making the surrounding land uninhabitable. The overall death toll from subsequent cancers is still highly debated.

In the 2011 Japanese accident, a 9.0 magnitude earthquake triggered a tsunami that flooded the coastal plant and caused the plant's cooling system pumps to fail. Three of the nuclear cores within the plant dramatically overheated and experienced a partial meltdown (in which the fuel became so hot that it melted). The accident was intensified by the loss of water in the fuel storage ponds (pools of water normally used to keep spent fuel as well as future fuel cool), which caused the fuel stored in the ponds to also overheat. The release of radiation into the environment, while significant, was much lower in Japan than at Chernobyl. Even several years after the accident, no radioactivity-related deaths have been reported at the Fukushima plant or the surrounding area. The cleanup of the site, however, will continue for many years.

▲ In 2011, the Fukushima Daiichi Nuclear Power Plant in Japan overheated as a result of a 9.0 magnitude earthquake triggered by a tsunami that flooded the coastal plant and caused the cooling system pumps to fail.

As serious as nuclear accidents are, a nuclear power plant *cannot* become a nuclear bomb. The uranium fuel used in electricity generation is not sufficiently enriched in U-235 to produce a nuclear detonation. Moreover, U.S. nuclear power plants have additional safety features designed to prevent similar accidents. For example, U.S. nuclear power plants have large containment structures designed to contain radioactive debris in the event of an accident.

A second problem associated with nuclear power is waste disposal. Although the amount of nuclear fuel used in electricity generation is small compared to other fuels, the products of the reaction are radioactive and have very long half-lives (thousands of years or more). What do we do with this waste? Currently, in the United States, nuclear waste is stored on site at the nuclear power plants. A permanent disposal site was being developed in Yucca Mountain, Nevada. The site had originally been scheduled to be operational in 2010, and that date was later delayed to 2017. However, the Obama administration determined that the Yucca Mountain site was untenable, and in 2010, the license application to develop this site was withdrawn. President Obama has formed a committee (called the Blue Ribbon Commission on America's Nuclear Future) that is charged with developing alternatives to Yucca Mountain. The committee has made several important recommendations. These include both the development of a temporary above-ground storage facility as well as a permanent underground facility. However, the committee has not made any decisions regarding Yucca Mountain as a potential site for the storage facility.

17.9   Nuclear Fusion: The Power of the Sun

▶ Compare and contrast nuclear fission and nuclear fusion.

As we have learned, nuclear fission is the splitting of a heavy nucleus to form two or more lighter ones. **Nuclear fusion**, by contrast, is the combination of two light nuclei to form a heavier one. Both fusion and fission emit large amounts of energy.

Nuclear fusion is the energy source of stars, including our sun. In stars, hydrogen atoms fuse together to form helium atoms, emitting energy in the process.

Nuclear fusion is also the basis of modern nuclear weapons called hydrogen bombs. A modern hydrogen bomb has up to 1000 times the explosive force of the first atomic bombs. These bombs employ the following fusion reaction:

$$^2_1H + {}^3_1H \longrightarrow {}^4_2He + {}^1_0n$$

In this reaction, deuterium (the isotope of hydrogen with one neutron) and tritium (the isotope of hydrogen with two neutrons) combine to form helium-4 and a neutron. Because fusion reactions require two positively charged nuclei (which repel each other) to fuse together, extremely high temperatures are required. In a hydrogen bomb, a small fission bomb is detonated first, providing temperatures high enough for fusion to proceed.

Nuclear fusion has been intensely investigated as a way to produce electricity. Because of the higher-energy production—fusion provides about ten times more energy per gram of fuel than fission—and because the products of the reaction are less dangerous than those of fission, fusion holds promise as a future energy source. However, in spite of intense efforts, fusion electricity generation remains elusive. One of the main problems is the high temperature required for fusion to occur—no material can withstand these temperatures. Whether fusion will ever be a viable energy source remains to be seen.

CONCEPTUAL ✔ **CHECKPOINT 17.5**

What is the main difference between nuclear fission and nuclear fusion?

(a) Fission is the combining of two nuclei to form one, while fusion is the splitting of one nucleus to form two.

(b) Fusion is the combining of two nuclei to form one, while fission is the splitting of one nucleus to form two.

(c) Fission gives off energy while fusion absorbs energy.

17.10   The Effects of Radiation on Life

▶ Describe how radiation exposure affects biological molecules and how exposure is measured.

Radiation can ionize atoms in biological molecules, thereby initiating reactions that can alter the molecules. When radiation damages important molecules in living cells, problems can develop. The ingestion of radioactive materials—especially alpha and beta emitters—is particularly dangerous because radioactive decay then occurs within the body and can do more damage than external radiation. The effects of radiation are divided into three different types: acute radiation damage, increased cancer risk, and genetic effects.

Acute Radiation Damage

Acute radiation damage results from exposures to large amounts of radiation in a short period of time. The main sources of this kind of exposure are nuclear bombs or exposed nuclear reactor cores. The resulting high levels of radiation kill large numbers of cells. Rapidly dividing cells, such as those in the immune system and the intestinal lining, are most susceptible. Consequently, people exposed to high levels of radiation have weakened immune systems and a lowered ability to absorb nutrients from food. In milder cases, recovery is possible with time. In more extreme cases, death results, often from unchecked infection.

Increased Cancer Risk

Lower doses of radiation over extended periods of time can increase cancer risk because radiation can damage DNA, the molecules in cells that carry instructions

DNA and its function in the body are explained in more detail in Chapter 19.

for cell growth and replication. When the DNA within a cell is damaged, the cell normally dies. Occasionally, however, changes in DNA cause cells to grow abnormally and to become cancerous. These cancerous cells grow into tumors that can spread and, in some cases, cause death. Cancer risk increases with increased radiation exposure. However, cancer is so prevalent and has so many convoluted causes that it is difficult to determine an exact threshold for increased cancer risk from radiation exposure.

Genetic Defects

Another possible effect of radiation exposure is genetic defects in offspring. If radiation damages the DNA of reproductive cells—such as eggs or sperm—the offspring that develop from those cells may have genetic abnormalities. Genetic defects of this type have been observed in laboratory animals exposed to high levels of radiation. However, such genetic defects—with a clear causal connection to radiation exposure—have yet to be observed in humans, even in studies of Hiroshima survivors.

Measuring Radiation Exposure

Common units of radioactivity include the *curie*, defined as 3.7×10^{10} decay events per second, and the *roentgen*, defined as the amount of radiation that produces 2.58×10^{-4} C of charge per kilogram of air. Human radiation exposure is often reported in a unit called the **rem**. The rem, which stands for *roentgen equivalent man*, is a weighted measure of radiation exposure that accounts for the ionizing power of the different types of radiation. On average, each person in the United States is exposed to approximately one-third of a rem of radiation per year from natural sources. It takes much more radiation than the natural amount to produce measurable health effects in humans. The first measurable effects, a decreased white blood cell count, occur at instantaneous exposures of approximately 20 rem (Table 17.4). Exposures of 100 rem show a definite increase in cancer risk, and exposures of more than 500 rem often result in death.

TABLE 17.4 Effects of Radiation Exposure

Dose (rem)	Probable Outcome
20–100	decreased white blood cell count; possible increase in cancer risk
100–400	radiation sickness; skin lesions; increase in cancer risk
500	death

17.11 Radioactivity in Medicine

▶ Describe how radioactivity is used in the diagnosis and treatment of disease.

Radioactivity is often perceived as dangerous; however, it is also enormously useful to physicians in the diagnosis and treatment of disease. We can broadly divide the use of radioactivity into **isotope scanning** and **radiotherapy**.

Isotope Scanning

Recall from Section 17.1 that in isotope scanning, a radioactive isotope is introduced into the body and the radiation emitted by the isotope is detected. Since different isotopes are taken up by different organs or tissues, isotope scanning has a variety of uses. For example, the radioactive isotope phosphorus-32 is preferentially taken up by cancerous tissue. A cancer patient is given this isotope so that physicians can locate and identify cancerous tumors. Other isotopes commonly used in medicine

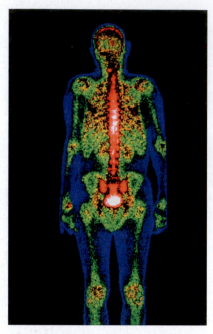

▲ **FIGURE 17.12 An isotope scan** Technetium-99 is often used as the radiation source for bone scans such as this one.

include iodine-131, used to diagnose thyroid disorders, and technetium-99, which can produce images of several different internal organs (◀ **FIGURE 17.12**).

Radiotherapy

Because radiation kills cells, and because it is particularly effective at killing rapidly dividing cells, it is often used as a therapy for cancer (cancer cells divide more quickly than normal cells). Gamma rays are focused on internal tumors to kill them (▼ **FIGURE 17.13**). The gamma-ray beam is usually aimed at the tumor from a number of different angles, maximizing the exposure of the tumor while minimizing the exposure of the healthy tissue around the tumor. (See *Chemistry and Health: Radiation Treatment for Cancer* in Chapter 9.) Nonetheless, cancer patients undergoing radiation therapy usually develop the symptoms of radiation sickness, which include vomiting, skin burns, and hair loss.

Some people wonder why radiation—which is known to cause cancer—is used to treat cancer. The answer lies in risk analysis. A cancer patient is normally exposed to radiation doses of about 100 rem. Such a dose increases cancer risk by about 1%. However, if the patient has a 100% chance of dying from the cancer that he or she already has, such a risk becomes acceptable, especially since there is often a significant chance of curing the cancer.

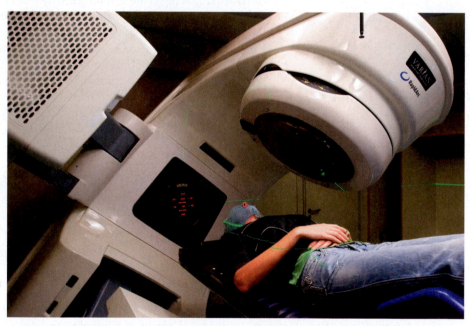

▲ **FIGURE 17.13 Radiotherapy for cancer** This treatment involves exposing a malignant tumor to gamma rays, typically from radioisotopes such as cobalt-60. The beam is moved in a circular pattern around the tumor to maximize exposure of the cancer cells while minimizing exposure of healthy tissues.

PEARSON
eText
2.0

CONCEPTUAL ✓ CHECKPOINT 17.6

Which type of radiation is most likely to be used for isotope scanning?

(a) alpha

(b) beta

(c) gamma

Chapter **17** in Review

MasteringChemistry™ provides end-of-chapter exercises, feedback-enriched tutorial problems, animations, and interactive activities to encourage problem solving practice and deeper understanding of key concepts and topics.

Self-Assessment Quiz

PEARSON
eText
2.0

Q1. Which daughter nuclide results from the alpha decay of bismuth-214?

(a) $^{210}_{81}Tl$

(b) $^{218}_{86}Rn$

(c) $^{214}_{85}At$

(d) $^{210}_{83}Bi$

Q2. Which nuclear equation accurately represents the beta decay of Xe-133?

(a) $^{133}_{54}Xe \longrightarrow ^{133}_{53}I + ^{0}_{-1}e$

(b) $^{133}_{54}Xe + ^{0}_{-1}e \longrightarrow ^{133}_{53}I$

(c) $^{133}_{54}Xe \longrightarrow ^{129}_{52}Cs + ^{4}_{2}He$

(d) $^{133}_{54}Xe \longrightarrow ^{133}_{55}Cs + ^{0}_{-1}e$

Q3. What is the missing particle in this nuclear equation?

$$^{22}_{12}Mg \longrightarrow ^{22}_{11}Na + \underline{\quad}$$

(a) $^{0}_{-1}e$

(b) $^{0}_{+1}e$

(c) $^{0}_{-1}\gamma$

(d) $^{0}_{+1}\gamma$

Q4. Which form of radioactive decay would you be most likely to detect if it was happening in the room next to the one you are currently in?

(a) alpha

(b) beta

(c) gamma

(d) positron emission

Q5. The chart shown in the right column above shows the mass of a decaying nuclide versus time. What is the half-life of the decay?

(a) 15 min

(b) 25 min

(c) 35 min

(d) 70 min

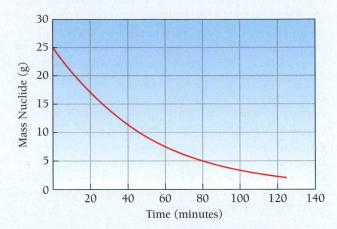

Q6. Iron-59 is a beta emitter with a half-life of 44.5 days. If a sample initially contains 16 mol of iron-59, how much iron-59 is left in the sample after 178 days?
(a) 0.0 mol (b) 1.0 mol (c) 2.0 mol (d) 4.0 mol

Q7. An artifact has a carbon-14 concentration that is 12.5% of that found in living organisms. How old is the artifact? (The half-life of carbon-14 is 5715 yr.)
(a) 5,715 yr (b) 11,430 yr
(c) 17,145 yr (d) 22,860 yr

Q8. Which issue is not associated with nuclear power generation?
(a) danger of overheated nuclear core
(b) waste disposal
(c) global warming
(d) none of the above (all of the above are problems associated with nuclear power generation)

Q9. Which reaction is a fission reaction?

(a) $^{238}_{92}U \longrightarrow ^{234}_{90}Th + ^{4}_{2}He$

(b) $^{2}_{1}H + ^{3}_{1}H \longrightarrow ^{4}_{2}He + ^{1}_{0}n$

(c) $^{14}_{7}N + ^{4}_{2}He \longrightarrow ^{17}_{8}O + ^{1}_{1}H$

(d) $^{235}_{92}U + ^{1}_{0}n \longrightarrow ^{142}_{56}Ba + ^{91}_{36}Kr + 3^{1}_{0}n$

Q10. Which medical procedure uses nuclear radiation?
(a) chemotherapy
(b) X-ray imaging
(c) MRI (magnetic resonance imaging)
(d) isotope scanning

Answers: 1:a, 2:d, 3:b, 4:c, 5:c, 6:b, 7:c, 8:c, 9:d, 10:d

Chemical Principles

Relevance

The Nature and Discovery of Radioactivity

Radioactivity is the emission of particles from unstable atomic nuclei. It can be divided into four types.

- **Alpha particles** are composed of 2 protons and 2 neutrons and have the symbol ^4_2He. Alpha particles have a high ionizing power but a low penetrating power.

- **Beta particles** are electrons emitted from atomic nuclei when a neutron changes into a proton. Beta particles have the symbol $^0_{-1}\text{e}$ and have intermediate ionizing power and intermediate penetrating power.

- **Gamma rays** are high-energy, short-wavelength photons. Gamma rays have the symbol $^0_0\gamma$ and have low ionizing power and high penetrating power.

- **Positrons** are emitted from atomic nuclei when protons change into neutrons. Positrons have the same mass as electrons but opposite charge and are represented by the symbol $^0_{+1}\text{e}$. They have intermediate ionizing power and intermediate penetrating power.

Radioactivity is a fundamental part of the behavior of some atoms, and it also has many applications. For example, radioactivity is used to diagnose and treat diseases, including cancer, thyroid diseases, abnormal kidney and bladder function, and heart disease. Natural radioactivity is part of our environment and can be used to date ancient objects. The discovery of radioactivity led to the discovery of fission, which in turn led to the development of nuclear bombs and nuclear energy.

Detecting Radioactivity

Radioactive emissions carry a large amount of energy and are therefore easily detected. The most common and inexpensive way to detect radioactivity is with photographic film, which is used in film-badge dosimeters to monitor exposure in people working with or near radioactive sources. Detection devices such as a Geiger-Müller counter or a scintillation counter give instantaneous readings of radiation levels.

Since radioactivity is invisible, it must be detected using film or instruments. The detection of radioactivity is important as both a scientific and a practical tool. Our understanding of what radioactivity is and our continuing research to understand it and its effects on living organisms require the ability to detect it. Our safety in areas where radioactive substances are used also depends on our ability to detect radiation.

Half-Life and Radiocarbon Dating

The half-life of a radioactive nuclide is the time it takes for half of the parent nuclides in a radioactive sample to decay. The presence of radioactive carbon-14 (with a half-life of 5715 years) in the environment provides a natural clock we can use to estimate the age of many artifacts and fossils. All living things contain carbon-14. When they die, the carbon-14 decays with its characteristic half-life. A measurement of the amount of carbon-14 remaining in a fossil or artifact reveals its age.

The half-life of a radioactive nuclide determines the activity of the nuclide and how long it will be radioactive. Nuclides with short half-lives are very active (many decay events per unit time) but are not radioactive for long. Nuclides with long half-lives are less active (fewer decays per unit time) but are radioactive for a long time.

Fission, the Atomic Bomb, and Nuclear Power

Fission—the splitting of the atom into smaller fragments—was discovered in 1939. Fission occurs when a U-235 nucleus absorbs a neutron. The nucleus becomes unstable, breaking apart to produce barium, krypton, neutrons, and a lot of energy.

The discovery of fission in 1939 changed the world. Within six years, the United States developed and tested fission nuclear bombs, which ended World War II. Fission can also be used to generate electricity. The fission reaction heats water to create steam, which turns the turbine on an electrical generator. Nuclear reactors generate about 20% of the electricity in the United States and up to 70% in some other nations.

Nuclear Fusion

Nuclear fusion is the combination of two light nuclei to form a heavier one. Nuclear fusion is the energy source of stars, including our sun.

Modern nuclear weapons are fusion bombs with 1000 times the power of the first fission bombs. Nuclear fusion is being explored as a way to generate electricity but has not yet proven successful.

The Effects of Radiation on Life and Nuclear Medicine

Radiation can damage molecules within living cells. Instantaneous high exposure to radiation can lead to radiation sickness and even death. Long-term lower exposure levels can increase cancer risk. Radiation is also used to attack cancerous tumors and to image internal organs through isotope scanning.

Radiation can be used for both good and harm. The destructive effects of radiation can be employed in a nuclear bomb. However, radiation can also be a precise tool in the physician's arsenal against disease.

Chemical Skills

Examples

LO: Write nuclear equations for alpha decay (Section 17.3).

Begin with the symbol for the isotope undergoing decay on the left side of the equation and the symbol for an alpha particle on the right side. Leave a space or a question mark for the unknown daughter nuclide.

Equalize the sum of the mass numbers and the sum of the atomic numbers on both sides of the equation by writing the appropriate mass number and atomic number for the unknown daughter nuclide.

Deduce the identity of the unknown daughter nuclide from the atomic number.

EXAMPLE **17.6** — **Writing Nuclear Equations for Alpha Decay**

Write the nuclear equation for the alpha decay of Po-214.

SOLUTION

$$^{214}_{84}Po \longrightarrow {}^{x}_{y}? + {}^{4}_{2}He$$
$$^{214}_{84}Po \longrightarrow {}^{210}_{82}? + {}^{4}_{2}He$$
$$^{214}_{84}Po \longrightarrow {}^{210}_{82}Pb + {}^{4}_{2}He$$

LO: Write nuclear equations for beta decay (Section 17.3).

Begin with the symbol for the isotope undergoing decay on the left side of the equation and the symbol for a beta particle on the right side.

Equalize the sum of the mass numbers and the sum of the atomic numbers on both sides of the equation by writing the appropriate mass number and atomic number for the unknown daughter nuclide.

Deduce the identity of the unknown daughter nuclide from the atomic number.

EXAMPLE **17.7** — **Writing Nuclear Equations for Beta Decay**

Write the nuclear equation for the beta decay of Bi-214.

SOLUTION

$$^{214}_{83}Bi \longrightarrow {}^{x}_{y}? + {}^{0}_{-1}e$$
$$^{214}_{83}Bi \longrightarrow {}^{214}_{84}? + {}^{0}_{-1}e$$
$$^{214}_{83}Bi \longrightarrow {}^{214}_{84}Po + {}^{0}_{-1}e$$

LO: Write nuclear equations for positron decay (Section 17.3).

Begin with the symbol for the isotope undergoing decay on the left side of the equation and the symbol for a positron on the right side.

Equalize the sum of the mass numbers and the sum of the atomic numbers on both sides of the equation by writing the appropriate mass number and atomic number for the unknown daughter nuclide.

Deduce the identity of the unknown daughter nuclide from the atomic number.

EXAMPLE **17.8** — **Writing Nuclear Equations for Positron Decay**

Write the nuclear equation for the positron decay of C-11.

SOLUTION

$$^{11}_{6}C \longrightarrow {}^{x}_{y}? + {}^{0}_{+1}e$$
$$^{11}_{6}C \longrightarrow {}^{11}_{5}? + {}^{0}_{+1}e$$
$$^{11}_{6}C \longrightarrow {}^{11}_{5}B + {}^{0}_{+1}e$$

LO: Use half-life (Section 17.5).

To use half-life to determine the time it takes for a sample to decay to a specified amount or the amount of a sample left after a specified time, draw a table showing the amount of the nuclide as a function of the number of half-lives. For each half-life, divide the amount of parent nuclide by 2.

EXAMPLE **17.9** — **Using Half-Life**

Po-210 is an alpha emitter with a half-life of 138 days. How many grams of Po-210 remain after 552 days if the sample initially contained 5.80 g of Po-210?

SOLUTION

Mass of Po-210 (g)	Number of Half-Lives	Time (days)
5.80	0	0
2.90	1	138
1.45	2	276
0.725	3	414
0.363	4	552

The amount of Po-210 left after 552 days is 0.363 g.

LO: Use carbon-14 content to determine the age of fossils or artifacts (Section 17.6).

To determine the age of an artifact or fossil based on its carbon-14 content, you can consult Table 17.3 or build your own table beginning with 100% carbon-14 (relative to living organisms) and reducing the amount by a factor of one-half for each half-life.

EXAMPLE **17.10**

Using Carbon-14 Content to Determine the Age of Fossils or Artifacts

Some wood ashes from a fire pit in the ruins of an ancient village have a carbon-14 content that is 25% of the amount found in living organisms. How old are the ashes and, by implication, the village?

SOLUTION

C-14(%)*	Number of Half-Lives	Time (yrs)
100	0	0
50.0	1	5715
25.0	2	11,430

*Percent relative to living organisms. The ashes are from wood that was living 11,430 years ago.

Key Terms

alpha (α) particle [17.3]
alpha (α) radiation [17.3]
beta (β) particle [17.3]
beta (β) radiation [17.3]
chain reaction [17.7]
critical mass [17.7]
daughter nuclide [17.3]

film-badge dosimeter [17.4]
gamma (γ) radiation [17.3]
gamma ray [17.3]
Geiger-Müller counter [17.4]
half-life [17.5]
ionizing power [17.3]
isotope scanning [17.11]

nuclear equation [17.3]
nuclear fission [17.7]
nuclear fusion [17.9]
parent nuclide [17.3]
penetrating power [17.3]
phosphorescence [17.2]
positron [17.3]

positron emission [17.3]
radioactive [17.1]
radioactivity [17.1]
radiocarbon dating [17.6]
radiotherapy [17.11]
rem [17.10]
scintillation counter [17.4]

Exercises

Questions

1. What is radioactivity? What does it mean for an atom to be radioactive?
2. How was radioactivity first discovered? By whom?
3. What are uranic rays?
4. What role did Marie Sklodowska Curie play in the discovery of radioactivity? How was she acknowledged for her work in radioactivity?
5. Explain what each symbol in the notation represents.

$$^{A}_{Z}X$$

6. Radioactivity originates from the _____ of radioactive atoms.
7. What is alpha radiation? What is the symbol for an alpha particle?
8. What happens to an atom when it emits an alpha particle?
9. How do the ionizing power and penetrating power of alpha particles compare to other types of radiation?

10. What is beta radiation? What is the symbol for a beta particle?
11. What happens to an atom when it emits a beta particle?
12. How do the ionizing power and penetrating power of beta particles compare to other types of radiation?
13. What is gamma radiation? What is the symbol for a gamma ray?
14. What happens to an atom when it emits a gamma ray?
15. How do the ionizing power and penetrating power of gamma particles compare to other types of radiation?
16. What is positron emission? What is the symbol for a positron?
17. What happens to an atom when it emits a positron?
18. How do the ionizing power and penetrating power of positrons compare to other types of radiation?
19. What is a nuclear equation? What does it mean for a nuclear equation to be balanced?

20. Identify the parent nuclides and daughter nuclides in the nuclear equation. Which kind of radioactive decay is involved?

$$^{231}_{91}\text{Pa} \longrightarrow ^{227}_{89}\text{Ac} + ^4_2\text{He}$$

21. What is a film-badge dosimeter, and how does it work?
22. How does a Geiger-Müller counter detect radioactivity?
23. Explain how a scintillation counter works.
24. What are some sources of natural radioactivity?
25. Explain the concept of half-life.
26. What is a radioactive decay series?
27. What is the source of radon in our environment? Why is radon problematic?
28. What is the source of carbon-14 in our environment? Why do all living organisms contain a uniform amount of carbon-14?
29. What happens to the carbon-14 in a living organism when it dies? How can this be used to establish how long ago the organism died?
30. How do we know that carbon-14 (or radiocarbon) dating is accurate? What is the age limit for which carbon-14 dating is useful?
31. Explain Fermi's experiment in which he bombarded uranium with neutrons. Include a nuclear equation in your answer.
32. What is nuclear fission? How and by whom was it discovered?
33. Why can nuclear fission be used in a bomb? Include the concept of a chain reaction in your explanation.
34. What is a critical mass?
35. What was the main goal of the Manhattan Project? Who was the project leader?
36. How can nuclear fission be used to generate electricity?
37. Explain the purpose of the control rods in a nuclear reactor core. How do they work?
38. What are the main advantages and problems associated with nuclear electricity generation?
39. Can a nuclear reactor detonate the way a nuclear bomb can? Why or why not?
40. What is nuclear fusion?
41. Do modern nuclear weapons use fission, fusion, or both? Explain.
42. Can nuclear fusion be used to generate electricity? What are the advantages of fusion over fission for electricity generation? What are the problems with fusion?
43. How does radiation affect the molecules within living organisms?
44. What is acute radiation damage to living organisms?
45. Explain how radiation can increase cancer risk.
46. Explain how radiation can cause genetic defects. Has this ever been observed in laboratory animals? In humans?
47. What is the main unit of radiation exposure? How much radiation is the average U.S. resident exposed to per year?
48. Describe the outcomes of radiation exposure at different doses (in rem).
49. Explain the medical use of isotope scanning.
50. How is radioactivity used to treat cancer?

Problems

ISOTOPIC AND NUCLEAR PARTICLE SYMBOLS

51. Draw the symbol for the isotope of lead that contains 128 neutrons.

52. Draw the symbol for the isotope of bismuth that contains 124 neutrons.

53. How many protons and neutrons are in this nuclide?

$$^{207}_{81}\text{Tl}$$

54. How many protons and neutrons are in this nuclide?

$$^{219}_{86}\text{Rn}$$

55. Identify the particle represented by each symbol as an alpha particle, a beta particle, a gamma ray, a positron, a neutron, or a proton.

(a) $^0_{-1}\text{e}$

(b) ^1_0n

(c) $^0_0\gamma$

56. Identify the particle represented by each symbol as an alpha particle, a beta particle, a gamma ray, a positron, a neutron, or a proton.

(a) ^1_1p

(b) ^4_2He

(c) $^0_{+1}\text{e}$

57. Complete the table.

Chemical Symbol	Atomic Number (Z)	Mass Number (A)	# Protons	# Neutrons
Tc	___	95	___	___
___	56	128	___	___
Eu	___	___	___	82
Fr	___	___	___	136

58. Complete the table.

Chemical Symbol	Atomic Number (Z)	Mass Number (A)	# Protons	# Neutrons
Pd	46	___	___	54
Ce	___	136	___	___
___	84	208	___	___
___	___	___	88	138

RADIOACTIVE DECAY

59. Write a nuclear equation for the alpha decay of each nuclide.
(a) U-234
(b) Th-230
(c) Ra-226
(d) Rn-222

60. Write a nuclear equation for the alpha decay of each nuclide.
(a) Po-218
(b) Po-214
(c) Po-210
(d) Th-227

61. Write a nuclear equation for the beta decay of each nuclide.
(a) Pb-214
(b) Bi-214
(c) Th-231
(d) Ac-227

62. Write a nuclear equation for the beta decay of each nuclide.
(a) Pb-211
(b) Tl-207
(c) Th-234
(d) Pa-234

63. Write a nuclear equation for positron emission by each nuclide.
(a) C-11
(b) N-13
(c) O-15

64. Write a nuclear equation for positron emission by each nuclide.
(a) Co-55
(b) Na-22
(c) F-18

65. Fill in the blanks in the partial decay series.

$$^{241}_{94}\text{Pu} \longrightarrow {}^{241}_{95}\text{Am} + \underline{\quad}$$
$$^{241}_{95}\text{Am} \longrightarrow {}^{237}_{93}\text{Np} + \underline{\quad}$$
$$^{237}_{93}\text{Np} \longrightarrow \underline{\quad} + {}^{4}_{2}\text{He}$$
$$\underline{\quad} \longrightarrow {}^{233}_{92}\text{U} + {}^{0}_{-1}\text{e}$$

66. Fill in the blanks in the partial decay series.

$$^{225}_{88}\text{Ra} \longrightarrow {}^{225}_{89}\text{Ac} + \underline{\quad}$$
$$^{225}_{89}\text{Ac} \longrightarrow \underline{\quad} + {}^{4}_{2}\text{He}$$
$$\underline{\quad} \longrightarrow {}^{217}_{85}\text{At} + {}^{4}_{2}\text{He}$$
$$^{217}_{85}\text{At} \longrightarrow \underline{\quad} + {}^{4}_{2}\text{He}$$

67. Write a partial decay series for Th-232 undergoing these sequential decays.

$$\alpha, \beta, \beta, \alpha$$

68. Write a partial decay series for Rn-220 undergoing these sequential decays.

$$\alpha, \alpha, \beta, \alpha$$

HALF-LIFE

69. Suppose you a have a 100,000-atom sample of a radioactive nuclide that decays with a half-life of 2.0 days. How many radioactive atoms are left after 10 days?

70. Iodine-131 is often used in nuclear medicine to obtain images of the thyroid. If you start with 4.0×10^{10} I-131 atoms, how many are left after approximately 1 month? I-131 has a half-life of 8.0 days.

71. A patient is given 0.050 mg of technetium-99m (where m means metastable—an unstable but long-lived state), a radioactive isotope with a half-life of about 6.0 hours. How long until the radioactive isotope decays to 6.3×10^{-3} mg?

72. Radium-223 decays with a half-life of 11.4 days. How long does it take for a 0.240-mol sample of radium to decay to 1.50×10^{-2} mol?

73. One of the nuclides in spent nuclear fuel is U-234, an alpha emitter with a half-life of 2.44×10^5 years. If a spent fuel assembly contains 2.80 kg of U-234, how long does it take for the amount of U-234 to decay to less than 0.10 kg?

74. One of the nuclides in spent nuclear fuel is U-235, an alpha emitter with a half-life of 703 million years. How long does it take for the amount of U-235 to reach one-eighth of its initial amount?

75. A radioactive sample contains 2.45 g of an isotope with a half-life of 3.8 days. How much of the isotope in grams remains after 11.4 days?

76. A 68-mg sample of a radioactive nuclide is administered to a patient to obtain an image of her thyroid. If the nuclide has a half-life of 12 hours, how much of the nuclide remains in the patient after 4.0 days?

77. Each of the tabulated nuclides is used in nuclear medicine. List them in order of most active (largest number of decay events per second) to least active (smallest number of decay events per second).

Nuclide	Half-Life
P-32	14.3 days
Cr-51	27.7 days
Ga-67	78.3 hours
Sr-89	50.0 days

78. Each of the tabulated nuclides is used in nuclear medicine. List them in order of most active (largest number of decay events per second) to least active (smallest number of decay events per second).

Nuclide	Half-Life
Y-90	64.1 hours
Tc-99m	6.02 hours
In-111	2.8 days
I-131	8.0 days

RADIOCARBON DATING

79. A wooden boat discovered just south of the Great Pyramid in Egypt had a carbon-14 content of approximately 50% of that found in living organisms. How old is the boat?

80. A layer of peat buried beneath the glacial sediments from the last ice age had a carbon-14 content of 25% of that found in living organisms. How long ago was this ice age?

81. An ancient skull has a carbon-14 content of 1.563% of that found in living organisms. How old is the skull?

82. A mammoth skeleton has a carbon-14 content of 12.50% of that found in living organisms. When did the mammoth live?

FISSION AND FUSION

83. Write the nuclear reaction for the neutron-induced fission of U-235 to form Xe-144 and Sr-90. How many neutrons are produced in the reaction?

84. Write the nuclear reaction for the neutron-induced fission of U-235 to produce Te-137 and Zr-97. How many neutrons are produced in the reaction?

85. Write the nuclear equation for the fusion of two H-2 atoms to form He-3 and one neutron.

86. Write the nuclear equation for the fusion of H-3 with H-1 to form He-4.

Cumulative Problems

87. Complete each nuclear equation.

(a) $^{1}_{1}p + ^{9}_{4}Be \longrightarrow$ ___ $+ ^{4}_{2}He$

(b) $^{209}_{83}Bi +$ ___ $\longrightarrow ^{272}_{111}Rg + ^{1}_{0}n$

(c) $^{179}_{74}W + ^{0}_{-1}e \longrightarrow$ ___

88. Complete each nuclear equation.

(a) $^{27}_{13}Al + ^{4}_{2}He \longrightarrow$ ___ $+ ^{1}_{0}n$

(b) ___ $+ ^{1}_{0}n \longrightarrow + ^{29}_{14}Si + ^{4}_{2}He$

(c) $^{241}_{95}Am \longrightarrow + ^{237}_{93}Np +$ ___

89. A *breeder nuclear reactor* is a reactor in which U-238 (which does not undergo fission) is converted into Pu-239 (which does undergo fission). The process involves bombardment of U-238 by neutrons to form U-239, which undergoes two sequential beta decays. Write nuclear equations to represent this process.

90. Write a series of nuclear equations in which Al-27 reacts with a neutron and the product undergoes an alpha decay followed by a beta decay.

91. The fission of U-235 produces 3.2×10^{-11} J/atom. How much energy does it produce per mole of U-235? Per kilogram of U-235?

92. The fusion of deuterium and tritium produces 2.8×10^{-12} J for every atom of deuterium and atom of tritium. How much energy is produced per mole of deuterium and mole of tritium?

93. Bi-210 is a beta emitter with a half-life of 5.0 days. If a sample contains 1.2 g of Bi-210, how many beta emissions occur in 5.0 days?

94. Po-218 is an alpha emitter with a half-life of 3.0 minutes. If a sample contains 55 mg of Po-218, how many alpha emissions occur in 6.0 minutes?

95. If a person living in a high-radon area is exposed to 0.400 rem of radiation from radon per year, and his total exposure is 0.585 rem, what percentage of his total exposure is due to radon?

96. An X-ray technician is exposed to 0.020 rem of radiation at work. If her total exposure is the national average (0.36 rem), what fraction of her exposure is due to on-the-job exposure?

97. Radium-226 (atomic mass 226.03 amu) decays to radon-224, a radioactive gas. The half-life of radium-226 is 1.6×10^3 years. If a 1.5-g sample of radium-226 decays for 45 days, what volume of radon gas (at 25.0 °C and 1.0 atm) is produced?

98. Consider the fission reaction.

$$^{235}_{92}\text{U} + {}^{1}_{0}\text{n} \longrightarrow {}^{142}_{56}\text{Ba} + {}^{91}_{36}\text{Kr} + 3{}^{1}_{0}\text{n} + \text{Energy}$$

What mass of Kr-91 (atomic mass 92.93 amu) is produced by the complete fission of 15 g of U-235 (atomic mass 235.04 amu)?

Highlight Problems

99. Closely examine the diagram representing the alpha decay of sodium-20 and draw the missing nucleus.

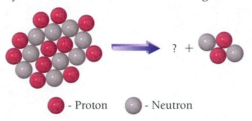

- Proton   - Neutron

100. Closely examine the diagram representing the beta decay of fluorine-21 and draw the missing nucleus.

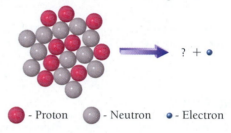

- Proton   - Neutron   - Electron

101. Closely examine the diagram representing the positron emission of carbon-10 and draw the missing nucleus.

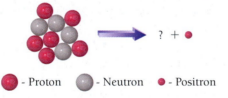

- Proton   - Neutron   - Positron

102. A radiometric dating technique uses the decay of U-238 to Pb-206 (the half-life for this process is 4.5 billion years) to determine the age of the oldest rocks on Earth and by implication the age of Earth itself. The oldest uranium-containing rocks on Earth contain approximately equal numbers of uranium atoms and lead atoms. Assuming the rocks were pure uranium when they were formed, how old are the rocks?

Questions for Group Work

Discuss these questions with the group and record your consensus answer.

103. Complete the table of particles involved in radioactive decay.

Particle Name	Symbol	Mass Number	Atomic Number or Charge
alpha particle	$^{4}_{2}\text{He}$	_____	_____
positron	_____	0	0
proton	$^{1}_{0}\text{n}$	_____	_____

104. Radon-220 undergoes alpha decay with a half-life of 55.6 s. If there are 16,000 atoms present initially, make a table showing how many atoms are present at 0 s, 55.6 s,

111.2 s, 166.8 s, 222.4 s, and 278.0 s. (Note that the times selected for observation are multiples of the half-life.) Make a graph of number of atoms present on the y-axis and total time on the x-axis.

105. Write all the balanced nuclear equations for each step of the nuclear decay sequence that starts with U-238 and ends with U-234. Refer to Figure 17.9 for the decay processes involved.

106. For each member in your group, suggest one thing that all types of nuclear reactions have in common and one way in which they are different from each other. Try to get one contribution from each group member. Record your answers as complete sentences.

Data Interpretation and Analysis

107. A common isotope used in medical imaging is technetium-99m, where the m stands for metastable. Metastable means that the technetium-99m isotope exists in a state that has excess energy for a time that is longer than normal. That excess energy is released over time as a gamma ray according to the following nuclear equation:

$$^{99m}_{43}\text{Tc} \longrightarrow\ ^{99}_{43}\text{Tc} + ^{0}_{0}\gamma$$

A sample initially containing 0.500 mg of technetium-99m is monitored as a function of time. Based on its rate of gamma-ray emission, a graph, showing the mass of technetium-99m as a function of time, is prepared. Study the graph and answer the questions that follow.

(a) What is the mass of technetium-99m present at 200 minutes? At 400 minutes?

(b) What is the half-life of technetium-99m in minutes? In hours?

(c) If a patient is given a 0.0500-mg dose of technetium-99m, how much of it is left in the patient's body after 1 day (24 hours)? (For this problem, assume that the technetium-99m is not biologically removed from the body.)

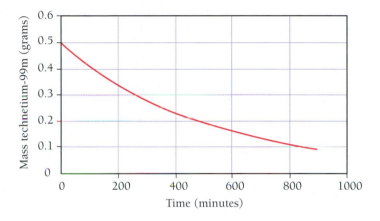

Answers to Skillbuilder Exercises

Skillbuilder 17.1 .... $^{216}_{84}\text{Po} \longrightarrow\ ^{212}_{82}\text{Pb} + ^{4}_{2}\text{He}$

Skillbuilder 17.2 .... $^{228}_{89}\text{Ac} \longrightarrow\ ^{228}_{90}\text{Th} + ^{0}_{-1}\text{e}$

Skillbuilder Plus, p. 615 .... $^{235}_{92}\text{U} \longrightarrow\ ^{231}_{90}\text{Th} + ^{4}_{2}\text{He};$
$^{231}_{90}\text{Th} \longrightarrow\ ^{231}_{91}\text{Pa} + ^{0}_{-1}\text{e};$
$^{231}_{91}\text{Pa} \longrightarrow\ ^{227}_{89}\text{Ac} + ^{4}_{2}\text{He}$

Skillbuilder 17.3 .... $^{22}_{11}\text{Na} \longrightarrow\ ^{22}_{10}\text{Ne} + ^{0}_{+1}\text{e}$

Skillbuilder 17.4 .... $7.00 \times 10^{-3}\,\text{mol}$

Skillbuilder 17.5 .... No, the carbon-14 content suggests that the scroll is from very recent times.

Answers to Conceptual Checkpoints

17.1 (c) The alpha decay decreases its atomic number by two to 82, and the beta decay increases it by one to 83.

17.2 (a) In alpha decay, the nucleus loses a helium nucleus (2 protons and 2 neutrons), reducing its mass number by 4. The other forms of decay listed involve electrons or positrons, which have negligible mass compared to that of nuclear particles, or gamma-ray photons, which have no mass.

17.3 The half-life is 100 years.

17.4 (a) If you divide 1,000,000 by 2, then divide the remainder by 2, and repeat this process eight more times, you are left with approximately 977 atoms.

17.5 (b) Fusion is combining; fission is splitting.

17.6 (c) Gamma radiation has the most penetrating power, and therefore it most readily penetrates through bodily tissues for easier detection.

Appendix: Mathematics Review

Basic Algebra

In chemistry, you often have to solve an equation for a particular variable. For example, suppose you want to solve the following equation for V:

$$PV = nRT$$

To solve an equation for a particular variable, you must isolate that variable on one side of the equation. The rest of the variables or numbers will then be on the other side of the equation. To solve the above equation for V, divide both sides by P.

$$\frac{PV}{P} = \frac{nRT}{P}$$

$$V = \frac{nRT}{P}$$

The Ps cancel, and you are left with an expression for V. For another example, consider solving the following equation for °F:

$$°C = \frac{(°F - 32)}{1.8}$$

First, eliminate the 1.8 in the denominator of the right side by multiplying both sides by 1.8.

$$(1.8)\,°C = \frac{(°F - 32)}{1.8}(1.8)$$

$$(1.8)\,°C = (°F - 32)$$

Then eliminate the -32 on the right by adding 32 to both sides.

$$(1.8)\,°C + 32 = (°F - 32) + 32$$

$$(1.8)\,°C + 32 = °F$$

You are now left with an expression for °F.

In general, solve equations by following these guidelines:

- Cancel numbers or symbols in the denominator (bottom part of a fraction) by multiplying by the number or symbol to be canceled.
- Cancel numbers or symbols in the numerator (upper part of a fraction) by dividing by the number or symbol to be canceled.
- Eliminate numbers or symbols that are added by subtracting the same number or symbol.
- Eliminate numbers or symbols that are subtracted by adding the same number or symbol.
- Whether you add, subtract, multiply, or divide, **always perform the same operation for both sides of a mathematical equation**. (Otherwise, the two sides will no longer be equal.)

For a final example, solve the following equation for x:

$$\frac{67x - y + 3}{6} = 2z$$

Cancel the 6 in the denominator by multiplying both sides by 6.

$$(6)\frac{67x - y + 3}{6} = (6)2z$$

$$67x - y + 3 = 12z$$

Eliminate the +3 by subtracting 3 from both sides.

$$67x - y + 3 - 3 = 12z - 3$$

$$67x - y = 12z - 3$$

Eliminate the $-y$ by adding y to both sides.

$$67x - y + y = 12z - 3 + y$$

$$67x = 12z - 3 + y$$

Cancel the 67 by dividing both sides by 67.

$$\frac{67x}{67} = \frac{12z - 3 + y}{67}$$

$$x = \frac{12z - 3 + y}{67}$$

▶ FOR PRACTICE **Using Algebra to Solve Equations**

Solve each of the following for the indicated variable:

(a) $P_1V_1 = P_2V_2$; solve for V_2

(b) $\dfrac{V_1}{T_1} = \dfrac{V_2}{T_2}$; solve for T_1

(c) $PV = nRT$; solve for n

(d) K = °C + 273; solve for °C

(e) $\dfrac{3x + 7}{2} = y$; solve for x

(f) $\dfrac{32}{y + 3} = 8$; solve for y

ANSWERS

(a) $V_2 = \dfrac{P_1V_1}{P_2}$

(b) $T_1 = \dfrac{V_1T_2}{V_2}$

(c) $n = \dfrac{PV}{RT}$

(d) °C = K − 273

(e) $x = \dfrac{2y - 7}{3}$

(f) $y = 1$

Mathematical Operations with Scientific Notation

Writing numbers in scientific notation is covered in detail in Section 2.2. Briefly, a number written in scientific notation consists of a **decimal part**, a number that is usually between 1 and 10, and an **exponential part**, 10 raised to an **exponent**, n.

Each of the following numbers is written in both scientific and decimal notation:

$$1.0 \times 10^5 = 100{,}000 \qquad 1.0 \times 10^{-6} = 0.000001$$

$$6.7 \times 10^3 = 6700 \qquad 6.7 \times 10^{-3} = 0.0067$$

Multiplication and Division

To multiply numbers expressed in scientific notation, multiply the decimal parts and add the exponents.

$$(A \times 10^m)(B \times 10^n) = (A \times B) \times 10^{m+n}$$

To divide numbers expressed in scientific notation, divide the decimal parts and subtract the exponent in the denominator from the exponent in the numerator.

$$\frac{(A \times 10^m)}{(B \times 10^n)} = \left(\frac{A}{B}\right) \times 10^{m-n}$$

Consider the following example involving multiplication:

$$(3.5 \times 10^4)(1.8 \times 10^6) = (3.5 \times 1.8) \times 10^{4+6}$$

$$= 6.3 \times 10^{10}$$

Consider the following example involving division:

$$\frac{(5.6 \times 10^7)}{(1.4 \times 10^3)} = \left(\frac{5.6}{1.4}\right) \times 10^{7-3}$$

$$= 4.0 \times 10^4$$

Addition and Subtraction

To add or subtract numbers expressed in scientific notation, rewrite all the numbers so that they have the same exponent, and then add or subtract the decimal parts of the numbers. The exponents remain unchanged.

$$A \times 10^n$$

$$\pm B \times 10^n$$

$$\overline{(A \pm B) \times 10^n}$$

Notice that the numbers *must have* the same exponent.
Consider the following example involving addition:

$$4.82 \times 10^7$$

$$+3.4 \times 10^6$$

First, express both numbers with the same exponent. In this case, rewrite the lower number and perform the addition as follows:

$$4.82 \times 10^7$$

$$\underline{+0.34 \times 10^7}$$

$$5.16 \times 10^7$$

Consider the following example involving subtraction:

$$7.33 \times 10^5$$

$$\underline{-1.9 \times 10^4}$$

First, express both numbers with the same exponent. In this case, rewrite the lower number and perform the subtraction as follows:

$$7.33 \times 10^5$$
$$-0.19 \times 10^5$$
$$\overline{7.14 \times 10^5}$$

▶ **FOR PRACTICE**   **Mathematical Operations with Scientific Notation**

Perform each of the following operations:

(a) $(2.1 \times 10^7)(9.3 \times 10^5)$

(b) $(5.58 \times 10^{12})(7.84 \times 10^{-8})$

(c) $\dfrac{(1.5 \times 10^{14})}{(5.9 \times 10^8)}$

(d) $\dfrac{(2.69 \times 10^7)}{(8.44 \times 10^{11})}$

(e) $\begin{array}{r} 1.823 \times 10^9 \\ +1.11 \times 10^7 \\ \hline \end{array}$

(f) $\begin{array}{r} 3.32 \times 10^{-5} \\ +3.400 \times 10^{-7} \\ \hline \end{array}$

(g) $\begin{array}{r} 6.893 \times 10^9 \\ -2.44 \times 10^8 \\ \hline \end{array}$

(h) $\begin{array}{r} 1.74 \times 10^4 \\ -2.9 \times 10^3 \\ \hline \end{array}$

ANSWERS

(a) 2.0×10^{13}

(b) 4.37×10^5

(c) 2.5×10^5

(d) 3.19×10^{-5}

(e) 1.834×10^9

(f) 3.35×10^{-5}

(g) 6.649×10^9

(h) 1.45×10^4

Logarithms

The logarithm (or log) of a number is the exponent to which 10 must be raised to obtain that number. For example, the log of 100 is 2 because 10 must be raised to the second power to get 100. Similarly, the log of 1000 is 3 because 10 must be raised to the third power to get 1000. The logs of several multiples of 10 are shown as follows:

$$\log 10 = 1$$
$$\log 100 = 2$$
$$\log 1000 = 3$$
$$\log 10{,}000 = 4$$

Because $10^0 = 1$ by definition, $\log 1 = 0$.

The log of a number smaller than 1 is negative because 10 must be raised to a negative exponent to get a number smaller than 1. For example, the log of 0.01 is -2 because 10 must be raised to the power of -2 to get 0.01. Similarly, the log of 0.001 is -3 because 10 must be raised to the power of -3 to get 0.001. The logs of several fractional numbers are as follows:

$$\log 0.1 = -1$$
$$\log 0.01 = -2$$
$$\log 0.001 = -3$$
$$\log 0.0001 = -4$$

The logs of numbers that are not multiples of 10 can be calculated on your calculator. See your calculator manual for specific instructions.

Inverse Logarithms

The inverse logarithm or invlog function (sometimes called antilog) is exactly the opposite of the log function. For example, the log of 100 is 2 and the inverse log of 2 is 100. The log function and the invlog function undo one another.

$$\log \ 100 = 3$$

$$\text{invlog} \ 3 = 1000$$

$$\text{invlog} \ (\log \ 1000) = 1000$$

The inverse log of a number is simply 10 raised to that number.

$$\text{invlog} \ x = 10^x$$

$$\text{invlog} \ 3 = 10^3 = 1000$$

The inverse logs of numbers can be calculated on your calculator. See your calculator manual for specific instructions.

▶ **FOR PRACTICE**   **Logarithms and Inverse Logarithms**

Perform each of the following operations:

(a) $\log \ 1.0 \times 10^5$ (f) invlog 1.44

(b) $\log \ 59$ (g) invlog -6.0

(c) $\log \ 1.0 \times 10^{-5}$ (h) invlog -0.250

(d) $\log 0.068$ (i) invlog (log 88)

(e) invlog 7.0

ANSWERS

(a) 5.00   (f) 28

(b) 1.77   (g) 1×10^{-6}

(c) -5.00 (h) 0.56

(d) -1.17 (i) 88

(e) 1×10^7

Answers to Odd-Numbered Exercises

NOTE: Answers in the Questions section are written as briefly as possible. Student answers may vary and still be correct.

CHAPTER 1
QUESTIONS

1. Soda fizzes due to the interactions between carbon dioxide and water under high pressure. At room temperature, carbon dioxide is a gas and water is a liquid. Through the use of pressure, the makers of soda force the carbon dioxide gas to dissolve in the water. When the can is sealed, the solution remains mixed. When the can is opened, the pressure is released and the carbon dioxide molecules escape in bubbles of gas.

3. Chemists study molecules and interactions at the molecular level to learn about and explain macroscopic events. Chemists attempt to explain why ordinary things are as they are.

5. Chemistry is the science that seeks to understand what matter does by studying what atoms and molecules do.

7. The scientific method is the way chemists investigate the chemical world. The first step consists of observing the natural world. Later observations can be combined to create a scientific law, which summarizes and predicts behavior. Theories are models that strive to explain the cause of the observed phenomenon. Theories are tested through experiment. When a theory is not well established, it is sometimes referred to as a hypothesis.

9. A law is simply a general statement that summarizes and predicts observed behavior. Theories seek to explain the causes of observed behavior.

11. To say "It is just a theory" makes it seem as if theories are easily discardable. However, many theories are very well established and are as close to truth as we get in science. Established theories are backed up with years of experimental evidence, and they are the pinnacle of scientific understanding.

13. The atomic theory states that all matter is composed of small, indestructible particles called atoms. John Dalton formulated this theory.

PROBLEMS

15. a. observation   b. theory
    c. law   d. observation

17.
Mass (g)	Volume (L)	Ratio (g/L)
22.5	1.6	14
35.8	2.55	14.0
70.2	5.00	14.0
98.5	7.01	14.1

The ratio of mass to volume is constant.

19. a. All atoms contain a degree of chemical reactivity. The larger the size of an atom, the higher the chemical reactivity of that atom.
    b. There are many correct answers. One example is: Conceivably, when the size of an atom is increased, the surface area of the atom is also increased; an atom with a greater surface area is more likely to react chemically.

25. a. 2.2 billion people
    c. 4.8 billion people
    e. 9 billion people

CHAPTER 2
QUESTIONS

1. Without units, the results are unclear and it is hard to keep track of what each separate measurement entails.

3. Often scientists work with very large or very small numbers that contain a lot of zeros. Scientific notation allows these numbers to be written more compactly, and the information is more organized.

5. Zeros count as significant digits when they are interior zeros (zeros between two numbers) and when they are trailing zeros (zeros after a decimal point). Zeros are **not** significant digits when they are leading zeros, which are zeros to the left of the first nonzero number.

7. For calculations involving only multiplication and division, the result carries the same number of significant figures as the factor with the fewest significant figures.

9. In calculations involving both multiplication/division and addition/subtraction, do the steps in parentheses first; next determine the correct number of significant figures in the intermediate answer; then do the remaining steps.

11. The basic SI unit of length is the meter. The kilogram is the SI unit of mass. Lastly, the second is the SI unit of time.

13. For measuring a Frisbee, the unit would be the meter and the prefix multiplier would be *centi-*. The final measurement would be in centimeters.

15. a. 2.42 cm   b. 1.79 cm
    c. 21.58 cm   d. 21.85 cm

17. Units act as a guide in the calculation and are able to show if the calculation is off track. The units must be followed in the calculation, so that the answer is correctly written and understood.

19. A conversion factor is a quantity used to relate two separate units. They are constructed from any two quantities known to be equivalent.

nversion factor is $\dfrac{1\ ft}{12\ in.}$. For a feet-to-inches conversion, the conversion factor must be inverted $\left(\dfrac{12\ in.}{1\ ft}\right)$.

23. **a.** Sort the information into the **given** information (the starting point for the problem) and the **find** information (the end point).

  b. Create a solution map to get from the given information to the information you are trying to find. This will likely include conversion factors or equations.

  c. Follow the solution map to solve the problem. Carry out mathematical operations and cancel units as needed.

  d. Ask, does this answer make physical sense? Are the units correct? Is the number of significant figures correct?

25. The solution map for converting grams to pounds is:

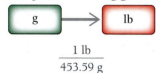

$$\dfrac{1\ lb}{453.59\ g}$$

27. The solution map for converting meters to feet is:

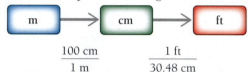

$$\dfrac{100\ cm}{1\ m} \qquad \dfrac{1\ ft}{30.48\ cm}$$

29. The density of a substance is the ratio of its mass to its volume. Density is a fundamental property of materials and differs from one substance to another. Density can be used to relate two separate units, thus working as a conversion factor. Density is a conversion factor between mass and volume.

PROBLEMS

31. **a.** 3.8802×10^7 **b.** 1.419×10^6
    c. 1.9746×10^7 **d.** 5.84×10^5

33. **a.** $7.461 \times 10^{-11}\ m$ **b.** $1.58 \times 10^{-5}\ mi$
    c. $6.32 \times 10^{-7}\ m$ **d.** $1.5 \times 10^{-5}\ m$

35. **a.** 602,200,000,000,000,000,000,000
    b. 0.00000000000000000016 C
    c. 299,000,000 m/s
    d. 344 m/s

37. **a.** 32,200,000
    b. 0.0072
    c. 118,000,000,000
    d. 0.00000943

39. 2,000,000,000 2×10^9
  1,211,000,000 1.211×10^9
  0.000874 8.74×10^{-4}
  320,000,000,000 3.2×10^{11}

41. **a.** 54.9 mL   **b.** 48.7 °C
    c. 46.83 °C   **d.** 64 mL

43. **a.** 0.005050   **b.** 0.0000000000000060
    c. 220,103   **d.** 0.00108

45. **a.** 4   **b.** 4
    c. 6   **d.** 5

47. **a.** correct   **b.** 3
    c. 7   **d.** correct

49. **a.** 256.0   **b.** 0.0004893
    c. 2.901×10^{-4} **d.** 2.231×10^6

51. **a.** 2.3   **b.** 2.4
    c. 2.3   **d.** 2.4

53. **a.** 42.3   **b.** correct
    c. correct   **d.** 0.0456

55.

8.32466	8.325	8.3	8
84.57225	84.57	85	8×10^1
132.5512	132.6	1.3×10^2	1×10^2

57. **a.** 0.054   **b.** 0.619
    c. 1.2×10^8 **d.** 6.6

59. **a.** 4.22×10^3 **b.** correct
    c. 3.9969   **d.** correct

61. **a.** 110.6   **b.** 41.4
    c. 183.3   **d.** 1.22

63. **a.** correct   **b.** 1.0982
    c. correct   **d.** 3.53

65. **a.** 3.9×10^3 **b.** 632
    c. 8.93×10^4 **d.** 6.34

67. **a.** 3.15×10^3 **b.** correct
    c. correct   **d.** correct

69. **a.** 3.55×10^3 g   **b.** 8.944 m
    c. 4.598×10^{-3} kg   **d.** 18.7 mL

71. **a.** 0.588 L   **b.** 34.1 μg
    c. 10.1 ns   **d.** 2.19×10^{-12} m

73. **a.** 57.2 cm   **b.** 38.4 m
    c. 0.754 km   **d.** 61 mm

75. **a.** 15.7 in.   **b.** 91.2 ft
    c. 6.21 mi   **d.** 8478 lb

77.

5.08×10^8 m	5.08×10^5 km	508 Mm
5.08×10^{-1} Gm	5.08×10^{-4} Tm	
2.7976×10^{10} m	2.7976×10^7 km	27,976 Mm
27976×10^1 Gm	2.7976×10^{-2} Tm	
1.77×10^{12} m	1.77×10^9 km	1.77×10^6 Mm
1.77×10^3 Gm	1.77 Tm	
1.5×10^8 m	1.5×10^5 m	1.5×10^2 Mm
0.15 Gm	1.5×10^{-4} Tm	
4.23×10^{11} m	4.23×10^8 km	4.23×10^5 Mm
423 Gm	0.423 Tm	

79. **a.** 2.255×10^7 kg   **b.** 2.255×10^4 Mg
    c. 2.255×10^{13} mg   **d.** 2.255×10^4 metric tons

81. 1.5×10^3 g

83. 5.0×10^1 min

85. 4.7×10^3 cm^3

87. **a.** 1.0×10^6 m^2 **b.** 1.0×10^{-6} m^3
    c. 1.0×10^{-9} m^3

89. **a.** 6.2×10^5 pm^3 **b.** 6.2×10^{-4} nm^3
    c. 6.2×10^{-1} Å^3

91. **a.** $2.15 \times 10^{-4} \, km^2$ **b.** $2.15 \times 10^4 \, dm^2$
    c. $2.15 \times 10^6 \, cm^2$

93. $1.49 \times 10^6 \, mi^2$

95. **a.** $2.5 \times 10^3 \, km/day$ **b.** 95 ft/s
    c. 29 m/s   **d.** $1.9 \times 10^3 \, yd/min$

97. $3.42 \times 10^{-3} \, g/lb$; 0.599 g

99. $11.4 \, g/cm^3$, lead

101. $1.26 \, g/cm^3$

103. Yes, the density of the crown is $19.3 \, g/cm^3$.

105. **a.** $4.30 \times 10^2 \, g$ **b.** 3.12 L

107. **a.** $3.38 \times 10^4 \, g$ (gold); $5.25 \times 10^3 \, g$ (sand)
    b. Yes, the mass of the bag of sand is different from the mass of the gold vase; thus, the weight-sensitive pedestal will sound the alarm.

109. $10.6 \, g/cm^3$

111. $2.7 \times 10^3 \, \dfrac{kg}{m^3}$

113. $2.5 \times 10^5 \, lb$

115. $1.19 \times 10^5 \, kg$

117. 18 km/L

119. 768 mi

121. Metal A is denser than metal B.

123. $2.26 \, g/cm^3$

125. 1.32 cm

127. 108 km; 47.2 km

129. $9.1 \times 10^{10} \, g/cm^3$

134. **a.** 8.2%   **c.** 24.4 million cubic kilometers

CHAPTER 3
QUESTIONS

1. Matter is defined as anything that occupies space and possesses mass. It can be thought of as the physical material that makes up the universe.

3. The three states of matter are solid, liquid, and gas.

5. In a crystalline solid, the atoms/molecules are arranged in geometric patterns with repeating order. In amorphous solids, the atoms/molecules do not have long-range order.

7. The atoms/molecules in gases are not in contact with each other and are free to move relative to one another. The spacing between separate atoms/molecules is very far apart. A gas has no fixed volume or shape; rather, it assumes both the shape and the volume of the container it occupies.

9. A mixture is two or more pure substances combined in variable proportions.

11. Pure substances are those composed of only one type of atom or molecule.

13. A mixture is formed when two or more pure substances are mixed together; however, a new substance is not formed. A compound is formed when two or more elements are bonded together and form a new substance.

15. In a physical change, the composition of the substance does not change, even though its appearance might change. However, in a chemical change, the substance undergoes a change in its composition.

17. Energy is defined as the capacity to do work.

19. Kinetic energy is the energy associated with the motion of an object. Potential energy is the energy associated with the position or composition of an object.

21. Three common units for energy are joules, calories, and kilowatt-hours.

23. An endothermic reaction is one that absorbs energy from the surroundings. The products have more energy than the reactants in an endothermic reaction.

25. Heat is the transfer of thermal energy caused by a temperature difference, whereas temperature is a measure of the thermal energy of matter.

27. Heat capacity is the quantity of heat energy required to change the temperature of a given amount of the substance by 1 °C.

29. $°F = \dfrac{9}{5}(°C) + 32$

PROBLEMS

31. **a.** element
    b. element
    c. compound
    d. compound

33. **a.** homogeneous
    b. heterogeneous
    c. homogeneous
    d. homogeneous

35. **a.** pure substance-element
    b. mixture-homogeneous
    c. mixture-heterogeneous
    d. mixture-heterogeneous

37. **a.** chemical   **b.** physical
    c. physical   **d.** chemical

39. physical–colorless; odorless; gas at room temperature; one liter has a mass of 1.260 g under standard conditions; mixes with acetone; chemical–flammable; polymerizes to form polyethylene

41. **a.** chemical   **b.** physical
    c. chemical   **d.** chemical

43. **a.** physical   **b.** chemical

45. $2.10 \times 10^2 \, kg$

47. **a.** Yes   **b.** No

49. 15.1 g of water

51. **a.** $2.46 \times 10^3 \, J$ **b.** $4.16 \times 10^{-3} \, Cal$
    c. 32.0 Cal   **d.** $2.35 \times 10^5 \, J$

53. **a.** $9.0 \times 10^7 \, J$ **b.** 0.249 Cal
    c. $1.31 \times 10^{-4} \, kWh$ **d.** $1.1 \times 10^4 \, cal$

55.

J	cal	Cal	kWh
225	53.8	5.38×10^{-2}	6.25×10^{-5}
3.44×10^6	8.21×10^5	8.21×10^2	9.54×10^{-1}
1.06×10^9	2.54×10^8	2.54×10^5	295
6.49×10^5	1.55×10^5	155	1.80×10^{-1}

57. $3.697 \times 10^9 \, J$

59. $8 \times 10^2 \, kJ$; 17 days

61. Exothermic.

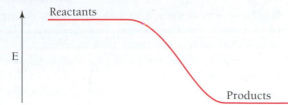

63. a. exothermic, $-\Delta H$ **b.** endothermic, $+\Delta H$
   c. exothermic, $-\Delta H$

65. a. $1.00 \times 10^2\ ^\circ\text{C}$ **b.** $-3.2 \times 10^2\ ^\circ\text{F}$
   c. $298\ \text{K}$ **d.** $3.10 \times 10^2\ \text{K}$

67. $-62\ ^\circ\text{C}$, $211\ \text{K}$

69. $159\ \text{K}$, $-173\ ^\circ\text{F}$

71. $-75.5\ ^\circ\text{F}$

73.

0.0 K	$-459.4\ ^\circ$F	$-273.0\ ^\circ$C
301 K	82.5 °F	28.1 °C
282 K	47 °F	8.5 °C

75. $9.0 \times 10^3\ \text{J}$

77. $8.7 \times 10^5\ \text{J}$

79. $58\ ^\circ\text{C}$

81. $31\ ^\circ\text{C}$

83. $1.0 \times 10^1\ ^\circ\text{C}$

85. $0.24\ \text{J/g}^\circ\text{C}$; silver

87. $2.2\ \text{J/g}^\circ\text{C}$

89. When warm drinks are placed into the ice, they release heat, which then melts the ice. The prechilled drinks, on the other hand, are already cold, so they do not release much heat.

91. 612 J

93. $49\ ^\circ\text{C}$

95. 70.2 J

97. $1.7 \times 10^4\ \text{kJ}$

99. $67\ ^\circ\text{C}$

101. 6.0 kWh

103. 22 g of fuel

105. 78 g

107. $27.2\ ^\circ\text{C}$

109. $5.96 \times 10^5\ \text{kJ}$; \$25

111. -40°

113. a. pure substance   **b.** pure substance
   c. pure substance   **d.** mixture

115. physical change

117. Small temperature changes in the ocean have a great impact on global weather because of the high heat capacity of water.

119. a. Sacramento is farther inland than San Francisco, so Sacramento is not as close to the ocean. The ocean water has a high heat capacity and will be able to keep San Francisco cooler in the hot days of summer. However, Sacramento is away from the ocean in a valley, so it will experience high temperatures in the summer.
   b. San Francisco is located right next to the ocean, so the high heat capacity of the seawater keeps the temperature in the city from dropping. In the winter, the ocean actually helps to keep the city warmer, compared to an inland city like Sacramento.

124. a. Petroleum, natural gas, and coal; 80%
   c. Coal and petroleum

CHAPTER 4
QUESTIONS

1. Democritus theorized that matter was ultimately composed of small, indivisible particles called atoms. Upon dividing matter, one would find tiny, indestructible atoms.

3. Rutherford's gold foil experiment involved sending positively charged alpha particles through a thin sheet of gold foil and detecting if there was any deflection of the particles. He found that most passed straight through, yet some particles showed some deflection. This result contradicts the plum-pudding model of the atom because the plum-pudding model does not explain the deflection of the alpha particles.

5.

Particle	Mass (kg)	Mass (amu)	Charge
Proton	1.67262×10^{-27}	1	+1
Neutron	1.67493×10^{-27}	1	0
Electron	0.00091×10^{-27}	0.00055	−1

7. Matter is usually charge-neutral due to protons and electrons having opposite charges. If matter were not charge-neutral, many unnatural things would occur, such as objects repelling or attracting each other.

9. A chemical symbol is a unique one- or two-letter abbreviation for an element. It is listed below the atomic number for that element on the periodic table.

11. Mendeleev noticed that many patterns were evident when elements were organized by increasing mass; from this observation he formulated the periodic law. He also organized the elements based on this law and created the basis for the periodic table being used today.

13. The periodic table is organized by listing the elements in order of increasing atomic number.

15. Nonmetals have varied properties (solid, liquid, or gas at room temperature); however, as a whole they tend to be poor conductors of heat and electricity, and they all tend to gain electrons when they undergo chemical changes. They are located toward the upper right side of the periodic table.

17. Each column within the main-group elements in the periodic table is labeled as a family or group of elements. The elements within a group usually have similar chemical properties.

19. An ion is an atom or group of atoms that has lost or gained electrons and has become charged.

21. a. ion charge = 1+
   b. ion charge = 2+
   c. ion charge = 3+
   d. ion charge = 2−
   e. ion charge = 1−

23. The percent natural abundance of isotopes is the relative amount of each different isotope in a naturally occurring sample of a given element.

25. Isotopes are noted in this manner: $^A_Z X$. X represents the chemical symbol, A represents the mass number, and Z represents the atomic number.

PROBLEMS

27. a. Correct.
b. False; different elements contain different types of atoms according to Dalton.
c. False; one cannot have 1.5 hydrogen atoms; combinations must be in simple, whole-number ratios.
d. Correct.

29. a. Correct.
b. False; most of the volume of the atom is empty space occupied by tiny, negatively charged electrons.
c. False; the number of negatively charged particles outside the nucleus equals the number of positively charged particles inside the nucleus.
d. False; the majority of the mass of an atom is found in the nucleus.

31. Solid matter seems to have no empty space within it because electromagnetic forces hold the atoms in a tight arrangement and the variation in density is too small to perceive with our eyes.

33. d
35. b
37. approximately 1.8×10^3 electrons
39. 5.4×10^{-4} g
41. a. 87   **b.** 36   **c.** 91   **d.** 32   **e.** 13
43. a. 18   **b.** 50   **c.** 54   **d.** 8   **e.** 81
45. a. C, 6   **b.** N, 7
c. Na, 11   **d.** K, 19
e. Cu, 29
47. a. manganese, 25   **b.** silver, 47
c. gold, 79   **d.** lead, 82
e. sulfur, 16

49.

Element Name	Element Symbol	Atomic Number
Gold	Au	79
Tin	Sn	50
Arsenic	As	33
Copper	Cu	29
Iron	Fe	26
Mercury	Hg	80

51. a. metal   **b.** metal
c. nonmetal   **d.** nonmetal
e. metalloid
53. a, d, e
55. a, b
57. c, d
59. b, e
61. a. halogen   **b.** noble gas
c. halogen   **d.** neither
e. noble gas
63. a. 6A   **b.** 3A   **c.** 4A   **d.** 4A   **e.** 5A
65. b, oxygen; it is in the same group or family.
67. b, chlorine and fluorine; they are in the same family or group.
69. d

71.

Chemical Symbol	Group Number	Group Name	Metal or Nonmetal
K	1A	Alkali Metals	Metals
Br	7A	Halogens	Nonmetal
Sr	2A	Alkaline Earth	Metal
He	8A	Noble Gas	Nonmetal
Ar	8A	Noble Gas	Nonmetal

73. a. e^- **b.** O^{2-} **c.** $2e^-$ **d.** Cl^-
75. a. 2−   **b.** 3+   **c.** 4+   **d.** 1−
77. a. 11 protons, 10 electrons
b. 56 protons, 54 electrons
c. 8 protons, 10 electrons
d. 27 protons, 24 electrons
79. a. False; Ti^{2+} has 22 protons and 20 electrons.
b. True
c. False; Mg^{2+} has 12 protons and 10 electrons.
d. True
81. a. Rb^+ **b.** K^+ **c.** Al^{3+} **d.** O^{2-}
83. a. 3 electrons lost   **b.** 1 electron lost
c. 1 electron gained   **d.** 2 electrons gained

85.

Symbol	Ion Commonly Formed	Number of Electrons in Ion	Number of Protons in Ion
Te	Te^{2-}	54	52
In	In^{3+}	46	49
Sr	Sr^{2+}	36	38
Mg	Mg^{2+}	10	12
Cl	Cl^-	18	17

87. a. $Z = 1, A = 3$ **b.** $Z = 24, A = 52$
c. $Z = 20, A = 42$ **d.** $Z = 73, A = 182$
89. a. $^{16}_{8}O$ **b.** $^{19}_{9}F$ **c.** $^{23}_{11}Na$ **d.** $^{27}_{13}Al$
91. a. $^{60}_{27}Co$ **b.** $^{22}_{10}Ne$ **c.** $^{131}_{53}I$ **d.** $^{244}_{94}Pu$
93. a. 11 protons, 12 neutrons
b. 88 protons, 178 neutrons
c. 82 protons, 126 neutrons
d. 7 protons, 7 neutrons
95. 6 protons, 8 neutrons, $^{14}_{6}C$
97. 85.47 amu
99. a. 49.31%   **b.** 78.91 amu
101. 121.8 amu, Sb
103. 7.8×10^{17} electrons
105. 4.2×10^{-45} m³; 6.2×10^{-31} m³; 6.7×10^{-13}%
107.

Number Symbol	Number of Protons	Number of Neutrons	A (Mass Number)	Natural Abundance
Sr-84 or $^{84}_{38}Sr$	38	46	84	0.56%
Sr-86 or $^{86}_{38}Sr$	38	48	86	9.86%
Sr-87 or $^{87}_{38}Sr$	38	49	87	7.00%
Sr-88 or $^{88}_{38}Sr$	38	50	88	82.58%

Atomic mass of Sr = 87.62 amu

109.

Symbol	Z	A	Number of Protons	Number of Electrons	Number of Neutrons	Charge
Zn^{2+}	30	64	30	28	34	2+
Mn^{3+}	25	55	25	22	30	3+
P	15	31	15	15	16	0
O^{2-}	8	16	8	10	8	2−
S^{2-}	16	34	16	18	18	2−

111. 153 amu, 52.2%

113. The atomic theory and nuclear model of the atom are both theories because they attempt to provide a broader understanding and model behavior of chemical systems.

115. Atomic mass is measured as the mean value of masses of all isotopes in a sample. In the case of fluorine, only the 19.00 amu isotope is naturally occurring. In the case of chlorine, about 76% of naturally occurring atoms are 35 amu, and 24% are 37 amu.

117. 69.3% Cu-63, 30.7% Cu-65

119. a. Nt-304 = 72%; Nt-305 = 4%; Nt-306 = 24%

b.

| 120 |
| Nt |
| 304.5 |

125. a. There is a periodic trend in which atomic radius increases abruptly and then decreases more gradually. The trend repeats itself with each row in the periodic table.

c. They all fall in the column 8A (the rightmost column) in the periodic table. They are all noble gases.

CHAPTER 5

QUESTIONS

1. Yes; when elements combine with other elements, a compound is created. Each compound is unique and contains properties different from those of the elements that compose it.

3. The law of constant composition states that all samples of a given compound have the same proportions of their constituent elements. Joseph Proust formulated this law.

5. The more metallic element is generally listed first in a chemical formula.

7. The empirical formula gives the relative number of atoms of each element in a compound. The molecular formula gives the actual number of atoms of each element in a molecule of the compound.

9. An atomic element is one that exists in nature, with a single atom as the basic unit. A molecular element is one that exists as a diatomic molecule as the basic unit. Molecular elements include H_2, N_2, O_2, F_2, Cl_2, Br_2, and I_2.

11. The systematic name can be directly derived by looking at the compound's formula. The common name for a compound acts like a nickname and can only be learned through familiarity.

13. The block that contains the elements for Type II compounds is known as the transition metals.

15. The basic form for the names of Type II ionic compounds is to have the name of the metal cation first, followed by the charge of the metal cation (in parentheses, using Roman numerals), and finally the base name of the nonmetal anion with -ide attached to the end.

17. For compounds containing a polyatomic anion, the name of the cation is first, followed by the name for the polyatomic anion. Also, if the compound contains both a polyatomic cation and a polyatomic anion, one would just use the names of both polyatomic ions.

19. The form for naming molecular compounds is to have the first element preceded by a prefix to indicate the number of atoms present. This is then followed by the second element with its corresponding prefix and -ide placed on the end of the second element.

21. To correctly name a binary acid, one must begin the first word with hydro-, which is followed by the base name of the nonmetal plus -ic added on the end. Finally, the word acid follows the first word.

23. To name an acid with oxyanions ending with -ite, one must take the base name of the oxyanion and attach -ous to it; the word acid follows this.

PROBLEMS

25. Yes; the ratios of sodium to chlorine in both samples were equal.

27. 2.06×10^3 g

29.

	Mass N_2O	Mass N	Mass O
Sample A	2.85	1.82	1.03
Sample B	4.55	2.91	1.64
Sample C	3.74	2.39	1.35
Sample D	1.74	1.11	0.63

31. NI_3

33. a. Fe_3O_4 **b.** PCl_3
    c. PCl_5 **d.** Ag_2O

35. a. 4   **b.** 4
    c. 6   **d.** 4

37. a. magnesium, 1; chlorine, 2
    b. sodium, 1; nitrogen, 1; oxygen, 3
    c. calcium, 1; nitrogen, 2; oxygen, 4
    d. strontium, 1; oxygen, 2; hydrogen, 2

39.

Formula	Number of $C_2H_3O_2$	Number of C Atoms	Number of H Atoms	Number of O Atoms	Number of Metal Atoms
$Mg(C_2H_3O_2)_2$	2	4	6	4	1
$NaC_2H_3O_2$	1	2	3	2	1
$Cr_2(C_2H_3O_2)_4$	4	8	12	8	2

41. a. CH_3 **b.** NO_2
    c. C_2H_3O **d.** NH_3

43. a. molecular   **b.** atomic
    c. atomic   **d.** molecular

45. a. molecular   **b.** ionic
    c. ionic   **d.** molecular

47. helium $\longrightarrow$ single atoms
$CCl_4 \longrightarrow$ molecules
$K_2SO_4 \longrightarrow$ formula units
bromine $\longrightarrow$ diatomic molecules

49. a. formula units **b.** single atoms
c. molecules **d.** molecules

51. a. ionic; forms only one type of ion
b. molecular
c. molecular
d. ionic; forms more than one type of ion

53. a. Na_2S **b.** SrO
c. Al_2S_3 **d.** $MgCl_2$

55. a. $KC_2H_3O_2$ **b.** K_2CrO_4
c. K_3PO_4 **d.** KCN

57. a. Li_3N, Li_2O, LiF **b.** Ba_3N_2, BaO, BaF_2
c. AlN, Al_2O_3, AlF_3

59. a. cesium chloride **b.** strontium bromide
c. potassium oxide **d.** lithium fluoride

61. a. chromium(II) chloride **b.** chromium(III) chloride
c. tin(IV) oxide **d.** lead(II) iodide

63. a. forms more than one type of ion, chromium(III) oxide
b. forms only one type of ion, sodium iodide
c. forms only one type of ion, calcium bromide
d. forms more than one type of ion, tin(II) oxide

65. a. barium nitrate **b.** lead(II) acetate
c. ammonium iodide **d.** potassium chlorate
e. cobalt(II) sulfate **f.** sodium perchlorate

67. a. hypobromite ion **b.** bromite ion
c. bromate ion **d.** perbromate ion

69. a. $CuBr_2$ **b.** $AgNO_3$
c. KOH **d.** Na_2SO_4
e. $KHSO_4$ **f.** $NaHCO_3$

71. a. sulfur dioxide **b.** nitrogen triiodide
c. bromine pentafluoride **d.** nitrogen monoxide
e. tetranitrogen tetraselenide

73. a. CO **b.** S_2F_4
c. Cl_2O **d.** PF_5
e. BBr_3 **f.** P_2S_5

75. a. PBr_5 phosphorus pentabromide
b. P_2O_3 diphosphorus trioxide
c. SF_4 sulfur tetraflouride
d. correct

77. a. oxyacid, nitrous acid, nitrite
b. binary acid, hydroiodic acid
c. oxyacid, sulfuric acid, sulfate
d. oxyacid, nitric acid, nitrate

79. a. hypochlorous acid
b. chlorous acid
c. chloric acid
d. perchloric acid

81. a. H_3PO_4 **b.** HBr **c.** H_2SO_3

83. a. 63.02 amu **b.** 199.88 amu
c. 153.81 amu **d.** 211.64 amu

85. PBr_3, Ag_2O, PtO_2, $Al(NO_3)_3$

87. a. CH_4 **b.** SO_3 **c.** NO_2

89. a. 12 **b.** 4 **c.** 12 **d.** 7

91. a. 8 **b.** 12 **c.** 12

93.

Formula	Type	Name
N_2H_4	molecular	dinitrogen tetrahydride
KCl	ionic	potassium chloride
H_2CrO_4	acid	chromic acid
$Co(CN)_3$	ionic	cobalt(III) cyanide

95. a. calcium nitrite
b. potassium oxide
c. phosphorus trichloride
d. correct
e. potassium iodite

97. a. $Sn(SO_4)_2$ 310.9 amu **b.** HNO_2 47.02 amu
c. $NaHCO_3$ 84.01 amu **d.** PF_5 125.97 amu

99. a. platinum(IV) oxide 227.08 amu
b. dinitrogen pentoxide 108.02 amu
c. aluminum chlorate 277.33 amu
d. phosphorus pentabromide 430.47 amu

101. C_2H_4

103. 10 different isotopes can exist. 151.88 amu, 152.88 amu, 153.88 amu, 154.88 amu, 155.88 amu, 156.88 amu, 157.88 amu, 158.88 amu, 159.88 amu, and 160.88 amu.

105. a. molecular element **b.** atomic element
c. ionic compound **d.** molecular compound

107. a. $NaOCl$; $NaOH$
b. $Al(OH)_3$; $Mg(OH)_2$
c. $CaCO_3$
d. $NaHCO_3$, $Ca_3(PO_4)_2$, $NaAl(SO_4)_2$

112. a. 1950: 314 ppm; 2000: 371 ppm; The carbon dioxide increased by 57 ppm between 1950 and 2000.
c. 436 ppm

CHAPTER 6
QUESTIONS

1. Chemical composition lets us determine how much of a particular element is contained within a particular compound.

3. There are 6.022×10^{23} atoms in 1 mole of atoms.

5. One mole of any element has a mass equal to its atomic mass in grams.

7. a. 30.97 g **b.** 195.08 g
c. 12.01 g **d.** 52.00 g

9. Each element has a different atomic mass number. So, the subscripts that represent mole ratios cannot be used to represent the ratios of grams of a compound. The grams per mole of one element always differ from the grams per mole of a different element.

11. a. 11.19 g $H \equiv 100$ g H_2O
b. 53.29 g $O \equiv 100$ g fructose
c. 84.12 g $C \equiv 100$ g octane
d. 52.14 g $C \equiv 100$ g ethanol

13. The empirical formula gives the smallest whole-number ratio of each type of atom. The molecular formula gives the specific number of each type of atom in the molecule. The molecular formula is always a multiple of the empirical formula.

15. The empirical formula mass of a compound is the sum of the masses of all the atoms in the empirical formula.

PROBLEMS

17. 3.5×10^{24} atoms

19. a. 2.0×10^{24} atoms **b.** 5.8×10^{21} atoms
 c. 1.38×10^{25} atoms **d.** 1.29×10^{23} atoms

21.

Element	Moles	Number of Atoms
Ne	0.552	3.32×10^{23}
Ar	5.40	3.25×10^{24}
Xe	1.78	1.07×10^{24}
He	1.79×10^{-4}	1.08×10^{20}

23. a. 72.7 dozen **b.** 6.06 gross
 c. 1.74 reams **d.** 1.45×10^{-21} mol

25. 0.321 mol

27. 28.6 g

29. a. 2.05×10^{-2} mol **b.** 0.623 mol
 c. 0.401 mol **d.** 3.21×10^{-3} mol

31.

Element	Moles	Mass
Ne	1.11	22.5
Ar	0.117	4.67 g
Xe	7.62	1.00
He	1.44×10^{-4}	5.76×10^{-4} g

33. 8.07×10^{18} atoms

35. 8.44×10^{22} atoms

37. a. 1.16×10^{23} atoms **b.** 2.81×10^{23} atoms
 c. 2.46×10^{22} atoms **d.** 7.43×10^{23} atoms

39. 1.9×10^{21} atoms

41. 1.61×10^{25} atoms

43.

Element	Mass	Moles	Number of Atoms
Na	38.5 mg	1.67×10^{-3}	1.01×10^{21}
C	13.5 g	1.12	6.74×10^{23}
V	1.81×10^{-20} g	3.55×10^{-22}	214
Hg	1.44 kg	7.18	4.32×10^{24}

45. b

47. a. 0.654 mol **b.** 1.22 mol
 c. 96.6 mol **d.** 1.76×10^{-5} mol

49.

Compound	Mass	Moles	Molecules
H_2O	112 kg	6.22×10^3	3.74×10^{27}
N_2O	6.33 g	0.144	8.66×10^{22}
SO_2	156	2.44	1.47×10^{24}
CH_2Cl_2	5.46	0.0643	3.87×10^{22}

51. 6.20×10^{21} molecules

53. a. 1.2×10^{23} molecules **b.** 1.21×10^{24} molecules
 c. 3.5×10^{23} molecules **d.** 6.4×10^{22} molecules

55. 0.10 mg

57. $\$6.022 \times 10^{21}$ total. $\$8.5 \times 10^{11}$ Each person would be a billionaire.

59. 5.4 mol Cl

61. d, 3 mol O

63. a. 2.5 mol C **b.** 0.230 mol C
 c. 22.7 mol C **d.** 201 mol C

65. a. 2 moles H per mole of molecules; 8 H atoms present
 b. 4 moles H per mole of molecules; 20 H atoms present
 c. 3 moles H per mole of molecules; 9 H atoms present

67. a. 22.3 g **b.** 29.4 g
 c. 21.6 g **d.** 12.9 g

69. a. 1.4×10^3 kg **b.** 1.4×10^3 kg
 c. 2.1×10^3 kg

71. 84.8% Sr

73. 36.1% Ca; 63.9% Cl

75. 10.7 g

77. 6.6 mg

79. a. 63.65% **b.** 46.68%
 c. 30.45% **d.** 25.94%

81. a. 39.99% C; 6.73% H; 53.28% O
 b. 26.09% C; 4.39% H; 69.52% O
 c. 60.93% C; 15.37% H; 23.69% N
 d. 54.48% C; 13.74% H; 31.78% N

83. a. 58.50% O **b.** 42.13% O **c.** 72.71% O

85. Fe_3O_4, 72.36% Fe; Fe_2O_3, 69.94% Fe; $FeCO_3$, 48.20% Fe; magnetite

87. NO_2

89. a. NiI_2 **b.** $SeBr_4$ **c.** $BeSO_4$

91. C_2H_6N

93. a. C_3H_6O **b.** $C_5H_{10}O_2$ **c.** $C_9H_{10}O_2$

95. P_2O_3

97. NCl_3

99. C_4H_8

101. a. C_6Cl_6 **b.** C_2HCl_3 **c.** $C_6H_3Cl_3$

103. 2.43×10^{23} atoms

105. 2×10^{21} molecules

107.

Substance	Mass	Moles	Number of Particles
Ar	0.018 g	4.5×10^{-4}	2.7×10^{20}
NO_2	8.33×10^{-3} g	1.81×10^{-4}	1.09×10^{20}
K	22.4 mg	5.73×10^{-4}	3.45×10^{20}
C_8H_{18}	3.76 kg	32.9	1.98×10^{25}

109. a. CuI_2: 20.03% Cu; 79.97% I
 b. $NaNO_3$: 27.05% Na; 16.48% N; 56.47% O
 c. $PbSO_4$: 68.32% Pb; 10.57% S; 21.10% O
 d. CaF_2: 51.33% Ca; 48.67% F

111. 1.8×10^3 kg rock

113. 59 kg Cl

115. 1.1×10^2 g H

117.

Formula	Molar Mass	%C (by mass)	%H (by mass)
C_2H_4	28.06	85.60%	14.40%
C_4H_{10}	58.12	82.66%	17.34%
C_4H_8	56.12	85.60%	14.40%
C_3H_8	44.09	81.71%	18.29%

119. $C_4H_6O_2$

121. $C_{10}H_{14}N_2$

123. 70.4% KBr, 29.6% KI

125. 29.6 g SO_2

127. 2.66 kg Fe

129. **a.** 1×10^{57} atoms per star

b. 1×10^{68} atoms per galaxy

c. 1×10^{79} atoms in the universe

131. $C_{16}H_{10}$

135. **a.** −20%

c. women, 4.4 L; men, 5.4 L

CHAPTER 7

QUESTIONS

1. A chemical reaction is the change of one or more substances into different substances, for example, burning wood, rusting iron, and protein synthesis.

3. The main evidence of a chemical reaction includes a color change, the formation of a solid, the formation of a gas, the emission of light, and the emission or absorption of heat.

5. **a.** gas   **b.** liquid

c. solid   **d.** aqueous

7. **a.** reactants: 4 Ag, 2 O, 1 C products: 4 Ag, 2 O, 1 C balanced: yes

b. reactants: 1 Pb, 2 N, 6 O, 2 Na, 2 Cl products: 1 Pb, 2 N, 6 O, 2 Na, 2 Cl balanced: yes

c. reactants: 3 C, 8 H, 2 O products: 3 C, 8 H, 10 O balanced: no

9. If a compound dissolves in water, then it is soluble. If it does not dissolve in water, it is insoluble.

11. When ionic compounds containing polyatomic ions dissolve in water, the polyatomic ions usually dissolve as intact units.

13. The solubility rules are a set of empirical rules for ionic compounds that were deduced from observations on many compounds. The rules help us determine whether particular compounds will be soluble or insoluble.

15. The precipitate will always be insoluble; it is the solid that forms upon mixing two aqueous solutions.

17. Acid–base reactions involve an acid and a base reacting to form water and an ionic compound. An example is the reaction between hydrobromic acid and sodium hydroxide: $HBr + NaOH \longrightarrow H_2O + NaBr$

19. Gas-evolution reactions are reactions that evolve a gas. An example is the reaction between hydrochloric acid and sodium bicarbonate:
$HCl + NaHCO_3 \longrightarrow H_2O + CO_2 + NaCl$

21. Combustion reactions are a type of redox reaction and are characterized by the exothermic reaction of a substance with O_2. An example is the reaction between methane and oxygen: $CH_4 + 2 O_2 \longrightarrow CO_2 + 2 H_2O$

23. A synthesis reaction combines simpler substances to form more complex substances. An example is the reaction between elemental potassium and chloride: $2 K + Cl_2 \longrightarrow 2 KCl$. A decomposition decomposes a more complex substance into simpler substances. An example is the decomposition of water: $2 H_2O \longrightarrow 2 H_2 + O_2$

PROBLEMS

25. **a.** Yes; there is a color change showing a chemical reaction.

b. No; the state of the compound changes, but no chemical reaction takes place.

c. Yes; there is a formation of a solid in a previously clear solution.

d. Yes; there is a formation of a gas when the yeast is added to the solution.

27. Yes; a chemical reaction has occurred, for the presence of the bubbles is evidence for the formation of a gas.

29. Yes; a chemical reaction has occurred. We know this due to the color change of the hair.

31. **a.** Reactants: 1 Pb, 6 O, 2 Na, 2 Cl Products: 1 Pb; 6 O; 2 Na; 2 Cl; Balanced

b. Reactants: 3 C, 8 H, 2 O; Products: 3 C, 8 H, 10 O; Not balanced

33. Placing a subscript 2 after H_2O would change the compound from water to hydrogen peroxide (H_2O_2). To balance chemical reactions, one must add coefficients, not subscripts.
$2 H_2O(l) \longrightarrow 2 H_2(g) + O_2(g)$

35. **a.** $PbS + 2 HCl \longrightarrow PbCl_2 + H_2S$

b. $CO + 3 H_2 \longrightarrow CH_4 + H_2O$

c. $Fe_2O_3 + 3 H_2 \longrightarrow 2 Fe + 3 H_2O$

d. $4 NH_3 + 5 O_2 \longrightarrow 4 NO + 6 H_2O$

37. **a.** $Mg(s) + 2 CuNO_3(aq) \longrightarrow 2 Cu(s) + Mg(NO_3)_2(aq)$

b. $2 N_2O_5(g) \longrightarrow 4 NO_2(g) + O_2(g)$

c. $Ca(s) + 2 HNO_3(aq) \longrightarrow H_2(g) + Ca(NO_3)_2(aq)$

d. $2 CH_3OH(l) + 3 O_2(g) \longrightarrow 2 CO_2(g) + 4 H_2O(g)$

39. $2 Na(s) + 2 H_2O(l) \longrightarrow H_2(g) + 2 NaOH(aq)$

41. $2 SO_2(g) + O_2(g) + 2 H_2O(l) \longrightarrow 2 H_2SO_4(aq)$

43. $V_2O_5(s) + 2 H_2(g) \longrightarrow V_2O_3(s) + 2 H_2O(l)$

45. $C_{12}H_{22}O_{11}(aq) + H_2O(l) \longrightarrow 4 CO_2(g) + 4 C_2H_5OH(aq)$

47. **a.** $Na_2S(aq) + Cu(NO_3)_2(aq) \longrightarrow 2 NaNO_3(aq) + CuS(s)$

b. $4 HCl(aq) + O_2(g) \longrightarrow 2 H_2O(l) + 2 Cl_2(g)$

c. $2 H_2(g) + O_2(g) \longrightarrow 2 H_2O(l)$

d. $FeS(s) + 2 HCl(aq) \longrightarrow FeCl_2(aq) + H_2S(g)$

49. **a.** $BaO_2(s) + H_2SO_4(aq) \longrightarrow BaSO_4(s) + H_2O_2(aq)$

b. $2 Co(NO_3)_3(aq) + 3 (NH_4)_2S(aq) \longrightarrow$
$Co_2S_3(s) + 6 NH_4NO_3(aq)$

c. $Li_2O(s) + H_2O(l) \longrightarrow 2 LiOH(aq)$

d. $Hg_2(C_2H_3O_2)_2(aq) + 2 KCl(aq) \longrightarrow$
$Hg_2Cl_2(s) + 2 KC_2H_3O_2(aq)$

51. **a.** $2 Rb(s) + 2 H_2O(l) \longrightarrow 2 RbOH(aq) + H_2(g)$

b. Equation is balanced.

c. $2 NiS(s) + 3 O_2(g) \longrightarrow 2 NiO(s) + 2 SO_2(g)$

d. $3 PbO(s) + 2 NH_3(g) \longrightarrow$
$3 Pb(s) + N_2(g) + 3 H_2O(l)$

53. $C_6H_{12}O_6(aq) + 6\,O_2(g) \longrightarrow 6\,CO_2(g) + 6\,H_2O(l)$

55. $2\,NO(g) + 2\,CO(g) \longrightarrow N_2(g) + 2\,CO_2(g)$

57. **a.** soluble; Na^+, $C_2H_3O_2^-$ **b.** soluble; Sn^{2+}, NO_3^-
c. insoluble **d.** soluble; Na^+, PO_4^{3-}

59. $AgCl$; $BaSO_4$; $CuCO_3$; Fe_2S_3

61.

Soluble	Insoluble
K_2S	Hg_2I_2
BaS	$Cu_3(PO_4)_2$
NH_4Cl	MgS
Na_2CO_3	$CaSO_4$
K_2SO_4	$PbSO_4$
SrS	$PbCl_2$
Li_2S	Hg_2Cl_2

63. **a.** NO REACTION
b. $K_2SO_4(aq) + BaBr_2(aq) \longrightarrow BaSO_4(s) + 2\,KBr(aq)$
c. $2\,NaCl(aq) + Hg_2(C_2H_3O_2)_2(aq) \longrightarrow$
$$Hg_2Cl_2(s) + 2\,NaC_2H_3O_2(aq)$$
d. NO REACTION

65. **a.** $Na_2CO_3(aq) + Pb(NO_3)_2(aq) \longrightarrow$
$$PbCO_3(s) + 2\,NaNO_3(aq)$$
b. $K_2SO_4(aq) + Pb(C_2H_3O_2)_2(aq) \longrightarrow$
$$PbSO_4(s) + 2\,KC_2H_3O_2(aq)$$
c. $Cu(NO_3)_2(aq) + BaS(aq) \longrightarrow$
$$CuS(s) + Ba(NO_3)_2(aq)$$
d. NO REACTION

67. **a.** correct
b. NO REACTION
c. correct
d. $Pb(NO_3)_2(aq) + 2\,LiCl(aq) \longrightarrow$
$$PbCl_2(s) + 2\,LiNO_3(aq)$$

69. K^+, NO_3^-

71. **a.** $Ag^+(aq) + NO_3^-(aq) + K^+(aq) + Cl^-(aq) \longrightarrow$
$$AgCl(s) + K^+(aq) + NO_3^-(aq)$$
$Ag^+(aq) + Cl^-(aq) \longrightarrow AgCl(s)$
b. $Ca^{2+}(aq) + S^{2-}(aq) + Cu^{2+}(aq) + 2\,Cl^-(aq) \longrightarrow$
$$CuS(s) + Ca^{2+}(aq) + 2\,Cl^-(aq)$$
$Cu^{2+}(aq) + S^{2-}(aq) \longrightarrow CuS(s)$
c. $Na^+(aq) + OH^-(aq) + H^+(aq) + NO_3^-(aq) \longrightarrow$
$$H_2O(l) + Na^+(aq) + NO_3^-(aq)$$
$H^+(aq) + OH^-(aq) \longrightarrow H_2O(l)$
d. $6\,K^+(aq) + 2\,PO_4^{3-}(aq) + 3\,Ni^{2+}(aq) + 6\,Cl^-(aq) \longrightarrow$
$$Ni_3(PO_4)_2(s) + 6\,K^+(aq) + 6\,Cl^-(aq)$$
$3\,Ni^{2+}(aq) + 2\,PO_4^{3-}(aq) \longrightarrow Ni_3(PO_4)_2(s)$

73. $Hg_2^{2+}(aq) + 2\,NO_3^-(aq) + 2\,Na^+(aq) + 2\,Cl^-(aq) \longrightarrow$
$$Hg_2Cl_2(s) + 2\,Na^+(aq) + 2\,NO_3^-(aq)$$
$Hg_2^{2+}(aq) + 2\,Cl^-(aq) \longrightarrow Hg_2Cl_2(s)$

75. **a.** $2\,Na^+(aq) + CO_3^{2-}(aq) + Pb^{2+}(aq) + 2\,NO_3^-(aq)$
$$\longrightarrow PbCO_3(s) + 2\,Na^+(aq) + 2\,NO_3^-(aq)$$
$Pb^{2+}(aq) + CO_3^{2-}(aq) \longrightarrow PbCO_3(s)$
b. $2\,K^+(aq) + SO_4^{2-}(aq) + Pb^{2+}(aq) + 2\,CH_3CO_2^-(aq)$
$$\longrightarrow PbSO_4(s) + 2\,K^+(aq) + 2\,CH_3CO_2^-(aq)$$
$Pb^{2+}(aq) + SO_4^{2-}(aq) \longrightarrow PbSO_4(s)$

c. $Cu^{2+}(aq) + 2\,NO_3^-(aq) + Ba^{2+}(aq) + S^{2-}(aq) \longrightarrow$
$$CuS(s) + Ba^{2+}(aq) + 2\,NO_3^-(aq)$$
$Cu^{2+}(aq) + S^{2-}(aq) \longrightarrow CuS(s)$
d. NO REACTION

77. $HCl(aq) + KOH(aq) \longrightarrow H_2O(l) + KCl(aq)$
$H^+(aq) + OH^-(aq) \longrightarrow H_2O(l)$

79. **a.** $2\,HCl(aq) + Ba(OH)_2(aq) \longrightarrow 2\,H_2O(l) + BaCl_2(aq)$
b. $H_2SO_4(aq) + 2\,KOH(aq) \longrightarrow 2\,H_2O(l) + K_2SO_4(aq)$
c. $HClO_4(aq) + NaOH(aq) \longrightarrow H_2O(l) + NaClO_4(aq)$

81. **a.** $HBr(aq) + NaHCO_3(aq) \longrightarrow$
$$H_2O(l) + CO_2(g) + NaBr(aq)$$
b. $NH_4I(aq) + KOH(aq) \longrightarrow$
$$H_2O(l) + NH_3(g) + KI(aq)$$
c. $2\,HNO_3(aq) + K_2SO_3(aq) \longrightarrow$
$$H_2O(l) + SO_2(g) + 2\,KNO_3(aq)$$
d. $2\,HI(aq) + Li_2S(aq) \longrightarrow H_2S(g) + 2\,LiI(aq)$

83. b and d are redox reactions; a and c are not.

85. **a.** $2\,C_2H_6(g) + 7\,O_2(g) \longrightarrow 4\,CO_2(g) + 6\,H_2O(g)$
b. $2\,Ca(s) + O_2(g) \longrightarrow 2\,CaO(s)$
c. $2\,C_3H_8O(l) + 9\,O_2(g) \longrightarrow 6\,CO_2(g) + 8\,H_2O(g)$
d. $2\,C_4H_{10}S(l) + 15\,O_2(g) \longrightarrow$
$$8\,CO_2(g) + 10\,H_2O(g) + 2\,SO_2(g)$$

87. **a.** $2\,Ag(s) + Br_2(g) \longrightarrow 2\,AgBr(s)$
b. $2\,K(s) + Br_2(g) \longrightarrow 2\,KBr(s)$
c. $2\,Al(s) + 3\,Br_2(g) \longrightarrow 2\,AlBr_3(s)$
d. $Ca(s) + Br_2(g) \longrightarrow CaBr_2(s)$

89. **a.** double displacement
b. synthesis or combination
c. single displacement
d. decomposition

91. **a.** synthesis **b.** decomposition
c. synthesis

93. **a.** $2\,Na^+(aq) + 2\,I^-(aq) + Hg_2^{2+}(aq) + 2\,NO_3^-(aq)$
$$\longrightarrow Hg_2I_2(s) + 2\,Na^+(aq) + 2\,NO_3^-(aq)$$
b. $2\,H^+(aq) + 2\,ClO_4^-(aq) + Ba^{2+}(aq) + 2\,OH^-(aq)$
$$\longrightarrow 2\,H_2O(l) + Ba^{2+}(aq) + 2\,ClO_4^-(aq)$$
$H^+(aq) + OH^-(aq) \longrightarrow H_2O(s)$
c. NO REACTION
d. $2\,H^+(aq) + 2\,Cl^-(aq) + 2\,Li^+(aq) + CO_3^{2-}(aq) \longrightarrow$
$$H_2O(l) + CO_2(g) + 2\,Li^+(aq) + 2\,Cl^-(aq)$$
$2\,H^+(aq) + CO_3^{2-}(aq) \longrightarrow H_2O(l) + CO_2(g)$

95. **a.** NO REACTION
b. NO REACTION
c. $K^+(aq) + HSO_3^-(aq) + H^+(aq) + NO_3^-(aq) \longrightarrow$
$$H_2O(l) + SO_2(g) + K^+(aq) + NO_3^-(aq)$$
$H^+(aq) + HSO_3^-(aq) \longrightarrow H_2O(l) + SO_2(g)$
d. $Mn^{3+}(aq) + 3\,Cl^-(aq) + 3\,K^+(aq) + PO_4^{3-}(aq) \longrightarrow$
$$MnPO_4(s) + 3\,K^+(aq) + 3\,Cl^-(aq)$$
$Mn^{3+}(aq) + PO_4^{3-}(aq) \longrightarrow MnPO_4(s)$

97. **a.** acid–base; $KOH(aq) + HC_2H_3O_2(aq) \longrightarrow$
$$H_2O(l) + KC_2H_3O_2(aq)$$
b. gas evolution; $2\,HBr(aq) + K_2CO_3(aq) \longrightarrow$
$$H_2O(l) + CO_2(g) + 2\,KBr(aq)$$

c. synthesis; $2\,H_2(g) + O_2(g) \longrightarrow 2\,H_2O(l)$

d. precipitation; $2\,NH_4Cl(aq) + Pb(NO_3)_2(aq) \longrightarrow$
$$PbCl_2(s) + 2\,NH_4NO_3(aq)$$

99. a. oxidation–reduction; single displacement

b. gas evolution; acid–base

c. gas evolution; double displacement

d. precipitation; double displacement

101. $3\,CaCl_2(aq) + 2\,Na_3PO_4(aq) \longrightarrow$
$$Ca_3(PO_4)_2(s) + 6\,NaCl(aq)$$
$3\,Ca^{2+}(aq) + 6\,Cl^-(aq) + 6\,Na^+(aq) + 2\,PO_4^{3-}(aq)$
$$\longrightarrow Ca_3(PO_4)_2(s) + 6\,Na^+(aq) + 6\,Cl^-(aq)$$
$3\,Ca^{2+}(aq) + 2\,PO_4^{3-}(aq) \longrightarrow Ca_3(PO_4)_2(s)$
$3\,Mg(NO_3)_2(aq) + 2\,Na_3PO_4(aq) \longrightarrow$
$$Mg_3(PO_4)_2(s) + 6\,NaNO_3(aq)$$
$3\,Mg^{2+}(aq) + 6\,NO_3^-(aq) + 6\,Na^+(aq) + 2\,PO_4^{3-}(aq)$
$$\longrightarrow Mg_3(PO_4)_2(s) + 6\,Na^+(aq) + 6\,NO_3^-(aq)$$
$3\,Mg^{2+}(aq) + 2\,PO_4^{3-}(aq) \longrightarrow Mg_3(PO_4)_2(s)$

103. *Correct answers may vary; representative correct answers are:

a. addition of a solution containing SO_4^{2-};
$Pb^{2+}(aq) + SO_4^{2-}(aq) \longrightarrow PbSO_4(s)$

b. addition of a solution containing SO_4^{2-};
$Ca^{2+}(aq) + SO_4^{2-}(aq) \longrightarrow CaSO_4(s)$

c. addition of a solution containing SO_4^{2-};
$Ba^{2+}(aq) + SO_4^{2-}(aq) \longrightarrow BaSO_4(s)$

d. addition of a solution containing Cl^-;
$Hg_2^{2+}(aq) + 2\,Cl^-(aq) \longrightarrow Hg_2Cl_2(s)$

105. 0.168 mol Ca; 6.73 g Ca

107. 0.00128 mol NaCl, 0.0750 g NaCl

109. a. chemical   **b.** physical

114. a. Water sample A contains Ca^{2+} and Cu^{2+}.

CHAPTER 8

QUESTIONS

1. Reaction stoichiometry is very important to chemistry. It gives us a numerical relationship between the reactants and products that allows chemists to plan and carry out chemical reactions to obtain products in the desired quantities.

For example, how much CO_2 is produced when a given amount of C_8H_{10} is burned?

How much $H_2(g)$ is produced when a given amount of water decomposes?

3. 1 mol $Cl_2 \equiv$ 2 mol NaCl

5. mass A $\longrightarrow$ moles A $\longrightarrow$ moles B $\longrightarrow$ mass B
(A = reactant, B = product)

7. The limiting reactant is the reactant that limits the amount of product in a chemical reaction.

9. The actual yield is the amount of product actually produced by a chemical reaction. The percent yield is the percentage of the theoretical yield that was actually attained.

11. d

13. The enthalpy of reaction is the total amount of heat generated or absorbed by a particular chemical reaction. The quantity is important because it quantifies the change in heat for the chemical reaction. It is useful for determining the necessary starting conditions and predicting the outcome of various reactions.

PROBLEMS

15. a. 2 mol C   **b.** 1 mol C

    c. 3 mol C   **d.** 1.5 mol C

17. a. 2.6 mol NO_2 **b.** 11.6 mol NO_2

    c. 8.90×10^3 mol NO_2 **d.** 2.012×10^{-3} mol NO_2

19. c

21. a. 3.50 mol HCl   **b.** 3.50 mol H_2O

    c. 0.875 mol Na_2O_2 **d.** 1.17 mol SO_3

23. a. 2.4 mol PbO(s), 2.4 mol $SO_2(g)$

    b. 1.6 mol PbO(s), 1.6 mol $SO_2(g)$

    c. 5.3 mol PbO(s), 5.3 mol $SO_2(g)$

    d. 3.5 mol PbO(s), 3.5 mol $SO_2(g)$

25.

mol N_2H_4	mol N_2O_4	mol N_2	mol H_2O
4	2	6	8
6	3	9	12
4	2	6	8
11	5.5	16.5	22
3	1.5	4.5	6
8.26	4.13	12.4	16.5

27. $2\,C_4H_{10}(g) + 13\,O_2(g) \longrightarrow$
$$8\,CO_2(g) + 10\,H_2O(g);\ 32\ \text{mol}\ O_2$$

29. a. $Pb(s) + 2\,AgNO_3(aq) \longrightarrow Pb(NO_3)_2(aq) + 2\,Ag(s)$

    b. 19 mol $AgNO_3$

    c. 56.8 mol Ag

31. a. 0.157 g O_2 **b.** 0.500 g O_2

    c. 114 g O_2 **d.** 2.86×10^{-4} g O_2

33. a. 4.0 g NaCl   **b.** 4.3 g $CaCO_3$

    c. 4.0 g MgO   **d.** 3.1 g NaOH

35. a. 8.9 g Al_2O_3, 9.7 g Fe   **b.** 3.0 g Al_2O_3, 3.3 g Fe

37.

Mass CH_4	Mass O_2	Mass CO_2	Mass H_2O
0.645 g	2.57 g	1.77 g	1.45 g
22.32 g	89.00 g	61.20 g	50.12 g
5.044 g	20.11 g	13.83 g	11.32 g
1.07 g	4.28 g	2.94 g	2.41 g
3.18 kg	12.7 kg	8.72 kg	7.14 kg
8.57×10^2 kg	3.42×10^3 kg	2.35×10^3 kg	1.92×10^3 kg

39. a. 2.3 g HCl   **b.** 4.3 g HNO_3

    c. 2.2 g H_2SO_4

41. 123 g H_2SO_4, 2.53 g H_2

43. a. 2 mol A   **b.** 1.8 mol A

    c. 4 mol B   **d.** 40 mol B

45. a. 1.5 mol C   **b.** 3 mol C

    c. 3 mol C   **d.** 96 mol C

47. a. 1 mol K   **b.** 1.8 mol K

    c. 1 mol Cl_2 **d.** 14.6 mol K

49. a. 1.3 mol MnO_3 **b.** 4.8 mol MnO_3

    c. 0.107 mol MnO_3 **d.** 27.5 mol MnO_3

51. 3 mol A, 0 mol B, 4 mol C

53. a. 2 Cl_2 **b.** 3 Cl_2

    c. 2 Cl_2

55. a. 1.0 g F_2 **b.** 10.5 g Li
c. 6.79×10^3 g F_2

57. a. 1.3 g $AlCl_3$ **b.** 24.8 g $AlCl_3$
c. 2.17 g $AlCl_3$

59. 74.6%

61. CaO; 25.7 g $CaCO_3$; 75.5%

63. O_2; 5.07 g NiO; 95.9%

65. Pb^{2+}; 262.7 g $PbCl_2$; 96.09%

67. TiO_2: 0 g, C: 7.0 g, Ti: 5.99 g, CO: 7.00 g

69. a. exothermic, $-\Delta H$ **b.** endothermic, $+\Delta H$
c. exothermic, $-\Delta H$

71. a. 55 kJ   **b.** 110 kJ   **c.** 28 kJ   **d.** 55 kJ

73. 4.78×10^3 kJ

75. 34.9 g C_8H_{18}

77. N_2

79. 0.152 g Ba^{2+}

81. 1.5 g HCl

83. 3.1 kg CO_2

85. 4.7 g Na_3PO_4

87. 469 g Zn

89. $2\ NH_4NO_3(s) \longrightarrow$
        $2\ N_2(g) + O_2(g) + 4\ H_2O(l)$; 2.00×10^2 g O_2

91. salicylic acid ($C_7H_6O_3$); 2.71 g $C_9H_8O_4$; 74.1%

93. NH_3; 120 kg (CH_4N_2O); 72.9%

95. 2.4 mg $C_4H_6O_4S_2$

97. 1.0×10^3 g CO_2

99. b; the loudest explosion will occur when the ratio is 2 hydrogen to 1 oxygen, for that is the ratio that occurs in water.

101. 2.8×10^{13} kg CO_2 per year; 1.1×10^2 years

106. a. Experiments 1, 2, and 3
c. 2 A + 1 B
e. 84.8%

CHAPTER 9

QUESTIONS

1. Both the Bohr model and the quantum-mechanical model for the atom were developed in the early 1900s. These models serve to explain how electrons are arranged within the atomic structure and how the electrons affect the chemical and physical properties of each element.

3. White light contains a spectrum of wavelengths and therefore a spectrum of color. Colored light is produced by a single wavelength and is therefore a single color.

5. Energy carried per photon is greater for shorter wavelengths than for longer wavelengths. Wavelength and frequency are inversely related—the shorter the wavelength, the higher the frequency.

7. X-rays pass through many substances that block visible light and are therefore used to image bones and organs.

9. Ultraviolet light contains enough energy to damage biological molecules, and excessive exposure increases the risk of skin cancer and cataracts.

11. Microwaves can only heat things containing water; therefore the food, which contains water, becomes hot, but the plate does not.

13. The Bohr model is a representation for the atom in which electrons travel around the nucleus in circular orbits with a fixed energy at specific, fixed distances from the nucleus.

15. The Bohr orbit describes the path of an electron as an orbit or trajectory (a specified path). A quantum-mechanical orbital describes the path of an electron using a probability map.

17. The e^- has wave–particle duality, which means the path of an electron is not predictable. The motion of a baseball is predictable. A probability map shows a statistical, reproducible pattern of where the electron is located.

19. The subshells are s (1 orbital, which contains a maximum of 2 electrons); p (3 orbitals, which contain a maximum of 6 electrons); d (5 orbitals, which contain a maximum of 10 electrons); and f (7 orbitals, which contain a maximum of 14 electrons).

21. The Pauli exclusion principle states that separate orbitals may hold no more than 2 electrons, and when 2 electrons are present in a single orbital, they must have opposite spins. When writing electron configurations, the principle means that no box can have more than 2 arrows, and the arrows will point in opposite directions.

23. [Ne] represents $1s^2 2s^2 2p^6$.
[Kr] represents $1s^2 2s^2 2p^6 3s^2 3p^6 4s^2 3d^{10} 4p^6$.

25.

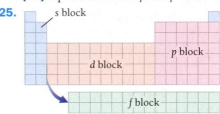

27. Group 1 elements form 1+ ions because they lose one valence electron in the outer s shell to obtain a noble gas configuration. Group 7 elements form 1− ions because they gain an electron to fill their outer p orbital to obtain a noble gas configuration.

PROBLEMS

29. a. 1.0 ns   **b.** 13.21 ms   **c.** 4 h 10 min

31. infrared

33. radiowaves < microwaves < infrared < ultraviolet

35. gamma, ultraviolet, or X-rays

37. a. radio waves < infrared < X-rays
b. radio waves < infrared < X-rays
c. X-rays < infrared < radio waves

39. energies, distances

41. $n = 6 \longrightarrow n = 2$: 410 nm
$n = 5 \longrightarrow n = 2$: 434 nm

43.

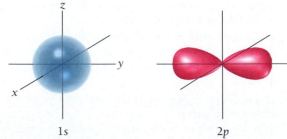

The 2s and 3p orbitals are bigger than the 1s and 2p orbitals.

45. Electron in the $2s$ orbital

47. $2p \longrightarrow 1s$

49. a. $1s^2 2s^2 2p^6 3s^2 3p^6 4s^2 3d^{10} 4p^6 5s^2$
b. $1s^2 2s^2 2p^6 3s^2 3p^6 4s^2 3d^{10} 4p^2$
c. $1s^2 2s^1$
d. $1s^2 2s^2 2p^6 3s^2 3p^6 4s^2 3d^{10} 4p^6$

51. a. He

    $1s$ $2s$

b. B

    $1s$ $2s$ $2p$

c. Li

    $1s$ $2s$ $2p$

d. N

    $1s$ $2s$ $2p$

53. a. $[Ar]4s^2 3d^{10} 4p^1$ b. $[Ar]4s^2 3d^{10} 4p^3$
c. $[Kr]5s^1$ d. $[Kr]5s^2 4d^{10} 5p^2$

55. a. $[Ar]4s^2 3d^{10}$ b. $[Ar]4s^1 3d^{10}$
c. $[Kr]5s^2 4d^2$ d. $[Ar]4s^2 3d^6$

57. Valence electrons are underlined.
a. $1s^2 2s^2 2p^6 3s^2 3p^6 \underline{4s^2} 3d^{10} \underline{4p^6}$
b. $1s^2 2s^2 2p^6 3s^2 3p^6 \underline{4s^2} 3d^{10} \underline{4p^2}$
c. $1s^2 2s^2 2p^6 \underline{3s^2} \underline{3p^5}$
d. $1s^2 2s^2 2p^6 3s^2 3p^6 4s^2 3d^{10} 4p^6 \underline{5s^2}$

59. a. Br

, 1 unpaired electron

    $4s$ $4p$

b. Kr

, 0 unpaired electron

    $4s$ $4p$

c. Na

, 1 unpaired electron

    $3s$

d. In

, 1 unpaired electron

    $5s$ $5p$

61. a. 6   b. 6   c. 7   d. 1

63. a. ns^1 b. ns^2 c. $ns^2 np^3$ d. $ns^2 np^5$

65. a. $[Ne]3s^2 3p^1$ b. $[He]2s^2$
c. $[Kr]5s^2 4d^{10} 5p^1$ d. $[Kr]5s^2 4d^2$

67. a. $[Kr]5s^2$ b. $[Kr]5s^2 4d^1$
c. $[Ar]4s^2 3d^2$ d. $[Kr]5s^2 4d^{10} 5p^4$

69. a. 2   b. 3   c. 5   d. 6

71. Period 1 has two elements. Period 2 has eight elements. The number of subshells is equal to the principal quantum number. For Period 1, $n = 1$ and the s subshell contains only two elements. For Period 2, $n = 2$ and contains s and p subshells that have a total of 8 elements.

73. a. Al   b. S
c. Ar   d. Mg

75. a. Cl   b. Ga
c. Fe   d. Rb

77. a. As   b. Br
c. cannot tell   d. S

79. Pb < Sn < Te < S < Cl

81. a. In   b. Si
c. Pb   d. C

83. F < S < Si < Ge < Ca < Rb

85. a. Sr   b. Bi
c. cannot tell   d. As

87. S < Se < Sb < In < Ba < Fr

89. 18 e⁻

91. Alkaline earth metals have the general electron configuration of ns^2. If they lose their two s electrons, they will obtain the stable electron configuration of a noble gas. This loss of electrons will give the metal a 2+ charge.

93. a. $1s^2 2s^2 2p^6 3s^2 3p^6$ b. $1s^2 2s^2 2p^6 3s^2 3p^6$
c. $1s^2 2s^2 2p^6 3s^2 3p^6$ d. $1s^2 2s^2 2p^6 3s^2 3p^6 4s^2 3d^{10} 4p^6$
They all have noble gas electron configurations.

95. Metals tend to form positive ions because they tend to lose electrons. Elements on the left side of the periodic table have only a few extra electrons, which they will lose to gain a noble gas configuration. Metalloids tend to be elements with 3 to 5 valence electrons; they could lose or gain electrons to obtain a noble gas configuration. Nonmetals tend to gain electrons to fill their almost full valence shell, so they tend to form negative ions and are on the right side of the table.

97. a. Can only have 2 in the s shell and 6 in the p shell: $1s^2 2s^2 2p^6 3s^2 3p^3$.
b. There is no $2d$ subshell: $1s^2 2s^2 2p^6 3s^2 3p^2$.
c. There is no $1p$ subshell: $1s^2 2s^2 2p^3$.
d. Can only have 6 in the p shell: $1s^2 2s^2 2p^6 3s^2 3p^3$.

99. Bromine is highly reactive because it reacts quickly to gain an electron and obtain a stable valence shell. Krypton is a noble gas because it already has a stable valence shell.

101. K

103. 660 nm

105. 8 min, 19 sec

107. a. 1.5×10^{-34} m
b. 1.88×10^{-10} m
Electrons have wave–particle duality, whereas golf balls do not.

109. The ionization energy dips at column 3A because removing an electron from one of those atoms leaves the atom with a fairly stable, filled s-orbital as its valence shell. For the group 6A elements, special stability occurs when those elements lose an electron and achieve a half-filled p-orbital as their valence shell.

111. Ultraviolet light is the only one of these three types of light that contains enough energy to break chemical bonds in biological molecules.

117. c. Al $[Ne]3s^2 3p^1$ P $[Ne]3s^2 3p^3$ S $[Ne]3s^2 3p^4$
The Al exception is due to the transition from the s orbitals to the p orbital.

The S exception is due to the pairing of 2 electrons in one orbital for S (compared to P in which all of the p orbitals are singly occupied).

    e. Si [Ne]$3s^2 3p^2$

       P [Ne]$3s^2 3p^3$

The electron affinity of Si is more exothermic because in Si, the incoming electron can singly occupy a p orbital. However, in P, the incoming electron must pair with another electron in a p orbital.

CHAPTER 10

QUESTIONS

1. Bonding theories predict how atoms bond together to form molecules, and they also predict what combinations of atoms form molecules and what combinations do not. Likewise, bonding theories explain the shapes of molecules, which in turn determine many of their physical and chemical properties.

3. Atoms with 8 valence electrons are particularly stable and are said to have an octet. Atoms such as hydrogen, helium, lithium, and beryllium are exceptions to the octet rule as they achieve stability when their outermost shell contains two electrons—a duet. A chemical bond is the sharing or transfer of electrons to attain stable electron configurations among the bonding atoms.

5. The Lewis structure for potassium has 1 valence electron, whereas the Lewis structure for monatomic chlorine has 7 valence electrons. From these structures we can determine that if potassium gives up its one valence electron to chlorine, K^+ and Cl^- are formed; therefore the formula must be KCl.

7. Double and triple bonds are shorter and stronger than single bonds.

9. You determine the number of electrons that go into the Lewis structure of a molecule by summing the valence electrons of each atom in the molecule.

11. The octet rule is not sophisticated enough to be correct every time. For example, some molecules that exist in nature have an odd number of valence electrons and thus will not have octets on all their constituent atoms. Some elements tend to form compounds in nature in which they have more (sulfur) or less (boron) than 8 valence electrons.

13. VSEPR theory predicts the shape of molecules using the idea that electron groups repel each other.

15. **a.** 180°   **b.** 120°   **c.** 109.5°

17. Electronegativity is the ability of an element to attract electrons within a covalent bond.

19. A polar covalent bond is a covalent bond that has a dipole moment.

21. If a polar liquid and a nonpolar liquid are mixed, they will separate into distinct regions because the polar molecules will be attracted to one another and will exclude the nonpolar molecules.

PROBLEMS

23. a. $1s^2 2s^2 2p^3$, ·N̈:   **b.** $1s^2 2s^2 2p^2$, ·Ç·

    c. $1s^2 2s^2 2p^6 3s^2 3p^5$, :Çl·   **d.** $1s^2 2s^2 2p^6 3s^2 3p^6$, :Är̈:

25. a. :Ï: **b.** ·S̈:

    c. ·G̈e·   **d.** ·Ca·

27. :Ẍ: Halogens tend to gain 1 electron in a chemical reaction.

29. M: Alkaline earth metals tend to lose 2 electrons in a chemical reaction.

31. a. Al^{3+} **b.** Mg^{2+}

    c. $[:S̈e:]^{2-}$ **d.** $[:N̈:]^{3-}$

33. a. Kr   **b.** Ne

    c. Kr   **d.** Xe

35. a. covalent   **b.** ionic

    c. covalent   **d.** ionic

37. a. $Na^+[:F̈:]^-$ **b.** $Ca^{2+}[:Ö:]^{2-}$

    c. $[:B̈r:]^- Sr^{2+}[:B̈r:]^-$ **d.** $K^+[:Ö:]^{2-} K^+$

39. a. CaS   **b.** $MgBr_2$

    c. CsI   **d.** Ca_3N_2

41. a. $[:F̈:]^- Mg^{2+}[:F̈:]^-$ **b.** $Mg^{2+}[:Ö:]^{2-}$

    c. $Mg^{2+}[:N̈:]^{3-} Mg^{2+}[:N̈:]^{3-} Mg^{2+}$

43. a. $Cs^+[:Çl:]^-$ **b.** $Ba^{2+}[:Ö:]^{2-}$

    c. $[:Ï:]^- Ca^{2+}[:Ï:]^-$

45. a. Hydrogen exists as a diatomic molecule because two hydrogen molecules achieve a stable duet when they share their electrons and form a single covalent bond.

    b. Iodine achieves a stable octet when two atoms share electrons and form a single bond.

    c. Nitrogen achieves a stable octet when two atoms share electrons and form a triple bond.

    d. Oxygen achieves a stable octet when two atoms share electrons and form a double bond.

47. a. H—P̈—H with H below P

    b. :Çl—S̈—Çl:

    c. :F̈—F̈:   **d.** H—Ï:

49. a. Ö=Ö   **b.** :C≡O:

    c. H—Ö—N̈=Ö   **d.** :Ö=S̈—Ö:

51. a. H—C≡C—H   **b.** H—C=C—H (with H, H below)

    c. H—N̈=N̈—H   **d.** H—N̈—N̈—H (with H, H below)

53. a. :N≡N: **b.** S̈=Si=S̈

    c. H—Ö—H   **d.** :Ï—N̈—Ï: with :Ï: below N

55. a. Ö=S̈e—Ö: ⟷ :Ö—S̈e=Ö

    b. $\left[\ddot{O}=C-\ddot{O}: \atop :\ddot{O}: \right]^{2-} ⟷ \left[:\ddot{O}-C=\ddot{O} \atop :\ddot{O}: \right]^{2-} ⟷ \left[:\ddot{O}-C-\ddot{O}: \atop :\ddot{O}: \right]^{2-}$

c. $[:\ddot{C}l-\ddot{O}:]^-$ d. $[:\ddot{O}-\ddot{C}l-\ddot{O}:]^-$

57. a. $\left[:\ddot{O}-\overset{\overset{\displaystyle :\ddot{O}:}{|}}{\underset{\underset{\displaystyle :\ddot{O}:}{|}}{P}}-\ddot{O}:\right]^{3-}$ **b.** $[:C\equiv N:]^-$

c. $[:\ddot{O}=\ddot{N}-\ddot{O}:]^- \longleftrightarrow [:\ddot{O}-\ddot{N}=\ddot{O}:]^-$

d. $\left[:\ddot{O}-\overset{\overset{\displaystyle :\ddot{O}:}{|}}{\underset{\underset{\displaystyle :\ddot{O}:}{|}}{S}}-\ddot{O}:\right]^{2-}$

59. a. $:\ddot{C}l-\overset{\overset{\displaystyle }{|}}{\underset{\underset{\displaystyle :\ddot{C}l:}{|}}{B}}-\ddot{C}l:$

b. $\ddot{O}=\dot{N}-\ddot{O}: \longleftrightarrow :\ddot{O}-\dot{N}=\ddot{O}$

c. $H-\overset{\overset{\displaystyle H}{|}}{\underset{\underset{\displaystyle H}{|}}{B}}-H$

61. a. 4   **b.** 4
   c. 2   **d.** 4
63. a. 2 bonding groups, 2 lone pairs
   b. 3 bonding groups, 1 lone pair
   c. 2 bonding groups, 0 lone pair
   d. 4 bonding groups, 0 lone pair
65. a. tetrahedral   **b.** trigonal planar
   c. linear   **d.** trigonal planar
67. a. 109.5°   **b.** 120°
   c. 180°   **d.** 120°
69. a. linear, linear
   b. trigonal planar, bent
   c. tetrahedral, bent
   d. tetrahedral, trigonal pyramidal
71. a. 180°   **b.** 120°
   c. 109.5°   **d.** 109.5°
73. a. linear, linear
   b. trigonal planar, bent (about both nitrogen atoms)
   c. tetrahedral, trigonal pyramidal (about both nitrogen atoms)
75. a. trigonal planar   **b.** bent
   c. trigonal planar   **d.** tetrahedral
77. a. 1.2   **b.** 1.8   **c.** 2.8
79. Cl > Si > Ga > Ca > Rb
81. a. polar covalent   **b.** ionic
   c. pure covalent   **d.** polar covalent
83. $H_2 < ICl < HBr < CO$
85. a. polar   **b.** nonpolar
   c. nonpolar   **d.** polar
87. a. $(+):C\equiv O:(-)$ **b.** nonpolar
   c. nonpolar   **d.** $(+)H-\ddot{B}r:(-)$
89. a. nonpolar   **b.** polar
   c. nonpolar   **d.** polar

91. a. nonpolar   **b.** polar
   c. nonpolar   **d.** polar
93. a. $1s^22s^22p^63s^23p^6\underline{4s^2}$, Ca: (underlined electrons are the ones included)
   b. $1s^22s^22p^33s^23p^6\underline{4s^2}3d^{10}\underline{4p^1}$, $\dot{G}a:$
   c. $[Ar]\underline{4s^2}3d^{10}\underline{4p^3}$, $\cdot\dot{A}s:$
   d. $[Kr]\underline{5s^2}4d^{10}\underline{5p^5}$, $:\ddot{I}:$
95. a. ionic, $K^+[:\ddot{S}:]^{2-}K^+$
   b. covalent, $H-\overset{\overset{\displaystyle \ddot{O}:}{\|}}{C}-\ddot{F}:$
   c. ionic, $Mg^{2+}[:\ddot{S}e:]^{2-}$
   d. covalent, $:\ddot{B}r-\overset{\overset{\displaystyle :\ddot{B}r:}{|}}{P}-\ddot{B}r:$
97. $:\ddot{C}l-\overset{\overset{\displaystyle \ddot{O}\cdot}{\|}}{C}-\ddot{C}l:$ polar,

$\overset{\overset{\displaystyle O}{\|}}{\underset{\underset{\displaystyle Cl \quad Cl}{}}{C}}$

99. $H-\overset{\overset{\displaystyle H}{|}}{\underset{\underset{\displaystyle H}{|}}{C}}-\overset{\overset{\displaystyle \ddot{O}:}{\|}}{C}-\ddot{O}H$ $H\underset{\underset{\displaystyle H}{}}{\overset{\overset{\displaystyle H}{}}{C}}-\overset{\overset{\displaystyle O}{\diagup\!\!\backslash}}{C}\underset{OH}{}$

101. $H:\ddot{C}l: + Na^+[:\ddot{O}:H]^- \longrightarrow H:\ddot{O}:H + Na^+[:\ddot{C}l:]^-$
103. $K\cdot$, $:\ddot{C}l-\ddot{C}l:$, $K^+[:\ddot{C}l:]^-$, Cl reduced, K oxidized
105. a. $K^+[:\ddot{O}:H]^-$

b. $K^+\left[:\ddot{O}=\overset{:\ddot{O}:}{N}-\ddot{O}:\right]^- \longleftrightarrow K^+\left[:\ddot{O}-\overset{\overset{\displaystyle \ddot{O}}{\|}}{N}-\ddot{O}:\right]^- \longleftrightarrow K^+\left[:\ddot{O}-\overset{:\ddot{O}:}{N}=\ddot{O}:\right]^-$

c. $Li^+[:\ddot{I}:\ddot{O}:]^-$

d. $Ba^{2+}\left[\overset{\overset{\displaystyle O}{\|}}{\underset{\underset{\displaystyle :\ddot{O}: \quad :\ddot{O}:}{}}{C}}\right]^{2-} \longleftrightarrow Ba^{2+}\left[\overset{:\ddot{O}:}{\underset{:\ddot{O}: \quad O}{C}}\right]^{2-} \longleftrightarrow Ba^{2+}\left[\overset{:\ddot{O}:}{\underset{O \quad :\ddot{O}:}{C}}\right]^{2-}$

107. a. $\overset{\overset{\displaystyle :\ddot{F}: \quad :\ddot{F}: \quad :\ddot{F}:}{}}{\underset{\underset{\displaystyle :\ddot{F}: \quad :\ddot{F}:}{}}{P}}$ **b.** $\overset{\overset{\displaystyle :\ddot{F}: \quad :\ddot{F}:}{}}{\underset{\underset{\displaystyle :\ddot{F}: \quad :\ddot{F}:}{}}{S}}$

c. $\overset{\overset{\displaystyle :\ddot{F}: \quad :\ddot{F}:}{}}{\underset{\underset{\displaystyle :\ddot{F}: \quad :\ddot{F}:}{}}{Se}}$

109. CH_2O_2 or $H-\overset{\overset{\displaystyle :\ddot{O}:}{\|}}{C}-\ddot{O}-H$

111. $H-\ddot{O}-\dot{\ddot{O}}\cdot$ HOO is not stable because one oxygen atom does not have an octet. The geometry for HOO is *bent*.

113. a. $\left[\ddot{\text{O}}-\ddot{\text{O}}\cdot\right]^{-}$

b. $\left[\ddot{\ddot{\text{O}}}\cdot\right]^{-}$

c. $\cdot\ddot{\text{O}}-\text{H}$

d. $\text{H}-\overset{\overset{\displaystyle H}{|}}{\underset{\underset{\displaystyle H}{|}}{\text{C}}}-\ddot{\text{O}}-\ddot{\text{O}}\cdot$

115. a. The structure has 2 bonding electron pairs and 2 lone pairs. The Lewis structure is analogous to that of water, and the molecular geometry is bent.

b. Correct

c. The structure has 3 bonding electron pairs and 1 lone pair. The Lewis structure is analogous to that of NH_3, and the geometry is trigonal pyramidal.

d. Correct

$$\ddot{\text{O}}-\dot{\text{N}}=\ddot{\text{O}}$$

119. a. $\left[\ddot{\text{O}}=\text{N}=\ddot{\text{O}}\right]^{+}$

$\left[\ddot{\text{O}}-\ddot{\text{N}}=\ddot{\text{O}}\right]^{-}$

CHAPTER 11

QUESTIONS

1. Pressure is the push (or force) exerted per unit area by gaseous molecules as they collide with the surfaces around them.

3. The kinetic molecular theory makes four main assumptions. The first is that the gas is a collection of molecules in constant motion. Second, there is no attraction or repulsion between the particles, and collisions are perfectly elastic. Third, there is a lot of space in between the particles relative to the particle size. Lastly, the speed of the particles increases with temperature.

5. The pain we experience in our ears during a change in altitude is due to a pressure difference between the cavities inside of our ears and the surrounding air.

7. Boyle's law states that the volume of a gas and its pressure are inversely proportional. This relationship can be explained by the kinetic molecular theory. If the volume of a sample is decreased, the same number of particles are crowded into a smaller space, causing more collisions with the walls of the container. This causes the pressure to increase.

9. When an individual is more than a couple of meters underwater, the air pressure in the lungs is greater than the air pressure at the water's surface. If a snorkel were used, it would move the air from the lungs to the surface, making it very difficult to breathe.

11. Increasing the temperature of the air in the balloon causes it to expand. As the volume of the air increases, the density decreases, allowing it to float in the cooler, more dense air surrounding it.

13. Avogadro's law states that the volume of a gas is directly proportional to the amount of gas in moles. Kinetic molecular theory predicts that if the number of gas particles increases at a constant pressure and temperature, the volume increases.

15. The ideal gas law is most accurate when the volume of gas particles is small compared to the space between them. It is also accurate when the forces between particles are not important. The ideal gas law breaks down at high pressures and low temperatures. This breakdown occurs because the gases are no longer acting according to the kinetic molecular theory.

17. Dalton's law states that the sum of the partial pressures in a gas mixture must equal the total pressure. $P_{\text{tot}} = P_A + P_B + P_C + \ldots$

19. Deep-sea divers breathe helium with oxygen because helium, unlike nitrogen, does not have physiological effects under high-pressure conditions. The oxygen concentration in the mixture is low to avoid oxygen toxicity.

21. Vapor pressure is the partial pressure of a gas above its liquid. Partial pressure increases with increasing temperature.

PROBLEMS

23. a. 1.680 atm   **b.** 2.35 atm

c. 8.64 atm   **d.** 0.599 atm

25. a. 1.7×10^3 torr   **b.** 36 mmHg

c. 1.28×10^3 mmHg   **d.** 834.2 torr

27.

Pascals	Atmospheres	mmHg	Torr	PSI
882	0.00871	6.62	6.62	0.128
5.65×10^4	0.558	424	424	8.20
1.71×10^5	1.69	1.28×10^3	1.28×10^3	24.8
1.02×10^5	1.01	764	764	14.8
3.32×10^4	0.328	249	249	4.82

29. a. 0.832 atm   **b.** 632 mmHg

c. 12.2 psi   **d.** 8.43×10^4 Pa

31. a. 809.0 mmHg   **b.** 1.065 atm

c. 809.0 torr   **d.** 107.9 kPa

33. 518 mmHg

35. 1.8 L

37.

P_1	V_1	P_2	V_2
755 mmHg	2.85 L	885 mmHg	2.43 L
9.35 atm	1.33 L	4.32 atm	2.88 L
192 mmHg	382 mL	152 mmHg	482 mL
211 atm	226 mL	3.82 atm	125 mL

39. 4.0 L

41. 58.9 mL

43.

V_1	T_1	V_2	T_2
1.08 L	25.4 °C	1.33 L	94.5 °C
58.9 mL	77 K	228 mL	298 K
115 cm³	12.5 °C	119 cm³	22.4 °C
232 L	18.5 °C	294 L	96.2 °C

45. 6.8 L

47. 4.33 L

49.

V_1	n_1	V_2	n_2
38.5 mL	1.55×10^{-3}	49.4 mL	1.99×10^{-3}
8.03 L	1.37	26.8 L	4.57
11.2 L	0.628	15.7 L	0.881
422 mL	0.0109	671 mL	0.0174

51. 1.71×10^3 mmHg

53. 0.76 L

55. 877 mmHg

57.

P_1	V_1	T_1	P_2	V_2	T_2
121 atm	1.58 L	12.2 °C	1.54 torr	1.33 L	32.3 °C
721 torr	141 mL	135 K	801 torr	152 mL	162 K
5.51 atm	0.879 L	22.1 °C	4.87 atm	1.05 L	38.3 °C

59. 5.11 L

61. 2.1 mol

63. 1.42 mol

65.

P	V	n	T
1.05 atm	1.19 L	0.112 mol	136 K
112 torr	40.8 L	0.241 mol	304 K
1.50 atm	28.5 mL	1.74×10^{-3} mol	25.4 °C
0.559 atm	0.439 L	0.0117 mol	255 K

67. 0.23 mol

69. 44.0 g/mol

71. 4.00 g/mol

73. 571 torr

75. 10.7 atm

77. 7.00×10^2 mmHg

79. 0.87 atm N_2; 0.25 atm O_2

81. 0.34 atm

83. a. 504 L **b.** 81 L
c. 49 L **d.** 6.0×10^2 L

85. a. 59.1 L **b.** 30.0 L
c. 72.1 L **d.** 0.125 L

87. a. 0.350 g **b.** 0.221 g
c. 8.15 g

89. 28 L

91. 33 L H_2; 16 L CO

93. 8.82 L

95. 12.6 g

97. 0.5611 g

99. $V = \dfrac{nRT}{P}$

$$= \dfrac{1.00 \text{ mol} \left(0.0821\dfrac{\text{L} \cdot \text{atm}}{\text{mol} \cdot \text{K}}\right)(273 \text{ K})}{1.00 \text{ atm}} = 22.4 \text{ L}$$

101. 27.8 g/mol

103. C_4H_{10}

105. 0.828 g

107. 0.128 g

109. 0.935 L

111. HCl + $NaHCO_3 \longrightarrow CO_2 + H_2O + NaCl$; 1.11×10^{-3} mol

113. a. SO_2, 0.0127 mol **b.** 65.6%

115. a. NO_2, 24.0 g **b.** 61.6%

117. 11.7 L

119. 356 torr

121. 0.15 atm

123. c. From the ideal gas law, we see that pressure is directly proportional to the number of moles of gas per unit volume (n/V). The gas in (c) contains the greatest concentration of particles and thus has the highest pressure.

125. 22.8 g

127. V_2 = 0.76 L, actual volume is 0.61 L. Difference is because air contains about 20% oxygen, which condenses into a liquid at −183 °C.

132. a. No **c.** The mole ratio is 1:1.

CHAPTER 12
QUESTIONS

1. Intermolecular forces are attractive forces that occur between molecules. Intermolecular forces are what living organisms depend on for many physiological processes. Intermolecular forces are also responsible for the existence of liquids and solids.

3. The magnitude of intermolecular forces relative to the amount of thermal energy in the sample determines the state of the matter.

5. Properties of solids:
   a. Solids have high densities in comparison to gases.
   b. Solids have a definite shape.
   c. Solids have a definite volume.
   d. Solids may be crystalline or amorphous.

7. Surface tension is the tendency of liquids to minimize their surface area. Molecules at the surface have few neighbors to interact with via intermolecular forces.

9. Evaporation is a physical change in which a substance is converted from its liquid form to its gaseous form. Condensation is a physical change in which a substance is converted from its gaseous form to its liquid form.

11. Evaporation below the boiling point occurs because molecules on the surface of the liquid experience fewer attractions to the neighboring molecules and can therefore break away. At the boiling point, evaporation occurs faster because more of the molecules have sufficient thermal energy to break away (including internal molecules).

13. Acetone has weaker intermolecular forces than water. Acetone is more volatile than water.

15. Vapor pressure is the partial pressure of a gas in dynamic equilibrium with its liquid. It increases with increasing temperature, and it also increases with decreasing strength of intermolecular forces.

17. A steam burn is worse than a water burn at the same temperature (100 °C) because when the steam condenses on the skin, it releases large amounts of additional heat.

19. As the first molecules freeze, they release heat, making it harder for other molecules to freeze without the aid of a refrigeration mechanism, which would draw heat out.

21. The melting of ice is endothermic. ΔH for melting is positive (+), whereas ΔH for freezing is negative (−).

23. Dispersion forces are the default intermolecular force present in all molecules and atoms. Dispersion forces are caused by fluctuations in the electron distribution within molecules or atoms. Dispersion forces are the weakest type of intermolecular force and increase with increasing molar mass.

25. Hydrogen bonding is an intermolecular force and is sort of a super dipole–dipole force. Hydrogen bonding occurs in compounds containing hydrogen atoms bonded directly to fluorine, oxygen, or nitrogen.

27. Dispersion force, dipole–dipole force, hydrogen bond, ion–dipole

29. Molecular solids as a whole tend to have low to moderately low melting points relative to other types of solids; however, strong molecular forces can increase their melting points relative to each other.

31. Ionic solids tend to have much higher melting points relative to the melting points of other types of solids.

33. Water is unique for a couple of reasons. Water has a low molar mass, yet it is still liquid at room temperature and has a relatively high boiling point. Unlike other substances, which contract upon freezing, water expands upon freezing.

PROBLEMS

35. The 55 mL of water in a dish with a diameter of 12 cm will evaporate more quickly because it has a larger surface area.

37. Acetone feels cooler while evaporating from one's hand, for it is more volatile than water and evaporates much faster.

39. The ice's temperature will increase from $-5\,°C$ to $0\,°C$, where it will then stay constant while the ice completely melts. After the melting process is complete, the water will continue to rise steadily in temperature until it reaches room temperature ($25\,°C$).

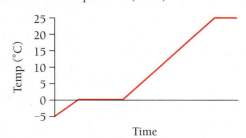

41. Allowing 0.50 g of $100\,°C$ steam to condense on your hand would cause a more severe burn than spilling 0.50 g of $100\,°C$ water on your hand. When the steam condenses on the skin, it releases large amounts of additional heat.

43. The $-8\,°C$ ice chest will cause the water in the watery bag of ice to freeze, which is an exothermic process. The freezing process of the water will release heat, and the temperature of the ice chest will increase.

45. The ice chest filled with ice at $0\,°C$ will be colder after a couple of hours than the ice chest filled with water at $0\,°C$. This is because the ice can absorb additional heat as it melts.

47. Denver is a mile above sea level and thus has a lower air pressure. Due to this low pressure, the point at which the vapor pressure of water equals the external pressure will occur at a lower temperature.

49. 76.3 kJ

51. 6.8 kJ

53. 10.4 kJ

55. 3.70×10^2 g

57. 12.5 kJ

59. 11.4 kJ

61. 6.53 kJ

63. a. dispersion
b. dispersion
c. dispersion, dipole–dipole
d. dispersion, dipole–dipole, hydrogen bonding

65. a. dispersion, dipole–dipole
b. dispersion, dipole–dipole, hydrogen bonding
c. dispersion
d. dispersion

67. Ion–dipole forces (between the ions and the water molecules); and hydrogen bonding (between the water molecules themselves)

69. d, because it has the highest molecular weight

71. CH_3OH, due to strong hydrogen bonding

73. NH_3, because it has the ability to form hydrogen bonds

75. These two substances are not miscible, for H_2O is polar and $CH_3CH_2CH_2CH_2CH_3$ is nonpolar.

77. a. no   b. yes
c. yes

79. a. atomic   b. molecular
c. ionic   d. atomic

81. a. molecular   b. ionic
c. molecular   d. molecular

83. c. LiCl(s); it is an ionic solid and possesses ionic bonds, resulting in a higher melting point.

85. a. Ti(s); Ti is a covalent atomic solid, and Ne is a nonbonding atomic solid.
b. $H_2O(s)$; while both are molecular solids, water has strong hydrogen bonding.
c. Xe(s); both are nonbonding atomic solids, but Xe has a higher molar mass.
d. NaCl(s); NaCl(s) is an ionic solid, and $CH_4(s)$ is a molecular solid.

87. $Ne < SO_2 < NH_3 < H_2O < NaF$

89. a. 2.6×10^4 J   b. 26 kJ
c. 6.2×10^3 cal   d. 6.2 Cal

91. $27\,°C$

93. 1.1×10^2 g

95. 57.5 kJ

97. a. H—S̈e—H Bent: Dispersion and dipole–dipole
b. Ö=S̈—Ö: Bent: Dispersion and dipole–dipole
c. H—C—C̈l: Tetrahedral: Dispersion and dipole–dipole (with :C̈l: above and :C̈l: below)
d. :Ö=C=Ö: Linear: Dispersion

99. $Na^+ \left[:\ddot{F}:\right]^-$; $Mg^{2+}\left[:\ddot{O}:\right]^{2-}$; MgO has a higher melting point because the magnitude of the charges on the ions is greater.

101. As the molecular weight increases from Cl to I, the greater the London dispersion forces present, which will increase the boiling point as observed. However, HF is the only compound listed that has the ability to form hydrogen bonds, which explains the anomaly in the trend.

103. 23.6 °C

105. The molecule in the interior has the most neighbors. The molecule on the surface is more likely to evaporate. In three dimensions, the molecules that make up the surface area have the fewest neighbors and are more likely to evaporate than the molecules in the interior.

107. **a.** Yes, if the focus is on the melting of icebergs. No, the melting of an ice cube in a cup of water will *not* raise the level of the liquid in the cup; when the ice cube melts, the volume of the water created will be less than the volume of the initial ice cube.

b. Yes, for the ice sheets that sit on the continent of Antarctica are above sea level, and if some melted, the water created would be added to the ocean without decreasing the amount of ice below the ocean's surface.

113. **a.** Pentane and hexane

c. Dispersion forces; the boiling point increases as the molar mass increases

e. 100–105 °C

CHAPTER 13

QUESTIONS

1. A solution is a homogeneous mixture of two or more substances. Some examples are air, seawater, soda water, and brass.

3. In a solution, the solvent is the majority component of the mixture, and the solute is the minority component. For example, in a seawater solution, the water is the solvent, and the salt content is the solute.

5. Solubility is the amount of the compound, usually in grams, that will dissolve in a specified amount of solvent.

7. In solutions with solids, soluble ionic solids form strong electrolyte solutions, while soluble molecular solids form nonelectrolyte solutions. Strong electrolyte solutions are solutions containing solutes that dissociate into ions, for example, $BaCl_2$ and NaOH.

9. Recrystallization is a common way to purify a solid. In recrystallization, enough solid is put into high-temperature water until a saturated solution is created. Then the solution cools slowly, and crystals result from the solution. The crystalline structure tends to reject impurities, resulting in a purer solid.

11. The bubbles formed on the bottom of a pot of heated water (before boiling) are dissolved air coming out of the solution. These gases come out of solution because the solubility of the dissolved nitrogen and oxygen decreases as the temperature of the water rises.

13. The solubility of gases increases with increasing pressure. When a soda can is opened, the pressure is lowered, decreasing the solubility of carbon dioxide. This causes bubbles of carbon dioxide to come out of the solution.

15. Mass percent is the number of grams of solute per 100 grams of solution. Molarity is defined as the number of moles of solute per liter of solution.

17. The boiling point of a solution containing a nonvolatile solute is higher than the boiling point of the pure solvent. The melting point of the solution, however, is lower.

19. Molality is a common unit of concentration of a solution expressed as number of moles of solute per kilogram of solvent.

21. Water tends to move from lower concentrations to higher concentrations, and when the saltwater is being passed through the human body, the salt content draws the water out of the body, causing dehydration.

PROBLEMS

23. **c** and **d** are solutions

25. **a.** solute: salt, solvent: water

b. solute: sugar, solvent: water

c. solute: CO_2, solvent: water

27. **a.** hexane **b.** water

c. ethyl ether **d.** water

29. ions, strong electrolyte solution

31. saturated

33. recrystallization

35. **a.** no **b.** yes

c. yes

37. At room temperature water contains some dissolved oxygen gas; however, the boiling of the water will remove dissolved gases.

39. Under higher pressure, the gas (nitrogen) will be more easily dissolved in the blood. To reverse this process, the diver should ascend to relieve the pressure.

41. **a.** 7.64% **b.** 3.50%

c. 4.64%

43. 12%

45.

Mass Solute (g)	Mass Solvent (g)	Mass Solution (g)	Mass%
15.5	238.1	253.6	6.11%
22.8	167.2	190.0	12.0%
28.8	183.3	212.1	13.6%
56.9	315.2	372.1	15.3%

47. 8.9 g NaCl

49. **a.** 1.8 g **b.** 10.5 mg

c. 0.46 kg

51. **a.** 2.6 kg **b.** 1.0×10^2 g

c. 18 g

53. 1.6×10^2 g

55. 1.3×10^3 g

57. 11 L

59. **a.** 0.194 M **b.** 0.234 M

c. 0.41 M

61. **a.** 0.149 M **b.** 0.461 M

c. 3.00×10^{-2} M

63. 0.57 M

65. a. 1.8 mol **b.** 0.38 mol
  c. 0.238 mol

67. a. 0.59 L **b.** 0.083 L
  c. 0.15 L

69.

Solute	Mass Solute	Mol Solute	Volume Solution	Molarity
KNO_3	22.5 g	0.223	125 mL	1.78 M
$NaHCO_3$	2.10 g	0.0250	250.0 mL	0.100 M
$C_{12}H_{22}O_{11}$	55.38 g	0.162	1.08 L	0.150 M

71. 2.7 g

73. 19 g

75. 8.8 L

77. a. 0.15 M **b.** 0.30 M
  c. 0.45 M

79. a. 0.24 M Na^+, 0.12 M SO_4^{2-}
  b. 0.50 M K^+, 0.25 M CO_3^{2-}
  c. 0.11 M Rb^+, 0.11 M Br^-

81. 0.29 M

83. Dilute 0.045 L of the stock solution to 2.5 L.

85. 6.0×10^2 mL

87. 17.7 mL

89. a. 0.025 L **b.** 0.020 L
  c. 1.03 L

91. 4.45 mL

93. 0.373 M

95. 1.2 L

97. a. 1.0 m **b.** 3.92 m
  c. 0.52 m

99. 1.49 m

101. a. $-1.6\,°C$ **b.** $-2.70\,°C$
  c. $-8.9\,°C$ **d.** $-4.37\,°C$

103. a. $100.060\,°C$ **b.** $100.993\,°C$
  c. $101.99\,°C$ **d.** $101.11\,°C$

105. $-1.27\,°C$, $100.348\,°C$

107. 2.28 M, 12.3%

109. 0.43 L

111. 319 mL

113. 1.03 M

115. 0.17 L

117. 9.0 mL

119. 8.0×10^1 mL

121. $-3.60\,°C$, $100.992\,°C$

123. $-1.86\,°C$, $100.508\,°C$

125. 1.8×10^2 g/mol

127. $101.8\,°C$

129. 39.8 g glucose, 85.2 g sucrose

131. a. Water will flow from left to right.
  b. Water will flow from right to left.
  c. Water won't flow between the two.

133. 3×10^4 L

139. a. 10.3 ppb; 3.81 ppb, 1.69 ppb
  c. If the water provider used first draw samples, they would have been required to take action. If they used 2 min flush samples, they would not have been required to take action. Residents probably don't flush their pipes before taking water, so the first draw technique is probably closer to actual practice.

CHAPTER 14

QUESTIONS

1. Sour gummy candies are coated with a white powder that is a mixture of citric acid and tartaric acid. The combination of these two acids creates the sour taste.

3. The main component of stomach acid is hydrochloric acid. Its purpose is to help break down food and kill harmful bacteria.

5. The properties of bases are bitter taste, slippery feel, and the ability to turn red litmus paper blue. Some common substances that contain bases are ammonia, Drano, baking soda, and antacids.

7. The Arrhenius definition of an acid is a substance that produces H^+ ions in aqueous solution. An example: $HCl(aq) \longrightarrow H^+(aq) + Cl^-(aq)$

9. The Brønsted–Lowry definition states that an acid is a proton donor and a base is a proton acceptor. The following is an example of a chemical equation demonstrating this definition:

$$\underset{\text{acid}}{HCl(aq)} + \underset{\text{base}}{H_2O(l)} \longrightarrow H_3O^+(aq) + Cl^-(aq)$$

11. An acid–base neutralization reaction occurs when an acid and a base are mixed and the $H^+(aq)$ from the acid combines with the $OH^-(aq)$ from the base to form $H_2O(l)$. An example follows.

$$HCl(aq) + KOH(aq) \longrightarrow H_2O(l) + KCl(aq)$$

13. $2\,HCl(aq) + K_2O(s) \longrightarrow H_2O(l) + 2\,KCl(aq)$

15. A titration is a laboratory procedure in which a reactant in a solution of known concentration is reacted with another reactant in a solution of unknown concentration until the reaction has reached the equivalence point. The equivalence point is the point at which the reactants are in exact stoichiometric proportions.

17. A strong acid is one that will completely dissociate in solution, while a weak acid does not completely dissociate in solution.

19. Monoprotic acids (such as HCl) contain only one hydrogen ion that will dissociate in solution, while diprotic acids (such as H_2SO_4) contain two hydrogen ions that will dissociate in solution.

21. Yes, pure water contains H_3O^+ ions. Through self-ionization, water acts as an acid and a base with itself; water is amphoteric.

23. a. $[H_3O^+] > 1.0 \times 10^{-7}$ M; $[OH^-] < 1.0 \times 10^{-7}$ M
  b. $[H_3O^+] < 1.0 \times 10^{-7}$ M; $[OH^-] > 1.0 \times 10^{-7}$ M
  c. $[H_3O^+] = 1.0 \times 10^{-7}$ M; $[OH^-] = 1.0 \times 10^{-7}$ M

25. The pOH of a solution is the negative base-10 logarithm of the concentration of OH^- ions ($-\log[OH^-]$). A change of 2.0 pOH units corresponds to a 100-fold change in $[OH^-]$.

27. A buffer is a solution that resists pH change by neutralizing added acid or added base.

PROBLEMS

29. a. acid; $H_2SO_4(aq) \longrightarrow H^+(aq) + HSO_4^-(aq)$

  b. base; $Sr(OH)_2(aq) \longrightarrow Sr^{2+}(aq) + 2\ OH^-(aq)$

  c. acid; $HBr(aq) \longrightarrow H^+(aq) + Br^-(aq)$

  d. base; $NaOH(aq) \longrightarrow Na^+(aq) + OH^-(aq)$

31.

	B-L Acid	B-L Base	Conjugate Acid	Conjugate Base
a.	HBr	H_2O	H_3O^+	Br^-
b.	H_2O	NH_3	NH_4^+	OH^-
c.	HNO_3	H_2O	H_3O^+	NO_3^-
d.	H_2O	C_5H_5N	$C_5H_5NH^+$	OH^-

33. a, c

35. a. Cl^- **b.** HSO_3^-

  c. CHO_2^- **d.** F^-

37. a. NH_4^+ **b.** $HClO_4$

  c. H_2SO_4 **d.** HCO_3^-

39. a. $HI(aq) + NaOH(aq) \longrightarrow H_2O(l) + NaI(aq)$

  b. $HBr(aq) + KOH(aq) \longrightarrow H_2O(l) + KBr(aq)$

  c. $2\ HNO_3(aq) + Ba(OH)_2(aq) \longrightarrow$
  $$2\ H_2O(l) + Ba(NO_3)_2(aq)$$

  d. $2\ HClO_4(aq) + Sr(OH)_2(aq) \longrightarrow$
  $$2\ H_2O(l) + Sr(ClO_4)_2(aq)$$

41. a. $2\ HBr(aq) + 2\ Rb(s) \longrightarrow 2\ RbBr(aq) + H_2(g)$

  b. $2\ HBr(aq) + Mg(s) \longrightarrow MgBr_2(aq) + H_2(g)$

  c. $2\ HBr(aq) + 2\ Ba(s) \longrightarrow 2\ BaBr(aq) + H_2(g)$

  d. $6\ HBr(aq) + 2\ Al(s) \longrightarrow 2\ AlBr_3(aq) + 3\ H_2(g)$

43. a. $MgO(aq) + 2\ HI(aq) \longrightarrow H_2O(l) + MgI_2(aq)$

  b. $K_2O(aq) + 2\ HI(aq) \longrightarrow H_2O(l) + 2\ KI(aq)$

  c. $Rb_2O(aq) + 2\ HI(aq) \longrightarrow H_2O(l) + 2\ RbI(aq)$

  d. $CaO(aq) + 2\ HI(aq) \longrightarrow H_2O(l) + CaI_2(aq)$

45. a. $6\ HClO_4(aq) + Fe_2O_3(s)$
  $$\longrightarrow 2\ Fe(ClO_4)_3(aq) + 3\ H_2O(l)$$

  b. $H_2SO_4(aq) + Sr(s) \longrightarrow SrSO_4(s) + H_2(g)$

  c. $H_3PO_4(aq) + 3\ KOH(aq) \longrightarrow 3\ H_2O(l) + K_3PO_4(aq)$

47. a. 0.1400 M **b.** 0.138 M

  c. 0.08764 M **d.** 0.182 M

49. 0.1090 M H_2SO_4

51. 16.9 mL

53. a. strong **b.** weak

  c. strong **d.** weak

55. a. $[H_3O^+] = 1.7$ M **b.** $[H_3O^+] = 1.5$ M

  c. $[H_3O^+] < 0.38$ M **d.** $[H_3O^+] < 1.75$ M

57. a. strong **b.** weak

  c. strong **d.** weak

59. a. $[OH^-] = 0.25$ M **b.** $[OH^-] < 0.25$ M

  c. $[OH^-] = 0.50$ M **d.** $[OH^-] = 1.25$ M

61. a. acidic **b.** acidic

  c. neutral **d.** basic

63. a. 6.7×10^{-6} M, basic **b.** 1.1×10^{-6} M, basic

  c. 4.5×10^{-9} M, acidic **d.** 1.4×10^{-11} M, acidic

65. a. 3.7×10^{-3} M, acidic **b.** 4.0×10^{-13} M, basic

  c. 9.1×10^{-5} M, acidic **d.** 3.0×10^{-11} M, basic

67. a. basic **b.** neutral

  c. acidic **d.** acidic

69. a. 7.77 **b.** 7.00

  c. 5.66 **d.** 3.13

71. a. 2.8×10^{-9} M **b.** 5.9×10^{-12} M

  c. 1.3×10^{-3} M **d.** 6.0×10^{-2} M

73. a. 7.28 **b.** 6.42

  c. 3.86 **d.** 12.98

75. a. 1.8×10^{-10} M **b.** 3.4×10^{-2} M

  c. 3.2×10^{-13} M **d.** 1.8×10^{-6} M

77. a. 1.810 **b.** 11.107

  c. 2.724 **d.** 10.489

79. a. pOH = 8.82, acidic **b.** pOH = 4.15, basic

  c. pOH = 7.00, neutral **d.** pOH = 2.06, basic

81. a. pOH = 6.08, basic **b.** pOH = 12.74, acidic

  c. pOH = 5.59, basic **d.** pOH = 1.274, basic

83. a. pH = 5.5, acidic **b.** pH = 9.8, basic

  c. pH = 12.3, basic **d.** pH = 7.0, neutral

85. c and d are buffers

87. $HCl(aq) + NaF(aq) \longrightarrow HF(aq) + NaCl(aq)$
  $HCl(aq) + KC_2H_3O_2(aq) \longrightarrow HC_2H_3O_2(aq) + KCl(aq)$

89. a. $HC_2H_3O_2$ **b.** NaH_2PO_4

  c. NaCHOO

91. 50.0 mL

93. 0.16 L

95. 65.2 g

97. 60.0 g/mol

99. 0.17 L

101. a. 2.60, acidic **b.** 11.75, basic

  c. 8.02, basic **d.** 1.710, acidic

103.

$[H_3O^+]$	$[OH^-]$	pH	pOH	Acidic or Basic
1.0×10^{-4}	1.0×10^{-10}	4.00	10.00	acidic
5.5×10^{-3}	1.8×10^{-12}	2.26	11.74	acidic
3.1×10^{-9}	3.2×10^{-6}	8.50	5.50	basic
4.8×10^{-9}	2.1×10^{-6}	8.32	5.68	basic
2.8×10^{-8}	3.5×10^{-7}	7.55	6.45	basic

105. a. $[H_3O^+] = 0.0088$ M
  $[OH^-] = 1.1 \times 10^{-12}$ M
  pH = 2.06

  b. $[H_3O^+] = 1.5 \times 10^{-3}$ M
  $[OH^-] = 6.7 \times 10^{-12}$ M
  pH = 2.82

  c. $[H_3O^+] = 9.77 \times 10^{-4}$ M
  $[OH^-] = 1.02 \times 10^{-11}$ M
  pH = 3.010

  d. $[H_3O^+] = 0.0878$ M
  $[OH^-] = 1.14 \times 10^{-13}$ M
  pH = 1.057

107. a. $[OH^-] = 0.15$ M

$[H_3O^+] = 6.7 \times 10^{-14}$ M

pH = 13.18

b. $[OH^-] = 3.0 \times 10^{-3}$ M

$[H_3O^+] = 3.3 \times 10^{-12}$ M

pH = 11.48

c. $[OH^-] = 9.6 \times 10^{-4}$ M

$[H_3O^+] = 1.0 \times 10^{-11}$ M

pH = 10.98

d. $[OH^-] = 8.7 \times 10^{-5}$ M

$[H_3O^+] = 1.1 \times 10^{-10}$ M

pH = 9.94

109. 1.49 L

111. 11.495

113. 3.0×10^{12} H^+ ions

115. 0.024 mol $Sr(OH)_2$, 0.076 mol NaOH

117. a. weak   **b.** strong

     c. weak   **d.** strong

119. approximately 8 times more concentrated

124. a. $HX(aq) + NaOH(aq) \longrightarrow H_2O(l) + NaX(aq)$

     c. pH = 8.9 at equivalence point; pH = 5 halfway to equivalence point

CHAPTER 15

QUESTIONS

1. The two general concepts involved in equilibrium are sameness and changelessness.

3. By controlling reaction rates, chemists can control the amount of a product that forms in a given period of time and have control of the outcome.

5. The two factors that influence reaction rates are concentration and temperature. The rate of a reaction increases with increasing concentration. The rate of a reaction increases with increasing temperature.

7. In a chemical reaction, dynamic equilibrium is the condition in which the rate of the forward reaction equals the rate of the reverse reaction.

9. Because the rate of the forward and reverse reactions is the same at equilibrium, the relative concentrations of reactants and products become constant.

11. The equilibrium constant is a measure of how far a reaction goes; it is significant because it is a way to quantify the concentrations of the reactants and products at equilibrium.

13. A small equilibrium constant shows that a reverse reaction is favored and that when equilibrium is reached, there will be more reactants than products. A large equilibrium constant shows that a forward reaction is favored and that when equilibrium is reached, there will be more products than reactants.

15. No, the particular concentrations of reactants and products at equilibrium will not always be the same for a given reaction—they will depend on the initial concentrations.

17. Various answers depending on the answer for Question 10.

19. Decreasing the concentration of a reactant in a reaction mixture at equilibrium causes the reaction to shift to the left.

21. Decreasing the concentration of a product in a reaction mixture at equilibrium causes the reaction to shift to the right.

23. Increasing the pressure of a reaction mixture at equilibrium if the product side has fewer moles of gas particles than the reactant side causes the reaction to shift to the right.

25. Decreasing the pressure of a reaction mixture at equilibrium if the product side has fewer moles of gas particles than the reactant side causes the reaction to shift to the left.

27. Increasing the temperature of an exothermic reaction mixture at equilibrium causes the reaction to shift left, absorbing some of the added heat. Decreasing the temperature of an exothermic reaction mixture at equilibrium causes the reaction to shift right, releasing heat.

29. $K_{sp} = [A^{2+}][B^-]^2$

31. The solubility of a compound is the amount of the compound that dissolves in a certain amount of liquid, and the molar solubility is the solubility in units of moles per liter.

33. Two reactants with a large K_{eq} for a particular reaction might not react immediately when combined because of a large activation energy, which is an energy hump that normally exists between the reactants and products. The activation energy must be overcome before the system will undergo a reaction.

35. No, a catalyst does not affect the value of the equilibrium constant; it simply lowers the activation energy and increases the rate of a chemical reaction.

PROBLEMS

37. Rate would decrease because the effective concentration of the reactants has been decreased, which lowers the rate of a reaction.

39. Reaction rates tend to decrease with decreasing temperature, so all life processes (chemical reactions) would have decreased rates.

41. The rate would be lower because the concentration of reactants decreases as they are consumed in the reaction.

43. a. $K_{eq} = \dfrac{[N_2O_4]}{[NO_2]^2}$ **b.** $K_{eq} = \dfrac{[NO]^2[Br_2]}{[BrNO]^2}$

     c. $K_{eq} = \dfrac{[H_2][CO_2]}{[H_2O][CO]}$ **d.** $K_{eq} = \dfrac{[CS_2][H_2]^4}{[CH_4][H_2S]^2}$

45. a. $K_{eq} = \dfrac{[Cl_2]}{[PCl_5]}$ **b.** $K_{eq} = [O_2]^3$

     c. $K_{eq} = \dfrac{[H_3O^+][F^-]}{[HF]}$ **d.** $K_{eq} = \dfrac{[NH_4^+][OH^-]}{[NH_3]}$

47. $K_{eq} = \dfrac{[H_2]^2[S_2]}{[H_2S]^2}$

49. a. products   **b.** both

     c. reactants   **d.** both

51. 0.0394

53. 1.79×10^{-5}

55. 0.0987

57. 7.0 M

59. 0.119 M

61.

$T(K)$	$[N_2]$	$[H_2]$	$[NH_3]$	K_{eq}
500	0.115	0.105	0.439	1.45×10^3
575	0.110	0.25	0.128	9.6
775	0.120	0.140	0.00439	0.0584

63. a. shift right   **b.** shift left
c. shift right
65. a. no effect   **b.** shift left
c. shift left   **d.** shift right
67. a. shift right   **b.** shift left
69. a. no effect   **b.** no effect
71. a. shift right   **b.** shift left
73. a. shift left   **b.** shift right
75. a. no effect   **b.** shift right
c. shift left   **d.** shift right
e. no effect
77. a. $CaSO_4(s) \longrightarrow Ca^{2+}(aq) + SO_4^{2-}(aq)$
  $K_{sp} = [Ca^{2+}][SO_4^{2-}]$
b. $AgCl(s) \longrightarrow Ag^+(aq) + Cl^-(aq)$
  $K_{sp} = [Ag^+][Cl^-]$
c. $CuS(s) \longrightarrow Cu^{2+}(aq) + S^{2-}(aq)$
  $K_{sp} = [Cu^{2+}][S^{2-}]$
d. $FeCO_3(s) \longrightarrow Fe^{2+}(aq) + CO_3^{2-}(aq)$
  $K_{sp} = [Fe^{2+}][CO_3^{2-}]$
79. $K_{sp} = [Fe^{2+}][OH^-]^2$
81. 7.0×10^{-11}
83. 1.35×10^{-4} M
85. 7.04×10^{-5} M
87. 2.61×10^{-3} M

89.

Compound	[Cation]	[Anion]	K_{sp}
$SrCO_3$	2.4×10^{-5}	2.4×10^{-5}	5.8×10^{-10}
SrF_2	1.0×10^{-3}	2.0×10^{-3}	4.0×10^{-9}
Ag_2CO_3	2.6×10^{-4}	1.3×10^{-4}	8.8×10^{-12}

91. 3.3×10^2
93. 5.34 g
95. b, c, d
97. 1.13×10^{-18} M, 1.62×10^{-15} g
99. Yes
101. 7.07×10^{-4}
103. 1.2×10^{-2} g
105. 0.021 g K_2CO_3
107. e
109. 35.5 L
114. a. $CaCO_3(s) \longrightarrow CaO(s) + CO_2(g)$
c. K = 0.00040 at 950 K; K = 0.0030 at 1050 K
e. 0.11 g

CHAPTER 16

QUESTIONS

1. A fuel-cell electric vehicle is an automobile running on an electric motor that is powered by hydrogen. The fuel cells use the electron-gaining tendency of oxygen and the electron-losing tendency of hydrogen to force electrons to move through a wire, creating the electricity that powers the car.

3. a. Oxidation is the gain of oxygen, and reduction is the loss of oxygen.

b. Oxidation is the loss of electrons, and reduction is the gain of electrons.

c. Oxidation is an increase in oxidation state, and reduction is a decrease in oxidation state.

5. gain

7. The oxidization state of a free element is zero. The oxidization state of a monoatomic ion equals its charge.

9. For an ion, the sum of the oxidation states of the individual atoms must add up to *the charge of the ion*.

11. In a redox reaction, an atom that undergoes an increase in oxidation state is *oxidized*. An atom that undergoes a decrease in oxidation state is *reduced*.

13. When balancing redox equations, the number of electrons lost in the oxidation half-reaction must *equal* the number of electrons gained in the reduction half-reaction.

15. When balancing aqueous redox reactions, charge is balanced using *electrons*.

17. The metals at the top of the activity series are the most reactive.

19. The metals at the bottom of the activity series are least likely to lose electrons.

21. If the metal is listed above H_2 on the activity series, it will dissolve in acids such as HCl or HBr.

23. Oxidation occurs at the *anode* of an electrochemical cell.

25. The salt bridge joins the two half-cells or completes the circuit; it allows the flow of ions between the two half-cells.

27. The common dry-cell battery does not contain large amounts of liquid water and is composed of a zinc case that acts as the anode. The cathode is a carbon rod immersed in a moist paste of MnO_2 that also contains NH_4Cl. The anode and cathode reactions that occur produce a voltage of about 1.5 volts.

anode reaction:

$Zn(s) \longrightarrow Zn^{2+}(aq) + 2 e^-$

cathode reaction:

$2 MnO_2(s) + 2 NH_2^+(aq) + 2 e^- \longrightarrow$
$\qquad\qquad 2 Mn_2O_3(s) + 2 NH_3(g) + H_2O(l)$

29. Fuel cells are like batteries, but the reactants are constantly replenished. The reactants constantly flow through the battery, generating electrical current as they undergo a redox reaction.

anode reaction:

$2 H_2(g) + 4 OH^-(aq) \longrightarrow 4 H_2O(g) + 4 e^-$

cathode reaction:

$O_2(g) + 2 H_2O(l) + 4 e^- \longrightarrow 4 OH^-(aq)$

31. Corrosion is the oxidation of metals; the most common example is rusting of iron.

oxidation:

$2 Fe(s) \longrightarrow 2 Fe^{2+}(aq) + 4 e^-$

reduction:

$O_2(g) + 2 H_2O(l) + 4 e^- \longrightarrow 4 OH^-(aq)$

overall:

$2 Fe(s) + O_2(g) + 2 H_2O(l) \longrightarrow 2 Fe(OH)_2(s)$

PROBLEMS

33. a. H_2 **b.** Al
   c. Al

35. a. Sr is oxidized; O_2 is reduced.
   b. Ca is oxidized; Cl_2 is reduced.
   c. Mg is oxidized; Ni^{2+} is reduced.

37. a. Sr is the reducing agent; O_2 is the oxidizing agent.
   b. Ca is the reducing agent; Cl_2 is the oxidizing agent.
   c. Mg is the reducing agent; Ni^{2+} is the oxidizing agent.

39. b (F_2), **d** (Cl_2)

41. a (K), **c** (Fe)

43. a. N_2 is oxidized and is the reducing agent.
     O_2 is reduced and is the oxidizing agent.
   b. C is oxidized and is the reducing agent.
     O_2 is reduced and is the oxidizing agent.
   c. Sb is oxidized and is the reducing agent.
     Cl_2 is reduced and is the oxidizing agent.

45. a. 0   **b.** +2
   c. +3   **d.** 0

47. a. Na: +1; Cl: −1   **b.** Ca: +2; F: −1
   c. S: +4; O: −2   **d.** H: +1; S: −2

49. a. +2   **b.** +4
   c. +1

51. a. C: +4; O: −2   **b.** O: −2; H: +1
   c. N: +5; O: −2   **d.** N: +3; O: −2

53. a. +1   **b.** +3
   c. +5   **d.** +7

55. a. Cu, +2; N, +5; O, −2
   b. Sr, +2; O, −2; H, +1
   c. K, +1; O, −2; Cr, +6
   d. Na, +1; H, +1; O, −2; C, +4

57. a. Sb + 5 ⟶ +3, reduced
     Cl − 1 ⟶ 0, oxidized
   b. C + 2 ⟶ +4, oxidized
     Cl 0 ⟶ −1, reduced
   c. N + 2 ⟶ +3, oxidized
     Br 0 ⟶ −1, reduced
   d. H 0 ⟶ +1, oxidized
     C + 4 ⟶ +2, reduced

59. Na is the reducing agent.
   H is the oxidizing agent.

61. a. $3 K(s) + Cr^{3+}(aq) \longrightarrow Cr(s) + 3 K^+(aq)$
   b. $Mg(s) + 2 Ag^+(aq) \longrightarrow Mg^{2+}(aq) + 2 Ag(s)$
   c. $2 Al(s) + 3 Fe^{2+}(aq) \longrightarrow 2 Al^{3+}(aq) + 3 Fe(s)$

63. a. reduction, $5 e^- + MnO_4^-(aq) + 8 H^+(aq) \longrightarrow$
$$Mn^{2+}(aq) + 4 H_2O(l)$$
   b. oxidation, $2 H_2O(l) + Pb^{2+}(aq) \longrightarrow$
$$PbO_2(s) + 4 H^+(aq) + 2 e^-$$
   c. reduction, $10 e^- + 2 IO_3^-(aq) + 12 H^+(aq) \longrightarrow$
$$I_2(s) + 6 H_2O(l)$$
   d. oxidation, $SO_2(g) + 2 H_2O(l) \longrightarrow$
$$SO_4^{2-}(aq) + 4 H^+(aq) + 2 e^-$$

65. a. $PbO_2(s) + 4 H^+(aq) + 2 I^-(aq) \longrightarrow$
$$I_2(s) + Pb^{2+}(aq) + 2 H_2O(l)$$
   b. $5 SO_3^{2-}(aq) + 6 H^+(aq) + 2 MnO_4^-(aq) \longrightarrow$
$$5 SO_4^{2-}(aq) + 2 Mn^{2+}(aq) + 3 H_2O(l)$$
   c. $S_2O_3^{2-}(aq) + 4 Cl_2(g) + 5 H_2O(l) \longrightarrow$
$$2 SO_4^{2-}(aq) + 8 Cl^-(aq) + 10 H^+(aq)$$

67. a. $ClO_4^-(aq) + 2 H^+(aq) + 2 Cl^-(aq) \longrightarrow$
$$ClO_3^-(aq) + Cl_2(aq) + H_2O(l)$$
   b. $3 MnO_4^-(aq) + 24 H^+(aq) + 5 Al(s) \longrightarrow$
$$3 Mn^{2+}(aq) + 5 Al^{3+}(aq) + 12 H_2O(l)$$
   c. $Br_2(aq) + Sn(s) \longrightarrow Sn^{2+}(aq) + 2 Br^-(aq)$

69. a. $3 ClO^-(aq) + 2 Cr(OH)_4^-(aq) + 2 OH^-(aq) \longrightarrow$
$$3 Cl^-(aq) + 2 CrO_4^{2-}(aq) + 5 H_2O(l)$$
   b. $2 MnO_4^-(aq) + Br(aq) + H_2O(l) \longrightarrow$
$$2 MnO_2(s) + BrO_3^-(aq) + 2 OH^-(aq)$$

71. a, Ag

73. b, Cu^{2+}

75. b, Al

77. **b** and **c** occur spontaneously in the forward direction.

79. Fe, Cr, Zn, Mn, Al, Mg, Na, Ca, K, Li

81. Mg

83. a. no reaction
   b. $2 HCl(aq) + Fe(s) \longrightarrow H_2(g) + FeCl_2(aq)$
   c. no reaction
   d. $6 HCl(aq) + 2 Al(s) \longrightarrow 3 H_2(g) + 2 AlCl_3(aq)$

85.

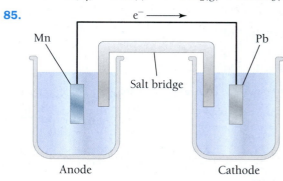

87. d

89. $Zn(s) + 2 MnO_2(s) + 2 H_2O(l) \longrightarrow$
$$Zn(OH)_2(s) + 2 MnO(OH)(s)$$

91.

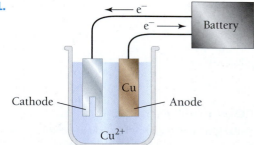

93. a, Zn; c, Mn

95. a. redox; Zn is oxidized; Co is reduced.
   b. not redox
   c. not redox
   d. redox; K is oxidized; Br is reduced.

97. $16 H^+(aq) + 2 MnO_4^-(aq) + 5 Zn(s) \longrightarrow$
$$2 Mn^{2+}(aq) + 5 Zn^{2+}(aq) + 8 H_2O(l); 34.9 \text{ mL}$$

99. Yes, the reaction will occur spontaneously.
$$Mg(s) \longrightarrow Mg^{2+}(aq) + 2 e^-$$
$$Ag^+(aq) + e^- \longrightarrow Ag(s)$$
$$2 Ag^+(aq) + Mg(s) \longrightarrow Mg^{2+}(aq) + 2 Ag(s)$$

101. 2.95%

103. 0.054 mol

105. **a.** $2 Cr(s) + 6 HI(aq) \longrightarrow$
$$2 Cr^{3+}(aq) + 6 I^-(aq) + 3 H_2(g), 98 \text{ mL HI}$$

b. $2 Al(s) + 6 HI(aq) \longrightarrow$
$$2 Al^{3+}(aq) + 6 I^-(aq) + 3 H_2(g), 68 \text{ mL HI}$$

c. no

d. no

107. 0.67 cm

109. 8.9×10^3 s or 2.5 h

111.

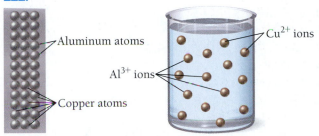

113. Many of the Zn atoms on the electrode would become Zn^{2+} ions in solution. Many Ni^{2+} ions in solution would become Ni atoms on the electrode.

116. a.

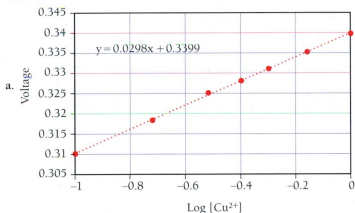

$y = 0.0298x + 0.3399$

Log [Cu²⁺] (x-axis), Voltage (y-axis)

Yes, the graph is linear.

c. i. 0.0573 M; ii. 0.857 M

CHAPTER 17

QUESTIONS

1. Radioactivity is the emission of tiny, invisible particles by disintegration of atomic nuclei. Many of these particles can pass right through matter. Atoms that emit these particles are radioactive.

3. Uranic rays were the name given by Henri Becquerel to the radiation emitted by crystals that contained uranium.

5. X: chemical symbol, used to identify the element.

A: mass number, which is the sum of the number of protons and number of neutrons in the nucleus.

Z: atomic number, which is the number of protons in the nucleus.

7. Alpha radiation occurs when an unstable nucleus emits a small piece of itself composed of 2 protons and 2 neutrons. The symbol for an alpha particle is $_2^4He$.

9. Alpha particles have high ionizing power and low penetrating power compared to beta and gamma particles.

11. When an atom emits a beta particle, its atomic number increases by one because it now has an additional proton. The mass of an atom does not change as a result of beta emission.

13. Gamma radiation is electromagnetic radiation, and the symbol for a gamma ray is $_0^0\gamma$.

15. Gamma particles have low ionizing power and high penetrating power compared to alpha and beta particles.

17. When an atom emits a positron, its atomic number decreases by one because it now has one less proton. The mass of an atom does not change when it emits a positron.

19. A nuclear equation represents the changes that occur during radioactivity and other nuclear processes. For a nuclear equation to be balanced, the sum of the atomic numbers on both sides of the equation must be equal, and the sum of the mass numbers on both sides of the equation must be equal.

21. A film-badge dosimeter is a badge that consists of photographic film held in a small case that is pinned to clothing. It is used to monitor a person's exposure to radiation. The more exposed the film has become in a given period of time, the more the person has been exposed to radioactivity.

23. In a scintillation counter, the radioactive particles pass through a material that emits ultraviolet or visible light in response to excitation by radioactive particles. The light is detected and turned into an electrical signal.

25. The half-life is the time it takes for one-half of the parent nuclides in a radioactive sample to decay to the daughter nuclides. One can relate the half-life of objects to find their radioactive decay rates.

27. The decaying of uranium in the ground is the source of radon in our environment. Radon increases the risk of lung cancer because it is a gas that can be inhaled.

29. When an organism dies, it stops incorporating carbon-14 into its tissues, and the amount present at its death will decay with a half-life of 5730 years. Using this information, one can determine the age of the organism by the amount of carbon-14 still present in the remains.

31. Fermi believed that if a neutron could be incorporated into the nucleus of an atom, the nucleus might undergo beta decay and convert a neutron into a proton. The nuclear equation for this process is:
$$_{92}^{238}U + _0^1n \longrightarrow _{92}^{239}U \longrightarrow _{93}^{239}X + _{-1}^0e$$

33. Fission can be used in a bomb because it is a self-amplifying reaction (fission of one atom induces fission of another) that can produce great amounts of energy.

35. The main goal of the Manhattan Project was to build an atomic bomb before the Germans did. Its project leader was J. R. Oppenheimer.

37. In nuclear reactors, control rods control the amount of fission that can occur. When the control rods are inserted into the fuel assembly, they absorb neutrons, preventing them from inducing fission in the fuel rods.

39. No, a nuclear reactor cannot detonate the way a nuclear bomb can because the uranium fuel used in electricity generation is not sufficiently enriched in U-235 to produce a nuclear detonation.

41. Modern nuclear weapons use both fission and fusion. In the hydrogen bomb, a small fission bomb is detonated first to create a high enough temperature for the fusion reaction to proceed.

43. Radiation can affect the molecules in living organisms by ionizing them.

45. Lower doses of radiation over extended periods of time can increase cancer risk by damaging DNA. Occasionally, a change in DNA can cause cells to grow abnormally and to become cancerous.

47. The main unit of radiation exposure is the rem, which stands for *roentgen equivalent man*. The average American is exposed to 1/3 of a rem of radiation per year.

49. Isotope scanning can be used in the medical community to detect and identify cancerous tumors. Likewise, isotope scanning can produce necessary images of several different internal organs.

PROBLEMS

51. $^{128}_{82}Pb$

53. 81 protons, 126 neutrons

55. a. beta particle   **b.** neutron

     c. gamma ray

57.

Chemical Symbol	Atomic Number (Z)	Mass Number (A)	Number of Protons	Number of Neutrons
Tc	43	95	43	52
Ba	56	128	56	72
Eu	63	145	63	82
Fr	87	223	87	136

59. a. $^{234}_{92}U \longrightarrow {}^{230}_{90}Th + {}^{4}_{2}He$

     b. $^{230}_{90}Th \longrightarrow {}^{226}_{88}Ra + {}^{4}_{2}He$

     c. $^{226}_{88}Ra \longrightarrow {}^{222}_{86}Rn + {}^{4}_{2}He$

     d. $^{222}_{86}Rn \longrightarrow {}^{218}_{84}Po + {}^{4}_{2}He$

61. a. $^{214}_{82}Pb \longrightarrow {}^{214}_{83}Bi + {}^{0}_{-1}e$

     b. $^{214}_{83}Bi \longrightarrow {}^{214}_{84}Po + {}^{0}_{-1}e$

     c. $^{231}_{90}Th \longrightarrow {}^{231}_{91}Pa + {}^{0}_{-1}e$

     d. $^{227}_{89}Ac \longrightarrow {}^{227}_{90}Th + {}^{0}_{-1}e$

63. a. $^{11}_{6}C \longrightarrow {}^{11}_{5}B + {}^{0}_{+1}e$

     b. $^{13}_{7}N \longrightarrow {}^{13}_{6}C + {}^{0}_{+1}e$

     c. $^{15}_{8}O \longrightarrow {}^{15}_{7}N + {}^{0}_{+1}e$

65. $^{241}_{94}Pu \longrightarrow {}^{241}_{95}Am + {}^{0}_{-1}e$

$^{241}_{95}Am \longrightarrow {}^{237}_{93}Np + {}^{4}_{2}He$

$^{237}_{93}Np \longrightarrow {}^{233}_{91}Pa + {}^{4}_{2}He$

$^{233}_{91}Pa \longrightarrow {}^{233}_{92}U + {}^{0}_{-1}e$

67. $^{232}_{90}Th \longrightarrow {}^{228}_{88}Ra + {}^{4}_{2}He$

$^{228}_{88}Ra \longrightarrow {}^{228}_{89}Ac + {}^{0}_{-1}e$

$^{228}_{89}Ac \longrightarrow {}^{228}_{90}Th + {}^{0}_{-1}e$

$^{228}_{90}Th \longrightarrow {}^{224}_{88}Ra + {}^{4}_{2}He$

69. 3.1×10^{3} atoms

71. 18 hr

73. 1.2×10^{6} yr

75. 0.31 g

77. Ga-67 > P-32 > Cr-51 > Sr-89

79. 5,715 yr

81. 34,290 yr

83. $^{235}_{92}U + {}^{1}_{0}n \longrightarrow {}^{144}_{54}Xe + {}^{90}_{38}Sr + 2{}^{1}_{0}n$; 2 neutrons

85. $^{2}_{1}H + {}^{2}_{1}H \longrightarrow {}^{3}_{2}He + {}^{1}_{0}n$

87. a. $^{6}_{3}Li$ **b.** $^{64}_{28}Ni$ **c.** $^{179}_{73}Ta$

89. $^{238}_{92}U + {}^{1}_{0}n \longrightarrow {}^{239}_{92}U$; ${}^{239}_{92}U \longrightarrow {}^{0}_{-1}e + {}^{239}_{93}Np$;

$^{239}_{93}Np \longrightarrow {}^{0}_{-1}e + {}^{239}_{94}Pu$

91. per mole = 1.9×10^{13} J

per kg = 8.1×10^{13} J

93. $1.7 \times 10^{21} \beta$ emissions

95. 68.4%

97. 8.7×10^{-6} L

99. nucleus with 9 protons and 7 neutrons

101. nucleus with 5 protons and 5 neutrons

107. a. 0.34 g at 200 minutes; 0.23 g at 400 minutes

     c. 0.003125 mg

Glossary

absolute zero The coldest temperature possible. Absolute zero (0 K or −273 °C or −459 °F) is the temperature at which molecular motion stops. Lower temperatures do not exist.

acid A molecular compound that dissolves in solution to form H^+ ions. Acids have the ability to dissolve some metals and will turn litmus paper red.

acid–base reaction A reaction that forms water and typically a salt.

acidic solution A solution containing a concentration of H_3O^+ ions greater than 1.0×10^{-7} M (pH < 7).

activation energy The amount of energy that must be absorbed by reactants before a reaction can occur; an energy hump that normally exists between the reactants and products.

activity series of metals A listing of metals (and hydrogen) in order of decreasing activity, decreasing ability to oxidize, and decreasing tendency to lose electrons.

actual yield The amount of product actually produced by a chemical reaction.

addition polymer A polymer formed by addition of monomers to one another without elimination of any atoms.

addition reaction A chemical reaction in which additional atoms or functional groups add to a molecule.

alcohol An organic compound containing an —OH functional group bonded to a carbon atom and having the general formula ROH.

aldehyde An organic compound with the general formula RCHO.

alkali metals The Group 1A elements, which are highly reactive metals.

alkaline battery A dry cell employing half-reactions that use a base.

alkaline earth metals The Group 2A elements, which are fairly reactive metals.

alkaloid Organic compound that acts as a base and is typically found in plants.

alkane A hydrocarbon in which all carbon atoms are connected by single bonds. Noncyclic alkanes have the general formula C_nH_{2n+2}.

alkene A hydrocarbon that contains at least one double bond between carbon atoms. Noncyclic alkenes have the general formula C_nH_{2n}.

alkyl group In an organic molecule, any group containing only singly bonded carbon atoms and hydrogen atoms.

alkyne A hydrocarbon that contains at least one triple bond between carbon atoms. Noncyclic alkynes have the general formula C_nH_{2n-2}.

alpha particle A particle consisting of two protons and two neutrons (a helium nucleus), represented by the symbol $_2^4He$.

alpha (α) radiation Radiation emitted by an unstable nucleus, consisting of alpha particles.

alpha (α)-helix The most common secondary protein structure. The amino acid chain is wrapped into a tight coil from which the side chains extend outward. The structure is maintained by hydrogen bonding interactions between NH and CO groups along the peptide backbone of the protein.

amine An organic compound that contains nitrogen and has the general formula NR_3, where R may be an alkyl group or a hydrogen atom.

amino acid A molecule containing an amine group, a carboxylic acid group, and an R group (also called a side chain). Amino acids are the building blocks of proteins.

amorphous A type of solid matter in which atoms or molecules do not have long-range order (e.g., glass and plastic).

amphoteric In Brønsted–Lowry terminology, able to act as either an acid or a base.

anion A negatively charged ion.

anode The electrode where oxidation occurs in an electrochemical cell.

aqueous solution A homogeneous mixture of a substance with water.

aromatic ring A ring of carbon atoms containing alternating single and double bonds.

Arrhenius acid A substance that produces H^+ ions in aqueous solution.

Arrhenius base A substance that produces OH^- ions in aqueous solution.

Arrhenius definitions (of acids and bases) The definitions of an acid as a substance that produces H^+ ions in aqueous solution and a base as a substance that produces OH^- ions in aqueous solution.

atmosphere (atm) The average pressure at sea level, 101,325 Pa (760 mmHg).

atom The smallest identifiable unit of an element.

atomic element An element that exists in nature with single atoms as the basic unit.

atomic mass A weighted average of the masses of each naturally occurring isotope of an element; atomic mass is the average mass of the atoms of an element.

atomic mass unit (amu) The unit commonly used to express the masses of protons, neutrons, and nuclei. 1 amu = 1.66×10^{-24} g.

atomic number (Z) The number of protons in the nucleus of an atom.

atomic size The size of an atom, which is determined by how far the outermost electrons are from the nucleus.

atomic solid A solid whose component units are individual atoms (e.g., diamond, C; iron, Fe).

atomic theory A theory stating that all matter is composed of tiny particles called atoms.

Avogadro's law A law stating that the volume (V) of a gas and the amount of the gas in moles (n) are directly proportional.

Avogadro's number The number of entities in a mole, 6.022×10^{23}.

balanced equation A chemical equation in which the numbers of each type of atom on both sides of the equation are equal.

ball-and-stick model   A way to represent molecules in which an atom is represented with a ball and a bond is represented with a stick.

base   A molecular compound that dissolves in solution to form OH^- ions. Bases have a slippery feel and turn litmus paper blue.

base chain   The longest continuous chain of carbon atoms in an organic compound.

basic solution   A solution containing a concentration of OH^- ions greater than 1.0×10^{-7} M (pH > 7).

bent   The molecular geometry in which 3 atoms are not in a straight line. This geometry occurs when the central atoms contain 4 electron groups (2 bonding and 2 nonbonding) or 3 electron groups (2 bonding and 1 nonbonding).

beta (β) particle   A form of radiation consisting of an energetic electron and represented by the symbol $_{-1}^{0}e$.

beta (β) radiation   Energetic electrons emitted by an unstable nucleus.

beta (β)-pleated sheet   A common pattern in the secondary structure of proteins. The protein chain is extended in a zigzag pattern, and the peptide backbones of adjacent strands interact with one another through hydrogen bonding to form sheets.

binary acid   An acid containing only hydrogen and a nonmetal.

binary compound   A compound containing only two different kinds of elements.

biochemistry   The study of the chemical substances and processes that occur in living organisms.

Bohr model   A model for the atom in which electrons travel around the nucleus in circular orbits at specific, fixed distances from the nucleus.

boiling point   The temperature at which the vapor pressure of a liquid is equal to the pressure above it.

boiling point elevation   The increase in the boiling point of a solution caused by the presence of the solute.

bonding pair   Electrons that are shared between two atoms in a chemical bond.

bonding theory   A model that predicts how atoms bond together to form molecules.

Boyle's law   A law maintaining that the volume (V) of a gas and its pressure (P) are inversely proportional.

branched alkane   An alkane composed of carbon atoms bonded in chains containing branches.

Brønsted–Lowry acid   A proton (H^+ ion) donor.

Brønsted–Lowry base   A proton (H^+ ion) acceptor.

buffer   A solution that resists pH change by neutralizing added acid or added base.

Calorie (Cal)   An energy unit equivalent to 1000 little-*c* calories.

calorie (cal)   The amount of energy required to raise the temperature of 1 g of water by 1 °C.

carbohydrate   A polyhydroxyl aldehyde or ketone, containing multiple —OH groups and often having the general formula $(CH_2O)_n$.

carbonyl group   A carbon atom double bonded to an oxygen atom.

carboxylic acid   An organic compound with the general formula RCOOH.

catalyst   A substance that increases the rate of a chemical reaction but is not consumed by the reaction.

cathode   The electrode where reduction occurs in an electrochemical cell.

cation   A positively charged ion.

cell   The smallest structural unit of living organisms that has the properties associated with life.

cell membrane   The structure that bounds the cell and holds the contents of the cell together.

cellulose   A common polysaccharide composed of repeating glucose units linked together.

Celsius (°C) scale   A temperature scale often used by scientists. On this scale, water freezes at 0 °C and boils at 100 °C at 1 atm pressure. Room temperature is approximately 22 °C.

chain reaction   A self-sustaining chemical or nuclear reaction yielding energy or products that cause further reactions of the same kind.

charge   A fundamental property of protons and electrons. Charged particles experience forces such that like charges repel and unlike charges attract.

Charles's law   A law stating that the volume (V) of a gas and its temperature (T) expressed in kelvins are directly proportional.

chemical bond   The sharing or transfer of electrons to attain stable electron configurations among the bonding atoms.

chemical change   A change in which matter changes its composition.

chemical energy   The energy associated with chemical changes.

chemical formula   A way to represent a compound. At a minimum, the chemical formula indicates the elements present in the compound and the relative number of atoms of each element.

chemical property   A property that a substance can display only through changing its composition.

chemical reaction   The process by which one or more substances transform into different substances via a chemical change. Chemical reactions often emit or absorb energy.

chemical symbol   A one- or two-letter abbreviation for an element. Chemical symbols are listed directly below the atomic number in the periodic table.

chemistry   The science that seeks to understand the behavior of matter by studying what atoms and molecules do.

chromosome   A biological structure containing genes, located within the nucleus of a cell.

climate change   Changes in the Earth's climate caused by human emission of gases (especially CO_2) into the atmosphere.

codon   A sequence of three bases in a nucleic acid that codes for one amino acid.

colligative properties   Physical properties of solutions that depend on the number of solute particles present but not the type of solute particles.

collision theory   A theory of reaction rates stating that effective collisions between reactant molecules must take place in order for the reaction to occur.

combination reaction   A chemical reaction in which simpler substances combine to form more complex substances.

combined gas law   A law that combines Boyle's law and Charles's law; it is used to calculate how a property of a gas (P, V, or T) changes when two other properties are changed at the same time.

combustion   The burning of a substance in the presence of oxygen.

combustion reaction   A reaction in which a substance reacts with oxygen, emitting heat and forming one or more oxygen-containing compounds.

complementary base   In DNA, a base capable of precise pairing with a specific other DNA base.

complete ionic equation   A chemical equation showing all the species as they are actually present in solution.

complex carbohydrate   A carbohydrate composed of many repeating saccharide units.

compound A substance composed of two or more elements in fixed, definite proportions.

compressible Able to occupy a smaller volume when subjected to increased pressure. Gases are compressible because, in the gas phase, atoms or molecules are widely separated.

concentrated solution A solution containing large amounts of solute.

condensation A physical change in which a substance is converted from its gaseous form to its liquid form.

condensation polymer A class of polymers that expel atoms, usually water, during their formation or polymerization.

condensed structural formula A shorthand way of writing a structural formula.

conjugate acid–base pair In Brønsted–Lowry terminology, two substances related to each other by the transfer of a proton.

conversion factor A factor used to convert between two separate units; a conversion factor is constructed from any two quantities known to be equivalent.

copolymers Polymers that are composed of two different kinds of monomers and result in chains composed of alternating units rather than a single repeating unit.

core electrons The electrons that are not in the outermost principal shell of an atom.

corrosion The oxidation of metals (e.g., rusting of iron).

covalent atomic solid An atomic solid, such as diamond, that is held together by covalent bonds.

covalent bond The bond that results when two nonmetals combine in a chemical reaction. In a covalent bond, the atoms share their electrons.

critical mass The mass of uranium or plutonium required for a nuclear reaction to be self-sustaining.

crystalline A type of solid matter with atoms or molecules arranged in a well-ordered, three-dimensional array with long-range, repeating order (e.g., salt and diamond).

cytoplasm In a cell, the region between the nucleus and the cell membrane.

Dalton's law of partial pressure A law stating that the sum of the partial pressures of each component in a gas mixture equals the total pressure.

daughter nuclide The nuclide product of a nuclear decay.

decanting A way to separate a mixture in which one layer is carefully poured off of another layer.

decimal part One part of a number expressed in scientific notation.

decomposition reaction A reaction in which a complex substance decomposes to form simpler substances; $AB \longrightarrow A + B$.

density (d) A fundamental property of materials that relates mass and volume and differs from one substance to another. The units of density are those of mass divided by volume, most commonly expressed in g/cm^3, g/mL, or g/L.

dilute solution A solution containing small amounts of solute.

dimer A molecule formed by the joining together of two smaller molecules.

dipeptide Two amino acids linked together via a peptide bond.

dipole–dipole force The interaction between two molecules having dipole–dipole moments (or between two polar molecules).

dipole moment A separation of charge within a chemical bond that produces a bond with a positive end and a negative end.

diprotic acid An acid containing two ionizable protons.

disaccharide A carbohydrate that can be decomposed into two simpler carbohydrates.

dispersion force The intermolecular force present in all molecules and atoms. Dispersion forces are caused by fluctuations in the electron distribution within molecules or atoms.

displacement reaction A reaction in which one element displaces another in a compound; $A + BC \longrightarrow AC + B$.

dissociation In aqueous solution, the process by which a solid ionic compound separates into its ions.

distillation A way to separate mixtures in which the mixture is heated to boil off the more volatile component.

disubstituted benzene A benzene in which two hydrogen atoms have been replaced by an atom or group of atoms.

DNA (deoxyribonucleic acid) Long chainlike molecules that occur in the nucleus of cells and act as blueprints for the construction of proteins.

dot structure A drawing that represents the valence electrons in atoms as dots; it shows a chemical bond as the sharing or transfer of electron dots.

double bond The bond that exists when two electron pairs are shared between two atoms. In general, double bonds are shorter and stronger than single bonds.

double displacement A reaction in which two elements or groups of elements in two different compounds exchange places to form two new compounds; $AB + CD \longrightarrow AD + CB$.

dry cell An ordinary battery (voltaic cell); it does not contain large amounts of liquid water.

duet The name for the two electrons corresponding to a stable Lewis structure in hydrogen and helium.

dynamic equilibrium In a chemical reaction, the condition in which the rate of the forward reaction equals the rate of the reverse reaction.

electrical current The flow of electric charge—for example, electrons flowing through a wire or ions through a solution.

electrical energy Energy associated with the flow of electric charge.

electrochemical cell A device that creates electrical current from a redox reaction.

electrolysis A process in which electrical current is used to drive an otherwise nonspontaneous redox reaction.

electrolyte solution A solution containing a solute that dissociates into ions.

electrolytic cell An electrochemical cell used for electrolysis.

electromagnetic radiation A type of energy that travels through space at a constant speed of 3.0×10^8 m/s (186,000 miles/s) and exhibits both wavelike and particlelike behavior. Light is a form of electromagnetic radiation.

electromagnetic spectrum A spectrum that includes all wavelengths of electromagnetic radiation.

electron A negatively charged particle that occupies most of the atom's volume but contributes almost none of its mass.

electron configuration A representation that shows the occupation of orbitals by electrons for a particular element.

electron geometry The geometrical arrangement of the electron groups in a molecule.

electron group A general term for a lone pair, single bond, or multiple bond in a molecule.

electron spin A fundamental property of all electrons that causes them to have magnetic fields associated with them. The spin of an electron can either be oriented up $\left(+\frac{1}{2}\right)$ or down $\left(-\frac{1}{2}\right)$.

electronegativity The ability of an element to attract electrons within a covalent bond.

element A substance that cannot be broken down into simpler substances.

emission spectrum (plural, emission spectra) A spectrum associated with the emission of electromagnetic radiation by elements or compounds.

empirical formula A formula for a compound that gives the smallest whole-number ratio of each type of atom.

empirical formula molar mass The sum of the molar masses of all the atoms in an empirical formula.

endothermic Describes a process that absorbs heat energy.

endothermic reaction A chemical reaction that absorbs energy from the surroundings.

energy The capacity to do work.

English system A unit system commonly used in the United States.

enthalpy The amount of thermal energy absorbed or emitted by a process under conditions of constant pressure.

enthalpy of reaction (ΔH_{rxn}) The amount of thermal energy absorbed or emitted by a chemical reaction under conditions of constant pressure.

enzyme A biological catalyst that increases the rates of biochemical reactions; enzymes are abundant in living organisms.

equilibrium constant (K_{eq}) The ratio, at equilibrium, of the concentrations of the products raised to their stoichiometric coefficients divided by the concentrations of the reactants raised to their stoichiometric coefficients.

equivalence point The point in a reaction at which the reactants are in exact stoichiometric proportions.

equivalent The stoichiometric proportions of elements and compounds in a chemical equation.

ester An organic compound with the general formula RCOOR.

ester linkage A type of bond with the general structure —COO—. Ester linkages join glycerol to fatty acids.

ether An organic compound with the general formula ROR.

evaporation A process in which molecules of a liquid, undergoing constant random motion, acquire enough energy to overcome attractions to neighbors and enter the gas phase.

excited state An unstable state for an atom or a molecule in which energy has been absorbed but not reemitted, raising an electron from the ground state into a higher energy orbital.

exothermic Describes a process that releases heat energy.

exothermic reaction A chemical reaction that releases energy to the surroundings.

experiment A procedure that attempts to measure observable predictions to test a theory or law.

exponent A number that represents the number of times a term is multiplied by itself. For example, in 2^4 the exponent is 4 and represents $2 \times 2 \times 2 \times 2$.

exponential part One part of a number expressed in scientific notation; it represents the number of places the decimal point has moved.

Fahrenheit (°F) scale The temperature scale that is most familiar in the United States; water freezes at 32 °F and boils at 212 °F at 1 atm prssure.

family (of elements) A group of elements that have similar outer electron configurations and therefore similar properties. Families occur in vertical columns in the periodic table.

family (of organic compounds) A group of organic compounds with the same functional group.

fatty acid A type of lipid consisting of a carboxylic acid with a long hydrocarbon tail.

fibrous protein Proteins with tertiary structures in which coiled amino acid chains align roughly parallel to one another, forming long, water-insoluble fibers.

film badge dosimeter Badge used to measure radiation exposure, consisting of photographic film held in a small case that is pinned to clothing.

filtration A method of separating a mixture composed of a solid and a liquid in which the mixture is poured through filter paper held in a funnel to capture the solid component.

formula mass The average mass of the molecules (or formula units) that compose a compound.

formula unit The basic unit of ionic compounds; the smallest electrically neutral collection of cations and anions that compose the compound.

fossil fuels Hydrocarbon-based fuels that originate from plant and animal life that existed on Earth in prehistoric times. The main types of fossil fuels are natural gas, petroleum, and coal.

freezing point depression The decrease in the freezing point of a solvent caused by the presence of a solute.

frequency The number of wave cycles or crests that pass through a stationary point in one second.

fuel cell A voltaic cell in which the reactants are constantly replenished.

functional group A set of atoms that characterizes a family of organic compounds.

gamma radiation High-energy, short-wavelength electromagnetic radiation emitted by an atomic nucleus.

gamma ray The shortest-wavelength, most energetic form of electromagnetic radiation. Gamma ray photons are represented by the symbol $^0_0\gamma$.

gas A state of matter in which atoms or molecules are widely separated and free to move relative to one another.

gas-evolution reaction A reaction that occurs in solution and forms a gas as one of the products.

Geiger-Müller counter A radioactivity detector consisting of a chamber filled with argon gas that discharges electrical signals when high-energy particles pass through it.

gene A sequence of codons within a DNA molecule that codes for a single protein. Genes vary in length from hundreds to thousands of codons.

genetic material The inheritable blueprint for making organisms.

globular protein Proteins with tertiary structures in which amino acid chains fold in on themselves, forming water-soluble globules.

glycogen A type of polysaccharide; it has a structure similar to that of starch, but the chain is highly branched.

glycolipid A biological molecule composed of a nonpolar fatty acid and hydrocarbon chain and a polar section composed of a sugar molecule such as glucose.

glycoside linkage The link between monosaccharides in a polysaccharide.

greenhouse gases Gases in the Earth's atmosphere that allow sunlight to enter the atmosphere but prevent heat from escaping.

ground state The state of an atom or molecule in which the electrons occupy the lowest possible energy orbitals available.

group (of elements) Elements that have similar outer electron configurations and therefore similar properties. Groups occur in vertical columns in the periodic table.

half-cell A compartment in which the oxidation or reduction half-reaction occurs in a galvanic or voltaic cell.

half-life The time it takes for one-half of the parent nuclides in a radioactive sample to decay to the daughter nuclides.

half-reaction Either the oxidation part or the reduction part of a redox reaction.

halogens The Group 7A elements, which are very reactive nonmetals.

heat The transfer or exchange of thermal energy caused by a temperature difference.

heat absorption One type of evidence of a chemical reaction, involving the intake of energy.

heat capacity The quantity of heat energy required to change the temperature of a given amount of a substance by 1 °C.

heat of fusion The amount of heat required to melt one mole of a solid at its melting point with no change in temperature.

heat of vaporization The amount of heat required to vaporize one mole of a liquid at its boiling point with no change in temperature.

heterogeneous mixture A mixture, such as oil and water, that has two or more regions with different compositions.

homogeneous mixture A mixture, such as salt water, that has the same composition throughout.

human genome All of the genetic material of a human being; the total DNA of a human cell.

Hund's rule A rule stating that when filling orbitals of equal energy, electrons will occupy empty orbitals singly before pairing with other electrons.

hydrocarbon A compound that contains only carbon and hydrogen atoms.

hydrogen bond A strong dipole–dipole interaction between molecules containing hydrogen directly bonded to a small, highly electronegative atom, such as N, O, or F.

hydrogenation The chemical addition of hydrogen to a compound.

hydronium ion The H_3O^+ ion. Chemists often use $H^+(aq)$ and $H_3O^+(aq)$ interchangeably to mean the same thing—a hydronium ion.

hypothesis A theory or law before it has become well established; a tentative explanation for an observation or a scientific problem that can be tested by further investigation.

hypoxia A shortage of oxygen in the tissues of the body.

ideal gas constant (R) The proportionality constant in the ideal gas law. $R = 0.0821 \text{ L} \cdot \text{atm}/\text{mol} \cdot \text{K}$

ideal gas law A law that combines the four properties of a gas—pressure (P), volume (V), temperature (T), and number of moles (n) in a single equation showing their interrelatedness: $PV = nRT$ (R = ideal gas constant).

indicator A substance that changes color with acidity level, often used to detect the end point of a titration.

infrared (IR) light The fraction of the electromagnetic spectrum between visible light and microwaves. Infrared light is invisible to the human eye.

insoluble Not soluble in water.

instantaneous (temporary) dipole A type of intermolecular force resulting from transient shifts in electron density within an atom or molecule.

intermolecular forces Attractive forces that exist between molecules.

International System (SI) The standard set of units for science measurements, based on the metric system.

ion An atom (or group of atoms) that has gained or lost one or more electrons, so that it has an electric charge.

ion–dipole force An intermolecular force that occurs between an ion and a polar molecule.

ion product constant (K_w) The product of the H_3O^+ ion concentration and the OH^- ion concentration in an aqueous solution. At room temperature, $K_w = 1.0 \times 10^{-4}$.

ionic bond The bond that results when a metal and a non-metal combine in a chemical reaction. In an ionic bond, the metal transfers one or more electrons to the nonmetal.

ionic compound A compound formed between a metal and one or more nonmetals.

ionic solid A solid compound composed of metals and non-metals joined by ionic bonds.

ionization energy The energy required to remove an electron from an atom in the gaseous state.

ionize Convert (an atom, molecule, or substance) into an ion or ions, typically by removing one or more electrons.

ionizing power The ability of radiation to ionize other molecules and atoms.

isomer A molecule with the same molecular formula but different structure as another molecule.

isoosmotic Describes solutions having equal osmotic pressure.

isotope One of two or more atoms with the same number of protons but different numbers of neutrons.

isotope scanning The use of radioactive isotopes to identify disease in the body.

Kelvin (K) scale The temperature scale that assigns 0 K to the coldest temperature possible, absolute zero (−273 °C or −459 °F), the temperature at which molecular motion stops. The size of the kelvin is identical to that of the Celsius degree.

ketone An organic compound with the general formula *RCOR*.

kilogram (kg) The SI standard unit of mass.

kilowatt-hour (kWh) A unit of energy equal to 3.6 million joules.

kinetic energy Energy associated with the motion of an object.

kinetic molecular theory A simple model for gases that predicts the behavior of most gases under many conditions.

law of conservation of energy A law stating that energy can be neither created nor destroyed. The total amount of energy is constant and cannot change; it can only be transferred from one object to another or converted from one form to another.

law of conservation of mass A law stating that in a chemical reaction, matter is neither created nor destroyed.

law of constant composition A law stating that all samples of a given compound have the same proportions of their constituent elements.

Le Châtelier's principle A principle stating that when a chemical system at equilibrium is disturbed, the system shifts in a direction that minimizes the disturbance.

lead-acid storage battery An automobile battery consisting of six electrochemical cells wired in series. Each cell produces 2 volts for a total of 12 volts.

Lewis model A simple model for a chemical bond in which atoms transfer or share electrons to attain a noble gas electron configuration (usually referred to as an octet).

Lewis structure A drawing that represents chemical bonds between atoms as shared or transferred electrons; the valence electrons of atoms are represented as dots.

light A form of electromagnetic radiation.

limiting reactant The reactant that determines the amount of product formed in a chemical reaction.

linear Describes the molecular geometry of a molecule containing two electron groups (two bonding groups and no lone pairs).

lipid A cellular component that is insoluble in water but soluble in nonpolar solvents.

lipid bilayer A structure formed by lipids in the cell membrane.

liquid A state of matter in which atoms or molecules are packed close to each other (about as closely as in a solid) but are free to move around and by each other.

liter (L) A unit of volume equal to 1.057 quarts.

logarithmic scale A scale involving logarithms. A logarithm entails an exponent that indicates the power to which a number is raised to produce a given number (e.g., the logarithm of 100 to the base 10 is 2).

lone pair Electrons that are only on one atom in a Lewis structure.

main-group elements Groups 1A–8A on the periodic table. These groups have properties that tend to be predictable based on their position in the periodic table.

mass A measure of the quantity of matter within an object.

mass number (A) The sum of the number of neutrons and protons in an atom.

mass percent composition (or mass percent) The percentage, by mass, of each element in a compound.

matter Anything that occupies space and has mass. Matter exists in three different states: solid, liquid, and gas.

melting point The temperature at which a solid turns into a liquid.

messenger RNA (mRNA) Long chainlike molecules that act as blueprints for the construction of proteins.

metallic atomic solid An atomic solid, such as iron, which is held together by metallic bonds that, in the simplest model, consist of positively charged ions in a sea of electrons.

metallic character The properties typical of a metal, especially the tendency to lose electrons in chemical reactions. Elements become more metallic as you move from right to left across the periodic table.

metalloids Those elements that fall along the boundary between the metals and the nonmetals in the periodic table; their properties are intermediate between those of metals and those of nonmetals.

metals Elements that tend to lose electrons in chemical reactions. They are found at the left side and in the center of the periodic table.

meter (m) The SI standard unit of length.

metric system The unit system commonly used throughout most of the world.

microwaves The part of the electromagnetic spectrum between the infrared region and the radio wave region. Microwaves are efficiently absorbed by water molecules and can therefore be used to heat water-containing substances.

millimeter of mercury (mmHg) A unit of pressure that originates from the method used to measure pressure with a barometer. Also called a *torr*.

miscibility The ability of two liquids to mix without separating into two phases, or the ability of one liquid to mix with (dissolve in) another liquid.

mixture A substance composed of two or more different types of atoms or molecules combined in variable proportions.

molality (m) A common unit of solution concentration, defined as the number of moles of solute per kilogram of solvent.

molar mass The mass of one mole of atoms of an element or one mole of molecules (or formula units) for a compound. An element's molar mass in grams per mole is numerically equivalent to the element's atomic mass in amu.

molar solubility The solubility of a substance in units of moles per liter (mol/L).

molar volume The volume occupied by one mole of gas. Under standard temperature and pressure conditions the molar volume of ideal gas is 22.5 L.

molarity (M) A common unit of solution concentration, defined as the number of moles of solute per liter of solution.

mole Avogadro's number (6.022×10^{23}) of particles— especially, of atoms, ions, or molecules. A mole of any element has a mass in grams that is numerically equivalent to its atomic mass in amu.

molecular compound A compound formed from two or more nonmetals. Molecular compounds have distinct molecules as their simplest identifiable units.

molecular element An element that does not normally exist in nature with single atoms as the basic unit. These elements usually exist as diatomic molecules—2 atoms of that element bonded together—as their basic units.

molecular equation A chemical equation showing the complete, neutral formulas for every compound in a reaction.

molecular formula A formula for a compound that gives the specific number of each type of atom in a molecule.

molecular geometry The geometrical arrangement of the atoms in a molecule.

molecular model A three-dimensional representation of a molecule.

molecular solid A solid whose composite units are molecules.

molecule Two or more atoms joined in a specific arrangement by chemical bonds. A molecule is the smallest identifiable unit of a molecular compound.

monomer An individual repeating unit that makes up a polymer.

monoprotic acid An acid containing only one ionizable proton.

monosaccharide A carbohydrate that cannot be decomposed into simpler carbohydrates.

monosubstituted benzene A benzene in which one of the hydrogen atoms has been replaced by another atom or group of atoms.

net ionic equation An equation that shows only the species that actually participate in a reaction.

neutral solution A solution in which the concentrations of H_3O^+ and OH^- are equal (pH = 7).

neutralization A chemical reaction in which an acid and a base react to form water.

neutralization reaction A reaction that takes place when an acid and a base are mixed; the $H^+(aq)$ from the acid combines with the $OH^-(aq)$ from the base to form $H_2O(l)$.

neutron A nuclear particle with no electrical charge and nearly the same mass as a proton.

nitrogen narcosis An increase in nitrogen concentration in bodily tissues and fluids that results in feelings of drunkenness.

noble gases The Group 8A elements, which are chemically unreactive.

nonbonding atomic solid An atomic solid that is held together by relatively weak dispersion forces.

nonelectrolyte solution A solution containing a solute that dissolves as molecules; therefore, the solution does not conduct electricity.

nonmetals Elements that tend to gain electrons in chemical reactions. They are found at the upper right side of the periodic table.

nonpolar Describes a molecule that does not have a dipole moment.

nonvolatile Describes a compound that does not vaporize easily.

normal alkane (or n-alkane) An alkane composed of carbon atoms bonded in a straight chain with no branches.

normal boiling point The boiling point of a liquid at a pressure of 1 atmosphere.

nuclear equation An equation that represents the changes that occur during radioactivity and other nuclear processes.

nuclear fission The process by which a heavy nucleus is split into nuclei of smaller masses and energy is emitted.

nuclear fusion The combination of light atomic nuclei to form heavier ones with emission of large amounts of energy.

nuclear radiation The energetic particles emitted from the nucleus of an atom when it is undergoing a nuclear process.

nuclear theory of the atom A theory stating that most of the atom's mass and all of its positive charge are contained in a small, dense nucleus. Most of the volume of the atom is empty space occupied by negatively charged electrons.

nucleic acids Biological molecules, such as deoxyribonucleic acid (DNA) and ribonucleic acid (RNA), that store and transmit genetic information.

nucleotide An individual unit of a nucleic acid. Nucleic acids are polymers of nucleotides.

nucleus (of a cell) The part of the cell that contains the genetic material.

nucleus (of an atom) The small core containing most of the atom's mass and all of its positive charge. The nucleus is made up of protons and neutrons.

observation Often the first step in the scientific method. An observation must measure or describe something about the physical world.

octet The number of electrons, 8, around atoms with stable Lewis structures.

octet rule A rule that states that an atom will give up, accept, or share electrons in order to achieve a filled outer electron shell, which usually consists of 8 electrons.

orbital The region around the nucleus of an atom where an electron is most likely to be found.

orbital diagram An electron configuration in which electrons are represented as arrows in boxes corresponding to orbitals of a particular atom.

organic chemistry The study of carbon-containing compounds and their reactions.

organic molecule A molecule whose main structural component is carbon.

osmosis The flow of solvent from a lower-concentration solution through a semipermeable membrane to a higher-concentration solution.

osmotic pressure The pressure produced on the surface of a semipermeable membrane by osmosis or the pressure required to stop osmotic flow.

oxidation The gain of oxygen, the loss of hydrogen, or the loss of electrons (the most fundamental definition).

oxidation state (or oxidation number) A number that can be used as an aid in writing formulas and balancing equations. It is computed for each element based on the number of electrons assigned to it in a scheme where the most electronegative element is assigned all of the bonding electrons.

oxidation–reduction (redox) reaction A reaction in which electrons are transferred from one substance to another.

oxidizing agent In a redox reaction, the substance being reduced. Oxidizing agents tend to gain electrons easily.

oxyacid An acid containing hydrogen, a nonmetal, and oxygen.

oxyanion An anion containing oxygen. Most polyatomic ions are oxyanions.

oxygen toxicity The result of increased oxygen concentration in bodily tissues.

parent nuclide The original nuclide in a nuclear decay.

partial pressure The pressure due to any individual component in a gas mixture.

pascal (Pa) The SI unit of pressure, defined as 1 newton per square meter.

Pauli exclusion principle A principle stating that no more than 2 electrons can occupy an orbital and that the 2 electrons must have opposite spins.

penetrating power The ability of a radioactive particle to penetrate matter.

peptide bond The bond between the amine end of one amino acid and the carboxylic acid end of another. Amino acids link together via peptide bonds to form proteins.

percent natural abundance The percentage amount of each isotope of an element in a naturally occurring sample of the element.

percent yield In a chemical reaction, the percentage of the theoretical yield that was actually attained.

periodic law A law that states that when the elements are arranged in order of increasing relative mass, certain sets of properties recur periodically.

periodic table An arrangement of the elements in which atomic number increases from left to right and elements with similar properties fall in columns called families or groups.

permanent dipole A separation of charge resulting from the unequal sharing of electrons between atoms.

pH The negative log of the concentration of H_3O^+ in a solution; the pH scale is a compact way to specify the acidity of a solution.

phenyl group The term for a benzene ring when other substituents are attached to it.

phospholipid A lipid with the same basic structure as a triglyceride, except that one of the fatty acid groups is replaced with a phosphate group.

phosphorescence The slow, long-lived emission of light that sometimes follows the absorption of light by some atoms and molecules.

photon A particle of light or a packet of light energy.

physical change A change in which matter does not change its composition, even though its appearance might change.

physical property A property that a substance displays without changing its composition.

pOH A scale used to measure basicity. $pOH = -\log[OH^-]$

polar covalent bond A covalent bond between atoms of different electronegativities. Polar covalent bonds have a dipole moment.

polar molecule A molecule with polar bonds that add together to create a net dipole moment.

polyatomic ion An ion composed of a group of atoms with an overall charge.

polymer A molecule with many similar units, called monomers, bonded together in a long chain.

polypeptide A short chain of amino acids joined by peptide bonds.

polysaccharide A long, chainlike molecule composed of many linked monosaccharide units. Polysaccharides are polymers of monosaccharides.

positron A nuclear particle that has the mass of an electron but carries a 1+ charge.

positron emission Expulsion of a positron from an unstable atomic nucleus. In positron emission, a proton is transformed into a neutron.

potential energy The energy of a body that is associated with its position or the arrangement of its parts.

pounds per square inch (psi) A unit of pressure. (1 atm = 14.7 psi)

precipitate An insoluble product formed through the reaction of two solutions containing soluble compounds.

precipitation reaction A reaction that forms a solid or precipitate when two aqueous solutions are mixed.

prefix multipliers Prefixes used by the SI system with the standard units. These multipliers change the value of the unit by powers of 10.

pressure The force exerted per unit area by gaseous molecules as they collide with the surfaces around them.

primary protein structure The sequence of amino acids in a protein's chain. Primary protein structure is maintained by the covalent peptide bonds between individual amino acids.

principal quantum number A number that indicates the shell that an electron occupies.

principal shell The shell indicated by the principal quantum number.

product A final substance produced in a chemical reaction; represented on the right side of a chemical equation.

property A characteristic we use to distinguish one substance from another.

protein A biological molecule composed of a long chain of amino acids joined by peptide bonds. In living organisms, proteins serve many varied and important functions.

proton A positively charged nuclear particle. A proton's mass is approximately 1 amu.

pure substance A substance composed of only one type of atom or molecule.

quantum (pl. quanta) The precise amount of energy possessed by a photon; the difference in energy between two atomic orbitals.

quantum number (n) An integer that specifies the energy of an orbital. The higher the quantum number n, the greater the distance between the electron and the nucleus and the higher its energy.

quantum-mechanical model The foundation of modern chemistry; explains how electrons exist in atoms and how they affect the chemical and physical properties of elements.

quaternary structure In a protein, the way that individual chains fit together to compose the protein. Quaternary structure is maintained by interactions between the R groups of amino acids on the different chains.

R group (side chain) An organic group attached to the central carbon atom of an amino acid.

radio waves The longest wavelength and least energetic form of electromagnetic radiation.

radioactive Describes a substance that emits tiny, invisible, energetic particles from the nuclei of its component atoms.

radioactivity The emission of tiny, invisible, energetic particles from the unstable nuclei of atoms. Many of these particles can penetrate matter.

radiocarbon dating A technique used to estimate the age of fossils and artifacts through the measurement of natural radioactivity of carbon atoms in the environment.

radiotherapy Treatment of disease with radiation, such as the use of gamma rays to kill rapidly dividing cancer cells.

random coil The name given to an irregular pattern of a secondary protein structure.

rate of a chemical reaction (reaction rate) The amount of reactant that changes to product in a given period of time. Also defined as the amount of a product that forms in a given period of time.

reactant An initial substance in a chemical reaction, represented on the left side of a chemical equation.

recrystallization A technique used to purify a solid; involves dissolving the solid in a solvent at high temperature, creating a saturated solution, then cooling the solution to cause the crystallization of the solid.

redox (oxidation–reduction) reaction A chemical reaction in which electrons are transferred from one reactant to another.

reducing agent In a redox reaction, the substance being oxidized. Reducing agents tend to lose electrons easily.

reduction The loss of oxygen, the gain of hydrogen, or the gain of electrons (the most fundamental definition).

rem Stands for *roentgen equivalent man;* a weighted measure of radiation exposure that accounts for the ionizing power of the different types of radiation.

resonance structures Two or more Lewis structures that are necessary to describe the bonding in a molecule or ion.

reversible reaction A reaction that is able to proceed in both the forward and reverse directions.

RNA (ribonucleic acid) Long chainlike molecules that occur throughout cells and take part in the construction of proteins.

salt An ionic compound that usually remains dissolved in a solution after an acid–base reaction has occurred.

salt bridge An inverted, U-shaped tube containing a strong electrolyte; completes the circuit in an electrochemical cell by allowing the flow of ions between the two half-cells.

saturated fat A triglyceride composed of saturated fatty acids. Saturated fat tends to be solid at room temperature.

saturated hydrocarbon A hydrocarbon that contains no double or triple bonds between the carbon atoms.

saturated solution A solution that holds the maximum amount of solute under the solution conditions. If additional solute is added to a saturated solution, it will not dissolve.

scientific law A statement that summarizes past observations and predicts future ones. Scientific laws are usually formulated from a series of related observations.

scientific method The way that scientists learn about the natural world. The scientific method involves observations, laws, hypotheses, theories, and experimentation.

scientific notation A system used to write very big or very small numbers, often containing many zeros, more compactly and precisely. A number written in scientific notation consists of a decimal part and an exponential part (10 raised to a particular exponent).

scintillation counter A device used to detect radioactivity in which energetic particles traverse a material that emits ultraviolet or visible light when excited by their passage. The light is detected and turned into an electrical signal.

second (s) The SI standard unit of time.

secondary protein structure Short-range periodic or repeating patterns often found in proteins. Secondary protein structure is maintained by interactions between amino acids that are fairly close together in the linear sequence of the protein chain or adjacent to each other on neighboring chains.

semiconductor A compound or element exhibiting intermediate electrical conductivity that can be changed and controlled.

semipermeable membrane A membrane that selectively allows some substances to pass through but not others.

SI units The most convenient system of units for science measurements, based on the metric system. The set of standard units agreed on by scientists throughout the world.

significant digits (figures) The non–place-holding digits in a reported measurement; they represent the precision of a measured quantity.

simple carbohydrate (simple sugar) A monosaccharide or disaccharide.

single-displacement reaction A reaction in which one element displaces another in a compound.

solid A state of matter in which atoms or molecules are packed close to each other in fixed locations.

solubility The amount of a compound, usually in grams, that will dissolve in a certain amount of solvent.

solubility rules A set of empirical rules used to determine whether an ionic compound is soluble.

solubility-product constant (K_{sp}) The equilibrium expression for a chemical equation that represents the dissolving of an ionic compound in solution.

soluble Dissolves in solution.

solute The minority component of a solution.

solution A homogeneous mixture of two or more substances.

solution map In this book, a solution map is a visual outline of the solution to a problem.

solvent The majority component of a solution.

space-filling model A way to represent molecules in which atoms are represented with spheres that overlap with one another.

specific heat capacity (or specific heat) The heat capacity of a substance in joules per gram degree Celsius (J/g °C).

spectator ions Ions that do not participate in a reaction; they appear unchanged on both sides of a chemical equation.

standard temperature and pressure (STP) Conditions often assumed in calculations involving gases: $T = 0$ °C (273 K) and $P = 1$ atm.

starch A common polysaccharide composed of repeating glucose units.

state of matter One of the three forms in which matter can exist: solid, liquid, and gas.

steroid A biological compound containing a 17-carbon 4-ring system.

stock solution A concentrated form in which solutions are often stored.

stoichiometry The numerical relationships among chemical quantities in a balanced chemical equation. Stoichiometry allows us to predict the amounts of products that form in a chemical reaction based on the amounts of reactants.

strong acid An acid that completely ionizes in solution.

strong base A base that completely dissociates in solution.

strong electrolyte A substance whose aqueous solutions are good conductors of electricity.

strong electrolyte solution A solution containing a solute that dissociates into ions; therefore, a solution that conducts electricity well.

structural formula A two-dimensional representation of molecules that not only shows the number and type of atoms, but also how the atoms are bonded together.

sublimation A physical change in which a substance is converted from its solid form directly into its gaseous form.

subshell In quantum mechanics, specifies the shape of the orbital and is represented by different letters (s, p, d, f).

substituent An atom or a group of atoms that has been substituted for a hydrogen atom in an organic compound.

substitution reaction A reaction in which one or more atoms are replaced by one or more different atoms.

supersaturated solution A solution holding more than the normal maximum amount of solute.

surface tension The tendency of liquids to minimize their surface area, resulting in a "skin" on the surface of the liquid.

synthesis reaction A reaction in which simpler substances combine to form more complex substances; A + B $\longrightarrow$ AB.

temperature A measure of the thermal energy in a sample of matter.

temporary dipole A type of intermolecular force resulting from transient shifts in electron density within an atom or molecule.

terminal atom An atom that is located at the end of a molecule or chain.

tertiary protein structure A protein's structure that consists of the large-scale bends and folds due to interactions between the *R* groups of amino acids that are separated by large distances in the linear sequence of the protein chain.

tetrahedral The molecular geometry of a molecule containing four electron groups (four bonding groups and no lone pairs).

theoretical yield The maximum amount of product that can be made in a chemical reaction based on the amount of limiting reactant.

theory A proposed explanation for observations and laws. A theory presents a model of the way nature works and predicts behavior that extends well beyond the observations and laws from which it was formed.

thermal energy A type of kinetic energy associated with the temperature-dependent random movement of atoms and molecules.

titration A laboratory procedure used to determine the amount of a substance in solution. In a titration, a reactant in a solution of known concentration is reacted with another reactant in a solution of unknown concentration until the reaction reaches the end point.

torr A unit of pressure named after the Italian physicist Evangelista Torricelli; also called a millimeter of mercury.

transition elements The elements in columns designated with B or in columns 3–12 of the periodic table.

transition metals The elements in the middle of the periodic table whose properties tend to be less predictable based simply on their position in the periodic table. Transition metals lose electrons in their chemical reactions, but do not necessarily acquire noble gas configurations.

triglyceride A fat or oil; a tryglyceride is a triester composed of glycerol with three fatty acids attached.

trigonal planar The molecular geometry of a molecule containing three electron groups, three bonding groups, and no lone pairs.

trigonal pyramidal The molecular geometry of a molecule containing four electron groups, three bonding groups, and one lone pair.

triple bond A chemical bond consisting of three electron pairs shared between two atoms. In general, triple bonds are shorter and stronger than double bonds.

ultraviolet (UV) light The fraction of the electromagnetic spectrum between the visible region and the X-ray region. UV light is invisible to the human eye.

units Previously agreed-on quantities used to report experimental measurements. Units are vital in chemistry.

unsaturated fat (or oil) A triglyceride composed of unsaturated fatty acids. Unsaturated fats tend to be liquids at room temperature.

unsaturated hydrocarbon A hydrocarbon that contains one or more double or triple bonds between its carbon atoms.

unsaturated solution A solution holding less than the maximum possible amount of solute under the solution conditions.

valence electrons The electrons in the outermost principal shell of an atom; they are involved in chemical bonding.

valence shell electron pair repulsion (VSEPR) A theory that allows prediction of the shapes of molecules based on the idea that electrons—either as lone pairs or as bonding pairs—repel one another.

vapor pressure The partial pressure of a vapor in dynamic equilibrium with its liquid.

vaporization The phase transition between a liquid and a gas.

viscosity The resistance of a liquid to flow; manifestation of intermolecular forces.

visible light The fraction of the electromagnetic spectrum that is visible to the human eye, bounded by wavelengths of 400 nm (violet) and 780 nm (red).

vital force A mystical or supernatural power that, it was once believed, was possessed only by living organisms and allowed them to produce organic compounds.

vitalism The belief that living things contain a nonphysical "force" that allows them to synthesize organic compounds.

volatile Tending to vaporize easily.

voltage The potential difference between two electrodes; the driving force that causes electrons to flow.

voltaic (galvanic) cell An electrochemical cell that creates electrical current from a spontaneous chemical reaction.

volume A measure of space. Any unit of length, when cubed, becomes a unit of volume.

wavelength The distance between adjacent wave crests in a wave.

weak acid An acid that does not completely ionize in solution.

weak base A base that does not completely dissociate in solution.

weak electrolyte A substance whose aqueous solutions are poor conductors of electricity.

work The result of a force acting on a distance.

X-rays The portion of the electromagnetic spectrum between the ultraviolet (UV) region and the gamma-ray region.

Credits

Index

Fundamental Physical Constants

Atomic mass unit	$1 \text{ amu} = 1.660539 \times 10^{-27} \text{ kg}$
	$1 \text{ g} = 6.022142 \times 10^{23} \text{ amu}$
Avogadro's number	$N_A = 6.022142 \times 10^{23}/\text{mol}$
Electron charge	$e = 1.602176 \times 10^{-19} \text{ C}$
Gas constant	$R = 8.314472 \text{ J}/(\text{mol} \cdot \text{K})$
	$= 0.0820582 \text{ (L} \cdot \text{atm)}/(\text{mol} \cdot \text{K})$
Mass of electron	$m_e = 5.485799 \times 10^{-4} \text{ amu}$
	$= 9.109382 \times 10^{-31} \text{ kg}$
Mass of neutron	$m_n = 1.008665 \text{ amu}$
	$= 1.674927 \times 10^{-27} \text{ kg}$
Mass of proton	$m_p = 1.007276 \text{ amu}$
	$= 1.672622 \times 10^{-27} \text{ kg}$
Pi	$\pi = 3.1415926536$
Planck's constant	$h = 6.626069 \times 10^{-34} \text{ J} \cdot \text{s}$
Speed of light in vacuum	$c = 2.99792458 \times 10^8 \text{ m/s}$

Useful Geometric Formulas

Perimeter of a rectangle $= 2l + 2w$

Circumference of a circle $= 2\pi r$

Area of a triangle $= (1/2)(\text{base} \times \text{height})$

Area of a circle $= \pi r^2$

Surface area of a sphere $= 4\pi r^2$

Volume of a sphere $= (4/3)\pi r^3$

Volume of a cylinder or prism $= \text{area of base} \times \text{height}$

Important Conversion Factors

Length: SI unit = meter (m)
- $1 \text{ m} = 39.37 \text{ in.}$
- $1 \text{ in.} = 2.54 \text{ cm (exactly)}$
- $1 \text{ mile} = 5280 \text{ ft} = 1.609 \text{ km}$
- $1 \text{ angstrom (Å)} = 10^{-10} \text{ m}$

Volume: SI unit = cubic meter (m³)
- $1 \text{ L} = 1000 \text{ cm}^3 = 1.057 \text{ qt (U.S.)}$
- $1 \text{ gal (U.S.)} = 4 \text{ qt} = 8 \text{ pt}$
  - $= 128 \text{ fluid ounces}$
  - $= 3.785 \text{ L}$

Mass: SI unit = kilogram (kg)
- $1 \text{ kg} = 2.205 \text{ lb}$
- $1 \text{ lb} = 16 \text{ oz} = 453.6 \text{ g}$
- $1 \text{ ton} = 2000 \text{ lb}$
- $1 \text{ metric ton} = 1000 \text{ kg} = 1.103 \text{ tons}$
- $1 \text{ g} = 6.022 \times 10^{23} \text{ atomic mass units (amu)}$

Pressure: SI unit = pascal (Pa)
- $1 \text{ Pa} = 1 \text{ N/m}^2$
- $1 \text{ bar} = 10^5 \text{ Pa}$
- $1 \text{ atm} = 1.01325 \times 10^5 \text{ Pa (exactly)}$
  - $= 1.01325 \text{ bar}$
  - $= 760 \text{ mmHg}$
  - $= 760 \text{ torr (exactly)}$

Energy: SI unit = joule (J)
- $1 \text{ J} = 1 \text{ N} \cdot \text{m}$
- $1 \text{ cal} = 4.184 \text{ J (exactly)}$
- $1 \text{ L} \cdot \text{atm} = 101.33 \text{ J}$

Temperature: SI unit = kelvin (K)
- $K = {}^\circ C + 273.15$
- ${}^\circ C = (5/9)({}^\circ F - 32^\circ)$
- ${}^\circ F = (9/5)({}^\circ C) + 32^\circ$